Texts & Monographs in Symbolic Computation

A Series of the Research Institute for Symbolic Computation, Johannes Kepler University, Linz, Austria

Mathematics is a key technology in modern society. Symbolic Computation is on its way to become a key technology in mathematics. "Texts and Monographs in Symbolic Computation" provides a platform devoted to reflect this evolution. In addition to reporting on developments in the field, the focus of the series also includes applications of computer algebra and symbolic methods in other subfields of mathematics and computer science, and, in particular, in the natural sciences. To provide a flexible frame, the series is open to texts of various kind, ranging from research compendia to textbooks for courses.

Indexed by zbMATH.

More information about this subseries at http://www.springer.com/series/3073

Johannes Blümlein • Carsten Schneider
Editors

Anti-Differentiation and the Calculation of Feynman Amplitudes

Springer

Editors
Johannes Blümlein
Deutsches Elektronen-Synchrotron, DESY
Zeuthen, Germany

Carsten Schneider
Research Institute for Symbolic Computation
Johannes Kepler University Linz
Linz, Austria

ISSN 0943-853X ISSN 2197-8409 (electronic)
Texts & Monographs in Symbolic Computation
ISBN 978-3-030-80218-9 ISBN 978-3-030-80219-6 (eBook)
https://doi.org/10.1007/978-3-030-80219-6

Mathematics Subject Classification: 34-xx, 33-xx, 39-xx, 44-xx, 45-xx

This Springer imprint is published by the registered company Springer Nature Switzerland AG.
The registered company address is: Gewerbestrasse 11, 6330 Cham, Switzerland

Preface

The analytic calculation of Feynman amplitudes, and in particular the evaluation of Feynman integrals, is a challenging task within elementary particle physics. In particular, the simplification of such often large-scale expressions in terms of special functions and constants plays a decisive role to improve the prediction and to obtain a deeper understanding of these predictions of renormalizable quantum field theories of elementary particle physics for experimental precision measurements. The main focus of the book *Anti-Differentiation and the Calculation of Feynman Amplitudes* is put on sophisticated and efficient symbolic methods and related tools from mathematics and theoretical physics to support this task. Special focus is put on the manipulation and simplification of non-trivial Feynman integrals, and more generally of rather general classes of multi-integrals and connected multi-sums that are relevant in many other research areas of natural and technical science.

Among many fascinating techniques in symbolic computation and special functions, the book addresses the following key aspects that become important in the up-to-date calculations of Feynman amplitudes:

- Advanced symbolic integration and summation tools are carved out that support in the task to explore and simplify multiple integrals and sums to special functions and constants.
- Sophisticated special function techniques are elaborated that explore important properties of major functions (e.g., integrals over elliptic functions and related sums).
- Efficient technologies, like Integration-By-Parts (IBP) methods, are worked out that enable one to represent physical expressions in terms of (coupled systems) of linear differential equations.
- Highly general computer algebra methods are introduced that support in the task to solve linear differential and difference equations efficiently.

This book brings together sophisticated tools from pure mathematics, computer algebra, and the theoretical particle physics community that are combined non-trivially or that have strong potentials for future interaction. In particular, we expect that the following tools presented in this book will be crucial for challenging calculations that will be fundamental for ongoing and future experiments at the Large Hardron Collider (LHC) and its planned successor at CERN, the FCC.

In *Analytic Integration Methods in QFT* (Johannes Blümlein), a general overview of the mathematical and symbolic computation methods for the treatment of Feynman integrals is elaborated. In particular, the chapters of this book are introduced accordingly in this general presentation.

In *Extensions of the AZ-algorithm and the Package MultiIntegrate* (Jakob Ablinger) an extension of the (continuous) multivariate Almkvist-Zeilberger algorithm is introduced that can compute linear difference and differential equations of Feynman integrals. Special focus is also put on solving the found equations in terms of nested sums and integrals.

In *Empirical Determinations of Feynman Integrals Using Integer Relation Algorithms* (Kevin Acres and David Broadhurst) efficient algorithms (in particular, PSLQ and LLL) are presented and applied to evaluate and explore sophisticated Feynman integrals in terms of special constants.

In *$N = 4$ SYM Gauge Theories: The $2 \to 6$ Amplitude in the Regge Limit* (Jochen Bartels) the Regge limit of scattering amplitudes is utilized, which is based upon unitarity, energy discontinuity, and the analytic structure of the arising Feynman amplitudes.

In *Direct Integration for Multi-Leg Amplitudes: Tips, Tricks, and When They Fail* (Jacob L. Bourjaily, Yang-Hui He, Andrew J. McLeod, Marcus Spradlin, Cristian Vergu, Matthias Volk, Matt von Hippel, and Matthias Wilhelm) it is demonstrated how the hyperlogarithmic integration method in interaction with new techniques can be used to transform certain Feynman integrals to hyperlogarithms.

In *A Geometrical Framework for Amplitude Recursions: Bridging Between Trees and Loops* (Johannes Brödel and André Kaderli) it is illustrated how scattering amplitudes, mostly coming from string theory, can be expressed in terms of (elliptic) multiple zeta values by manipulating iterated integrals on Riemann surfaces with boundaries.

In *Differential Galois Theory and Integration* (Thomas Dreyfus and Jacques-Arthur Weil) an important simplification step of coupled systems of linear differential equations is presented that supports the solving task and may deliver extra knowledge of algebraic relations between integrals.

In *Top-Down Decomposition: A Cut-Based Approach to Integral Reductions* (Hjalte Frellesvig) an alternative approach of the well-established IBP methods is introduced to find linear relations between Feynman integrals.

In *Hypergeometric Functions and Feynman Diagrams* (Mikhail Kalmykov, Vladimir Bytev, Bernd Kniehl, Sven-Olaf Moch, Bennie F. L. Ward, and Scott A. Yost) various relevant tools, such as the calculation of ϵ-expansions, for hypergeometric function and more generally Appell function representations of certain classes of Feynman integrals are elaborated.

In *Differential Equations and Feynman Integrals* (Anatoly V. Kotikov) it is shown how linear differential equations for Feynman integrals can be extracted by using IBP methods. In addition, methods are presented that are currently used to solve such equations in the setting of Feynman integrals.

In *Holonomic Anti-Differentiation and Feynman Amplitudes* (Christoph Koutschan) the holonomic system approach is applied to compute linear differential equations of Feynman integrals. In particular, hypergeometric functions and sunrise integrals are considered in more details.

In *Outer Space as a Combinatorial Backbone for Cutkosky Rules and Coactions* (Dirk Kreimer) a co-action of bridge-free graphs (that extends the known definition for simple graphs) is connected non-trivially to Feynman integrals.

In *Integration-by-Parts: A Survey* (Peter Marquard) an overview of IBP methods in particular of Laporta's algorithm is given. Special emphasis is put on complexity aspects and different tactics how one might handle them.

In *Calculating Four-Loop Corrections in QCD* (Sven-Olaf Moch and Vitaly Magerya) the main working flow is presented how the calculations of perturbative corrections in QCD at four loops can be accomplished by using computer algebra programs such as `Form`.

In *Contiguous Relations and Creative Telescoping* (Peter Paule) a complete list of constructive theorems is carved out, which predict whether a finite set of hypergeometric series whose summands are similar satisfies a contiguous relation. In particular, an enhanced version of the package `fastZeil` is presented that can compute the predicted relations explicitly.

In *Nested Integrals and Rationalizing Transformations* (Clemens G. Raab) a general overview of symbolic methods for the treatment of nested integrals is given. In particular, a comprehensive list of univariate rationalizing transformations is explored.

In *Term Algebras, Canonical Representations and Difference Ring Theory for Symbolic Summation* (Carsten Schneider) a general summation framework for the existing difference ring and field algorithms is developed that is relevant for the simplification of Feynman integrals.

In *Expansion by Regions: An Overview* (Vladimir A. Smirnov) different techniques are addressed that can calculate expansions of a given multi-loop Feynman integral in a given limit where some kinematic invariants and/or masses have certain scaling measured in powers of a given small parameter.

In *Some Steps Towards Improving IBP Calculations and Related Topics* (J. A. M. Vermaseren) interesting aspects of IBP reductions and their major bottlenecks are discussed with potential future techniques how they could be overcome with computer algebra.

In *Iterated Integrals Related to Feynman Integrals Associated to Elliptic Curves* (Stefan Weinzierl) the mathematical background of elliptic curves (and the connected elliptic integrals and modular forms) is introduced in the context of Feynman integrals.

The volume reflects and summarizes the state of the art having been obtained so far. As it is shown, intense developments have taken place in this field during the last 25 years. Many more extensions lay ahead and are necessary to be performed to reach the level of ultimate precision at the theory side to cope with the accuracies to be reached at the FCC collider.

Zeuthen, Germany
Linz, Austria
April 2020

Johannes Blümlein
Carsten Schneider

Contents

Contributors

Jakob Ablinger Research Institute for Symbolic Computation (RISC), Johannes Kepler University Linz, Linz, Austria

Kevin Acres Cryptography Research Centre, Technology Innovation Institute, Abu Dhabi, UAE

Jochen Bartels II. Institut für Theoretische Physik, Universität Hamburg, Hamburg, Germany

Johannes Blümlein Deutsches Elektronen-Synchrotron, DESY, Zeuthen, Germany

Jacob L. Bourjaily Niels Bohr Institute, University of Copenhagen, Copenhagen, Denmark
Institute for Gravitation and the Cosmos, Department of Physics, Pennsylvania State University, University Park, PA, USA

David Broadhurst School of Physical Sciences, Open University, Milton Keynes, UK

Johannes Brödel Humboldt University, Berlin, Germany

Vladimir Bytev BLTP JINR, Dubna, Moscow Region, Russia

Thomas Dreyfus Institut de Recherche Mathématique Avancée, U.M.R. 7501 Université de Strasbourg et C.N.R.S., Strasbourg, France

Hjalte Frellesvig Niels Bohr International Academy, Copenhagen, Denmark

Yang-Hui He Department of Mathematics, University of London, London, UK
Merton College, University of Oxford, Oxford, UK
School of Physics, NanKai University, Tianjin, P.R. China

André Kaderli Humboldt University, Berlin, Germany

Mikhail Kalmykov JINR, Dubna, Russia

Bernd A. Kniehl II. Institut fuer Theoretische Physik, Universität Hamburg, Hamburg, Germany

Anatoly V. Kotikov Bogoliubov Laboratory of Theoretical Physics Joint Institute for Nuclear Research, Dubna, Russia

Christoph Koutschan Johann Radon Institute for Computational and Applied Mathematics, Austrian Academy of Sciences, Linz, Austria

Dirk Kreimer Department of Mathematics, Humboldt University, Berlin, Germany
Department of Physics, Humboldt University, Berlin, Germany

Vitaly Magerya Institut für Theoretische Physik, Campus Süd Karlsruher Institut für Technologie (KIT), Karlsruhe, Germany

Peter Marquard Deutsches Elektronen-Synchrotron, DESY, Zeuthen, Germany

Andrew J. McLeod Niels Bohr Institute, University of Copenhagen, Copenhagen, Denmark

Sven-Olaf Moch II. Institut für Theoretische Physik, Universität Hamburg, Hamburg, Germany

Peter Paule Research Institute for Symbolic Computation (RISC), Johannes Kepler University Linz, Linz, Austria

Clemens G. Raab Institut für Algebra, Johannes Kepler Universität Linz, Linz, Austria

Carsten Schneider Research Institute for Symbolic Computation (RISC), Johannes Kepler University Linz, Linz, Austria

Vladimir A. Smirnov Skobeltsyn Institute of Nuclear Physics of Moscow State University, Moscow, Russia

Marcus Spradlin Department of Physics, Brown University, Providence, RI, USA

Cristian Vergu Niels Bohr Institute, University of Copenhagen, Copenhagen, Denmark

J. A. M. Vermaseren Nikhef, Amsterdam, The Netherlands

Matthias Volk Niels Bohr Institute, University of Copenhagen, Copenhagen, Denmark

Matt von Hippel Niels Bohr Institute, University of Copenhagen, Copenhagen, Denmark

Bennie F. L. Ward Department of Physics, Baylor University, Waco, TX, USA

Jacques-Arthur Weil XLIM, U.M.R 7252, Université de Limoges et C.N.R.S, Limoges Cedex, France

Stefan Weinzierl PRISMA Cluster of Excellence, Institut für Physik, Johannes Gutenberg-Universität Mainz, Mainz, Germany

Matthias Wilhelm Niels Bohr Institute, University of Copenhagen, Copenhagen, Denmark

Scott A. Yost Department of Physics, The Citadel, Charleston, SC, USA

Analytic Integration Methods in Quantum Field Theory: An Introduction

Johannes Blümlein

Abstract A survey is given on the present status of analytic calculation methods and the mathematical structures of zero- and single scale Feynman amplitudes which emerge in higher order perturbative calculations in the Standard Model of elementary particles, its extensions and associated model field theories, including effective field theories of different kind.

1 Introduction

Analytic precision calculations for the observables in renormalizable quantum field theories have developed during the last 70 years significantly. These methods have helped to put the Standard Model of elementary particles to tests of an unprecedented accuracy, requested by the scientific method [1]. Present and future high luminosity experiments [2, 3] will demand even higher precision predictions at the theory side. This goes along with mastering large sets of analytic data by methods of computer algebra and special mathematical methods to perform the corresponding integrals analytically.

Here analytic integration is understood as antidifferentiation. In this context the first question going to arise is: which is the space to represent a certain class of integrals in an irreducible manner. As history showed, this question is usually answered in an iterative way. Often not all the existing relations in a given class of functions can be revealed right from the beginning. It is even so that in some cases it remained unclear over many decades whether all relations are already found

Contribution to the Volume "Antidifferentiation and the Calculation of Feynman Amplitudes", Springer, Berlin, 2021.

J. Blümlein (✉)
Deutsches Elektronen-Synchrotron, DESY, Zeuthen, Germany
e mail: Johannes.Bluemlein@desy.de

J. Blümlein, C. Schneider (eds.), *Anti-Differentiation and the Calculation of Feynman Amplitudes*, Texts & Monographs in Symbolic Computation,
https://doi.org/10.1007/978-3-030-80219-6_1

or not. One example for this are the multiple zeta values [4]. The development of physics applications is of course not stopped by this, as also partial solutions are of great help in reducing the large complexity to be dealt with. The application of quite different techniques has often to be combined to finally tackle these cutting edge problems. Moreover, it is this process which delivers new insights and is able to produce even more refined technologies.

The problem of analytic integration of Feynman diagrams is nowadays a field of research requesting to join fundamental ideas from theoretical physics, computer algebra and computing technology, as well as of a growing number of fields in pure mathematics. The topics are so challenging that the experts in all these different fields are attracted by them in the solution of the different problems.

The present interdisciplinary workshop, organized by the Wolfgang-Pauli Center, arose from the idea to get scientists working in the field of Quantum Field Theory, computer algebra and pure mathematics together to review the status of the analytic solution of Feynman integrals reached and to prepare for further developments.

The topics of the workshop included both techniques to reduce the number of Feynman diagrams by physical relations, such as the integration by parts relations [5–7] as well as the mathematical methods to compute these integrals analytically. The latter include the method of generalized hypergeometric functions [8] and the general theory of contiguous relations [9], the methods of integer relations [10], guessing methods of one-dimensional quantities, hyperlogarithms [11], the solution of master-integrals using difference and differential equations [12–15], Risch algorithms on nested integrals and rationalization algorithms [16], holonomic integration [17], the multivalued Almkvist-Zeilberger algorithm [18], expansion by regions [19], elliptic integrals and related topics [20, 21], cutting techniques [22], and special multi-leg applications [23–25]. In different precision calculations these technologies are applied.[1]

In this paper we give a brief introduction into the topic,[2] covering the main steps in multi-loop perturbative calculations in Sect. 2. Then we turn to the different symbolic integration techniques of Feynman parameter integrals in Sect. 3 and describe the associated function spaces in Sect. 4. All these technologies serve the purpose to reach a higher theoretical precision for many observables in Quantum Field Theory to cope with the experimental precision data and to either confirm the Standard Model of elementary particles to higher accuracy or to find signals of new physics. Some aspects of this are discussed in Sects. 5 and 6 contains the conclusions.

[1] For a summary on recent massless calculations, see [26].

[2] For other recent surveys on integration methods for Feynman integrals see [27–32].

2 Principle Computation Steps for Feynman Diagrams

In any large scale calculation there is the need to generate the Feynman diagrams in an automated way. One of the important packages to provide this is `QGRAF` [33], for which the corresponding physics file containing the Feynman rules has to be provided. Necessary group-theoretic calculations, as e.g. the color algebra in QCD, can be carried out using the package `Color` [34]. Finally, there is the necessity to perform the Dirac- and Lorentz-algebra, in an efficient manner, which is provided by `Form` [35].

Next, the integration by parts reduction [36] has to be performed. For many processes up to three loops in QCD the Laporta algorithm [37] in its different implementations [5, 38–43] is sufficient. At even higher orders the complexity becomes larger and larger and it is necessary to combine different methods or to device algorithms tailored to the particular problem to be solved [5–7]. There will be certainly more developments in this field in the future.

After this reduction one obtains the master integrals, which represent the quantity to be finally calculated, and it has to be decided in which way the computation shall be put forward. One way, if it can be pursued, is to calculate the individual master integrals to the necessary depth in the dimensional parameter $\varepsilon = D - 4$, with D the dimension of the space-time using the analytic calculation technologies described in Sect. 3. At lower orders in the coupling constant up to moderate complexity in the involved mass scales this is possible. It may even be that a single technology, like the calculation of the master integrals by solving the associated differential equations provides the full solution.

However, in particular in massive calculations it is possible, that the master integrals contain elliptic parts at 3-loop order, but the quantity to be calculated is known to be free of these contributions. In such a case one may use the method of arbitrarily high Mellin moments for single scale quantities [44] to express the moments of the master integrals. Here the elliptic structures are fully encoded in just rational coefficients. One forms the observable to be calculated which are given as series of Mellin moments weighted by ζ- and color factors and here the elliptic or hyperelliptic contributions cancel. The method of guessing [45, 46] will then enable one to find the recurrence relation for the complete result. This method requires a large number of moments, which, however, can be algorithmically provided [44]. Recent applications are [47–50]. The solution of the recurrences provided by guessing is using difference ring theory [51] implemented in the package `Sigma` [52, 53] in the case that the recurrences factorize to first order. Other cases are discussed in Sects. 3 and 4.

Let us now turn to the specific antidifferentiation methods for Feynman integrals.

3 Symbolic Integration of Feynman Parameter Integrals

Most of the analytic methods described in this section have more general applications than just to be used for the evaluation of Feynman integrals and were developed even without knowing of this particular application. Still the challenges to integrate also involved Feynman diagrams have refined many of these methods significantly. Non of these methods is universal and it is often an appropriate combination of these methods leading to optimal solutions of a project at hand, w.r.t. the necessary computational requests such as memory, storage and computational time in the case of the high end calculations.

Many of the problems can be cast into discrete formulations, allowing to make use of methods of difference ring theory. Here the packages `Sigma` [52, 53], `EvaluateMultiSums` and `SumProduction` [54–56], see also [57], can be used.

In the following we describe the PSLQ method for zero-dimensional quantities, hypergeometric functions and their generalizations, analytic solutions using Mellin–Barnes integrals, hyperlogarithms, guessing techniques, the method of difference and differential equations, and the Almkvist-Zeilberger algorithm.

3.1 PSLQ: Zero-Dimensional Integrals

In expanding perturbatively in the coupling constant several physical quantities are zero-dimensional, i.e. they can be represented by numbers only. A recent example consists in the QCD β-function, now known to 5-loop order [58]. The respective expressions are given by the color factors of the gauge group, rational terms and special numbers, as e.g. multiple zeta values [4, 59]. If one knows the potential pool of all the contributing special numbers one may try to determine the rational coefficients of the whole problem by providing enough numerical digits for the corresponding quantity. One method to obtain such an experimental result is PSLQ [60].

One example is given by determining the integral

$$I_1 = \int_0^1 dx \frac{\text{Li}_2(x)}{1+x} \approx 0.38889584616810632909974350804769310098 85, \quad (1)$$

where $\text{Li}_2(x)$ denotes the classical dilogarithm [61]. I_1 is a weight $\mathsf{w} = 3$ multiple zeta value [4] for which the basis is known. It is spanned by

$$\{\ln^3(2), \zeta_2 \ln(2), \zeta_3\}. \quad (2)$$

PSLQ delivers the following representation

$$I_1 = \ln^3(2) - \frac{5}{8}\zeta_3, \tag{3}$$

which is also obtained by a direct analytic calculation. In any case it is important to have enough digits available. If a result has been obtained, it should be verified by an even larger number of digits. The results usually remain experimental. In many complex applications it is difficult to prove the result analytically. Advanced applications of these and similar methods are discussed in [10]. PSLQ methods can be used also to determine the constants discussed in Sect. 4.5.

3.2 *Generalized Hypergeometric Functions and Their Extensions*

The integrands of multi-dimensional Feynman parameter integrals are hyperexponential, i.e. given by products of multivariate polynomial expressions raised to real powers, implied by the dimensional parameter ε. These types of functions correspond to the integrands defining the (generalized) hypergeometric functions [62–64] and their generalizations such as the Appell-, Kampe-De-Feriet- and related functions [65–77]. The advantage of these integral representations is that they usually have a lower dimensional series representation compared to their integral representations and a part of the original integrals can be performed in this way. The simplest function is Euler's Beta-function implying the series of ${}_{p+1}F_p$ functions

$$B(a_1, a_2) = \int_0^1 dt\, t^{a_1-1}(1-t)^{a_2-1} \tag{4}$$

$$\begin{aligned} {}_3F_2(a_1, a_2, a_3; b_1, b_2; x) = {} & \frac{\Gamma(b_2)}{\Gamma(a_3)\Gamma(b_2-a_3)} \int_0^1 dt\, t^{a_3-1}(1-t)^{-a_3+b_2-1} \\ & \times {}_2F_1(a_1, a_2; b_1; tx). \end{aligned} \tag{5}$$

Up to the level of the massless and massive two-loop calculations for single-scale quantities in QCD these representations are usually sufficient [78–81]. In the case of three-loop ladder graphs also Appell-functions [82, 83] contribute. A survey on the status of this method has been given in [8]. In relating the different special functions of this kind contiguous relations play an essential role, which has been discussed in [9] in detail. One ends up with a series of infinite sum representations, for which the ε-expansion is performed. These sums have to be further dealt with by using summation methods, cf. Sect. 3.6.

3.3 The Analytic Mellin–Barnes Technique

Only the simpler hyperexponential integrands can be represented by the higher transcendental functions described in Sect. 3.2. One major problem to proceed are the structures of some of the hyperexponential factors, for which the contributing variables cannot be cast into a form required in the previous case. Here the use of Mellin–Barnes integrals [84, 85] is of help, which are defined by

$$\frac{1}{(a+b)^\alpha} = \frac{1}{\Gamma(\alpha)} \frac{1}{2\pi i} \int_{-i\infty}^{i\infty} dz \Gamma(\alpha+z)\Gamma(-z) \frac{b^z}{a^{\alpha+z}}, \quad \alpha \in \mathbb{R}, \alpha > 0, \quad (6)$$

cf. e.g. [86]. Here the contour integral is understood to be either closed to the left or the right surrounding the corresponding singularities. Note that also the functions in Sect. 3.2 have representations in terms of Pochhammer–Umlauf integrals [62, 87, 88] and therefore Mellin–Barnes integral representations. The different Mellin–Barnes integrals can be turned into a number of infinite series by the residue theorem, leading to nested sums to be dealt with further by the summation technologies implemented in the package `Sigma` [52, 53]. Here it is not a priori clear that all the sums can be solved, which will turn out by working through the algorithm. One is advised therefore not to use the Mellin–Barnes splitting of the integrands extensively, although being possible [89]. On the other hand, one may apply the packages for Mellin–Barnes integrals [90–93] to obtain numerical results for comparisons, to check the final analytic results. One reason that a summation problem cannot be solved completely is related to the fact, that the associated recurrences are not first order factorizing and other technologies have to be applied. Mellin–Barnes integrals do significantly extend the methods described in Sect. 3.2 and may lead to new higher transcendental functions not known yet in the literature.

3.4 Hyperlogarithms

The idea behind the method of hyperlogarithms is that for certain multivariate Feynman parameter integrals an order of integrations can be found in which the respective parameters to be integrated over always occur in linear form (Fubini sequence) [94]. In this way the corresponding integrals are cast into iterative Kummer–Poincaré integrals [95–99]. Originally the method could only be applied to non-singular integrals in the dimensional parameter ε and an extension has been worked out in [100] to the singular case. An implementation of the algorithm has been given in [101]. The method has first been applied to massless Feynman integrals. A generalization for massive integrals, also containing local operator insertions, has been given in [102, 103], where multi-linearity is broken in part, still yielding analytic results. The method is interesting but of limited use, since it applies to structurally simple cases only and requires more than just first order factorization of the associated differential equations or the related nested sum representations, through the application of which much more general cases can be solved.

3.5 *The Method of Guessing*

The integral transform

$$\mathbf{M}[f(x)](N) = \int_0^1 dx x^{N-1} f(x) \tag{7}$$

defines the Mellin transform, which will often appear in the following. If single-variate multiple Feynman parameter integrals $f(x)$ in QCD processes are viewed in terms of their Mellin moments $\{a(N)|_{N=1}^{\infty}\}$ for fixed values of N, one obtains series of rational numbers weighted by color factors and multiple zeta values or other special numbers, cf. [104–107]. It turns out in very many practical cases that the general N solution generating the individual moments $a(N)$ obey recursion relations. This is the case e.g. for (massive) operator matrix elements [108] but also for single-scale Wilson coefficients, [109].

One would like now to determine this recurrence on the basis of a (large) number of these moments algorithmically. The corresponding algorithms are called guessing methods [46], which are also available in `Sage` [110], exploiting the fast integer algorithms available there. The method returns the wanted difference equation, and tests it by a larger series of further moments. This method has been applied in [45] to obtain from more than 5000 moments the massless unpolarized three-loop anomalous dimensions and Wilson coefficients in deep-inelastic scattering [109, 111]. More recently, the method has been applied *ab initio* in the calculation of three-loop splitting functions [47, 49, 112] and the massive two- and three-loop form factor [113, 114]. The largest systems solved in this context were massive operator matrix elements needing $\sim$ 8000 moments [48] to derive the corresponding recurrences.

One then tries to solve these recurrences with the package `Sigma` [52, 53], which will either find the solution or does at least factor off all the first order factors, separating the remaining part to be solved using other techniques. The large number of moments needed is generated using the method described in [44]. This algorithm will play a central role in many upcoming calculations in the singlevariate case.

Other algorithms such as `Mincer` [115], `MATAD` [116] or `Q2E` [117, 118] do also provide Mellin moments. However, the number of moments which can be obtained with these formalisms is rather low. Still these packages play a very essential role in higher order calculations, since they provide independent tests and they are used both for predicting intermediary and final results for indispensable comparisons.

3.6 Difference Equations and Summation Methods

Many of the problems occurring in analytic Feynman integral calculation can be mapped to summation problems and the solution of difference equations. Infinite and finite sums appear in binomial and Mellin–Barnes decompositions and also in the expansion of Pochhammer symbols depending on the dimensional parameter ε into the associated Laurent series. Moreover, ordinary differential equations in a variable x can be transformed into recursions in the variable N by a Mellin transform [119]. Furthermore, structures in x-space can be expanded into formal Taylor series, the Nth coefficient of which, $a(N)$, also obeys a certain recurrence.

Nested sums over hypergeometric terms $h(k)$, with $h(k+1)/h(k) \in \mathbb{K}(k)$, for some field $\mathbb{K}$ that contains the rational numbers $\mathbb{Q}$, will form the basis of the summation problems we briefly consider in the following. One first considers finite sums, i.e. those terminating at an upper integer for all summation quantifiers. The corresponding sums are then cast into the form

$$S_{b,\mathbf{a}}(N;\mathbf{c}) = \sum_{k=1}^{N} h_b(k;\mathbf{c}) S_{\mathbf{a}}(k;\mathbf{c}). \tag{8}$$

Here $\{\mathbf{c}\}$ denotes a finite set of constants, which has to be added to the ground field of the difference ring and $h_b(k;\mathbf{c})$ is a hypergeometric term. The c_i's are also given by certain physical parameters in multi-scale processes. Infinite sums can be dealt with by considering limiting procedures implemented in the package `HarmonicSums` [120–128]. In solving a summation problem at hand the associated sum-algebra is built and the corresponding sums appearing in the final result are simplified accordingly.

All summation problems which lead to first order factorizing recurrences can be solved using difference ring theory [51] implemented in the package `Sigma` [52, 53]. This concerns a rather wide class of cases. By the systematic use of these techniques, harmonic sums [120, 121] generalized harmonic sums [122, 129], cyclotomic harmonic sums [123], and finite binomial and inverse-binomial sums [124] can be dealt with. A part of these function spaces has been found and systematically explored by these techniques. Recent developments in this field are summarized in [12].

These methods also apply to cases, which are not factorizing to first order, as e.g. in [48]. Here all first order factors are separated from a remainder non-factorizing recurrence. The latter one can be further dealt with using different techniques.

3.7 Differential Equations

The IBP-relations [36] do naturally imply systems of differential equations for the master integrals. In the case of single-scale quantities these are systems of ordinary differential equations, Eq. (9), which have to be solved, providing the necessary boundary conditions. Early investigations following this approach were [130–133].

$$\frac{d}{dx}\begin{pmatrix} f_1 \\ \vdots \\ f_n \end{pmatrix} = \begin{pmatrix} A_{11} & \dots & A_{1,n} \\ \vdots & & \vdots \\ A_{n1} & \dots & A_{n,n} \end{pmatrix}\begin{pmatrix} f_1 \\ \vdots \\ f_n \end{pmatrix} + \begin{pmatrix} g_1 \\ \vdots \\ g_n \end{pmatrix}, \tag{9}$$

One way to solve the system (9) consists in decoupling it using the methods of [134, 135] encoded in `Oresys` [136]. In this way one obtains one scalar differential equation of higher order

$$\sum_{k=0}^{n} p_{n-k}(x)\frac{d^{n-k}}{dx^{n-k}} f_1(x) = \overline{g}(x), \tag{10}$$

with $p_n \neq 0$, and $(n-1)$ equations for the remaining solutions, which are fully determined by the solution $f_1(x)$. One also may transform Eq. (9) into Mellin space, decouple there and solve using the efficient methods of the package `Sigma`, cf. [83].

In the case of first order factorization the decoupled differential operator of (10) can be written in form of a combination of iterative integrals, cf. Sect. 4.2,

$$f_1(x) = \sum_{k=1}^{n+1} \gamma_k g_k(x), \ \gamma_k \in \mathbb{C}, \tag{11}$$

$$g_k(x) = h_0(x) \int_0^x dy_1 h_1(y_1) \int_0^{y_1} dy_2 h_2(y_2) \dots \int_0^{y_{k-2}} dy_{k-1} h_{k-1}(y_{k-1}) \times \int_0^{y_{k-1}} dy_k q_k(y_k), \tag{12}$$

with $q_k(x) = 0$ for $1 \leq k \leq m$. Further, $\gamma_{m+1} = 0$ if $\bar{g}(x) = 0$ in (10), and $\gamma_{m+1} = 1$ and $q_{m+1}(x)$ being a mild variation of $\bar{g}(x)$ if $\bar{g}(x) \neq 0$.

Starting with master integrals appearing in quantum field theories one obtains linear differential equations with rational coefficients. Thus looking for d'Alembertian solutions [137] by factorizing the differential operator into linear factors, one finds all solutions that are expressible in terms of iterative integrals of the form (12) where the letters $h_i(x)$ are hyperexponential, i.e., $\frac{d}{dx}h_i(x)/h(x) \in \mathbb{K}$ for some field $\mathbb{K}$ containing $\mathbb{Q}$. Such solutions can be computed, e.g., with the package HarmonicSums.

Liouvillian solutions [138] can also be calculated with `HarmonicSums` utilizing Kovacic's algorithm [139]. This algorithm has been applied in various massive three-loop calculations so far, cf. [48, 83, 140]. A solution algorithm for first order systems has also been presented in [113]. In these algorithms no specific choice of a basis is necessary and the different contributions in the ε-expansion are obtained straightforwardly.

In the multivariate case, the so-called ε-representation of a linear system of partial differential equations

$$\partial_m f(\varepsilon, x_n) = A_m(\varepsilon, x_n) f(\varepsilon, x_n) \tag{13}$$

is important, as has been recognized in [141, 142], see also [143]. The matrices A_n can now be transformed in the non-Abelian case by

$$A'_m = B^{-1} A_m B - B^{-1}(\partial_m B), \tag{14}$$

as well-known [144, 145]. One then intends to find a matrix B to transform (13) into the form

$$\partial_m f(\varepsilon, x_n) = \varepsilon A_m(x_n) f(\varepsilon, x_n), \tag{15}$$

if possible. This then yields solutions in terms of iterative integrals. Here a formalism for the basis change to the ε-basis has been proposed in [146] and implemented in the singlevariate case in [147, 148] and in the multivariate case in [149]. All these methods apply only if the systems are first order factorizable.

In the solution of differential equations it is often important to rationalize roots as much as possible [16, 150, 151]. The corresponding algorithms are important for the derivation of iterated root-valued integrals also in the multivariate case [152, 153]. The corresponding structures do then allow expansions in small parameters to obtain even more compact analytic results, since normally the iterated integrals with various involved root-valued letters turn out to form very large expressions. Recent developments have been summarized in [16]. General aspects on the solution of differential equation systems were summarized in [14], while aspects of differential Galois theory were discussed in [13]. In the solution of ordinary differential equations emerging in the context of Feynman diagrams holonomic integration often provides a powerful tool, see [17].

3.8 *Multivalued Almkvist-Zeilberger Algorithm*

Singlevariate Feynman parameter integrals $I(x, \varepsilon)$ are integrals over $\{x_i|_{1=1}^n\} \in [0, 1]^n$ with one more free parameter $x \in [0, 1]$ and the dimensional parameter ε. A Mellin transform leads to the function $\hat{I}(N, \varepsilon)$. The Almkvist–Zeilberger algorithm [154, 155] provides a method to either find an associated differential equation for $I(x)$ or a difference equation for $I(N)$, the coefficients of which are either polynomials in $\{x, \varepsilon\}$ or $\{N, \varepsilon\}$,

$$\sum_{l=0}^{m} P_l(x, \varepsilon) \frac{d^l}{dx^l} I(x, \varepsilon) = N(x, \varepsilon) \tag{16}$$

$$\sum_{l=0}^{m} R_l(N, \varepsilon) I(N + l, \varepsilon) = M(N, \varepsilon). \tag{17}$$

Both equations may be inhomogeneous, where the inhomogeneities emerge as known functions from lower order problems. An optimized and improved algorithm for the input class of Feynman integrals has been implemented in the `MultiIntegrate` package [83, 127]. It can either produce homogeneous equations of the form (16, 17) or equations with an inhomogeneity formed out of already known functions. The method extends successively the structural form of the difference or differential equation for the functions $\hat{I}$ or I unless a solution is found.

This algorithm is of great use in specific cases in which either direct summation problems or the solution of associated differential equations becomes to voluminous or in the case of very long computation times. Like also in the case of guessing the corresponding recurrences turn out to be well homogeneized, which makes their solution easier.

4 The Function Spaces

The solution of the different massless and massive higher loop calculations for zero-, single-, and multiscale problems induce specific function spaces, which also form algebras. These structures have been revealed in more detail after 1997 along with performing more and more involved computations. Before it has been known that specific numbers play a role in Feynman integral calculations, see e.g. [156], and in the one-dimensional case Nielsen integrals [157], generalizing the classical polylogarithms [61], were in use.

The first generalizations of these functions led to nested sum structures on the one hand [120, 121], and iterative integrals over certain alphabets on the other hand [128]. These structures do also apply to not too involved multi-scale problems. Iterative non-iterative integrals occurred with the advent of complete elliptic integrals in letters, or more generally higher transcendental functions for whose integral representation the variable to be integrated over cannot be transformed in one of the integration boundaries only. Synonymous objects appear in the associated sums.

In the following we describe the hierarchies of spaces for iterated integrals and nested sums, which contribute in Feynman diagram calculations. Beyond these structures there are also problems leading to non first order factorizable recurrences and differential operators. In all calculations special numbers occur, which have representations by iterated integrals at $x = 1$ or through nested sums in the limit $N \to \infty$. Even others appear in the context of the quantities discussed in Sect. 4.4. Finally, we will also discuss numerical representations of all these functions, including the analytic continuation of nested sums to $N \in \mathbb{C}$.

4.1 Nested Sums

Considering the singlevariate case sum representations have the form given in Eq. (8). Very often finite sums of another type have first to be brought into this representation using the algorithms encoded in the package `Sigma` [52, 53]. Furthermore, also infinite sums have to be handled, which are usually considered as the limit $N \to \infty$ of the associated finite sums.

The sums obey quasi-shuffle relations [158, 159], see Sect. 4.3. The simplest structures are the finite harmonic sums [120, 121], where $g_b(k) = (\text{sign}(b))^k/k^{|b|}$, $b \in \mathbb{N}\backslash\{0\}$. A generalization is obtained in the cyclotomic case [123]. Here the characteristic summands are $g_{a,b,c}(k) = (\pm 1)^k/(ak+b)^c$, with $a, b, c \in \mathbb{N}\backslash\{0\}$. Further, the generalized harmonic sums have letters of the type b^k/k^c, with $c \in \mathbb{N}\backslash\{0\}, b \neq 0, b \in \mathbb{R}$, [122]. A generalization of the last two classes of sums are those generated by the Mellin transform of iterative integrals with letters induced by quadratic forms [160], see Sect. 4.2. Another generalization are nested finite binomial and inverse-binomial sums, containing also other sums discussed before. An example is given by

$$\sum_{i=1}^{N} \frac{1}{(2i+1)\binom{2i}{i}} \sum_{j=1}^{i} \binom{2j}{j} \frac{(-1)^j}{j^3} = \frac{1}{2} \int_0^1 \frac{(-x)^N - 1}{x+1} \frac{x}{\sqrt{x+\frac{1}{4}}} \mathrm{H}^*_{\mathsf{w}_{14},0,0}(x)$$
$$- \frac{\mathrm{H}^*_{-\frac{1}{4},0,0}(0)}{2\pi} \int_0^1 dx \frac{\left(\frac{x}{4}\right)^N - 1}{x-4} \frac{x}{\sqrt{1-x}}, \tag{18}$$

see [124]. Here the indices w_k label specific letters given in [124] and the iterated integrals H^* are defined over the support $[x, 1]$. Infinite binomial and inverse-binomial sums have been considered in [161, 162]. Given the general structure of (8) many more iterated sums can be envisaged and may still appear in even higher order calculations.

4.2 Iterated Integrals

Iterated integrals are of the form

$$\mathrm{H}_{b,\mathbf{a}}(x) = \int_0^x dy f_b(y) \mathrm{H}_{\mathbf{a}}(y), \quad \mathrm{H}_\emptyset = 1,\, f_c \in \mathfrak{A},\; x \in [0,1], \tag{19}$$

where f_c are real or complex-valued functions and are the letters of the alphabet $\mathfrak{A}$. For certain letters regularizations are required since otherwise the corresponding integrals do not exist. This occurs if the letters have poles in $x \in [0,1]$. Iterated integrals obey shuffle relations [159, 163] which allows one to represent them over a basis of fewer integrals; for further details the reader is referred to Sect. 4.3.

The simplest iterative integrals having been considered in quantum field theory are the Nielsen integrals for the two-letter alphabets $\{1/x, 1/(1-x)\}$ or $\{1/x, 1/(1+x)\}$ [157], covering also the polylogarithms [61]. This class has later been extended to the harmonic polylogarithms [128] built over the alphabet $\{1/x, 1/(1-x), 1/(1+x)\}$.

A further extension is to the real representations of the cyclotomic polylogarithms, with $\{1/x, 1/\Phi_k(x)\}$ [123], where $\Phi_k(x)$ denotes the kth cyclotomic polynomial. Another extension is given by Kummer–Poincaré iterative integrals over the alphabet $\{1/(x-a_i),\; a_i \in \mathbb{C}\}$, [95–99]. Properties of these functions have been studied in [122, 129]. One can avoid integrals defined over complex numbers by allowing the more flexible definition of integrals with denominator polynomials $P(x)$, which one can factor into

$$P(x) = \prod_{k=1}^{n}(x-a_k)\prod_{l=1}^{m}(x^2+b_l x + c_l), \quad a_k, b_l, c_l \in \mathbb{R} \tag{20}$$

in real representations. One then performs partial fractioning for $1/P(x)$ and forms iterative integrals out of the obtained letters in

$$\mathfrak{A}_R = \left\{ \frac{1}{x-a_i}, \frac{1}{x^2+b_i x + c_i}, \frac{x}{x^2+b_i x+c_i} \middle| a_i, b_i, c_i \in \mathbb{R},\; 4c_i \geq b_i \right\}, \tag{21}$$

cf. [160].

The iterated integrals (19) can be analytically continued from $x \in [0,1]$ to the complex plane by observing their respective cuts. For the harmonic polylogarithms this has been described in [164]. For the other cases the corresponding algorithm is implemented in `HarmonicSums`, see also [165].

Further classes are found for square-root valued letters as studied e.g. in [124]. In multi-scale problems, cf. e.g. [152, 153, 166], further root-valued letters appear, like also the Kummer-elliptic integrals [153], which are iterative integrals, because the elliptic structure is due to incomplete elliptic integrals.

The occurrence of several masses or additional external non-factorizing scales in higher order loop- and phase-space integrals leads in general to rational and root-valued letters with real parameter letters in the contributing alphabet, cf. [152, 153, 166]. In the case of the loop integrals one obtains letters of the kind

$$\frac{1}{1-x(1-\eta)}, \frac{\sqrt{x(1-x)}}{\eta+x(1-\eta)}, \sqrt{x(1-\eta(1-x)}, \quad \eta \in [0,1]. \tag{22}$$

The iterative integrals and constants which appeared in [166, 167] could finally be all integrated to harmonic polylogarithms containing complicated arguments, at least up to one remaining integration, which allows their straightforward numerical evaluation.

In the case of phase space integrals with more scales, e.g. [152], also letters contribute, which may imply incomplete elliptic integrals and iterated structures thereof. The integrands could not be rationalized completely by variable transformations, see also [150]. Contributing letters are e.g.

$$\frac{x}{\sqrt{1-x^2}\sqrt{1-k^2x^2}}, \frac{x}{\sqrt{1-x^2}\sqrt{1-k^2x^2(k^2(1-x^2(1-z^2))-z^2)}}, \tag{23}$$

with $k, z \in [0, 1]$. The corresponding iterative integrals are called Kummer-elliptic integrals. They are derived using the techniques described in [124, 168, 169].

4.3 General Properties of Nested Sums and Iterated Integrals

Iterated integrals obey shuffle relations

$$\begin{aligned} \mathrm{H}_{a_1,\ldots a_m}(z) \cdot \mathrm{H}_{b_1,\ldots b_n}(z) &= \mathrm{H}_{a_1,\ldots a_n}(z) ⧢ \mathrm{H}_{b_1,\ldots b_n}(z) \\ &= \sum_{c_i \in \{a_1,\ldots a_m ⧢ b_1,\ldots b_n\}} \mathrm{H}_{c_i}(z). \end{aligned} \tag{24}$$

Here the order of the letter sequences of the quantities to be shuffled is preserved. The associated algebras are called shuffle algebras [159, 163]. The counting of the basis elements in the respective class [170] may be done by counting its Lyndon words [171] or using the Witt-formulae [172].

Likewise, nested sums over hypergeometric terms form quasi-shuffle or stuffle [59] algebras [158]. The stuffling relations are obtained by [129, 159]

$$\begin{aligned} S_{a_1,\ldots,a_n}(N) \cdot S_{b_1,\ldots,b_m}(N) &= \sum_{l_1=1}^{N} \frac{\mathrm{sign}(a_1)^{l_1}}{l_1^{|a_1|}} S_{a_2,\ldots,a_n}(l_1) \cdot S_{b_1,\ldots,b_m}(l_1) \\ &+ \sum_{l_2=1}^{N} \frac{\mathrm{sign}(b_1)^{l_2}}{l_2^{|b_1|}} S_{a_1,\ldots,a_n}(l_2) \cdot S_{b_2,\ldots,b_m}(l_2) \\ &- \sum_{l=1}^{N} \frac{(\mathrm{sign}(a_1)\mathrm{sign}(b_1))^{l_2}}{l^{|a_1|+|b_1|}} S_{a_2,\ldots,a_n}(l) \cdot S_{b_2,\ldots,b_m}(l) \end{aligned} \tag{25}$$

for harmonic sums and similar for the sums in extended spaces, see [122–124, 126, 173, 174]. These algebraic relations allow one to reduce the number of contributing functions already significantly.

In the case of the iterated integrals different classes of mappings of the main argument may be used, which are helpful in many cases. The most important ones are

$$k \cdot z \to z, \ \ k \in \mathbb{Q}, \quad 1 - z \to z, \quad \frac{1}{z} \to z, \quad \frac{1-z}{1+z} \to z. \tag{26}$$

Depending on the class of iterative integrals to be considered, not all of these relations map inside this class, but can lead to functions in respective extensions. In these cases one just considers the wider space. Iterated integrals also obey the differentiation relation

$$\frac{d}{dz}\mathrm{H}_{b,\mathbf{a}}(z) = f_b(z)\mathrm{H}_{\mathbf{a}}(z). \tag{27}$$

Beyond the quasi-shuffle relations, also nested sums obey further relations if considering their analytic continuation to $N \in \mathbb{Q}, \mathbb{R}$ or $\mathbb{C}$, cf. [122–124, 126, 173, 174]. These relations are called structural relations, cf. [126]. The double- [4] and multiple arguments relations and the differential relations, applied to the associated Mellin transforms, belong to this class. The simplest double argument relation reads

$$S_{n_1,\ldots,n_p}(N) = 2^{n_1+n_2+\ldots n_p-p} \sum_{\pm} S_{\pm n_1,\ldots,\pm n_p}(2N)\,, \tag{28}$$

for the harmonic sums. Examples for the differential relation are

$$\frac{d}{dN}S_k(N) = \frac{(-1)^{k-1}}{(k-1)!}\psi^{(k)}(N+1) = -k(S_{k+1}(N) - \zeta_{k+1}) \tag{29}$$

$$\frac{d}{dN}S_{-k}(N) = \frac{(-1)^{k-1}}{(k-1)!}\beta^{(k)}(N+1) = -k\left[S_{-(k+1)}(N) + \left(1 - \frac{\zeta_{k+1}}{2^k}\right)\right], \tag{30}$$

with $\beta(N) = [\psi((N+1)/2) - \psi(N/2)]/2$. Therefore all single harmonic sums fall into a single equivalence class under differentiation for N, which is represented by the harmonic sum $S_1(N)$. The number of elements in the respective classes after applying the structural relations can also be counted by Witt-like formulae.

4.4 *Solutions in the Case of Non First Order Factorizable Recurrences and Differential Operators*

Non-first order factorizing systems of differential or difference equations for the master integrals, cf. Sect. 3.6, occur at a certain order in massive Feynman diagram calculations. Well-known examples for this are the sun-rise integral, cf. e.g. [175–181], the kite integral [182–184], the three-loop QCD-corrections to the ρ-parameter [185–187], and the three-loop QCD corrections to the massive operator matrix element A_{Qg} [48]. In the case of the ρ-parameter a Heun equation [188] remains after separating the first order factorizing terms. Its solution can be given in terms of ${}_2F_1$-functions with a certain rational argument [187, 189] and rational parameters. These structures will later turn out not to occur accidentally. Next one may investigate whether these solutions can be expressed in terms of complete elliptic integrals. This can be checked algorithmically using the triangle group [190].

In the examples mentioned one can find representations in terms of complete elliptic integrals of the first and second kind, **K** and **E**, cf. [191, 192]. Here the question arises whether an argument transformation allows for a representation through only **K**. It turns out that this in not possible in the present case according to the criteria given in [193, 194].

The homogeneous solution of the Heun equations are given by ${}_2F_1$-solutions $\psi_k^{(0)}(x), k = 1, 2$, at a specific rational argument. However, these integrals cannot be represented such that the variable x just appears in the boundaries of the integral. The inhomogeneous solution reads

$$\psi(x) = \psi_1^{(0)}(x)\left[C_1 - \int dx \psi_2^{(0)}(x)\frac{N(x)}{W(x)}\right] + \{1 \to 2\}, \tag{31}$$

with $N(x)$ and $W(x)$ the inhomogeneity and the Wronskian. $C_{1,2}$ are the integration constants. Through partial integration the ratio $N(x)/W(x)$ can be transformed into an iterative integral. Since $\psi_k^{(0)}(x)$ cannot be written as iterative integrals, $\psi(x)$ is obtained as an *iterative non-iterative integral* [187, 195] of the type

$$\mathbb{H}_{a_1,\ldots,a_{m-1};a_m,F_m(r(y_m)),a_{m+1},\ldots a_q}(x) = \tag{32}$$
$$\int_0^x dy_1 f_{a_1}(y_1) \int_0^{y_1} dy_2 \ldots \int_0^{y_{m-1}} dy_m f_{a_m}(y_m) F_m[r(y_m)] H_{a_{m+1},\ldots,a_q}(y_m),$$

with $r(x)$ a rational function and F_m a *non-iterative integral.* In general, usually more non-iterative integrals will appear in (32). F_m denotes *any* non-iterative integral, implying a very general representation, cf. [187].[3] In [197] an ε-form for the Feynman diagrams of elliptic cases has been found recently. Here transcendental letters contribute. This is in accordance with our earlier finding, Eq. (32), which, as well is an iterative integral over all objects between the individual iterations and to which now also the non-iterative higher transcendental functions $F_m[r(y_m)]$ contribute. One may obtain fast convergent representations of $\mathbb{H}(x)$ by overlapping series expansions around $x = x_0$ outside possible singularities, see [187] for details.

Now we return to the elliptic case. Here one one may transform the kinematic variable x occurring as $\mathbf{K}(k^2) = \mathbf{K}(r(x))$ into the variable $q = \exp[i\pi\tau]$ analytically with

$$k^2 = r(x) = \frac{\vartheta_2^4(q)}{\vartheta_3^4(q)}, \tag{33}$$

by applying a cubic order Legendre–Jacobi transformation, where $\vartheta_l, l = 1, \ldots, 4$ denote Jacobi's ϑ-functions and $\mathsf{Im}(\tau) > 0$. In this way Eq. (31) is rewritten in terms of the new variable. The integrands are given by products of meromorphic modular forms, cf. [198–200], which can be written as a linear combination of ratios of Dedekind's η-function

$$\eta(\tau) = q^{\frac{1}{12}} \prod_{k=1}^{\infty} (1 - q^{2k}) \,. \tag{34}$$

Depending on the largest multiplier $k \in \mathbb{N}$, k_m, of τ in the argument of the η-function, the solution transforms under the congruence subgroup $\Gamma_0(k_m)$ and one can perform Fourier expansions in q around the different cusps of the problem, cf. [201, 202].

For holomorphic modular forms, one obtains representations in Eisenstein series with character, while in the meromorphic case additional η-factors in the denominators are present. In the former case the q-integrands can be written in

[3] This representation has been used in a more specific form also in [196] later.

terms of elliptic polylogarithms in the representation [179, 180]

$$\mathrm{ELi}_{n,m}(x, y) = \sum_{k=1}^{\infty}\sum_{l=1}^{\infty} \frac{x^k}{k^n}\frac{y^l}{l^m} q^{kl} \tag{35}$$

and products thereof, cf. [180]. The corresponding q-integrals can be directly performed. The solution (31) usually appears for single master integrals. Other master integrals are obtained integrating further other letters, so that finally representations by $\mathbb{H}(x)$ occur. Iterated modular forms, resp. Eisenstein series, have been also discussed recently in [203–205].

Returning to the example of the ρ-parameter we find that it cannot be represented in terms of elliptic polylogarithms only because of the emergence of the complete elliptic integral **E**, for which the singularity in q implies Dedekind η-functions appearing in the denominator. These factors have no (known) closed form q-expansion, cf. [206]. Let us also remark that the corresponding non-iterative solutions are sometimes found mapping first into the non-physical region. In the end one has to perform an analytic continuation back to the physical case, which requires to have closed form expressions. Recent developments in the field of Feynman integrals and elliptic structures are discussed in [20, 21].

Let us mention that in some applications also non-factorizable differential equations of 3rd order and higher can occur. The higher the order the less is known about the analytic structure of the solutions in the general case. In the future one will be confronted with these cases and practical solutions for them have to be found, including highly precise numerical representations in the physical cases. This issue is presently under study.

4.5 Spaces of Special Numbers

For the sums of Sect. 4.1 which are convergent in the limit $N \to \infty$ and the iterated integrals of Sect. 4.2 which can be evaluated at $x = 1$ one obtains two sets of special numbers. They span the solution spaces for zero-scale quantities and appear as boundary values for single-scale problems. Examples for these special numbers are the multiple zeta values [4], associated to the harmonic sums and harmonic polylogarithms, special generalized numbers [122] like $\mathrm{Li}_2(1/3)$ and $\mathrm{Li}_k(-1/2)$, cf. [160], associated to generalized sums and to Kummer–Poincaré iterated integrals, special cyclotomic numbers [123] like Catalan's number, special binomial numbers [124], as e.g. $\mathrm{arccot}(\sqrt{7})$, and special constants in the elliptic case [187, 207]. The latter numbers are given by integrals involving complete elliptic integrals at special rational arguments and related functions. In general these numbers obey more relations than the finite sums and iterated integrals. One may use the PSLQ-method to get a first information on relations between these numbers occurring in a given problem and prove the conjectured relations afterwards.

4.6 Numerical Representations

Physical observables based on single scale quantities can either be represented in Mellin N-space or x-space. Many of these representations are given in terms of either nested sums or iterative integrals. However, there are also contributions due to iterative non-iterative integrals.

Representations in Mellin N-space allow the exact analytic solution of evolution equations [208] and scheme-invariant evolution equations can be derived in this way [209, 210]. The x-space representation is then obtained by a single numerical integral around the singularities of the respective quantity for $N \in \mathbb{C}$, cf. [208], requiring to know the complex representation of the integrand in N-space.

In the case of harmonic sums semi-numerical representations were given in [211, 212]. Furthermore, it is known that the basic harmonic sums, except of $S_1(N)$, which is represented by the Digamma function, and its polynomials, have a representation by factorial series [213, 214], which has been used in [126, 173] for their asymptotic representation, see also [215].

The asymptotic representation of these quantities is thus given to arbitrary precision and one may use the recurrence relations of these quantities to analytically continue (8) from integer values of N to $N \in \mathbb{C}$. Here it is important to observe the crossing relations for the respective process [216, 217] which either implies the analytic continuation from the even or from the odd integers. These steps also apply to the other types of sums which were described in [122–124, 160] analogously, appearing in certain physical problems, cf. [83, 102].

In the case that the corresponding relations are not given in tabulated form, they can be calculated using the package `HarmonicSums`. Relations for harmonic sums are also implemented in `summer` [120], and for generalized harmonic sums in `nestedsums` [218], `Xsummer` [219], and `PolyLogTools` [220].

In other applications one may want to work in x-space directly. Here numerical representations are available for the Nielsen integrals [157], the harmonic polylogarithms [164, 165, 221, 222], the Kummer–Poincaré iterative integrals [165], the cyclotomic harmonic polylogarithms [113], and those implied by quadratic forms [160]. These representations are also useful to lower the number of numerical integrations for more general problems, e.g. in the multivariate case. The relations for the corresponding quantities are implemented for the harmonic polylogarithms in [128, 221] and for all iterative integrals mentioned, including general iterative integrals, in the package `HarmonicSums`. Moreover, the packages `summer` [120], the multiple zeta values data mine [4], and `HarmonicSums` also provide extensive lists of special numbers in various tabulated basis representations allowing quick numerical evaluation. Dynamical numerical evaluations are provided by the package described in [165] and by `HarmonicSums` for non tabulated cases.

Finally, we remark that numerical evaluations of a series of elliptic integral solutions were given in [223, 224].

5 Precision Goals in Testing the Standard Model

I would like to finally discuss the application of the mathematical methods described to precision predictions for dedicated observables in Quantum Field Theory, which are measured at high precision both in low energy experiments and at present and future colliders.

At low energies central quantities are $(g-2)_{e,\mu}$ [225], for which the $O(\alpha^5)$ contributions have been computed numerically [226] and the $O(\alpha^4)$ contributions and parts of $O(\alpha^5)$ terms have been calculated analytically [207, 227–231]. For various years there is a discrepancy between the experimental data and the theoretical prediction. For massive calculations also the on-shell renormalization and decoupling constants are important. At present highest loop order they were given in [232].

In massless QCD the level of 3-loop corrections for the anomalous dimensions and Wilson coefficients has been reached [48, 49, 109, 111, 112]. The corrections to the β-function [58] are available at five loop order. The heavy flavor corrections to deep-inelastic structure functions reached the level of 3-loop corrections [48, 108, 166, 167, 233] and are on the way to be completed. Furthermore, there are also analytic 3-loop corrections to the inclusive Higgs production rate [234] and the Drell-Yan process [235, 236], while the NNLO corrections for the $t\bar{t}$-production in hadronic collisions [237] has been computed numerically, because of the presence of more involved integrals, still to be solved analytically.

All these processes are essential to pin down the accuracy of the parton distribution functions in the region of a clear twist-2 dominance [238], also accounting for jet production cross sections in $pp \rightarrow Z+$jet at NNLO [239, 240] and in ep two-jet production [241].

The final goal is here the precision measurement of the strong coupling constant $\alpha_s(M_Z^2)$ in a widely unique manner. This can also be achieved using the method of scheme-invariant evolution equations [242] for which the initial conditions are measured directly. Furthermore, one would like to determine at least the charm quark mass, m_c [243], in a correlated way with the parton densities and $\alpha_s(M_Z^2)$. Here one wants to reach relative accuracies of the order of 1% and better. It finally may be necessary to study QCD evolution at the level of N^{3}LO [244], in particular if one wants to include small x effects and check the analytic predictions of the BFKL formalism [245].

Facing future colliders such as the FCC_ee [3] a measurement of the fine structure constant $\alpha(M_Z^2)$ is possible at very high accuracy by using the forward–backward asymmetry [246] and one needs to know precision predictions on the QED radiative corrections [247]. This measurement may yield an independent access to the size of the hadronic contributions to α. In studying the Z resonance at the FCC_ee one expects precisions of $\sim$ 100 keV for M_Z and the width of the Z boson, Γ_Z, which requires refined QED corrections. Those for the initial state radiation have been calculated in [152] using a wide host of methods described in this article. There are more goals, as e.g. the precision understanding of the top-threshold, cf. e.g. [248–254], the measurement of $\sin^2\theta_W$ [255], and the precision measurement of the W-boson mass.

Beyond the more inclusive measurements we have described, there is a large list of hard exclusive reactions needed in the analysis of the experimental data at the LHC and at future colliders. These corrections require a lot more numerical technologies, because of the number of different scales present. For a recent summary of the status see [256].

Effective field methods can also be applied to classical gravity to derive higher order post-Newtonian corrections for the inspiraling process of two massive objects. With these methods currently the level of the 5th post-Newtonian order has been reached [257].

6 Conclusions

With the progress in analytic precision calculations in Quantum Field Theory more and more mathematical technologies are used to solve the corresponding integrals analytically. The classical and Nielsen polylogarithms turned out to be not sufficient any more to represent intermediary and the final results in the late 1990ies. Moreover, the method of hypergeometric functions, which has fully provided the corresponding integral representations up this point failed to cover more involved structures. This applied already to massless and massive calculations for single scale quantities in QCD at 3-loop order. The first indication for this was that the arguments appearing in the 2-loop Nielsen integral representations became more and more complicated. At this time it has also been discovered that Mellin-space representations lead to essential compactifications [120, 121] and later it turned out that all the single-scale 2-loop results can be written in terms of just six harmonic sums [174, 258, 259].

The iterative integral structure has been known from the classical polylogarithms [61] and Nielsen integrals [157] and led to the harmonic polylogarithms [128]. During the following years more and more of these structures have been revealed. Here the difference ring techniques [12, 51–53] played an essential role, since the corresponding structures were found constructively, mostly in massive calculations. The Mellin transform of these quantities allowed then to find the associated iterated integrals.

For some years now also iterative non-iterative integrals are known and were widely studied in the case of complete elliptic integrals. However, we expect more involved structures to emerge. Possible new structures of this kind could be Abel-integrals [260], integrals related to K3-surfaces [261], and Calabi–Yau structures [262].

In massive calculations [48, 50] we observe a growing number of non first order factorizable recurrences, probably already containing structures beyond the elliptic level. Their solution calls for a general method, which might provide semi-analytic numerical representations in the end, which can be tuned to any precision. Yet one is also interested in the concrete mathematical structures of these cases. Global methods like the recurrences or the method of differential equations will have a hard time to reveal those. It is rather important here to analyze the multidimensional integrands first, which is provided by applying cutting techniques in a systematic manner, performing various Hilbert-transforms [263–265]. This has been successfully practiced at one-loop order, see e.g. [266], and also revealed in a nice manner the emergence of elliptic integrals.[4] The method has been advocated early by M. Veltman in his PhD thesis [267], see also [268].

In the future we will see an intense cooperation of theoretical physicists, mathematicians and computer scientists working on large scale computer algebra on the topic of the calculation of Feynman integrals by antidifferentiation. The field will conquer new horizons, one never thought of. All the participating fields will enormously profit from this work and new masterpieces of the esprit humain will be seen.

Acknowledgments I would like to thank to all who have contributed to the present volume and all my colleagues with whom I have had countless fruitful discussions on the present topic during the last 30 years. This project has received funding from the European Union's Horizon 2020 research and innovation programme under the Marie Skłodowska–Curie grant agreement No. 764850, SAGEX.

References

1. M. Veltman, Facts and Mysteries in Elementary Particle Physics (World Scientific, Singapore, 2003)
2. The ILC: https://en.wikipedia.org/wiki/International_Linear_Collider
   J.A. Aguilar-Saavedra et al. [ECFA/DESY LC Physics Working Group], *TESLA: The Superconducting electron positron linear collider with an integrated x-ray laser laboratory. Technical design report. Part 3. Physics at an e^+e^- linear collider*, hep-ph/0106315;
   E. Accomando et al. [ECFA/DESY LC Physics Working Group], Phys. Rept. **299**, 1–78 (1998). [hep-ph/9705442];
   The Future Circular Collider,
   https://en.wikipedia.org/wikiFuture_Circular_Collider;
   TH FCC-ee design study, http://tlep.web.cern.ch;
   A. Abada et al. [FCC], Eur. Phys. J. ST **228**(4), 755–1107 (2019);
   J.L. Abelleira Fernandez et al. [LHeC Study Group], J. Phys. G **39**, 075001 (2012). [arXiv:1206.2913 [physics.acc-ph]];
   D. Boer, M. Diehl, R. Milner, R. Venugopalan, W. Vogelsang, D. Kaplan, H. Montgomery, S. Vigdor, A. Accardi, E.C. Aschenauer, et al., *Gluons and the quark sea at high energies: Distributions, polarization, tomography*. [arXiv:1108.1713 [nucl-th]]
3. A. Abada et al. [FCC], Eur. Phys. J. ST **228**(2), 261–623 (2019)
4. J. Blümlein, D. Broadhurst, J.A.M. Vermaseren, Comput. Phys. Commun. **181**, 582–625 (2010). [arXiv: 0907.2557 [math-ph]]

[4]To see the same on the basis of a Heun differential equation is much more difficult [187].

5. P. Marquard, *Integration-by-Parts: A Survey*, contribution to this volume
6. J. Vermaseren, *Some Steps Towards Improving IBP Calculations and Related Topics*, contribution to this volume
7. H. Frellesvig, *Top-Down Decomposition: A Cut-Based Approach to Integral Reductions*, contribution to this volume
8. M. Kalmykov, V. Bytev, B.A. Kniehl, S.O. Moch, B.F.L. Ward, S.A. Yost, *Hypergeometric Functions and Feynman Diagrams*. arXiv:2012.14492 [hep-th]
9. P. Paule, *Contiguous Relations and Creative Telescoping*, contribution to this volume
10. K. Acres, D. Broadhurst, *Empirical determinations of Feynman integrals using integer relation algorithms*. [arXiv:2103.06345 [hep-ph]]
11. E. Panzer, talk at this workshop, contribution not received
12. C. Schneider, *Term Algebras, Canonical Representations and Difference Ring Theory for Symbolic Summation*, contribution to this volume
13. T. Dreyfus, J.-A. Weil, *Differential Galois Theory and Integration*, contribution to this volume
14. A.V. Kotikov, *Differential equations and Feynman integrals*. [arXiv:2102.07424 [hep-ph]]
15. J. Henn, talk at this workshop, contribution not received
16. C.G. Raab, *Nested Integrals and Rationalizing Transformations*, contribution to this volume
17. C. Koutschan, *Holonomic Anti-Differentiation and Feynman Amplitudes*, contribution to this volume
18. J. Ablinger, *Extensions of the AZ-algorithm and the Package MultiIntegrate*, [arXiv:2101.11385 [cs.SC]], contribution to this volume
19. V.A. Smirnov, *Expansion by Regions: An Overview*, contribution to this volume
20. S. Weinzierl, *Iterated integrals related to Feynman integrals associated to elliptic curves*, [arXiv:2012.08429 [hep-th]]
21. J. Broedel, A. Kaderli, *A Geometrical Framework for Amplitude Recursions: Bridging Between Trees and Loops*, contribution to this volume
22. D. Kreimer, *Outer Space as a Combinatorial Backbone for Cutkosky Rules and Coactions*, [arXiv:2010.11781 [hep-th]], contribution to this volume
23. J.L. Bourjaily, Y.-H. He, A.J. McLeod, M. Spradlin, C. Vergu, M. Volk, M. von Hippel, M. Wilhelm, *Direct Integration for Multi-leg Amplitudes: Tips, Tricks, and When They Fail*, contribution to this volume
24. J. Bartels, *N=4 SYM Gauge Theories: the* $2 \to 6$ *Amplitude in the Regge Limit*, contribution to this volume
25. G. Papathanasiou, talk at this workshop, contribution not received
26. S. Moch, V. Magerya, *Calculating Four-Loop Corrections in QCD*, contribution to this volume
27. S. Weinzierl, *Introduction to Feynman Integrals*, arXiv:1005.1855 [hep-ph]
28. J. Ablinger, J. Blümlein, C. Schneider, J. Phys. Conf. Ser. **523**, 012060 (2014). [arXiv: 1310.5645 [math-ph]]
29. J. Ablinger, J. Blümlein, in *Computer Algebra in Quantum Field Theory. Integration, Summation and Special Functions*, ed. by C. Schneider, J. Blümlein (Springer, Wien, 2012), pp. 1–32 [arXiv: 1304.7071 [math-ph]]
30. S. Weinzierl, in *Computer Algebra in Quantum Field Theory. Integration, Summation and Special Functions*, ed. by C. Schneider, J. Blümlein (Springer, Wien, 2012), pp. 381–406 [arXiv:1301.6918 [hep-ph]]
31. C. Duhr, in *Journeys Through the Precision Frontier: Amplitudes for Colliders*, ed. by L. Dixon, F. Petriello. 2014 TASI Lectures (World Scientific, Singapore, 2015), pp. 419–476. [arXiv:1411.7538 [hep-ph]]
32. J. Blümlein, C. Schneider, Int. J. Mod. Phys. A **33**(17), 1830015 (2018). [arXiv:1809.02889 [hep-ph]].
33. P. Nogueira, J. Comput. Phys. **105**, 279–289 (1993)
34. T. van Ritbergen, A. Schellekens, J.A.M. Vermaseren, Int. J. Mod. Phys. **A14**, 41–96 (1999). [hep-ph/9802376]

35. J.A.M. Vermaseren, *New features of FORM*. math-ph/0010025;
M. Tentyukov, D. Fliegner, M. Frank, A. Onischenko, A. Retey, H.M. Staudenmaier, J.A.M. Vermaseren, AIP Conf. Proc. **583**(1), 202 (2002). [cs/0407066 [cs-sc]];
M. Tentyukov, J.A.M. Vermaseren, Comput. Phys. Commun. **181**, 1419–1427 (2010). [hep-ph/0702279];
B. Ruijl, U. Takahiro, J.A.M. Vermaseren, *FORM version 4.2*. arXiv:1707.06453[hep-ph]
36. J. Lagrange, *Nouvelles recherches sur la nature et la propagation du son*, Miscellanea Taurinensis, t. II, 1760-61; Oeuvres t. I, p. 263;
C.F. Gauss, Theoria attractionis corporum sphaeroidicorum ellipticorum homogeneorum methodo novo tractate, Commentationes societas scientiarum Gottingensis recentiores, Vol III, 1813, Werke Bd. **V** pp. 5–7;
G. Green, *Essay on the Mathematical Theory of Electricity and Magnetism*, Nottingham, 1828 [Green Papers, pp. 1–115];
M. Ostrogradski, Mem. Ac. Sci. St. Peters., **6**, 129–133 (1831);
K.G. Chetyrkin, F.V. Tkachov, Nucl. Phys. **B192**, 159–204 (1981)
37. S. Laporta, Int. J. Mod. Phys. **A15**, 5087–5159 (2000). [hep-ph/0102033]
38. A. Smirnov, JHEP **10**, 107 (2008). [arXiv: 0807.3243[hep-ph]]
39. A.V. Smirnov, F.S. Chuharev, *FIRE6: Feynman Integral REduction with Modular Arithmetic*. arXiv:1901.07808 [hep-ph]
40. C. Studerus, Comput. Phys. Commun. **181**, 1293–1300 (2010). [arXiv: 0912.2546 [physics.comp-ph]]
41. A. von Manteuffel, C. Studerus, *Reduze 2 - Distributed Feynman Integral Reduction*. arXiv:1201.4330 [hep-ph]
42. P. Marquard, D. Seidel, *The* `Crusher` *algorithm* (unpublished)
43. P. Maierhöfer, J. Usovitsch, P. Uwer, Comput. Phys. Commun. **230**, 99–112 (2018). [arXiv:1705.05610 [hep-ph]];
J. Klappert, F. Lange, P. Maierhöfer, J. Usovitsch, [arXiv:2008.06494 [hep-ph]]
44. J. Blümlein, C. Schneider, Phys. Lett. **B771**, 31–36 (2017). [arXiv: 1701.04614 [hep-ph]]
45. J. Blümlein, M. Kauers, S. Klein, C. Schneider, Comput. Phys. Commun. **180**, 2143–2165 (2009). [arXiv:0902.4091[hep-ph]]
46. M. Kauers, M. Jaroschek, F. Johansson, in *Computer Algebra and Polynomials*, ed. by J. Gutierrez, J. Schicho, Josef, M. Weimann. Lecture Notes in Computer Science, vol. 8942 (Springer, Berlin, 2015), pp. 105–125. [arXiv:1306.4263 [cs.SC]]
47. J. Ablinger, A. Behring, J. Blümlein, A. De Freitas, A. von Manteuffel, C. Schneider, Nucl. Phys. B **922**, 1–40 (2017). [arXiv:1705.01508 [hep-ph]]
48. J. Blümlein, J. Ablinger, A. Behring, A. De Freitas, A. von Manteuffel, C. Schneider, C. Schneider, in *PoS (QCDEV2017)*, 031 (2017). [arXiv:1711.07957 [hep-ph]]
49. A. Behring, J. Blümlein, A. De Freitas, A. Goedicke, S. Klein, A. von Manteuffel, C. Schneider, K. Schönwald, Nucl. Phys. B **948**, 114753 (2019). [arXiv:1908.03779 [hep-ph]]
50. J. Blümlein, P. Marquard, N. Rana, C. Schneider, Nucl. Phys. B **949**, 114751 (2019). [arXiv:1908.00357 [hep-ph]]
51. M. Karr, J. ACM **28**, 305–350 (1981);
M. Bronstein, J. Symbolic Comput. **29**(6), 841–877 (2000);
C. Schneider, *Symbolic Summation in Difference Fields*, Ph.D. Thesis RISC, Johannes Kepler University, Linz technical report 01–17 (2001);
C. Schneider, An. Univ. Timisoara Ser. Mat.-Inform. **42**, 163–179 (2004);
C. Schneider, J. Differ. Equ. Appl. **11**, 799–821 (2005);
C. Schneider, Appl. Algebra Engrg. Comm. Comput. **16**, 1–32 (2005);
C. Schneider, J. Algebra Appl. **6**, 415–441 (2007);
C. Schneider, Clay Math. Proc. **12**, 285–308 (2010). [arXiv:0904.2323 [cs.SC]];
C. Schneider, Ann. Comb. **14**, 533–552 (2010). [arXiv:0808.2596];
C. Schneider, in *Computer Algebra and Polynomials, Applications of Algebra and Number Theory*, ed. by J. Gutierrez, J. Schicho, M. Weimann. Lecture Notes in Computer Science (LNCS) **8942**, 157–191 (2015). [arXiv:1307.7887 [cs.SC]];

C. Schneider, J. Symb. Comput. **72**, 82–127 (2016). [arXiv:1408.2776 [cs.SC]]. C. Schneider, J. Symb. Comput. **80**, 616–664 (2017). [arXiv:1603.04285 [cs.SC]]. S.A. Abramov, M. Bronstein, M. Petkovšek, Carsten Schneider, J. Symb. Comput. **107**, 23–66 (2021). [arXiv:2005.04944 [cs.SC]]
52. C. Schneider, Sém. Lothar. Combin. **56**, 1–36 (2007) article B56b.
53. C. Schneider, in *Computer Algebra in Quantum Field Theory: Integration, Summation and Special Functions*, ed. by C. Schneider, J. Blümlein. Texts and Monographs in Symbolic Computation (Springer, Wien, 2013), pp. 325–360. [arXiv:1304.4134 [cs.SC]]
54. J. Ablinger, J. Blümlein, S. Klein, C. Schneider, Nucl. Phys. Proc. Suppl. **205-206**, 110–115 (2010). [arXiv:1006.4797[math-ph]]
55. J. Blümlein, A. Hasselhuhn, C. Schneider, in *PoS (RADCOR2011)*, 032 (2012). [arXiv: 1202.4303 [math-ph]]
56. C. Schneider, J. Phys. Conf. Ser. **523**, 012037 (2014). [arXiv:1310.0160[cs.Sc]]
57. C. Krattenthaler, C. Schneider, in *Algorithmic Combinatorics: Enumerative Combinatorics, Special Functions and Computer Algebra*, ed. by V. Pillwein, C. Schneider (Springer, Wien, 2020), pp. 249–296
58. P.A. Baikov, K.G. Chetyrkin, J.H. Kühn, Phys. Rev. Lett. **118**(8), 082002 (2017). [arXiv:1606.08659 [hep-ph]];
F. Herzog, B. Ruijl, T. Ueda, J.A.M. Vermaseren, A. Vogt, JHEP **02**, 090 (2017). [arXiv:1701.01404 [hep-ph]];
T. Luthe, A. Maier, P. Marquard, Y. Schröder, JHEP **10**, 166 (2017). [arXiv:1709.07718 [hep-ph]];
K. G. Chetyrkin, G. Falcioni, F. Herzog, J. A. M. Vermaseren, JHEP **10**, 179 (2017). arXiv:1709.08541 [hep-ph]
59. J.M. Borwein, D.M. Bradley, D.J. Broadhurst, and P. Lisonek, Trans. Am. Math. Soc. **353**, 907–941 (2001). [math/9910045]
60. H.R.P. Ferguson, D.H. Bailey, *A Polynomial Time, Numerically Stable Integer Relation Algorithm*, RNR Techn. Rept. RNR-91-032, Jul. 14, 1992
61. A. Devoto, D.W. Duke, Riv. Nuovo Cim. **7N6**, 1–39 (1984);
L. Lewin, *Dilogarithms and Associated Functions*, (Macdonald, London, 1958); L. Lewin, *Polylogarithms and Associated Functions* (North Holland, New York, 1981)
62. F. Klein, *Vorlesungen über die hypergeometrische Funktion*, Wintersemester 1893/94, Die Grundlehren der Mathematischen Wissenschaften, vol. 39 (Springer, Berlin, 1933)
63. W.N. Bailey, *Generalized Hypergeometric Series* (Cambridge University Press, Cambridge, 1935)
64. L.J. Slater, *Generalized Hypergeometric Functions* (Cambridge University Press, Cambridge, 1966)
65. P. Appell, J. Kampé de Fériet, *Fonctions Hypergéométriques et Hypersphériques, Polynomes D' Hermite* (Gauthier-Villars, Paris, 1926)
66. P. Appell, *Les Fonctions Hypergëométriques de Plusieur Variables* (Gauthier-Villars, Paris, 1925)
67. J. Kampé de Fériet, *La fonction hypergëométrique* (Gauthier-Villars, Paris, 1937)
68. H. Exton, *Multiple Hypergeometric Functions and Applications* (Ellis Horwood, Chichester, 1976)
69. H. Exton, *Handbook of Hypergeometric Integrals* (Ellis Horwood, Chichester, 1978)
70. M.J. Schlosser, in *Computer Algebra in Quantum Field Theory: Integration, Summation and Special Functions*, ed. by C. Schneider, J. Blümlein, (Springer, Wien, 2013), pp. 305–324. arXiv:1305.1966 [math.CA]
71. C. Anastasiou, E.W.N. Glover, C. Oleari, Nucl. Phys. **B572**, 307–360 (2000). [hep--ph/9907494]
72. C. Anastasiou, E.W.N. Glover, C. Oleari, Nucl. Phys. **B565**, 445–467 (2000). [hep--ph/9907523]
73. H.M. Srivastava, P.W. Karlsson, *Multiple Gaussian Hypergeometric Series* (Ellis Horwood, Chicester, 1985)

74. G. Lauricella, Rediconti del Circolo Matematico di Palermo **7**(S1), 111–158 (1893)
75. S. Saran, Ganita **5**, 77–91 (1954)
76. S. Saran, Acta Math. **93**, 293–312 (1955)
77. J. Blümlein, M. Saragnese, C. Schneider, *Hypergeometric Structures in Feynman Integrals*, DESY 21–071
78. R. Hamberg, W.L. van Neerven, T. Matsuura, Nucl. Phys. **B359**, 343–405 (1991); [Erratum: Nucl. Phys. **B644**, 403–404 (2002)]
79. R. Hamberg, *Second order gluonic contributions to physical quantities*, Ph.D. Thesis, Leiden University, 1991
80. M. Buza, Y. Matiounine, J. Smith, R. Migneron, W.L. van Neerven, Nucl. Phys. **B472**, 611–658 (1996). [hep-ph/9601302]
81. I. Bierenbaum, J. Blümlein, S. Klein, Nucl. Phys. **B780**, 40–75 (2007). [hep-ph/0703285]
82. J. Ablinger, J. Blümlein, A. Hasselhuhn, S. Klein, C. Schneider, F. Wißbrock, Nucl. Phys. **B864**, 52–84 (2012). [arXiv:1206.2252[hep-ph]]
83. J. Ablinger, A. Behring, J. Blümlein, A. De Freitas, A. von Manteuffel, C. Schneider, Comput. Phys. Commun. **202**, 33–112 (2016). [arXiv:1509.08324 [hep-ph]]
84. E.W. Barnes, Q. J. Math. **41**, 136–140 (1910)
85. H. Mellin, Math. Ann. **68**(3), 305–337 (1910)
86. V.A. Smirnov, *Feynman Integral Calculus* (Springer, Berlin, 2006)
87. L. Pochhammer, Math. Ann. **35**, 495–526 (1890)
88. A. Kratzer, W. Franz, *Transzendente Funktionen* (Geest & Portig, Leipzig, 1960)
89. J. Blümlein, S. Klein, C. Schneider, F. Stan, J. Symb. Comput. **47**, 1267–1289 (2012). arXiv:1011.2656 [cs.SC]
90. M. Czakon, Comput. Phys. Commun. **175**, 559–571 (2006). [hep-ph/0511200]
91. A. Smirnov, V. Smirnov, Eur. Phys. J. **C62**, 445–449 (2009). [arXiv:0901.0386 [hep-ph]]
92. J. Gluza, K. Kajda, T. Riemann, Comput. Phys. Commun. **177**, 879–893 (2007). [arXiv:0704.2423[hep-ph]]
93. J. Gluza, K. Kajda, T. Riemann, V. Yundin, Eur. Phys. J. **C71**, 1516 (2011). [arXiv:1010.1667[hep-ph]]
94. F. Brown, Commun. Math. Phys. **287**, 925–958 (2009). arXiv:0804.1660 [math.AG]
95. E.E. Kummer, J. Reine Angew. Math. (Crelle) **21**, 74–90 (1840); 193–225; 328–371
96. H. Poincaré, Acta Math. **4**, 201–312 (1884)
97. J.A. Lappo-Danilevsky, *Mémoirs sur la Théorie des Systèmes Différentielles Linéaires* (Chelsea Publ. Co, New York, 1953)
98. K.T. Chen, Trans. A.M.S. **156**(3), 359–379 (1971)
99. A.B. Goncharov, Math. Res. Lett. **5**, 497–516 (1998)
100. A. von Manteuffel, E. Panzer, R.M. Schabinger, JHEP **02**, 120 (2015). [arXiv:1411.7392[hep-ph]]
101. E. Panzer, Comput. Phys. Commun. **188**, 148–166 (2015). [arXiv:1403.3385[hep-th]]
102. J. Ablinger, J. Blümlein, C. Raab, C. Schneider, F. Wißbrock, Nucl. Phys. **B885**, 409–447 (2014). [arXiv:1403.1137[hep-ph]]
103. F.Ph. Wißbrock, *$O(\alpha_s^3)$ contributions to the heavy flavor Wilson coefficients of the structure function $F_2(x, Q^2)$ at $Q^2 \gg m^2$*, Ph.D. Thesis, TU Dortmund, 2015
104. S.A. Larin, T. van Ritbergen, J.A.M. Vermaseren, Nucl. Phys. **B427**, 41–52 (1994)
105. S.A. Larin, P. Nogueira, T. van Ritbergen, J.A.M. Vermaseren, Nucl. Phys. **B492**, 338–378 (1997). [hep-ph/9605317]
106. A. Retey, J.A.M. Vermaseren, Nucl. Phys. **B604**, 281–311 (2001). [hep-ph/0007294]
107. J. Blümlein, J.A.M. Vermaseren, Phys. Lett. **B606**, 130–138 (2005). [hep-ph/0411111]
108. I. Bierenbaum, J. Blümlein, S. Klein, Nucl. Phys. B **820**, 417–482 (2009). arXiv:0904.3563 [hep-ph]
109. J.A.M Vermaseren, A. Vogt, S. Moch, Nucl. Phys. **B724**, 3–182 (2005). [hep-ph/0504242]
110. `Sage`, http://www.sagemath.org/
111. S. Moch, J.A.M. Vermaseren, A. Vogt, Nucl. Phys. **B688**, 101–134 (2004). [hep--ph/0403192];

A. Vogt, S. Moch, J.A.M. Vermaseren, Nucl. Phys. **B691**, 129–181 (2004). [hep--ph/0404111];
S. Moch, J.A.M. Vermaseren, A. Vogt, Nucl. Phys. B **889**, 351–400 (2014). arXiv:1409.5131 [hep-ph]
112. J. Blümlein, P. Marquard, C. Schneider, K. Schönwald, *The three-loop unpolarized and polarized non-singlet anomalous dimensions from off shell operator matrix elements*. Nucl. Phys. B, in print [arXiv:2107.06267 [hep-ph]]
113. J. Ablinger, J. Blümlein, P. Marquard, N. Rana, C. Schneider, Nucl. Phys. B **939**, 253–291 (2019). arXiv:1810.12261 [hep-ph]
114. J. Ablinger, J. Blümlein, P. Marquard, N. Rana, C. Schneider, Phys. Lett. B **782**, 528–532 (2018). arXiv:1804.07313 [hep-ph]
115. S.G. Gorishnii, S.A. Larin, L.R. Surguladze, F.V. Tkachov, Comput. Phys. Commun. **55**, 381–408 (1989);
S.A. Larin, F.V. Tkachov, J.A.M. Vermaseren, *The FORM version of MINCER* NIKHEF-H-91-18
116. M. Steinhauser, Comput. Phys. Commun. **134**, 335–364 (2001). arXiv:hep-ph/0009029
117. R. Harlander, T. Seidensticker, M. Steinhauser, Phys. Lett. **B426**, 125–132 (1998). [hep--ph/9712228]
118. T. Seidensticker, *Automatic application of successive asymptotic expansions of Feynman diagrams*, in *Proc. 6th International Workshop on New Computing Techniques in Physics Research*, Crete, Greece, April 1999. arXiv:hep-ph/9905298
119. N.E. Nörlund, *Vorlesungen über Differenzenrechnung* (Springer, Berlin, 1924); reprinted by (Chelsea Publishing Company, New York, 1954)
120. J.A.M. Vermaseren, Int. J. Mod. Phys. A **14**, 2037–2076 (1999). arXiv:hep-ph/9806280 [hep-ph]
121. J. Blümlein, S. Kurth, Phys. Rev. D **60**, 014018 (1999). arXiv:hep-ph/9810241 [hep-ph]
122. J. Ablinger, J. Blümlein, C. Schneider, J. Math. Phys. **54**, 082301 (2013). arXiv: 1302.0378 [math-ph]
123. J. Ablinger, J. Blümlein, C. Schneider, J. Math. Phys. **52**, 102301 (2011). arXiv: 1105.6063 [math-ph]
124. J. Ablinger, J. Blümlein, C.G. Raab, C. Schneider, J. Math. Phys. **55**, 112301 (2014). arXiv:1407.1822 [hep-th]
125. J. Ablinger, J. Blümlein, C. Schneider, J. Phys. Conf. Ser. **523**, 012060 (2014). arXiv:1310.5645 [math-ph];
J. Ablinger, PoS (LL2014) 019. arXiv:1407.6180[cs.SC]; *A Computer Algebra Toolbox for Harmonic Sums Related to Particle Physics*, Diploma Thesis, JKU Linz, 2009, arXiv:1011.1176[math-ph]; PoS (LL2016) 067; Experimental Mathematics 26 (2017). arXiv:1507.01703 [math.CO]; PoS (RADCOR2017) 001, arXiv:1801.01039 [cs.SC]; PoS (LL2018) 063;
J. Ablinger, Experimental Mathematics 26 (2017). arXiv:1507.01703 [math.CO]; *Discovering and Proving Infinite Pochhammer Sum Identities*, arXiv:1902.11001 [math.CO]
126. J. Blümlein, Comput. Phys. Commun. **180**, 2218–2249 (2009). arXiv:0901.3106 [hep-ph]
127. J. Ablinger, *Computer Algebra Algorithms for Special Functions in Particle Physics*, Ph.D. Thesis, Linz U. (2012). arXiv:1305.0687[math-ph]
128. E. Remiddi, J.A.M. Vermaseren, Int. J. Mod. Phys. **A15**, 725–754 (2000). [hep-ph/9905237]
129. S. Moch, P. Uwer, S. Weinzierl, J. Math. Phys. **43**, 3363–3386 (2002). arXiv:hep-ph/0110083 [hep-ph]
130. A.V. Kotikov, Phys. Lett. **B254**, 158–164 (1991)
131. Z. Bern, L.J. Dixon, D.A. Kosower, Phys. Lett. **B302**, 299–308 (1993). [Erratum: Phys. Lett. B318, 649 (1993)] [hep-ph/9212308]
132. E. Remiddi, Nuovo Cim. **A110**, 1435–1452 (1997). [hep-th/9711188]
133. T. Gehrmann, E. Remiddi, Nucl. Phys. **B580**, 485–518 (2000). [hep-ph/9912329]
134. A. Bostan, F. Chyzak, É. de Panafieu, *Complexity Estimates for Two Uncoupling Algorithms*. Proceedings of ISSAC'13, Boston, June 2013

135. B. Zürcher, *Rationale Normalformen von pseudo-linearen Abbildungen*. Master's thesis, Mathematik, ETH Zürich (1994)
136. S. Gerhold, *Uncoupling systems of linear Ore operator equations*, Master's thesis, RISC, J. Kepler University, Linz (2002)
137. S.A. Abramov and M. Petkovšek, *D'Alembertian solutions of linear differential and difference equations*, in *Proc. ISSAC'94*, ed. by J. von zur Gathen (ACM Press, 1994), pp. 169–174
138. M.F. Singer, Am. J. Math. **103**(4), 661–682 (1981)
139. J.J. Kovacic, J. Symb. Comput. **2**, 3–43 (1986)
140. J. Ablinger, A. Behring, J. Blümlein, G. Falcioni, A. De Freitas, P. Marquard, N. Rana, C. Schneider, Phys. Rev. D **97**(9), 094022 (2018). arXiv:1712.09889 [hep-ph]
141. A.V. Kotikov, The Property of maximal transcendentality in the N=4 Supersymmetric Yang-Mills, in *Subtleties in quantum field theory*, ed. by D. Diakonov, pp. 150–174. arXiv:1005.5029 [hep-th]
142. J.M. Henn, Phys. Rev. Lett. **110**, 251601 (2013). [arXiv:1304.1806[hep-th]]
143. J.M. Henn, J. Phys. **A48**, 153001 (2015). [arXiv:1412.2296[hep-ph]]
144. V.E. Zakharov, S.V. Manakov, S.P. Novikov, L.P. Pitaevskii, *Teoria Solitonov: metod obratnoi zadatschi* (Nauka, Moskva, 1980)
145. S.Yu. Sakovich, J. Phys. A Math. Gen. **28**, 2861–2869 (1995)
146. R.N. Lee, JHEP **04**, 108 (2015). [arXiv:1411.0911[hep-ph]]
147. M. Prausa, Comput. Phys. Commun. **219**, 361–376 (2017). [arXiv:1701.00725 [hep-ph]]
148. O. Gituliar, V. Magerya, Comput. Phys. Commun. **219**, 329–338 (2017). [arXiv:1701.04269 [hep-ph]]
149. C. Meyer, Comput. Phys. Commun. **222**, 295–312 (2018). [arXiv:1705.06252 [hep-ph]]
150. M. Besier, D. Van Straten, S. Weinzierl, Commun. Num. Theor. Phys. **13**, 253–297 (2019). arXiv:1809.10983 [hep-th]
151. M. Besier, P. Wasser, S. Weinzierl, Comput. Phys. Commun. **253**, 107197 (2020). arXiv:1910.13251 [cs.MS]
152. J. Ablinger, J. Blümlein, A. De Freitas, K. Schönwald, Nucl. Phys. B **955**, 115045 (2020). arXiv:2004.04287 [hep-ph];
J. Blümlein, A. De Freitas, C. Raab, K. Schönwald, Nucl. Phys. B **956**, 115055 (2020). arXiv:2003.14289 [hep-ph]; Phys. Lett. B **801**, 135196 (2020). arXiv:1910.05759 [hep-ph]; Phys. Lett. B **791**, 206–209 (2019). arXiv:1901.08018 [hep-ph]
153. J. Blümlein, A. De Freitas, C.G. Raab, K. Schönwald, Nucl. Phys. B **945**, 114659 (2019). arXiv:1903.06155 [hep-ph];
J. Blümlein, C. Raab, K. Schönwald, Nucl. Phys. B **948**, 114736 (2019). arXiv:1904.08911 [hep-ph]
154. G. Almkvist, D. Zeilberger, J. Symb. Comp. **10**, 571–591 (1990)
155. M. Apagodu, D. Zeilberger, Adv. Appl. Math. (Special Regev Issue) **37**, 139–152 (2006)
156. D.J. Broadhurst, D. Kreimer, Phys. Lett. B **393**, 403–412 (1997). arXiv:hep-th/9609128 [hep-th]
157. N. Nielsen Nova Acta Leopold. **XC**(Nr. 3), 125–211 (1909);
K.S. Kölbig, J.A. Mignoco, E. Remiddi, BIT **10**, 38–74 (1970);
K.S. Kölbig, SIAM J. Math. Anal. **17**, 1232–1258 (1986)
158. M.E. Hoffman, J. Algebraic Combin. **11**, 49–68 (2000). arXiv:math/9907173 [math.QA]
159. J. Blümlein, Comput. Phys. Commun. **159**, 19–54 (2004). [hep-ph/0311046]
160. J. Ablinger, J. Blümlein, C. Schneider, Phys. Rev. D **103**(9), 096025 (2021) [hep-th]
161. A.I. Davydychev, M.Yu. Kalmykov, Nucl. Phys. **B699**, 3–64 (2004). [hep-th/0303162]
162. S. Weinzierl, J. Math. Phys. **45**, 2656–2673 (2004). [hep-ph/0402131]
163. C. Reutenauer, *Free Lie Algebras* (Calendron Press, Oxford, 1993)
164. T. Gehrmann, E. Remiddi, Comput. Phys. Commun. **141**, 296–312 (2001). [hep-ph/0107173]
165. J. Vollinga, S. Weinzierl, Comput. Phys. Commun. **167**, 177–194 (2005). arXiv:hep-ph/0410259 [hep-ph]
166. J. Ablinger, J. Blümlein, A. De Freitas, C. Schneider, K. Schönwald, Nucl. Phys. B **927**, 339–367 (2018). arXiv:1711.06717 [hep-ph]

167. J. Ablinger, J. Blümlein, A. De Freitas, A. Goedicke, C. Schneider, K. Schönwald, Nucl. Phys. B **932**, 129–240 (2018). arXiv:1804.02226 [hep-ph]
168. C.G. Raab. *On the arithmetic of d'Alembertian functions*, presentation at the 19th Conference on Applications of Computer Algebra (ACA 2013), Málaga, Spain, 2–6 July 2013, in preparation
169. Li Guo, G. Regensburger, M. Rosenkranz, J. Pure Appl. Algebra **218**, 456–473 (2014)
170. D.E. Radford, J. Algebra **58**, 432–454 (1979)
171. R.C. Lyndon, Trans. Am. Math. Soc. **77**, 202–215 (1954); **78**, 329–332 (1955)
172. E. Witt, Journ. Reine und Angew. Mathematik **177**, 152–160 (1937); Math. Zeitschr. **64**, 195–216 (1956)
173. J. Blümlein, Clay Math. Proc. **12**, 167–188 (2010). [arXiv:0901.0837[math-ph]]
174. J. Blümlein, Nucl. Phys. B Proc. Suppl. **135**, 225–231 (2004). arXiv:hep-ph/0407044 [hep-ph]
175. D.J. Broadhurst, J. Fleischer, O. V. Tarasov, Z. Phys. **C60**, 287–302 (1993). [hep-ph/9304303]
176. S. Bloch, P. Vanhove, J. Number Theor. **148**, 328–364 (2015). [hep-th/ 1309.5865]
177. S. Laporta, E. Remiddi, Nucl. Phys. **B704**, 349–386 (2005). [hep-ph/0406160]
178. L. Adams, C. Bogner, S. Weinzierl, J. Math. Phys. **54**, 052303 (2013). [hep-ph/1302.7004]
179. L. Adams, C. Bogner, S. Weinzierl, J. Math. Phys. **55**(10), 102301 (2014). [hep--ph/1405.5640]
180. L. Adams, C. Bogner, S. Weinzierl, J. Math. Phys. **56**(7), 072303 (2015). [hep--ph/1504.03255]
181. L. Adams, C. Bogner, S. Weinzierl, J. Math. Phys. **57**(3), 032304 (2016). [hep--ph/1512.05630]
182. A. Sabry, Nucl. Phys. **33**, 401–430 (1962)
183. E. Remiddi, L. Tancredi, Nucl. Phys. **B907**, 400–444 (2016). [arXiv:1602.01481[hep-ph]]
184. L. Adams, C. Bogner, A. Schweitzer, S. Weinzierl, J. Math. Phys. **57**(12), 122302 (2016). [hep-ph/1607.01571]
185. J. Grigo, J. Hoff, P. Marquard, M. Steinhauser, Nucl. Phys. B **864**, 580–596 (2012). arXiv:1206.3418 [hep-ph]
186. J. Blümlein, A. De Freitas, M. Van Hoeij, E. Imamoglu, P. Marquard, C. Schneider, in *PoS (LL2018)*, 017. arXiv:1807.05287 [hep-ph]
187. J. Ablinger, J. Blümlein, A. De Freitas, M. van Hoeij, E. Imamoglu, C.G. Raab, C.S. Radu, C. Schneider, J. Math. Phys. **59**(6), 062305 (2018). arXiv: 1706.01299 [hep-th]
188. A. Ronveaux, (ed.), *Heun's Differential Equations* (The Clarendon Press Oxford, Oxford, 1995)
189. E. Imamoglu, M. van Hoeij, J. Symb. Comput. **83**, 245–271 (2017). arXiv:1606.01576 [cs.SC]
190. K. Takeuchi, J. Fac. Sci, Univ. Tokyo, Sect. 1A **24**, 201–272 (1977)
191. F.G. Tricomi, *Elliptische Funktionen* (Geest & Portig, Leipzig, 1948); übersetzt und bearbeitet von M. Krafft
192. E.T. Whittaker, G.N. Watson, *A Course of Modern Analysis* (Cambridge University Press, Cambridge, 1996); reprint of 4th edition (1927)
193. S. Herfurtner, Math. Ann. **291**, 319–342 (1991)
194. H. Movasati, S. Reiter, Bull. Braz. Math Soc. **43**, 423–442 (2012). arXiv: 0902.0760[math.AG]
195. J. Blümlein, talks at: The 5th International Congress on Mathematical Software ZIB Berlin from July 11 to July 14, 2016, Session: Symbolic computation and elementary particle physics. https://www.risc.jku.at/conferences/ICMS2016/; and QCD@LHC2016, U. Zürich, August 22 to August 26, 2016. https://indico.cern.ch/event/516210/timetable/#all.detailed
196. E. Remiddi, L. Tancredi, Nucl. Phys. **B925**, 212–251 (2017). [arXiv: 1709.03622[hep-ph]]
197. L. Adams, S. Weinzierl, Phys. Lett. **B781**, 270–278 (2018). [arXiv: 1802. 05020[hep-ph]]
198. J.-P. Serre, *A Course in Arithmetic* (Springer, Berlin, 1973)
199. H. Cohen, F. Strömberg, *Modular Forms, A Classical Approach*. Graduate Studies in Mathematics, vol. 179 (AMS, providence, RI, 2017)

200. K. Ono, *The Web of Modularity: Arithmetic of the Coefficients of Modular Forms and q-Series*. CBMS Regional Conference Series in Mathematics, vol. 102 (AMS, Providence, RI, 2004)
201. H.H. Chan, W. Zudilin, Mathematika **56**, 107–117 (2010)
202. D.J. Broadhurst, *Eta quotients, Eichler integrals and L-series*, talk at HMI Bonn, February 2018
203. J. Broedel, C. Duhr, F. Dulat, B. Penante, L. Tancredi, JHEP **1808**, 014 (2018). arXiv:1803.10256 [hep-th]
204. L. Adams, S. Weinzierl, Commun. Num. Theor. Phys. **12**, 193–251 (2018). arXiv:1704.08895 [hep-ph]
205. J. Broedel, C. Duhr, F. Dulat, B. Penante, L. Tancredi, JHEP **05**, 120 (2019). arXiv:1902.09971 [hep-ph]; JHEP **01**, 023 (2019). arXiv:1809.10698 [hep-th]; J. Broedel, C. Duhr, F. Dulat, L. Tancredi, JHEP **05**, 093 (2018). arXiv:1712.07089 [hep-th]; Phys. Rev. D **97**(11), 116009 (2018). arXiv:1712.07095 [hep-ph]
206. K. Martin, *Modular Forms*. Lecture Notes (2016) U. Oklahoma, http://www2.math.ou.edu/~kmartin/mfs/;
G.H. Hardy, S. Ramanujan, Proc. Lond. Math. Soc. (2) **17**, 75–115 (1918);
H. Rademacher, Proc. Lond. Math. Soc. (2) **43**, 241–254 (1937); Ann. Math. **44**, 416–422 (1943);
J.H. Brunier, K. Ono, Adv. Math. **246**, 198–219 (2013)
207. S. Laporta, Phys. Lett. B **772**, 232–238 (2017). arXiv:1704.06996 [hep-ph]
208. J. Blümlein, A. Vogt, Phys. Rev. D **58**, 014020 (1998). arXiv:hep-ph/9712546
209. J. Blümlein, V. Ravindran, W.L. van Neerven, Nucl. Phys. B **586**, 349–381 (2000). arXiv:hep-ph/0004172
210. J. Blümlein, A. Guffanti, Nucl. Phys. Proc. Suppl. **152**, 87–91 (2006). arXiv:hep-ph/0411110
211. J. Blümlein, Comput. Phys. Commun. **133**, 76–104 (2000). [hep-ph/0003100]
212. J. Blümlein, S.-O. Moch, Phys. Lett. **B614**, 53–61 (2005). [hep-ph/0503188]
213. N. Nielsen, *Handbuch der Theorie der Gammafunktion* (Teubner, Leipzig, 1906); reprinted by (Chelsea Publishing Company, Bronx, New York, 1965)
214. E. Landau, *Über die Grundlagen der Theorie der Fakultätenreihen*. S.-Ber. math.-naturw. Kl. Bayerische Akad. Wiss. München **36**, 151–218 (1906)
215. A.V. Kotikov, V.N. Velizhanin, *Analytic continuation of the Mellin moments of deep inelastic structure functions*. arXiv:hep-ph/0501274
216. H.D. Politzer, Phys. Rept. **14**, 129–180 (1974)
217. J. Blümlein, N. Kochelev, Nucl. Phys. B **498**, 285–309 (1997). arXiv:hep-ph/9612318 [hep-ph]
218. S. Weinzierl, Comput. Phys. Commun. **145**, 357–370 (2002). [math-ph/0201011]
219. S. Moch, P. Uwer, Comput. Phys. Commun. **174**, 759–770 (2006). [math-ph/0508008]
220. C. Duhr, F. Dulat, *PolyLogTools - Polylogs for the Masses*. arXiv:1904.07279 [hep-th]
221. D. Maitre, Comput. Phys. Commun. **174**, 222–240 (2006). [hep-ph/0507152]
222. J. Ablinger, J. Blümlein, M. Round, C. Schneider, Comput. Phys. Commun. **240**, 189–201 (2019). arXiv:1809.07084 [hep-ph]
223. C. Bogner, A. Schweitzer, S. Weinzierl, Nucl. Phys. B **922**, 528–550 (2017). arXiv:1705.08952 [hep-ph]
224. G. Passarino, Eur. Phys. J. C **77**(2), 77 (2017). arXiv:1610.06207 [math-ph];
M. Walden, S. Weinzierl, *Numerical evaluation of iterated integrals related to elliptic Feynman integrals*. arXiv:2010.05271 [hep-ph]
225. K. Melnikov, A. Vainshtein, Springer Tracts Mod. Phys. **216**, 1–176 (2006);
F. Jegerlehner, A. Nyffeler, Phys. Rept. **477**, 1–110 (2009). arXiv:0902.3360 [hep-ph];
F. Jegerlehner, Springer Tracts Mod. Phys. **274**, 1–711 (2018)
226. T. Aoyama, M. Hayakawa, T. Kinoshita, M. Nio, Phys. Rev. Lett. **109**, 111807 (2012). arXiv:1205.5368 [hep-ph]; Phys. Rev. Lett. **109**, 111808 (2012). arXiv:1205.5370 [hep-ph]
227. P.A. Baikov, A. Maier, P. Marquard, Nucl. Phys. B **877**, 647–661 (2013). arXiv:1307.6105 [hep-ph]

228. A. Kurz, T. Liu, P. Marquard, A.V. Smirnov, V.A. Smirnov, M. Steinhauser, EPJ Web Conf. **118**, 01033 (2016). arXiv:1511.08222 [hep-ph]
229. A. Kurz, T. Liu, P. Marquard, A. Smirnov, V. Smirnov, M. Steinhauser, Phys. Rev. D **93**(5), 053017 (2016). arXiv:1602.02785 [hep-ph]
230. P. Marquard, A.V. Smirnov, V.A. Smirnov, M. Steinhauser, D. Wellmann, EPJ Web Conf. **218**, 01004 (2019). arXiv:1708.07138 [hep-ph]
231. S. Volkov, Phys. Rev. D **100**(9), 096004 (2019). arXiv:1909.08015 [hep-ph]
232. P. Marquard, A.V. Smirnov, V.A. Smirnov, M. Steinhauser, Phys. Rev. Lett. **114**(14), 142002 (2015). arXiv:1502.01030 [hep-ph];
P. Marquard, A.V. Smirnov, V.A. Smirnov, M. Steinhauser, D. Wellmann, Phys. Rev. D **94**(7), 074025 (2016). arXiv:1606.06754 [hep-ph];
S. Laporta, Phys. Lett. B **802**, 135264 (2020). arXiv:2001.02739 [hep-ph];
Y. Schröder, M. Steinhauser, JHEP **01**, 051 (2006). arXiv:hep-ph/0512058 [hep-ph];
K.G. Chetyrkin, J.H. Kühn, C. Sturm, Nucl. Phys. B **744**, 121–135 (2006). arXiv:hep-ph/0512060 [hep-ph];
T. Liu, M. Steinhauser, Phys. Lett. B **746**, 330–334 (2015). arXiv:1502.04719 [hep-ph];
M. Fael, K. Schönwald, M. Steinhauser, Phys. Rev. Lett. **125**(5), 052003 (2020). arXiv:2005.06487 [hep-ph]; JHEP **10**, 087 (2020). arXiv:2008.01102 [hep-ph]; *Third order corrections to the semi-leptonic $b \to c$ and the muon decays*. arXiv:2011.13654 [hep-ph]
233. J. Ablinger, J. Blümlein, S. Klein, C. Schneider, F. Wißbrock, Nucl. Phys. B **844**, 26–54 (2011). arXiv:1008.3347 [hep-ph];
J. Ablinger, J. Blümlein, A. De Freitas, A. Hasselhuhn, A. von Manteuffel, M. Round, C. Schneider, F. Wißbrock, Nucl. Phys. B **882**, 263–288 (2014). arXiv:1402.0359 [hep-ph];
J. Ablinger, A. Behring, J. Blümlein, A. De Freitas, A. Hasselhuhn, A. von Manteuffel, M. Round, C. Schneider, F. Wißbrock, Nucl. Phys. **B886**, 733–823 (2014). [arXiv:1406.4654[hep-ph]];
J. Ablinger, A. Behring, J. Blümlein, A. De Freitas, A. von Manteuffel, C. Schneider, Nucl. Phys. B **890**, 48–151 (2014). arXiv:1409.1135 [hep-ph];
A. Behring, J. Blümlein, A. De Freitas, A. Hasselhuhn, A. von Manteuffel, C. Schneider, Phys. Rev. D **92**(11), 114005 (2015). arXiv:1508.01449 [hep-ph];
A. Behring, J. Blümlein, A. De Freitas, A. von Manteuffel, C. Schneider, Nucl. Phys. B **897**, 612–644 (2015). arXiv:1504.08217 [hep-ph];
J. Ablinger, J. Blümlein, A. De Freitas, A. Hasselhuhn, C. Schneider, F. Wißbrock, Nucl. Phys. B **921**, 585–688 (2017). arXiv:1705.07030 [hep-ph];
A. Behring, J. Blümlein, G. Falcioni, A. De Freitas, A. von Manteuffel, C. Schneider, Phys. Rev. D **94**(11), 114006 (2016). arXiv:1609.06255 [hep-ph];
A. Behring, J. Blümlein, A. De Freitas, A. von Manteuffel, K. Schönwald, C. Schneider, Nucl. Phys. B **964**, 115331 (2021). arXiv:2101.05733 [hep-ph];
J. Ablinger, J. Blümlein, A. De Freitas, A. Goedicke, M. Saragnese, C. Schneider, K. Schönwald, Nucl. Phys. B **955**, 115059 (2020). arXiv:2004.08916 [hep-ph];
J. Ablinger, J. Blümlein, A. De Freitas, M. Saragnese, C. Schneider, K. Schönwald, Nucl. Phys. B **952**, 114916 (2020). arXiv:1911.11630 [hep-ph]
234. C. Anastasiou, C. Duhr, F. Dulat, F. Herzog, B. Mistlberger, Phys. Rev. Lett. **114**, 212001 (2015). arXiv:1503.06056 [hep-ph]
235. C. Duhr, F. Dulat, B. Mistlberger, Phys. Rev. Lett. **125**(17), 172001 (2020). arXiv:2001.07717 [hep-ph]
236. C. Duhr, F. Dulat, B. Mistlberger, JHEP **11**, 143 (2020). arXiv:2007.13313 [hep-ph]
237. M. Czakon, P. Fiedler, A. Mitov, Phys. Rev. Lett. **110**, 252004 (2013). arXiv:1303.6254 [hep-ph]
238. S. Alekhin, J. Blümlein, S. Moch, Phys. Rev. D **86**, 054009 (2012). arXiv:1202.2281 [hep-ph];
A. Accardi, et al. Eur. Phys. J. C **76**(8), 471 (2016). arXiv:1603.08906 [hep-ph];
S. Alekhin, J. Blümlein, S. Moch, R. Placakyte, Phys. Rev. D **96**(1), 014011 (2017). arXiv:1701.05838 [hep-ph]

239. A. Gehrmann-De Ridder, T. Gehrmann, E.W.N. Glover, A. Huss, T.A. Morgan, Phys. Rev. Lett. **117**(2), 022001 (2016). arXiv:1507.02850 [hep-ph]
240. R. Boughezal, J.M. Campbell, R.K. Ellis, C. Focke, W.T. Giele, X. Liu, F. Petriello, Phys. Rev. Lett. **116**(15), 152001 (2016). arXiv:1512.01291 [hep-ph]
241. J. Currie, T. Gehrmann, A. Huss, J. Niehues, JHEP **07**, 018 (2017). [Erratum: JHEP **12**, 042 (2020)] arXiv:1703.05977 [hep-ph];
V. Andreev et al. [H1], Eur. Phys. J. C **77**(11), 791 (2017). arXiv:1709.07251 [hep-ex]
242. J. Blümlein, A. Guffanti, Nucl. Phys. B Proc. Suppl. **152**, 87–91 (2006). arXiv:hep-ph/0411110 [hep-ph];
J. Blümlein, H. Böttcher, A. Guffanti, Nucl. Phys. B **774**, 182–207 (2007). arXiv:hep-ph/0607200 [hep-ph];
J. Blümlein, M. Saragnese, Phys. Lett. B **820**, 136589 (2021) [arXiv:2107.01293 [hep-ph]]
243. S. Alekhin, J. Blümlein, K. Daum, K. Lipka, S. Moch, Phys. Lett. B **720**, 172–176 (2013). arXiv:1212.2355 [hep-ph]
244. S. Moch, B. Ruijl, T. Ueda, J.A.M. Vermaseren, A. Vogt, JHEP **10**, 041 (2017). arXiv:1707.08315 [hep-ph]
245. V.S. Fadin, E.A. Kuraev, L.N. Lipatov, Phys. Lett. B **60**, 50–52 (1975);
S. Catani, F. Hautmann, Nucl. Phys. B **427**, 475–524 (1994). arXiv:hep-ph/9405388 [hep-ph];
V.S. Fadin, L.N. Lipatov, Phys. Lett. B **429**, 127–134 (1998). arXiv:hep-ph/9802290 [hep-ph]
246. P. Janot, JHEP **02**, 053 (2016). [Erratum: JHEP **11**, 164 (2017)] arXiv:1512.05544 [hep-ph]
247. J. Blümlein, A. De Freitas, K. Schönwald, Phys. Lett. B **816**, 136250 (2021) [arXiv:2102.12237 [hep-ph]]
248. M. Beneke, Y. Kiyo, A. Maier, J. Piclum, Comput. Phys. Commun. **209**, 96–115 (2016). arXiv:1605.03010 [hep-ph]
249. M. Beneke, A. Maier, T. Rauh, P. Ruiz-Femenia, JHEP **02**, 125 (2018). arXiv:1711.10429 [hep-ph]
250. M. Beneke, Y. Kiyo, P. Marquard, A. Penin, J. Piclum, M. Steinhauser, Phys. Rev. Lett. **115**(19), 192001 (2015). arXiv:1506.06864 [hep-ph]
251. F. Bach, B.C. Nejad, A. Hoang, W. Kilian, J. Reuter, M. Stahlhofen, T. Teubner, C. Weiss, JHEP **03**, 184 (2018). arXiv:1712.02220 [hep-ph]
252. A.H. Hoang, M. Stahlhofen, JHEP **05**, 121 (2014). arXiv:1309.6323 [hep-ph]
253. K. Seidel, F. Simon, M. Tesar, S. Poss, Eur. Phys. J. C **73**(8), 2530 (2013). arXiv:1303.3758 [hep-ex]
254. F. Simon, in *PoS (ICHEP2016)*, 872. arXiv:1611.03399 [hep-ex]
255. I. Dubovyk, A. Freitas, J. Gluza, T. Riemann, J. Usovitsch, JHEP **08**, 113 (2019). arXiv:1906.08815 [hep-ph]; Phys. Lett. B **783**, 86–94 (2018). arXiv:1804.10236 [hep-ph]
256. G. Heinrich, *Collider Physics at the Precision Frontier*. arXiv:2009.00516 [hep-ph]
257. S. Foffa, P. Mastrolia, R. Sturani, C. Sturm, W.J. Torres Bobadilla, Phys. Rev. Lett. **122**(24), 241605 (2019). arXiv:1902.10571 [gr-qc];
J. Blümlein, A. Maier, P. Marquard, Phys. Lett. B **800**, 135100 (2020). arXiv:1902.11180 [gr-qc];
J. Blümlein, A. Maier, P. Marquard, G. Schäfer, Nucl. Phys. B **965**, 115352 (2021). arXiv:2010.13672 [gr-qc]
258. J. Blümlein, V. Ravindran, Nucl. Phys. B **716**, 128–172 (2005). arXiv:hep-ph/0501178 [hep-ph]
259. J. Blümlein, V. Ravindran, Nucl. Phys. B **749**, 1–24 (2006). arXiv:hep-ph/0604019 [hep-ph]
260. C. Neumann, *Vorlesungen über Riemann's Theorie der Abel'schen Integrale*, 2nd edn. (Teubner, Leipzig, 1884)
261. F. Brown, O. Schnetz, Duke Math. J. **161**(10), 1817–1862 (2012)
262. K. Bönisch, C. Duhr, F. Fischbach, A. Klemm, C. Nega, *Feynman integrals in dimensional regularization and extensions of Calabi-Yau motives* [arXiv:2108.05310 [hep-th]].
263. D. Hilbert, *Grundzüge einer allgemeinen Theorie der linearen Integralgleichungen* (Teubner, Leipzig, 1912)
264. R. de L. Kronig, J. Opt. Soc. Am. **12**, 547–557 (1926)

265. H.A. Kramers, Atti Cong. Intern. Fisici (Transactions of Volta Centenary Congress) Como 2 (1927)
266. S. Abreu, R. Britto, C. Duhr, E. Gardi, JHEP **12**, 090 (2017). [arXiv:1704.07931[hep-th]]
267. M.J.G. Veltman, Physica **29**, 186–207 (1963)
268. E. Remiddi, Helv. Phys. Acta **54**, 364–382 (1982);
E. Remiddi, *Differential Equations and Dispersion Relations for Feynman Amplitudes*, in *Elliptic Integrals, Elliptic Functions and Modular Forms in Quantum Field Theory*, ed. by J. Blümlein, C. Schneider, P. Paule (Springer, Wien, 2019), pp. 391–414

Extensions of the AZ-Algorithm and the Package MultiIntegrate

Jakob Ablinger

Abstract We extend the (continuous) multivariate Almkvist-Zeilberger algorithm in order to apply it for instance to special Feynman integrals emerging in renormalizable Quantum field Theories. We will consider multidimensional integrals over hyperexponential integrals and try to find closed form representations in terms of nested sums and products or iterated integrals. In addition, if we fail to compute a closed form solution in full generality, we may succeed in computing the first coefficients of the Laurent series expansions of such integrals in terms of indefinite nested sums and products or iterated integrals. In this article we present the corresponding methods and algorithms. Our Mathematica package `MultiIntegrate`, can be considered as an enhanced implementation of the (continuous) multivariate Almkvist Zeilberger algorithm to compute recurrences or differential equations for hyperexponential integrands and integrals. Together with the summation package `Sigma` and the package `HarmonicSums` our package provides methods to compute closed form representations (or coefficients of the Laurent series expansions) of multidimensional integrals over hyperexponential integrands in terms of nested sums or iterated integrals.

1 Introduction

The Almkvist-Zeilberger was first formulated by Apagodu and Zeilberger [14, 31] and has later been refined and generalized [1, 23–25, 28]. It attracted attention in renormalizable Quantum Field Theory in the frame of the calculation of Feynman integrals.

In the following we briefly want to summarize the structure of those integrals. The very general class of Feynman integrals which are, for instance, considered

J. Ablinger (✉)
Johannes Kepler University Linz, Research Institute for Symbolic Computation (RISC), Linz, Austria
e-mail: Jakob.Ablinger@risc.jku.at

J. Blümlein, C. Schneider (eds.), *Anti-Differentiation and the Calculation of Feynman Amplitudes*, Texts & Monographs in Symbolic Computation,
https://doi.org/10.1007/978-3-030-80219-6_2

in [18] are of relevance for many physical processes at high energy colliders, such as the Large Hadron Collider and others. The considered integrals are two-point Feynman integrals in D-dimensional Minkowski space with one time- and $(D-1)$ Euclidean space dimensions, $\varepsilon = D - 4$ and $\varepsilon \in \mathbb{R}$ with $|\varepsilon| \ll 1$ of the following structure:

$$\mathcal{I}(\varepsilon, N, p) = \int \frac{d^D p_1}{(2\pi)^D} \cdots \int \frac{d^D p_k}{(2\pi)^D} \frac{\mathcal{N}(p_1, \ldots p_k; p; m_1 \ldots m_k; \Delta, N)}{(-p_1^2 + m_1^2)^{l_1} \ldots (-p_k^2 + m_k^2)^{l_k}} \prod_V \delta_V \,. \tag{1}$$

They can be shown to obey difference equations with respect to N, see, e.g., [17]. In (1) the external momentum p and the loop momenta p_i denote D-dimensional vectors, $m_i > 0, m_i \in \mathbb{R}$ are scalars (masses), $m_i \in \{0, M\}, k, l_i \in \mathbb{N}, k \geq 2, l_i \geq 1$, and Δ is a light-like D-vector, $\Delta.\Delta = 0$. The numerator function $\mathcal{N}$ is a polynomial in the scalar products $p.p_i,\ p_i.p_k$ and of monomials $(\Delta.p_{(i)})^{n_i}, n_i \in \mathbb{N}, n_i \geq 0$. $N \in \mathbb{N}$ denotes the spin of a local operator stemming from the light cone expansion, see, e.g., [26] and references therein, which contributes to the numerator function $\mathcal{N}$ with a polynomial in $\Delta.p_i$ of maximal degree N, cf. [15]. Furthermore it is assumed for simplicity that only one of the loops is formed of massive lines. The δ_V occurring in (1) are shortcuts for Dirac delta distributions in D dimensions $\delta_V = \delta^{(D)}\left(\sum_{l=1}^{k} a_{V,l} p_l\right), a_{V,l} \in \mathbb{Q}$.

These integrals are mathematically well defined and in [18] it is shown how they can be mapped onto integrals on the m-dimensional unit cube with the following structure:

$$\mathcal{I}(\varepsilon, N) = C(\varepsilon, N, M) \int_0^1 dy_1 \ldots \int_0^1 dy_m \frac{\sum_{i=1}^{k} \prod_{l=1}^{r_i} [P_{i,l}(y)]^{\alpha_{i,l}(\varepsilon, N)}}{[Q(y)]^{\beta(\varepsilon)}}\,, \tag{2}$$

with $k \in \mathbb{N}, r_1, \ldots, r_k \in \mathbb{N}$ and where $\beta(\varepsilon)$ is given by a rational function in ε, i.e., $\beta(\varepsilon) \in \mathbb{R}(\varepsilon)$, and similarly $\alpha_{i,l}(\varepsilon, N) = n_{i,l}N + \overline{\alpha}_{i,l}$ for some $n_{i,l} \in \{0, 1\}$ and $\overline{\alpha}_{i,l} \in \mathbb{R}(\varepsilon)$, see also [20] in the case no local operator insertions are present. $C(\varepsilon, N, M)$ is a factor, which depends on the dimensional parameter ε, the integer parameter N and the mass M. $P_i(y), Q(y)$ are polynomials in the remaining Feynman parameters $y = (y_1, \ldots, y_m)$ written in multi-index notation. In (2) all terms which stem from local operator insertions were geometrically resumed; see [15]. In [18] it was already mentioned that after splitting the integral (2), the integrands fit into the input class of the multivariate Almkvist-Zeilberger algorithm. Hence, if the split integrals are properly defined, they obey homogeneous recurrence relations in N due to the theorems in [14]. In [18] the integrals of (2) are transformed further to a multi-sum representation, while in this article (and also in [1, 9, 11]) we want to tackle them directly by looking on integrals of the form

$$\mathcal{I}(n) = \mathcal{I}(\varepsilon, n) = \int_{u_d}^{o_d} \cdots \int_{u_1}^{o_1} F(n; x_1, \ldots, x_d; \varepsilon)\, dx_1 \cdots dx_d, \tag{3}$$

with $d, n \in \mathbb{N}$, $F(n; x_1, \ldots, x_d; \varepsilon)$ a hyperexponential term, $\varepsilon > 0$ a real parameter and $u_i, o_i \in \mathbb{R} \cup \{-\infty, \infty\}$.
In [1] we only considered a discrete variable n but here we will also consider continuous variables, i.e., we will also deal with integrals of the form

$$\mathcal{I}(x) = \mathcal{I}(\varepsilon, x) = \int_{u_d}^{o_d} \ldots \int_{u_1}^{o_1} F(x; x_1, \ldots, x_d; \varepsilon)\, dx_1 \cdots dx_d, \tag{4}$$

with $d \in \mathbb{N}$, $x \in \mathbb{R}$, $F(x; x_1, \ldots, x_d; \varepsilon)$ a hyperexponential term, $\varepsilon > 0$ a real parameter and $u_i, o_i \in \mathbb{R} \cup \{-\infty, \infty\}$. We will use our package `MultiIntegrate`[1] [1, 11] that can be considered as an enhanced implementation of the multivariate Almkvist Zeilberger algorithm to compute recurrences/differential equations for the integrands and integrals. For solving recurrences `MultiIntegrate` relies on the solver implemented in the packages `Sigma`[1] [37–39] and `EvaluateMultiSums`[1] [38, 39], while for solving differential equation it relies on the solver implemented in the package `HarmonicSums`[1] [1, 2, 6, 8, 10].

Throughout this article $\mathbb{K}$ denotes a field with $\mathbb{Q} \subseteq \mathbb{K}$ (e.g., $\mathbb{K} = \mathbb{Q}(\varepsilon)$ forms a rational function field) in which the usual operations can be computed.

The remainder of this article is structured in two main sections. In Sect. 2 we will recall the multi-variate Almkvist-Zeilberger algorithm and its modifications as presented in [1] to solve integrals of the form (3), while in Sect. 3 we will present a method based on the continuous Almkvist-Zeilberger algorithm to solve integrals of the form (4). The reason for Sect. 2 is its similarity to the continuous case in Sect. 3, in addition it allows us to present a complete picture of the functionality of the package `MultiIntegrate`. However, in order to keep Sect. 2 short, we omit examples in this section and refer to [1, 9, 11] for further illustrations.

2 A Fine-Tuned Multi-Variate Almkvist-Zeilberger Algorithm

In this section we will recall a method (presented in [1] and [11]) to compute integrals of the form (3), that is based on slight modifications of the multi-variate Almkvist-Zeilberger algorithm [14] and implemented in the package `MultiIntegrate`. The method relies on finding and solving recurrences.
In general, consider the integrand

$$F(n; x_1, \ldots, x_d) = P(n; x_1, \ldots, x_d) \cdot H(n; x_1, \ldots, x_d), \tag{5}$$

[1]The `Mathematica` packages `MultiIntegrate`, `HarmonicSums`, `Sigma` and `EvaluateMultiSums` can be downloaded at https://risc.jku.at/software.

with a multivariate polynomial $P(n; x_1, \dots, x_d) \in \mathbb{K}[n, x_1, \dots, x_d]$ and

$$H(n; x_1, \dots, x_d) = e^{\frac{a(x_1,\dots,x_d)}{b(x_1,\dots,x_d)}} \cdot \left(\prod_{p=1}^{P} S_p(x_1, \dots, x_d)^{\alpha_p}\right) \cdot \left(\frac{s(x_1, \dots, x_d)}{t(x_1, \dots, x_d)}\right)^n,$$

where $a(x_1, \dots, x_d)$ and $b(x_1, \dots, x_d) \neq 0$, $s(x_1, \dots, x_d)$ and $t(x_1, \dots, x_d) \neq 0$ and $S_p(x_1, \dots, x_d) \in \mathbb{K}[x_1, \dots, x_d]$, and $\alpha_p \in \mathbb{K}$. Such integrands have the property that the logarithmic derivatives are rational, i.e.,

$$\frac{D_{x_i} \overline{H}(n; x_1, \dots, x_d)}{\overline{H}(n; x_1, \dots, x_d)} = \frac{q_i(x_1, \dots, x_d)}{r_i(x_1, \dots, x_d)}$$

for some $q_i(n, x_1, \dots, x_d), r_i(n, x_1, \dots, x_d) \in \mathbb{K}[n, x_1, \dots, x_d]$ and are called *hyperexponential* in x_i. Note that this class of integrands covers a big class of Feynman integrals (by choosing the rational function field $\mathbb{K} = \mathbb{Q}(\varepsilon)$) that contains at most one mass [18, 42].
Then due to [14] there exists a non-negative integer L, there exist $e_0(n), \dots, e_L(n) \in \mathbb{K}[n]$ (or equivalently from $\mathbb{K}(n)$), *not all zero*, and there also exist $R_i(n; x_1, \dots, x_d) \in \mathbb{K}(n, x_1, \dots, x_d)$ such that

$$G_i(n; x_1, \dots, x_d) := R_i(n; x_1, \dots, x_d) F(n; x_1, \dots, x_d) \tag{6}$$

satisfy the integrand recurrence

$$\sum_{i=0}^{L} e_i(n) F(n+i; x_1, \dots, x_d) = \sum_{i=1}^{d} D_{x_i} G_i(n; x_1, \dots, x_d), \tag{7}$$

where D_{x_i} stands for the derivative w.r.t x_i.

2.1 The General Method

The proof of the existence, and in particular a method to compute such an integrand recurrence (7), is based on the following observation [14]. Fix a non-negative integer L (with the role given above) and define

$$\overline{H}(n; x_1, \dots, x_d) := e^{\frac{a(x_1,\dots,x_d)}{b(x_1,\dots,x_d)}} \cdot \left(\prod_{p=1}^{P} S_p(x_1, \dots, x_d)^{\alpha_p}\right) \cdot \frac{s(x_1, \dots, x_d)^n}{t(x_1, \dots, x_d)^{n+L}},$$

Then we have

$$\sum_{i=0}^{L} e_i(n)F(n+i; x_1, \ldots, x_d) = h(x_1, \ldots, x_d)\overline{H}(n; x_1, \ldots, x_d),$$

where $h(x_1, \ldots, x_d)$ is a polynomial i.e.,

$$h(x_1, \ldots, x_d) := \sum_{i=1}^{L} e_i(n)P(n+i, x_1, \ldots, x_d)\frac{s(x_1, \ldots, x_d)^i}{t(x_1, \ldots, x_d)^{i-L}}.$$

and, by construction, the logarithmic derivatives of $\overline{H}(n; x_1, \ldots, x_d)$ are a rational functions in the x_i, i.e., we have that

$$\frac{D_{x_i}\overline{H}(n; x_1, \ldots, x_d)}{\overline{H}(n; x_1, \ldots, x_d)} = \frac{q_i(x_1, \ldots, x_d)}{r_i(x_1, \ldots, x_d)}$$

for explicitly given $q_i(n, x_1, \ldots, x_d), r_i(n, x_1, \ldots, x_d) \in \mathbb{K}[n, x_1, \ldots, x_d]$. For $i = 1, \ldots, d$ we make the general ansatz

$$G_i(n; x_1, \ldots, x_d) = \overline{H}(n; x_1, \ldots, x_d) \cdot r_i(n, x_1, \ldots, x_d) \cdot X_i(n; x_1, \ldots, x_d). \quad (8)$$

Then it turns out that for L chosen sufficiently large[2] there exist polynomials $X_i(n; x_1, \ldots, x_d) \in \mathbb{K}[n][x_1, \ldots, x_d]$ with $1 \leq i \leq L$ and polynomials $e_i(n) \in \mathbb{K}[n]$ (not all zero) such that (7) holds. Motivated by this fact, one searches for these unknowns X_i and e_i as follows. Note that the ansatz (7) is equivalent to (see [14])

$$\begin{aligned}\sum_{i=1}^{d}[D_{x_i}r_i(x_1, \ldots, x_d) + q_i(x_1, \ldots, x_d)] \cdot X_i(n; x_1, \ldots, x_d)\\ + r_i(x_1, \ldots, x_d) \cdot D_{x_i}X_i(n; x_1, \ldots, x_d)\\ = \sum_{i=0}^{L} e_i(n)\, P(n; x_1, \ldots, x_d)\, s(x_1, \ldots, x_d)^i\, t(x_1, \ldots, x_d)^{L-i}. \quad (9)\end{aligned}$$

We choose appropriate degree bounds w.r.t. the $x_1, \ldots, x_d$ for the X_i ($1 \leq i \leq d$) and plug the polynomials with unknown coefficients from $\mathbb{K}[n]$ (from $\mathbb{K}(n)$) into (9). By coefficient comparison this yields a linear system in $\mathbb{K}(n)$ with the unknowns $e_i(n)$ and the unknown coefficients of the polynomials X_i. Finally,

[2]There exist upper bounds for a particular input. But usually, these bounds are too high and one tries smaller values.

we can seek for a non-trivial solution for (9) and thus for (7). To optimize the search for a non-trivial solution we make use of homomorphic image computations in our implementation. More precisely, we plug in some concrete integers for the parameters and reduce all integer coefficients modulo a prime. If there is no solution in the homomorphic setting, there is no solution in the general setting. By choosing these values sufficiently generically we can also minimize the risk of obtaining a homomorphic solution that does not extend to a general solution. In the end, we clear denominators in n such that the $e_i(n)$ turn to polynomials.

If $F(n; \ldots, x_{i-1}, u_i, x_{i+1}, \ldots) = 0$ and $F(n; \ldots, x_{i-1}, o_i, x_{i+1}, \ldots) = 0$ then

$$\mathcal{I}(n) := \int_{u_d}^{o_d} \ldots \int_{u_1}^{o_1} F(n; x_1, \ldots, x_d) dx_1 \ldots dx_d,$$

satisfies the homogeneous linear recurrence equation with polynomial coefficients

$$\sum_{i=0}^{L} e_i(n)\mathcal{I}(n+i) = 0. \tag{10}$$

The general method now is straightforward: Given an integrand of the form (5), we can set $L = 0$, look for degree bounds for $X_i(x_1, \ldots, x_d)$ and try to find a solution of (10) by coefficient comparison. If we do not find a solution of (10) with not all $e_i(n)$'s equal to zero (we stop the calculation if the homomorphic image check fails), we increase L by one, look for new degree bounds for $X_i(x_1, \ldots, x_d)$ and try again to find a solution of (10). Again, if we do not find a solution with not all $e_i(n)$'s equal to zero, we increase L by one and repeat the process.

The discrete multiple Almkvist-Zeilberger algorithm is implemented in the command `mAZ` of `MultiIntegrate`.

Once we found a recurrence we exploit algorithms from [12, 32, 34, 36] which can constructively decide if a solution with certain initial values is expressible in terms of indefinite nested products and sums. This covers harmonic sums [16, 40], S-sums [8, 30], cyclotomic sums [6] and binomial sums [10, 27] as special cases. In our implementation we make use of the algorithms implemented in the summation package `Sigma`. For details on which solutions can be found using `Sigma`, we refer to [18].

2.2 *Dealing with Non-Standard Boundary Conditions*

Unfortunately, in many cases the integrand (5) does not vanish at the integration bounds and we end up in a linear recurrence with a non-trivial inhomogeneous part which can be written as a linear combination of integrals with at least one integral operator less. In the following we will deal with non-standard boundary conditions in two different ways, see [1].

2.2.1 Dealing with Inhomogeneous Recurrences

In [1] a method that deals with the inhomogeneous recurrence similar to [18] can be found. It gives rise to a recursive method. To be more precise, we consider the integral

$$\mathcal{I}(n) := \int_{u_d}^{o_d} \cdots \int_{u_1}^{o_1} F(n; x_1, \ldots, x_d) dx_1 \ldots dx_d.$$

Suppose that we found

$$\sum_{i=0}^{L} e_i(n) F(n+i; x_1, \ldots, x_d) = \sum_{i=1}^{d} D_{x_i} G_i(n; x_1, \ldots, x_d) \tag{11}$$

where at least one $G_i(n; x_1, \ldots, x_d)$ does not vanish at the integration limits. By integration with respect to $x_1, \ldots, x_d$ we can deduce that $\mathcal{I}(n)$ satisfies the inhomogeneous linear recurrence equation

$$\sum_{i=0}^{L} e_i(n) \mathcal{I}(n+i) =$$
$$\sum_{i=1}^{d} \int_{u_d}^{o_d} \cdots \int_{u_{i-1}}^{o_{i-1}} \int_{u_{i+1}}^{o_{i+1}} \cdots \int_{u_1}^{o_1} O_i(n) dx_1 \ldots dx_{i-1} dx_{i+1} \ldots dx_d$$
$$- \sum_{i=1}^{d} \int_{u_d}^{o_d} \cdots \int_{u_{i-1}}^{o_{i-1}} \int_{u_{i+1}}^{o_{i+1}} \cdots \int_{u_1}^{o_1} U_i(n) dx_1 \ldots dx_{i-1} dx_{i+1} \ldots dx_d$$

with

$$U_i(n) := G_i(n; x_1, \ldots, x_{i-1}, o_i, x_{i+1} \ldots, x_d)$$
$$O_i(n) := G_i(n; x_1, \ldots, x_{i-1}, u_i, x_{i+1} \ldots, x_d).$$

Note that the inhomogeneous part of the above recurrence equation is a sum of $2 \cdot d$ integrals of dimension $d - 1$, which fit again into the input class of the multiple Almkvist-Zeilberger algorithm. Hence we can apply the algorithms to the $2 \cdot d$ integrals recursively until we arrive at the base case of one-dimensional integrals for which we have to solve an inhomogeneous linear recurrence relation where the inhomogeneous part is free of integrals. Given the solutions for the one-dimensional integrals we can step by step find the solutions of higher dimensional integrals until we finally find the solution for $\mathcal{I}(n)$ by solving again an inhomogeneous linear recurrence equation and combining it with the initial values. Note that we have to calculate initial values with respect to n for all the integrals arising in this process.

Summarizing, with these algorithms we use the following strategy (note that we assume that we are able to compute the initial values for the arising integrals); compare [1, 18]:

Divide and Conquer Strategy

1. BASE CASE: If $\mathcal{I}(n)$ has no integration quantifiers, return $\mathcal{I}(n)$.
2. DIVIDE: As worked out above, compute a recurrence relation

$$a_0(n)\mathcal{I}(n) + \cdots + a_d(n)\mathcal{I}(n+d) = h(n) \tag{12}$$

   with polynomial coefficients $a_i(n) \in \mathbb{K}[n]$, $a_m(n) \neq 0$ and the right side $h(n)$ containing a linear combination of hyperexponential multi-integrals each with less than d integration quantifiers.
3. CONQUER: Apply the strategy recursively to the simpler integrals in $h(n)$. This results in an indefinite nested product-sum expressions $\tilde{h}(n)$ with

$$\tilde{h}(n) = h(n), \quad \forall n \geq \delta \text{ for some } \delta \in \mathbb{N}. \tag{13}$$

   If the method fails to find the $\tilde{h}(n)$ in terms of indefinite nested product-sum expressions, STOP.
4. COMBINE: Given (12) with (13), compute, if possible, $\tilde{\mathcal{I}}(n)$ in terms of nested product-sum expressions such that

$$\tilde{\mathcal{I}}(n) = \mathcal{I}(n), \quad \forall n \geq \delta \text{ for some } \delta \in \mathbb{N}. \tag{14}$$

   by solving the recurrence.

This divide and conquer strategy is implemented in the command `mAZIntegrate` of `MultiIntegrate`.

Remark 1 We remark that this approach works nicely, if the initial values of the integrals in the inhomogeneous part can be calculated efficiently. Further details on this approach are given in [1, 7]. We remark further that similar approaches have been explored in [18, 19] and [7] based on [41, 43] and [35], respectively, in order to derive recurrences for hypergeometric multi-sums.

2.2.2 Adapting the Ansatz to Find Homogeneous Recurrences

In order to avoid the difficulties of inhomogeneous recurrences we adapt the ansatz. Namely, we can always obtain a homogeneous recurrence of the form (10) by changing (8) to

$$G_i(n; x_1, \ldots, x_d) = \overline{H}(n; x_1, \ldots, x_d) \cdot r_i(x_1, \ldots, x_d) \cdot X_i(x_1, \ldots, x_d)(x_i - u_i)(x_i - o_i), \tag{15}$$

i.e., the G_i are forced to vanish at the integration bounds. Then with this Ansatz (9) the underlying linear system turns into

$$\sum_{i=1}^{d} [D_{x_i} r_i(x_1, \ldots, x_d) + q_i(x_1, \ldots, x_d)] \cdot X_i(x_1, \ldots, x_d)(x_i - u_i)(x_i - o_i) + r_i(x_1, \ldots, x_d) \cdot D_{x_i} X_i(x_1, \ldots, x_d)(x_i - u_i)(x_i - o_i) = \sum_{i=0}^{L} e_i(N)\, P(N; x_1, \ldots, x_d)\, s(x_1, \ldots, x_d)^i\, t(x_1, \ldots, x_d)^{L-i}. \tag{16}$$

The general method now is straightforward: Given an integrand of the form (5), we can set $L = 0$, look for degree bounds for $X_i(x_1, \ldots, x_d)$ and try to find a solution of (16) by coefficient comparison. If we do not find a solution of (16) with not all $e_i(n)$'s equal to zero, we increase L by one, look for new degree bounds for $X_i(x_1, \ldots, x_d)$ and try again to find a solution of (16). Again, if we do not find a solution with not all $e_i(n)$'s equal to zero, we increase L by one and repeat the process.
Once we found a recurrence we can use the recurrence solver implemented in the summation package `Sigma` to try to solve it.

This strategy is implemented in the command `mAZDirectIntegrate` of `MultiIntegrate`.

Remark 2 The advantage of this approach is, that we do not have to deal with integrals (and initial conditions) recursively, since the recurrence is homogenous, however the additional conditions on the ansatz might increase the order of the recurrence drastically. In particular, the routine is more robust: no abortion can occur due to problematic integral arising from the recursion.

2.3 Computing Series Expansions of the Integrals

Due to time and memory limitations, not finding all solutions of the recurrences or due to missing initial values (in full generality) we might fail to process certain integrals using the methods described in the previous subsection. Therefore, inspired by [18], a method which computes ε-expansions of integrals of the form (3) was developed in [1]. In the following we recall this method. We assume that the integral $\mathcal{I}(\varepsilon, n)$ from (3) has a Laurent expansion in ε for each $n \in \mathbb{N}$ with $n \geq \lambda$ for some $\lambda \in \mathbb{N}$ and thus it is an analytic function in ε throughout an annular region centered by 0 where the pole at $\varepsilon = 0$ has some order $K \in \mathbb{Z}$. Hence we can write it in the form

$$\mathcal{I}(\varepsilon, n) = \sum_{k=-K}^{\infty} \varepsilon^k I_k(n). \tag{17}$$

In the following we try to find the first coefficients $I_t(n), I_{t+1}(n), \ldots, I_u(n)$ in terms of indefinite nested product-sum expressions of the expansion

$$\mathcal{I}(\varepsilon, n) = I_t(n)\varepsilon^t + I_{t+1}(n)\varepsilon^{t+1} + I_{t+2}(n)\varepsilon^{t+2} + \ldots \tag{18}$$

with $t = -K \in \mathbb{Z}$. We start by computing a recurrence for $\mathcal{I}(\varepsilon, n)$ in the form

$$\begin{aligned} a_0(\varepsilon, n)J(\varepsilon, n) + a_1(\varepsilon, n)J(\varepsilon, n+1) + \cdots + a_d(\varepsilon, n)J(\varepsilon, n+d) \\ = h_{-K}(n)\varepsilon^{-K} + h_{-K+1}(n)\varepsilon^{-K+1} + \cdots + h_u(n)\varepsilon^u + \ldots \end{aligned} \tag{19}$$

In order to accomplish this task, we can use the methods presented in the previous section. Given the recurrence we exploit an algorithm from [18] which can constructively decide if a formal Laurent series solution with certain initial values is expressible (up to a certain order) in terms of indefinite nested products and sums.

This algorithm is implemented in the package `Sigma` and can be summarized as follows (see [18] and compare [1, 11]).

Suppose we are given the linear recurrence (19) of order d where the $a_i(\varepsilon, n)$ are polynomials in n and ε and where the inhomogeneous part can be expanded in ε up to order u. Consider a function which has a Laurent series expansion

$$\mathcal{I}(\varepsilon, n) = F_t(n)\varepsilon^t + F_{t+1}(n)\varepsilon^{t+1} + \dots \tag{20}$$

and which is a solution of the given recurrence for all $n \geq n_0$ for some $n_0 \in \mathbb{N}$. Then together with the d initial values $F_j(n_0), \dots, F_j(n_0+d-1)$ with $t \leq j \leq u$, all values $F_t(n), \dots, F_u(n)$ with $n \geq n_0$ can be computed provided that the values $h_i(n)$ for all i with $t \leq i \leq u$ and all integers n with $n \geq n_0$ can be computed. In addition, if the $h_t(n), \dots, h_u(n)$ are given explicitly in terms of indefinite nested product-sum expressions, there is an algorithm which decides constructively if the $F_t(n), \dots, F_u(n)$ can be given in terms of indefinite nested product-sum expressions.

Having such a Laurent series recurrence solver in hand we can combine it with the methods from the previous sections. Let $\mathcal{I}(\varepsilon, n)$ be a multi-integral of the form (3) and assume that $\mathcal{I}(\varepsilon, n)$ has a series expansion (18) for all $n \geq \lambda$ for some $\lambda \in \mathbb{N}$. If we succeed in finding a homogeneous differential equation, for instance by using the method form Sect. 2.2.2 we can directly apply the Laurent series recurrence solver, supposing that we can handle the initial values. This has been exploited in the frame of [11].

This strategy is implemented in the command `mAZExpandedDirect-Integrate` of `MultiIntegrate`.

Of course we can again think of a recursive method to compute the first coefficients (compare [1, 18]), say $F_t(n), \dots, F_u(n)$ of (18). Note that we have the same advantages and disadvantages as mentioned for the recursive method in Sect. 2.2.1, but if we assume that we can handle the initial values we can use the following strategy.

Divide and Conquer Strategy

1. BASE CASE: If $\mathcal{I}(\varepsilon, n)$ has no integration quantifiers, compute the expansion by standard methods.
2. DIVIDE: As worked out before, compute a recurrence relation

$$a_0(\varepsilon, n)\mathcal{I}(\varepsilon, n) + \dots + a_d(\varepsilon, n)\mathcal{I}(\varepsilon, n+d) = h(\varepsilon, n) \tag{21}$$

with polynomial coefficients $a_i(\varepsilon, n) \in \mathbb{K}[\varepsilon, n]$, $a_m(\varepsilon, n) \neq 0$ and the right side $h(\varepsilon, n)$ containing a linear combination of hyperexponential multi-integrals each with less than d integration quantifiers.
3. CONQUER: Apply the strategy recursively to the simpler integrals in $h(\varepsilon, n)$. This results in an expansion of the form

$$h(\varepsilon, n) = h_t(n) + h_1(n)\varepsilon + \cdots + h_u(n)\varepsilon^u + O(\varepsilon^{u+1}); \tag{22}$$

if the method fails to find the $h_t(n), \ldots, h_u(n)$ in terms of indefinite nested product-sum expressions, STOP.
4. COMBINE: Given (21) with (22), compute, if possible, the $F_t(n), \ldots, F_u(n)$ of (18) in terms of nested product-sum expressions by using `Sigma`.

This divide and conquer strategy is implemented in the command `mAZ-ExpandedIntegrate` of `MultiIntegrate`.

3 A Fine-Tuned Continuous Multi-Variate Almkvist-Zeilberger Algorithm

In this section we present a method to compute integrals of the form (4), that is based on slight modifications of the continuous multi-variate Almkvist-Zeilberger algorithm [14] and implemented in the package `MultiIntegrate`. Unlike in the discrete case, this method relies on finding and solving differential equations. In general, consider the hyperexponential integrand

$$F(x; x_1, \ldots, x_d) = P(x; x_1, \ldots, x_d) \cdot H(x; x_1, \ldots, x_d), \tag{23}$$

with a multivariate polynomial $P(x; x_1, \ldots, x_d) \in \mathbb{K}[x, x_1, \ldots, x_d]$ and

$$H(x; x_1, \ldots, x_d) = e^{\frac{a(x,x_1,\ldots,x_d)}{b(x,x_1,\ldots,x_d)}} \cdot \left(\prod_{p=1}^{P} S_p(x, x_1, \ldots, x_d)^{\alpha_p} \right),$$

where $a(x, x_1, \ldots, x_d)$, $b(x, x_1, \ldots, x_d)$ and $S_p(x, x_1, \ldots, x_d) \in \mathbb{K}[x, x_1, \ldots, x_d]$, with $b(x, x_1, \ldots, x_d) \neq 0$, and $\alpha_p \in \mathbb{K}$. Then due to [14] there exists a non-negative integer L, there exist $e_0(x), e_1(x), \ldots, e_L(x) \in \mathbb{K}[x]$ (or equivalently from $\mathbb{K}(x)$), *not all zero*, and there also exist $R_i(x; x_1, \ldots, x_d) \in \mathbb{K}(x, x_1, \ldots, x_d)$ such that

$$G_i(x; x_1, \ldots, x_d) := R_i(x; x_1, \ldots, x_d) F(x; x_1, \ldots, x_d) \tag{24}$$

satisfy the integrand differential equation

$$\sum_{i=0}^{L} e_i(x) D_x^i F(x; x_1, \ldots, x_d) = \sum_{i=1}^{d} D_{x_i} G_i(x; x_1, \ldots, x_d). \tag{25}$$

3.1 The General Method

The proof of the existence, and in particular a method to compute such a differential equation (25), is based on the following observation [14]. Fix a non-negative integer L (with the role given above), define

$$\overline{H}(x; x_1, \ldots, x_d) := \frac{e^{\frac{a(x,x_1,\ldots,x_d)}{b(x,x_1,\ldots,x_d)}}}{b(x, x_1, \ldots, x_d)^{2L}} \cdot \left(\prod_{p=1}^{P} S_p(x, x_1, \ldots, x_d)^{\alpha_p} \right),$$

Then we have

$$\sum_{i=0}^{L} e_i(x) D_x^i F(x; x_1, \ldots, x_d) = h(x, x_1, \ldots, x_d) \overline{H}(x; x_1, \ldots, x_d).$$

for some polynomial $h(x, x_1, \ldots, x_d)$ that can be determined and, by construction, the logarithmic derivatives of $\overline{H}(x; x_1, \ldots, x_d)$ are rational functions in the x_i, i.e., we have that

$$\frac{D_{x_i} \overline{H}(x; x_1, \ldots, x_d)}{\overline{H}(x; x_1, \ldots, x_d)} = \frac{q_i(x, x_1, \ldots, x_d)}{r_i(x, x_1, \ldots, x_d)}$$

for explicitly given $q_i(x, x_1, \ldots, x_d), r_i(x, x_1, \ldots, x_d) \in \mathbb{K}[x, x_1, \ldots, x_d]$.
For $i = 1, \ldots, d$ we make the general ansatz

$$G_i(x; x_1, \ldots, x_d) = \overline{H}(x; x_1, \ldots, x_d) \cdot r_i(x, x_1, \ldots, x_d) \cdot X_i(x; x_1, \ldots, x_d). \tag{26}$$

Then it turns out that for L chosen sufficiently large[3] there exist polynomials $X_i(x; x_1, \ldots, x_d) \in \mathbb{K}[x][x_1, \ldots, x_d]$ with $1 \leq i \leq L$ and polynomials $e_i(x) \in \mathbb{K}[x]$ (not all zero) such that (25) holds. Motivated by this fact, one searches for

[3]There exist upper bounds for a particular input. But usually, these bounds are too high and one tries smaller values.

these unknowns X_i and e_i as follows. Note that the ansatz (25) is equivalent to (see [14])

$$\sum_{i=1}^{d}[D_{x_i} r_i(x, x_1, \ldots, x_d) + q_i(x, x_1, \ldots, x_d)] \cdot X_i(x; x_1, \ldots, x_d)$$
$$+ r_i(x, x_1, \ldots, x_d) \cdot D_{x_i} X_i(x, x_1, \ldots, x_d) = h(x, x_1, \ldots, x_d). \quad (27)$$

Finally, we choose appropriate degree bounds w.r.t. the $x_1, \ldots, x_d$ for the X_i $(1 \leq i \leq d)$ and plug the polynomials with unknown coefficients from $\mathbb{K}[x]$ (from $\mathbb{K}(x)$) into (27). By coefficient comparison this yields a linear system in $\mathbb{K}(x)$ with the unknowns $e_i(x)$ and the unknown coefficients of the polynomials X_i. Finally, we can seek a non-trivial solution for (27) and thus for (25). In the end, we clear denominators in x such that the $e_i(x)$ turn to polynomials.
If $F(x; \ldots, x_{i-1}, u_i, x_{i+1}, \ldots) = 0$ and $F(x; \ldots, x_{i-1}, o_i, x_{i+1}, \ldots) = 0$ then

$$\mathcal{I}(x) := \int_{u_d}^{o_d} \ldots \int_{u_1}^{o_1} F(x; x_1, \ldots, x_d) dx_1 \ldots dx_d,$$

satisfies the homogeneous linear differential equation with polynomial coefficients

$$\sum_{i=0}^{L} e_i(x) D_x^i \mathcal{I}(x) = 0. \quad (28)$$

The general method now is straightforward: Given an integrand of the form (23), we can set $L = 0$, look for degree bounds for $X_i(x, x_1, \ldots, x_d)$ and try to find a solution of (28) by coefficient comparison. If we do not find a solution of (28) with not all $e_i(x)$'s equal to zero (with homomorphic image testing to decide non-existence efficiently), we increase L by one, look for new degree bounds for $X_i(x, x_1, \ldots, x_d)$ and try again to find a solution of (28). Again, if we do not find a solution with not all $e_i(x)$'s equal to zero, we increase L by one and repeat the process.

> The continuous Almkvist-Zeilberger algorithm is implemented in the command `cmAZ` of `MultiIntegrate`.

Once we found a differential equation we can make use of the differential equation solver implemented in `HarmonicSums`. This solver finds all solutions of holonomic differential equations that can be expressed in terms of iterated integrals over hyperexponential alphabets [3, 12, 13, 22, 32] (with harmonic polylogarithms [33], cyclotomic polylogarithms [6] and iterated integrals over root-valued alphabets [10] as special cases); these solutions are called d'Alembertian solutions [12], in addition for differential equations of order two it finds all solutions that are Liouvillian [4, 29].

Example 1 (cmAZ) The following problem, which was already solved in [21], was communicated to us by D. Broadhurst. The goal is to find a differential equation satisfied by

$$Y(h) = \int_0^1 \int_u^1 \frac{1}{\sqrt{uv(1-u)(1-v)(1-uh)(1-(1-v)h)}} dv du. \tag{29}$$

In order to fit (29) to the requirements of the AZ-algorithm we transform it using the substitution $v \to u/(1+(u-1)z)$, which leads to

$$\int_0^1 \int_0^1 \frac{1}{\sqrt{(1-hu)(z-1)(1+(u-1)z)(h(u-1)(z-1)+z-uz-1)}} dz du. \tag{30}$$

Now we can apply our implementation:

In[1]:= $\mathrm{cmAZ}\left[\frac{1}{\sqrt{(1-hu)(z-1)(1+(u-1)z)(h(u-1)(z-1)+z-uz-1)}}, h, \{u, z\},\right.$
$\left.\mathrm{AddFactors} \to \left\{(1-u)^3(1-z)^6, (1-z)^3\right\}\right]$

Out[1]= $-1 + 2h + 2\left(1 - 7h + 7h^2\right) D_h + 6(-1+h)h(-1+2h)D_h^2 + 2(-1+h)^2h^2D_h^3$

Note that in this example the integrand is not vanishing at the integration bounds, still we could derive a homogeneous differential equation, for details we refer to the next session. However, here the right hand side can be computed easily and we find the following differential equation, which is equivalent to the one found in [21]:

$$\left((h-1)^2h^2D_h^3 + 3(h-1)h(2h-1)D_h^2 + (1-7h+7h^2)D_h + h - \tfrac{1}{2}\right)$$
$$Y(h) = \frac{(h^2+4h-4)}{\sqrt{1-h}(2-h)^2}.$$

In a similar way this was already proven by D. van Straten.

3.2 Dealing with Non-Standard Boundary Conditions

Unfortunately, in many cases the integrand (23) does not vanish at the integration bounds and we end up in a linear differential equation with a non-trivial inhomogeneous part which can be written as a linear combination of integrals with at least one integral operator less. In the following we will deal with non-standard boundary conditions in two different ways, similar to the discrete case of Sect. 2.

3.2.1 Dealing with Inhomogeneous Differential Equations

In the previous section a method that deals with the inhomogeneous recurrences was stated, here we will use similar considerations that will give rise to a recursive method. To be more precise, we consider the integral

$$\mathcal{I}(x) := \int_{u_d}^{o_d} \cdots \int_{u_1}^{o_1} F(x; x_1, \ldots, x_d) dx_1 \ldots dx_d.$$

Suppose that we found

$$\sum_{i=0}^{L} e_i(x) D_x^i F(x; x_1, \ldots, x_d) = \sum_{i=1}^{d} D_{x_i} G_i(x; x_1, \ldots, x_d) \tag{31}$$

where at least one $G_i(x; x_1, \ldots, x_d)$ does not vanish at the integration limits. By integration with respect to $x_1, \ldots, x_d$ we can deduce that $\mathcal{I}(x)$ satisfies the inhomogeneous linear differential equation

$$\sum_{i=0}^{L} e_i(x) D_x^i \mathcal{I}(x) =$$

$$\sum_{i=1}^{d} \int_{u_d}^{o_d} \cdots \int_{u_{i-1}}^{o_{i-1}} \int_{u_{i+1}}^{o_{i+1}} \cdots \int_{u_1}^{o_1} O_i(x) dx_1 \ldots dx_{i-1} dx_{i+1} \ldots dx_d$$

$$- \sum_{i=1}^{d} \int_{u_d}^{o_d} \cdots \int_{u_{i-1}}^{o_{i-1}} \int_{u_{i+1}}^{o_{i+1}} \cdots \int_{u_1}^{o_1} U_i(x) dx_1 \ldots dx_{i-1} dx_{i+1} \ldots dx_d$$

with

$$U_i(x) := G_i(x; x_1, \ldots, x_{i-1}, o_i, x_{i+1} \ldots, x_d)$$
$$O_i(x) := G_i(x; x_1, \ldots, x_{i-1}, u_i, x_{i+1} \ldots, x_d).$$

Note that the inhomogeneous part of the above differential equation is a sum of $2 \cdot d$ integrals of dimension $d - 1$, which fit again into the input class of the continuous multiple Almkvist-Zeilberger algorithm. Hence we can apply the algorithms to the $2{\cdot}d$ integrals recursively until we arrive at the base case of one-dimensional integrals for which we have to solve an inhomogeneous differential equation where the inhomogeneous part is free of integrals. Given the solutions for the one-dimensional

integrals we can step by step find the solutions of higher dimensional integrals until we finally find the solution for $\mathcal{I}(x)$ by solving again an inhomogeneous linear differential equation and combining it with the initial conditions. Note that we have to calculate the initial conditions with respect to x for all the integrals arising in this process.
Summarizing, we use the following strategy (note that we assume that we are able to compute the initial conditions for the arising integrals):

Divide and Conquer Strategy

1. BASE CASE: If $\mathcal{I}(x)$ has no integration quantifiers, return $\mathcal{I}(x)$.
2. DIVIDE: As worked out above, compute a differential equation

$$a_0(x)\mathcal{I}(x) + a_1(x)D_x\mathcal{I}(x) + \cdots + a_d(x)D_x^d\mathcal{I}(x) = h(x) \tag{32}$$

   with polynomial coefficients $a_i(x) \in \mathbb{K}[x]$, $a_m(x) \neq 0$ and the right side $h(x)$ containing a linear combination of hyperexponential multi-integrals each with less than d integration quantifiers.
3. CONQUER: Apply the strategy recursively to the simpler integrals in $h(x)$. This results in an iterated integral expressions $\tilde{h}(x)$ with

$$\tilde{h}(x) = h(x). \tag{33}$$

   If the method fails to find the $\tilde{h}(x)$ in terms of iterated integral expressions, STOP.
4. COMBINE: Given (32) with (33), compute, if possible, $\tilde{\mathcal{I}}(x)$ in terms of iterated integral expressions such that

$$\tilde{\mathcal{I}}(x) = \mathcal{I}(x) \tag{34}$$

   by solving the differential equation.

This divide and conquer strategy is implemented in the command `cmAZ-Integrate` of `MultiIntegrate`.

Remark 3 We remark that this approach works nicely, if the initial conditions of the integrals in the inhomogeneous part can be calculated efficiently.

Example 2 (cmAZIntegrate) We consider the integral

$$\int_{-1}^{1}\int_{-1}^{1}\int_{-1}^{1}\int_{-1}^{1} e^{-x(w_1 w_2+w_3 w_4)} dw_4 dw_3 dw_2 dw_1 : \tag{35}$$

In[2]:= **cmAZIntegrate[e^{−x(w1 w2+w3 w4)}, x, {{w1, −1, 1}, {w2, −1, 1}, {w3, −1, 1}, {w4, −1, 1}}]**

Out[2]= $$\frac{8\left(\mathrm{G}\left(\frac{e^{-\tau}}{\tau},\frac{e^{-\tau}}{\tau};x\right)-\mathrm{G}\left(\frac{e^{-\tau}}{\tau},\frac{e^{\tau}}{\tau};x\right)-\mathrm{G}\left(\frac{e^{\tau}}{\tau},\frac{e^{-\tau}}{\tau};x\right)+\mathrm{G}\left(\frac{e^{\tau}}{\tau},\frac{e^{\tau}}{\tau};x\right)\right)}{x^2}$$

Note that the iterated integrals are defined recursively by

$$\mathrm{G}\left(f_1(\tau), f_2(\tau), \cdots, f_k(\tau); x\right) = \int_0^x f_1(\tau_1)\mathrm{G}\left(f_2(\tau), \cdots, f_k(\tau); \tau_1\right) d\tau_1,$$

with the special case $\mathrm{G}(x) = 1$, compare e.g., [5].
Here, in a first step the differential equation

$$2f(x) + xD_x f(x) =$$
$$\int_{-1}^{1}\int_{-1}^{1}\int_{-1}^{1} e^{x(-w_1 w_2+w_4)} dw_4 dw_2 dw_1 + \int_{-1}^{1}\int_{-1}^{1}\int_{-1}^{1} e^{-x(w_1 w_2+w_4)} dw_4 dw_2 dw_1$$
$$+\int_{-1}^{1}\int_{-1}^{1}\int_{-1}^{1} e^{-x(w_2+w_3 w_4)} dw_4 dw_3 dw_2 + \int_{-1}^{1}\int_{-1}^{1}\int_{-1}^{1} e^{x(w_2-w_3 w_4)} dw_4 dw_3 dw_2$$

is computed. The procedure is applied recursively to all the integrals on the right hand side, which leads to

$$2f(x) + xf'(x) = -\frac{8e^{-x}\left(e^{2x}-1\right)\left(\mathrm{G}\left(\frac{e^{-\tau}}{\tau};x\right)-\mathrm{G}\left(\frac{e^{\tau}}{\tau};x\right)\right)}{x^2}$$

Finally, solving this differential equation and combining with initial conditions yields the result.

3.2.2 Adapting the Ansatz to Find Homogeneous Differential Equations

In order to avoid the difficulties of inhomogeneous differential equations we adapt the ansatz. Namely, we can always obtain a homogeneous differential equation of the form (28) by changing (26) to

$$G_i(x; x_1, \ldots, x_d) = \overline{H}(x; x_1, \ldots, x_d) \cdot r_i(x_1, \ldots, x_d) \cdot X_i(x_1, \ldots, x_d)(x_i - u_i)(x_i - o_i), \tag{36}$$

i.e., the G_i are forced to vanish at the integration bounds. Then with this ansatz (27) the underlying linear system turns into

$$\sum_{i=1}^{d} [D_{x_i} r_i(x_1, \ldots, x_d) + q_i(x_1, \ldots, x_d)] \cdot X_i(x_1, \ldots, x_d)(x_i - u_i)(x_i - o_i) \tag{37}$$
$$+ r_i(x, x_1, \ldots, x_d) \cdot D_{x_i} X_i(x_1, \ldots, x_d)(x_i - u_i)(x_i - o_i) = h(x, x_1, \ldots, x_d).$$

The general method now is straightforward: Given an integrand of the form (23), we can set $L = 0$, look for degree bounds for $X_i(x, x_1, \ldots, x_d)$ and try to find a solution of (37) by coefficient comparison. If we do not find a solution of (37) with not all $e_i(x)$'s equal to zero (again homomorphic image testing is used for speedups), we increase L by one, look for new degree bounds for $X_i(x, x_1, \ldots, x_d)$ and try again to find a solution of (37). Again, if we do not find a solution with not all $e_i(x)$'s equal to zero, we increase L by one and repeat the process.
Once we found a differential equation we can use the differential equation solver implemented in the package `HarmonicSums` to try to find a closed form solution.

This strategy implemented in the command `cmAZDirectIntegrate` of `MultiIntegrate`.

Remark 4 The advantage of this approach is, that we do not have to deal with integrals (and initial conditions) recursively, since the differential equation is homogenous, however the additional conditions on the ansatz might increase the order of the differential equation drastically.

Example 3 (cmAZDirectIntegrate) We consider again the integral given in (35):

In[3]:= **cmAZDirectIntegrate[$e^{-x(w_1*w_2+w_3*w_4)}$, x, {{w_1, −1, 1}, {w_2, −1, 1}, {w_3, −1, 1}, {w_4, −1, 1}}]**

Out[3]= $$\frac{8\left(\mathrm{G}\left(\frac{e^{-\tau}}{\tau}, \frac{e^{-\tau}}{\tau}; x\right) - \mathrm{G}\left(\frac{e^{-\tau}}{\tau}, \frac{e^{\tau}}{\tau}; x\right) - \mathrm{G}\left(\frac{e^{\tau}}{\tau}, \frac{e^{-\tau}}{\tau}; x\right) + \mathrm{G}\left(\frac{e^{\tau}}{\tau}, \frac{e^{\tau}}{\tau}; x\right)\right)}{x^2}$$

Here the differential equation

$$0 = 32\left(4x - 16x^3 + 9x^5\right) f(x) - 4\left(27 - 148x^2 + 598x^4 - 63x^6\right) f'(x)$$
$$- 4\left(117x - 568x^3 + 556x^5 - 9x^7\right) f''(x) - \left(478x^2 - 2919x^4 + 603x^6\right) f^{(3)}(x)$$
$$- 5\left(34x^3 - 247x^5 + 9x^7\right) f^{(4)}(x) - \left(23x^4 - 189x^6\right) f^{(5)}(x) - \left(x^5 - 9x^7\right) f^{(6)}(x)$$

is derived. Solving and combing it with the initial condition yields the result given in `Out[3]`.

3.3 Computing Series Expansions of the Integrals

Due to time and memory limitations, not finding all solutions of the differential equations or due to missing initial conditions (in full generality) we might fail to process certain integrals using the methods described in the previous subsection. Therefore, inspired by the previous section we are seeking a method which computes ε-expansions of integrals of the form (4).
Again we assume that the integral $\mathcal{I}(\varepsilon, x)$ from (4) has a Laurent expansion in ε for $x \in \mathbb{R}$ with $x_\alpha < x < x_\beta$ for some $x_\alpha < 0, x_\beta > 0 \in \mathbb{R}$ and thus it is an analytic function in ε throughout an annular region centered by 0 where the pole at $\varepsilon = 0$ has some order $K \in \mathbb{Z}$. Hence we can write it in the form

$$\mathcal{I}(\varepsilon, x) = \sum_{k=-K}^{\infty} \varepsilon^k I_k(x). \tag{38}$$

In the following we try to find the first coefficients $I_t(x), I_{t+1}(x), \ldots, I_u(x)$ in terms of iterated integral expressions of the expansion

$$\mathcal{I}(\varepsilon, x) = I_t(x)\varepsilon^t + I_{t+1}(x)\varepsilon^{t+1} + I_{t+2}(x)\varepsilon^{t+2} + \ldots \tag{39}$$

with $t = -K \in \mathbb{Z}$. Assume that we managed to compute a differential equation satisfied by $\mathcal{I}(\varepsilon, x)$ in the form

$$a_0(\varepsilon, x)J(\varepsilon, x) + a_1(\varepsilon, x)D_x J(\varepsilon, x) + \cdots + a_d(\varepsilon, x)D_x^d J(\varepsilon, x)$$
$$= h_{-K}(x)\varepsilon^{-K} + h_{-K+1}(x)\varepsilon^{-K+1} + \cdots + h_u(x)\varepsilon^u + \ldots. \tag{40}$$

In order to find such a differential equation we can use the methods presented in the previous subsections. In the package `HarmonicSums` we implemented an algorithm that tries to find (39), given a differential equation (40) and suitable initial conditions given as power series expansions about $x = 0$ starting from some $s \in \mathbb{Z}$:

$$
\begin{aligned}
I_t(x) &= I_{t,s}x^s + I_{t,s+1}x^{s+1} + \cdots + I_{t,s+2}x^{s+d-1} + O(x^{s+d}) \\
I_{t+1}(x) &= I_{t+1,s}x^s + I_{t+1,s+1}x^{s+1} + \cdots + I_{t+1,s+2}x^{s+d-1} + O(x^{s+d}) \\
&\vdots \\
I_u(x) &= I_{u,s}x^s + I_{u,s+1}x^{s+1} + \cdots + I_{u,s+2}x^{s+d-1} + O(x^{s+d})
\end{aligned}
\tag{41}
$$

In the following we will illustrate the basic calculation steps of this algorithm, which can be considered as the continuous version of the algorithm presented in [18]; see Sect. 2.3. Inserting the ansatz (39) into (40) yields

$$
\begin{aligned}
&a_0(\varepsilon, x)\Big[I_t(x)\varepsilon^t + I_{t+1}(x)\varepsilon^{t+1} + I_{t+2}(x)\varepsilon^{t+2} + \ldots\Big] + \\
&a_1(\varepsilon, x)\Big[D_x I_t(x)\varepsilon^t + D_x I_{t+1}(x)\varepsilon^{t+1} + D_x I_{t+2}(x)\varepsilon^{t+2} + \ldots\Big] \\
&+ \cdots + \\
&a_d(\varepsilon, x)\Big[D_x^d I_t(x)\varepsilon^t + D_x^d I_{t+1}(x)\varepsilon^{t+1} + D_x^d I_{t+2}(x)\varepsilon^{t+2} + \ldots\Big] \\
&\qquad = h_t(x)\varepsilon^t + h_{t+1}(x)\varepsilon^{t+1} + \cdots + h_u(x)\varepsilon^u + \ldots.
\end{aligned}
\tag{42}
$$

Since two Laurent series agree if they agree coefficient-wise, we obtain the following constraint for $I_t(x)$ by coefficient comparison:

$$
\sum_{k=0}^{d} a_k(0, x) D_x^k I_t(x) = h_t(x), \tag{43}
$$

with the initial condition given in (41). We are now in the position to try to find an explicit representation using `HarmonicSums`'s differential equation solver [3, 4]. We assume that we could find an iterated integral representation $\tilde{I}_t(x)$ such that $\tilde{I}_t(x) = I_t(x)$ for all $x \in (x_\alpha, x_\beta)$. In order to obtain the next coefficient of the Laurent series in ε, we insert $\tilde{I}_t(x)$ into (42), which yields

$$
\begin{aligned}
&a_0(\varepsilon, x)\Big[I_{t+1}(x)\varepsilon^{t+1} + I_{t+2}(x)\varepsilon^{t+2} + I_{t+3}(x)\varepsilon^{t+3} + \ldots\Big] + \\
&a_1(\varepsilon, x)\Big[D_x I_{t+1}(x)\varepsilon^{t+1} + D_x I_{t+2}(x)\varepsilon^{t+2} + D_x I_{t+3}(x)\varepsilon^{t+3} + \ldots\Big] \\
&+ \cdots + \\
&a_d(\varepsilon, x)\Big[D_x^d I_{t+1}(x)\varepsilon^{t+1} + D_x^d I_{t+2}(x)\varepsilon^{t+2} + D_x^d I_{t+3}(x)\varepsilon^{t+3} + \ldots\Big] \\
&\qquad = \tilde{h}_{t+1}(x)\varepsilon^{t+1} + \tilde{h}_{t+2}(x)\varepsilon^{t+1} + \cdots + \tilde{h}_u(x)\varepsilon^u + \ldots
\end{aligned}
\tag{44}
$$

with $\tilde{h}_i(x) = h_i(x) - g_i(x)$, where the $g_i(x)$ satisfy

$$a_0(\varepsilon, x)\tilde{I}_t(x) + a_1(\varepsilon, x)D_x\tilde{I}_t(x) + \cdots + a_d(\varepsilon, x)D_x^d\tilde{I}_t(x) \\ = h_t(x)\varepsilon^t + g_{t+1}(x)\varepsilon^{t+1} + \cdots + g_u(x)\varepsilon^u + \ldots.$$

Now we repeat the above procedure: by coefficient comparison we obtain the following constraint for $I_{t+1}(x)$:

$$\sum_{k=0}^{d} a_k(0, x)D_x^k I_{t+1}(x) = \tilde{h}_{t+1}(x). \tag{45}$$

Assuming that we can find a solution $\tilde{I}_{t+1}(x)$ of (45) in terms of iterated integrals that satisfy the initial condition from (41) such that $\tilde{I}_{t+1}(x) = I_{t+1}(x)$ for all $x \in (x_\alpha, x_\beta)$ we can update the ansatz (44):

$$\begin{aligned} &a_0(\varepsilon, x)\Big[I_{t+2}(x)\varepsilon^{t+2} + I_{t+3}(x)\varepsilon^{t+3} + I_{t+4}(x)\varepsilon^{t+4} + \ldots\Big]+ \\ &a_1(\varepsilon, x)\Big[D_x I_{t+2}(x)\varepsilon^{t+2} + D_x I_{t+3}(x)\varepsilon^{t+3} + D_x I_{t+4}(x)\varepsilon^{t+4} + \ldots\Big] \\ &+\cdots+ \\ &a_d(\varepsilon, x)\Big[D_x^d I_{t+2}(x)\varepsilon^{t+2} + D_x^d I_{t+3}(x)\varepsilon^{t+3} + D_x^d I_{t+4}(x)\varepsilon^{t+4} + \ldots\Big] \\ &\qquad = \tilde{\tilde{h}}_{t+2}(x)\varepsilon^{t+2} + \tilde{\tilde{h}}_{t+3}(x)\varepsilon^{t+3} + \cdots + \tilde{\tilde{h}}_u(x)\varepsilon^u + \ldots \end{aligned} \tag{46}$$

with $\tilde{\tilde{h}}_i(x) = \tilde{h}_i(x) - \tilde{g}_i(x)$, where the $\tilde{g}_i(x)$ satisfy

$$a_0(\varepsilon, x)\tilde{I}_{t+1}(x) + a_1(\varepsilon, x)D_x\tilde{I}_{t+1}(x) + \cdots + a_d(\varepsilon, x)D_x^d\tilde{I}_{t+1}(x) \\ = \tilde{h}_{t+1}(x)\varepsilon^t + \tilde{g}_{t+2}(x)\varepsilon^{t+2} + \cdots + \tilde{g}_u(x)\varepsilon^u + \ldots.$$

We can repeat this process as long as we can compute solutions and as long as needed. The illustrated calculation steps can be summarized with the following theorem.

Theorem 1 *Suppose we are given a linear differential equation*

$$a_0(\varepsilon, x)J(\varepsilon, x) + a_1(\varepsilon, x)D_x J(\varepsilon, x) + \cdots + a_d(\varepsilon, x)D_x^d J(\varepsilon, x) \\ = h_k(x)\varepsilon^k + h_{k+1}(x)\varepsilon^{k+1} + \cdots + h_u(x)\varepsilon^u + \ldots$$

of order d where the $a_i(\varepsilon, x)$ are polynomials in x and ε and where the inhomogeneous part can be expanded in ε up to order u in terms of expressions in iterated integrals over hyperexponential alphabets. Consider a function which has a Laurent series expansion

$$J(\varepsilon, x) = F_k(x)\varepsilon^k + F_{k+1}(x)\varepsilon^{k+1} + \dots$$

and which is a solution of the given differential equation for all $x \in \mathbb{R}$ with $x_\alpha < x < x_\beta$ for some $x_\alpha < 0, x_\beta > 0 \in \mathbb{R}$. Then together with the initial conditions

$$F_j(x) = F_{j,s}x^s + F_{j,s+1}x^{s+1} + \dots + F_{j,s+2}x^{s+d-1} + O(x^{s+d})$$

with $k \le j \le u$, all $F_k(x), \dots, F_u(x)$ with $x_\alpha < x < x_\beta$ can be computed in terms of expressions in iterated integrals over hyperexponential alphabets provided that the values $h_j(x)$ for all j with $k \le i \le u$ and $x_\alpha < x < x_\beta$ can be computed in terms of expressions in iterated integrals over hyperexponential alphabets.

This algorithm is implemented in the package `HarmonicSums` and with this implementation in hand we can try to find Laurent series solutions of integrals of the form (4). Let $I(\varepsilon, n)$ be a multi-integral of the form (4) and assume that $I(\varepsilon, x)$ has a series expansion (39) for all $x \in \mathbb{R}$ with $x_\alpha < x < x_\beta$ for some $x_\alpha < 0, x_\beta > 0 \in \mathbb{R}$. If we succeed in finding a homogeneous differential equation, for instance by using the method form Sect. 3.2.1 we can directly apply the Laurent series differential equation solver, supposing that we can handle the initial conditions.

This strategy is implemented in the command `cmAZExpandedDirect-Integrate` of `MultiIntegrate`.

Example 4 (cmAZExpandedDirectIntegrate) We consider the integral

$$I(\varepsilon, w) = \int_0^1 \int_0^1 \int_0^1 e^{xyw}((1-w)x(1-y))^{\frac{\varepsilon}{2}}((1-w)y(1-x)z(1-z))dzdydx \tag{47}$$

with the given initial condition

$$I(\varepsilon, w) = \underbrace{\frac{8}{3(2+\varepsilon)^2(4+\varepsilon)^2} - \frac{4w(28+\varepsilon(12+\varepsilon))}{3(2+\varepsilon)(4+\varepsilon)^2(6+\varepsilon)^2} + \frac{w^2(-1664+\varepsilon(12+\varepsilon(12+\varepsilon))(72+\varepsilon(16+\varepsilon)))}{3(2+\varepsilon)(4+\varepsilon)^2(6+\varepsilon)^2(8+\varepsilon)^2}}_{init:=} + O(w^3).$$

We want to find the first two terms of the ε−expansion of $I(\varepsilon, w)$, i.e., we want to compute $I_0(w)$ and $I_1(w)$ such that $I(\varepsilon, w) = I_0(w) + \varepsilon I_1(w) + O(\varepsilon^2)$. This can be achieved by using our implementation:

In[4]:= **cmAZExpandedDirectIntegrate[I(ε, w), w, {ε, 0, 1}, {{x, 0, 1}, {y, 0, 1}, {z, 0, 1}}, InitValues → init]**

Out[4]=
$$\Bigg\{\Bigg\{\frac{1}{6}-\frac{1}{6w}-\frac{\mathrm{G}\left(\frac{1-\varepsilon^{\tau}}{\tau};w\right)}{6w^2}+\frac{\mathrm{G}\left(\frac{1-\varepsilon^{\tau}}{\tau};w\right)}{6w},-\frac{1}{12w^2}+\frac{\varepsilon^{w}}{12w^2}+\frac{1}{12w}-\frac{\varepsilon^{w}}{12w}$$
$$-\frac{\mathrm{G}\left(\frac{1}{1-\tau};w\right)}{12}+\frac{\mathrm{G}\left(\frac{1}{1-\tau};w\right)}{12w}-\frac{\mathrm{G}\left(\frac{1-\varepsilon^{\tau}}{\tau};w\right)}{12}-\frac{\mathrm{G}\left(\frac{1-\varepsilon^{\tau}}{\tau};w\right)}{12w^2}+\frac{\mathrm{G}\left(\frac{1-\varepsilon^{\tau}}{\tau};w\right)}{6w}$$
$$+\frac{\mathrm{G}\left(\frac{1}{1-\tau},\frac{1-\varepsilon^{\tau}}{\tau};w\right)}{12w^2}-\frac{\mathrm{G}\left(\frac{1}{1-\tau},\frac{1-\varepsilon^{\tau}}{\tau};w\right)}{12w}-\frac{\mathrm{G}\left(\frac{1}{\tau},\frac{1-\varepsilon^{-\tau}}{\tau};w\right)}{12w^2}+\frac{\mathrm{G}\left(\frac{1}{\tau},\frac{1-\varepsilon^{-\tau}}{\tau};w\right)}{12w}$$
$$+\frac{\mathrm{G}\left(\frac{1}{\tau},\frac{1-\varepsilon^{\tau}}{\tau};w\right)}{12w^2}-\frac{\mathrm{G}\left(\frac{1}{\tau},\frac{1-\varepsilon^{\tau}}{\tau};w\right)}{12w}+\frac{\mathrm{G}\left(\frac{1-\varepsilon^{\tau}}{\tau},\frac{1}{1-\tau};w\right)}{12w^2}-\frac{\mathrm{G}\left(\frac{1-\varepsilon^{\tau}}{\tau},\frac{1}{1-\tau};w\right)}{12w}$$
$$+\frac{\mathrm{G}\left(\frac{1-\varepsilon^{\tau}}{\tau},\frac{1-\varepsilon^{-\tau}}{\tau};w\right)}{12w^2}-\frac{\mathrm{G}\left(\frac{1-\varepsilon^{\tau}}{\tau},\frac{1-\varepsilon^{-\tau}}{\tau};w\right)}{12w}\Bigg\},\{0,1\}\Bigg\}$$

Of course we can again think of a recursive method to compute the first coefficients, say $F_t(n), \ldots, F_u(n)$ of (39). Note that we have the same advantages and disadvantages as mentioned for the recursive method in Sect. 3.2.2, but if we assume that we can handle the initial conditions we can use to following strategy.

Divide and Conquer Strategy

1. BASE CASE: If $\mathcal{I}(\varepsilon, x)$ has no integration quantifiers, compute the expansion by standard methods.
2. DIVIDE: As worked out before, compute a differential equation
$$a_0(\varepsilon, x)\mathcal{I}(\varepsilon, x) + \cdots + a_d(\varepsilon, x)D_x^d\mathcal{I}(\varepsilon, x) = h(\varepsilon, x) \tag{48}$$
with polynomial coefficients $a_i(\varepsilon, x) \in \mathbb{K}[\varepsilon, x]$, $a_m(\varepsilon, x) \neq 0$ and the right side $h(\varepsilon, x)$ containing a linear combination of hyperexponential multi-integrals each with less than d integration quantifiers.
3. CONQUER: Apply the strategy recursively to the simpler integrals in $h(\varepsilon, x)$. This results in an expansion of the form
$$h(\varepsilon, x) = h_t(x) + h_1(x)\varepsilon + \cdots + h_u(x)\varepsilon^u + O(\varepsilon^{u+1}); \tag{49}$$
if the method fails to find the $h_t(x), \ldots, h_u(x)$ in terms of iterated integral expressions, STOP.

4. COMBINE: Given (48) with (49), compute, if possible, the $F_t(n), \ldots, F_u(n)$ of (39) in terms of iterated integral expressions by using `HarmonicSums`.

This divide and conquer strategy is implemented in the command `cmAZ-ExpandedIntegrate` of `MultiIntegrate`.

Example 5 (cmAZExpandedIntegrate) Again we consider the integral given in (47) with the same initial condition. In order to compute the first two terms of the ε-expansion of $I(\varepsilon, w)$, we can also use the following function call:

In[5]:= **cmAZExpandedIntegrate[I(ε, w), w,{ε, 0, 1}, {{x, 0, 1}, {y, 0, 1}, {z, 0, 1}}, InitValues → init]**

Out[5]=
$$\Bigg\{\Bigg\{\frac{1}{6}-\frac{1}{6w}-\frac{\mathrm{G}\left(\frac{1-\varepsilon^{\tau}}{\tau};w\right)}{6w^2}+\frac{\mathrm{G}\left(\frac{1-\varepsilon^{\tau}}{\tau};w\right)}{6w},-\frac{1}{12w^2}+\frac{\varepsilon^{w}}{12w^2}+\frac{1}{12w}-\frac{\varepsilon^{w}}{12w}$$
$$-\frac{\mathrm{G}\left(\frac{1}{1-\tau};w\right)}{12}+\frac{\mathrm{G}\left(\frac{1}{1-\tau};w\right)}{12w}-\frac{\mathrm{G}\left(\frac{1-\varepsilon^{\tau}}{\tau};w\right)}{12}-\frac{\mathrm{G}\left(\frac{1-\varepsilon^{\tau}}{\tau};w\right)}{12w^2}+\frac{\mathrm{G}\left(\frac{1-\varepsilon^{\tau}}{\tau};w\right)}{6w}$$
$$+\frac{\mathrm{G}\left(\frac{1}{1-\tau},\frac{1-\varepsilon^{\tau}}{\tau};w\right)}{12w^2}-\frac{\mathrm{G}\left(\frac{1}{1-\tau},\frac{1-\varepsilon^{\tau}}{\tau};w\right)}{12w}-\frac{\mathrm{G}\left(\frac{1}{\tau},\frac{1-\varepsilon^{-\tau}}{\tau};w\right)}{12w^2}+\frac{\mathrm{G}\left(\frac{1}{\tau},\frac{1-\varepsilon^{-\tau}}{\tau};w\right)}{12w}$$
$$+\frac{\mathrm{G}\left(\frac{1}{\tau},\frac{1-\varepsilon^{\tau}}{\tau};w\right)}{12w^2}-\frac{\mathrm{G}\left(\frac{1}{\tau},\frac{1-\varepsilon^{\tau}}{\tau};w\right)}{12w}+\frac{\mathrm{G}\left(\frac{1-\varepsilon^{\tau}}{\tau},\frac{1}{1-\tau};w\right)}{12w^2}-\frac{\mathrm{G}\left(\frac{1-\varepsilon^{\tau}}{\tau},\frac{1}{1-\tau};w\right)}{12w}$$
$$+\frac{\mathrm{G}\left(\frac{1-\varepsilon^{\tau}}{\tau},\frac{1-\varepsilon^{-\tau}}{\tau};w\right)}{12w^2}-\frac{\mathrm{G}\left(\frac{1-\varepsilon^{\tau}}{\tau},\frac{1-\varepsilon^{-\tau}}{\tau};w\right)}{12w}\Bigg\},\{0,1\}\Bigg\}$$

4 Conclusion

In this paper we summarize the theoretical background of our package `Multi-Integrate` which can be downloaded at https://risc.jku.at/software and which provides several methods to deal with multiple integrals over hyperexponential integrands.

Acknowledgments This work was supported by the Austrian Science Fund (FWF) grant SFB F50 (F5009-N15) and by the bilateral project WTZ BG 03/2019 (KP-06-Austria/8/2019), funded by OeAD (Austria) and Bulgarian National Science Fund. The author would like to thank C. Schneider for useful discussions.

References

1. J. Ablinger, Computer Algebra Algorithms for Special Functions in Particle Physics. PhD thesis, J. Kepler University Linz, 2012
2. J. Ablinger, The package HarmonicSums: Computer algebra and analytic aspects of nested sums, in *PoS LL2014*, 019 (2014)
3. J. Ablinger, Inverse mellin transform of holonomic sequences, in *PoS LL2016*, 067 (2016)
4. J. Ablinger, Computing the inverse mellin transform of holonomic sequences using Kovacic's algorithm, in *PoS RADCOR2017*, 069 (2018)
5. J. Ablinger, Discovering and proving infinite pochhammer sum identities. Experimental Mathematics (2019)
6. J. Ablinger, J. Blümlein, C. Schneider, Harmonic sums and polylogarithms generated by cyclotomic polynomials. J. Math. Phys. **52**(10), 102301 (2011)
7. J. Ablinger, J. Blümlein, M. Round, C. Schneider, Advanced computer algebra algorithms for the expansion of Feynman integrals, in *PoS LL2012*, 050 (2012)
8. J. Ablinger, J. Blümlein, C. Schneider, Analytic and algorithmic aspects of generalized harmonic sums and polylogarithms. J. Math. Phys. **54**(8), 082301 (2013)
9. J. Ablinger, J. Blümlein, A. De Freitas, A. Hasselhuhn, A. von Manteuffel, M. Round, C. Schneider, F. Wißbrock, The transition matrix element $A_{gq}(N)$ of the variable flavor number scheme at $O(\alpha_s^3)$. Nucl. Phys. B **882**, 263–288 (2014)
10. J. Ablinger, J. Blümlein, C.G. Raab, C. Schneider, Iterated binomial sums and their associated iterated integrals. J. Math. Phys. **55**(11), 112301 (2014)
11. J. Ablinger, A. Behring, J. Blümlein, A. De Freitas, A. von Manteuffel, C. Schneider, Calculating three loop ladder and V-topologies for massive operator matrix elements by computer algebra. Comput. Phys. Commun. **202**, 33–112 (2016)
12. S. Abramov, M. Petkovšek, D'Alembertian solutions of linear differential and difference equations, in *Proceedings of ISSAC'94*, ed. by J. von zur Gathen, pp. 169–174 (1994)
13. S. Abramov, E. Zima, D'Alembertian solutions of inhomogeneous linear equations (differential, difference, and some other), in *Proceedings of ISSAC'96*, ed. by Y.N. Lakshman, pp. 232–240 (1996)
14. M. Apagodu, D. Zeilberger. Multi-variable Zeilberger and Almkvist-Zeilberger algorithms and the sharpening of Wilf-Zeilberger theory. Adv. Appl. Math. **37**(2), 139–152 (2006)
15. I. Bierenbaum, J. Blümlein, S. Klein, Mellin moments of the $o(\alpha_s^3)$ heavy flavor contributions to unpolarized deep-inelastic scattering at $q^2 \gg m^2$ and anomalous dimensions. Nucl. Phys. B **820**(1), 417–482 (2009)
16. J. Blümlein, S. Kurth, Harmonic sums and Mellin transforms up to two-loop order. Phys. Rev. D **60**(1), 014018 (1999)
17. J. Blümlein, M. Kauers, S. Klein, C. Schneider, Determining the closed forms of the $O(a_s^3)$ anomalous dimensions and Wilson coefficients from Mellin moments by means of computer algebra. Comput. Phys. Commun. **180**(11), 2143–2165 (2009)
18. J. Blümlein, S. Klein, C. Schneider, F. Stan, A symbolic summation approach to Feynman integral calculus. J. Symb. Comput. **47**(10), 1267–1289 (2011)
19. J. Blümlein, M. Round, C. Schneider, Refined holonomic summation algorithms in particle physics. Adv. Comput. Algebra WWCA 2016 **226**, 51–91 (2018)
20. C. Bogner, S. Weinzierl, Feynman graph polynomials. Int. J. Modern Phys. A **25**, 2585–2618 (2010)
21. D. Broadhurst, W. Zudilin, A magnetic double integral. J. Aust. Math. Soc. **1**, 9–25 (2019)
22. M. Bronstein, Linear ordinary differential equations: breaking through the order 2 barrier, in *Proceedings of ISSAC'92*, pp. 42–48 (1992)
23. S. Chen, M. Kauers, Trading order for degree in creative telescoping. J. Symb. Comput. **47**(8), 968–995 (2012)
24. S. Chen, M. Kauers, Order-degree curves for hypergeometric creative telescoping, in *Proceedings of ISSAC'12*, pp. 122–129 (2012)

25. S. Chen, M. Kauers, C. Koutschan, A generalized apagodu-zeilberger algorithm, in *Proceedings of ISSAC'14*, pp. 107–114 (2014)
26. Y. Frishman, Operator products at almost light like distances. Ann. Phys. **66**, 373–389 (1971)
27. M.Y. Kalmykov, O. Veretin, Single scale diagrams and multiple binomial sums. Phys. Lett. B **483**, 315–323 (2000)
28. C. Koutschan, A fast approach to creative telescoping. Math. Comput. Sci. **4**(2–3), 259–266 (2010)
29. J.J. Kovacic, An algorithm for solving second order linear homogeneous differential equations. J. Symb. Comput. **2**, 3–43 (1986)
30. S.-O. Moch, P. Uwer, S. Weinzierl, Nested sums, expansion of transcendental functions, and multiscale multiloop integrals. J. Math. Phys. **43**(6), 3363–3386 (2002)
31. M. Mohammed, D. Zeilberger, Sharp upper bounds for the orders of the recurrences outputted by the Zeilberger and q-Zeilberger algorithms. J. Symb. Comput. **39**(2), 201–207 (2005)
32. M. Petkovšek, Hypergeometric solutions of linear recurrences with polynomial coefficients. J. Symb. Comput. **14**, 243–264 (1992)
33. E. Remiddi, J.A.M. Vermaseren, Harmonic polylogarithms. Int. J. Modern Phys. A **15**, 725–754 (2000)
34. C. Schneider, *Symbolic summation in difference fields*. PhD thesis, RISC, J. Kepler University Linz, May 2001
35. C. Schneider, A new Sigma approach to multi-summation. Adv. Appl. Math. **34**, 740–767 (2005)
36. C. Schneider, Solving parameterized linear difference equations in terms of indefinite nested sums and products. J. Differ. Equ. Appl. **11**(9), 799–821 (2006)
37. C. Schneider, Symbolic summation assists combinatorics. SÃminaire Lotharingien de Combinatoire **56**, 1–36 (2007)
38. C. Schneider, Simplifying multiple sums in difference fields, in *Computer Algebra in Quantum Field Theory: Integration, Summation and Special Functions*, ed. by C. Schneider, J. Blümlein. Texts & Monographs in Symbolic Computation (Springer, Wien, 2013), pp. 325–360
39. C. Schneider, Modern summation methods for loop integrals in quantum field theory: The packages sigma, EvaluateMultiSums and SumProduction. J. Phys. Conf. Ser. **523**, 012037 (2014)
40. J.A.M. Vermaseren, Harmonic sums, Mellin transforms and integrals. Int. J. Modern Phys. A **14**(13), 2037–2076 (1999)
41. K. Wegschaider, Computer generated proofs of binomial multi-sum identities. Master's thesis, RISC, J. Kepler University, May 1997
42. S. Weinzierl, Feynman graphs, in *Computer Algebra in Quantum Field Theory: Integration, Summation and Special Functions*, ed. by C. Schneider, J. Blümlein. Texts & Monographs in Symbolic Computation (Springer, Wien, 2013), pp. 381–406
43. H.S. Wilf, D. Zeilberger, An algorithmic proof theory for hypergeometric (ordinary and "q") multisum/integral identities. Inventiones Mathematicae **108**, 575–633 (1992)

Empirical Determinations of Feynman Integrals Using Integer Relation Algorithms

Kevin Acres and David Broadhurst

Abstract Integer relation algorithms can convert numerical results for Feynman integrals to exact evaluations, when one has reason to suspect the existence of reductions to linear combinations of a basis, with rational or algebraic coefficients. Once a tentative reduction is obtained, confidence in its validity is greatly increased by computing more decimal digits of the terms and verifying the stability of the result. Here we give examples of how the PSLQ and LLL algorithms have yielded remarkable reductions of Feynman integrals to multiple polylogarithms and to the periods and quasi-periods of modular forms. Moreover, these algorithms have revealed quadratic relations between Feynman integrals. A recent application concerning black holes involves quadratic relations between combinations of Feynman integrals with algebraic coefficients.

1 Introduction

The mathematical problem at hand is easy to state: given numerical approximations to $n > 2$ real numbers, x_k, is there at least one *probable* relation $\sum_{k=1}^{n} z_k x_k = 0$ with integer coefficients z_k, at least two of which are non-zero? If so, produce such a relation.

By way of example, in 1985 Broadhurst studied periods coming from 6-loop counterterms [7] in ϕ^4 theory and found, with good confidence, the relations

$$P_{6,1} = 168\zeta_9, \quad P_{6,2} = \frac{1063}{9}\zeta_9 + 8\zeta_3^3, \quad 16P_{6,3} + P_{6,4} = 1440\zeta_5\zeta_3 \tag{1}$$

K. Acres
Cryptography Research Centre, Technology Innovation Institute, Abu Dhabi, UAE
e-mail: kevin@tii.ae

D. Broadhurst (✉)
School of Physical Sciences, Open University, Milton Keynes, UK
e-mail: david.broadhurst@open.ac.uk

J. Blümlein, C. Schneider (eds.), *Anti-Differentiation and the Calculation of Feynman Amplitudes*, Texts & Monographs in Symbolic Computation,
https://doi.org/10.1007/978-3-030-80219-6_3

between the periods $P_{6,k}$, as labeled in the later census by Schnetz [25], and Riemann zeta values $\zeta_s = \sum_{n>0} 1/n^s$. There was a strong intuition that $P_{6,3}$ and $P_{6,4}$ might involve the multiple zeta value (MZV)

$$\zeta_{5,3} = \sum_{m>n>0} \frac{1}{m^5 n^3} = 0.037707672984847544011304 78\ldots \qquad (2)$$

along with ζ_8 and the product $\zeta_5\zeta_3$. Yet such relations were not discovered, at the level of accuracy then attainable.

A decade later, Broadhurst and Kreimer [8] solved this problem, by improving accuracy for the periods and using the PSLQ algorithm developed by Ferguson and Bailey [18], which identified

$$P_{6,3} = \frac{36}{5}\left(12\zeta_{5,3} - 29\zeta_8\right) + 252\zeta_5\zeta_3. \qquad (3)$$

Moreover, they found $\zeta_{3,5,3}$, with weight 11 and depth 3, in some 7-loop periods.

Much experimenting with PSLQ led to the Broadhurst-Kreimer conjecture [9] that the number $N(w,d)$ of *primitive* MZVs of weight w and depth d is generated by

$$\prod_{w>2}\prod_{d>0}(1 - x^w y^d)^{N(w,d)} = 1 - \frac{x^3 y}{1-x^2} + \frac{x^{12}y^2(1-y^2)}{(1-x^4)(1-x^6)} \qquad (4)$$

with a final term inferred by relating MZVs to *alternating* sums.

2 PSLQ and LLL

PSLQ came from work by Helaman Ferguson and Rodney Forcade [17] in 1977, implemented in *multiple-precision* FORTRAN by David Bailey in 1992, improved and parallelized by Bailey and Broadhurst [2] in 1999, and used extensively in the study of Feynman integrals since then.

The algorithm proceeds as follows. First we *initialize*:

1. For $j := 1$ to n: for $i := 1$ to n: if $i = j$ then set $A_{ij} := 1$ and $B_{ij} := 1$ else set $A_{ij} := 0$ and $B_{ij} := 0$; endfor; endfor.
2. For $k := 1$ to n: set $s_k :=$ **sqrt** $\left(\sum_{j=k}^{n} x_j^2\right)$; endfor. Set $t = 1/s_1$.
   For $k := 1$ to n: set $y_k := t x_k$; $s_k := t s_k$; endfor.
3. For $j := 1$ to $n-1$: for $i := 1$ to $j-1$: set $H_{ij} := 0$; endfor;
   set $H_{jj} := s_{j+1}/s_j$; for $i := j+1$ to n: set $H_{ij} := -y_i y_j/(s_j s_{j+1})$; endfor; endfor.

4. For $i := 2$ to n: for $j := i - 1$ to 1 step -1: set $t :=$ **round**(H_{ij}/H_{jj}); $y_j := y_j + t y_i$; for $k := 1$ to j: set $H_{ik} := H_{ik} - t H_{jk}$; endfor; for $k := 1$ to n: set $A_{ik} := A_{ik} - t A_{jk}$, $B_{kj} := B_{kj} + t B_{ki}$; endfor; endfor; endfor.

Thus the numerical data, in the vector x, is converted to a vector y, by taking square roots of partial sums of squares. Then the matrices H, A and B are created, with a crucial *rounding* in Step 4.

Then we proceed by *iteration*:

1. Select m such that $(\mathbf{4/3})^{i/2}|H_{ii}|$ is maximal when $i = m$. **Swap** the entries of y indexed m and $m+1$, the corresponding rows of A and H, and the corresponding columns of B.
2. If $m \leq n-2$ then set $t_0 :=$ **sqrt**$(H_{mm}^2 + H_{m,m+1}^2)$, $t_1 := H_{mm}/t_0$ and $t_2 := H_{m,m+1}/t_0$; for $i := m$ to n: set $t_3 := H_{im}$, $t_4 := H_{i,m+1}$, $H_{im} := t_1 t_3 + t_2 t_4$ and $H_{i,m+1} := -t_2 t_3 + t_1 t_4$; endfor; endif.
3. For $i := m+1$ to n: for $j := \min(i-1, m+1)$ to 1 step -1: set $t :=$ **round**(H_{ij}/H_{jj}) and $y_j := y_j + t y_i$; for $k := 1$ to j: set $H_{ik} := H_{ik} - t H_{jk}$; endfor; for $k := 1$ to n: set $A_{ik} := A_{ik} - t A_{jk}$ and $B_{kj} := B_{kj} + t B_{ki}$; endfor; endfor; endfor.
4. If the largest entry of A exceeds the precision, then **fail**, else if a component of the y vector is very small, then output the **relation** from the corresponding column of B, else go back to Step 1.

The constant $4/3$ in the *swap* of Step 1 of the iteration ensures that the algorithm will find a relation, provided that one exists and the data have been specified to sufficient accuracy. In practice, reliable results may be obtained with a smaller constant.

For big problems, *parallelized* PSLQ [2] has been vital, especially for the magnetic moment of the electron [21]. For smaller problems, there is an alternative.

2.1 LLL

In 1982, Arjen Lenstra, Hendrik Lenstra and László Lovász gave the LLL algorithm [22] for lattice reduction to a basis with short and almost orthogonal components. An extension of this underlies the `lindep` procedure in Pari-GP [24], which we here apply to the problem of determining the period $P_{6,3}$.

```
P63=107.71102484102;\\ only 14 digits specified
V=[P63,zetamult([5,3]),zeta(8),zeta(5)*zeta(3)];
for(d=11,16,U=lindep(V,d);U*=sign(U[1]);print([d,U~]));
[11, [4, -827, 173, -460]]
[12, [4, -827, 173, -460]]
[13, [4, -827, 173, -460]]
[14, [5, -432, 1044, -1260]]
[15, [5, -432, 1044, -1260]]
[16, [196, 1652, -9045, -9701]]
```

In this case, using 14 good digits of $P_{6,3}$, we happen to obtain the correct result (3). In practice, one would need several more digits, for better confidence.

In what follows, all claimed relations have been checked using at least 100 more digits than were needed for the discoveries, making the probability of a mistake less than $1/10^{100}$.

2.2 Improvement and Parallelization of PSLQ

Multi-level improvement: perform most operations at 64-bit precision, some at intermediate precision (Bailey and Broadhurst [2] chose 125 digits) and only the bare minimum of the most delicate operations at full precision (more than 10,000 digits, for some big problems).

Multi-pair improvement: swap up to $0.4n$ disjoint pairs of the n indices at each iteration. In this case, it is not proven that the algorithm will succeed, but it has not yet been found to fail.

Parallelization: distribute the disjoint-pair jobs; for each pair, distribute the full-precision matrix multiplication in the outermost loop.

2.3 Examples

Bailey and Broadhurst [2], working at 10,000 digits, found that the constant associated with the fourth bifurcation of the logistic map is the root of a polynomial of degree 240.

They tested a conjecture on alternating sums of the form

$$\zeta\begin{pmatrix}\sigma_1, \sigma_2 \cdots \sigma_d \\ s_1, s_2 \cdots s_d\end{pmatrix} = \sum_{k_1>k_2>\cdots>k_d>0} \frac{\sigma_1^{k_1}}{k_1^{s_1}} \frac{\sigma_2^{k_2}}{k_2^{s_2}} \cdots \frac{\sigma_d^{k_d}}{k_d^{s_d}} \tag{5}$$

where $\sigma_j = \pm 1$ are signs and $s_j > 0$ are integers, namely that at weight $w = \sum_j s_j$ every convergent alternating sum is a rational linear combination of elements of a basis of size $F_{w+1} = F_w + F_{w-1}$, i.e. the Fibonacci number with index $w + 1$. At $w = 11$, many integer relations of size $F_{12} + 1 = 145$ were found, at 5000-digit precision.

For weights $w \leq 20$, inverse binomial sums [6] of the form

$$S(w) = \sum_{n=1}^{\infty} \frac{1}{n^w \binom{2n}{n}} \tag{6}$$

were reduced to multiple polylogarithms of the sixth root of unity [10], with $S(20)$ given by 106 terms.

2.4 Relations in the Multiple Zeta Value Data Mine

The Broadhurst-Kreimer conjecture (4) came from the PSLQ discovery that

$$
\begin{aligned}
2^5 \cdot 3^3 \zeta_{4,4,2,2} = 2^{14} \sum_{m>n>0} \frac{(-1)^{m+n}}{(m^3 n)^3} &+ 2^5 \cdot 3^2\, \zeta_3^4 + 2^6 \cdot 3^3 \cdot 5 \cdot 13\, \zeta_9\, \zeta_3 \\
&+ 2^6 \cdot 3^3 \cdot 7 \cdot 13\, \zeta_7\, \zeta_5 + 2^7 \cdot 3^5\, \zeta_7\, \zeta_3\, \zeta_2 + 2^6 \cdot 3^5\, \zeta_5^2\, \zeta_2 \\
&- 2^6 \cdot 3^3 \cdot 5 \cdot 7\, \zeta_5\, \zeta_4\, \zeta_3 - 2^8 \cdot 3^2\, \zeta_6\, \zeta_3^2 - \frac{13177 \times 15991}{691} \zeta_{12} \\
&+ 2^4 \cdot 3^3 \cdot 5 \cdot 7\, \zeta_{6,2}\, \zeta_4 - 2^7 \cdot 3^3\, \zeta_{8,2}\, \zeta_2 - 2^6 \cdot 3^2 \cdot 11^2 \zeta_{10,2} \qquad (7)
\end{aligned}
$$

which shows that, at weight 12, a depth 4 MZV is reducible to terms of depth $d \leq 2$, and their products, if one allows an *alternating* double sum in the basis.

When constructing the MZV data mine, Blümlein, Broadhurst and Vermaseren [4] proved this, by massive use of computer algebra. It would be much harder to prove an LLL discovery at weight 21 and depth 7, where

$$
81 \zeta_{6,2,3,3,5,1,1} + 326 \sum_{j>k>l>m>n>0} \frac{(-1)^{k+m}}{(jk^2 l m^2 n)^3} \qquad (8)
$$

was reduced to 150 terms containing MZVs of depths $d \leq 5$.

3 Counterterms at 7 Loops

Broadhurst found reductions to MZVs for a pair of 7-loop periods [11]

$$
\begin{aligned}
P_{7,8} &= \frac{22383}{20} \zeta_{11} + \frac{4572}{5} \left(\zeta_{3,5,3} - \zeta_3 \zeta_{5,3} \right) - 700 \zeta_3^2 \zeta_5 \\
&\quad + 1792 \zeta_3 \left(\frac{9}{320} \left(12 \zeta_{5,3} - 29 \zeta_8 \right) + \frac{45}{64} \zeta_5 \zeta_3 \right) \qquad (9) \\
P_{7,9} &= \frac{92943}{160} \zeta_{11} + \frac{3381}{20} \left(\zeta_{3,5,3} - \zeta_3 \zeta_{5,3} \right) - \frac{1155}{4} \zeta_3^2 \zeta_5 \\
&\quad + 896 \zeta_3 \left(\frac{9}{320} \left(12 \zeta_{5,3} - 29 \zeta_8 \right) + \frac{45}{64} \zeta_5 \zeta_3 \right) \qquad (10)
\end{aligned}
$$

that had been expected to involve alternating sums.

These results were later proven, one by the methods of Erik Panzer [23] and the other by the methods of Oliver Schnetz [26]. Their methods yielded complicated combinations of *alternating* sums, which were then reduced to the MZV formulas (9, 10) by use of proven results in the MZV data mine [4].

The period $P_{7,11}$ in the census of Schnetz [25] is much more demanding. All other periods up to 7 loops reduce to MZVs; only $P_{7,11}$ requires multiple polylogarithms of *sixth* roots of unity, of the form (5) with $\sigma_j^6 = 1$.

Panzer evaluated $\sqrt{3}P_{7,11}$ in terms of 4589 such sums, each of which he evaluated to 5000 digits. Then he found an empirical reduction to a 72-dimensional basis. The rational coefficient of π^{11} in his result was [23]

$$C_{11} = -\frac{964259961464176555529722140887}{2733669078108291387021448260000} \tag{11}$$

whose denominator contains 8 primes greater than 11, namely 19, 31, 37, 43, 71, 73, 50909 and 121577.

Using LLL, it was possible to find a much better basis, with no prime greater 3 in the denominator of any coefficient. Let $A = \mathrm{d}\log(x)$, $B = -\mathrm{d}\log(1-x)$ and $D = -\mathrm{d}\log(1-\exp(2\pi\mathrm{i}/6)x)$ be letters, forming words W that define iterated integrals $Z(W)$. Let

$$W_{m,n} = \sum_{k=0}^{n-1} \frac{\zeta_3^k}{k!} A^{m-2k} D^{n-k}, \tag{12}$$

$P_n = (\pi/3)^n/n!$, $I_n = \mathrm{Cl}_n(2\pi/3)$ and $I_{a,b} = \Im Z(A^{b-a-1} D A^{2a-1} B)$. Using

$$I_{2,9} = 91(11T_{2,9}) - 898T_{3,8} + 11I_{4,7} - 292P_{11} \tag{13}$$

$$I_{3,8} = 24(11T_{2,9}) + 841T_{3,8} - 190I_{4,7} - 255P_{11} \tag{14}$$

to transform to $T_{2,9}$ and $T_{3,8}$, the result becomes

$$\begin{aligned}\sqrt{3}P_{7,11} &= -10080\Im Z(W_{7,4} + W_{7,2}P_2) + 50400\zeta_3\zeta_5 P_3 \\ &+ \left(35280\Re Z(W_{8,2}) + \frac{46130}{9}\zeta_3\zeta_7 + 17640\zeta_5^2\right) P_1 \\ &- 13277952T_{2,9} - 7799049T_{3,8} + \frac{6765337}{2} I_{4,7} - \frac{583765}{6} I_{5,6} \\ &- \frac{121905}{4}\zeta_3 I_8 - 93555\zeta_5 I_6 - 102060\zeta_7 I_4 - 141120\zeta_9 I_2 \\ &+ \frac{42452687872649}{6} P_{11}.\end{aligned} \tag{15}$$

4 Periods and Quasi-Periods in Electrodynamics

The magnetic moment of the electron, in Bohr magnetons, has quantum electrodynamic contributions $\sum_{L=0}^{4} a_L(\alpha/\pi)^L$ given up to $L = 4$ loops by [21]

$$a_0 = \quad 1 \quad \text{[Dirac, 1928]} \tag{16}$$

$$a_1 = \quad 0.5 \quad \text{[Schwinger, 1947]} \tag{17}$$

$$a_2 = -0.32847896557919378458217281696489239241111929867962\ldots \tag{18}$$

$$a_3 = \quad 1.18124145658720000627475398221287785336878939093213\ldots \tag{19}$$

$$a_4 = -1.91224576492644557415264716743983005406087339065872\ldots \tag{20}$$

In 1957, corrections by Petermann and by Sommerfield resulted in

$$a_2 = \frac{197}{144} + \frac{\zeta_2}{2} + \frac{3\zeta_3 - 2\pi^2 \log 2}{4}. \tag{21}$$

In 1996, Laporta and Remiddi obtained

$$\begin{aligned} a_3 = {} & \frac{28259}{5184} + \frac{17101\zeta_2}{135} + \frac{139\zeta_3 - 596\pi^2 \log 2}{18} \\ & - \frac{39\zeta_4 + 400U_{3,1}}{24} - \frac{215\zeta_5 - 166\zeta_3\zeta_2}{24}. \end{aligned} \tag{22}$$

The 3-loop contribution contains a weight-4 depth-2 alternating sum

$$U_{3,1} = \sum_{m>n>0} \frac{(-1)^{m+n}}{m^3 n} = \frac{\zeta_4}{2} + \frac{(\pi^2 - \log^2 2)\log^2 2}{12} - 2\sum_{n>0} \frac{1}{2^n n^4}. \tag{23}$$

Equally fascinating is the Bessel moment [14]

$$B = -\int_0^\infty \frac{27550138t + 35725423t^3}{48600} I_0(t) K_0^5(t) \mathrm{d}t \tag{24}$$

in the evaluation by Laporta [21], at 4800 digits, of

$$a_4 = P + B + E + U \approx 2650.565 - 1483.685 - 1036.765 - 132.027 \approx -1.912 \tag{25}$$

where P comprises multiple polylogs, E comprises integrals whose integrands contain logs and products of elliptic integrals and U comes from 6 light-by-light integrals, still under investigation.

4.1 Bessel Moments and Modular Forms

Gauss noted on 30 May 1799 that the lemniscate constant

$$\int_0^1 \frac{\mathrm{d}x}{\sqrt{1-x^4}} = \frac{(\Gamma(1/4))^2}{4\sqrt{2\pi}} = \frac{\pi/2}{\mathrm{agm}(1,\sqrt{2})} \tag{26}$$

is given by the reciprocal of an arithmetic-geometric mean. This is an example of the Chowla-Selberg formula at the first singular value [5]. In 1939, Watson [27] encountered the sixth singular value, in work on integrals from condensed matter physics. Here, $\left(\sum_{n\in\mathbf{Z}} \exp(-\sqrt{6}\pi n^2)\right)^4$ gives the product of $\Gamma(k/24)$ with $k = 1, 5, 7, 11$, as observed by Glasser and Zucker [20] in 1977. In 2007, Broadhurst and Laporta identified a Feynman period at the *fifteenth* singular value [3], where $\left(\sum_{n\in\mathbf{Z}} \exp(-\sqrt{15}\pi n^2)\right)^4$ gives the product of $\Gamma(k/15)$ with $k = 1, 2, 4, 8$.

With $N = a + b$ Bessel functions and $c \geq 0$, we define moments

$$M(a, b, c) = \int_0^\infty I_0^a(t) K_0^b(t) t^c \mathrm{d}t \tag{27}$$

that converge for $b > a \geq 0$. Then the 5-Bessel matrix

$$\begin{bmatrix} M(1,4,1) & M(1,4,3) \\ M(2,3,1) & M(2,3,3) \end{bmatrix} = \begin{bmatrix} \pi^2 C & \pi^2 \left(\frac{2}{15}\right)^2 \left(13C - \frac{1}{10C}\right) \\ \frac{\sqrt{15}\pi}{2} C & \frac{\sqrt{15}\pi}{2} \left(\frac{2}{15}\right)^2 \left(13C + \frac{1}{10C}\right) \end{bmatrix} \tag{28}$$

involves a single new constant

$$C = \frac{\pi}{16}\left(1 - \frac{1}{\sqrt{5}}\right)\left(\sum_{n=-\infty}^{\infty} \exp(-\sqrt{15}\pi n^2)\right)^4 = \frac{1}{240\sqrt{5}\pi^2} \prod_{k=0}^{3} \Gamma\left(\frac{2^k}{15}\right) \tag{29}$$

and its *reciprocal*. The determinant $2\pi^3/\sqrt{3^3 5^5}$ of matrix (28) is an algebraic multiple of π. This is an example of an all-loop result discovered by Broadhurst and Mellit [12] and proven by Yajun Zhou [28].

The L-series for $N = 5$ Bessel functions comes from a *modular form* of weight 3 and level 15:

$$\eta_n = q^{n/24} \prod_{k>0} (1 - q^{nk}), \quad q = \exp(2\pi i \tau), \tag{30}$$

$$f_{3,15}(\tau) = (\eta_3\eta_5)^3 + (\eta_1\eta_{15})^3 = \sum_{n>0} A_5(n) q^n \tag{31}$$

$$L_5(s) = \sum_{n>0} \frac{A_5(n)}{n^s} \quad \text{for } s > 2 \tag{32}$$

$$L_5(1) = \sum_{n>0} \frac{A_5(n)}{n} \left(2 + \frac{\sqrt{15}}{2\pi n}\right) \exp\left(-\frac{2\pi n}{\sqrt{15}}\right) \tag{33}$$

$$= 5C = \frac{5}{\pi^2} \int_0^\infty I_0(t) K_0^4(t) t \,\mathrm{d}t\,. \tag{34}$$

4.2 Periods and Quasi-Periods for the Laporta Problem

Laporta's 4-loop work [21] engages the first row of the 6-Bessel determinant

$$\det \begin{bmatrix} M(1,5,1) & M(1,5,3) \\ M(2,4,1) & M(2,4,3) \end{bmatrix} = \frac{5\zeta_4}{32} \tag{35}$$

associated to a modular form $f_{4,6}(\tau) = (\eta_1\eta_2\eta_3\eta_6)^2$ with weight 4 and level 6 [11, 29]. At top left we have $M(1,5,1)$, from the on-shell 4-loop sunrise diagram, in two spacetime dimensions. Below it, $M(2,4,1)$ comes from cutting an internal line. The second column comes from differentiating the first, with respect to the external momentum, to produce quasi-periods associated with a *weakly* holomorphic modular form

$$\widehat{f}_{4,6}(\tau) = \mu f_{4,6}(\tau), \quad \mu = \frac{1}{32}\left(w + \frac{3}{w}\right)^4 - \frac{9}{16}\left(w + \frac{3}{w}\right)^2, \quad w = \frac{3\eta_3^4\eta_2^2}{\eta_1^4\eta_6^2}. \tag{36}$$

With $s = 1, 2$, we computed compute 10,000 digits of the Eichler integrals

$$\frac{\Omega_s}{(2\pi)^s} = \int_{1/\sqrt{3}}^\infty f_{4,6}\left(\frac{1+iy}{2}\right) y^{s-1} dy, \quad \frac{\widehat{\Omega}_s}{(2\pi)^s} = \int_{1/\sqrt{3}}^\infty \widehat{f}_{4,6}\left(\frac{1+iy}{2}\right) y^{s-1} dy. \tag{37}$$

4.3 Laporta's Intersection Number

LLL readily gave 4 linear relations

$$\frac{2}{\pi^2}\begin{bmatrix} 4M(1,5,1) & \frac{36}{5}\,(M(1,5,1)+M(1,5,3)) \\ \frac{5}{3}M(2,4,1) & 3\,(M(2,4,1)+M(2,4,3)) \end{bmatrix} = \begin{bmatrix} -\Omega_2 & \widehat{\Omega}_2 \\ -\Omega_1 & \widehat{\Omega}_1 \end{bmatrix} \tag{38}$$

between Feynman integrals, the periods $\Omega_{1,2}$ and the quasi-periods $\widehat{\Omega}_{1,2}$.

The intersection number is the determinant of this matrix, namely $1/12$. Broadhurst and Roberts [14] converted this into a quadratic relation between 4 hypergeometric series:

$$\begin{aligned} F_a &= {}_4F_3(\ \ 1/2,\ \ 2/3,\ \ 2/3, 5/6;\ \ 7/6, 7/6, 4/3;\ 1) \\ F_b &= {}_4F_3(-1/2,\ \ 1/6,\ \ 1/3, 4/3; -1/6, 5/6, 5/3;\ 1) \\ F_c &= {}_4F_3(\ \ 1/6,\ \ 1/3,\ \ 1/3, 1/2;\ \ 2/3, 5/6, 5/6;\ 1) \\ F_d &= {}_4F_3(-7/6, -1/2, -1/3, 2/3; -5/6, 1/6, 1/3;\ 1) \end{aligned} \tag{39}$$

namely

$$7F_aF_b + 10F_cF_d = 40, \tag{40}$$

which was later proven by Yajun Zhou [30].

5 Quadratic Relations

In this section, we give quadratic relations between Feynman integrals, recently discovered by using the LLL algorithm. If one has n integrals, there are $n(n+1)/2$ products to consider, in linear combinations with rational or algebraic coefficients, which may give a rational or algebraic multiple of a power of π. This problem soon explodes. We begin with a conjecture obtained after intensive use of LLL and tested with up to $n = 100$ Feynman integrals.

Conjecture (Broadhurst and Roberts [14]) *With the Feynman, de Rham and Betti matrices below,*

$$F_N D_N F_N^{\mathtt{tr}} = B_N. \tag{41}$$

The elements of the Feynman matrices F_N are the Bessel moments

$$F_{2k+1}(u,a) = \frac{(-1)^{a-1}}{\pi^u} M(k+1-u, k+u, 2a-1) \tag{42}$$

$$F_{2k+2}(u,a) = \frac{(-1)^{a-1}}{\pi^{u+1/2}} M(k+1-u, k+1+u, 2a-1) \tag{43}$$

with u and a, as well as later indices v and b, running from 1 to k. $F_N^{\mathtt{tr}}$ is the transpose of F_N.

The Betti matrices B_N have rational elements given by

$$B_{2k+1}(u,v) = (-1)^{u+k}2^{-2k-2}(k+u)!(k+v)!Z(u+v) \tag{44}$$

$$B_{2k+2}(u,v) = (-1)^{u+k}2^{-2k-3}(k+u+1)!(k+v+1)!Z(u+v+1) \tag{45}$$

$$Z(m) = \frac{1+(-1)^m}{(2\pi)^m}\zeta_m. \tag{46}$$

For the de Rham matrices D_N, let v_k and w_k be the rationals generated by

$$\frac{J_0^2(t)}{C(t)} = \sum_{k\ge0}\frac{v_k}{k!}\left(\frac{t}{2}\right)^{2k} = 1 - \frac{17t^2}{54} + \frac{3781t^4}{186624} + \ldots \tag{47}$$

$$\frac{2J_0(t)J_1(t)}{tC(t)} = \sum_{k\ge0}\frac{w_k}{k!}\left(\frac{t}{2}\right)^{2k} = 1 - \frac{41t^2}{216} + \frac{325t^4}{186624} + \ldots \tag{48}$$

where $J_0(t) = I_0(it)$, $J_1(t) = -J_0'(t)$ and

$$C(t) = \frac{32(1-J_0^2(t) - tJ_0(t)J_1(t))}{3t^4} = 1 - \frac{5t^2}{27} + \frac{35t^4}{2304} - \frac{7t^6}{9600} + \ldots \tag{49}$$

Construct rational bivariate polynomials $H_s = H_s(y,z)$ by the recursion

$$H_s = zH_{s-1} - (s-1)yH_{s-2} - \sum_{k=1}^{s-1}\binom{s-1}{k}(v_kH_{s-k} - w_kzH_{s-k-1}) \tag{50}$$

for $s > 0$, with $H_0 = 1$ and $H_{-1} = 0$. Use these to define

$$d_s(N,c) = \frac{H_s(3c/2, N+2-2c)}{4^s s!}. \tag{51}$$

Finally, construct de Rham matrices with the rational elements

$$D_N(a,b) = \sum_{c=-b}^{a} d_{a-c}(N,-c)d_{b+c}(N,c)c^{N+1}. \tag{52}$$

The discovery of formula (52) for the coefficients of these quadratic relations involved intensive use of LLL, at high numerical precision. At 20 loops, there are 100 Feynman integrals to consider, with 5050 products. Javier Fresán, Claude Sabbah and Jeng-Daw Yu [19] have verified that our formulas hold up to 20 loops, after which they ran out of computing power. They encountered subtleties when N

is divisible by 4. These are entirely avoided by our uniform formula (52). Yajun Zhou [31] has given an illuminating classical proof, with generalizations.

5.1 Quadratic Relations at Weight 6 and Level 24

Here we establish relations between Bessel moments (27) with $a + b = 6$, $c = 0, 2, 4$, and Eichler integrals of modular forms of weight 6 and level 24, constructed from eta quotients. This connection was suggested by the discovery of the linear and quadratic relations

$$\frac{M(0,6,0)}{M(2,4,0)} = \frac{3M(0,6,2) - 8M(0,6,4)}{3M(2,4,2) - 8M(2,4,4)} = 3\pi^2, \tag{53}$$

$$\det\begin{bmatrix} M(0,6,0) & 3M(0,6,2) - 8M(0,6,4) \\ M(1,5,0) & 3M(1,5,2) - 8M(1,5,4) \end{bmatrix} = \frac{5\pi^6}{16}. \tag{54}$$

We begin by defining three eta quotients, subject to two algebraic relations:

$$r = \left(\frac{\eta_2\eta_{12}}{\eta_4\eta_6}\right)^6 = \frac{s-t}{st} = 9t - 8s = q - 6q^3 + 15q^5 + O(q^7), \tag{55}$$

$$s = \left(\frac{\eta_4\eta_{12}}{\eta_2\eta_6}\right)^3 = q + 3q^3 + 6q^5 + O(q^7), \tag{56}$$

$$t = \left(\frac{\eta_6\eta_{12}}{\eta_2\eta_4}\right)^2 = q + 2q^3 + 7q^5 + O(q^7). \tag{57}$$

The moments $M(0,6,0)$ and $M(1,5,0)$ are related to periods of the modular form

$$f_1(\tau) = (\eta_2\eta_4\eta_6\eta_{12})^3\left(\frac{1}{s^2} - 64s^2\right) = q - 9q^3 - 34q^5 - 240q^7 + 81q^9 + O(q^{11}). \tag{58}$$

To find its quasi-periods, we form a column vector of 5 cusp forms

$$\mathbf{f}(\tau) = (\eta_2\eta_4\eta_6\eta_{12})^3\begin{bmatrix} 1/s^2 - 64s^2 \\ 1/s^2 + 64s^2 \\ 1/r^2 - r^2 \\ 1/t^2 - 81t^2 \\ 1/t^2 + 81t^2 + 54 \end{bmatrix} = \mathbf{T}\begin{bmatrix} q \\ q^3 \\ q^5 \\ q^7 \\ q^9 \end{bmatrix} + O(q^{11}), \tag{59}$$

$$\mathbf{T} = \begin{bmatrix} 1 & -9 & -34 & -240 & 81 \\ 1 & -9 & 94 & 144 & 81 \\ 1 & 9 & 38 & 120 & 81 \\ 1 & -7 & -74 & -24 & -383 \\ 1 & 47 & -74 & -24 & 697 \end{bmatrix}, \qquad (60)$$

with the Hecke matrix **T** recording the first 5 non-vanishing Fourier coefficients of the 5 modular forms. The first three components of **f** are new forms, while the remaining two are old forms.

Since there are 16 cusp forms of weight 6 and level 24, we began our investigation with a far more fearsome 16-dimensional problem. After intensive study of the relationship between Rademacher sums [1] and the determinants and permanents [15] of matrices of periods and quasi-periods of modular forms, we were able to reduce the Bessel-moment problem to the 5-dimensional problem presented here.

For each of the 5 cusp forms f_k, we seek a weakly holomorphic form $\widehat{f_k}$, such that the periods of f_k and quasi-periods of $\widehat{f_k}$ yield a determinant that is a rational multiple of a power of π. For $k = 1$, this will solve the Bessel-moment problem.

To construct 5 weakly homomorphic modular forms, we define a column vector

$$\mathbf{g}(\tau) = \left(\frac{\eta_{12}^5}{\eta_4\eta_6^2}\right)^6 \begin{bmatrix} 1 \\ a+35 \\ a^2+40a+646 \\ a^3+45a^2+840a+8352 \\ a^4+50a^3+1059a^2+12308a+84817 \end{bmatrix}, \quad a = 72\frac{\eta_4\eta_{12}^5}{\eta_2^5\eta_6}, \qquad (61)$$

with monic polynomials in $a = O(q^2)$, determined by the requirement that

$$\frac{-27}{\tau^6} g_k\left(-\frac{1}{24\tau}\right) = \frac{1}{q^{2k-1}} + O(q), \qquad (62)$$

which records the singular behaviour near the cusp of $g_k(\tau)$ at $\tau = 0$. We avoid this singularity by taking Eichler integrals from $\tau = \frac{\mathrm{i}}{4}$ to $\tau = \mathrm{i}\infty$, with extremely good behaviour of $g_k(\tau)$ at the end-points.

The Fourier expansion of g_k begins at q^{11}. Thus we may add, to any combination of the weakly holomorphic forms in **g**, a combination of the cusp forms in **f**, since the latter are determined by their expansions up to q^9, recorded in the Hecke matrix **T**. Our Ansatz for the weakly homomorphic partners in $\widehat{\mathbf{f}}$ has the form

$$\widehat{\mathbf{f}} = \widetilde{\mathbf{T}}^{-1}(\mathbf{U}\mathbf{g} + \mathbf{V}\mathbf{T}^{-1}\mathbf{f}), \quad \mathbf{U} = 18^2 \begin{bmatrix} 1 & 0 & 0 & 0 & 0 \\ 0 & 3^5 & 0 & 0 & 0 \\ 0 & 0 & 5^5 & 0 & 0 \\ 0 & 0 & 0 & 7^5 & 0 \\ 0 & 0 & 0 & 0 & 9^5 \end{bmatrix}, \qquad (63)$$

which uses inverses of $\mathbf{T}$ and its transpose $\widetilde{\mathbf{T}}$. The diagonal matrix $\mathbf{U}$ reflects the singular behaviour (62) of the weakly holomorphic modular forms g_k near $\tau = 0$.

Our final challenge is to determine the matrix $\mathbf{V}$, which we expect, from previous work, to be symmetric, with a vanishing first row and column. Here we give our eventual result

$$\mathbf{V} = 6^2 \begin{bmatrix} 0 & 0 & 0 & 0 & 0 \\ 0 & -4 & 54 & 648 & 5995 \\ 0 & 54 & 2916 & 77508 & 1150848 \\ 0 & 648 & 77508 & 3039444 & 64431936 \\ 0 & 5995 & 1150848 & 64431936 & 1865595908 \end{bmatrix}, \tag{64}$$

with explanation of how it was obtained, empirically, by integer-relation searches.

We define periods, quasi-periods and determinants as follows:

$$\begin{bmatrix} P_k(s) \\ \widehat{P}_k(s) \end{bmatrix} = -\mathrm{i} \int_0^\infty \begin{bmatrix} f_k((1+\mathrm{i}y)/4) \\ \widehat{f}_k((1+\mathrm{i}y)/4) \end{bmatrix} y^{s-1} \mathrm{d}y,$$
$$D_k(s,t) = P_k(s)\widehat{P}_k(t) - P_k(t)\widehat{P}_k(s), \tag{65}$$

with $k = 1, 2, 3, 4, 5$ and $s = 1, 2, 3, 4, 5$. The Eichler integrals are from $\tau = \frac{1}{4}$ to $\tau = \mathrm{i}\infty$, along the vertical line where the real part of τ is $\frac{1}{4}$ and hence $q = \exp(2\pi\mathrm{i}\tau)$ is pure imaginary. Since f_k and $\widehat{f}_k$ have Fourier expansions in odd powers of q, they too are pure imaginary. Hence the periods and quasi-periods are real.

Our criteria for the elements of $\mathbf{V}$ are the determinant conditions

$$D_k(1,3) = D_k(1,5) = D_k(2,4) = 0, \quad D_k(1,2) = \frac{d_k}{\pi^5}, \tag{66}$$

where d_k is a rational number. The first three conditions ensure the matching of the period polynomial of f_k with the quasi-period polynomial of $\widehat{f}_k$. The fourth condition encodes a non-trivial quadratic relation between periods and quasi-periods. Moreover, we were able to use Rademacher sums [1] and Petersson inner products [24] to show that $d_k = \frac{21}{2}, -\frac{135}{14}, -\frac{15}{2}, \frac{15}{2}, -\frac{6}{7}$, for $k = 1, 2, 3, 4, 5$.

Thus we have $5 \times 4 = 20$ integer-relation conditions with which to determine $4 + 3 + 2 + 1 = 10$ independent elements of $\mathbf{V}$, giving us great confidence in our results.

Here we record the relations between periods, quasi-periods and Bessel moments at $k = 1$, which have been checked at 1000-digit precision:

$$P_1(1) = 21 P_1(3) = 9 P_1(5) = 28 \frac{M(0,6,0)}{\pi^6}, \tag{67}$$

$$P_1(2) = 3 P_1(4) = 8 \frac{M(1,5,0)}{\pi^5}, \tag{68}$$

$$\widehat{P}_1(1) = 21\widehat{P}_1(3) = 9\widehat{P}_1(5) = 7\,\frac{17M(0,6,0) + 48(3M(0,6,2) - 8M(0,6,4))}{80\pi^6}, \tag{69}$$

$$\widehat{P}_1(2) = 3\widehat{P}_1(4) = 2\,\frac{17M(1,5,0) + 48(3M(1,5,2) - 8M(1,5,4))}{80\pi^5}. \tag{70}$$

5.2 Quadratic Relations at Levels 14 and 34

In 2019, Philip Candelas, Xenia de la Ossa, Mohamed Elmi and Duco van Straten announced a discovery of a family of Calabi-Yau manifolds with rank-2 attractor points [16].

They compactified a 10-dimensional supergravity theory on a Calabi-Yau threefold with complex structure, to obtain 4-dimensional black holes, with event horizons whose areas are determined by their electric and magnetic charges and by ratios of periods of modular forms of weight 4 and levels 14 or 34.

Hearing of this on a visit to Oxford in November 2019, Broadhurst observed that their Calabi-Yau periods come from solutions to a homogeneous differential equation associated with 4 loop sunrise integrals, namely

$$M_{m,n}(z) = \int_0^\infty I_0(xz)[I_0(x)]^m[K_0(x)]^{5-m}x^{2n+1}\mathrm{d}x \tag{71}$$

$$N_{m,n}(z) = z\int_0^\infty I_1(xz)[I_0(x)]^m[K_0(x)]^{5-m}x^{2n+2}\mathrm{d}x \tag{72}$$

with $m \in \{0, 1, 2\}$, integers $n \geq 0$ and real $z^2 < (5 - 2m)^2$. The uncut diagram $M_{0,0}(z)$ satisfies an *inhomogeneous* differential equation.

The external mass is z. At $z = 1$ we obtain Laporta's on-shell periods, for the magnetic moment of the electron at 4 loops, coming from the modular form $f_{4,6}(\tau) = (\eta_1\eta_2\eta_3\eta_6)^2$ with level 6. With mass $z = \sqrt{17} - 4$, we obtain level-34 periods. At the space-like point $z = \sqrt{-7}$, we obtain level-14 periods.

At each of the levels 14 and 34, Candelas et al. considered 16 Calabi-Yau periods, coming from 4 solutions to a homogeneous fourth-order differential equation, together with the first 3 derivatives of each solution. They were unable to identify all of these 16 periods.

Using LLL we found that 8 Feynman integrals, at each level, suffice to solve their problem, completely. These 8 integrals determine a pair of periods and a pair of quasi-periods, at each of the weights 2 and 4. Hence they satisfy two quadratic relations. At level 34, the coefficients in these relations are algebraic numbers in $\mathbf{Q}(\sqrt{17})$.

5.3 Level 14, at Space-Like Momentum

At level 14, with $z = \sqrt{-7}$, we identified

$$f_{4,14}(\tau) = \frac{(\eta_2\eta_7)^6}{(\eta_1\eta_{14})^2} - 4(\eta_1\eta_2\eta_7\eta_{14})^2 + \frac{(\eta_1\eta_{14})^6}{(\eta_2\eta_7)^2} \tag{73}$$

as the relevant modular form of weight 4. Its periods are critical values of the L-function $L(f_{4,14}, s) = ((2\pi)^s/\Gamma(s)) \int_0^\infty f_{4,14}(\mathrm{i}y)y^{s-1}\mathrm{d}y$, with

$$L(f_{4,14}, 3) = M_{1,0}(\sqrt{-7}) = \int_0^\infty J_0(\sqrt{7}x)I_0(x)K_0^4(x)x\mathrm{d}x = \frac{\pi^2}{7}L(f_{4,14}, 1) \tag{74}$$

$$\frac{1}{2}L(f_{4,14}, 2) = M_{2,0}(\sqrt{-7}) = \int_0^\infty J_0(\sqrt{7}x)I_0^2(x)K_0^3(x)x\mathrm{d}x. \tag{75}$$

There is also a modular form of weight 2 to consider, $f_{2,14}(\tau) = \eta_1\eta_2\eta_7\eta_{14}$. This provides a modular parametrization of a quartic elliptic curve, namely

$$d^2 = (1+h)(1+8h)(1+5h+8h^2), \tag{76}$$

$$h = \left(\frac{\eta_2\eta_{14}}{\eta_1\eta_7}\right)^3 = q + 3q^2 + 6q^3 + 13q^4 + O(q^5), \tag{77}$$

$$d = \frac{q}{f_{2,14}}\frac{\mathrm{d}h}{\mathrm{d}q} = 1 + 7q + 27q^2 + 92q^3 + 259q^4 + O(q^5), \tag{78}$$

yielding an L-value and a j-invariant:

$$L(f_{2,14}, 1) = \frac{\omega_+}{3}, \quad j\left(\frac{\omega_+ + \mathrm{i}\omega_-}{2\omega_+}\right) = \left(\frac{5\times 43}{28}\right)^3, \tag{79}$$

with periods determined by arithmetic-geometric means

$$\omega_\pm = \frac{2\pi}{\mathrm{agm}\left(\sqrt{2^{9/2} \pm 13},\, 2^{11/4}\right)} \tag{80}$$

and also by Feynman integrals:

$$\frac{\omega_+}{2} = 3M_{2,0}(\sqrt{-7}) + 4N_{2,0}(\sqrt{-7}), \tag{81}$$

$$\frac{\pi\omega_-}{2} = 3M_{1,0}(\sqrt{-7}) + 4N_{1,0}(\sqrt{-7}). \tag{82}$$

The quasi-periods at weight 2 are $\widehat{\omega}_{\pm}$, with

$$\frac{3\widehat{\omega}_+}{16} = 7M_{2,0}(\sqrt{-7}) + 8N_{2,0}(\sqrt{-7}) + 28M_{2,1}(\sqrt{-7}), \tag{83}$$

$$\frac{3\pi\widehat{\omega}_-}{16} = 7M_{1,0}(\sqrt{-7}) + 8N_{1,0}(\sqrt{-7}) + 28M_{1,1}(\sqrt{-7}). \tag{84}$$

Suppressing the argument $z = \sqrt{-7}$, we obtain the quadratic relation

$$\det\begin{bmatrix} 3M_{2,0} + 4N_{2,0} & M_{2,0} + 28M_{2,1} \\ 3M_{1,0} + 4N_{1,0} & M_{1,0} + 28M_{1,1} \end{bmatrix} = -\frac{3\pi^2}{32} \tag{85}$$

from Legendre's relation for complete elliptic integrals.

At weight 4 we found

$$\det\begin{bmatrix} M_{2,0} & 39N_{2,0} - 427M_{2,1} - 112N_{2,1} \\ M_{1,0} & 39N_{1,0} - 427M_{1,1} - 112N_{1,1} \end{bmatrix} = \frac{3\pi^2}{32} \tag{86}$$

as the quadratic relation between the periods and quasi-periods of $f_{4,14}$.

We define the weight-4 periods and quasi-periods as

$$\mathcal{G}_m = M_{m,0}, \quad \widehat{\mathcal{G}}_m = 7(35M_{m,0} - 122M_{m,1}) + 2(39N_{m,0} - 112N_{m,1}), \tag{87}$$

for $m = 1, 2$ and $z = \sqrt{-7}$. The quasi-periods come from Eichler integrals of a weakly holomorphic form obtained by multiplying $f_{2,14}^2$ by a polynomial that is quartic in h and linear in d/h. For one of the quasi-periods, the dependence on d/h is irrelevant. We used LLL to determine that

$$\widehat{\mathcal{G}}_2 = 5\pi^2 \int_{1/\sqrt{7}}^{\infty} g\left(\frac{1+\mathrm{i}y}{2}\right) y\mathrm{d}y \tag{88}$$

$$g(\tau) = (253 + 645h + 1446h^2 + 2064h^3 + 1024h^4) f_{2,14}^2. \tag{89}$$

Then the other quasi-period comes from the determinant $\mathcal{G}_2\widehat{\mathcal{G}}_1 - \mathcal{G}_1\widehat{\mathcal{G}}_2 = 3(\pi/4)^2$.

5.4 Level 34, with Mass $\sqrt{17} - 4$

At level 34, with $z = u = \sqrt{17} - 4$, we used Pari-GP to identify the modular form of weight 4. Let $\chi(n)$ be the Dirichlet character defined for prime p by $\chi(17) = 0$ and otherwise by $\chi(p) = \pm 1$ according as whether p is or is not a square modulo 17. Pari-GP declares that there are 12 cusp forms of level 34 and weight 4 with this character. Feynman integrals choose a pair of new forms whose Fourier coefficients, $A_4(n)$ and $\overline{A}_4(n)$, are Gaussian integers, related by complex conjugation.

Let $L_4(s)$ be the analytic continuation of

$$L_4(s)=\sum_{n>0}\frac{A_4(n)}{n^s}=\frac{1}{1+2^{1-s}}\prod_{p>2}\frac{1}{1-A_4(p)p^{-s}+\chi(p)p^{3-2s}} \tag{90}$$

with the choice of sign $A_4(3)=2\mathrm{i}$. For prime p, $A_4(p)$ is real if $\chi(p)=+1$ and imaginary if $\chi(p)=-1$, while $A_4(17)/17=1-4\mathrm{i}$ is truly complex.

Feynman integrals determine the critical L-values at weight 4:

$$L_4(3)=\left(\frac{13-u+(1+13u)\mathrm{i}}{17}\right)M_{1,0}(u), \tag{91}$$

$$L_4(2)=4\left(\frac{5-3u+(3+5u)\mathrm{i}}{17}\right)M_{2,0}(u), \tag{92}$$

$$L_4(1)=\left(\frac{7-11u+(11+7u)\mathrm{i}}{\pi^2}\right)M_{1,0}(u). \tag{93}$$

At weight 2 they determine the periods and quasi-periods of the elliptic curve

$$y^2=\left(x+\frac{5-u}{8}\right)\left(x+\frac{5+u}{8}\right)\left(x+\frac{3+u}{2}\right) \tag{94}$$

whose real and imaginary periods are

$$\omega_1=\frac{4\pi}{\mathrm{agm}\left(\sqrt{4u},\sqrt{14+10u}\right)},\quad \omega_2=\frac{-4\pi\mathrm{i}}{\mathrm{agm}\left(\sqrt{14+6u},\sqrt{14+10u}\right)}. \tag{95}$$

The elliptic periods $\omega_{1,2}$ and quasi-periods $\widehat{\omega}_{1,2}$ are determined by

$$\frac{\omega_1}{4}=\mathcal{P}_2=(2+3u)M_{2,0}(u)+4(4+u)N_{2,0}(u) \tag{96}$$

$$\frac{\pi\mathrm{i}\omega_2}{4}=\mathcal{P}_1=(2+3u)M_{1,0}(u)+4(4+u)N_{1,0}(u) \tag{97}$$

$$\frac{3\widehat{\omega}_1}{8(1+u)}=\widehat{\mathcal{P}}_2=M_{2,0}(u)+2(5+u)N_{2,0}(u)+2u(3+u)(4+u)M_{2,1}(u) \tag{98}$$

$$\frac{3\pi\mathrm{i}\widehat{\omega}_2}{8(1+u)}=\widehat{\mathcal{P}}_1=M_{1,0}(u)+2(5+u)N_{1,0}(u)+2u(3+u)(4+u)M_{1,1}(u) \tag{99}$$

with Legendre's condition giving $\mathcal{P}_1\widehat{\mathcal{P}}_2-\mathcal{P}_2\widehat{\mathcal{P}}_1=3(\pi/4)^2/(1+u)$.

At weight 4, the periods $\mathcal{H}_m = M_{m,0}(u)$ and quasi-periods

$$\begin{aligned}\widehat{\mathcal{H}}_m &= 81M_{m,0}(u) + 3(2+u)(u-6)N_{m,0}(u)\\ &\quad + u^2(2+u)(4+u)(96+11u)M_{m,1}(u) + 136(1-u)N_{m,1}(u) \qquad (100)\end{aligned}$$

yield the intersection number $\mathcal{H}_1\widehat{\mathcal{H}}_2 - \mathcal{H}_2\widehat{\mathcal{H}}_1 = 3(\pi/8)^2/u$.

The numbers which remained unidentified in [16] are now easy to determine. They involve the permanents [15] of the matrices of Feynman integrals whose determinants yield intersection numbers that are algebraic multiples of powers of π.

6 Summary

PSLQ and LLL have enlivened quests for analytical results, provided strong tests on conjectures and condensed huge expressions. Parallel PSLQ was of the essence in Laporta's work in electrodynamics. LLL led to a conjecture on quadratic relations for all loops, to determinations of quasi-periods at weight 6 and level 24 and to exact results for black-hole problems that involve modular forms of levels 14 and 34. Our new results at levels 14, 24 and 34 were obtained from extending methods developed in our work on eta quotients [1]. The permanents which were lacking in [16] yield Rademacher sums [1] that are sums of products of Bessel functions and Kloosterman sums [13] .

References

1. K. Acres, D. Broadhurst, *Eta Quotients and Rademacher Sums*, in *Elliptic Integrals, Elliptic Functions and Modular Forms in Quantum Field Theory*, ed. by J. Blümlein, C. Schneider, P. Paule (Springer, 2019), pp. 1–27. https://arxiv.org/abs/1810.07478
2. D.H. Bailey, D.J. Broadhurst, Parallel integer relation detection: techniques and applications. Math. Comp. **70**, 1719–1736 (2001). https://arxiv.org/abs/math/9905048
3. D. Bailey, J. Borwein, D. Broadhurst, M. Larry Glasser, Elliptic integral evaluations of Bessel moments. J. Phys. **A41**, 205203 (2008). https://arxiv.org/abs/0801.0891
4. J. Blümlein, D.J. Broadhurst, J.A.M. Vermaseren, The multiple zeta value data mine. Comput. Phys .Commun. **181**, 582–625 (2010). https://arxiv.org/abs/0907.2557
5. J.M. Borwein, P.B. Borwein, *Pi and the AGM: a study in analytic number theory and computational complexity*, Monographies et Études de la Société Mathématique du Canada (Wiley, Toronto, 1987)
6. J.M. Borwein, D.J. Broadhurst, J. Kamnitzer, Central binomial sums, multiple Clausen values and zeta values. Exp. Math. **10**, 25–34 (2001). https://arxiv.org/abs/hep-th/0004153
7. D.J. Broadhurst, *Massless scalar Feynman diagrams: five loops and beyond*. Open University report OUT-4102-18, 1985. https://arxiv.org/abs/1604.08027
8. D.J. Broadhurst, D. Kreimer, Knots and numbers in ϕ^4 theory to 7 loops and beyond. Int. J. Mod. Phys. **C6**, 519–524 (1995). https://arxiv.org/abs/hep-ph/9504352

9. D.J. Broadhurst, D. Kreimer, Association of multiple zeta values with positive knots via Feynman diagrams up to 9 loops. Phys. Lett. **B393**, 403–412 (1997). https://arxiv.org/abs/hep-th/9609128
10. D.J. Broadhurst, Massive 3-loop Feynman diagrams reducible to SC^* primitives of algebras of the sixth root of unity. Eur. Phys. J. **C8**, 311–333 (1999). https://arxiv.org/abs/hep-th/9803091
11. D. Broadhurst, O. Schnetz, *Algebraic geometry informs perturbative quantum field theory*, in *Loops and Legs in Quantum Field Theory, PoS (LL2014)*, 078 (2014). https://pos.sissa.it/211/078/pdf
12. D. Broadhurst, A. Mellit, *Perturbative quantum field theory informs algebraic geometry*, in *Loops and Legs in Quantum Field Theory, PoS (LL2016)*, 079 (2016). https://pos.sissa.it/260/079/pdf
13. D. Broadhurst, Feynman integrals, L-series and Kloosterman moments. Commun. Number Theory Phys. **10**, 527–569 (2016). https://arxiv.org/abs/1604.03057
14. D. Broadhurst, D.P. Roberts, *Quadratic relations between Feynman integrals*, in *Loops and Legs in Quantum Field Theory, PoS (LL2018)*, 053 (2018). https://pos.sissa.it/303/053/pdf
15. F. Brown, A class of non-holomorphic modular forms III: real analytic cusp forms for $SL_2(Z)$. Res. Math. Sci. **5**, 34 (2018). https://doi.org/10.1007/s40687-018-0151-3
16. P. Candelas, X. de la Ossa, M. Elmi, D. van Straten, A one parameter family of Calabi-Yau manifolds with attractor points of rank two. J. High Energy Phys. **2020**, 202 (2020). https://arxiv.org/abs/1912.06146
17. H.R.P. Ferguson, R.W. Forcade, Generalization of the euclidean algorithm for real numbers to all dimensions higher than two. Bull. Am. Math. Soc. **1**, 912–914 (1979). https://doi.org/10.1090/S0273-0979-1979-14691-3
18. H.R.P. Ferguson, D.H. Bailey, *A polynomial time, numerically stable integer relation algorithm*. RNR technical report RNR-91-032, 1992. https://www.davidhbailey.com/dhbpapers/pslq.pdf
19. J. Fresán, C. Sabbah, J.-D. Yu, *Quadratic relations between Bessel moments*. https://arxiv.org/abs/2006.02702
20. M.L. Glasser, I.J. Zucker, Extended Watson integrals for the cubic lattices. PNAS **74**, 1800–1801 (1977). https://doi.org/10.1073/pnas.74.5.1800
21. S. Laporta, High-precision calculation of the 4-loop contribution to the electron $g-2$ in QED. Phys. Lett. **B772**, 232–238 (2017). https://arxiv.org/abs/1704.06996
22. A.K. Lenstra, H.W. Lenstra Jr., L. Lovász, Factoring polynomials with rational coefficients. Mathematische Annalen **261**, 515–534 (1982). https://doi.org/10.1007/BF01457454
23. E. Panzer, *Feynman integrals and hyperlogarithms*. Thesis, Humboldt-Universität zu Berlin, 2015. https://arxiv.org/abs/1506.07243
24. Pari-GP version 2.13.1, Univ. Bordeaux, 2021. http://pari.math.u-bordeaux.fr
25. O. Schnetz, Quantum periods: a census of ϕ^4 transcendentals. Commun. Number Theory Phys. **4**, 1–47 (2010). https://arxiv.org/abs/0801.2856
26. O. Schnetz, *Graphical functions and single-valued multiple polylogarithms*. https://arxiv.org/abs/1302.6445
27. G.N. Watson, Three triple integrals. Q. J. Math. **10**, 266–276 (1939). https://doi.org/10.1093/qmath/os-10.1.266
28. Y. Zhou, Wronskian factorizations and Broadhurst-Mellit determinant formulae. Commun. Number Theory Phys. **12**, 355–407 (2018). https://arxiv.org/abs/1711.01829
29. Y. Zhou, *Some algebraic and arithmetic properties of Feynman diagrams*, in *Elliptic Integrals, Elliptic Functions and Modular Forms in Quantum Field Theory*, ed. by J. Blümlein, C. Schneider, P. Paule (Springer, 2019), pp. 485–509. https://arxiv.org/abs/1801.05555
30. Y. Zhou, On Laporta's 4-loop sunrise formulae. Ramanujan J. **50**, 465–503 (2019). https://arxiv.org/abs/1801.02182
31. Y. Zhou, *Wrońskian algebra and Broadhurst-Roberts quadratic relations*. https://arxiv.org/abs/2012.03523

$\mathrm{N} = 4$ SYM Gauge Theories: The $2 \to 6$ Amplitude in the Regge Limit

Jochen Bartels

Abstract In this contribution we discuss the Regge limit of scattering amplitudes in $\mathrm{N} = 4$ SYM in the planar approximation. The analysis is based upon unitarity and energy discontinuities, and the analytic structure plays a vital role. We first summarize the lessons learned from the study of the remainder functions of the $2 \to 4$ and the $2 \to 5$ scattering amplitudes and then present new results for the $2 \to 6$ amplitude.

1 Introduction

Interest in $\mathrm{N} = 4$ supersymmetric gauge theory (SYM) [1–6] is connected with the hope that one can find a simple connection between weak and strong coupling which, in particular, allows to understand high order perturbation theory corrections [7–14]. Ultimately this may allow to find also higher order corrections in Quantum Chromodynamics (QCD). In recent years progress has been made in computing, in the planar approximation and for the Regge limit, next-to-leading order (NLO) corrections and, in a few cases, even all order corrections for parts of the remainder functions [12], and new methods are being developed. In the present contribution we also focus on the Regge limit, and we present new results which are based upon energy discontinuities and unitarity relations. We limit ourselves to the leading-order (LO) approximation, and particular attention will be given to the $2 \to 6$ amplitude.

J. Bartels (✉)
Universität Hamburg, II. Institut für Theoretische Physik, Hamburg, Germany
e-mail: joachim.bartels@desy.de

J. Blümlein, C. Schneider (eds.), *Anti-Differentiation and the Calculation of Feynman Amplitudes*, Texts & Monographs in Symbolic Computation,
https://doi.org/10.1007/978-3-030-80219-6_4

2 General Remarks

Let us start with a remark on the role of the Regge limit. Experience from studies of quantum field theories (QED, QCD) shows that, in the Regge limit, scattering amplitudes exhibit a simple factorizing structure which is valid to all orders in perturbation theory. As an example, for the 2 → 3 process in Fig. 1 the amplitude in the Regge limit has the form

$$T(s_1, s_2, t_1, t_2, \kappa) = \Gamma(t_1)\frac{\xi_1 s_1^{\alpha(t_1)}}{t_1} V(t_1, t_2, \kappa)\frac{\xi_2 s_2^{\alpha(t_2)}}{t_2}\Gamma(t_2), \tag{1}$$

where each element ($\Gamma(t_i)$, $V(t_1, t_2, \kappa)$ and $\alpha(t_i)$) can be written as a power expansion in the coupling constant (here t_i are the squares of the exchange momenta, $\xi_i = e^{-i\pi\alpha(t_i)} + \tau_i$ the signature factors). Therefore, if one computes LO, NLO, NLLO ...corrections to the scattering amplitude, these corrections have to be decomposed into corrections to $\Gamma(t)$ etc. This type of factorization holds for general $2 \to n$ amplitudes, and can easily be generalized also to contributions with Regge cuts in the complex angular momentum plane.

So far much of the analysis of the scattering amplitudes has been done in the planar approximation (large-N_c limit, where N_c denotes the number of colors). In general, Regge theory is defined for signatured amplitudes which are linear combinations of different kinematic regions. In the planar approximation, on the other hand, each kinematic region has to be studied separately. An example is given in Fig. 1. In the left diagram all energies are positive, in the right one the two produced particles have been crossed, i.e. the subenergies s_1 and s_3 have become negative. Equivalently one can say that the t_1 and t_2 channels have been twisted. In the following we will use the notation were the kinematic region for the right-hand side (rhs) will be noted as $\tau_1\tau_3$, i.e. the kinematic region will be labeled by the twisted t channels. The use of this planar approximation leads to a few special features. First, planar amplitudes have a quite different structure of Regge cuts. In Fig. 2, the left-hand side (lhs) amplitude is completely planar and has only Regge poles, whereas the right one has a Regge cut in the t_2-channel. For a more detailed discussion of this structure we refer to the appendix of [15]. Another feature of the planar approximation is the appearance of singularities in the Regge pole contributions: in exactly those kinematic regions where (in the planar approximation) also Regge cuts appear, the Regge pole contributions contain

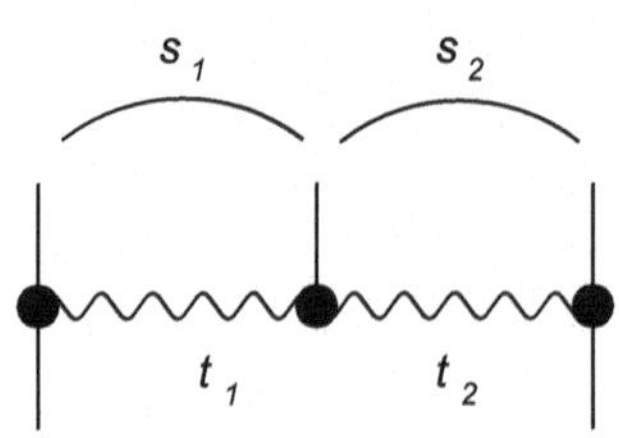

Fig. 1 The 2 → 3 amplitude: factorization

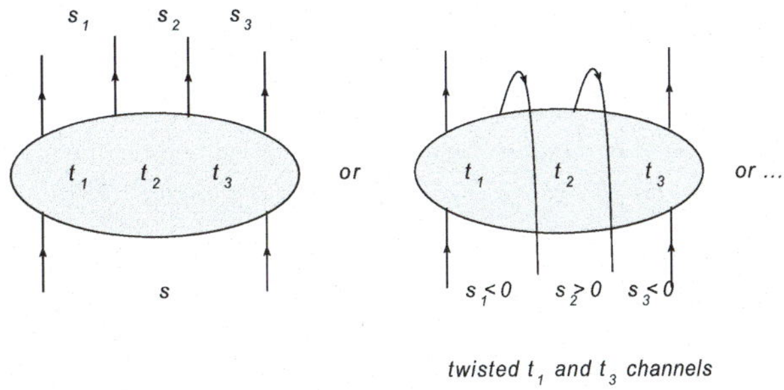

Fig. 2 Two different kinematic regions of the $2 \to 4$ amplitude

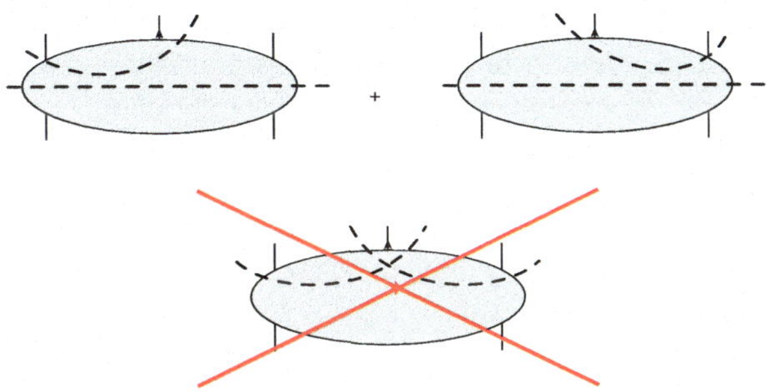

Fig. 3 The analytic structure of the $2 \to 3$ amplitude in the Regge limit: the upper line shows the two allowed terms, the lower one a forbidden double discontinuity

unphysical singularities which have to be removed by subtractions of the Regge cut contributions. This will be discussed in more detail in the following section.

As it has already been said, the calculations presented in this paper are based upon energy discontinuities and unitarity. One of the most important constraints comes from the Steinmann relations [16]: they exclude 'simultaneous discontinuities in overlapping channels'. A simple example is given in Fig. 3. In the upper line we have the two allowed double discontinuities, below a double discontinuity which is forbidden. It is straightforward to generalize this, for the maximal number of simultaneous energy discontinuities, to $2 \to n$ scattering amplitudes. The total number of terms equals C_{n-1} where C_n are the Catalan numbers with $C_n = 1, 1, 2, 5, 14, 42, \ldots$ $(n = 1, 2, 3, \ldots)$ and

$$C_{n+1} = \sum_{i=0}^{n} C_i C_{n-i}. \tag{2}$$

So far our remarks have been general. Now let us turn to the planar approximation of the Regge limit of N = 4 SYM gauge theories. What has been found is that for all energies being positive the amplitude is described by the all order BDS formula [1]:

$$T = T_{BDS}. \tag{3}$$

In the Regge limit there are only Regge pole contributions. However, if some t-channels are twisted (and energies become negative) the BDS formula is incomplete and requires corrections, the so-called remainder functions:

$$T = T_{BDS} \cdot R. \tag{4}$$

Here the remainder function R depends upon the kinematic region, is infrared finite and conformal invariant. In the Regge limit, R consists of Regge cut contributions. To find this remainder function is the main task: which type of Regge cuts appear in which kinematic region? One of the most promising features of these Regge cuts is integrability: the kernels for a Regge cut being a bound state of n reggeized gluons are equivalent to an integrable string consisting of n sites [15].

So far, existing studies of the $2 \to 4$, $2 \to 5$ cases have discussed only Regge cuts consisting of two reggeized gluons. For $2 \to n$ amplitudes the analysis [12] has been restricted to a special kinematic region which is also described only by this type of Regge cuts. The case $2 \to 6$ is of particular interest since it contains, for the first time, also a three reggeon cut. It requires the calculation of multiple energy discontinuities. Results for this process will be one of the main points of the present contribution.

3 A Few Features of the $2 \to 4$ and $2 \to 5$ Amplitudes

In order to explain in more detail some technical aspects which appear in the Regge limit in the planar approximation, we first briefly summarize aspects of the $2 \to 4$ and the $2 \to 5$ amplitudes.

3.1 The $2 \to 4$ Amplitude

Starting with the analytic structure of the $2 \to 4$ scattering amplitude (Fig. 4) we have five terms which have the maximal number of non-overlapping energy discontinuities and are in agreement with the Steinmann relations. Here the letters RR etc. refer to the two produced particles: the first letter stands for the left produced particle, the second one to the right one. R or L indicate on which side of the produced particle the energy cut enters.

In [2, 3, 17] in was found that the BDS-amplitude suggested by Bern, Dixon and Smirnov [1] in certain kinematic regions needs corrections which are given by Regge cuts. In the region where all energies are positive, only Regge poles contribute and the BDS-amplitude suggested by Bern, Dixon and Smirnov is complete. When analytically continuing into the region where the two produced particles have negative energies, the expression derived from the BDS amplitude is incomplete and needs a remainder function consisting of a two-reggeon Regge cut in the t_2-channel:

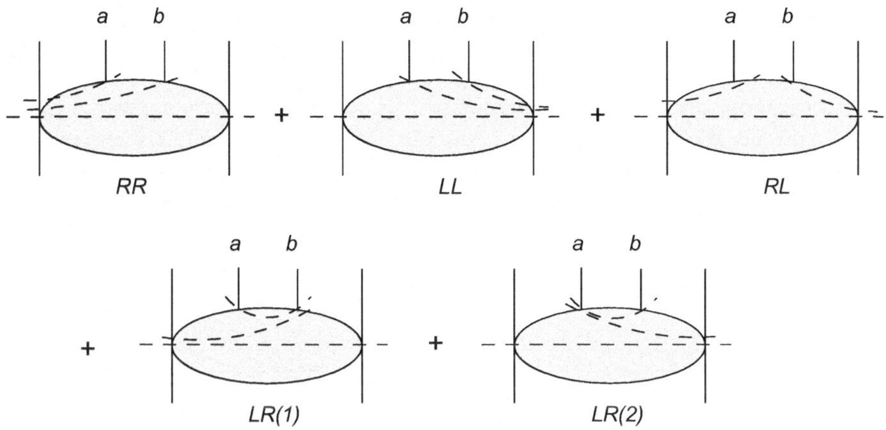

Fig. 4 The five terms of the 2 → 4 scattering amplitude: the dashed lines indicate the energy variables with a non-vanishing discontinuity

This cut is obtained from the discontinuity in s_2 and therefore can contribute only to the last two terms in Fig. 4, $LR(1)$ and $LR(2)$. For the result it is convenient to use the conformal spin representation. In leading order we have

$$f_{\omega_2} = \tag{5}$$
$$\frac{\alpha_s N_c}{4\pi} \sum_n (-1)^n \int d\nu \Phi_L \left(\frac{q_3^* k_a^*}{k_b^* q_1^*}\right)^{i\nu - \frac{n}{2}} \left(\frac{1}{\omega_2' - \omega(\nu, n)} - \frac{1}{\omega_2'}\right) \left(\frac{q_3 k_a}{k_b q_1}\right)^{i\nu + \frac{n}{2}} \Phi_R,$$

where ω_2' denotes the angular momentum of the t_2-channel, k_a and k_b are the momenta of the produced particles a and b, q_i the momenta of the exchange channels, and we have used the complex notation of the two dimensional momentum $k = (k_x, k_y)$: $k = k_x + ik_y$, $k^* = k_x - ik_y$. Furthermore

$$\omega(\nu, n) = \frac{g^2 N_c}{8\pi^2} \left(2\psi(1) - \mathcal{R}\psi(1 + i\nu + \frac{n}{2}) - \mathcal{R}\psi(1 + i\nu - \frac{n}{2}\right), \tag{6}$$

where $\psi(x)$ denotes the ψ-function, and $\mathcal{R}$ stands for 'real part'. In (5) we see the factorization pattern with the two impact factors

$$\Phi_L = \frac{1}{i\nu + \frac{n}{2}}, \quad \Phi_R = \frac{1}{i\nu - \frac{n}{2}}. \tag{7}$$

Next we address the issue of subtractions which we have mentioned before. As it was already said in Sect. 2, in contrast to signatured Regge amplitudes which are linear combinations of all different kinematic regions, the planar approximation considers each different region separately. As shown in [4, 17], starting from the factorizing form of the signatured Regge pole amplitude and then expanding into the sum of all different regions one finds that certain regions have unphysical singularities, and it is the appearance of these singularities which have to be removed by subtractions of the Regge cut amplitudes. As the simplest example, we start from Fig. 2: the planar kinematic region on the lhs of Fig. 2 is completely regular, and there are no Regge cuts in this region. On the other hand, in the twisted region $\tau_1\tau_3$ on the rhs of Fig. 2 the Regge pole contribution contains a singularity:

$$s_1^{\omega_1}(-s_2)^{\omega_2}s_3^{\omega_3}\left[e^{i\pi(\omega_a+\omega_b)} - 2ie^{i\pi\omega_2}\frac{\Omega_a\Omega_b}{\Omega_2}\right], \tag{8}$$

which we can also write as

$$s_1^{\omega_1}(-s_2)^{\omega_2}s_3^{\omega_3}\left[\cos\pi\omega_{ab} - 2i\left(-\frac{1}{2}\sin\pi(\omega_a+\omega_b) + \cos\pi\omega_2\frac{\Omega_a\Omega_b}{\Omega_2}\right)\right]. \tag{9}$$

Here

$$\begin{aligned} \omega_i &= --\frac{\gamma_K}{4}\ln\frac{|q_i|^2}{\lambda^2}, \quad i = 1,2,3 \\ \gamma_K &= 4a, \quad a = \frac{\alpha_s N_c}{2\pi} = \frac{g^2 N_c}{8\pi^2} \\ \Omega_2 &= \sin\pi\omega_2, \end{aligned} \tag{10}$$

i.e. ω_i denotes the gluon trajectory function in the t_i-channel, and

$$\Omega_a = -\frac{\gamma_K}{8}\ln\frac{|q_1|^2|q_2|^2}{|k_a|^2\lambda^2} \tag{11}$$

stands for the production vertex of particle a, and λ^2 denotes the square of the infrared cutoff. In (9) it is the round brackets, in particular the terms proportional to $1/\Omega_2$, which are unphysical and have to be removed.

At the same time it is known that there exist Regge cut contributions, and one easily sees that, in the planar approximation, the two reggeon cut in the t_2 channel appears just in the twisted kinematic region on the rhs of Fig. 2. It has the general form:

$$2is_1^{\omega_1}s_3^{\omega_3}\int\frac{d\omega_2'}{2\pi i}(-s_2)^{\omega_2'}W(\omega_2'), \tag{12}$$

where $W(\omega_2')$ has a left hand cut in the ω_2' plane. In particular, the singular part of the pole term in (9) has the same phase structure and thus can be removed by adding

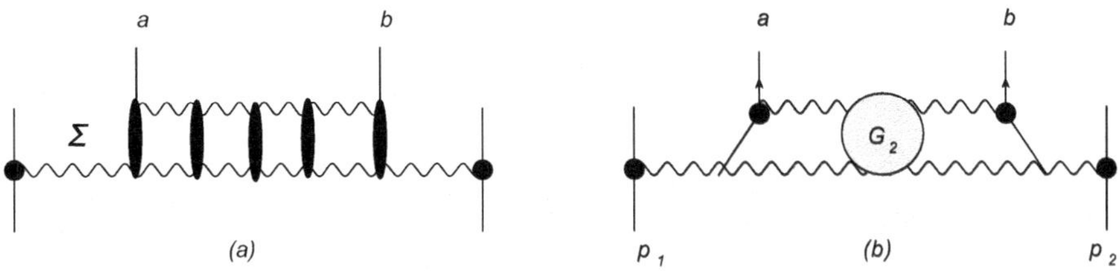

Fig. 5 The two reggeon cut in the $2 \to 4$ scattering amplitude. (**a**) Illustration of the energy discontinuity obtained from the unitarity integral. (**b**) Graphical illustration of the leading order remainder function f_{ω_2} in momentum space: the two production vertices and the blob which stands for the sum of the two gluon ladder graphs

the subtraction term

$$\delta W = 2is_1^{\omega_1}(-s_2)^{\omega_2}s_3^{\omega_3}\left(-\frac{1}{2}\sin\pi(\omega_a+\omega_b)+\cos\pi\omega_2\frac{\Omega_a\Omega_b}{\Omega_2}\right). \tag{13}$$

A similar analysis applies to the region $\tau_1\tau_2\tau_3$ where also the two reggeon cut contributes. As we have seen, for the $2 \to 4$ amplitude it is rather straightforward to determine the necessary subtraction of the Regge cut. For the $2 \to 5$ and $2 \to 6$ cases this is a much more difficult task.

This completes our brief summary of the simplest case, the $2 \to 4$ amplitude in leading order. All aspects discussed so far are relevant also for all higher order amplitudes, $2 \to 5$ etc.

Several comments are to be added:

(1) the impact factors Φ_L and Φ_R in (7) are given in the conformal (ν, n) representation. When going back to the momentum representation (see Fig. 5b) these impact factors are not pointlike. For this reason, for the attached two gluon state in the color octet representation the bootstrap property of the BFKL ladder does not apply, and the two reggeon cut survives.
(2) As we have said before, in the planar approximation (large N_c limit) the two gluon Regge cut contributes only in the kinematic regions $\tau_1\tau_3$ and $\tau_1\tau_2\tau_3$. The first one corresponds to the region where the produced particles a and b have negative energies. The absence of the Regge cuts in all other regions can be explained also in another way, namely by the planarity of Feynman diagrams [15]. Consider in Fig. 5b the loop on the left hand side of the two reggeon Green's function, and introduce Sudakov variables:

$$k' = \alpha' p_1 + \beta' p_2 + \mathbf{k}'. \tag{14}$$

Then the α' integration picks up poles from the propagator of the left impact factor and from the energy factor of the reggeon in the t_1 channel: as long as this reggeon is untwisted, both the energy cut singularity and the pole of the propagator lie on the same side of the α'-integration contour, and the integral vanishes. If on the other hand we twist the t_1 channel, the α'-integration is nonzero.

Since the appearance of [2, 3] where only the leading log approximation had been used several refinements have been made, e.g. [7, 11, 17, 18], First, higher order corrections of the impact factors have been calculated and NLO corrections and NNLO corrections of the BFKL kernel have been obtained. Also, the generalization to the 3 $\rightarrow$ 3 has been found [19]. The connection between Regge and collinear limits has been addressed in [20].

For the BFKL kernel of the two gluon Regge cut it has been shown [15] that it is part of an integrable open spin chain: the two gluon system belongs to a chain consisting two spins, a three reggeon cut (see below) provides the three spin chain etc.

3.2 The 2 → 5 Amplitude

The investigation of the 2 $\rightarrow$ 4 amplitude described in the previous section has been extended to the 2 $\rightarrow$ 5 amplitude [4–6]. Compared to the 2 $\rightarrow$ 4 scattering amplitude, the new piece of the remainder function is a 'long' Regge cut extending over the two t-channels (Fig. 6): t_2 and t_3.

Beginning again with the analytic structure, we now have 14 terms. In Fig. 7 we show those four terms which contain this new long cut: We note that $LLR(1)$ in the upper line and $LRR(1)$ in the lower line have the same discontinuity structure except for the shortest discontinuities in $s_3 = s_{bc}$ and $s_2 = s_{ab}$. This indicates that the production vertex for particle b consists of two distinct pieces, V_L and V_R. The same argument applies to $LLR(2)$ in the upper line and $LRR(2)$ in the lower line of Fig. 7.

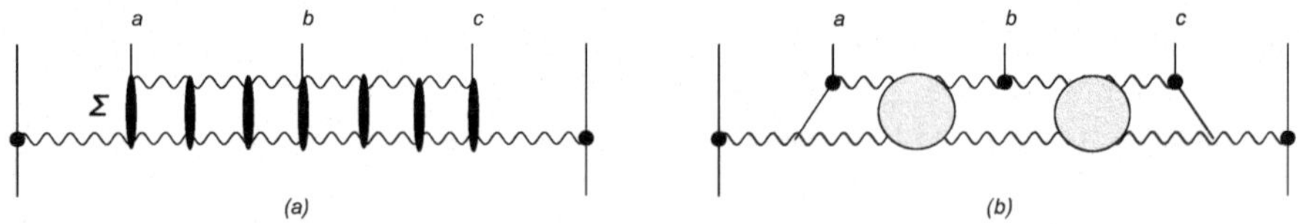

Fig. 6 Long Regge cut extending over the t_2 and t_3 channel. (**a**) unitarity integral and (**b**) leading order remainder function

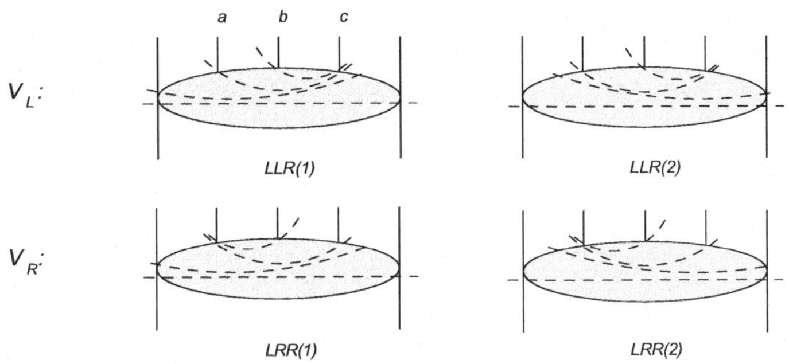

Fig. 7 Terms containing the long cut

As to the kinematic regions, in the region where all produced particles have positive energies again only Regge poles contribute, and the amplitude is fully described by the BDS formula. Now let us continue this pole contribution to other kinematic regions and describe in more detail how the singularities appear. As we have already seen for the 2 → 4 amplitude, this analysis is necessary for finding the subtractions of the Regge cut amplitudes. As discussed in detail in [4], the signatured 2 → $n+1$ production amplitude can be written in the factorized form:

$$\frac{A_{2\to n+1}^{\tau_i\tau_j\ldots\tau_n}}{\Gamma(t_1)\Gamma(t_n)} = |s_1|^{\omega_1}\xi_1 V^{\tau_1\tau_2;a_1}|s_2|^{\omega_2}\xi_2 V^{\tau_2\tau_3;a_2}|s_3|^{\omega_3}\xi_3 \times \ldots$$
$$\times |s_{n-1}|^{\omega_{n-1}}\xi_{n-1}V^{\tau_{n-1}\tau_n;a_{n-1}}|s_n|^{\omega_n}\xi_n, \tag{15}$$

where

$$\xi_i = e^{-i\pi\omega_i} - \tau_i \;\; ; \;\; \xi_{ij} = e^{-i\pi\omega_{ij}} + \tau_i\tau_j \;\; ; \;\; \xi_{ji} = e^{-i\pi\omega_{ji}} + \tau_i\tau_j \tag{16}$$

with

$$\omega_{ij} = \omega_i - \omega_j, \;\; \tau_i = \pm 1 \tag{17}$$

denote the given signature factors, and

$$V^{\tau_i\tau_j;a_j} = \frac{\xi_{ij}}{\xi_i}c_R^{ij;a_i} + \frac{\xi_{ji}}{\xi_j}c^{ij;a_i} \tag{18}$$

stands for the complex-valued production vertex of the particle a_j.

As it has been said before, the planar approximation is obtained as an expansion in monomials of signatures τ_i, and in this expansion each terms belongs to the planar approximation continued to a particular kinematic region. For this expansion it is convenient to write

$$\tilde{V}^{\tau_1\tau_2;a_1} = \xi_1 V^{\tau_1\tau_2;a_1}\xi_2 \tag{19}$$

and

$$\frac{A_{2\to n+1}^{\tau_i\tau_j\ldots\tau_n}}{\Gamma(t_1)|s_1|^{\omega_1}|s_2|^{\omega_2}\ldots|s_n|^{\omega_n}\Gamma(t_n)} = \tilde{V}^{\tau_1\tau_2;a_1}\frac{1}{\xi_2}\tilde{V}^{\tau_2\tau_3;a_2}\frac{1}{\xi_3}\cdots\frac{1}{\xi_{n-1}}\tilde{V}^{\tau_{n-1}\tau_n;a_{n-1}}. \tag{20}$$

In order to obtain this representation we observe that the production vertex can be expanded as:

$$\begin{aligned}\tilde{V}^{\tau_1\tau_2;a} &= e^{-i\pi\omega_1}c_R^{12;a_1} + e^{-i\pi\omega_2}c_L^{12;a_1} - \tau_1 e^{-i\pi\omega_1}\left(e^{-i\pi\omega_1}c_R^{12;a_1} + e^{-i\pi\omega_2}c_L^{12;a_1}\right)\\ &-\tau_2 e^{-i\pi\omega_2}\left(e^{-i\pi\omega_1}c_R^{12;a_1} + e^{-i\pi\omega_2}c_L^{12;a_1}\right) + \tau_1\tau_2\left(e^{-i\pi\omega_2}c_R^{12;a_1} + e^{-i\pi\omega_1}c_L^{12;a_1}\right),\end{aligned} \tag{21}$$

and the propagator can be written in the form:

$$\frac{1}{\xi_2} = \frac{1}{e^{-i\pi\omega_2} - \tau_2} = \frac{e^{-i\pi\omega_2} + \tau_2}{-2i\,\sin(\pi\omega_2)e^{-i\pi\omega_2}}. \tag{22}$$

Note the appearance of the nonphysical poles $\sim\ 1/\sin\pi\omega_2$ which should be canceled by the Regge cut contributions.

In this way one can determine the Regge pole contributions in all different kinematic regions, labeled by products of τ_i factors. In particular one finds that in all regions containing the product $\tau_1\tau_4$ the Regge pole terms are most singular and contain double poles of the form $\sim 1/(\sin\pi\omega_2\sin\pi\omega_3)$. In particular, in the region $\tau_1\tau_4$ the Regge pole term has the form

$$\begin{aligned}&|s_1|^{\omega_1}(-s_2)^{\omega_2}(-s_3)^{\omega_3}|s_4|^{\omega_4}\\ &\cdot\left[e^{i\pi(\omega_a+\omega_b+\omega_c)} - 2i e^{i\pi(\omega_2+\omega_3)}\frac{\sin(\pi\omega_a)\sin(\pi\omega_b)\sin(\pi\omega_c)}{\sin(\pi\omega_2)\sin(\pi\omega_3)}\right].\end{aligned} \tag{23}$$

A complete list of all kinematic regions can be found in [4].

The same result can also be obtained in a different way. Instead of (15), we can start from the analytic representation consisting of 14 terms. For each term the Regge pole contribution has a particular form, e.g. for the first term LLR(1) in Fig. 7 in the region where all energies are positive:

$$(-s_1)^{\omega_1}(-s_2)^{\omega_2}(-s_3)^{\omega_3}(-s_4)^{\omega_4} F^{pole}_{LLR(1)}$$
$$= (-s_1)^{\omega_1}(-s_2)^{\omega_2}(-s_3)^{\omega_3}(-s_4)^{\omega_4} \frac{\Omega_1 \, \Omega_{34}}{\Omega_3 \, \Omega_{14}} \frac{V_L(a)}{\Omega_{21}} \frac{V_L(b)}{\Omega_{32}} \frac{V_R(c)}{\Omega_{34}}, \tag{24}$$

where

$$\Omega_i = \sin \pi \omega_i, \quad \Omega_{ij} = \sin \pi(\omega_i - \omega_j), \quad \omega_{ij} = \omega_i - \omega_j. \tag{25}$$

When inserting, for each kinematic region, the corresponding phases of the energy factor, one simply computes the sum of the 14 terms. For the kinematic region $\tau_1 \tau_4$ one obtains, after some algebra, the above result (23).

Turning to the Regge cut contributions we start with the ansatz for the long cut in the region of all positive energies, e.g. for the first term in Fig. 7:

$$\int d\omega_2' \int d\omega_3' \int d\omega_4' (-s_3)^{\omega_3'-\omega_2'} (-s_{123})^{\omega_2'-\omega_1} (-s_{0123})^{\omega_1-\omega_4} (-s)^{\omega_4} F_{LLR(1)} \tag{26}$$

with

$$F_{LLR(1)} = \frac{W_{\omega_2\omega_3;R}}{\Omega_{32}\Omega_{14}}. \tag{27}$$

With this ansatz (and analogous ones for the 3 remaining terms) one finds that this Regge cut appears in all regions where the t_1 and t_4 channels are twisted, i.e. in the regions $\tau_1\tau_4$, $\tau_1\tau_2\tau_4$, $\tau_1\tau_3\tau_4$ and $\tau_1\tau_2\tau_3\tau_4$. These are exactly the same regions where also the pole terms are most singular, with double poles as in (23).

Based upon this Regge pole structure it is possible to determine the necessary subtraction terms for the Regge cut contributions, in analogy with the $2 \to 4$ case described above. Without going into details which are described in [5] we only list the results for the long cut. Introducing the trigonometric factors of the long cut, generalizing (27) we find, as expected, that the long cut appears in all regions containing the product $\tau_1\tau_4$, i.e. just in those regions where the Regge poles are most singular. From the general analytic representation we find, for the Regge cut contribution in the region $\tau_1\tau_4$,

$$2i|s_1|^{\omega_1}|s_2|^{\omega_2}|s_3|^{\omega_3}|s_4|^{\omega_4}\left(e^{-i\pi\omega_2}\widetilde{W}_{\omega_2\omega_3;R} + e^{-i\pi\omega_3}\widetilde{W}_{\omega_2\omega_3;L}\right), \tag{28}$$

where we have introduced the following combinations of long cut and short cuts:

$$\widetilde{W}_{\omega_2\omega_3;L} = \frac{W_{\omega_2\omega_3;L}}{\Omega_{32}} + \frac{\Omega_a}{\Omega_2} W_{\omega_3} \tag{29}$$

$$\widetilde{W}_{\omega_2\omega_3;R} = \frac{W_{\omega_2\omega_3;R}}{\Omega_{23}} + W_{\omega_2} \frac{\Omega_c}{\Omega_3}. \tag{30}$$

Here $W_{\omega_2\omega_3;L}$ and $W_{\omega_2\omega_3;R}$ denote the cut terms of the upper and lower lines of Fig. 7, resp. We define

$$\begin{aligned} W_{\omega_2\omega_3;L} &= \delta W_{\omega_2\omega_3;L} + W^{reg}_{\omega_2\omega_3;L} \\ W_{\omega_2\omega_3;R} &= \delta W_{\omega_2\omega_3;R} + W^{reg}_{\omega_2\omega_3;R}, \end{aligned} \tag{31}$$

where the subtraction terms are found to be

$$\begin{aligned} \Omega_{23}\delta \widetilde{W}_{\omega_2\omega_3;L} = & \frac{\Omega_a\Omega_b\Omega_c}{\Omega_3} \\ & -\frac{1}{2}\Big[\cos\pi(\omega_3-\omega_b)\cos\pi(\omega_a-\omega_c) - \cos\pi(\omega_3-\omega_a-\omega_b-\omega_c)\Big] \end{aligned} \tag{32}$$

$$\begin{aligned} \Omega_{32}\delta \widetilde{W}_{\omega_2\omega_3;R} = & \frac{\Omega_a\Omega_b\Omega_c}{\Omega_2} \\ & -\frac{1}{2}\Big[\cos\pi(\omega_2-\omega_b)\cos\pi(\omega_a-\omega_c) - \cos\pi(\omega_2-\omega_a-\omega_b-\omega_c)\Big]. \end{aligned} \tag{33}$$

With these subtractions we find for the scattering amplitude in the region $\tau_1\tau_4$ the regular expression:

$$\begin{aligned} & |s_1|^{\omega_1}|s_4|^{\omega_4}\Big[(-s_2)^{\omega_2}(-s_3)^{\omega_3}e^{i\pi\omega_b}\cos\pi\omega_{ac} \\ & +2i\left(e^{-i\pi\omega_3}\widetilde{W}^{reg}_{\omega_2\omega_3;L} + e^{-i\pi\omega_2}\widetilde{W}^{reg}_{\omega_2\omega_3;R}\right)\Big]. \end{aligned} \tag{34}$$

So far the given expressions are valid to all orders of the coupling constant. However, when computing, from unitarity, the partial waves $W_{\omega_2\omega_3;L}$ and $W_{\omega_2\omega_3;R}$, we first will restrict ourselves to the leading approximation. This implies that, in (34), we will approximate the phase factors by unity and hence deal with the sum $W^{reg}_{\omega_2\omega_3;L} + W^{reg}_{\omega_2\omega_3;R}$. The energy discontinuity Δ_{123} in leading order is given by:

$$\Delta_{123} = \widetilde{W}^{reg}_{\omega_2\omega_3;L} + \widetilde{W}^{reg}_{\omega_2\omega_3;R} - W^{reg}_{\omega_3} - W^{reg}_{\omega_2}, \tag{35}$$

where $W^{reg}_{\omega_2}$, $W^{reg}_{\omega_3}$ denote the regular parts of the short cuts of the t_2 and t_3 channels, and the unitarity integral which determines the energy discontinuity Δ_{123} in leading order has the form:

$$\Delta_{123} = |s_1|^{\omega_1}|s_2|^{\omega_2}|s_3|^{\omega_3}|s_4|^{\omega_4}\left(f_{\omega_2\omega_3} + \frac{\pi}{2}\delta_{14} - f_{\omega_2} - \frac{\pi}{2}\delta_{13} - f_{\omega_3} - \frac{\pi}{2}\delta_{24}\right), \tag{36}$$

$$f_{\omega_2\omega_3} = \frac{a}{2}\sum_{n_1,n_2}(-1)^{n_1+n_2} \tag{37}$$

$$\cdot \int\int \frac{d\nu_1 d\nu_2}{(2\pi)^2}\Phi_L\left(\frac{k_a^* q_3^*}{q_1^* k_b^*}\right)^{i\nu_1+\frac{n_1}{2}}\left(\frac{k_a q_3}{q_1 k_b}\right)^{i\nu_1-\frac{n_1}{2}}\left(\frac{s_{12}}{s_{02}}\right)^{\omega(\nu_1,n_1)}$$

$$\cdot B(\nu_1,\nu_2,n_1.n_2)\left(\frac{s_{23}}{s_{03}}\right)^{\omega(\nu_2,n_2)}\left(\frac{k_b^* q_4^*}{q_2^* k_c^*}\right)^{i\nu_2+\frac{n_2}{2}}\left(\frac{k_b q_4}{q_2 k_c}\right)^{i\nu_2-\frac{n_2}{2}}\Phi_R|_{\text{sub}}$$

and

$$\delta_{14} = V_{14} + \pi(\omega_a + \omega_c),\ \ V_{ik} = \frac{\gamma_k}{4}\ln\frac{|q_i|^2|q_k|^2}{|q_i - q_k|^2\lambda^2}. \tag{38}$$

V_{14} denotes the one loop approximation of the long cut which is contained already in the BDS formula. $f_{\omega_2\omega_3}$ is illustrated in Fig. 6. In (37) we have carried out the ω-integrals. The impact factors for particles a and c are the same as in the $2 \to 4$ amplitude. $B(\nu_1, \nu_2, n_1, n_2)$ denotes the new (leading order) production vertex of particle b (see Fig. 6b). The subscript $|_{\text{sub}}$ indicates that we have subtracted the one loop contribution.

Combining (35) with (36) we find

$$W^{reg}_{\omega_2\omega_3;L} + W^{reg}_{\omega_2\omega_3;R} = \left(f_{\omega_2\omega_3} + \frac{\pi}{2}\delta_{14}\right), \tag{39}$$

and the amplitude becomes:

$$|s_1|^{\omega_1}(-s_2)^{\omega_2}(-s_3)^{\omega_3}|s_4|^{\omega_4}\left[e^{i\pi\omega_b}\cos\pi\omega_{ac} + 2i\left(f_{\omega_2\omega_3} + \frac{\pi}{2}\delta_{14}\right)\right]. \tag{40}$$

Here the energy factors together with the phase $e^{i\pi(\delta_{14}+\omega_b)}$ are contained in the BDS amplitude.

Finally we have to discuss the production vertex of particle b. These results are new, and details are given in [21, 22]. In (37) $B(\nu_1, \nu_2, n_1, n_2)$ denotes the leading order vertex illustrated in Fig. 6b. This result was obtained from the (single) discontinuity in s_{123} (Fig. 8a) which, in leading order, leads to the sum of the two partial waves (35). To see the full structure we need to compute $W^{reg}_{\omega_2\omega_3;L}$ and

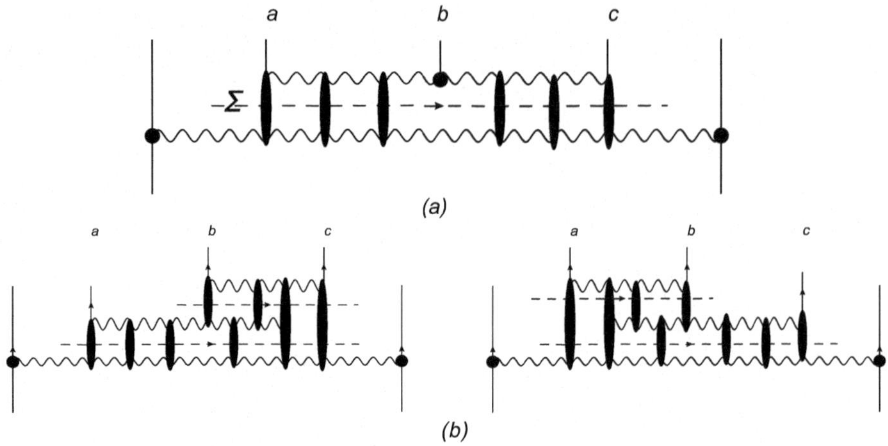

Fig. 8 Single and double discontinuities

$W^{reg}_{\omega_2\omega_3;R}$ separately, not only the sum. As it can be seen from Fig. 7, in order to discriminate between the first and the second line, we need to compute the double discontinuities $\Delta_{12}\Delta_{123}$ and $\Delta_{23}\Delta_{123}$ (Fig. 8b): This has been done recently in [21, 22], and we summarize the results. Writing down the leading order equations for the double discontinuities and evaluating the corresponding unitarity equations, one finds

$$\widetilde{W}^{\rm reg}_{\omega_2\omega_3;\rm R} = \frac{\pi\omega'_2 f_{\omega_2\omega_3} - f^{(b)}_{\omega_2\omega_3}}{\pi(\omega'_2 - \omega'_3)} \tag{41}$$

and

$$\widetilde{W}^{\rm reg}_{\omega_2\omega_3;\rm L} = \frac{\pi\omega'_3 f_{\omega_2\omega_3} - f^{(b)}_{\omega_2\omega_3}}{\pi(\omega'_3 - \omega'_2)} \tag{42}$$

(for simplicity, we have disregarded the one and two loop terms). Both partial waves, $\widetilde{W}^{\rm reg}_{\omega_2\omega_3;\rm R}$ and $\widetilde{W}^{\rm reg}_{\omega_2\omega_3;\rm L}$, consist of two terms. If we would disregard the second terms and take the sum of the first ones we simply are back to the previous result of the single discontinuity. So it is the second piece which is new and can be obtained only via the double discontinuities. The first equation (41) is illustrated in Fig. 9.

The new vertex contained in $f^{(b)}_{\omega_2\omega_3}$ can be found in [21, 22] and is illustrated in Fig. 10.

It is important to stress that in (41) and (42) both terms are of the same order in g^2. As we have already said, in the leading term of the scattering amplitude where we disregard all phases, in (39) only the sum $W_{\omega_2\omega_3;L} + W_{\omega_2\omega_3;R}$ appears which

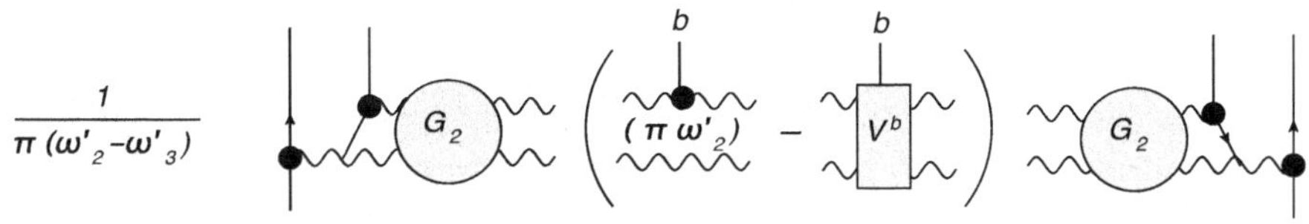

Fig. 9 Graphical illustration of the two terms of $\widetilde{W}_{23;R}$

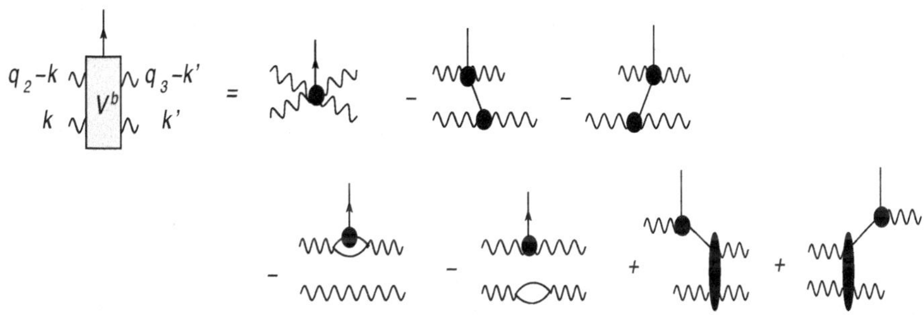

Fig. 10 The new piece of the production vertex

equals the sum of the two first terms only and cancels the second terms. Only if we expand the phases contained in the energy factors also these second terms show up:

$$
\begin{aligned}
&2ie^{-i\pi(\omega_2+\omega_3)} \int_2 \int_3 s_{12}^{\omega_2'} s_{23}^{\omega_3'} \left(e^{i\pi\omega_2'} \widetilde{W}^{\text{reg}}_{\omega_2\omega_3;\text{R}} + e^{i\pi\omega_3'} \widetilde{W}^{\text{reg}}_{\omega_2\omega_3;\text{L}} \right) \\
&\approx e^{-i\pi\omega_5} e^{i\pi\omega_d} \int_2 \int_3 s_{12}^{\omega_2'} s_{23}^{\omega_3'} \\
&\quad \cdot \left(\widetilde{W}^{reg}_{\omega_2\omega_3;R} + \widetilde{W}^{reg}_{\omega_2\omega_3;L} - i\pi\omega_2' \widetilde{W}^{reg}_{\omega_2\omega_3;R} - i\pi\omega_3' \widetilde{W}^{reg}_{\omega_2\omega_3;L} \right) \\
&\approx 2ie^{-i\pi(\omega_2+\omega_3+\omega_5)} e^{i\pi\omega_d} \int_2 \int_3 s_{12}^{\omega_2'} s_{23}^{\omega_3'} \left(f_{\omega_2\omega_3} + i f^{(b)}_{\omega_2\omega_3} \right) . \qquad (43)
\end{aligned}
$$

The '≈' sign indicates that we have expanded the phase factors, and our equations are valid only up to the first order in $i\pi$. It should be emphasized that the second terms in (41) and (42) still belong to the same leading approximation of the partial waves as the first ones, although in the scattering amplitude they appear proportional to $\sim i\pi$ as 'next-to-leading-order'.

As illustrated in Fig. 9, in both $\widetilde{W}^{\text{reg}}_{\omega_2\omega_3;\text{L}}$ and $\widetilde{W}^{\text{reg}}_{\omega_2\omega_3;\text{R}}$ the production vertex for particle b factorizes; in $f_{\omega_2\omega_3}$ the vertex is given by $B(\nu_1\nu_2, n_1, n_2)$, in $f^{(b)}_{\omega_2\omega_3}$ by V^b. When writing in (43) the scattering amplitude, the combination $f_{\omega_2\omega_3} + i f^{(b)}_{\omega_2\omega_3}$ appears and the production vertex is given by the complex valued combination $B + iV^b$. We therefore still have factorization, but the production vertex has become complex-valued.

It will be important to compare these results with those of [8, 12]: in these papers the $2 \to 5$ amplitude beyond LO (and even the generalization to $2 \to n$) has been studied in the kinematic regions $\tau_1 \tau_4$, and expressions for the production vertex have been derived. So far, a comparison with our results cannot be carried out: our expression have been derived im momentum space (as it is natural when using energy discontinuities and unitarity), and it will be necessary to perform the transformation to the conformal (ν, n) representation. This will be done in a separate paper. Also in [7, 11] the multiregge limit of the $2 \to n$ amplitude has been studied. Here a comparison is even more complicated and requires a closer investigation.

4 The 8 Point Function

We now turn to the next step, the generalization to the $2 \to 6$ production process. This process is of special interest since it contains new Regge cut pieces. We illustrate these new Regge cuts by their unitarity integrals in Figs. 11 and 12. First, there is a longer cut extending over 3 different t-channels (Fig. 11). More important, for the first time it also contains the product of two short two-reggeon cuts (Fig. 12a) and a Regge cut consisting of three reggeized gluons (Fig. 12b).

The analytic decomposition of the $2 \to 6$ amplitude now contains 42 terms. In Fig. 13 we show those terms which contain the double cuts.

For the 3 reggeon cut we have the four terms shown in Fig. 14.

The kinematic regions where these cuts appear are, again, determined by combining the trigonometric factors of the Regge poles and of the Regge cut amplitudes with the phases of the energy factors (the results are in accordance with the argument given in the appendix of [15]). In particular, the very long Regge cut extending over the t_2, t_3, and t_4 channels appears in all those kinematic regions

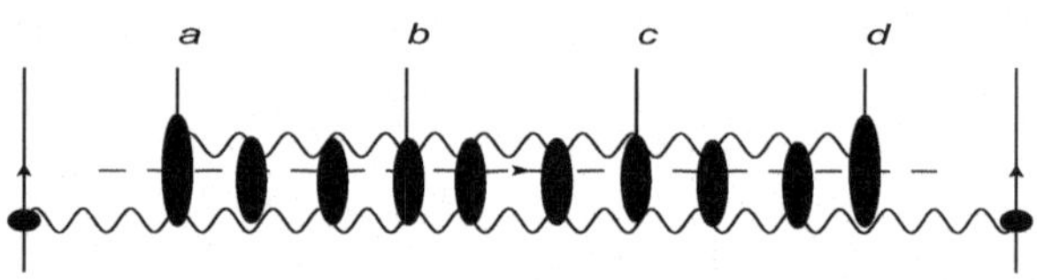

Fig. 11 The single discontinuity in s_{1234}

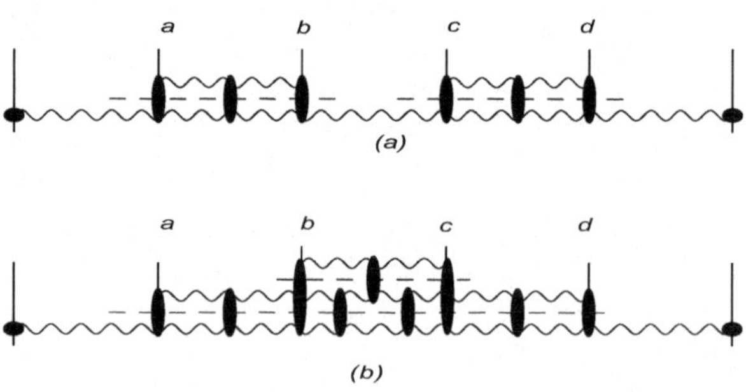

Fig. 12 Double discontinuities (**a**) in s_{12} and s_{34}, (**b**) in s_{23} and s_{1234}

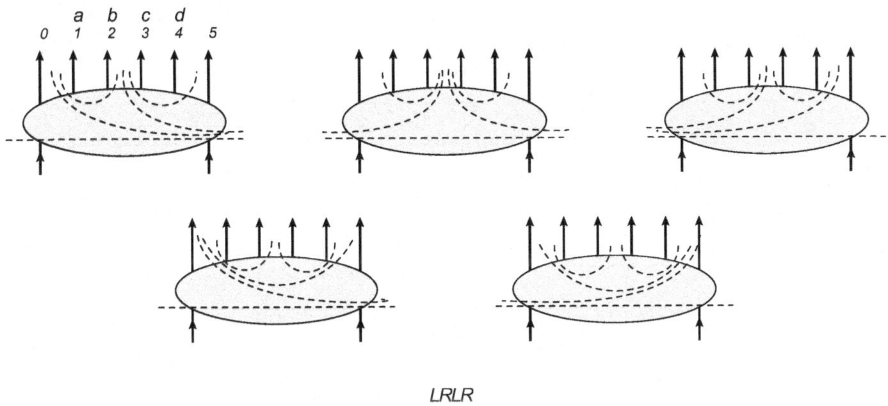

Fig. 13 The five terms containing the double cut

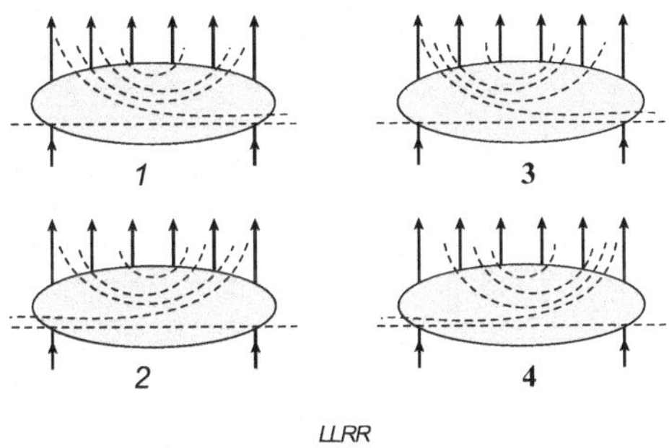

Fig. 14 The four terms containing the 3 reggeon cut

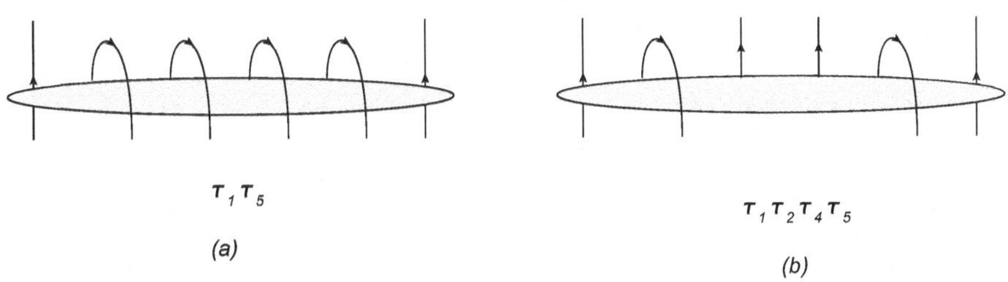

Fig. 15 Kinematic regions for (**a**) the very long cut and (**b**) for the three reggeon cut

where the t_1 and t_5 channels are twisted; in our notation, all regions containing the product of τ_1 and τ_5. This includes, in particular, the region $\tau_1\tau_5$, where the four produced particles have negative energies (Fig. 15a). The same applies to the product of the two short cuts. The three reggeon cut appears only in the region where particles a and d have negative energies, particles b and c positive energy (Fig. 15b): $\tau_1\tau_2\tau_4\tau_5$ and $\tau_1\tau_2\tau_3\tau_4\tau_5$.

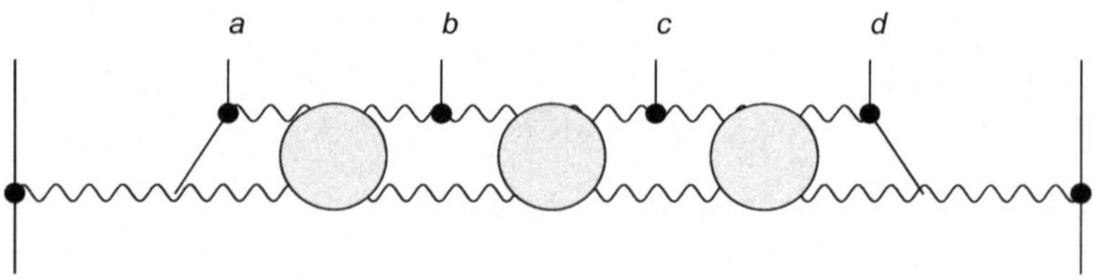

Fig. 16 The very long cut in the leading approximation

Details of this study are given in [21], and here we only quote the main results. In order to find the subtractions we, again, first have to determine the Regge pole contributions in all kinematic regions. This follows the procedure described in the previous section, and the results are listed in one of the appendices of [21]. In the next step one has to find the subtractions: this turns out to be a rather complicated task. As an important test, we have to show that not only the scattering amplitudes but also their single and multiple discontinuities are free from the Regge pole singularities. This is described in detail in [21], and the derivation is too lengthy to be described here.

From Fig. 12 we immediately see that for the two new contributions we need to consider double discontinuities. This provides a new test of the correct analytic structure in all energy variables, in accordance with the Steinmann relations. Furthermore, both the double cut and the three reggeon cut contributions are of the order $(i\pi)^2$, in contrast to the two reggeon cut contributions in $2 \to 4$, $2 \to 5$ and the very long cut in Fig. 11, for which the leading order is of the order $(i\pi)$. However, in our discussion of the production vertex of the $2 \to 5$ amplitude we also had to extend our analysis to double discontinuities of the order $(i\pi)^2$. We therefore conclude that, starting with $2 \to 5$, double and even higher order energy discontinuities need to be taken into account..

Below we list a few results. Beginning with the very long cut, we have to generalize our analysis of the long cut obtained in the $2 \to 5$ case. There are four terms which we denote by W_{LL}, W_{LR}, W_{RL}, and W_{RR}. Starting with the leading contribution for the scattering amplitude where all phases of the energy factors are neglected, only the sum appears: $W_{LL} + W_{LR} + W_{RL} + W_{RR}$, and for this sum we can use the single energy discontinuity in s_{1234}. The result is illustrated in Fig. 16 which nicely generalizes Fig. 6b. The scattering amplitude in the region $\tau_1\tau_5$ then has the form:

$$
\begin{aligned}
&e^{-i\pi(\omega_2+\omega_3+\omega_4)}\Big[e^{i\pi(\omega_b+\omega_c)}\cos\pi(\omega_a-\omega_d) \\
&+2i\Big[f_{\omega_2\omega_3\omega_4}+\frac{\pi}{2}\delta_{15} \\
&-i\left(f_{\omega_2}+\frac{\pi}{2}\delta_{13}-\frac{\pi(\omega_a-\omega_b)}{2}\right)\left(f_{\omega_4}+\frac{\pi}{2}\delta_{35}-\frac{\pi(\omega_d-\omega_c)}{2}\right)\Big]\Big], \quad (44)
\end{aligned}
$$

where the long cut is contained in $f_{\omega_2\omega_3\omega_4}$. In addition to the very long cut we also have the double cut which comes as a product of the short cuts in the t_2 and t_4 channels. The remainder function for the very long cut, $f_{\omega_2\omega_3\omega_4}$, has the form:

$$f_{\omega_2\omega_3\omega_4} = \frac{a}{2} \sum_{n_2,n_3,n_4} (-1)^{n_2+n_3-n_4} \tag{45}$$

$$\cdot \int\int \frac{d\nu_2 d\nu_3 d\nu_4}{(2\pi)^3} \Big[\Phi_L \left(\frac{k_a^* q_3^*}{q_1^* k_b^*}\right)^{i\nu_2+\frac{n_2}{2}} \left(\frac{k_a q_3}{q_1 k_b}\right)^{i\nu_2-\frac{n_2}{2}}$$

$$\cdot \left(\frac{s_{12}}{s_{02}}\right)^{\omega(\nu_2,n_2)} \cdot B(\nu_2, \nu_2, n_3.n_3) \left(\frac{k_b^* q_4^*}{q_2^* k_c^*}\right)^{i\nu_3+\frac{n_3}{2}} \left(\frac{k_b q_4}{q_2 k_c}\right)^{i\nu_3-\frac{n_3}{2}} \left(\frac{s_{23}}{s_{03}}\right)^{\omega(\nu_3,n_3)}$$

$$\cdot B(\nu_3, \nu_4, n_3, n_4) \left(\frac{k_c^* q_5^*}{q_3^* k_d^*}\right)^{i\nu_4+\frac{n_4}{2}} \left(\frac{k_c q_5}{q_3 k_d}\right)^{i\nu_4-\frac{n_4}{2}} \left(\frac{s_{34}}{s_{04}}\right)^{\omega(\nu_4,n_4)} \Phi_R \Big]|_{\text{sub}}.$$

The vertex functions Φ_L and Φ_R for the produced particles a and d are the same as those in the $2 \to 4$ and $2 \to 5$ amplitudes. The production vertex functions $B(\nu.n)$ belong to the pointlike vertices shown in Fig. 16, and they are the same as in (37). In (44) we have not yet written the complete answer, i.e. we have not yet included terms proportional to $(i\pi)^2$.

From the $2 \to 5$ amplitude we know that this is not the yet the final answer. Each production vertex for particles b and c consists of two pieces, V_L and V_R. In order to determine the partial waves W_{LL}, W_{LR}, W_{RL}, and W_{RR} separately, we need to compute multiple discontinuities. An example, $\Delta_{12}\Delta_{1234}$, is shown in Fig. 17.

The simplest way to find W_{LL} etc. is the following: the double discontinuity $\Delta_{12}\Delta_{1234}$ determines the sum $W_{RL} + W_{RR}$; similarly, the double discontinuity $\Delta_{34}\Delta_{1234}$ gives the sum $W_{RL} + W_{LL}$. Finally, the triple discontinuity $\Delta_{12}\Delta_{34}\Delta_{1234}$ determines W_{RL}, and from the single discontinuity Δ_{1234} we have the sum of all four terms. This then allows to find the four terms W_{LL}, W_{LR}, W_{RL}, and W_{RR} separately (in leading order). Without going into detail we illustrate the result for $\widetilde{W}_{RR}^{reg}$ in Fig. 18. Analogous results hold for the other terms, $\widetilde{W}_{RL}^{reg}$ etc.

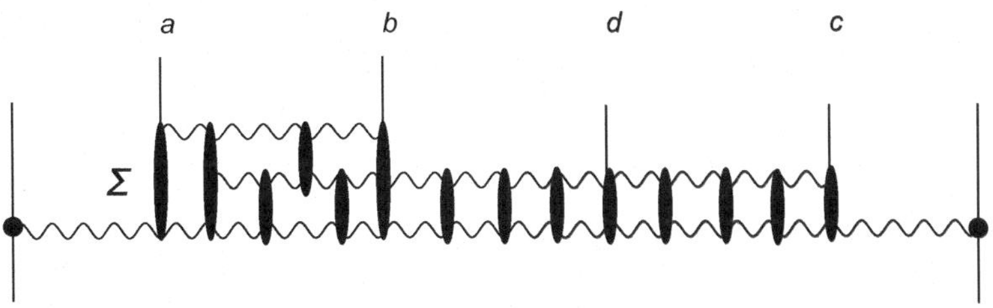

Fig. 17 Illustration of the double discontinuity $\Delta_{12}\Delta_{1234}$

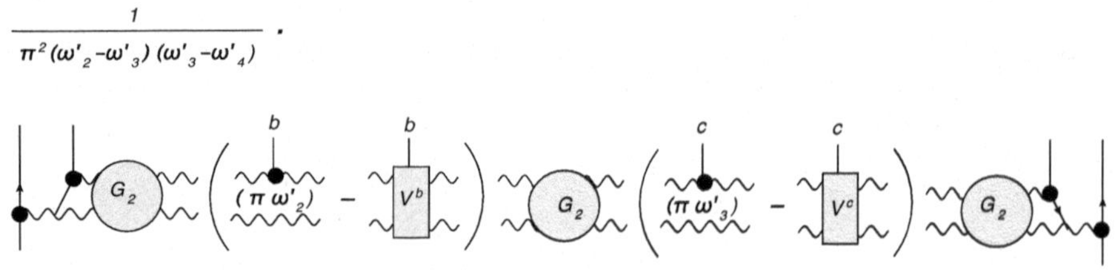

Fig. 18 Illustration of $\widetilde{W}_{RR}^{reg}$: factorization of the production vertices

Similar to the $2 \to 5$ case, we still have factorization of the production vertex: in $\widetilde{W}_{RR}^{reg}$ etc we have combinations of B and V^b (Fig. 18) for both produced particles. When writing down the amplitude and expanding the phases of the energy factors up to the order $i\pi$, we find—for the region $\tau_1\tau_5$—the combination $f_{\omega_2\omega_3\omega_4} + if^b_{\omega_2\omega_3\omega_4} + if^c_{\omega_2\omega_3\omega_4}$, in agreement with the product of the two complex valued production vertices which were found for $2 \to 5$. This demonstrates factorization: in the product of the two production vertices b and c each term agrees with the structure of the V_R vertex of the $2 \to 5$ amplitude shown Fig. 9. Further details can be found in [21]. The factorization of production vertices for the general $2 \to n$ case has also been established in [12]. However, for a closer comparison we refer to our discussion at the end of Sect. 3.

As to the product of the two short reggeon cuts in the t_2 and t_4 channels we only mention that they are obtained from the double discontinuity $\Delta_{12}\Delta_{34}$ (Fig. 12a). Each factor agrees with the partial wave obtained for the $2 \to 4$ vertex.

Finally we come to the 3 reggeon cut appearing in the region $\tau_1\tau_2\tau_4\tau_5$ (Fig. 12b): it is obtained from the double discontinuity $\Delta_{23}\Delta_{1234}$. The full amplitude in the region $\tau_1\tau_2\tau_4\tau_5$ consists of several pieces which in leading order are given by:

$$
\begin{aligned}
&e^{-i\pi\omega_3}\left(e^{i\pi(\omega_b+\omega_c-\omega_a-\omega_d)} + 2i\left[\frac{\pi}{2}\delta_{1245}\right.\right.\\
&+f_{\omega_2\omega_3\omega_4} - if^b_{\omega_2\omega_3\omega_4} - if^c\omega_2\omega_3\omega_4) - \left(f_{\omega_2\omega_3} + if^b_{\omega_2\omega_3}\right) - \left(f_{\omega_3\omega_4} + if^b_{\omega_3\omega_4}\right)\\
&+f_{\omega_3} - i\left(f_{\omega_2} + \frac{\pi}{2}\delta_{13} - \frac{\pi(\omega_a-\omega_b)}{2}\right)\left(f_{\omega_4} + \frac{\pi}{2}\delta_{35} - \frac{\pi(\omega_d-\omega_c)}{2}\right)\\
&\left.\left.+2if_{3\text{-reggeon}}\right]\right)\,. \qquad (46)
\end{aligned}
$$

The three reggeon cut term is given by

$$
\begin{aligned}
f_{3\text{-reggeon}} &= \Phi^d\frac{1}{\omega_2''-K^{octet}}V^b\frac{1}{\omega_3''-K^{3;octet}}V^c\frac{1}{\omega_4''-K^{octet}}\Phi^d|_{\text{more than two loops}}\\
&+\text{two loop terms}\,. \qquad (47)
\end{aligned}
$$

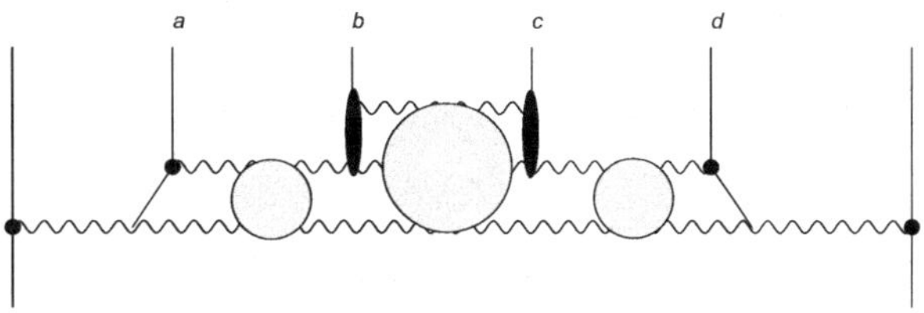

Fig. 19 The 3 reggeon cut in the leading approximation

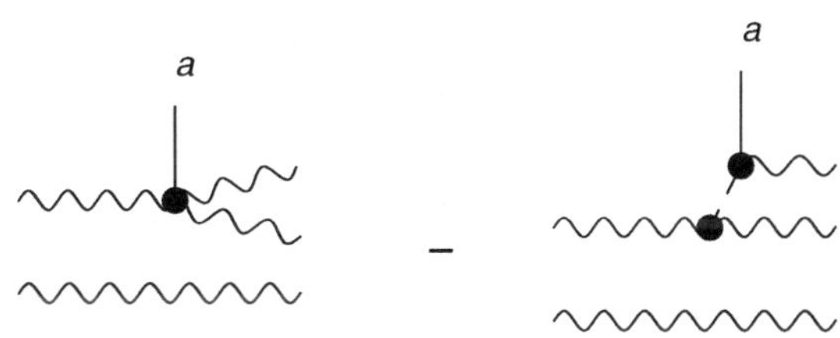

Fig. 20 The 2-reggeon → 3-reggeon transition

Here $f_{3\text{-reggeon}}$ in the momentum representation is illustrated in Fig. 19. For the produced particles a and d the impact factors $\Phi_{L,R}$ are the same as before (e.g. in the long cut or very long cut amplitudes). A new element is the transition 2 reggeons → 3 reggeons at the production of particle b. It is illustrated in Fig. 20. We note that it consists of two terms. In order to obtain this vertex in the conformal (ν, n) representation one has to convolute it with BFKL eigenfunctions: from the left with the two reggeon wavefunction, from the right with the three reggeon wavefunction. The result will be published in a separate paper.

For completeness we illustrate in momentum space the leading order 3 → 3 BFKL kernel for the 3 reggeon cut:

$$
\begin{aligned}
&K^{3-octet}(\mathbf{k}_1, \mathbf{k}_2, \mathbf{q}_3 - \mathbf{k}_1 - \mathbf{k}_2; \mathbf{k}_1', \mathbf{k}_2', \mathbf{q}_3 - \mathbf{k}_1' - \mathbf{k}_2') \\
&= -\frac{a}{2} \ln \frac{\mathbf{k}_1^2 (\mathbf{q}_3 - \mathbf{k}_1 - \mathbf{k}_2)^2}{\mathbf{q}_3^2 \mathbf{q}_3^2} + \frac{1}{2} \Big(K^{(1)}(\mathbf{k}_1, \mathbf{k}_2; (\mathbf{k}_1', \mathbf{k}_2') \\
&\quad + K^{(1)}(\mathbf{k}_2, \mathbf{q}_3 - \mathbf{k}_1 - \mathbf{k}_2; \mathbf{k}_2', \mathbf{q}_3 - \mathbf{k}_1' - \mathbf{k}_2') \Big) . \qquad (48)
\end{aligned}
$$

In Fig. 21a it is a sum of two infrared finite color singlet BFKL kernels $K^{(1)}$. It is this 3-gluon kernel which defines the open string Hamiltonian consisting of three sites. This result has been obtained so far only in the leading approximation. The BFKL kernel for two reggeized gluons is known also in higher order, but beyond leading order there exist also another higher order kernel, the interactions of three reggeized gluons [23] (Fig. 21b).

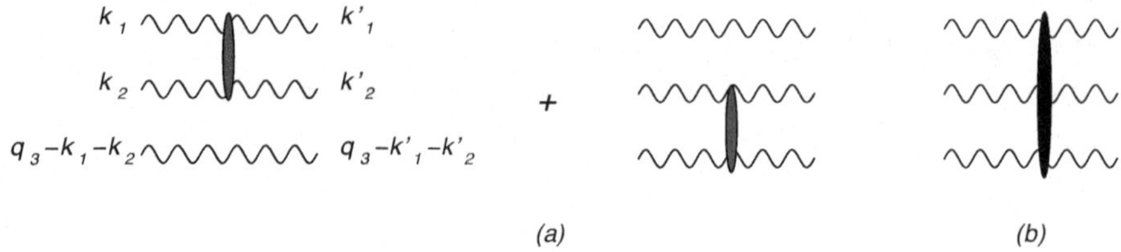

Fig. 21 Kernel of the 3-gluon octet state

The 3 reggeon cut contains a first generalization of the open spin chain [15] mentioned before: whereas the BFKL kernel of the two reggeon cut represents the spin chain of two sites, the kernel of the three reggeon cut has three sites. The diagram in Fig. 21b thus provides a new building block of the integrable structure contained in the 2 → n scattering amplitudes in the multi-Regge limit and, to our knowledge, needs further investigation.

5 Summary and Future Steps

In this paper we have outlined how unitarity and energy discontinuities can be used to compute, for the Regge limit, multiparticle amplitudes. An essential ingredient is the analytic structure: the amplitude is written as a sum of several terms, and each term has the maximal number of non-overlapping energy discontinuities. For the study of N = 4 SYM gauge theories one usually deals with the planar approximation (large N_c limit): this approximation, however, requires a few special considerations.

After the simpler cases of the 2 → 4 and 2 → 5 amplitudes new results have been obtained for the 2 → 6 amplitude. There are two novel features: the product of shorter Regge cuts and, for the first time, the appearance of a three reggeon cut.

Based upon the results known so far, it is tempting to make more general predictions for the planar approximation of the 2 → n case. First, for sufficiently large n, we expect to find more general products of Regge cuts. An example is shown in Fig. 22.

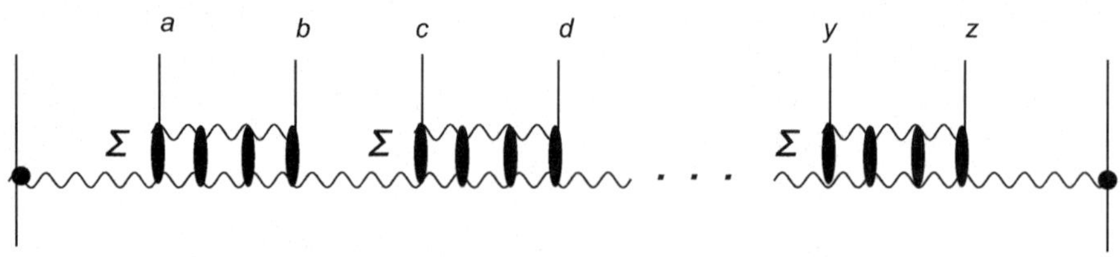

Fig. 22 Product of short cuts

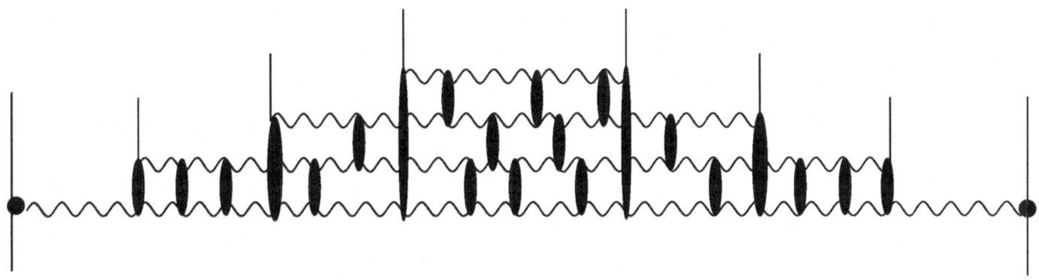

Fig. 23 The 4-reggeon cut

Next we expect Regge cuts consisting of n reggeized gluons (Fig. 23): as an example, the bound state of four reggeized gluons is expected to appear first in the $2 \to 8$ amplitude.

In these examples all elements are known, at least in leading order in momentum space. Using the methods outlined in this paper, also higher order corrections can be obtained by inserting, in the unitarity equations which are used for the discontinuities, higher order production amplitudes. An important task, however remains, namely the transformation to conformal variables ν and n. Also, the generalization from $2 \to n$ to $m \to n$ amplitudes is an interesting future step.

As we have mentioned before, in recent years new and elegant methods are being developed which may allow to compute more easily higher order corrections. We consider this as an important development. Our results described in this paper may be seen as being complementary to this. In particular, they may help to find also the explicit connection with scattering amplitudes and cross sections. A comparison of the results obtained by different methods will be an important task.

References

1. Z. Bern, L.J. Dixon, V.A. Smirnov, Phys. Rev. D **72** (2005) 085001. [hep-th/0505205]
2. J. Bartels, L.N. Lipatov, A. Sabio Vera, Phys. Rev. D **80** (2009) 045002. [arXiv:0802.2065 [hep-th]]
3. J. Bartels, L.N. Lipatov, A. Sabio Vera, Eur. Phys. J. C **65** (2010) 587. [arXiv:0807.0894 [hep-th]]
4. J. Bartels, A. Kormilitzin, L. Lipatov, Phys. Rev. D **89** (2014) 065002. [arXiv:1311.2061 [hep-th]]
5. J. Bartels, A. Kormilitzin, L.N. Lipatov, Phys. Rev. D 91, 045005 (2015). arXiv:1411.2294 [hep-th]
6. J. Bartels, A. Kormilitzin, L.N. Lipatov, A. Prygarin, Phys. Rev. D **86**, 065026 (2012). https://doi.org/10.1103/PhysRevD.86.065026. [arXiv:1112.6366 [hep-th]]
7. T. Bargheer, G. Papathanasiou, V. Schomerus, JHEP **1605**, 012 (2016). [arXiv:1512.07620 [hep-th]]
8. V. Del Duca, S. Druc, J. Drummond, C. Duhr, F. Dulat, R. Marzucca, G. Papathanasiou, B. Verbeek, JHEP **1806**, 116 (2018). [arXiv:1801.10605 [hep-th]]
9. V. Del Duca, C. Duhr, F. Dulat, B. Penante, JHEP **1901**, 162 (2019). [arXiv:1811.10398 [hep-th]]

10. S. Caron-Huot, L.J. Dixon, F. Dulat, M. von Hippel, A.J. McLeod, G. Papathanasiou, JHEP **1908**, 016 (2019). [arXiv:1903.10890 [hep-th]]
11. T. Bargheer, V. Chestnov, V. Schomerus, JHEP **05**, 002 (2020). [arXiv:1906.00990 [hep-th]]
12. V. Del Duca, S. Druc, J.M. Drummond, C. Duhr, F. Dulat, R. Marzucca, G. Papathanasiou, B. Verbeek, Phys. Rev. Lett. **124**(16), 161602 (2020). [arXiv:1912.00188 [hep-th]]
13. S. Caron-Huot, L.J. Dixon, J.M. Drummond, F. Dulat, J. Foster, Ö. Gürdoğan, M. von Hippel, A.J. McLeod, G. Papathanasiou, PoS **CORFU2019**, 003 (2020). [arXiv:2005.06735 [hep-th]]
14. B. Basso, L.J. Dixon, G. Papathanasiou, Phys. Rev. Lett. **124**(16), 161603 (2020). [arXiv:2001.05460 [hep-th]]
15. L.N. Lipatov, J. Phys. A **42**, 304020 (2009). [arXiv:0902.1444 [hep-th]]
16. O. Steinmann, Helv. Physica. Acta **33**, 257 (1960); Helv. Physica. Acta **33**, 347 (1960)
17. L.N. Lipatov, Theor. Math. Phys. **170**, 166 (2012) [arXiv:1008.1015 [hep-th]]
18. L.N. Lipatov, A. Prygarin, Phys. Rev. D **83**, 125001 (2011). [arXiv:1011.2673 [hep-th]]
19. J. Bartels, L.N. Lipatov, A. Prygarin, Phys. Lett. B **705**, 507–512 (2011). [arXiv:1012.3178 [hep-th]]
20. M. Sprenger, JHEP **01**, 035 (2017). https://doi.org/10.1007/JHEP01(2017)035. [arXiv:1610.07640 [hep-th]]
21. J. Bartels (2020). [arXiv:2005.08818 [hep-th]]
22. T. Bargheer, J. Bartels, in preparation
23. J. Bartels, V.S. Fadin, L.N. Lipatov, G.P. Vacca, Nucl. Phys. B **867**, 827–854 (2013). [arXiv:1210.0797 [hep-ph]]

Direct Integration for Multi-Leg Amplitudes: Tips, Tricks, and When They Fail

Jacob L. Bourjaily, Yang-Hui He, Andrew J. McLeod, Marcus Spradlin, Cristian Vergu, Matthias Volk, Matt von Hippel, and Matthias Wilhelm

Abstract Direct hyperlogarithmic integration offers a strong alternative to differential equation methods for Feynman integration, particularly for multi-particle diagrams. We review a variety of results by the authors in which this method, employed with some care, can compute diagrams of up to eight particles and four loops. We also highlight situations in which this method fails due to an algebraic obstruction. In a large number of cases the obstruction can be associated with a Calabi-Yau manifold.

J. L. Bourjaily
Niels Bohr Institute, University of Copenhagen, Copenhagen, Denmark

Institute for Gravitation and the Cosmos, Department of Physics, Pennsylvania State University, University Park, PA, USA
e-mail: bourjaily@psu.edu

A. J. McLeod · C. Vergu · M. Volk · M. von Hippel (✉) · M. Wilhelm
Niels Bohr Institute, University of Copenhagen, Copenhagen, Denmark
e-mail: amcleod@nbi.ku.dk; c.vergu@nbi.ku.dk; mvolk@nbi.ku.dk; mvonhippel@nbi.ku.dk; matthias.wilhelm@nbi.ku.dk

Y.-H. He
Department of Mathematics, University of London, London, UK

Merton College, University of Oxford, Oxford, UK

School of Physics, NanKai University, Tianjin, P.R. China
e-mail: hey@maths.ox.ac.uk

M. Spradlin
Department of Physics, Brown University, Providence, RI, USA
e-mail: marcus_spradlin@brown.edu

J. Blümlein, C. Schneider (eds.), *Anti-Differentiation and the Calculation of Feynman Amplitudes*, Texts & Monographs in Symbolic Computation,
https://doi.org/10.1007/978-3-030-80219-6_5

1 Introduction

Several methods are available for evaluating Feynman integrals in terms of hyperlogarithms. Of these, direct hyperlogarithmic integration is perhaps surprisingly a bit of a dark horse. The method, which in rough outline consists of partial-fractioning rational functions and re-expressing hyperlogarithms in the integration variable, applying the definition of the hyperlogarithm, and careful treatment of boundary values [1, 2], has been implemented in computer packages [3, 4], but remains less popular than differential equation methods [5–10] or methods based on the Mellin-Barnes representation [11–13].

In part, this lack of popularity is due to the method's dependence on linear reducibility. At each integration step, it must be possible to express the integrand in terms of rational functions and hyperlogarithms in the integration parameter. If there is no integration order such that this is possible then we say that the integral is not linearly reducible. This can happen, for example, if partial-fractioning in a previous integration step gives rise to irreducible algebraic roots in a later integration variable. If this happens then direct hyperlogarithmic integration is obstructed.

Despite this potential for obstruction, direct hyperlogarithmic integration has several advantages. Differential equation and Mellin-Barnes methods both have difficulty in problems with a large number of scales. In contrast, provided linear reducibility is preserved the number of scales has little impact on the difficulty of direct hyperlogarithmic integration. As such, it is particularly well-suited for multi-leg scattering amplitudes. These are especially relevant in the context of planar $\mathcal{N} = 4$ super Yang-Mills, where amplitudes with five particles and fewer are fully captured by the BDS ansatz [14].

In this talk, we present several direct integrations of multi-leg Feynman integrals by the authors, mostly in the context of planar $\mathcal{N} = 4$ super Yang-Mills [15–21]. We observe that obstructions to linear reducibility can be postponed or avoided entirely via a variety of techniques and best-practices. In some cases, integration can be performed to completion, resulting in an expression in terms of hyperlogarithms with rational arguments [17]. In others, integration can still be completed, but the resulting hyperlogarithms depend manifestly on algebraic roots in the kinematics. In several cases presented here, it is possible to show that this dependence is spurious, and the singularities of the result are in fact all rational in the kinematics [20, 21]. In still other cases, integration cannot be completed, and there is unavoidable algebraic dependence on the integration parameters at an intermediate stage. These cases have an intriguing commonality of structure: in each case, the obstructions to direct integration can be characterized in terms of Calabi-Yaus [15, 16, 18, 19].

2 Tips and Tricks for a Rational Result

As an illustrative example, consider the following direct integration, which should be thought of as the first step in a longer calculation:

$$\begin{aligned}\int_0^\infty &\frac{d\alpha}{\alpha^2+2f\alpha+g}\\ &=\int_0^\infty \frac{d\alpha}{2\sqrt{f^2-g}}\left(\frac{1}{\alpha+f-\sqrt{f^2-g}}-\frac{1}{\alpha+f+\sqrt{f^2-g}}\right)\\ &=\frac{1}{2\sqrt{f^2-g}}\log\left(\frac{f+\sqrt{f^2-g}}{f-\sqrt{f^2-g}}\right). \end{aligned}\tag{1}$$

If $\sqrt{f^2-g}$ happens to be a perfect square, the result will contain no irreducible roots and direct integration can proceed happily. If it is not, direct integration is obstructed. Our goal then is to avoid obstructions of this kind. Sometimes this can be accomplished simply by choosing a different order of variables in which to integrate: these orders can be found systematically via compatibility graph reduction [1, 2, 22, 23]. In other cases there is no integration order free of these obstructions and they must be dealt with in another way. Sometimes this can be done via a change of variables which rationalizes the square root, which have been studied systematically in [24].

These systematic methods are often useful, but they do not suffice to avoid every possible obstruction. Going beyond them requires re-thinking our starting assumptions. If we begin with a different representation for the integrand, integrals can be linearly reducible that were not in the original parametrization. This freedom of reparametrization is much less understood. In the following we discuss a few approaches which have been particularly helpful in our calculations of multi-leg diagrams, in which a different framing of a problem can make a seemingly obstructed integral reducible.

2.1 Loop-by-Loop Parametrization

It is a widely believed conjecture that, for L-loop Feynman integrals in four dimensions, any hyperlogarithms that appear are of maximum weight $2L$. With this in mind, one would expect it to be possible to write any such integral as a $2L$-fold integral over a rational function. This is not typically true of the standard Feynman parameter representation, which has a variable for each propagator, potentially leading to many more than $2L$ integration variables. In practice, this dependence on extra variables can obscure linear reducibility. Heuristically, the more integrations need to be performed the greater the chance that one will introduce a spurious algebraic root. As such, it is wise to begin with a representation depending on as few integration variables as possible, preferably only $2L$.

A method that can achieve this goal in many cases, and approach it in others, is loop-by-loop Feynman parametrization. This method is in some ways analogous to the loop-by-loop approach to the Baikov representation [25], but less general: it is particularly applicable to planar diagrams with massless propagators. In these cases it is often possible to Feynman parametrize one loop at a time, treating the other loop momenta as external. As the diagram is planar, it can be written so that only one propagator depends on a given one of the remaining loop momenta. Then after Feynman parametrizing the first loop, one can integrate in the Feynman parameter corresponding to that propagator, isolating the dependence on the next loop momentum in a "propagator-like" form. This then allows the next loop to be integrated in the same way, while at the same time reducing the number of integration parameters in the final result. This method was used more or less straightforwardly to obtain $2L$- or $2L+1$-parameter integrands for L-loop diagrams in [15–17, 20, 21]. It was discussed in a bit more detail in [26]. More complicated cases requiring more involved changes of variables were considered in [19], including a six-parameter representation of a three-loop integral and a nine-parameter representation of a four-loop integral.

2.2 *Momentum Twistors*

Much as it is wise to use as few integration parameters as possible, it is also wise to use as few kinematic parameters as possible. For planar integrals, a particularly natural way to do so is by employing momentum twistor space [27]. To define this space for an n-point diagram, we can first consider the dual space defined by dual x-coordinates $p_a = x_{a+1} - x_a$, with $x_{n+1} = x_1$. These coordinates automatically enforce momentum conservation. For massless external momenta, we then have the additional constraint that $(x_{a+1} - x_a)^2 = 0$, so the dual points are light-like separated.[1] To make this manifest, we go to momentum twistor space, in which each dual point x_a is assigned to a line span$\{z_{a-1}, z_a\}$ in $\mathbb{P}^3$.

In addition to being a natural minimal set of kinematic parameters, momentum twistors also have an additional advantage: they rationalize many of the kinematic square roots that would otherwise occur in scattering amplitudes. Construct the following $n \times n$ matrix of kinematic invariants:

$$G = \{G^a_b = (x_a - x_b)^2\}\,. \tag{2}$$

[1]For massive external momenta, we can represent each in terms of a pair of massless external momenta, so this discussion still applies.

In momentum twistors, the invariant $(x_a - x_b)^2$ is proportional to the determinant $\det\{z_{a-1}, z_a, z_{b-1}, z_b\}$. Thus, this matrix is linear in each of the momentum twistors. One can represent each pair $z_{a-1} \wedge z_a$ and $z_{b-1} \wedge z_b$ as a six-component vector, which shows that this matrix is in fact a Gramian matrix, with rank at most six. This means that all 7×7 minors of the matrix should vanish. Expressed in terms of $(x_a - x_b)^2$ or conformal cross-ratios, this will give rise to relations which are quadratic in each such variable, with solutions involving algebraic roots of 7×7 determinants of G. These appear quite generically when attempting direct integration of seven- and higher-point amplitudes in x_a variables or in cross-ratios. Because G is manifestly linear in the pairs $z_{a-1} \wedge z_a$ and $z_{b-1} \wedge z_b$, in these variables these algebraic roots are rationalized. Thus momentum twistors serve to rationalize a particularly common class of algebraic roots, permitting direct integration smoothly in more cases.

The combination of loop-by-loop parametrization and momentum twistors is already quite powerful. In [17], these methods sufficed to compute hyperlogarithmic representations for several classes of integrals, including the six-point "double penta-ladder" integrals of [28], seven-point integrals referred to as "Heptagon A" and "Heptagon B", and an eight-point family of integrals referred to as "Octagon A", all through four loops. Another eight-point integral in [17], referred to there as "Octagon B", could be computed at two loops through the same methods, resulting in hyperlogarithms which depend on an algebraic square root, arising from a kinematic configuration related to the four-mass box which will be discussed in Sect. 3. Finally, of the sixteen two-loop six-point integrals computed in [26] and the five seven-point two-loop integrals computed in [21], all but one were computed using essentially the above methods, augmented by a few minor tricks.

2.3 Splitting the Integration Path

While loop-by-loop Feynman parametrization and momentum twistors make many planar integrals linearly reducible, they do not fix all obstructions to linear reducibility. In cases when they fail, it is sometimes possible to perform direct integration regardless, via the expedient of splitting the integration path.

There are two distinct versions of this trick, corresponding to two distinct situations. In the first, an integrand may be linearly irreducible due to the occurrence of two polynomials (either in denominators of rational functions or singularities of hyperlogarithms) that do not have a common order in which they can be integrated. In such a situation, it is sometimes the case that the terms containing these polynomials can be separated: the integrand can be written as a sum of terms depending on one polynomial, terms depending on the other, and terms depending on neither. In these situations one can then split the integration, integrating one set of terms with one integration order and the other set of terms with another, so that each set of terms continues to be linearly reducible.

In the second situation, the integrand already contains algebraic roots. A single square root of a quadratic polynomial can always be rationalized via a rational change of variables. If there are multiple such square roots then simultaneously rationalizing them is a nontrivial task. The methods of [24] provide criteria to distinguish cases which can be rationalized from cases which cannot. However, it may not always be necessary to simultaneously rationalize all of the algebraic roots occurring in an integrand. This is because these roots, much like the mutually incompatible polynomials of the first situation, may appear only in separate terms of the integrand. Then by separating the integrand into terms each containing only one distinct square root of a quadratic polynomial, it is possible to group the integrand into pieces that can individually be rationalized by distinct changes of variables, thus allowing integration to continue.

The first of these situations held in the fifth integral considered in [21], which was successfully integrated using this method. The integrals considered in [20] were more complicated: these required both methods at different stages of the calculation, with the integrand first split into two pieces that integrate rationally along different orders and then split again into pieces that could be rationalized by distinct changes of variables.

3 Kinematic Square Roots at Symbol Level

The methods described in the preceding section allowed in several cases for integration to proceed all the way to a hyperlogarithmic expression. However, these expressions themselves may be expressed in terms of algebraic roots in kinematic parameters, even when written with rational parametrizations of momentum twistors.

In general, one expects some algebraic roots that appear in this way to be spurious, artifacts of our integration procedure that are not required to express the integral, while others may be unavoidable. The latter should be "physically meaningful" in some sense: it should be possible to characterize them in terms of Landau singularities [29], for example. In the case of planar diagrams with massless internal lines, these roots can originate from Gramian determinants smaller than the 6×6 determinants that momentum twistors rationalize, such as 4×4 Gramian determinants. These are the origin of the square root appearing in the four-mass box [30–32], which indeed is not rationalized with momentum twistors alone. This same kinematic behavior is responsible for the square root observed in the integral referred to as Octagon B in [17], as well as for the square roots that were expected for the integrals considered in [20].

Ideally we would like to represent an integral using whichever algebraic roots are necessary and no more, eliminating any spurious arguments in our hyperlogarithms. If our hyperlogarithms depended on rational arguments we could do this by going to a uniquely specifiable basis called a fibration basis [2, 33]. However, this is not possible when the arguments of the hyperlogarithms contain algebraic roots. When a root can be rationalized by another change of variables it is sometimes possible to go to a fibration basis after doing so and then transform back to show that the root does not contribute. This method was used in [26]. When this is not possible, sometimes one can propose a plausible ansatz of hyperlogarithms without algebraic roots, then match series expansions at a sufficient depth to be convincing. If neither of those are possible, then one generally cannot find an explicit hyperlogarithmic form without roots, but one may still be able to identify some roots as spurious at the level of the symbol [34].

Before discussing this, we should clear up some common misunderstandings regarding the symbol map. In particular, there are two slogans that are often repeated: **"The symbol trivializes all identities"** and **"The symbol of a constant vanishes"**. Both of these slogans are useful in the proper context, but neither is strictly true.

The symbol results from maximal iteration of the coaction on a hyperlogarithm, resulting in a tensor product of logarithms. If these are all logarithms of rational numbers and functions then the symbol will indeed trivialize all identities: one simply needs to factor the argument of each logarithm and expand. However, if any of the numbers or functions involved are algebraic then this procedure will not typically be unique, and thus will not trivialize all identities. This is because algebraic extensions of the rationals are in general not unique factorization domains.

To give a simple example, consider the integers extended by $\sqrt{-5}$. The number nine can be factorized in this ring in two different ways:

$$9 = 3 \times 3 = (2 + \sqrt{-5}) \times (2 - \sqrt{-5})\,. \tag{3}$$

As neither 3, $(2 + \sqrt{-5})$, or $(2 - \sqrt{-5})$ can be factorized further in this ring (more precisely they are irreducible elements), this shows that there are numbers which cannot be uniquely factorized.

In order to make the symbol useful in the presence of algebraic roots, then, we need to find a basis of symbol letters that is truly linearly independent, preferably where as few letters as possible are algebraic. If there are few enough algebraic letters this can be done by inspection, or almost as easily. This was the case for the fifth integral considered in [21]: its symbol depended on only one algebraic root, which appeared in 22 distinct letters. It was reasonably straightforward to find relations between these letters, expressing them in a basis of just five algebraic letters (as well as assorted rational letters). When the symbol was expanded in this basis all dependence on the five remaining algebraic letters dropped out, resulting in a purely rational symbol alphabet.

If there are many algebraic letters, especially of higher degree, then a more systematic approach is desirable. We will discuss such an approach below, employing software implementations of algebraic extensions of the integers. In order to do this we will have to consider constant symbol letters, which brings us to the second misleading slogan, the claim that the symbol of a constant vanishes. This slogan may seem plausible to readers used to calculations in planar $\mathcal{N} = 4$ super Yang-Mills, where the constants of interest at well-behaved kinematic points are typically multiple zeta values. As each term in the symbol of a multiple zeta value contains at least one entry equal to one, and $\ln 1 = 0$, it is true that the symbol of a multiple zeta value vanishes. However, we do not need to choose a well-behaved kinematic point. Choosing instead a generic kinematic point results in a non-vanishing constant symbol, with all of the properties that make symbols useful for non-constant functions.

With the above in mind, we integrated the eight-point integrals investigated in [20] at a particular, generic kinematic point, computed their symbols. These symbols were initially particularly complicated: one of the two integrals considered had a symbol with 8,367,616 terms in 2,024 letters, while the other had 9,941,483 terms in 2,156 letters. These initially involved very complicated algebraic numbers, in some cases up to degree 16. The most complicated letters had a common structure: they were of the form $\rho - \sigma$, where σ was a root of a fourth-order polynomial and ρ was a linear combination of at most two square roots and an integer. By grouping these letters according to the roots appearing in ρ and σ, it was possible to search for combinations that do not involve higher than square roots. This search was accomplished with the use of `SageMath` [35], in particular its `Pari` [36] functionality. The relations found in this way were sufficient to remove all higher roots from the symbols, leaving letters that were linear combinations of at most two square roots.

These letters still satisfy many nontrivial relations. To find them, we employ factorization in prime ideals. We sketch the method below:

The ideal generated by a number is defined as the set of its integer multiples. We use the following notation:

$$(p) \equiv \{mp|m \in \mathbb{Z}\} . \tag{4}$$

An ideal generated by a single element is called a *principal ideal.* We can also consider ideals generated by more than one element:

$$(a, b) \equiv \{ma + nb|m, n \in \mathbb{Z}\} . \tag{5}$$

Ideals of this kind can be multiplied, with $(a, b)(c, d) = (ac, ad, bc, bd)$. With these concepts in place, we can return to our earlier example. Suppose we wish to factor, not the number 9, but the ideal (9). Then the factorization in equation 3 can be further refined, by factoring principal ideals into ideals generated by two elements. We have,

$$(9) = (3) \times (3) = (2+\sqrt{-5}) \times (2-\sqrt{-5}) = (3, 1+\sqrt{-5})^2(3, 1-\sqrt{-5})^2 \tag{6}$$

This factorization is now unique: the ideals $(3, 1 + \sqrt{-5})$ and $(3, 1 - \sqrt{-5})$ are not merely irreducible, but prime.

Taking into account some subtleties regarding the unit element, and a generalization to fractional ideals (both of which we will not discuss here), factorization into prime ideals allowed for all remaining symbol letters in these integrals to be represented in terms of a truly multiplicatively independent basis. Writing the symbol in this basis, we found that all roots that were expected to be spurious cancel: the only surviving roots in each integral are those identified as "physically meaningful". Out of the original over 2000 symbol letters for each integral we find both integrals can be expressed in a common basis of just 35 symbol letters, leading to symbols that are 5216 and 5245 terms in length, three orders of magnitude smaller than our initial results. Remarkably, we find that in the combination that these two integrals appear in the physical amplitude the remaining algebraic letters actually cancel, and only integer letters remain. This result was later confirmed via other methods [37].

4 Parametric Square Roots: Elliptic and Beyond

Even with the methods of Sect. 2, some integrals are not linearly reducible. This happens when partial-fractioning or fibration gives rise to an algebraic root in the remaining integration variables that cannot be rationalized. As a square root of a quadratic polynomial can always be rationalized by a rational change of variables, the first nontrivial case involves cubic or quartic polynomials in a single variable. These polynomials define elliptic curves, and there is a growing literature on the Feynman integrals that contain them [38–60]. Via direct integration, we found that the two-loop ten-particle N^3MHV amplitude in planar $\mathcal{N} = 4$ super Yang-Mills has a supercomponent with this property: linear reducibility is obstructed by an elliptic curve [15].

In other cases, linear reducibility is obstructed by an algebraic root in more than one variable, for example $\sqrt{Q(x_1, x_2, \ldots)}$. There has been much less progress on Feynman integrals of this kind, but what progress exists has focused on analyzing the geometric properties of the algebraic varieties defined by these roots (for example, by the equation $y^2 = Q(x_1, x_2, \ldots)$). In particular, the most productive cases thus far have involved varieties that define Calabi-Yaus [1, 16, 18, 19, 60–70].

One way a variety may define a Calabi-Yau is if it can be embedded consistently in a weighted projective space, such that the sum of the coordinate weights is equal to the overall degree of the polynomial [71]. By "embedded consistently" we mean that it must scale uniformly under weighted rescaling of the coordinates. In the following subsections, we will describe several diagrams and classes of diagram that give rise to varieties with this property under direct integration. These examples will have a common structure: they can all be embedded in k-dimensional weighted projective space $\mathbb{WP}^{1,\ldots,1,k}$ (where all coordinate weights except one have weight 1, and the remaining coordinate has weight k). The origin of this property will be clear for the first class of diagrams we discuss, and more mysterious for the second.

4.1 Scalar Marginal Integrals

For our first set of examples, we consider scalar diagrams, writing them in the well-known Symanzik representation. For an L-loop Feynman diagram $\mathcal{I}$ with E internal edges $\{e_i\}$ with masses m_i in D dimensions, we write:

$$\mathcal{I} = \Gamma(E - LD/2) \int_{x_i \geq 0} [d^{E-1}x_i] \frac{\mathfrak{U}^{E-(L+1)D/2}}{\mathfrak{F}^{E-LD/2}}, \tag{7}$$

where the graph polynomials $\mathfrak{U}$ and $\mathfrak{F}$ are defined by:

$$\mathfrak{U} \equiv \sum_{\{T\} \in \mathfrak{T}_1} \prod_{e_i \notin T} x_i, \quad \mathfrak{F} \equiv \Bigg[\sum_{\{T_1, T_2\} \in \mathfrak{T}_2} s_{T_1} \Big(\prod_{e_i \notin T_1 \cup T_2} x_i \Big) \Bigg] + \mathfrak{U} \sum_{e_i} x_i m_i^2 . \tag{8}$$

The $\mathfrak{U}$ polynomial is a sum over spanning trees $\mathfrak{T}_1$ of the graph, while the $\mathfrak{F}$ polynomial includes a sum over disconnected pairs of trees that together span the graph (denoted $\mathfrak{T}_2$). s_{T_1} is the square of the sum of momenta flowing in to tree T_1, while $[d^{E-1}x_i]$ denotes projective integration over the E variables $\{x_i\}$.

There are two cases where this representation simplifies dramatically, making it easier to probe its geometry. In the first, consider a case where $E = LD/2$. The function $\Gamma(E - LD/2)$ diverges in this case, but if there are no subdivergences then the rest of the integral can be convergent in integer dimensions:

$$\bar{\mathcal{I}} = \int_{x_i \geq 0} [d^{E-1}x_i] \frac{1}{\mathfrak{U}^{D/2}} . \tag{9}$$

Integrals of this form have no kinematic dependence, so they are simply numbers. They are referred to in the mathematical literature as Feynman periods. These diagrams give rise to Calabi-Yaus upon direct integration [1].

Next, consider a case where $E = (L + 1)D/2$. We considered cases of this form in [18], this subsection reviews our discussion there. Here $\Gamma(E - LD/2) = \Gamma(D/2)$ is finite. Provided the integral is otherwise convergent we refer to this class as marginally convergent, or "marginal". In integer dimensions we find:

$$\mathcal{I} = \Gamma(D/2) \int_{x_i \geq 0} [d^{E-1}x_i] \frac{1}{\mathfrak{F}^{D/2}} . \tag{10}$$

Because of the kinematic dependence of $\mathfrak{F}$ this is now not merely a number, but a function.

In two dimensions, scalar marginal integrals consist of the comparatively well-studied higher-loop sunrise diagrams [60, 61, 63–65, 68–70]. In higher dimensions there are many more such diagrams.

All marginal integrals share common features, which ensure that if linear reducibility is obstructed the obstruction will define a Calabi-Yau. The Symanzik representation of these integrals depends on only the $\mathfrak{F}$ polynomial. This polynomial is homogeneous and has degree $L + 1$, so the denominator in the Symanzik representation $\mathfrak{F}^{D/2}$ has degree $(L + 1)D/2 = E$, the same as the number of variables. Direct integration preserves this property: each integration will decrease the number of variables and the overall degree of the denominator by one. If at some stage partial-fractioning introduces a square root of a polynomial in the remaining m integration parameters $\sqrt{Q(x_i)}$, the polynomial $Q(x_i)$ will be homogeneous and have overall degree $2m$. The equation $y^2 = Q(x_i)$ defines a variety. By assigning weight m to y and weight 1 to each of the x_i we may consistently embed this variety in a weighted projective space. A quick count shows that the sum of these coordinate weights is equal to the degree of our variety, showing that all such varieties will define Calabi-Yaus.

Specializing to graphs with massless particles in four dimensions, we can further prove a bound on the dimension of these spaces. Starting once again with the Symanzik representation,

$$\mathcal{I} = \int_{x_i \geq 0} [d^{2L+1}x_i] \frac{1}{\mathfrak{F}^2}, \tag{11}$$

as $\mathfrak{F}$ is linear in each variable when all propagators are massless we can integrate in any variable. Integrating in x_j and writing $\mathfrak{F} = \mathfrak{F}_0^{(j)} + x_j \mathfrak{F}_1^{(j)}$, we obtain

$$\mathcal{I} = \int_{x_i \geq 0} [d^{2L}x_i] \frac{1}{\mathfrak{F}_0^{(j)} \mathfrak{F}_1^{(j)}}. \tag{12}$$

Each of $\mathfrak{F}_0^{(j)}$ and $\mathfrak{F}_1^{(j)}$ is separately linear in each remaining integration variable, so we can once again integrate in any remaining x_k. Writing $\mathfrak{F}_i^{(j)} = \mathfrak{F}_{i,0}^{(j,k)} + x_k \mathfrak{F}_{i,1}^{(j,k)}$, we have

$$\mathcal{I} = \int_{x_i \geq 0} [d^{2L-1}x_i] \frac{\log\left(\mathfrak{F}_{0,0}^{(j,k)} \mathfrak{F}_{1,1}^{(j,k)}\right) - \log\left(\mathfrak{F}_{0,1}^{(j,k)} \mathfrak{F}_{1,0}^{(j,k)}\right)}{\mathfrak{F}_{0,0}^{(j,k)} \mathfrak{F}_{1,1}^{(j,k)} - \mathfrak{F}_{0,1}^{(j,k)} \mathfrak{F}_{1,0}^{(j,k)}}. \tag{13}$$

The denominator of the integrand is now at most quadratic in each variable, while the arguments of the logs are products of polynomials which are linear in each variable. We are thus able to integrate once more, but this time potentially at a cost of introducing a square root of a polynomial in the remaining $2L-1$ variables. If this polynomial is irreducibly quartic or cubic in each remaining integration parameter then the root cannot be rationalized by a rational change of variables. This is thus the first integration step at which direct integration can potentially be obstructed, in cases when there is no integration order that avoids an irreducibly quartic or cubic root. As such, the varieties that characterize this obstruction represent the highest-dimension Calabi-Yaus that can occur for this class of diagrams, demonstrating that Calabi-Yau dimension is bounded with loop order, at a maximum of $2L - 2$.

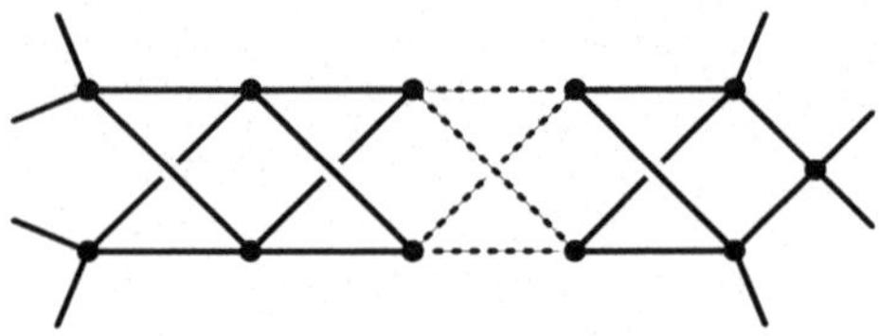

Fig. 1 The tardigrade diagrams, defined at even loops

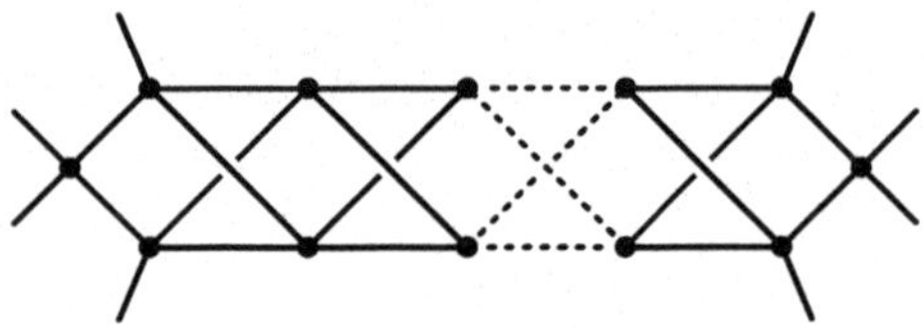

Fig. 2 The paramecium diagrams, defined at odd loops

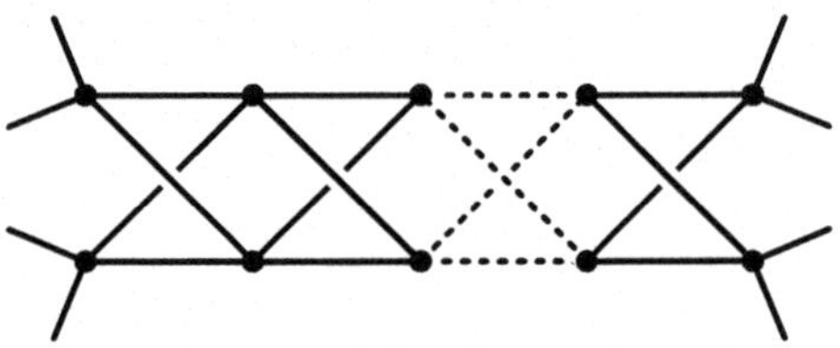

Fig. 3 The amoeba diagrams, defined at odd loops ≥ 5

It turns out that this bound is saturated. There are Feynman diagrams at each loop order for which every integration order is obstructed, and the obstruction defines a Calabi-Yau of dimension $2L - 2$. We characterize three infinite families of such diagrams. For even loops we find what we refer to as the tardigrade diagrams, depicted in Fig. 1. For odd loops we find two infinite families, which we refer to as paramecia and amoebas, depicted in Figs. 2 and 3 respectively. The amoeba diagrams are not maximally obstructed in this fashion at three loops, but otherwise saturate the bound at each higher order.

We have investigated a broader set of marginal diagrams in four dimensions. We find as the loop order increases, the majority of diagrams have maximal dimension obstructions of this kind.

4.2 More General Examples

The previous argument clarifies why Calabi-Yaus appear during direct integration of marginal diagrams. We have also found several examples of non-marginal diagrams that also give rise to Calabi-Yaus.

In each case discussed here, our starting point will be a $2L$-fold integral over a rational function. While such a representation is easy to obtain from the Symanzik form for marginal integrals (see Eq. (12) where this is manifest), for non-marginal integrals it is easier to find these representations via loop-by-loop parametrization (see Sect. 2.1). Heuristically, we believe that starting with a $2L$-fold will ensure that no integration is in any sense spurious, in analogy with the polylogarithmic case of transcendental weight $2L$ at L loops. In particular, since the integrals we consider occur in planar $\mathcal{N} = 4$ super Yang-Mills we expect uniform transcendentality: the integral should require exactly $2L$ parameters to express the integrand rationally, and no fewer.

The marginal integrals which were maximally obstructed manifested their obstructions after three integrations, at which point their integrand contained a dilogarithm. The integrals considered here will typically be less obstructed than this, involving Calabi-Yaus of dimension lower than $2L - 2$ at L loops. As such, it is in many cases quite cumbersome to perform a full direct integration up to the point it becomes obstructed, particularly for the higher-loop cases. As a proxy for this integration we instead typically took maximal codimension residues, which in rough terms probes whether the integrands can be iteratively partial-fractioned in the integration variables, but may not be sensitive to whether the resulting polylogarithms can at each stage be written in an appropriate fibration basis.

The first non-marginal diagrams which we found to give rise to Calabi-Yaus were the traintrack diagrams, higher-loop analogues to the elliptic double-box [16]. Depicted in Fig. 4, these diagrams, much like the higher-loop sunrise diagrams, increase in Calabi-Yau dimension at each loop order, with dimension $L - 1$ at L loops. In [19] we showed that the Calabi-Yau arising from the three-loop traintrack can be written, much like the marginal integrals, as a variety embedded in $\mathbb{WP}^{1,\ldots,1,k}$ (where there $k = 3$). Later, [72] analyzed the leading singularity structure of these integrals to all orders in twistor space, finding Calabi-Yau geometry at each order.

The three-loop wheel diagram (depicted in Fig. 5) was also analyzed in [19], and gives rise to a variety which can be embedded in $\mathbb{WP}^{1,1,1,1,4}$, corresponding to a Calabi-Yau threefold. In this case the embedding is slightly more subtle to derive, involving a deprojectivization and a particular choice of reprojectivization.

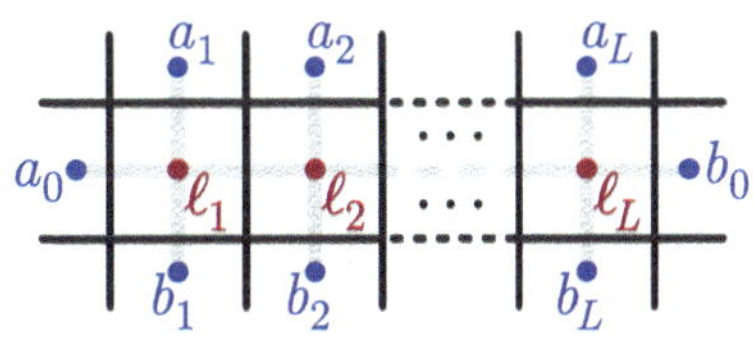

Fig. 4 The traintrack diagrams. These diagrams are planar, and we label here their dual coordinates

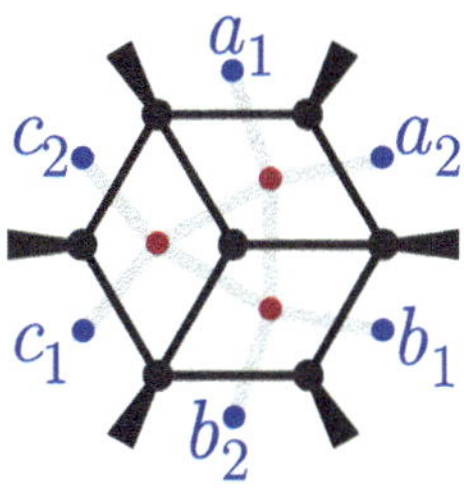

Fig. 5 The three-loop wheel diagram. This diagram is planar, and we label here its dual coordinates

There thus appears to be a common structure in each Calabi-Yau that has been observed in a Feynman diagram in the literature: all cases can be described with varieties embedded in $\mathbb{WP}^{1,\ldots,1,k}$. In [19] we went into some detail analyzing the properties of a generic Calabi-Yau defined in this space, including Hodge numbers and Euler characteristics. It is an open question whether this structure is universal, and if so what it can teach us about these diagrams.

5 Conclusions

We reviewed a variety of applications of direct hyperlogarithmic integration in the context of multi-leg scattering amplitudes. We presented tricks that allow one to avoid certain obstructions, but we also highlighted the structure of the obstructions that remain once these tricks are employed: obstructions which in a surprisingly varied set of cases define Calabi-Yaus.

It would be extremely interesting to go beyond the tips and tricks discussed here, and find a systematic algorithm that can find a hyperlogarithmic expression for any Feynman integral for which such an expression exists. One possibility is that such a method might arise from a motivic understanding of these integrals [73].

Along related lines, it will be important to understand which aspects of the Calabi-Yau geometries characterized in this work are universal, and which are specific to particular representations. The results of [72] suggest that such common features exist, but it still may be the case that one can express a single integral with multiple geometries (see for example the role of isogeny in [58]), or that the obstructions found here need to be augmented with other information for a full characterization.

Finally, it is worth investigating under what conditions Calabi-Yaus can arise from Feynman integrals. As the work reviewed here shows, they are more common than one might naively assume.

References

1. F.C.S. Brown, arXiv:0910.0114 [math.AG]
2. E. Panzer, https://doi.org/10.18452/17157. arXiv:1506.07243 [math-ph]
3. E. Panzer, Comput. Phys. Commun. **188**, 148–166 (2015). https://doi.org/10.1016/j.cpc.2014.10.019. arXiv:1403.3385 [hep-th]
4. C. Bogner, F. Brown, Commun. Num. Theor. Phys. **09**, 189–238 (2015). https://doi.org/10.4310/CNTP.2015.v9.n1.a3. arXiv:1408.1862 [hep-th]
5. A.V. Kotikov, Phys. Lett. B **254**, 158–164 (1991). https://doi.org/10.1016/0370-2693(91)90413-K
6. E. Remiddi, Nuovo Cim. A **110**, 1435–1452 (1997). arXiv:hep-th/9711188 [hep-th]
7. S. Müller-Stach, S. Weinzierl, R. Zayadeh, Commun. Math. Phys. **326**, 237–249 (2014). https://doi.org/10.1007/s00220-013-1838-3. arXiv:1212.4389 [hep-ph]

8. J.M. Henn, Phys. Rev. Lett. **110**, 251601 (2013). https://doi.org/10.1103/PhysRevLett.110.251601. arXiv:1304.1806 [hep-th]
9. R.N. Lee, JHEP **04**, 108 (2015). https://doi.org/10.1007/JHEP04(2015)108. arXiv:1411.0911 [hep-ph]
10. C. Meyer, JHEP **04**, 006 (2017). https://doi.org/10.1007/JHEP04(2017)006. arXiv:1611.01087 [hep-ph]
11. J. Gluza, K. Kajda, T. Riemann, Comput. Phys. Commun. **177**, 879–893 (2007). https://doi.org/10.1016/j.cpc.2007.07.001. arXiv:0704.2423 [hep-ph]
12. J. Blümlein, I. Dubovyk, J. Gluza, M. Ochman, C.G. Raab, T. Riemann, C. Schneider, PoS **LL2014**, 052 (2014). https://doi.org/10.22323/1.211.0052. arXiv:1407.7832 [hep-ph]
13. M. Ochman, T. Riemann, Acta Phys. Polon. B **46**(11), 2117 (2015). https://doi.org/10.5506/APhysPolB.46.2117. arXiv:1511.01323 [hep-ph]
14. Z. Bern, L.J. Dixon, V.A. Smirnov, Phys. Rev. D **72**, 085001 (2005). https://doi.org/10.1103/PhysRevD.72.085001. arXiv:hep-th/0505205 [hep-th]
15. J.L. Bourjaily, A.J. McLeod, M. Spradlin, M. von Hippel, M. Wilhelm, Phys. Rev. Lett. **120**(12), 121603 (2018). https://doi.org/10.1103/PhysRevLett.120.121603. arXiv:1712.02785 [hep-th]
16. J.L. Bourjaily, Y.H. He, A.J. Mcleod, M. Von Hippel, M. Wilhelm, Phys. Rev. Lett. **121**(7), 071603 (2018). https://doi.org/10.1103/PhysRevLett.121.071603. arXiv:1805.09326 [hep-th]
17. J.L. Bourjaily, A.J. McLeod, M. von Hippel, M. Wilhelm, JHEP **08**, 184 (2018). https://doi.org/10.1007/JHEP08(2018)184. arXiv:1805.10281 [hep-th]
18. J.L. Bourjaily, A.J. McLeod, M. von Hippel, M. Wilhelm, Phys. Rev. Lett. **122**(3), 031601 (2019). https://doi.org/10.1103/PhysRevLett.122.031601. arXiv:1810.07689 [hep-th]
19. J.L. Bourjaily, A.J. McLeod, C. Vergu, M. Volk, M. Von Hippel, M. Wilhelm, JHEP **01**, 078 (2020). https://doi.org/10.1007/JHEP01(2020)078. arXiv:1910.01534 [hep-th]
20. J.L. Bourjaily, A.J. McLeod, C. Vergu, M. Volk, M. Von Hippel, M. Wilhelm, JHEP **02**, 025 (2020). https://doi.org/10.1007/JHEP02(2020)025. arXiv:1910.14224 [hep-th]
21. J.L. Bourjaily, M. Volk, M. Von Hippel, JHEP **02**, 095 (2020). https://doi.org/10.1007/JHEP02(2020)095. arXiv:1912.05690 [hep-th]
22. F. Brown, Commun. Math. Phys. **287**, 925–958 (2009). https://doi.org/10.1007/s00220-009-0740-5. arXiv:0804.1660 [math.AG]
23. C. Bogner, M. Lüders, Contemp. Math. **648**, 11–28 (2015). arXiv:1302.6215 [hep-ph]
24. M. Besier, D. Van Straten, S. Weinzierl, Commun. Num. Theor. Phys. **13**, 253–297 (2019). https://doi.org/10.4310/CNTP.2019.v13.n2.a1. arXiv:1809.10983 [hep-th]
25. H. Frellesvig, C.G. Papadopoulos, JHEP **04**, 083 (2017). https://doi.org/10.1007/JHEP04(2017)083. arXiv:1701.07356 [hep-ph]
26. J.L. Bourjaily, F. Dulat, E. Panzer, Nucl. Phys. B **942**, 251–302 (2019). https://doi.org/10.1016/j.nuclphysb.2019.03.022. arXiv:1901.02887 [hep-th]
27. A. Hodges, JHEP **05**, 135 (2013). https://doi.org/10.1007/JHEP05(2013)135. arXiv:0905.1473 [hep-th]
28. S. Caron-Huot, L.J. Dixon, M. von Hippel, A.J. McLeod, G. Papathanasiou, JHEP **07**, 170 (2018). https://doi.org/10.1007/JHEP07(2018)170. arXiv:1806.01361 [hep-th]
29. L.D. Landau, Nucl. Phys. **13**(1), 181–192 (1960). https://doi.org/10.1016/B978-0-08-010586-4.50103-6
30. A.P. Hodges, Crossing and twistor diagrams. Twistor Newsletter **5**, 4 (1977)
31. A. Hodges, JHEP **08**, 051 (2013). https://doi.org/10.1007/JHEP08(2013)051. arXiv:1004.3323 [hep-th]
32. L. Mason, D. Skinner, J. Phys. A **44**, 135401 (2011). https://doi.org/10.1088/1751-8113/44/13/135401. arXiv:1004.3498 [hep-th]
33. F.C.S. Brown, Annales Sci. Ecole Norm. Sup. **42**, 371 (2009). arXiv:math/0606419 [math.AG]
34. A.B. Goncharov, M. Spradlin, C. Vergu, A. Volovich, Phys. Rev. Lett. **105**, 151605 (2010). https://doi.org/10.1103/PhysRevLett.105.151605. arXiv:1006.5703 [hep-th]
35. The Sage Developers. *`SageMath`, the Sage Mathematics Software System (Version 8.8)* (2019). Available at: www.sagemath.org

36. The PARI Group, Univ. Bordeaux. *PARI/GP version 2.11.0* (2018). Available at: pari.math.u-bordeaux.fr
37. S. He, Z. Li, C. Zhang, arXiv:2009.11471 [hep-th]
38. S. Laporta, E. Remiddi, Nucl. Phys. B **704**, 349–386 (2005). https://doi.org/10.1016/j.nuclphysb.2004.10.044. arXiv:hep-ph/0406160 [hep-ph]
39. S. Muller-Stach, S. Weinzierl, R. Zayadeh, PoS **LL2012**, 005 (2012). https://doi.org/10.22323/1.151.0005. arXiv:1209.3714 [hep-ph]
40. F. Brown, A. Levin, Preprint (2011). arXiv:1110.6917
41. S. Bloch, P. Vanhove, J. Number Theor. **148**, 328–364 (2015). https://doi.org/10.1016/j.jnt.2014.09.032. arXiv:1309.5865 [hep-th]
42. L. Adams, C. Bogner, S. Weinzierl, J. Math. Phys. **54**, 052303 (2013). https://doi.org/10.1063/1.4804996. arXiv:1302.7004 [hep-ph]
43. L. Adams, C. Bogner, S. Weinzierl, J. Math. Phys. **55**(10), 102301 (2014). https://doi.org/10.1063/1.4896563. arXiv:1405.5640 [hep-ph]
44. L. Adams, C. Bogner, S. Weinzierl, J. Math. Phys. **56**(7), 072303 (2015). https://doi.org/10.1063/1.4926985. arXiv:1504.03255 [hep-ph]
45. L. Adams, C. Bogner, S. Weinzierl, J. Math. Phys. **57**(3), 032304 (2016). https://doi.org/10.1063/1.4944722. arXiv:1512.05630 [hep-ph]
46. L. Adams, C. Bogner, A. Schweitzer, S. Weinzierl, J. Math. Phys. **57**(12), 122302 (2016). https://doi.org/10.1063/1.4969060. arXiv:1607.01571 [hep-ph]
47. L. Adams, S. Weinzierl, Commun. Num. Theor. Phys. **12**, 193–251 (2018). https://doi.org/10.4310/CNTP.2018.v12.n2.a1. arXiv:1704.08895 [hep-ph]
48. L. Adams, E. Chaubey, S. Weinzierl, Phys. Rev. Lett. **118**(14), 141602 (2017). https://doi.org/10.1103/PhysRevLett.118.141602. arXiv:1702.04279 [hep-ph]
49. C. Bogner, A. Schweitzer, S. Weinzierl, Nucl. Phys. B **922**, 528–550 (2017). https://doi.org/10.1016/j.nuclphysb.2017.07.008. arXiv:1705.08952 [hep-ph]
50. J. Broedel, C. Duhr, F. Dulat, L. Tancredi, JHEP **05**, 093 (2018). https://doi.org/10.1007/JHEP05(2018)093. arXiv:1712.07089 [hep-th]
51. J. Broedel, C. Duhr, F. Dulat, L. Tancredi, Phys. Rev. D **97**(11), 116009 (2018). https://doi.org/10.1103/PhysRevD.97.116009. arXiv:1712.07095 [hep-ph]
52. L. Adams, S. Weinzierl, Phys. Lett. B **781**, 270–278 (2018). https://doi.org/10.1016/j.physletb.2018.04.002. arXiv:1802.05020 [hep-ph]
53. J. Broedel, C. Duhr, F. Dulat, B. Penante, L. Tancredi, JHEP **08**, 014 (2018). https://doi.org/10.1007/JHEP08(2018)014. arXiv:1803.10256 [hep-th]
54. L. Adams, E. Chaubey, S. Weinzierl, Phys. Rev. Lett. **121**(14), 142001 (2018). https://doi.org/10.1103/PhysRevLett.121.142001. arXiv:1804.11144 [hep-ph]
55. J. Broedel, C. Duhr, F. Dulat, B. Penante, L. Tancredi, JHEP **01**, 023 (2019). https://doi.org/10.1007/JHEP01(2019)023. arXiv:1809.10698 [hep-th]
56. L. Adams, E. Chaubey, S. Weinzierl, JHEP **10**, 206 (2018). https://doi.org/10.1007/JHEP10(2018)206. arXiv:1806.04981 [hep-ph]
57. I. Hönemann, K. Tempest, S. Weinzierl, Phys. Rev. D **98**(11), 113008 (2018). https://doi.org/10.1103/PhysRevD.98.113008. arXiv:1811.09308 [hep-ph]
58. C. Bogner, S. Müller-Stach, S. Weinzierl, Nucl. Phys. B **954**, 114991 (2020). https://doi.org/10.1016/j.nuclphysb.2020.114991. arXiv:1907.01251 [hep-th]
59. J. Broedel, C. Duhr, F. Dulat, B. Penante, L. Tancredi, JHEP **05**, 120 (2019). https://doi.org/10.1007/JHEP05(2019)120. arXiv:1902.09971 [hep-ph]
60. J. Broedel, C. Duhr, F. Dulat, R. Marzucca, B. Penante, L. Tancredi, JHEP **09**, 112 (2019). https://doi.org/10.1007/JHEP09(2019)112. arXiv:1907.03787 [hep-th]
61. S. Groote, J.G. Korner, A.A. Pivovarov, Annals Phys. **322**, 2374–2445 (2007). https://doi.org/10.1016/j.aop.2006.11.001. arXiv:hep-ph/0506286 [hep-ph]
62. F. Brown, O. Schnetz, Duke Math. J. **161**(10), 1817–1862 (2012). https://doi.org/10.1215/00127094-1644201. arXiv:1006.4064 [math.AG]
63. S. Bloch, M. Kerr, P. Vanhove, Compos. Math. **151**(12), 2329–2375 (2015). https://doi.org/10.1112/S0010437X15007472. arXiv:1406.2664 [hep-th]

64. S. Bloch, M. Kerr, P. Vanhove, Adv. Theor. Math. Phys. **21**, 1373–1453 (2017). https://doi.org/10.4310/ATMP.2017.v21.n6.a1. arXiv:1601.08181 [hep-th]
65. D. Broadhurst, private communication
66. D. Festi, D. van Straten, Commun. Num. Theor. Phys. **13**, 463–485 (2019). https://doi.org/10.4310/CNTP.2019.v13.n2.a4. arXiv:1809.04970 [math.AG]
67. M. Besier, D. Festi, M. Harrison, B. Naskrecki, Commun. Num. Theor. Phys. **14**(4), 863–911 (2020). https://doi.org/10.4310/CNTP.2020.v14.n4.a4. arXiv:1908.01079 [math.AG]
68. C. Doran, A. Novoseltsev, P. Vanhove, To appear
69. A. Klemm, C. Nega, R. Safari, JHEP **04**, 088 (2020). https://doi.org/10.1007/JHEP04(2020)088. arXiv:1912.06201 [hep-th]
70. K. Bönisch, F. Fischbach, A. Klemm, C. Nega, R. Safari, arXiv:2008.10574 [hep-th]
71. T. Hubsch, *Calabi-Yau manifolds: A Bestiary for physicists* (World Scientific, 1992)
72. C. Vergu, M. Volk, JHEP **07**, 160 (2020). https://doi.org/10.1007/JHEP07(2020)160. arXiv:2005.08771 [hep-th]
73. F. Brown, C. Duhr, arXiv:2006.09413 [hep-th]

A Geometrical Framework for Amplitude Recursions: Bridging Between Trees and Loops

Johannes Broedel and André Kaderli

Abstract Various methods for the recursive evaluation of scattering amplitudes in quantum field theory and string theory have been put forward during the last couple of years. In these proceedings we describe a geometrical framework, which is believed to be capable of treating many of these recursions in a unified way. Our recursive framework is based on manipulating iterated integrals on Riemann surfaces with boundaries. A geometric parameter appears as variable of a differential equation of KZ or KZB type. The parameter interpolates between two associated regularized boundary values, which contain iterated integrals closely related to scattering amplitudes defined on two different geometries.

1 Introduction

The calculation of scattering amplitudes in perturbative quantum field theories relies on the evaluation of Feynman integrals associated to Feynman graphs, which in turn are a combinatorial representation of Feynmann's path integral formalism. A typical Feynman integral associated to an ℓ-loop process reads

$$(\mu^2)^{\nu-\frac{\ell D}{2}} \int \prod_{r=1}^{\ell} \frac{d^D k_r}{i\pi^{D/2}} \prod_{j=1}^{n} \frac{1}{(-q_j{}^2+m_j^2)^{\nu_j}}, \qquad \nu = \sum_{j=1}^{n} \nu_j \,. \tag{1}$$

where k_i are the loop momenta, q_j and m_j label the momenta and masses along the n (internal and external) propagators. The quantity D is the (spacetime) dimension and the integral shall usually be evaluated in four dimensions.

From the Feynman formalism, a multitude of different types of integrals can arise [1–3]. A first step towards treating the integrals in a uniform way is to introduce

J. Broedel (✉) · A. Kaderli
Humboldt University, Berlin, Germany
e-mail: jbroedel@physik.hu-berlin.de; kaderlia@physik.hu-berlin.de

J. Blümlein, C. Schneider (eds.), *Anti-Differentiation and the Calculation of Feynman Amplitudes*, Texts & Monographs in Symbolic Computation,
https://doi.org/10.1007/978-3-030-80219-6_6

Feynman parameters x_i, which amounts to a clever substitution of the momentum integrations in the Feynman integral (1). Leaving out a constant prefactor, one obtains an integral of the following type:

$$\int_{x_j \geq 0} \delta\Big(1 - \sum_{j=1}^{n} x_j\Big)\Big(\prod_{j=1}^{n} dx_j\Big)\, \mathcal{I}(x_1, \ldots, x_n, D)\,. \tag{2}$$

Integration is constrained to a simplex by the condition $x_j \geq 0$ and the δ distribution. Every simplex, however, can be parametrized iteratively. Consequently, each Feynman integral can be rewritten as a linear combination of iterated integrals.

The type of iterated integral, on the other hand, is determined by the differential form $\mathcal{I}$, which shall to be integrated over. This differential form can have singularities in the Feynman parameters. The singularity structure can be explored by writing the integrand as

$$\mathcal{I}(x_1, \ldots, x_n) = \frac{\mathcal{N}}{\mathcal{D}} \tag{3}$$

where the integrand becomes singular, whenever the denominator polynomial $\mathcal{D}$ has a zero.[1] The zero locus of this polynomial defines an algebraic curve, which can be taken as starting point for the definition of suitable differentials incorporating the symmetries of the Feynman diagram. From those differentials, one can then build iterated integrals. Once the differential forms and associated iterated integrals are known, it is usually possible to write down a differential equation for a set of master integrals. The resulting (matrix) equation should hopefully be translatable into typical differential equations for a set of master integrals as used heavily in modern Feynman as well as string-theoretic calculations [4–9].

While this mathematical account sounds very straightforward, it is peppered with practical difficulties: identification of suitable differential forms—that is, a cohomology—for a given algebraic curve is for example possible only for the simplest Feynman graphs.

Therefore, in these proceedings, we will take the two simplest algebraic curves, Riemann surfaces with boundary of genus zero and genus one, as examples. The corresponding differential forms generate iterated integrals, which are polylogarithms (genus zero) or elliptic analogues thereof (genus one). Whereas actual Feynman integrals might imply more complicated differential forms, almost all final results turn out to be expressible in terms of these simple iterated integrals and special values thereof: multiple zeta values (MZVs) and elliptic multiple zeta values (eMZVs). The only structural ingredient we need to add is an extra parameter, with respect to which a differential equation governing the recursion is established. For Feynman integrals, this would be an additional Feynman parameter, while for string

[1] The polynomials $\mathcal{N}$ and $\mathcal{D}$ are very closely related to the Symanzik polynomials.

configuration-space integrals the parameter describes an additional vertex insertion point.

So the recursive algorithms discussed in these proceedings are to be seen as prototypes for more complicated recursions. They are simple and thus mathematically very clean: they turn out to precisely describe recursions for amplitudes in open string theory at tree level (genus zero) and one-loop level (genus one).

While Feynman integrals have to be regularized case-by-case (which is usually done with dimensional regularization), for the classes of iterated integrals considered in these proceedings a standard way of regularization is available. Thus, one will not have to find a suitable regularization for each Feynman integral separately, but can rather rely on a general regularization scheme for all integrals occurring.

Pursuing this line of thoughts further, result for a scattering amplitude evaluated in the Feynman formalism is usually provided as loop expansion in the parameter ℓ and as expansion in the parameter ϵ of dimensional regularization. On the other hand, string scattering amplitudes defined as iterated integrals on Riemann surfaces are result in a genus expansion (parameter g) and an expansion in α', the inverse string tension. This suggestive correspondence might be substantiated by understanding "stringyness" once again as a simple regulating mechanism, which after all is a very old idea.

The whole subject is comparably involved algebraically, that is, there is a price to pay for the formalization. As a reward, the formalism is applicable to many different situations and is expected to lead to recursion relations for various types of iterated integrals and thus scattering amplitudes during the next couple of years.

2 Genus Zero

2.1 Iterated Integrals and Multiple Zeta Values

Let us review the most straightforward implementation of polylogarithms on a genus-zero Riemann surface. Consider the one-form[2]

$$\omega_a = \frac{dx}{x-a}, \quad x, a \in \mathbb{R} \tag{4}$$

[2]For simplicity, we consider real integration paths here exclusively. More general quantities a, for example complex functions of complex parameters, will lead to the hyperlogarithms discussed in Erik Panzer's talk.

and define iterated integrals [10]

$$G(a_1,\ldots,a_r;x)=\int_0^x \frac{dt}{t-a_1}G(a_2,\ldots,a_r;t)=\int_0^z \omega_{a_r}\cdots\omega_{a_1}\,,\quad G(;z)=1\,,\tag{5}$$

where $a_1 \neq z$ and $a_r \neq 0$. Given the iterated structure above, the integrals are subject to shuffle relations:

$$\begin{aligned}&G(a_1,a_2,\ldots,a_j;x)G(b_1,b_2,\ldots,b_k;x)\\&\quad=G\big((a_1,a_2,\ldots,a_j)\sqcup\!\sqcup(b_1,b_2,\ldots,b_k);x\big)\end{aligned}\tag{6}$$

and the differential forms in Eq. (4) imply that different integrands in Eq. (5) are related by partial fraction:

$$\frac{1}{x_i-x_k}\frac{1}{x_j-x_k}=\frac{1}{x_i-x_j}\frac{1}{x_j-x_k}+\frac{1}{x_j-x_i}\frac{1}{x_i-x_k}\,.\tag{7}$$

If the case $a_r=0$ was allowed, the integrals $G(a_1,\ldots,a_r;x)$ would not be well-defined, since they would diverge due to the simple pole at the lower integration boundary. This can be regularized by the convention

$$G(\underbrace{0,\ldots,0}_{n};x)=\frac{\log^n x}{n!}\,,\tag{8}$$

such that integrals $G(a_1,\ldots,a_r;x)$ associated to arbitrary labels $a_1,\ldots,a_r\neq x$, including $a_r=0$, can be defined as follows: the shuffle identity (6) is used to formally write and define the (a priori ill-defined) integral $G(a_1,\ldots,a_r;x)$ as an expansion in the well-defined integrals Eq. (5) and powers of logarithms $G(0,\ldots,0;x)$. This regularization scheme is called *shuffle regularization* and is compatible with (e.g. it preserves) the shuffle product. The integrals $G(a_1,\ldots,a_r;z)$ are called *multiple polylogarithms* (MPLs).

For the purpose of exploring open-string amplitudes at tree level, it is sufficient to confine ourselves to labels $a_i\in\{0,1\}$. Choosing $x=1$ leads to representation of *multiple zeta values* (MZVs) in terms of iterated integrals:

$$\begin{aligned}\zeta(n_1,\ldots,n_r)&=(-1)^r G(\underbrace{0\ldots01}_{n_r}\ldots\underbrace{0\ldots01}_{n_1};1)\\&=(-1)^r\int_0^1\omega_1\omega_0^{n_1-1}\omega_1\omega_0^{n_2-1}\ldots\omega_r\omega_0^{n_r-1}\\&=\sum_{1\le k_1<\cdots<k_r}\frac{1}{k_1^{n_1}\ldots k_r^{n_r}}\,,\end{aligned}\tag{9}$$

for $n_r > 1$. As before, this definition can be extended to arbitrary labels $n_1, \ldots, n_r \geq 1$ and integrals $G(a_1, \ldots, a_r; 1)$ with $a_1 = 1$, respectively, by a shuffle regularization with the conventions

$$\zeta(1) = -G(1, 1) = 0 \quad \text{as well as} \quad G(0; 1) = \log 1 = 0\,. \tag{10}$$

Multiple zeta values inherit shuffle relations from the polylogarithms; in addition there are the shuffle relations (best palpable in the sum representation in the last line of (9)). After considering all relations, a basis of MZVs at each conjectured transcendentality can be chosen, the mathematically most beautiful being the Hoffman basis [11].

2.2 Selberg Integrals and Open-String Configuration-Space Integrals

Selberg integrals serve as generating series for the iterated integrals introduced in the previous subsection. At the same time they contain the configuration-space integrals appearing in open-string tree-level amplitudes. The full scattering amplitude in open superstring theories at tree level can be calculated as correlation function of vertex operators inserted on the boundary of a genus-zero Riemann surface. Upon evaluation, those correlators separate into a polarization part (which can be calculated straightforwardly) and the so-called configuration-space integrals [12–14]. The best known example is the four-point Veneziano amplitude [15], which reads

$$\int_0^1 dx_3\, x_3^{s_{13}} (1 - x_3)^{s_{23}} \frac{s_{13}}{x_3} = \frac{\Gamma(1 + s_{13})\Gamma(1 + s_{23})}{\Gamma(1 + s_{13} + s_{23})}\,. \tag{11}$$

The complex parameters

$$s_{i_1 \ldots i_r} = \alpha'(k_{i_1} + \ldots + k_{i_r})^2 \tag{12}$$

are Mandelstam variables built from the momenta of the external particles. In these proceedings, these variables shall be assumed to be chosen such that all integrals considered are convergent [16, 17]. In contrast to the usual Mandelstam variables, a parameter α' is supplemented here, which serves as counting parameter and will be identified with the inverse string tension only later on.

The N-point configuration-space integrals in genus-zero open-string amplitudes are examples of Selberg integrals [18], which can be constructed as follows: consider the $(L+1)$-punctured Riemann sphere with fixed points

$$(x_1, x_2, x_{L+1}) = (0, 1, \infty)\,. \tag{13}$$

Writing

$$x_{ij} = x_{i,j} = x_i - x_j \,, \tag{14}$$

the corresponding Selberg integrals are iteratively defined by

$$\mathrm{S}[i_{k+1}, \dots, i_L](x_1, \dots, x_k) = \int_0^{x_k} \frac{dx_{k+1}}{x_{k+1,i_{k+1}}} \mathrm{S}[i_{k+2}, \dots, i_L](x_1, \dots, x_{k+1}) \,, \tag{15}$$

and the empty Selberg integral (or Selberg seed) is defined as[3]

$$\mathrm{S}[](x_1, \dots, x_L) = \prod_{0 \le x_i < x_j \le 1} |x_{ij}|^{s_{ij}} \,. \tag{16}$$

Definition (15) presumes that the so-called admissibility condition

$$1 \le i_p < p \qquad \forall p \in \{k+1, \dots, L\} \tag{17}$$

is met. The integral in Eq. (15) is referred to as of type $(k, L+1)$. It is, for fixed s_{ij}, defined on $\mathcal{M}_{0,k+1}$. Accordingly, these integrals form a basis of the twisted de Rham cohomology of the configuration space of $(L+1)$-punctured Riemann spheres with $k+1$ fixed punctures with respect to the pull-back of the connection $d + d\log \mathrm{S}$ [19]. Integrals with at least one label $i_p = 1$ may be reduced to this basis using integration by parts and partial fractioning.

2.3 Recursion for Open-String Amplitudes at Genus Zero

Aomoto [19] and Terasoma [20] showed that Selberg integrals of type $(2, L)$ can be obtained algebraically from those of type $(2, L-1)$: one starts from a basis vector $\mathbf{S}(x_3)$ for Selberg integrals of type $(3, L+1)$, which contain an auxiliary point x_3 in contrast to the integrals of type $(2, L)$ and $(2, L-1)$, respectively. Taking the derivative with respect to x_3 leads to an equation of Knizhnik–Zamolodchikov (KZ) type [21]

$$\frac{d}{dx_3}\mathbf{S}(x_3) = \left(\frac{e_0}{x_3} + \frac{e_1}{x_3 - 1}\right)\mathbf{S}(x_3) \,, \tag{18}$$

[3]We use the notation $\prod_{x_a \le x_i < x_j \le x_b} = \prod_{i,j \in \{1,2,\dots,L\}:\, x_a \le x_i < x_j \le x_b}$.

where the (braid) matrices e_0 and e_1 have entries which are homogeneous polynomials of degree one in the parameters s_{ij}. The regularized boundary values

$$\mathbf{C}_0 = \lim_{x_3 \to 0} x^{-e_0} \mathbf{S}(x_3)\,, \quad \mathbf{C}_1 = \lim_{x_3 \to 1} (1-x_3)^{-e_1} \mathbf{S}(x_3)\,. \tag{19}$$

of the differential equation (18) are Selberg integrals of type $(2, L-1)$ and $(2, L)$, respectively. They can be shown to be related by the Drinfeld associator [22, 23]

$$\mathbf{C}_1 = \mathbf{\Phi}(e_0, e_1)\; \mathbf{C}_0, \tag{20}$$

which is the generating series of multiple zeta values [24],

$$\begin{aligned}\mathbf{\Phi}(e_0, e_1) &= \sum_{w \geq 0} \sum_{k_1,\ldots,k_w \geq 1} e_0^{k_w-1} e_1 \ldots e_0^{k_2-1} e_1 e_0^{k_1-1} e_1 \zeta(k_1, k_2, \ldots, k_w) \\ &= 1 - \zeta(2)[e_0, e_1] - \zeta(3)\left([e_0, [e_0, e_1]] - [[e_0, e_1], e_1]\right) + \ldots\,. \end{aligned} \tag{21}$$

What makes this construction useful for physicists is the fact that the $(N-1)$-point and the N-point configuration-space integrals at genus zero can be identified (upon proper assignment of the Mandelstam variables) as linear combinations of the components of $\mathbf{C}_0$ and $\mathbf{C}_1$ respectively, where $N{=}L$. This relationship has been used to derive a recursive construction for all configuration-space integrals on genus zero: it provides an analogue of the Parke–Taylor formula [25] for string theory [26, 27]. The precise relation of the above formalism to open-string configuration-space integrals at genus zero has been discussed thoroughly in ref. [28] (Fig. 1).

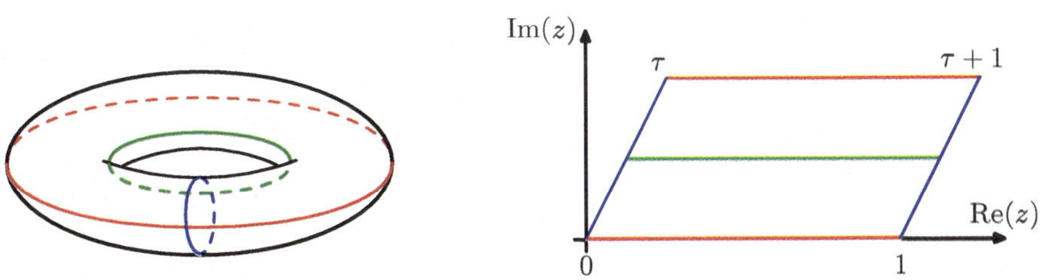

Fig. 1 The torus and its fundamental domain. The ratio of the lengths ω_B and ω_A of the B- and A-cycle respectively yields the modular parameter: $\tau = \omega_B/\omega_A$

3 Genus One

3.1 Iterated Integrals at Genus One and Elliptic Multiple Zeta Values

While there is a large collection of literature [29–31] on how to define homotopy-invariant integrals on an elliptic curve (or torus), we are going to focus here on what is one of the simplest approaches and simultaneously the best fit for a canonical generalization of the iterated integrals on a genus-zero surface introduced in Eq. (5). We will parametrize the elliptic curve by the modular parameter τ (or its exponentiated version $q = \exp(2\pi i\tau)$) and name the red and blue boundaries of the fundamental domain A- and B-cycle respectively. An (infinite) set of differential forms on the elliptic curve can be defined starting from the Kronecker series

$$F(z,\eta,\tau) = \frac{\theta_1'(0,\tau)\theta_1(z+\eta,\tau)}{\theta_1(z,\tau)\theta_1(\eta,\tau)}\,, \tag{22}$$

where θ_1 is the odd Jacobi function and the tick denotes a derivative with respect to the first argument. The Kronecker series is symmetric in z and η, but only quasiperiodic in the variable z:

$$F(z+1,\eta,\tau) = F(z,\eta,\tau), \qquad F(z+\tau,\eta,\tau) = e^{-2\pi i\eta}F(z,\eta,\tau)\,. \tag{23}$$

In addition, Fay's trisecant equation [32] implies the Fay identity

$$\begin{aligned} &F(z_1,\eta_1,\tau)F(z_2,\eta_2,\tau) \\ &\quad = F(z_1,\eta_1+\eta_2,\tau)F(z_2-z_1,\eta_2,\tau) + F(z_2,\eta_1+\eta_2,\tau)F(z_1-z_2,\eta_1,\tau)\,. \end{aligned} \tag{24}$$

Expanding the Kronecker form in the second argument, one obtains an infinite set of differential forms

$$\eta F(z,\eta,\tau)dz = \sum_{n=0}^{\infty} g^{(n)}(z,\tau)\,\eta^n dz, \tag{25}$$

satisfying

$$g^{(n)}(-z,\tau)dz = (-1)^n g^{(n)}(z,\tau)dz. \tag{26}$$

They are only quasi-periodic

$$g^{(n)}(z+1,\tau) = g^{(n)}(z,\tau) \tag{27a}$$

$$g^{(1)}(z+\tau,\tau) = g^{(1)}(z,\tau) - 2\pi i \tag{27b}$$

$$g^{(2)}(z+\tau,\tau) = g^{(2)}(z,\tau) - 2\pi i g^{(1)}(z,\tau) - \tfrac{1}{2}(2\pi i)^2 \tag{27c}$$

$$\vdots$$

and satisfy the (expanded) form of Fay relations:

$$\begin{aligned} g^{(n_1)}(t-z,\tau)g^{(n_2)}(t,\tau) = &-(-1)^{n_1}g^{(n_1+n_2)}(z,\tau) \\ &+\sum_{j=0}^{n_2}\binom{n_1-1+j}{j}g^{(n_2-j)}(z,\tau)g^{(n_1+j)}(t-z,\tau) \\ &+\sum_{j=0}^{n_1}\binom{n_2-1+j}{j}(-1)^{n_1+j}g^{(n_1-j)}(z,\tau)g^{(n_2+j)}(t,\tau). \end{aligned} \tag{28}$$

Taking the differential forms $g^{(n)}(z,\tau)$ as starting point, one defines the following iterated integrals:

$$\tilde{\Gamma}\left(\begin{smallmatrix} n_1, & n_2, & \ldots, & n_r \\ a_1, & a_2, & \ldots, & a_r \end{smallmatrix}; z, \tau\right) = \int_0^z dz'\, g^{(n_1)}(z'-a_1,\tau)\tilde{\Gamma}\left(\begin{smallmatrix} n_2, & \ldots, & n_r \\ a_2, & \ldots, & a_r \end{smallmatrix}; z, \tau\right), \tag{29}$$

for $(n_1, a_1) \neq (1, z)$ and $(n_r, a_r) \neq (1, 0)$, which naturally obey shuffle relations

$$\begin{aligned} &\tilde{\Gamma}(A_1, A_2, \ldots, A_j; z, \tau)\tilde{\Gamma}(B_1, B_2, \ldots, B_k; z, \tau) \\ &\quad = \tilde{\Gamma}\big((A_1, A_2, \ldots, A_j) ⧢ (B_1, B_2, \ldots, B_k); z, \tau\big) \end{aligned} \tag{30}$$

in terms of the combined letters $A_i = {n_i \atop a_i}$.

The function $g^{(1)}$ has a simple pole at zero: it thus qualifies as genus-one generalization of $\frac{1}{z}$ at genus zero. The integral over $g^{(1)}$ will be of particular interest below: it is the (A-cycle) generalization of the natural logarithm. Similar to the prescription in Eq. (8), the integral $\tilde{\Gamma}\left({1 \atop 0}; z, \tau\right)$ is a priori not well-defined and requires regularization because it exhibits an endpoint divergence at the lower integration boundary. Throughout the article, we are going to employ *tangential basepoint regularization* [33, 34]. In short, this amounts to subtracting the endpoint divergence by defining[4]

$$\tilde{\Gamma}(\begin{smallmatrix}1\\0\end{smallmatrix}; z, \tau) = \lim_{\epsilon\to 0}\int_{\epsilon}^{z} dz\, g^{(1)}(z,\tau) + \log(1 - e^{2\pi i\epsilon})$$

$$= \log(1 - e^{2\pi i z}) - \pi i z + 4\pi \sum_{k,l>0} \frac{1}{2\pi k}\,(1 - \cos(2\pi k z))\, q^{kl}\,. \tag{31}$$

In particular, when placing the branch cut of the logarithm such that $\log(-1) = \pi i$, one finds the following asymptotic behavior for $z \to 0$

$$\tilde{\Gamma}(\begin{smallmatrix}1\\0\end{smallmatrix}; z, \tau) \sim \log(-2\pi i z) \tag{32}$$

and $z \to 1$

$$\tilde{\Gamma}(\begin{smallmatrix}1\\0\end{smallmatrix}; z, \tau) \sim \log(-2\pi i(1 - z))\,. \tag{33}$$

The remaining integrals $\tilde{\Gamma}\left(\begin{smallmatrix}n_1, & n_2, & \dots, & n_r\\ a_1, & a_2, & \dots, & a_r\end{smallmatrix}; z, \tau\right)$ with $(n_r, a_r) = (1, 0)$ are then defined by shuffle regularization, similar to the genus-zero integrals from Eq. (5): they are defined by the well-defined iterated integrals from Eq. (29), the shuffle identity (30) and the regularized integral $\tilde{\Gamma}(\begin{smallmatrix}1\\0\end{smallmatrix}; z, \tau)$. This regularization procedure is compatible with the shuffle product, i.e. an algebra homomorphism. For the remainder of those proceedings, the τ-dependence will be mostly kept implicit for all integration kernels $g^{(n)}$ and all iterated elliptic integrals $\tilde{\Gamma}$. The latter are called *elliptic multiple polylogarithms* (eMPLs).

3.2 Elliptic Multiple Zeta Values

In the same way, as multiple zeta values can be represented as values of a special class of MPLs at one, so-called A-cycle *elliptic multiple zeta values* (eMZVs) [35–37] are defined as values of regularized eMPLs at one:

$$\omega(n_1, n_2, \dots, n_r) = \tilde{\Gamma}\left(\begin{smallmatrix}n_r, & \dots, & n_1\\ 0, & \dots, & 0\end{smallmatrix}; 1\right), \quad n_1 \neq 1. \tag{34}$$

In order to extend the definition (34) to the cases $n_1 = 1$, eMZVs need to be further regularized in a shuffle-compatible way. While a thorough discussion of the regularization procedure starting from the regulated integral in (31) can be found in ref. [40], regularization of A-cycle eMZVs practically amounts to defining

$$\omega(1) = 0 \tag{35}$$

[4] The limit $\epsilon \to 0$ is assumed to be taken within the unit interval.

and using shuffle relations (inherited from (30))

$$\omega(n_1, n_2, \ldots, n_r)\omega(k_1, k_2, \ldots, k_s) = \omega\big((n_1, n_2, \ldots, n_r) \sqcup\!\sqcup (k_1, k_2, \ldots, k_s)\big) , \tag{36}$$

to identify and isolate all those contributions. Furthermore, (26) implies

$$\omega(n_1, n_2, \ldots, n_{r-1}, n_r) = (-1)^{n_1+n_2+\ldots+n_r}\omega(n_r, n_{r-1}, \ldots, n_2, n_1) . \tag{37}$$

The two types of relations above by far do not exhaust all relations between elliptic multiple zeta values; in particular does the Fay identity (28) imply many more relations. A thorough discussion can be found in ref. [38] and a list of relations on the associated website [39].

3.3 Generalized Selberg Integrals at Genus One

In order to investigate a genus-one analogue of the genus-zero recursive construction in Sect. 2, we need a suitable analogue of genus-zero Selberg integrals (15): let $L \geq 2$, $0 = z_1 < z_L < \ldots < z_2 < 1$ and τ the modular parameter of the torus $\mathbb{C}/(\mathbb{Z} + \tau\mathbb{Z})$. Let the empty genus-one Selberg integral (or genus-one Selberg seed) be

$$\mathrm{S}^\tau = \mathrm{S}^\tau\big[\ \big](z_1, \ldots, z_L) = \prod_{0=z_1 \leq z_i < z_j \leq z_2} \exp\left(s_{ij}\tilde{\Gamma}_{ji}\right) . \tag{38}$$

where $\tilde{\Gamma}_{ji} = \tilde{\Gamma}(\begin{smallmatrix}1\\0\end{smallmatrix}; z_j - z_i, \tau)$. Genus-one Selberg integrals of weight $w = \sum_{i=k+1}^{L} n_i$ and type (k, L) are then defined recursively by

$$\begin{aligned}\mathrm{S}^\tau&\left[\begin{smallmatrix} n_{k+1}, & \ldots, & n_L \\ i_{k+1}, & \ldots, & i_L \end{smallmatrix}\right](z_1, \ldots, z_k)\\ &= \int_0^{z_k} dz_{k+1}\, g^{(n_{k+1})}_{k+1,i_{k+1}} \mathrm{S}^\tau\left[\begin{smallmatrix} n_{k+2}, & \ldots, & n_L \\ i_{k+2}, & \ldots, & i_L \end{smallmatrix}\right](z_1, \ldots, z_{k+1}) .\end{aligned} \tag{39}$$

where we use the shorthand notation

$$g^{(n)}_{ij} = g^{(n)}_{i,j} = g^{(n)}(z_i - z_j, \tau) . \tag{40}$$

Moreover, the so-called admissibility condition $1 \leq i_p < p$ is required for all $p \in \{k + 1, \ldots, L\}$, which is the genus-one analogue of Eq. (17).

To build a recursion following the structure of the genus-zero recursion reviewed in Sect. 2.3, we need to find a suitable class of genus-one Selberg integrals: to achieve this, we fix the symmetries of the torus by $z_1 = 0$, supplement one unintegrated auxiliary point z_2, such that $k = 2$ punctures are fixed and integrate over the remaining $L - 2$ punctures, but keep the number L of insertion points variable. The resulting class of genus-one Selberg integrals reads

$$\mathrm{S}^{\tau}\left[\begin{smallmatrix} n_3, & \ldots, & n_L \\ i_3, & \ldots, & i_L \end{smallmatrix}\right](z_1 = 0, z_2) = \int_{0=z_1<z_L<z_{L-1}<\cdots<z_2} \prod_{i=3}^{L} dz_i \; \mathrm{S}^{\tau} \prod_{k=3}^{L} g_{k,i_k}^{(n_k)} . \tag{41}$$

Again, we would like to identify a basis in the above class of integrals with respect to integration by parts and partial fractioning. While there was only one type of differential form in the genus-zero situation (which one could have assigned weight one), we have an infinite number here: all combinations of $n_3, \ldots, n_L$ can appear and for each of those combinations (almost) all admissible values can occur. This combinatorial problem can be solved [40] and we collect all basis elements in a vector $\mathbf{S}^{\tau}_{w}(z_2)$ of definite weight w and combine all those vectors into an infinitely large vector:

$$\mathbf{S}^{\tau}(z_2) = \begin{pmatrix} \mathbf{S}^{\tau}_{0}(z_2) \\ \mathbf{S}^{\tau}_{1}(z_2) \\ \mathbf{S}^{\tau}_{2}(z_2) \\ \vdots \end{pmatrix} . \tag{42}$$

The resulting vector $\mathbf{S}^{\tau}(z_2)$ is the analogue if the genus-zero Selberg vector $\mathbf{S}(x_3)$, which satisfies the KZ Eq. (18).

3.4 Selberg Recursion at Genus One

In this subsection, it will be argued that the derivative of the vector $\mathbf{S}^{\tau}(z_2)$ defined in Eq. (42) with respect to the auxiliary point z_2 can be written in the form

$$\frac{\partial}{\partial z_2}\mathbf{S}^{\tau}(z_2) = \sum_{n\geq 0} g_{21}^{(n)} x^{(n)} \mathbf{S}^{\tau}(z_2) , \tag{43}$$

where the non-vanishing entries of the matrices $x^{(n)}$ turn out to be homogeneous polynomials of degree one in the parameters s_{ij}. The resulting system is of elliptic KZB-type, whose solution will be described below.

Proving the above statement is elaborate and is spelt out in detail in ref. [40]. The proof is constructive and relies on formal explicit evaluation of the derivative for each entry of the vector $\mathbf{S}^{\tau}(z_2)$. Performing the derivative on the Selberg seed will bring down various terms of the form

$$\frac{\partial}{\partial z_i}\mathbf{S}^\tau = \sum_{k\neq i} s_{ik}\, g_{ik}^{(1)}\mathbf{S}^\tau \,, \tag{44}$$

whereas all other derivatives can be rewritten using integration by parts as to act on the Selberg seed exclusively. Thus one is left with Selberg integrals of definite length containing products of functions $g_{ij}^{(n)}$. Organizing these products in so-called *chains* (e.g. $g_{ij}^{(n_1)} g_{jk}^{(n_2)} g_{kl}^{(n_3)}$) allows to translate the application of Fay identities into graphical operations. Employing the (graphical analogue of) Fay identities algorithmically, one can show that in each of those integrands a factor $g_{21}^{(n)}$ can be isolated. Pulling this factor out of the integral (as neither the point z_1 or z_2 are integrated over) renders the remaining integral a basis integral, that is, a component of the original vector $\mathbf{S}^\tau(z_2)$. Accordingly, the Mandelstam variables arising from Eq. (44) can then be collected in the matrices $x^{(n)}$, yielding the closed system in Eq. (43) [40].

What remains, is to solve Eq. (43). In the same way as this has been done for the KZ-system in Sect. 2.3, one can solve the system by considering regularized boundary values, which are related by the elliptic KZB associator [41, 42]. Regularized boundary values for the KZB system in Eq. (43) are defined as

$$\mathbf{C}_1^\tau = \lim_{z_2\to 1}(-2\pi i(1-z_2))^{-x^{(1)}}\mathbf{S}^\tau(z_2) \text{ and } \mathbf{C}_0^\tau = \lim_{z_2\to 0}(-2\pi i z_2)^{-x^{(1)}}\mathbf{S}^\tau(z_2)\,. \tag{45}$$

These two boundary values are related (see e.g. [40]) by the elliptic KZB associator $\Phi(x^{(0)}, x^{(1)}, x^{(2)}, \ldots)$ via

$$\mathbf{C}_1^\tau = \Phi(x^{(0)}, x^{(1)}, x^{(2)}, \ldots)\,\mathbf{C}_0^\tau \,, \tag{46}$$

whereas the KZB associator is—in analogy to the KZ associator in Eq. (21)—a generating series for A-cycle eMZVs:

$$\begin{aligned}\Phi^\tau &= 1 + x^{(0)} - 2\zeta(2)x^{(2)} \\ &\quad + \frac{1}{2}x^{(0)}x^{(0)} - (x^{(0)}x^{(1)} - x^{(1)}x^{(0)})\omega(0,1;\tau) - \zeta(2)(x^{(0)}x^{(2)} + x^{(2)}x^{(0)}) \\ &\quad + \big(x^{(1)}x^{(2)} - x^{(2)}x^{(1)}\big)\big(\omega(0,3;\tau) - 2\zeta(2)\omega(0,1;\tau)\big) + 5\zeta(4)x^{(2)}x^{(2)} + \cdots\end{aligned} \tag{47}$$

Equation (46) is the main tool in the recursive construction at genus one. What remains to be done before it can be applied, is the investigation of the boundary values $\mathbf{C}_0^\tau$ and $\mathbf{C}_1^\tau$ pictured in Fig. 2. Careful evaluation of the boundary values is beyond the scope of these proceedings, but is performed in detail in [40]. The analysis relies on evaluating the matrix exponential in Eq. (45), and thus requires

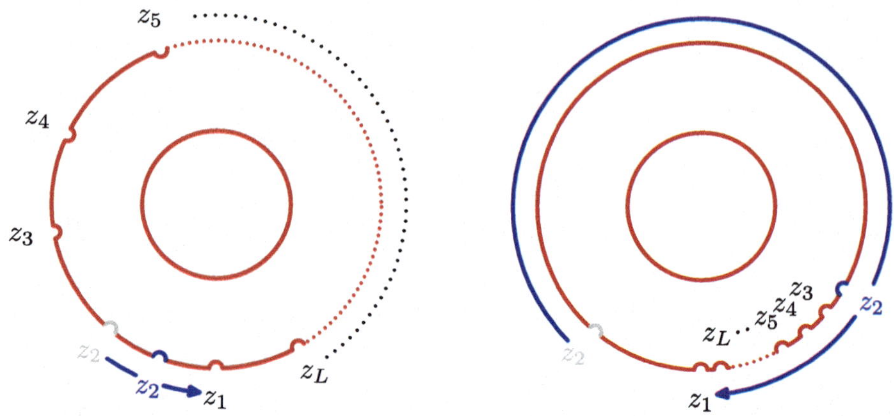

Fig. 2 Limits of the auxiliary point z_2 correspond to approaching the point $z_1 = 0 \equiv 1$—the origin of the fundamental domain—along the real line from the left and from the right. While the limit $z_2 \to 1$ describes a smooth merging of z_2 with the point 1, in the limit $z_2 \to 0$ the other points z_i are squeezed in the vanishing interval close to zero

consideration of eigenspaces and eigenvalues of the braid matrix $x^{(1)}$. Furthermore, when calculating results for Selberg integrals using Eq. (46), one has to limit the size of the system: the infinitely long vector $\mathbf{S}^\tau(z_2)$ has to be cut to finite length, i.e. one needs to consider entries up to a certain weight $w_{\max}$ only. The maximal weight, in turn depends on the order in α' the expansion of the Selberg integral shall be calculated. Dependencies and the process of cutting the system to finite size is again carefully examined in ref. [40], but will be used in the example in Sect. 3.5 below.

Similarly to the previous genus-zero section, the regularized boundary value $\mathbf{C}_1^\tau$ can be shown to contain $(L-1)$-point configuration-space integrals at genus one [37, 43–45] whereas $\mathbf{C}_0^\tau$ contains $(L+1)$-point configuration-space integrals at genus zero. Accordingly, the N-point configuration-space integrals appearing in open-string amplitudes at genus one can be calculated from the $(N+2)$-point integrals at genus zero via Eq. (46), with $N=L-1$. This is going to be exemplified in the next subsection.

3.5 *Recursive Evaluation of Two-Point Open-String Integrals at Genus One*

The successful concept for the calculation of genus-zero string-integrals from Selberg integrals, will be extended to genus one here. One-loop open-string amplitudes are calculated on a genus-one Riemann surface with boundary. Setting up the string correlation function, the problem can be divided in a polarization part and configuration-space integrals. Omitting the (rather straightforward) polarization part, we will furthermore limit our attention to those configuration-space integrals where points are inserted on one boundary only.

Instead of developing the full theory here, let us present the easiest nontrivial example: the two-point case, which we would like to calculate to second order in α'. The two-loop correction yields non-trivial results only, if the Mandelstam variables s_{ij} are treated as independent parameters of the integrals, which do not satisfy any constraints like momentum conservation.

The two-point configuration-space integral reads [45]

$$S^{\text{1-loop}}_{\text{2-point}}(\tilde{s}_{13}) = \int_0^1 dz_3 \exp\left(\tilde{s}_{13}\tilde{\Gamma}_{31}\right) = \sum_{n\geq 0} \tilde{s}_{13}^n\, \omega(\underbrace{1,\ldots,1}_{n},0)\,, \tag{48}$$

where the Mandelstam variable $\tilde{s}_{13}$ is associated to the loop momentum. Requiring two vertex insertion, the appropriate genus-one Selberg integral with an extra insertion point z_2 has length $L=3$ and the insertion points on the cylinder boundary are ordered as

$$0 = z_1 < z_3 < z_2 < 1 \equiv z_1 \bmod \mathbb{Z}\,. \tag{49}$$

In the limit $z_2 \to 1$, the punctures z_2 and z_1 merge, leaving us with two punctures for the one-loop string corrections. Accordingly, we are advised to consider the integrals

$$\begin{aligned} \mathrm{S}^\tau\!\left[\begin{smallmatrix} n_3 \\ i_3 \end{smallmatrix}\right]\!(0,z_2) &= \int_0^{z_2} dz_3\, \mathrm{S}^\tau\, g^{(n_3)}_{3i_3}\,, \qquad 1 \leq i_3 < 3\,, \\ \mathrm{S}^\tau &= \exp\left(s_{13}\tilde{\Gamma}_{31} + s_{12}\tilde{\Gamma}_{21} + s_{23}\tilde{\Gamma}_{23}\right)\,. \end{aligned} \tag{50}$$

In the same way as the components of the vector $\mathbf{S}^\tau(z_2)$ are ordered by weight, so are the vectors $\mathbf{C}_1^\tau$ and $\mathbf{C}_0^\tau$. While the two-point one-loop correction is contained in the weight-zero entry, the tree-level correction can be found in the weight-one component. Sorting out the details, the goal of calculating up to second order in α' implies maximal weight two in the KZB system.

At this point, we would like to refer the reader to ref. [40] for (actually a lot of) careful derivation and write down the explicit two-point realization of Eq. (46) right away:

$$\begin{pmatrix} S^{\text{1-loop}}_{\text{2-point}}(\tilde{s}_{13}) \\ * \\ * \\ * \end{pmatrix} + O\big((\alpha')^3\big) = \Phi_3^\tau(x^{(n)}_{\leq 2}) \begin{pmatrix} 0 \\ \frac{1}{s_{13}} \frac{\Gamma(1+s_{13})\Gamma(1+s_{23})}{\Gamma(1+s_{13}+s_{23})} \\ 0 \\ 0 \end{pmatrix}, \tag{51}$$

where $\tilde{s}_{13} = s_{13} + s_{23}$ and only a finite part of the associator (cf. Eq. (47)) has to be determined. As indicated by the subscript of the associator in Eq. (51), to calculate the one-loop configuration-space integral up to the second order in α', products of at most three (cut) matrices $x^{(n)}$ have to be included, which are given by:

$$x^{(0)}_{\leq 2} = \begin{pmatrix} 0 & s_{13} & 0 & 0 \\ 0 & 0 & -s_{23} & -s_{23} \\ 0 & 0 & 0 & 0 \\ 0 & 0 & 0 & 0 \end{pmatrix}, \quad x^{(1)}_{\leq 2} = \begin{pmatrix} s_{12} & 0 & 0 & 0 \\ 0 & s_{123} & 0 & 0 \\ 0 & 0 & s_{12}+s_{23} & -s_{23} \\ 0 & 0 & -s_{13} & s_{12}+s_{13} \end{pmatrix} \tag{52}$$

and

$$x^{(2)}_{\leq 2} = \begin{pmatrix} 0 & 0 & 0 & 0 \\ -s_{23} & 0 & 0 & 0 \\ 0 & s_{13} & 0 & 0 \\ 0 & s_{13} & 0 & 0 \end{pmatrix}. \tag{53}$$

Putting everything together, the relevant subpart of the matrix Eq. (51) reads

$$\begin{aligned} S^{\text{1-loop}}_{\text{2-point}}(\tilde{s}_{13}) + O\Big((\alpha')^3\Big) &= \Phi^\tau_3\Big(x^{(n)}_{\leq 2}\Big)_{0,1} \frac{1}{s_{13}} \frac{\Gamma(1+s_{13})\Gamma(1+s_{23})}{\Gamma(1+s_{13}+s_{23})} \\ &= 1 + (s_{13}+s_{23})\omega(1,0) + (s_{13}+s_{23})^2\omega(1,1,0) + O\Big((\alpha')^3\Big), \end{aligned} \tag{54}$$

where the subindex on the associator Φ specifies the appropriate matrix component. Nicely enough, this reproduces indeed the two-point one-loop string correction $S^{\text{1-loop}}_{\text{2-point}}(\tilde{s}_{13})$ given in Eq. (48) with the effective Mandelstam variable $\tilde{s}_{13} = s_{13}+s_{23}$ up to second order in α'.

3.6 Geometric Interpretation

What is the geometric meaning of the two limits $z_2 \to 0$ and $z_2 \to 1$ in the genus-one case? The latter limit has an easy explanation: the merging of the point z_2 with the point $0 \equiv 1$ happens in exactly the right way as to yield a finite result from two competing processes: the regularization of the boundary value and the behavior of the function $\tilde{\Gamma}$ for z_2 close to one. The resulting geometry is just the same as one has been starting with: just a point less. More involved is the other limit: when z_2 tends to zero, all other insertion points z_3 to z_L are squeezed in the (infinitesimal) interval $(0, z_2)$. Effectively, this limit amounts to *shortening* the elliptic A-cycle: this implies that the modular parameter τ, which is the ratio of the lengths of B- and A-cycle becomes very large (cf. Fig. 1). Simultaneously, as can be justified by

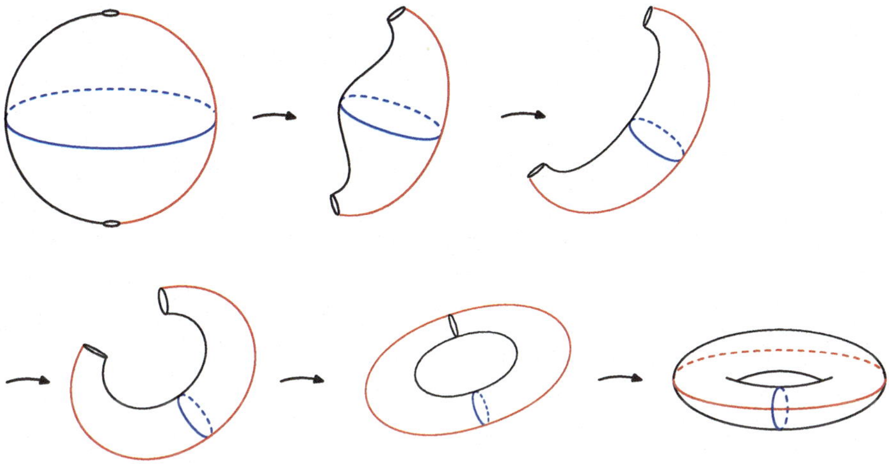

Fig. 3 Step by step morphing of the Riemann sphere to the torus by joining infinitesimal circles at north- and south pole. The reverse process is modeled by the limit $\tau \to i\infty$

an integral transformation [40], from the perspective of the Riemann sphere the red line (which used to be the A-cycle before) becomes infinitely long, resulting in half a great circle (the positive real axis) on the Riemann sphere (see Fig. 3).

4 General Framework and Outlook

4.1 What Does It Need for a General Recursion?

The two recursive algorithms reviewed above have several structural commonalities, however, are rather different when considering limits of the respective associated differential equations.

In both formalisms, an algebraic variety is taken as starting point and an integrable connection is associated. This connection is built from differential forms with at most simple poles, thus leading to logarithmic singularities after iterated integration. The differential forms incorporate the periodicities/cycles of the algebraic variety in question. Symmetries of the variety, for example the choice of origin, are implemented by fixing a couple of positions in the Selberg integral, which simultaneously singles out a canonical path for the integration using homotopy invariance.

In a next step, a differential equation with respect to an auxiliary point shall be established. For simplicity, let us assume the iterated integration to happen in the interval [0, 1], which is divided by several insertion points. (This is the case for both recursions discussed above.) In both scenarios, the auxiliary point is placed between the largest insertion point and one.

Once the auxiliary point, which is a parameter not to be integrated over, has been supplemented in a set of integrals, one can now identify a basis set of integrals and determine the derivative. It is not yet clear, what a necessary or sufficient condition for closure of this system of differential equations is: in the two scenarios above we have just been lucky (or standard enough).

To this end, one shall consider the boundary values. In an intricate interplay between regularization of the integrals, the regularization of the boundary values one can relate iterated integrals without auxiliary point featuring different numbers of insertion points and thus integrations.

While all of the above considerations have been fairly general, the geometric interpretation finally depends on the particular surface in question, on particular on its cycles. The geometric picture incorporated by taking the two limits of the differential equation in the genus-one case are discussed in Sect. 3.6 above.

While there are several further examples, where a similar approach has been successful, let us here mention the recent calculation of the maximal cut of multiloop banana amplitudes in refs. [46, 47]. The ingredients here are very similar: there is a (slightly more complicated) algebraic variety: a Calabi-Yau manifold, a Picard-Fuchs type differential equation (this time without auxiliary point), a basis set of integrals determined from the cohomology of the Calabi-Yau manifold. As turns out, the ideal of this Picard-Fuchs is a Gelfand–Kapranov–Zelevinsky(GKZ)-system, which delivers the desired result.

A final remark is in place here: Feynman integrals are associated to graphs with edges, while string amplitudes are expressed as correlation functions on two-dimensional worldsheets. Considering the results, however, there is always a way to replace the Feynman expression with a set of iterated integrals naturally defined on a Riemann surface. Even more: when taking (dimensional) regularization into account, the result of calculating a particular scattering process using the Feynman formalism will be a double expansion: the topological expansion in the number of loops ℓ and the expansion in the parameter ϵ of dimensional regularization. On the contrary, evaluating a string correlator in order to model string scattering, the result will be again a double-expansion: the topological expansion parametrized by the string coupling g_S and the expansion in the inverse string tension α'. It remains to be explored throughout the next years, whether those two double expansions can be related. Clearly, individual Feynman diagrams lead to divergent integrals, whose divergences cancel in the final, physical result only. This not being the case for string amplitudes points into the direction of a singular transformation. However, the idea of interpreting/identifying "stringyness" simply as a regulating mechanism, which comes across very naturally, is rather appealing.

Acknowledgments Both authors would like to thank the Kolleg Mathematik und Physik Berlin for supporting the conference "Antidifferentiation and the Calculation of Feynman Amplitudes". AK would like to thank the IMPRS for Mathematical and Physical Aspects of Gravitation, Cosmology and Quantum Field Theory, of which he is a member and which renders his studies possible.

References

1. A.B. Goncharov, M. Spradlin, C. Vergu, A. Volovich, Phys. Rev. Lett. **105**, 151605 (2010)
2. J. Ablinger, J. Blümlein, A. De Freitas, M. van Hoeij, E. Imamoglu, C. Raab, C. Radu, C. Schneider, J. Math. Phys. **59**(6), 062305 (2018).
3. J. Brödel, C. Duhr, F. Dulat, L. Tancredi, JHEP **05**, 093 (2018)
4. A. Kotikov, Phys. Lett. B **254**, 158 (1991)
5. E. Remiddi, Nuovo Cim. A **110**, 1435 (1997)
6. T. Gehrmann, E. Remiddi, Nucl. Phys. B **580**, 485 (2000)
7. J.M. Henn, Phys. Rev. Lett. **110**, 251601 (2013)
8. G. Puhlfürst, S. Stieberger, Nucl. Phys. B **902**, 186 (2016)
9. G. Puhlfürst, S. Stieberger, Nucl. Phys. B **906**, 168 (2016)
10. A. Goncharov, *Multiple Polylogarithms and Mixed Tate Motives*. math/0103059[math.AG]
11. D. Zagier, Ann. Math. **175**(2), 977 (2012)
12. C.R. Mafra, O. Schlotterer, S. Stieberger, Nucl. Phys. B **873**, 419 (2013)
13. C.R. Mafra, O. Schlotterer, S. Stieberger, Nucl. Phys. B **873**, 461 (2013)
14. J. Brödel, O. Schlotterer, S. Stieberger, Fortsch. Phys. **61**, 812 (2013)
15. G. Veneziano, Nuovo Cim. A **57**, 190 (1968)
16. S. Mandelstam, Phys. Rept. **13**, 259 (1974)
17. F. Brown, C. Dupont, *Single-Valued Integration and Superstring Amplitudes in Genus Zero*. arXiv:1910.01107[math.NT]
18. A. Selberg, Norsk Mat. Tidsskrift **26**, 71 (1944)
19. K. Aomoto, J. Math. Soc. Japan **39**(2), 191 (1987)
20. T. Terasoma, Compos. Math. **133**, 1 (2002)
21. V.G. Knizhnik, A.B. Zamolodchikov, Nucl. Phys. B **247**, 83 (1984)
22. V.G. Drinfeld, Algebra i Analiz **1**(6), 114 (1989)
23. V. Drinfeld, Leningrad Math. J. **2**(4), 829 (1991)
24. T. Le, J. Murakami, Nagoya Math J. **142**, 93 (1996)
25. S.J. Parke, T. Taylor, Phys. Rev. Lett. **56**, 2459 (1986).
26. J. Brödel, O. Schlotterer, S. Stieberger, T. Terasoma, Phys. Rev. D **89**(6), 066014 (2014)
27. C.R. Mafra, O. Schlotterer, JHEP **01**, 031 (2017)
28. A. Kaderli, J. Phys. A **53**(41), 415401 (2020)
29. A. Levin, Compos. Math. **106**, 267 (1997)
30. F. Brown, A. Levin, *Multiple Elliptic Polylogarithms*. arXiv:1110.6917v2 [math.NT]
31. A. Levin, G. Racinet, (2007). *Towards Multiple Elliptic Polylogarithms*. arXiv:math/0703237
32. D. Mumford, M. Nori, P. Norman, *Tata Lectures on Theta I, II*. Bd. **2** (Birkhäuser, Basel, 1983, 1984)
33. P. Deligne, Le Groupe Fondamental de la Droite Projective Moins Trois Points, in *Galois Groups Over*, vol. 16, ed. by Y. Ihara, K. Ribet, J.P. Serre. Mathematical Sciences Research Institute Publications (Springer, New York, 1989), pp. 79–297
34. F. Brown, Motivic periods and the projective line minus three points, in *Proceedings of the ICM 2014*, vol. 2, ed. by S.Y. Jang Y.R. Kim, D.-W. Lee, I. Yie (2014), pp. 295–318
35. B. Enriquez, Bull. Soc. Math. France **144**(3), 395 (2016)
36. N. Matthes, Elliptic multiple zeta values. Ph.D. thesis, Universität Hamburg, 2016
37. J. Brödel, C.R. Mafra, N. Matthes, O. Schlotterer, JHEP **07**, 112 (2015)
38. J. Brödel, N. Matthes, O. Schlotterer, J. Phys. A **49**(15), 155203 (2016)
39. J. Brödel, N. Matthes, O. Schlotterer, https://tools.aei.mpg.de/emzv
40. J. Brödel, A. Kaderli, *Amplitude Recursions with an Extra Marked Point*. arXiv:1912.09927[hep-th]
41. D. Bernard, Nucl. Phys. B **303**, 77 (1988)
42. D. Bernard, Nucl. Phys. B **309**, 145 (1988)
43. M.B. Green, J.H. Schwarz, L. Brink, Nucl. Phys. B **198**, 474 (1982)
44. L. Dolan, P. Goddard, Commun. Math. Phys. **285**, 219 (2009)

45. C.R. Mafra, O. Schlotterer, JHEP **03**, 007 (2020)
46. A. Klemm, C. Nega, R. Safari, JHEP **04**, 088 (2020)
47. K. Bönisch, F. Fischbach, A. Klemm, C. Nega, R. Safari, *Analytic Structure of all Loop Banana Amplitudes*. arXiv:2008.10574[hep-th]

Differential Galois Theory and Integration

Thomas Dreyfus and Jacques-Arthur Weil

Abstract In this chapter, we present methods to simplify reducible linear differential systems before solving. Classical integrals appear naturally as solutions of such systems. We will illustrate the methods developed in Dreyfus and Weil (Computing the Lie algebra of the differential Galois group: The reducible case, ArXiv **1904.07925** 2019) on several examples to reduce the differential system. This will give information on potential algebraic relations between integrals.

Keywords Ordinary differential equations · Differential Galois theory · Computer algebra · Integrals · Lie algebras · D-finite functions

1 Introduction

In this chapter, we will review properties of block triangular linear differential systems and their use to compute properties of integrals.

Let $\mathbf{k} = \mathbb{C}(x)$ and $A \in \mathrm{Mat}(n, \mathbf{k})$. We will study the corresponding *linear differential system* $[A] : \partial_x Y = AY$. More generally, we might consider linear differential systems over a differential field $(\mathbf{k}, \partial)$ of characteristic zero, that is a field $\mathbf{k}$ equipped with an additive morphism ∂ satisfying the Leibniz rule $\partial(ab) = a\partial(b) + \partial(a)b$.

The Galois theory of linear differential equations aims at understanding what are the algebraic relations between the solutions of $[A]$. We attach to $[A]$ a group that measures the relations. The computation of this differential Galois group is a hard

T. Dreyfus
Institut de Recherche Mathématique Avancée, U.M.R. 7501 Université de Strasbourg et C.N.R.S., Strasbourg, France
e-mail: dreyfus@math.unistra.fr

J.-A. Weil (✉)
XLIM, U.M.R 7252, Université de Limoges et C.N.R.S, Limoges Cedex, France
e-mail: Jacques-Arthur.Weil@unilim.fr

J. Blümlein, C. Schneider (eds.), *Anti-Differentiation and the Calculation of Feynman Amplitudes*, Texts & Monographs in Symbolic Computation,
https://doi.org/10.1007/978-3-030-80219-6_7

task in full generality. The goal of this chapter is to illustrate on examples the method described in [1] that focuses on the reduction of block-triangular linear differential systems. This approach is powerful enough to understand the desired relations on the solutions.

Given an invertible matrix $P \in \mathrm{GL}(n, \mathbf{k})$, the linear change of variables $Y = PZ$ produces a new differential system denoted $Z' = P[A]Z$ where

$$P[A] := P^{-1}AP - P^{-1}P'.$$

Two linear differential systems $[A]$ and $[B]$ are called *(gauge) equivalent* over $\mathbf{k}$ if there exists a gauge transformation, an invertible matrix $P \in \mathrm{GL}(n, \mathbf{k})$, such that $B = P[A]$. A linear differential system is called *reducible* (over $\mathbf{k}$) when it is gauge equivalent to a linear differential system in block triangular form:

$$[A] : \partial Y = AY, \text{ with } A = \left(\begin{array}{c|c} A_1 & 0 \\ \hline S & A_2 \end{array}\right) \in \mathrm{Mat}(\mathbf{k}). \tag{1}$$

It turns out that computing properties of integrals of D-finite functions[1] or of some types of iterated integrals may be reduced to computing solutions of such reducible linear differential systems. Computing the differential Galois groups of block triangular systems gives, in turn, information on properties of their solutions. This idea was promoted by Bertrand [2] or Berman and Singer [3] who showed how to compute Galois groups of some reducible systems and how this would reveal algebraic properties of integrals.

Our aim, in this chapter, is to show how to compute such algebraic properties. The underlying theory is developed in [1] and [4]; general references for differential Galois theory are for example [5–7]; general references for constructive theory of reduced forms of differential systems are [8–12]. We will rely on many examples rather than on cumbersome theory and provide references for interested readers. For example, we will consider the dilogarithm function Li_2 defined by $\mathrm{Li}_2'(x) = -\frac{\ln(1-x)}{x}$ and our 4-dimensional example (in the next sections) will provide a simple algorithmic proof that it is not only transcendent but algebraically independent of e^x, $\ln(x)$ and $\ln(1-x)$; the proof will only use rational solutions of linear first order differential equations.

The chapter is organized as follows. We begin by some examples to illustrate what the method will provide. Then we give a review of differential Galois theory and reduced forms of differential systems. We finish by explaining the strategy of [1] on two examples that are chosen so that almost all calculations can be reproduced easily.

[1] A function is *D-finite* when its is a solution of a linear differential equations with coefficients in $\mathbf{k}$.

2 Examples

2.1 A First Toy Example

We consider two confluent Heun[2] functions

$$f_1(x) = \exp\left(\frac{\sqrt{3}}{12}x\right)\mathrm{HeunC}\left(\frac{\sqrt{3}}{6}, -\frac{1}{3}, -\frac{1}{3}, \frac{1}{48}, \frac{11}{48}; x\right)$$
$$= 1 - \frac{7}{96}x - \frac{719}{46080}x^2 - \frac{127307}{10616832}x^3 - \frac{82319293}{10192158720}x^4 + O\left(x^5\right)$$

and

$$f_2(x) = \exp\left(\frac{\sqrt{3}}{12}x\right)\sqrt[3]{x}\mathrm{HeunC}\left(\frac{\sqrt{3}}{6}, \frac{1}{3}, -\frac{1}{3}, \frac{1}{48}, \frac{11}{48}; x\right)$$
$$= x^{\frac{1}{3}}\left(1 + \frac{25}{192}x + \frac{8977}{129024}x^2 + \frac{1099183}{26542080}x^3 + O(x^4)\right).$$

They form a basis of solutions of the second order linear differential equation

$$L(y) := \frac{\mathrm{d}^2}{\mathrm{d}x^2}y(x) + \frac{2}{3}\left(\frac{1}{x} + \frac{1}{x-1}\right)\frac{\mathrm{d}}{\mathrm{d}x}y(x) - \frac{(3x^2 - 6x + 7)}{144x(x-1)}y(x) = 0$$

The Wronskian relation gives us the algebraic relation $(f_1 f_2' - f_1' f_2)^3 = x^2(x-1)^2$. The equation has order two so the Kovacic algorithm [14–16] can be used to compute the differential Galois group and we find that no other algebraic relations exist between f_1, f_2, f_1' and f_2'. Now let $F_i(x) := \int^x f_i(t)dt$ be a primitive. We want to determine whether F_1 and F_2 are algebraically independent of the f_i and f_i' or not. The techniques explained below will show that this question reduces to asking whether there is a *rational* solution to the linear differential system

$$Z' = \begin{pmatrix} 0 & \frac{-3x^2+6x-7}{144x(x-1)} \\ -1 & \frac{1}{3}\frac{4x-2}{x(x-1)} \end{pmatrix} \cdot Z + \begin{pmatrix} 1 \\ 0 \end{pmatrix}$$

or, equivalently, whether the following linear differential equation (the left hand side turns out to be the adjoint operator of L) has a rational solution:

$$-g''(x) + \frac{2}{3}\frac{(2x-1)}{x(x-1)}g'(x) + \frac{(3x^4 - 9x^3 - 179x^2 + 185x - 96)}{144x^2(x-1)^2}g(x) = 1.$$

[2]See https://dlmf.nist.gov/31.12 or [13]. The notation *HeunC* is the syntax in Maple.

It can be seen directly or with a computer algebra system that this equation has no rational solution. The underlying theoretical tools are from the constructive differential Galois theory. However, in operational terms, it is rather easy to compute and check: no hard theory is required for the calculation. Let us unveil a corner of the underlying tools.

We have in fact studied the differential Galois group of the differential system $[A]$ with

$$A = \begin{pmatrix} 0 & 1 & 0 \\ \frac{3x^2-6x+7}{144(x-1)^3x^2} & -\frac{2}{3}\left(\frac{1}{x-1}+\frac{1}{x}\right) & 0 \\ 1 & 0 & 0 \end{pmatrix}$$

which admits the fundamental solution matrix

$$U := \begin{pmatrix} f_1(x) & f_2(x) & 0 \\ f_1'(x) & f_2'(x) & 0 \\ \int^x f_1(t)\mathrm{d}t & \int^x f_2(t)\mathrm{d}t & 1 \end{pmatrix}.$$

Our calculation of rational solution above shows (with the tools displayed below) that the differential Galois group of the system $[A]$ will have dimension 5. This in turn shows that the integrals $\int^x f_j(t)\mathrm{d}t$ are algebraically independent of f_1, f_2 and their derivatives. In fact, it even shows that both integrals are algebraically independent.

2.2 A Second Toy Example

Recall that the hypergeometric function is given by the formula

$${}_2F_1([a,b],[c])(x) = \sum_{n=0}^{\infty} \frac{(a)_n(b)_n}{(c)_n}\frac{x^n}{n!}, \quad \text{with} \quad (a)_0 = 1, \ (a)_n = a(a+1)\dots(a+n-1).$$

We now take two hypergeometric functions (with nice modular properties)

$$f_1(x) = {}_2F_1\left(\left[-\frac{1}{3}, \frac{1}{12}\right], \left[\frac{7}{12}\right]\right)(x)$$

and

$$f_2(x) = x^{5/12}{}_2F_1\left(\left[\frac{1}{12}, \frac{1}{2}\right], \left[\frac{17}{12}\right]\right)(x).$$

The vectors $Y_i := (f_i, f_i')^T$ are solutions of $Y' = A_1 Y$ below. To study properties of their integrals, we set $Y := (f, f', \int^x f(t)dt)^T$ and we have $Y' = AY$ where

$$A_1 = \begin{pmatrix} 0 & 1 \\ \frac{1}{36}\frac{1}{x(x-1)} & -\frac{7}{12x} - \frac{1}{6(x-1)} \end{pmatrix} \text{ and } A = \begin{pmatrix} A_1 & 0 \\ 1\,0 & 0 \end{pmatrix}.$$

We will see in the sequel how we may find a suitable change of variables

$$Q := \begin{pmatrix} 1 & 0 & 0 \\ 0 & 1 & 0 \\ -\frac{15}{44} + \frac{45x}{44} & -\frac{9x(x-1)}{11} & 1 \end{pmatrix}$$

such that

$$Q[A] = \begin{pmatrix} A_1 & 0 \\ 0\,0 & 0 \end{pmatrix}.$$

This shows that

$$\int^x f_i(t)\,\mathrm{dt} = -\frac{9}{11}x(x-1)f_i'(x) + \frac{15}{44}(3x-1)f_i(x) + c_i, \quad c_i \in \mathbb{C},$$

and the differential Galois group of $[A]$ has dimension 3. We note that none of these two examples is new and that efficient methods to handle these questions have been developed by Abramov and van Hoeij in [17, 18].

2.3 Integrals via Reducible Systems

These two examples have shown how, by augmenting the dimension of a linear differential system, we can study integrals of its solutions. Conversely, block triangular systems give rise to integrals via variation of constants; we review this for clarification. For a factorized reducible system $[A]$ of the form

$$A = \left(\begin{array}{c|c} A_1 & 0 \\ \hline S & A_2 \end{array}\right) = A_{diag} + A_{sub},$$

with $A_{diag} := \left(\begin{array}{c|c} A_1 & 0 \\ \hline 0 & A_2 \end{array}\right)$ and $A_{sub} := \left(\begin{array}{c|c} 0 & 0 \\ \hline S & 0 \end{array}\right)$, we have a fundamental solution matrix of the form

$$U = \left(\begin{array}{c|c} U_1 & 0 \\ \hline U_2 V & U_2 \end{array}\right) = \left(\begin{array}{c|c} U_1 & 0 \\ \hline 0 & U_2 \end{array}\right)\left(\begin{array}{c|c} \mathrm{Id}_{n_1} & 0 \\ \hline V & \mathrm{Id}_{n_2} \end{array}\right).$$

Once U_1 and U_2 are known, V is given by integrals: $\partial V = U_2^{-1} S U_1$. So it may seem that no further theory is required. However, U_1 and U_2 are fundamental solution matrices for systems $\partial U_i = A_i U_i$ so the relation $\partial V = U_2^{-1} S U_1$ may involve integrals of complicated D-finite functions.

Our approach will be to first "reduce" S as much as possible, using manipulations with rational functions, to prepare the system for easier solving. In return, we will obtain algebraic information on all (possibly iterated) integrals that may occur in the relation $\partial V = U_2^{-1} S U_1$.

We note, for the record, that factorized differential operators also correspond to block triangular differential systems.

2.4 *Example of Situations Involving Reducible Linear Differential Systems*

2.4.1 Operators from Statistical Physics and Combinatorics

In many models attached to statistical mechanics, quantities are expressed as multiple integrals depending on a parameter. They are generally holonomic in this parameter, meaning that they are D-finite, i.e. they are solutions to linear differential operators (for example, the so-called Ising operators). A description of this setting may be found in the books of Baxter [19] and McCoy [20] (notably Chapters 10 and 12) or in the surveys [21, 22] (and references therein). Similarly, in combinatorics, sequences that satisfy recurrence relations may be studied through their generating series, which are often D-finite.

Experimentally (see, for example, [23–28] or [29–32]), it turns out that many differential operators coming from these processes are factored into a product of smaller order factors—and the corresponding companion systems are reducible.

Other cases of reducible systems are the ones admitting reducible monodromies such as [33] or in the works of Kalmykov on Feynman integrals via Mellin-Barnes integrals [34–38]; about differential equations for Feynman integrals, one may consult the works of Smirnov [39, 40].

Finally, we mention the paper [41] with other examples of integrals where the techniques presented below may offer alternative approaches for some of the computations.

2.4.2 Variational Equations of Nonlinear Differential Systems

Another natural source of reducible systems is the old method of *variational equations* of nonlinear differential systems along a given particular solution ϕ. One can form the linear differential equation describing perturbations along this solution ϕ. The general principle is that obstructions to integrability of the nonlinear system can be read on this linear differential system. Ziglin [42] linked non-integrability to non-commutations in the monodromy group with concrete versions given e.g. in [43] and [44]. It was generalized in the theory of Morales-Ruiz and Ramis and extended with Simó [45, 46] for Hamiltonian systems; they prove that a Hamiltonian system is completely integrable only if all its variational equations have a virtually abelian differential Galois group. Extensions to nonhamiltonian differential equations can be found in [4, 47]. Applications to specific problems have been occasions to establish effective criteria. For example: the three body problem [48, 49], n-body problems [50, 51], Hill systems (movements of the moon) [52, 53] or a swinging Atwood machine [54].

These variational equations can be written in the form of reducible linear differential systems of big dimension. The simplification techniques outlined below are hence particularly relevant to make computation on such systems practical (see [10]).

We note that the Morales-Ramis theory has had a spectacular recent development, initiated in [55] where this variational approach is applied to path integrals thus establishing a beautiful and unexpected bridge with the previous subsection.

Reducible linear differential systems appear naturally in another type of perturbative approach: in ϵ-expansions of solutions for perturbed systems $\frac{d}{dx}Y = B(x, \epsilon)Y$ like the ones that appear in works of J. Blümlein, C. Raab, C. Schneider, J. Henn and others for example.

3 Reduced Forms of Linear Differential Systems

3.1 Ingredient #1: Differential Galois-Lie Algebra

In what follows, $(\mathbf{k}, \partial)$ is a differential field of characteristic 0. We outline a brief exposition of the Galois theory of linear differential equations, see [5–7] for expositions with proofs. We consider a linear differential system

$$[A]: \quad \partial Y = AY, \quad A \in \mathrm{Mat}(n, \mathbf{k}). \tag{2}$$

Let C be the field of constants of the differential field $\mathbf{k}$, that is $C := \{a \in \mathbf{k} | \partial(a) = 0\}$. We will assume that C is algebraically closed, i.e, every non constant polynomial equation has a solution in C.

A *Picard-Vessiot extension* is a field $K = \mathbf{k}(U)$, where U is a fundamental solution matrix[3] of $[A]$, such that the field of constants of K is still C. This can be constructed algebraically; alternatively, when $\mathbf{k}$ is a field of meromorphic functions, one may consider a local matrix of power series solutions at a regular point and this gives a Picard-Vessiot extension. A Picard-Vessiot extension is unique modulo differential field isomorphisms.

[3] An invertible $n \times n$ matrix U such that $U' = AU$.

The *differential Galois group* $G := \mathrm{Aut}_\partial(K/\mathbf{k})$ is the set of automorphisms of K which leave the base field $\mathbf{k}$ fixed and commute with the derivation. Let $\sigma \in G$. By construction, $\sigma(U)$ is also a fundamental solution matrix in K and we find that there exists a matrix $[\sigma] \in \mathrm{GL}(n, C)$ such that $\sigma(U) = U \cdot [\sigma]$. The map $\sigma \mapsto [\sigma]$ provides a faithful representation of G as a subgroup of $\mathrm{GL}(n, C)$, actually a linear algebraic group. If we change the fundamental solution, we obtain a conjugate representation.

We recall that, given an invertible matrix $P \in \mathrm{GL}(n, \mathbf{k})$, the linear change of variables $Y = PZ$ produces a new differential system denoted $\partial Z = P[A]Z$ where

$$P[A] := P^{-1}AP - P^{-1}\partial P.$$

We note that such a gauge transformation $P[A]$, with $P \in \mathrm{GL}(n, \mathbf{k})$, does not change the Galois group.

Given a Picard-Vessiot extension $K = \mathbf{k}(U)$, the polynomial relations among all entries of U (over $\mathbf{k}$) form an ideal I. The Galois group G can then be viewed as the set of matrices which stabilize this ideal I of relations. Thus the computation of G is strongly related to the understanding of the algebraic relations among the solutions.

The *Galois-Lie algebra of* $[A]$ is the Lie algebra $\mathfrak{g}$ of the differential Galois group G. It is defined as the tangent space of G at the identity Id. The dimension of $\mathfrak{g}$ measures the transcendence degree of K over $\mathbf{k}$, that is

$$\dim_C \mathfrak{g} = \mathrm{trdeg}(K/\mathbf{k}).$$

One way of computing the Lie algebra [5] is the following: $\mathfrak{g}$ is the set of matrices N such that $\mathrm{Id} + \epsilon N \in G(C[\epsilon])$ with $\epsilon^2 = 0$. In other terms, $\mathrm{Id} + \epsilon N$ satisfies the defining equations of the group modulo ϵ^2. For example, $\mathrm{SL}(n, C)$ (set of M such that $\det(M) = 1$) gives the Lie algebra $\mathfrak{sl}(n, C)$ of matrices N such that $\mathrm{Tr}(N) = 0$. The symplectic group $\mathrm{Sp}(2n, C)$ is the set of M such that $M^T \cdot J \cdot M = J$; its Lie algebra $\mathfrak{sp}(n, C)$ is found to the set of N such that $N^T J + JN = 0$, with $J = \begin{pmatrix} 0 & \mathrm{Id} \\ -\mathrm{Id} & 0 \end{pmatrix}$. The additive group $\mathbb{G}_a := \left\{ \begin{pmatrix} 1 & a \\ 0 & 1 \end{pmatrix}, a \in C \right\}$ admits the Lie algebra $\mathfrak{g}_a = Span_C \left\{ \begin{pmatrix} 0 & 1 \\ 0 & 0 \end{pmatrix} \right\}$; the multiplicative group $\mathbb{G}_m = \left\{ \begin{pmatrix} a & 0 \\ 0 & \frac{1}{a} \end{pmatrix}, a \in C^* \right\}$ admits the Lie algebra $\mathfrak{g}_m = Span_C \left\{ \begin{pmatrix} 1 & 0 \\ 0 & -1 \end{pmatrix} \right\}$.

The reduction technique exposed in this chapter aims at computing directly the Galois-Lie algebra $\mathfrak{g}$ before computing G itself. Although the theory is not obvious, the resulting calculations are reasonably simple.

3.2 *Ingredient #2: Lie Algebra* Lie(A) *Associated to* A

The Lie algebra $\mathrm{Lie}(A)$ associated to the matrix A is defined as follows. Let $a_1, \ldots, a_s \in \mathbf{k}$ be a basis of the C-vector space generated by the coefficients of

A. We can then decompose A as $A = \sum_{i=1}^{s} a_i M_i$ where the M_i are constant matrices. Now, we consider the smallest Lie algebra containing all the M_i: this is the vector space generated by the M_i and all their iterated Lie brackets ($[M, N] := MN - NM$); then we take its algebraic envelope.

Definition 1 The Lie algebra Lie(A) associated to the matrix A is the smallest *algebraic*[4] Lie algebra containing all the M_i.

The decomposition $A = \sum_{i=1}^{s} a_i M_i$ is not unique but the vector space generated by the M_i is unique. Thus, the associated Lie algebra Lie(A) does not depend on the chosen decomposition.

This Lie algebra Lie(A) appears in works of Magnus [57] or Feynman who use the Baker-Campbell-Hausdorff formula to write solutions of $\partial Y = AY$ as (infinite) products of exponentials constructed with Lie brackets. Wei and Norman give in [58, 59] a finite formula to solve the system when Lie(A) is solvable. This formula is well-known in physics and control theory but not as well among mathematicians. The terminology of Lie algebra associated to A appears[5] in [58, 59] (in there, it is defined as the Lie algebra generated by all values of $A(z_0)$ for z_0 spanning all constants minus singularities and the algebraic envelope is missing). In the sequel, we will study a 4-dimensional example and an 8-dimensional example where our technique will have some relations to the Wei-Norman approach; namely, we will change A to obtain an associated Lie algebra Lie(A) of minimal dimension so that solving formulas become optimal in some sense.

Example 1 (A 4-Dimensional Example) Let

$$A := \begin{pmatrix} 1 & \frac{1}{x} & \frac{1}{x-1} & 0 \\ 0 & 1 & 0 & \frac{1}{x-1} \\ 0 & 0 & 1 & -\frac{1}{x} \\ 0 & 0 & 0 & 1 \end{pmatrix}.$$

Note that this system is upper triangular, contrary to (1). This will illustrate that our method can be equivalently applied to upper and lower triangular systems. We obtain a Wei-Norman decomposition $A = M_1 + \frac{1}{x}M_2 + \frac{1}{x-1}M_3$, where

$$M_1 = \begin{pmatrix} 1&0&0&0 \\ 0&1&0&0 \\ 0&0&1&0 \\ 0&0&0&1 \end{pmatrix}, \quad M_2 := \begin{pmatrix} 0&1&0&0 \\ 0&0&0&0 \\ 0&0&0&-1 \\ 0&0&0&0 \end{pmatrix}, \quad M_3 := \begin{pmatrix} 0&0&1&0 \\ 0&0&0&1 \\ 0&0&0&0 \\ 0&0&0&0 \end{pmatrix}.$$

[4] A Lie algebra is called algebraic when it is the Lie algebra of a linear algebraic group. When the M_i are given, this can be computed, see [56], Section 3, or [1], Section 6.

[5] For this reason, some authors, including ourselves, call the decomposition $A = \sum_{i=1}^{s} a_i M_i$ a *Wei-Norman decomposition of* A.

We have

$$M_4 := [M_2, M_3] = \begin{pmatrix} 0 & 0 & 0 & 2 \\ 0 & 0 & 0 & 0 \\ 0 & 0 & 0 & 0 \\ 0 & 0 & 0 & 0 \end{pmatrix}.$$

All the other brackets in $\langle M_1, M_2, M_3, M_4 \rangle$ are zero and we find that $\mathrm{Lie}(A) = \langle M_1, M_2, M_3, M_4 \rangle$. It is solvable of depth 2: the first derived algebra (the set of all matrices in Lie(A) which can be written as a Lie bracket) is $\langle M_4 \rangle$ and the second derived algebra is $\{0\}$. We will continue below with this example. ◇

3.3 Linear Differential Systems in Reduced Form

We now turn to the link between $\mathrm{Lie}(A)$ and differential Galois theory, based on two important results of Kolchin and Kovacic. Proofs can be found in [5], Proposition 1.31 and Corollary 1.32; see also [60], Theorem 5.8, and [8], § 5.3 after Remark 31. Let $A \in \mathrm{Mat}(n, \mathbf{k})$; let G be the differential Galois group of $[A]$ and $\mathfrak{g}$ its Lie algebra, the Galois-Lie algebra of the system $[A]$. The first result is

$$\mathfrak{g} \subset \mathrm{Lie}(A).$$

So, the Lie algebra associated to A, an object which is very easy to compute, provides an "upper bound" on $\mathfrak{g}$. When we perform a gauge transformation $P \in \mathrm{GL}(n, \mathbf{k})$ to obtain a new system $P[A]$, G and $\mathfrak{g}$ are invariant while $\mathrm{Lie}(P[A])$ may vary. This shows that $\mathfrak{g}$ is a lower bound on all $\mathrm{Lie}(P[A])$, for all gauge transformations $P \in \mathrm{GL}(n, \mathbf{k})$. The second result of Kolchin and Kovacic is that this lower bound is reached. By definition, $\mathrm{Lie}(A)$ is the Lie algebra of an algebraic connected group H. Then there exists a gauge transformation[6] $P \in H(\overline{\mathbf{k}})$ such that $\mathfrak{g} = \mathrm{Lie}(P[A])$. Furthermore, if G is connected and under the very mild additional condition that $\mathbf{k}$ is a C^1-field[7] then we may choose $P \in H(\mathbf{k})$ (no algebraic extension).

Definition 2 A system $\partial Y = AY$ is in *reduced form* when the Lie algebra $\mathrm{Lie}(A)$ associated to A is equal to the Lie algebra of the differential Galois group of $\partial Y = AY$.

---

[6]The notation $H(\overline{\mathbf{k}})$ denotes matrices whose entries are in $\overline{\mathbf{k}}$ and satisfy all the equations defining the algebraic group H.

[7] A field $\mathbf{k}$ is a C^1-field when every non-constant homogeneous polynomial P over $\mathbf{k}$ has a non-trivial zero provided that the number of its variables is more than its degree. For example, $C(x)$ is a C^1-field and any algebraic extension of a C^1-field is a C^1-field (Tsen's theorem).

The results of Kolchin and Kovacic show that a reduced form always exists. We provide, in the sequel, constructive methods to obtain them when the systems are in block-triangular form. They will be illustrated on our 4-dimensional example in the next section.

Example 2 (4-Dimensional Example, Continued) We will show, in the next section, that $[A]$ is in reduced form. This system is easily integrated step by step and we find a fundamental solution matrix

$$U = e^x \begin{pmatrix} 1 & \ln(x) & \ln(x-1) & 2\,\mathrm{dilog}(x) + \ln(x-1)\ln(x) \\ 0 & 1 & 0 & \ln(x-1) \\ 0 & 0 & 1 & -\ln(x) \\ 0 & 0 & 0 & 1 \end{pmatrix}$$

where dilog is defined by $\mathrm{dilog}'(x) = \frac{\ln(x)}{1-x}$. We may take $\mathrm{dilog}(x) = \mathrm{Li}_2(1-x)$. When computing U, the terms requiring one integration ($\ln(x)$ and $\ln(x-1)$) correspond to the terms in M_2 and M_3 in the Wei-Norman decomposition of A. The term dilog comes from the existence of M_4, the Lie bracket of M_2 and M_3 in $\mathrm{Lie}(A)$.

The corresponding Galois group is a semi-direct product of a 1-dimensional torus (giving rise to the $\exp(x)$ in the solution) and of a vector group generated by M_2, M_3 (giving rise to the terms in ln) and M_4 (giving rise to the dilog). Its Lie algebra is $\mathrm{Lie}(A)$. ◇

The ideas behind this notion of reduced form have been used for inverse problems in differential Galois theory: given an algebraic group G, construct a differential system $[A]$ having G as its differential Galois group. It is also a technique known in differential geometry. Its use for direct problems in differential Galois theory is more recent. A remark in [5] suggests that this would be a good idea. In the context of Lie-Vessiot systems, Blazquez and Morales exploit this idea in [60]. It is developed in [10, 53, 61] in order to study variational equations in the context of integrability of Hamiltonian systems and the Morales-Ramis-Simó theory (and later in [4] to study algebraic properties of Painlevé equations). For irreducible systems (or systems in block diagonal form), a criterion for reduced forms is established in [8] with a decision procedure. Another, much more efficient approach is given in [11, 12] together with generalizations of the criterion of [8].

The approach described below allows, given the above results, to compute a reduced form of a block triangular linear differential system (the last case remaining after all the above contributions). It is based upon these works, notably [4], and is constructed in [1].

4 How to Compute a Reduced Form of a Reducible System

Assume now that $\mathbf{k} = C(x)$, where C is algebraically closed of characteristic zero where the derivation ∂ acts trivially. It is in particular a C^1-field. We consider a *block triangular* system over the differential field $C(x)$ in the same form as (1), that is

$$[A] : \partial Y = AY, \text{ with } A = \left(\begin{array}{c|c} A_1 & 0 \\ \hline S & A_2 \end{array}\right) \in \mathrm{Mat}(n, \mathbf{k}).$$

Let $A_{\mathrm{diag}} := \left(\begin{array}{c|c} A_1 & 0 \\ \hline 0 & A_2 \end{array}\right)$. In what follows, we will assume that the block diagonal part A_{diag} is in *reduced form* and we will show how to find a gauge transformation P such that $P[A]$ is in reduced form. By [1], Lemma 2.7, the differential Galois group is connected and the reduction matrix we are looking for has coefficients in $\mathbf{k}$. Instead of reproving all the theory (which, in this case, can be mostly found in [1]), we will work out in details a simple example where most of the required algorithmic elements appear. This may help convince the reader of how the method works (the details in [1] may be technical, at least in a first reading).

4.1 Shape of the Gauge Transformation

Let $\mathfrak{h}_{sub}$ be the set of *off-diagonal* constant matrices of the form $\left(\begin{array}{c|c} 0 & 0 \\ \hline S & 0 \end{array}\right)$ (same sizes as in relation (1)). We will extend the scalars to $\mathfrak{h}_{sub}(\mathbf{k}) := \mathfrak{h}_{sub} \otimes_C \mathbf{k}$, the off-diagonal matrices with coefficients in $\mathbf{k}$. Our first step is that we may find a reduction matrix in a very particular shape.

Lemma 1 ([10], Lemma 3.4) *There exists a gauge transformation* $P \in \left\{\mathrm{Id} + B, B \in \mathfrak{h}_{sub}(\mathbf{k})\right\}$ *such that* $\partial Y = P[A]Y$ *is in reduced form.*

The following is based on an observation from [53, 61]. Let $P = \mathrm{Id} + B$, $B \in \mathfrak{h}_{sub}(\mathbf{k})$. Suppose that, for all $Q \in \{\mathrm{Id} + B, B \in \mathfrak{h}_{sub}(\mathbf{k})\}$, we have $\mathrm{Lie}(P[A]) \subseteq \mathrm{Lie}(Q[P[A]])$; then, $P[A]$ is in reduced form. In other terms, no rational gauge transformation can turn it into a system with a smaller associated Lie algebra. In this case, $\mathrm{Lie}(P[A])$ will be the Lie algebra of the differential Galois group and this will give us transcendence relations and algebraic relations on the solutions; this will be seen on the main example of this section.

More generally, as we can see in [1], Section 5, if our method can reduce a system with two diagonal blocks then we can iterate this method to obtain a reduced form of a block-triangular system with an arbitrary number of blocks on the diagonal. Let us illustrate this iteration on a system with three diagonal blocks of the form

$$\left(\begin{array}{c|c|c} A_1 & 0 & 0 \\ \hline S_{2,1} & A_2 & 0 \\ \hline S_{3,1} & S_{3,2} & A_3 \end{array}\right)$$

where the block diagonal part is in reduced form (see [11, 12] for this). We will first reduce the south-east part (which is of the same form as (1)) into a form

$$\left(\begin{array}{c|c} A_2 & 0 \\ \hline S & A_3 \end{array}\right)$$

Let P_1 be the reduction matrix. By [1], Lemma 5.1, the following system is automatically in reduced form

$$A_d := \left(\begin{array}{c|c|c} A_1 & 0 & 0 \\ \hline 0 & A_2 & 0 \\ \hline 0 & S & A_3 \end{array}\right).$$

Now we perform the gauge transformation $\left(\begin{array}{c|c} \mathrm{Id} & 0 \\ \hline 0 & P_1 \end{array}\right)$ to obtain a system of the form

$$\left(\begin{array}{c|c|c} A_1 & 0 & 0 \\ \hline \tilde{S}_{2,1} & A_2 & 0 \\ \hline \tilde{S}_{3,1} & S & A_3 \end{array}\right)$$

(the $S_{i,j}$ may have changed after the first reduction step). We now see that this system in the same form as (1) with A_d as the block diagonal matrix. So a second reduction of a two-blocks triangular system allows to reduce the initial three-blocks triangular system.

This iteration is well seen in our 4-dimensional example below.

Example 3 (4-Dimensional Example, Continued) Let

$$A := \begin{pmatrix} 1 & 0 & \frac{1}{x} & 0 \\ \frac{1}{x-1} & 1 & 0 & -\frac{1}{x} \\ 0 & 0 & 1 & 0 \\ 0 & 0 & \frac{1}{x-1} & 1 \end{pmatrix}.$$

A simple application of a factorization algorithm shows that it is reducible. Indeed, letting

$$P := \begin{pmatrix} 0 & 0 & -1 & 0 \\ -1 & 0 & 0 & 0 \\ 0 & 0 & 0 & 1 \\ 0 & 1 & 0 & 0 \end{pmatrix},$$

we have

$$P[A] = \left(\begin{array}{cc|c|c} 1 & \frac{1}{x} & \frac{1}{x-1} & 0 \\ 0 & 1 & 0 & \frac{1}{x-1} \\ \hline 0 & 0 & 1 & -\frac{1}{x} \\ \hline 0 & 0 & 0 & 1 \end{array}\right).$$

This example is, of course, particularly simple. We use it to show how to apply the iteration procedure and Lemma 1 to simplify the system or prove that it cannot be simplified further.

Since we consider an upper triangular system, we start with the "north-west" corner. We let

$$B := \begin{pmatrix} 1 & \frac{1}{x} \\ 0 & 1 \end{pmatrix}.$$

The diagonal part is in reduced form (solutions are e^x and cannot be simplified using rational functions). The associated Lie algebra Lie(B) has dimension 2. Reduction would imply to have dimension 1. By Lemma 1, a reduction matrix would have the form

$$P := \begin{pmatrix} 1 & f(x) \\ 0 & 1 \end{pmatrix}.$$

The north-east coefficient of $P[B]$ is $\frac{1}{x} - f'(x)$. The coefficient $\frac{1}{x} - f'(x)$ could never be constant (the equation $f'(x) = \frac{1}{x}$ has no rational solution, the simple pole $\frac{1}{x}$ cannot be canceled by the derivative of a rational function). For any choice of f, Lie$(P[B])$ will have dimension 2. It follows that $[B]$ is in reduced form. So we iterate.

We now pick a bigger matrix B:

$$B := \left(\begin{array}{cc|c} 1 & \frac{1}{x} & \frac{1}{x-1} \\ 0 & 1 & 0 \\ \hline 0 & 0 & 1 \end{array}\right) \quad \text{and } B_{\text{diag}} := \left(\begin{array}{cc|c} 1 & \frac{1}{x} & 0 \\ 0 & 1 & 0 \\ \hline 0 & 0 & 1 \end{array}\right).$$

By [1], Lemma 5.1, and by the above calculation, we find that $[B_{\text{diag}}]$ is in reduced form. Lemma 1 thus shows that a reduction matrix would have the simple form

$$P := \begin{pmatrix} 1 & 0 & f(x) \\ 0 & 1 & g(x) \\ 0 & 0 & 1 \end{pmatrix}.$$

Now Lie(B_{diag}) has dimension 2 and Lie(B) has dimension 3. A reduction matrix should therefore map B to a matrix with an associated Lie algebra of dimension 2.

We have $P[B]_{2,3} = -g'(x)$ so there should exist constants g_1, g_2 and a rational function $g(x)$ such that $-g'(x) = g_1 + g_2\frac{1}{x}$. A necessary condition is $g_2 = 0$ and $g(x) = g_0 - g_1 x$. Similarly, there should exist constants f_1, f_2 such that $P[B]_{1,3} = f_1 + f_2\frac{1}{x}$. We now plug our condition on g into this relation and find that there should be a rational function f such that

$$f'(x) = -f_1 - g_1 + (g_0 - f_2)\frac{1}{x} + \frac{1}{x-1}.$$

Now, because of the pole of order 1 at $x = 1$, this can never have a rational solution (whatever the values of the unknown constants). It follows that, for any choice of f, $\mathrm{Lie}(P[B])$ will have dimension 3. So P is in reduced form.

Note that our main ingredient here has been to look for a rational solution of an inhomogeneous linear differential equation whose right-hand side contains parameters. There exist algorithms to compute conditions on the parameters (from the right-hand side) so that such an equation has rational solutions, see Sect. 4.2.2 below or [62], and this will be the key to what follows.

We continue iterating the reduction process. Now we will have

$$B := \begin{pmatrix} 1 & \frac{1}{x} & \frac{1}{x-1} & 0 \\ 0 & 1 & 0 & \frac{1}{x-1} \\ 0 & 0 & 1 & -\frac{1}{x} \\ 0 & 0 & 0 & 1 \end{pmatrix}, \quad B_{\mathrm{diag}} := \begin{pmatrix} 1 & \frac{1}{x} & \frac{1}{x-1} & 0 \\ 0 & 1 & 0 & 0 \\ 0 & 0 & 1 & 0 \\ 0 & 0 & 0 & 1 \end{pmatrix} \text{ and } P := \begin{pmatrix} 1 & 0 & 0 & f_1(x) \\ 0 & 1 & 0 & f_2(x) \\ 0 & 0 & 1 & f_3(x) \\ 0 & 0 & 0 & 1 \end{pmatrix}.$$

Using again [1], Lemma 5.1, we see that B_{diag} is in reduced form. Furthermore, $\mathrm{Lie}(B_{\mathrm{diag}})$ has dimension 3 and $\mathrm{Lie}(B)$ has dimension 4. We compute $P[B]$.

$$P[B] = \begin{pmatrix} 1 & \frac{1}{x} & \frac{1}{x-1} & \frac{f_2(x)}{x} + \frac{f_3(x)}{x-1} - \frac{\mathrm{d}}{\mathrm{d}x} f_1(x) \\ 0 & 1 & 0 & \frac{1}{x-1} - \frac{\mathrm{d}}{\mathrm{d}x} f_2(x) \\ 0 & 0 & 1 & -\frac{1}{x} - \frac{\mathrm{d}}{\mathrm{d}x} f_3(x) \\ 0 & 0 & 0 & 1 \end{pmatrix}.$$

The relation $P[B]_{3,4} = c_{1,3} + c_{2,3}\frac{1}{x} + c_{3,3}\frac{1}{x-1}$ gives conditions $c_{3,3} = 0$, $c_{2,3} = -1$ and $f_3(x) = -c_{1,3}x + c_{0,3}$. The same study on $P[B]_{2,4}$ gives us $f_2(x) = -c_{1,2}x + c_{0,2}$. Without finishing with the last coefficient, we see that $\mathrm{Lie}(P[B])$ contains matrices of the following forms (respectively because of terms in $\frac{1}{x}$ and $\frac{1}{x-1}$):

$$M_2 := \begin{pmatrix} 0 & 1 & 0 & \star \\ 0 & 0 & 0 & 0 \\ 0 & 0 & 0 & -1 \\ 0 & 0 & 0 & 0 \end{pmatrix} \text{ and } M_3 := \begin{pmatrix} 0 & 0 & 1 & \star \\ 0 & 0 & 0 & 1 \\ 0 & 0 & 0 & 0 \\ 0 & 0 & 0 & 0 \end{pmatrix}$$

whose Lie bracket is

$$M_4 := [M_2, M_3] = \begin{pmatrix} 0&0&0&2 \\ 0&0&0&0 \\ 0&0&0&0 \\ 0&0&0&0 \end{pmatrix}.$$

So we see that, whatever our future choices may be, Lie($P[B]$) will contain M_4 and hence have dimension 4. This shows that our system cannot be reduced so it is in reduced form. Furthermore, this suggests that our reduction conditions might have been stronger: requiring $P[B]_{3,4} = c_{1,3} + c_{2,3}\frac{1}{x} + c_{3,3}\frac{1}{x}$ was not enough. We could have imposed $P[B]_{3,4} = 0$ and $P[B]_{2,4} = 0$. Both these relations are easily seen to be impossible to fulfill with rational functions so our system is again seen to be in reduced form. ◇

To summarize what this example suggests: we need to "cancel" terms in the purely triangular part; this reduces to finding rational solutions of linear differential equations with parametrized right-hand sides. And the order of the computations matters: here, one needs to study the relations on f_2 and f_3 before studying relations on f_1. We will show, in the sequel, how to systematize these ideas, using an isotypical decomposition and an adapted flag structure, and how to make them algorithmic so that a computer algebra system may perform the calculations.

In this example, we had seen that a fundamental solution matrix could be written using e^x, $\ln(x)$, $\ln(x-1)$ and dilog(x). As $[B]$ is in reduced form, Lie(B) is the Lie algebra of the Galois group and it has dimension 4. This shows that these four functions are transcendent and algebraically independent. So our calculation above (long but not hard) gave us a simple proof that dilog(x) is algebraically independent of e^x, $\ln(x)$, $\ln(x-1)$.

4.2 The Adjoint Action of the Diagonal

We recall our notations so far. We have $n_i \times n_i$ matrices A_i with coefficients in $\mathbf{k}$ and

$$A = \left(\begin{array}{c|c} A_1 & 0 \\ \hline S & A_2 \end{array}\right) \in \mathrm{Mat}(n, \mathbf{k}),\ A_{\mathrm{diag}} := \left(\begin{array}{c|c} A_1 & 0 \\ \hline 0 & A_2 \end{array}\right).$$

If we take two off-diagonal matrices B_1 and B_2 in $\mathfrak{h}_{sub}$, we have $B_1.B_2 = 0$. This allows two simple calculations. First, let $P := \mathrm{Id} + \sum_i f_i B_i$, with $f_i \in \mathbf{k}$, $B_i \in \mathfrak{h}_{sub}$. Then

$$P[A] = A + \sum_i f_i [A_{\mathrm{diag}}, B_i] - \sum_i \partial(f_i) B_i. \quad (3)$$

Furthermore, $[A_{\text{diag}}, B_i] \in \mathfrak{h}_{sub}(\mathbf{k})$. These two calculations show that reduction will be governed by the *adjoint action* $\Psi : X \mapsto [A_{\text{diag}}, X]$ of the block diagonal part A_{diag} on $\mathfrak{h}_{sub}(\mathbf{k})$. This adjoint action Ψ is a linear map. Its matrix, on the canonical basis of $\mathfrak{h}_{sub}$, is

$$\Psi = A_2 \otimes \text{Id}_{\text{n}_1} - \text{Id}_{\text{n}_2} \otimes A_1^T.$$

When $\partial Y = A_{\text{diag}} Y$ has an abelian Lie algebra we may easily compute a Jordan normal form of $\Psi : X \mapsto [A_{\text{diag}}, X]$. Furthermore the eigenvalues of Ψ belong to $\mathbf{k}$. This is the idea behind [10]. In our case, we will need a more subtle structure, an isotypical decomposition into Ψ-invariant subspaces of $\mathfrak{h}_{sub}$.

Example 4 (An 8-Dimensional Example) We consider a matrix A given by

$$A := \left(\begin{array}{cccc|cccc} 1 & 0 & \frac{1}{x} & 0 & 0 & 0 & 0 & 0 \\ \frac{1}{x-1} & 1 & 0 & -\frac{1}{x} & 0 & 0 & 0 & 0 \\ 0 & 0 & 1 & 0 & 0 & 0 & 0 & 0 \\ 0 & 0 & \frac{1}{x-1} & 1 & 0 & 0 & 0 & 0 \\ \hline \star & \star & \star & \star & 1 & 0 & \frac{1}{x} & 0 \\ \star & \star & \star & \star & \frac{1}{x-1} & 1 & 0 & -\frac{1}{x} \\ \star & \star & \star & \star & 0 & 0 & 1 & 0 \\ \star & \star & \star & \star & 0 & 0 & \frac{1}{x-1} & 1 \end{array}\right).$$

The block diagonal part is given by two copies of our 4-dimensional example A_1 (here, $A_2 = A_1$) and we have shown that it was in reduced form. The off-diagonal part is given by

$$\begin{pmatrix} \frac{-3x+4}{4x^2} & \frac{x-4}{4x^2} & -\frac{1}{2(x-1)} + \frac{2x-2}{x^2} & -\frac{1}{x} \\ \frac{1}{2(x-1)} + \frac{-2x+5}{x^2} & \frac{x-4}{4x^2} & \frac{2}{x-1} + \frac{4}{x^2} & \frac{1}{2(x-1)} + \frac{2x-2}{x^2} \\ \frac{-1}{4(x-1)} & 0 & \frac{3x-4}{4x^2} & \frac{x+4}{4x^2} \\ -\frac{1}{2(x-1)} & \frac{1}{4(x-1)} & \frac{1}{2(x-1)} + \frac{2x-7}{x^2} & \frac{-x+4}{4x^2} \end{pmatrix}.$$

As $A_2 = A_1$, the matrix of the adjoint action of the diagonal on $\mathfrak{h}_{sub}$ is a (sparse) 16×16 matrix given by $\Psi = A_1 \otimes \text{Id}_4 - \text{Id}_4 \otimes A_1^T$. ◇

4.2.1 Isotypical Decomposition

Recall that $\mathfrak{h}_{sub}$ is the C-vector space of off-diagonal matrices. We now show how the adjoint action Ψ of the diagonal will govern the reduction strategy on $\mathfrak{h}_{sub}$.

Definition 3 A vector space $W \subset \mathfrak{h}_{sub}$ will be called a Ψ-*space* if $\Psi(W) \subset W \otimes_C \mathbf{k}$.

The importance of these Ψ-spaces is stated in the following lemma.

Lemma 2 ([1], Lemma 2.11) *Let* $A := \left(\begin{array}{c|c} A_1 & 0 \\ \hline S & A_2 \end{array}\right)$ *and assume that* $\partial Y = AY$ *is in reduced form. Then,* $\mathrm{Lie}(A) \cap \mathfrak{h}_{sub}$ *is a* Ψ*-space.*

So our reduction strategy will be to try to project onto the smallest possible Ψ-space using rational gauge transformations. In [1], we provide references to algorithms to decompose and factor into Ψ-spaces. This is obtained using an isotypical decomposition (eigenring methods) and a flag structure.

Lemma 3 (Krull-Schmidt) *The C-vector space* $\mathfrak{h}_{sub}$ *admits a unique isotypical decomposition*

$$\mathfrak{h}_{sub} = \bigoplus_{i=1}^{\kappa} W_i$$

where

- *each* W_i *is a* Ψ*-space;*
- $W_i \simeq \nu_i V_i$*, a direct sum of* ν_i Ψ*-spaces that are all isomorphic to an indecomposable* Ψ*-space* V_i *which admits a* flag decomposition

  $$V_i = V_i^{[\mu]} \supsetneq V_i^{[\mu-1]} \supsetneq \cdots \supsetneq V_i^{[1]} \supsetneq V_i^{[0]} = \{0\}$$

  and $V_i^{[j]}/V_i^{[j-1]}$ *is a sum of isomorphic irreducible* Ψ*-spaces for* $1 \le j \le \mu$*;*
- *For* $i \neq j$*, the* Ψ*-spaces* $V_i \subset W_i$ *and* $V_j \subset W_j$ *are not isomorphic.*

Once this decomposition and flag structure are computed, we perform, at each stage, a projection on a minimal Ψ-subspace in $V_i^{[j]}$. For some vectors $b_i \in \mathbf{k}^N$ and a matrix $E_{i,j}$ with coefficients in $\mathbf{k}$ (obtained by linear algebra), this reduces to computing all tuples $(F, c_1, \ldots, c_s)$, with $F \in \mathbf{k}^N$ and c_i constants, such that

$$F' = E_{i,j} \cdot F + \sum_i c_i b_i .$$

The resulting system $P[A]$ will be "minimal": it will be in reduced form. The proof of this result is technical and can be found in [1]. We will illustrate the process on our main example.

Example 5 (8-Dimensional Example, Continued) In this example, $\mathfrak{h}_{sub}$ decomposes as a direct sum $\mathfrak{h}_{sub} = \mathfrak{h}_1 \oplus \mathfrak{h}_5 \oplus \mathfrak{h}_{10}$ of three indecomposable Ψ-spaces.
We first study the adjoint action $\Psi = [A_{\mathrm{diag}}, \bullet]$ of A_{diag} on $\mathfrak{h}_5$. We find (see [63]) an adapted basis given by off-diagonal matrices $N_2, \ldots, N_6$ with south-west blocks

$$\begin{bmatrix} 0&0&0&0 \\ 2&0&0&0 \\ 0&0&0&0 \\ 0&0&-2&0 \end{bmatrix}, \begin{bmatrix} 0&0&-2&0 \\ 0&0&0&-2 \\ 0&0&0&0 \\ 0&0&0&0 \end{bmatrix}, \begin{bmatrix} 1&0&0&0 \\ 0&-1&0&0 \\ 0&0&-1&0 \\ 0&0&0&1 \end{bmatrix}, \begin{bmatrix} 0&0&0&0 \\ 0&0&0&0 \\ 1&0&0&0 \\ 0&1&0&0 \end{bmatrix}, \begin{bmatrix} 0&-1&0&0 \\ 0&0&0&0 \\ 0&0&0&1 \\ 0&0&0&0 \end{bmatrix}.$$

The matrix of the adjoint action Ψ on this basis of $\mathfrak{h}_5$ is

$$\Psi_5 := \left(\begin{array}{cc|c|cc} 0 & 0 & \frac{1}{x-1} & 0 & 0 \\ 0 & 0 & \frac{1}{x} & 0 & 0 \\ \hline 0 & 0 & 0 & \frac{1}{x} & \frac{1}{x-1} \\ \hline 0 & 0 & 0 & 0 & 0 \\ 0 & 0 & 0 & 0 & 0 \end{array}\right).$$

The flag structure on $\mathfrak{h}_5$ suggests the following reduction path. Try to remove elements in $\langle N_5, N_6\rangle$ if possible; then in $\langle N_4\rangle$; then in $\langle N_2, N_3\rangle$. How to do this will be made clear in the next section; the flag structure guides the order in which computations should be handled.

We turn to $\mathfrak{h}_{10}$. We find (see [63]) a basis adapted to the flag structure given by off-diagonal matrices $N_7, \ldots, N_{16}$ whose south-west blocks are:

$$\begin{bmatrix} 0&0&0&0 \\ 0&0&2&0 \\ 0&0&0&0 \\ 0&0&0&0 \end{bmatrix}, \begin{bmatrix} 0&0&0&0 \\ -1&0&0&0 \\ 0&0&0&0 \\ 0&0&-1&0 \end{bmatrix}, \begin{bmatrix} 0&0&1&0 \\ 0&0&0&-1 \\ 0&0&0&0 \\ 0&0&0&0 \end{bmatrix},$$

$$\begin{bmatrix} 0&0&0&0 \\ 0&0&0&0 \\ 0&0&0&0 \\ 1&0&0&0 \end{bmatrix}, \begin{bmatrix} \frac{1}{2}&0&0&0 \\ 0&-\frac{1}{2}&0&0 \\ 0&0&\frac{1}{2}&0 \\ 0&0&0&-\frac{1}{2} \end{bmatrix}, \begin{bmatrix} \frac{1}{2}&0&0&0 \\ 0&\frac{1}{2}&0&0 \\ 0&0&-\frac{1}{2}&0 \\ 0&0&0&-\frac{1}{2} \end{bmatrix}, \begin{bmatrix} 0&0&0&-1 \\ 0&0&0&0 \\ 0&0&0&0 \\ 0&0&0&0 \end{bmatrix},$$

$$\begin{bmatrix} 0&0&0&0 \\ 0&0&0&0 \\ \frac{1}{2}&0&0&0 \\ 0&-\frac{1}{2}&0&0 \end{bmatrix}, \begin{bmatrix} 0&-\frac{1}{2}&0&0 \\ 0&0&0&0 \\ 0&0&0&-\frac{1}{2} \\ 0&0&0&0 \end{bmatrix}, \begin{bmatrix} 0&0&0&0 \\ 0&0&0&0 \\ 0&-\frac{1}{2}&0&0 \\ 0&0&0&0 \end{bmatrix}$$

The matrix of the adjoint action Ψ on this adapted basis $N_7, \ldots, N_{16}$ is:

$$\Psi_{10} := \left(\begin{array}{c|cc|cccc|cc|c} 0 & \frac{1}{x} & \frac{1}{x-1} & 0 & 0 & 0 & 0 & 0 & 0 & 0 \\ \hline 0 & 0 & 0 & \frac{1}{x} & -\frac{1}{x-1} & 0 & 0 & 0 & 0 & 0 \\ 0 & 0 & 0 & 0 & 0 & -\frac{1}{x} & \frac{1}{x-1} & 0 & 0 & 0 \\ \hline 0 & 0 & 0 & 0 & 0 & 0 & 0 & \frac{1}{x-1} & 0 & 0 \\ 0 & 0 & 0 & 0 & 0 & 0 & 0 & 0 & \frac{1}{x-1} & 0 \\ 0 & 0 & 0 & 0 & 0 & 0 & 0 & \frac{1}{x} & 0 & 0 \\ 0 & 0 & 0 & 0 & 0 & 0 & 0 & 0 & \frac{1}{x} & 0 \\ \hline 0 & 0 & 0 & 0 & 0 & 0 & 0 & 0 & 0 & \frac{1}{x-1} \\ 0 & 0 & 0 & 0 & 0 & 0 & 0 & 0 & 0 & \frac{1}{x} \\ \hline 0 & 0 & 0 & 0 & 0 & 0 & 0 & 0 & 0 & 0 \end{array}\right).$$

4.2.2 Intermezzo: Reduction and Rational Solutions

Before we continue, let us make a quick excursion into our main algorithmic toolbox. we start with a simple case. We look for a condition on $P := \mathrm{Id} + \left(\begin{array}{c|c} 0 & 0 \\ \hline \beta & 0 \end{array}\right)$ to have

$$A = \left(\begin{array}{c|c} A_1 & 0 \\ \hline S & A_2 \end{array}\right) \longrightarrow P[A] = \left(\begin{array}{c|c} A_1 & 0 \\ \hline 0 & A_2 \end{array}\right).$$

A simple calculation shows that β should be a rational solution of the matrix linear differential system $\beta' = A_2\beta - \beta A_1 + S$. If we let $\mathtt{vec}$ denote the operator transforming a matrix into a vector by stacking its rows, we find (see [1]) that $\mathtt{vec}(\beta)' = \Psi \cdot \mathtt{vec}(\beta) - \mathtt{vec}(S)$, where Ψ is again the adjoint action of the diagonal defined above. So reduction will be governed by computing rational solutions of linear differential systems. When $\mathbf{k} = C(x)$, a computer algebra algorithm for this task has been given by Barkatou in [64], see [65] for a generalization to linear partial differential systems and a Maple implementation.

Now, our general tool (also found in the above references) will be an apparently more complicated problem. Given a matrix Ψ and vectors $b_1, \ldots, b_s$, we will look for tuples $(F, c_1, \ldots, c_s)$, with $F \in \mathbf{k}^N$ and c_i constant, such that $F' = \Psi \cdot F + \sum_i c_i b_i$. Such tuples form a *computable* vector space and the algorithms in [64, 65] provide this when $\mathbf{k} = C(x)$. Results and algorithms for general fields $\mathbf{k}$ can be found in [62].

We now pick concrete coefficients to show how to perform the reduction on our 8-dimensional example. A Maple worksheet[8] with this example and the chosen coefficients may be found at [63].

4.2.3 Reduction on $\mathfrak{h}_5$ (8-Dimensional Example)

To remove all of $\mathfrak{h}_5$, it would be enough to have a rational solution to the system

$$Y' = \Psi_5 . Y + b \text{ with } b = \begin{pmatrix} \frac{3}{x^2} - \frac{1}{x} \\ \frac{1}{x^2} - \frac{1}{x} \\ \frac{1}{x^2} - \frac{1}{2x} \\ 0 \\ \frac{1}{x^2} \end{pmatrix}$$

and Ψ_5 is given in Example 5 (page 162). This gives us reduction equations

[8] The reader may also find a pdf version at
http://www.unilim.fr/pages_perso/jacques-arthur.weil/DreyfusWeilReductionExamples.pdf.

$$\begin{array}{ll} (W^{[3]}): & \begin{cases} f_{3,1}'(x) = \frac{1}{x^2} \\ f_{3,2}'(x) = 0 \end{cases} \\ (W^{[2]}): & \left\{ f_{2,1}'(x) = \frac{1}{x-1} f_{3,1}(x) + \frac{1}{x} f_{3,2}(x) + \frac{1}{x^2} - \frac{1}{2x} \right. \\ (W^{[1]}): & \begin{cases} f_{1,1}'(x) = \frac{1}{x} f_{2,1}(x) + \frac{1}{x^2} - \frac{1}{x} \\ f_{1,2}'(x) = \frac{1}{x-1} f_{2,1}(x) + \frac{3}{x^2} - \frac{1}{x} \end{cases} \end{array}$$

The first two equations correspond to the highest level $W^{[3]}$ of the flag. To remove an element from $W^{[3]}$, there should be a rational solution to the equation $y' = c_1.\frac{1}{x^2} + c_2.0$. The C-vector space of pairs $(c_1, c_2) \in C^2$ such that there exists $f \in \mathbf{k}$ with $f' = c_1.\frac{1}{x^2} + c_2.0$ is found to be 2-dimensional; for $\underline{c} = (1, 0)$, we have $f_{3,1} := -\frac{1}{x} + c_{3,1}$; for $\underline{c} = (0, 1)$, we have $f_{3,2} := c_{3,2}$, where the $c_{3,i}$ are arbitrary constants (their importance will soon be visible). Our gauge transformation is $P^{[3]} = \mathrm{Id} + f_{3,1} N_6 + f_{3,2} N_5$ and $A^{[2]} := P^{[3]}[A]$ does not contain any terms from $W^{[3]}$.

Now $W^{[2]}$ is 1-dimensional. The equation for the reduction on $W_2^{[2]}$ is now

$$\begin{aligned} y' &= \frac{1}{x-1} f_{3,1}(x) + \frac{1}{x} f_{3,2}(x) + \frac{1}{x^2} - \frac{1}{2x} \\ &= \frac{1}{2} \frac{2c_{3,2}+1}{x} + \frac{c_{3,1}-1}{x-1} + \frac{1}{x^2}. \end{aligned}$$

We have necessary and sufficient conditions on the parameters $c_{3,i}$ to have a rational solution, namely $c_{3,1} = 1$, $c_{3,2} = -\frac{1}{2}$ and then a general rational solution $f_{2,1} := \frac{-1}{x} + c_{2,1}$. Our new gauge transformation is $P^{[2]} = \mathrm{Id} + (-\frac{1}{x} + c_{2,1}) N_4$ and $A^{[1]} := P^{[2]}[A^{[2]}]$ does not contain any term from $W^{[2]}$ any more.

Finally, we look for all $(c_1, c_2) \in C^2$ such that $c_1 f_{1,1} + c_2 f_{2,2}$ is rational: we look for non-zero pairs $(c_1, c_2) \in C^2$ such that there exists a rational solution $f \in \mathbf{k}$ of

$$\begin{aligned} y' &= c_1 \left(\frac{1}{x} \left(-\frac{1}{x} + c_{2,1} \right) + \frac{1}{x^2} - \frac{1}{x} \right) + c_2 \left(\frac{1}{x-1} \left(-\frac{1}{x} + c_{2,1} \right) + \frac{3}{x^2} - \frac{1}{x} \right) \\ &= \frac{3\,c_2}{x^2} + \frac{c_1\,(c_{2,1} - 1)}{x} + \frac{c_2\,(c_{2,1} - 1)}{x-1}. \end{aligned}$$

This integral is rational if and only if both residues are zero. As the solution $c_1 = c_2 = 0$ is not admissible, we see that a necessary and sufficient condition is $c_{2,1} = 1$. The set of desired pairs (c_1, c_2) is of dimension 2. For $\underline{c} = (1, 0)$, we have $f_{1,1} := c_{1,1}$, for $\underline{c} = (0, 1)$, we have $f_{1,2} := -\frac{3}{x} + c_{1,2}$, where the $c_{1,i}$ are constants and can be chosen arbitrarily. So our last gauge transformation matrix will be $P^{[1]} = \mathrm{Id} - \frac{3}{x} N_2$ and the reduction matrix on $\mathfrak{h}_5$ is

$$P_5 := P^{[3]} P^{[2]} P^{[1]} = \mathrm{Id} - \frac{3}{x} N_2 + \left(1 - \frac{1}{x}\right) N_4 - \frac{1}{2} N_5 + \left(1 - \frac{1}{x}\right) N_6.$$

The resulting matrix $\tilde{A} := P_5[A]$ contains no terms from $\mathfrak{h}_5$.

4.2.4 Reduction on $\mathfrak{h}_{10}$ (8-Dimensional Example)

The matrix Ψ_{10} is given in Example 5 page 162. The reduction equations are now

$$
\begin{aligned}
&(W^{[5]}) : \left\{ f'_{5,1}(x) = 0 \right. \\
&(W^{[4]}) : \begin{cases} f'_{4,1}(x) = \frac{1}{x} f_{5,1}(x) - \frac{1}{2x} \\ f'_{4,2}(x) = \frac{1}{x-1} f_{5,1}(x) - \frac{1}{2(x-1)} \end{cases} \\
&(W^{[3]}) : \begin{cases} f'_{3,1}(x) = \frac{1}{x} f_{4,1}(x) + \frac{1}{x} \\ f'_{3,2}(x) = \frac{1}{x} f_{4,2}(x) - \frac{1}{2x} \\ f'_{3,3}(x) = \frac{1}{x-1} f_{4,1}(x) \\ f'_{3,4}(x) = \frac{1}{x-1} f_{4,2}(x) - \frac{1}{2(x-1)} \end{cases} \\
&(W^{[2]}) : \begin{cases} f'_{2,1}(x) = \frac{1}{x-1} f_{3,1}(x) - \frac{1}{x} f_{3,2}(x) - \frac{1}{2(x-1)} \\ f'_{2,2}(x) = -\frac{1}{x-1} f_{3,3}(x) + \frac{1}{x} f_{3,4}(x) + \frac{1}{x^2} - \frac{1}{2(x-1)} \end{cases} \\
&(W^{[1]}) : \left\{ f'_{1,1}(x) = \frac{1}{x-1} f_{2,1}(x) + \frac{1}{x} f_{2,2}(x) + \frac{2}{x^2} + \frac{1}{(x-1)}. \right.
\end{aligned}
$$

We will let the reader solve this iteratively following the method from the previous section. This will give the following successive reductions

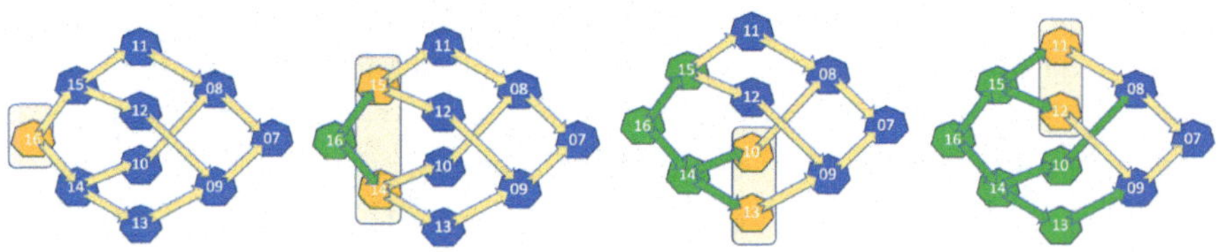

where the green denotes parts that have been successfully removed. However, we reach an obstruction when trying to remove N_{11} (once the equation for $f_{3,1}$ has a rational solution, the equation for $f_{3,3}(x)$ cannot have a rational solution).

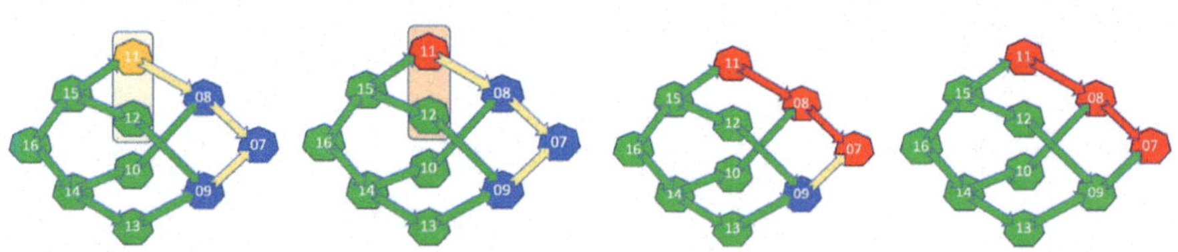

The reduction matrix is

$$
P_{10} := \mathrm{Id} + \left(c_{1,1} - \frac{1}{x} \right) N_7 - \frac{1}{x} N_8 - N_9 - \frac{1}{2} N_{11} + \frac{1}{2} N_{13} + \frac{1}{2} N_{14} - N_{15} + \frac{1}{2} N_{16}
$$

and we obtain the reduced form $A_{\mathrm{red}} := P_{10}[P_5[A]]$:

$$A_{\mathrm{red}} := \left(\begin{array}{cccc|cccc} 1 & 0 & \frac{1}{x} & 0 & 0 & 0 & 0 & 0 \\ \frac{1}{x-1} & 1 & 0 & -\frac{1}{x} & 0 & 0 & 0 & 0 \\ 0 & 0 & 1 & 0 & 0 & 0 & 0 & 0 \\ 0 & 0 & \frac{1}{x-1} & 1 & 0 & 0 & 0 & 0 \\ \hline -\frac{1}{2\,(x-1)} & 0 & 0 & 0 & 1 & 0 & \frac{1}{x} & 0 \\ 0 & \frac{1}{2\,(x-1)} & 0 & 0 & \frac{1}{x-1} & 1 & 0 & -\frac{1}{x} \\ 0 & 0 & -\frac{1}{2\,(x-1)} & 0 & 0 & 0 & 1 & 0 \\ 0 & 0 & 0 & \frac{1}{2\,(x-1)} & 0 & 0 & \frac{1}{x-1} & 1 \end{array}\right).$$

The associated Lie algebra is spanned by

$$\left(\begin{array}{cccc|cccc} 1&0&0&0&0&0&0&0 \\ 0&1&0&0&0&0&0&0 \\ 0&0&1&0&0&0&0&0 \\ 0&0&0&1&0&0&0&0 \\ \hline 0&0&0&0&1&0&0&0 \\ 0&0&0&0&0&1&0&0 \\ 0&0&0&0&0&0&1&0 \\ 0&0&0&0&0&0&0&1 \end{array}\right), \left(\begin{array}{cccc|cccc} 0&0&0&0&0&0&0&0 \\ 1&0&0&0&0&0&0&0 \\ 0&0&0&0&0&0&0&0 \\ 0&0&1&0&0&0&0&0 \\ \hline -\frac{1}{2}&0&0&0&0&0&0&0 \\ 0&\frac{1}{2}&0&0&1&0&0&0 \\ 0&0&-\frac{1}{2}&0&0&0&0&0 \\ 0&0&0&\frac{1}{2}&0&0&1&0 \end{array}\right), \left(\begin{array}{cccc|cccc} 0&0&1&0&0&0&0&0 \\ 0&0&0&-1&0&0&0&0 \\ 0&0&0&0&0&0&0&0 \\ 0&0&0&0&0&0&0&0 \\ \hline 0&0&0&0&0&0&1&0 \\ 0&0&0&0&0&0&0&-1 \\ 0&0&0&0&0&0&0&0 \\ 0&0&0&0&0&0&0&0 \end{array}\right),$$

$$\left(\begin{array}{cccc|cccc} 0&0&0&0&0&0&0&0 \\ 0&0&1&0&0&0&0&0 \\ 0&0&0&0&0&0&0&0 \\ 0&0&0&0&0&0&0&0 \\ \hline 0&0&0&0&0&0&0&0 \\ 0&0&0&0&0&0&1&0 \\ 0&0&0&0&0&0&0&0 \\ 0&0&0&0&0&0&0&0 \end{array}\right), \left(\begin{array}{cccc|cccc} 0&0&0&0&0&0&0&0 \\ 0&0&0&0&0&0&0&0 \\ 0&0&0&0&0&0&0&0 \\ 0&0&0&0&0&0&0&0 \\ \hline 0&0&0&0&0&0&0&0 \\ 0&0&1&0&0&0&0&0 \\ 0&0&0&0&0&0&0&0 \\ 0&0&0&0&0&0&0&0 \end{array}\right).$$

This gives us the Lie algebra $\mathfrak{g} = \mathrm{Lie}(A_{\mathrm{red}})$ of the differential Galois group. Note that, during the reduction process, we found the two incompatible equations $f'_{3,1}(x) = \frac{c_{4,1}+1}{x}$ and $f'_{3,3}(x) = \frac{c_{4,1}}{x-1}$, where $c_{4,1}$ is a constant. There were two mutually exclusive paths: either remove N_{11} or remove N_{13}. We removed N_{13} here by setting $c_{4,1} = -1$; the choice of removing N_{11} (by setting $c_{4,1} = 0$) gives a different reduced form whose associated Lie algebra is conjugated to the one we just found. We refer to [1] for the computations in that other path. We also remark that two of the matrices that could not be removed from $\mathfrak{h}_{sub}$ are "absorbed" as lower triangular parts of matrices coming from A_{diag}. It is 5-dimensional, whereas the Lie algebra $\mathrm{Lie}(A)$ associated to the original matrix A had dimension 14. This shows that the Picard-Vessiot extension is obtained from the Picard-Vessiot extension K_{diag}

for $[A_{\text{diag}}]$ by adding only one integral and the system has indeed been transformed into a form where solving is much simpler than before—and we also have proofs of transcendence properties for the remaining objects.

Acknowledgments This project has received funding from the ANR project *De Rerum Natura* ANR-19-CE40-0018. We warmly thank the organizers of the conference *Antidifferentiation and the Calculation of Feynman Amplitudes* for stimulating exchanges and lectures. We specially thank the referee for many observations and precisions that have enhanced the quality of this text.

References

1. T. Dreyfus, J.-A. Weil, Computing the Lie algebra of the differential Galois group: The reducible case (2019). ArXiv **1904.07925**
2. D. Bertrand, Unipotent radicals of differential Galois group and integrals of solutions of inhomogeneous equations. Math. Ann. **321**(3), 645–666 (2001)
3. P.H. Berman, M.F. Singer, Calculating the Galois group of $L_1(L_2(y)) = 0$, L_1, L_2 completely reducible operators. J. Pure Appl. Algebra **139**(1–3), 3–23 (1999). Effective methods in algebraic geometry (Saint-Malo, 1998)
4. G. Casale, J.-A. Weil, Galoisian methods for testing irreducibility of order two nonlinear differential equations. Pacific J. Math. **297**(2), 299–337 (2018)
5. M. van der Put, M.F. Singer, *Galois Theory of Linear Differential Equations*. Grundlehren der Mathematischen Wissenschaften [Fundamental Principles of Mathematical Sciences], vol. 328 (Springer, Berlin, 2003)
6. T. Crespo, Z. Hajto, *Algebraic Groups and Differential Galois Theory*. Graduate Studies in Mathematics, vol. 122 (American Mathematical Society, Providence, 2011)
7. M.F. Singer, Introduction to the Galois theory of linear differential equations, in *Algebraic Theory of Differential Equations*. London Mathematical Society Lecture Note Series, vol. 357 (Cambridge University Press, Cambridge, 2009), pp. 1–82
8. A. Aparicio-Monforte, É. Compoint, J.-A. Weil, A characterization of reduced forms of linear differential systems. J. Pure Appl. Algeb.**217**(8), 1504–1516 (2013)
9. C. Mitschi, M.F. Singer, Solvable-by-finite groups as differential Galois groups. Ann. Fac. Sci. Toulouse Math. **11**(3), 403–423 (2002)
10. A. Aparicio-Monforte, T. Dreyfus, J.-A. Weil, Liouville integrability: an effective Morales–Ramis–Simó theorem. J. Symb. Comput. **74**, 537–560 (2016)
11. M. Barkatou, T. Cluzeau, L. Di Vizio, J.-A. Weil, Reduced forms of linear differential systems and the intrinsic Galois-Lie algebra of Katz. SIGMA Symmetry Integra. Geom. Methods Appl. **16**(054), 13 (2020)
12. M. Barkatou, T. Cluzeau, J.-A. Weil, L. Di Vizio, Computing the lie algebra of the differential galois group of a linear differential system, in *Proceedings of the ACM on International Symposium on Symbolic and Algebraic Computation* (2016), pp. 63–70
13. O.V. Motygin, On evaluation of the confluent heun functions, in *2018 Days on Diffraction (DD)* (IEEE, Piscataway, 2018), pp. 223-229
14. J.J. Kovacic, An algorithm for solving second order linear homogeneous differential equations. J. Symbolic Comput. **2**(1), 3–43 (1986)
15. F. Ulmer, J.-A. Weil, Note on Kovacic's algorithm. J. Symbolic Comput. **22**(2), 179–200 (1996)
16. M. van Hoeij, J.-A. Weil, Solving second order differential equations with Klein's theorem, in *ISSAC '05: Proceedings of the 2005 International Symposium on Symbolic and Algebraic Computation (Beijing)* (ACM, New York, 2005)
17. S.A. Abramov, M. van Hoeij, Integration of solutions of linear functional equations. Integral Transform. Spec. Funct. **8**(1–2), 3–12 (1999)

18. S.A. Abramov, M. van Hoeij, A method for the integration of solutions of Ore equations, in *Proceedings of the 1997 International Symposium on Symbolic and Algebraic Computation (Kihei, HI)* (ACM, New York, 1997), pp. 172–175
19. R.J. Baxter, *Exactly Solved Models in Statistical Mechanics* (Academic [Harcourt Brace Jovanovich Publishers], London, 1982)
20. B.M. McCoy, *Advanced Statistical Mechanics*. International Series of Monographs on Physics, vol. 146 (Oxford University Press, Oxford, 2010)
21. B.M. McCoy, J-M. Maillard, The importance of the ising model. Prog. Theor. Phys. 127 (2012), 791–817 (2012)
22. B.M. McCoy, M. Assis, S. Boukraa, S. Hassani, J.-M. Maillard, W.P. Orrick, N. Zenine, *The saga of the Ising susceptibility*, in *New Trends in Quantum Integrable Systems* (World Scientific Publishing, Hackensack, 2011), pp. 287–306
23. S. Boukraa, S. Hassani, J.-M. Maillard, B.M. McCoy, J.-A. Weil, N. Zenine, Fuchs versus Painlevé. J. Phys. A **40**(42), 12589–12605 (2007)
24. A. Bostan, S. Boukraa, S. Hassani, J.-M. Maillard, J.-A. Weil, N. Zenine, Globally nilpotent differential operators and the square Ising model. J. Phys. A **42**(12), 125206, 50 (2009)
25. A. Bostan, S. Boukraa, S. Hassani, J.-M. Maillard, J.-A. Weil, N. Zenine, N. Abarenkova, Renormalization, isogenies, and rational symmetries of differential equations. Adv. Math. Phys. **2010**, 44p (2010)
26. A. Bostan, S. Boukraa, S. Hassani, M. van Hoeij, J.-M. Maillard, J.-A. Weil, N. Zenine, The Ising model: from elliptic curves to modular forms and Calabi-Yau equations. J. Phys. A Math. Theor. **44**(4), 045204, 44 (2011)
27. S. Boukraa, S. Hassani, J.-M. Maillard, J.-A. Weil, Differential algebra on lattice green and calabi-yau operators. J. Phys. A: Math. Theor. **47**(9), 095203 (2014)
28. S. Boukraa, S. Hassani, J.-M. Maillard, J.-A. Weil, Canonical decomposition of irreducible linear differential operators with symplectic or orthogonal differential galois groups. J. Phys. A: Math. Theor. **48**(10), 105202 (2015)
29. S. Hassani, Ch. Koutschan, J.-M. Maillard, N. Zenine, Lattice Green functions: the d-dimensional face-centered cubic lattice, $d = 8, 9, 10, 11, 12$. J. Phys. A **49**(16), 164003, 30 (2016)
30. C. Koutschan, Lattice Green functions of the higher-dimensional face-centered cubic lattices. J. Phys. A **46**(12), 125005, 14 (2013)
31. Y. Abdelaziz, S. Boukraa, C. Koutschan, J.-M. Maillard, Heun functions and diagonals of rational functions. J. Phys. A **53**(7), 075206, 24 (2020)
32. Y. Abdelaziz, S. Boukraa, C. Koutschan, J.-M. Maillard, Diagonals of rational functions, pullbacked ${}_2F_1$ hypergeometric functions and modular forms. J. Phys. A **51**(45), 455201, 30 (2018)
33. T.M. Sadykov, S. Tanabé, Maximally reducible monodromy of bivariate hypergeometric systems. Izv. Ross. Akad. Nauk Ser. Mat. **80**(1), 235–280 (2016)
34. M.Yu. Kalmykov, Gauss hypergeometric function: reduction, ϵ-expansion for integer/half-integer parameters and Feynman diagrams. J. High Energy Phys. **2006**(4), 056, 21 (2006)
35. V.V. Bytev, M.Yu. Kalmykov, B.A. Kniehl, Differential reduction of generalized hypergeometric functions from Feynman diagrams: one-variable case. Nuclear Phys. B **836**(3), 129–170 (2010)
36. M.Yu. Kalmykov, B.A. Kniehl, Mellin-Barnes representations of Feynman diagrams, linear systems of differential equations, and polynomial solutions. Phys. Lett. B **714**(1), 103–109 (2012)
37. M.Yu. Kalmykov, B.A. Kniehl, Counting the number of master integrals for sunrise diagrams via the Mellin-Barnes representation. J. High Energy Phys. **2017**(7), 031 (2017). front matter+27
38. M.Yu. Kalmykov, B.A. Kniehl, Counting master integrals: integration by parts vs. differential reduction. Phys. Lett. B **702**(4), 268–271 (2011)
39. V.A. Smirnov, Analytic tools for Feynman integrals, in *Springer Tracts in Modern Physics*, vol. 250 (Springer, Heidelberg, 2012)

40. R.N. Lee, A.V. Smirnov, V.A. Smirnov, Solving differential equations for Feynman integrals by expansions near singular points. J. High Energy Phys. **2018**(3), 008 (2018). front matter+14
41. J. Ablinger, J. Blümlein, C.G. Raab, C. Schneider, Iterated binomial sums and their associated iterated integrals. J. Math. Phys. **55**(11), 112301, 57 (2014)
42. S.L. Ziglin, Branching of solutions and nonexistence of first integrals in hamiltonian mechanics. I, Function. Analy. Appl. **16**(3), 181–189 (1982)
43. R.C. Churchill, D.L. Rod, On the determination of Ziglin monodromy groups. SIAM J. Math. Anal. **22**(6), 1790–1802 (1991)
44. V. Salnikov, Effective algorithm of analysis of integrability via the Ziglin's method. J. Dyn. Control Syst. **20**(4), 465–474 (English) (2014)
45. J.-J. Morales-Ruiz, J.-P. Ramis, Galoisian obstructions to integrability of Hamiltonian systems. I, II. Methods Appl. Anal. **8**(1), 33–95, 97–111 (2001)
46. J.-J. Morales-Ruiz, J.-P. Ramis, C. Simo, Integrability of Hamiltonian systems and differential Galois groups of higher variational equations. Ann. Sci. École Norm. Sup. **40**(6), 845–884 (2007)
47. M. Ayoul, N. Tien Zung, Galoisian obstructions to non-Hamiltonian integrability. C. R. Math. Acad. Sci. Paris **348**(23–24), 1323–1326 (2010)
48. A. Tsygvintsev, The meromorphic non-integrability of the three-body problem. J. Reine Angew. Math. **537**, 127–149 (2001)
49. D. Boucher, J.-A. Weil, Application of J.-J. Morales and J.-P. Ramis' theorem to test the non-complete integrability of the planar three-body problem, in *From Combinatorics to Dynamical Systems*. EMS IRMA Lectures in Mathematics and Theoretical Physics, vol. 3 (de Gruyter, Berlin, 2003), pp. 163–177
50. J.-J. Morales-Ruiz, S. Simon, On the meromorphic non-integrability of some N-body problems. Discrete Contin. Dyn. Syst. **24**(4), 1225–1273 (2009)
51. T. Combot, Non-integrability of the equal mass n-body problem with non-zero angular momentum. Celestial Mech. Dynam. Astronom. **114**(4), 319–340 (2012)
52. J.-J. Morales-Ruiz, C. Simó, S. Simon, Algebraic proof of the non-integrability of Hill's problem. Ergodic Theory Dynam. Syst. **25**(4), 1237–1256 (2005)
53. A. Aparicio-Monforte, J.-A. Weil, A reduced form for linear differential systems and its application to integrability of Hamiltonian systems. J. Symb. Comput. **47**(2), 192– 213 (2012)
54. O. Pujol, J.-P. Pérez, J.-P. Ramis, C. Simó, S. Simon, J.-A. Weil, Swinging Atwood Machine: experimental and numerical results, and a theoretical study. Physica D: Nonlinear Phenomena **239**(12), 1067–1081 (2010)
55. J.-J. Morales-Ruiz, A differential Galois approach to path integrals. J. Math. Phys. **61**(5), 052103, 12 (2020)
56. C. Fieker, W.A. de Graaf, Finding integral linear dependencies of algebraic numbers and algebraic Lie algebras. LMS J. Comput. Math. **10**, 271–287 (2007)
57. W. Magnus, On the exponential solution of differential equations for a linear operator. Comm. Pure Appl. Math. **7**, 649–673 (1954)
58. J. Wei, E. Norman, Lie algebraic solution of linear differential equations. J. Math. Phys. **4**, 575–581 (1963)
59. J. Wei, E. Norman, On global representations of the solutions of linear differential equations as a product of exponentials. Proc. Amer. Math. Soc. **15**, 327–334 (1964)
60. D. Blázquez-Sanz, J.-J. Morales-Ruiz, Differential Galois theory of algebraic Lie-Vessiot systems, in *Differential Algebra, Complex Analysis and Orthogonal Polynomials*. Contemporary Mathematics, vol. 509 (American Mathematical Society, Providence, 2010), pp. 1–58
61. A. Aparicio-Monforte, J.-A. Weil, A reduction method for higher order variational equations of Hamiltonian systems, in *Symmetries and Related Topics in Differential and Difference Equations, Contemporary Mathematics*, vol. 549 (American Mathematical Society, Providence, 2011), pp. 1–15
62. M.F. Singer, Liouvillian solutions of linear differential equations with Liouvillian coefficients. J. Symb. Comput. **11**(3), 251–273 (1991)

63. T. Dreyfus, J.-A. Weil, Maple worksheet with the examples for this paper: http://www.unilim.fr/pages_perso/jacques-arthur.weil/DreyfusWeilReductionExamples.mw, (2020)
64. M.A. Barkatou, On rational solutions of systems of linear differential equations. J. Symb. Comput. **28**(4–5), 547–567 (1999)
65. M.A. Barkatou, T. Cluzeau, C. El Bacha, J.-A. Weil, Computing closed form solutions of integrable connections, in *Proceedings of the 36th International Symposium on Symbolic and Algebraic computation* (New York, NY), ISSAC '12 (ACM, New York, 2012)

Top-Down Decomposition: A Cut-Based Approach to Integral Reductions

Hjalte Frellesvig

Abstract In this contribution we will discuss a new approach to the derivation of linear relations between Feynman integrals. This new approach uses the mathematical object known as the intersection number to define what amounts to an inner product between Feynman integrals, which can be used to project directly unto the basis of master integrals. In particular we will discuss one perspective to this intersection-based method, which we name the top-down approach. This approach can be seen as an integral level version of the algorithms based on integrand reduction and generalized unitarity cuts, that were revolutionizing NLO scattering computations in the early 2000s.

1 Introduction

The derivation of linear relations between Feynman integrals is a major bottleneck in state of the art two-loop scattering amplitude calculations in the Standard Model. Following the text-book approach of writing down the Feynman diagrams, and performing the Dirac and color algebra, will give a set of $\mathcal{O}(10000)$ scalar Feynman integrals to evaluate. Thankfully such integrals are related through linear relations, that allow them all to be written in terms of a minimal set of linearly independent objects called master integrals, which rather number a more manageable $\mathcal{O}(100)$. The traditional way of deriving such relations uses Integration-By-Parts (IBP) relations [2] systematized by Laporta's algorithm [3]. This algorithm has been implemented in a number of public codes, with names such as AIR [4], FIRE [5], Reduze [6], LiteRed [7], and Kira [8]. These codes are all highly optimized, and

Based on [1].

H. Frellesvig (✉)
Niels Bohr International Academy, Copenhagen, Denmark
e-mail: hjalte.frellesvig@nbi.ku.dk

J. Blümlein, C. Schneider (eds.), *Anti-Differentiation and the Calculation of Feynman Amplitudes*, Texts & Monographs in Symbolic Computation,
https://doi.org/10.1007/978-3-030-80219-6_8

the community is benefiting from the competition between the research groups concerning the speed with which the integral reductions are done. Without these codes (or their private counterparts) multi-loop scattering amplitude computations with number of loops and legs that is the standard today, would not be possible. Yet no matter the amount of optimization that is put in, approaches based on Laporta's algorithm have as an intermediate step the solution of a very large linear system relating the various Feynman integrals, so an approach to the derivation of the linear relations that omits this step would be highly desirable. Such an approach was found in ref. [9], and consists of a projection in the vector space formed by the Feynman integrals, with the *intersection number* playing the role of an inner product between them. At one-loop (corresponding to the next-to leading order (NLO) in the perturbative expansion) the situation is different, particularly because the Feynman integrals all are known. Historically, the main bottleneck there has been the step of the Feynman diagram generation and Dirac algebra, which even at one-loop is hard to do in "real time" in the context of cross-section computations. This problem has however been solved using methods of integrand reduction and generalized unitarity cuts [10–22] which together form what is occasionally known as the OPP method after the authors of ref. [13]. The introduction of these techniques lead to the "NLO revolution" which took place around 2012, after which all one-loop scattering amplitudes may be considered known. The purpose of this document is to discuss an approach to integral decomposition using intersection numbers, which utilize insights from the OPP method, with the aim of combining these two aspects of a scattering amplitude computation, the integrand decomposition and the derivation of the integral relations, into one unified step.

2 Integrand Decomposition and the OPP Method

In this section we will discuss the OPP method and integrand reduction. For the one-loop case, these developments were made during the 1990s and early 2000s in a series of papers, of which refs. [10–22] probably is a non-exhaustive list.

For one-loop scattering amplitudes, the numerator may schematically[1] be written as a sum of numerators of one-loop diagrams with 4, 3, 2, and 1 propagators respectively. That is

(1)

[1]Equation (1) is brushing a lot under the rug. Besides the spurious terms discussed below, it does not include the rational term $\mathcal{R}$ containing the contributions not captured by four-dimensional cuts, nor does it include the extraction of pentagon-terms.

From Feynman diagrams an expression of this form can be obtained by identifying terms in the numerator of the LHS as combinations of propagators and objects that integrate to zero, but a better approach that does not require the computation of loop-level Feynman diagrams, consists of applying generalized unitarity cuts. In this context a generalized unitary cut may be seen as a δ-function insertion, or alternatively as a contour deformation into a complex loop around the pole formed by the propagator, that is

$$\frac{1}{D} \to 2\pi i \delta(D) \qquad \text{or} \qquad \int_C \frac{f(k)\mathrm{d}k}{D(k)} \to \oint \frac{f(k)\mathrm{d}k}{D(k)} \tag{2}$$

Let us start by extracting the box-coefficients d_i of Eq. (1). For each d_i this can be done by performing the quadruple-cut (that is four generalized unitarity cuts) corresponding to the four propagators of the corresponding box integral:

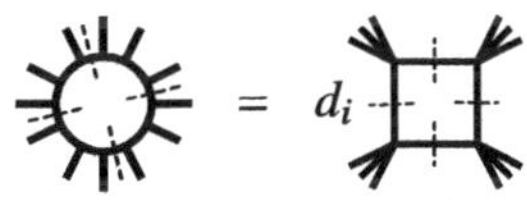

(3)

since the cut puts the lower terms to zero. The cut of the box on the RHS is 1, while the cut of the general integrand on the LHS may be identified as a product of tree-level amplitudes:

$$d_i = \sum \mathcal{A}_{i1}^{(0)} \mathcal{A}_{i2}^{(0)} \mathcal{A}_{i3}^{(0)} \mathcal{A}_{i4}^{(0)} \tag{4}$$

where the four tree-level amplitudes correspond to the four "pieces" the LHS of Eq. (3) gets cut into, and the sum goes over the various spin and flavour states the propagating particles may carry. When all the d_i-coefficients have been found, the triangle-coefficients c_j are next. They may be isolated on triple cuts:

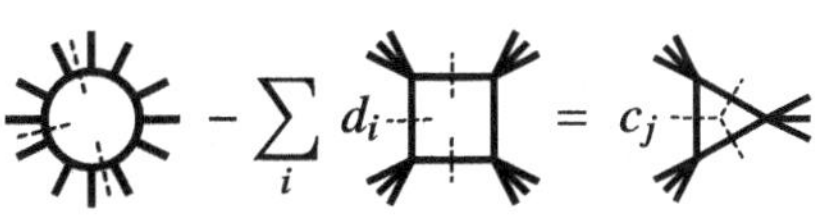

(5)

where again the LHS may be identified as a product of tree level amplitudes. The same procedure may be repeated for the bubble and tadpole coefficients b and a.

Yet writing the expansion as in Eq. (1) swipes something under the rug. There is indeed only one coefficient contributing to the physical amplitude for each term on the RHS. But at the integrand level there are additional terms, which, however, all integrate to zero. For instance a fully reduced box-integral is in full generality written as

$$\int_C \frac{d + \tilde{d}\, k\cdot\eta}{P_1(k)P_2(k)P_3(k)P_4(k)} \mathrm{d}^D k \tag{6}$$

with

$$\eta^{\mu} \propto \varepsilon^{\mu\nu_1\nu_2\nu_3} p_{1\nu_1} p_{2\nu_2} p_{3\nu_3} \tag{7}$$

defined to be perpendicular to the external momenta of the box (of which only p_1,p_2,p_3 are independent) such that the corresponding integral integrates to zero. This means that the coefficient $\tilde{d}$ doesn't contribute to any physics, and such a term is known as a spurious term. Yet it is necessary to extract it, since it will contribute to the subtraction needed to extract the triangle-coefficients c_j. For the triangles the situation is even worse since the most general irreducible integral is

$$\int_C \frac{\Delta_{\text{tri}}(k)}{P_1(k)P_2(k)P_3(k)P_4(k)} \mathrm{d}^D k \tag{8}$$

with

$$\begin{aligned}\Delta_{\text{tri}} = c \,+\, \tilde{c}_{10}\,(k\cdot\eta_1) \,+\, \tilde{c}_{01}\,(k\cdot\eta_2) \,+\, \tilde{c}_{11}\,(k\cdot\eta_1)(k\cdot\eta_2) \,+\, \tilde{c}_{12}\,(k\cdot\eta_1)(k\cdot\eta_2)^2 \\ +\, \tilde{c}_{21}\,(k\cdot\eta_1)^2(k\cdot\eta_2) \,+\, \tilde{c}_{20;02}\,((k\cdot\eta_1)^2 - (k\cdot\eta_2)^2)\end{aligned} \tag{9}$$

where η_1 and η_2 are defined to be perpendicular to the external momenta of the triangle, and to each other. Again the tilded c-coefficients multiply terms that integrate to zero, so this means that for the triangles there are 6 spurious term for each genuine triangle coefficient. A similar explosion of terms take place for the bubbles and the tadpoles.

Another potential issue of the expansion of Eq. (1) is that we may end up computing coefficients of integrals that are not independent. For instance in the fully mass-less case the triangle and bubble integrals are related through the well-known relation

 $= \frac{2(D-3)}{s(D-4)} \times$  (10)

meaning that it is redundant to discuss triangle-coefficients and bubble-coefficients as distinct objects.

Yet of course none of these issues—the presence of spurious terms and the potential over-counting—are real obstacles at the one-loop level and none of it prevented the NLO revolution.

At the two-loop level the situation is however different. The computation of Feynman integrals is so much harder at the multi-loop level, that this is where the main obstacles are for such computations rather than at the integrand reduction step. Yet still there have been attempts as generalizing the procedure outlined above to the multi-loop case [23–43]. And while such an approach has proven useful for actual NNLO scattering amplitude computations, it has not brought about any "NNLO revolution". The reason for this is that the two problems outlined above get a lot worse at the multiloop level. Let us limit ourselves to the fully massless double-box. There an integrand reduction similar to Eq. (1) may be written as

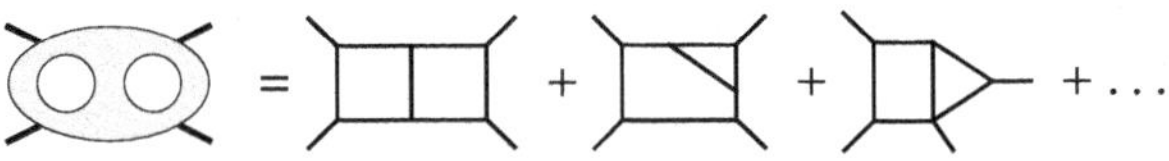

(11)

We have not written prefactors in front of the terms on the RHS, and that is because in the multiloop case there is not just one term per integral sector. For the double-box the irreducible numerator may be written as

$$\Delta_{\text{d-box}} = c_1 + c_2(k_2 \cdot p_1) + \ldots + c_{32}(k_1 \cdot p_4)(k_2 \cdot p_1)^3(k_2 \cdot \eta) \tag{12}$$

with 32 terms in total of which half are spurious and the other half not. This means there will be 16 double-box terms contributing to the physics. Compare this with the fact that the double-box has two master-integrals in its highest sector, meaning we end up with eight times the minimal number of coefficients. For the next two terms on the RHS of Eq. (11) the situation is even worse [44]. The second integral has $10 + 10$ terms in its irreducible numerator, and the third has $35 + 34$, but none of these two integrals have any master integrals at all in their highest sector, so an ideal approach would not be considering them at all.

These examples should motivate why it is worthwhile to search for a different approach that reduce the integrand directly unto master integrals, omitting the two issues of spurious and redundant terms—an approach that might be thought of as integral level OPP.

3 Integral Reduction and Intersection Theory

As discussed in the introduction, the reduction of Feynman integrals unto a minimal basis of master integrals, is a major bottleneck in modern day scattering amplitude computations. Writing a Feynman integral in terms of master integrals

$$I = \sum_{i}^{\nu} c_i I_i \tag{13}$$

with ν denoting the number of master integrals, seems reminiscent of writing a vector in terms of as set of basis-vectors. And indeed that correspondence is not merely metaphorical, it is not hard to realize that Feynman integrals through their linear relations do form a vectorspace. For vector spaces endowed with an inner product, extracting the coefficients of the basis vectors can be done with a projection:

$$\begin{aligned} \langle v| &= \langle v|v_j\rangle \left(\mathbf{C}^{-1}\right)_{ji} \langle v_i| && \text{with} && \mathbf{C}_{ij} = \langle v_i|v_j\rangle \\ &= \sum_i c_i \langle v_i| && \text{with} && c_i = \langle v|v_j\rangle \left(\mathbf{C}^{-1}\right)_{ji} \end{aligned} \tag{14}$$

The case of an orthonormal basis may be more familiar, there

$$\langle v_i | v_j \rangle = \delta_{ij} \qquad \Rightarrow \qquad c_i = \langle v | v_i \rangle \tag{15}$$

If this was possible to do for Feynman integrals, it would be possible to get around the IBP relations and the associated large linear system that have to be inverted in the traditional method of integral reduction. But to do this requires the introduction of an object which may play the role of an inner product between Feynman integrals.

To get there we have to introduce a parametric representation for Feynman integrals. We will focus on the Baikov representation [45] (see also [46–51]), but many of the following steps would be possible also with more traditional parametrizations such as Schwinger or Feynman parameters. In the Baikov representation, a Feynman integral is expressed as

$$I = \int \frac{\mathrm{d}^D k_1}{\pi^{D/2}} \cdots \frac{\mathrm{d}^D k_L}{\pi^{D/2}} \frac{N(\mathbf{k})}{P_1(\mathbf{k})^{a_1} \cdots P_{\mathrm{P}}(\mathbf{k})^{a_{\mathrm{P}}}} = K \int_C \frac{N(\mathbf{x})\, \mathcal{B}^\gamma(\mathbf{x})\, \mathrm{d}^n x}{x_1^{a_1} \cdots x_{\mathrm{P}}^{a_{\mathrm{P}}}} \tag{16}$$

Here $\gamma = d - E - L - 1$ where E is the number of independent external momenta, and K (which we leave out in the following) additionally a function of the kinematics. $\mathcal{B}$ is the Baikov polynomial defined as

$$\mathcal{B} = \det G \left(\left\{ k_1, \cdots, k_L, p_1, \cdots, p_E \right\} \right) \tag{17}$$

with the Gram matrix G being defined as the matrix of scalar product of its argument vectors with themselves. $\mathcal{B}$ will be a polynomial function of the Baikov variables x_i which equal the propagators. The integration contour C equals the area in which the Baikov polynomial is positive, and the a_i are integers. There is also a loop-by-loop version of Baikov parametrization [50] where the Baikov polynomial of Eq. (16) is replaced with a product of polynomials raised to different powers $\prod_{i=1}^{2L-1} \mathcal{B}_i^{\gamma_i}$ trading some simplicity for a smaller number of integration variables.

We may write the integral of Eq. (16) as

$$I = \int_C u\, \phi \qquad \text{with} \qquad u = \mathcal{B}^\gamma \qquad \text{and} \qquad \phi = \frac{N(\mathbf{x})\, \mathrm{d}^n x}{x_1^{a_1} \cdots x_{\mathrm{P}}^{a_{\mathrm{P}}}} \tag{18}$$

where the u is a multivalued function and ϕ a differential form with rational prefactor. In fact ϕ and C may be thought of as not just a form and a contour, but as a representative of an equivalence-class of forms and contours respectively which share the property that the integral of Eq. (18) gives the same result. In this context C is known as a twisted cycle, and ϕ as a twisted cocycle. With this notation we may further write I as a pairing between a twisted cycle and cocycle, $I = \langle \phi | C]$ with the multivalued function u being implicit. It is likewise possible to define a dual integral $[C|\varphi\rangle = \int_C u^{-1} \varphi$ showcasing the definition of the dual cycle and cocycle. And from this we have all the ingredients needed to define the *intersection number* as a different kind of pairing, not between a cycle and a cocycle but rather between a cocycle and a dual cocycle $\langle\phi|\varphi\rangle$. This object plays exactly the role of an inner product between Feynman integrals needed in order to be able to use Eq. (14) to extract coefficients of master integrals. Writing all the integrals as $I_i = \langle\phi_i|C]$ the relation becomes

$$I = \sum_i^{\nu} c_i I_i \quad \text{with} \quad c_i = \langle\phi|\varphi_j\rangle \left(\mathbf{C}^{-1}\right)_{ji} \quad \text{with} \qquad \mathbf{C}_{ij} = \langle\phi_i|\varphi_j\rangle \tag{19}$$

The exact mathematical definition of the intersection number will not be given here. For this the reader should regard the mathematical literature [53–55]. Rather we will discuss how to compute it. In the univariate case (that is when the ϕ_i are univariate differential forms) the intersection number is given as

$$\langle\phi|\varphi\rangle = \sum_{p\in\mathcal{P}} \text{Res}_{z=p}(\psi\varphi) \qquad \text{with} \qquad (\mathrm{d}+\omega)\psi = \phi \tag{20}$$

Here $\omega \equiv \mathrm{dlog}(u)$ is known as the twist, and $\mathcal{P}$ is the set of poles of ω. ψ has to be found by solving the differential equation on the right of Eq. (20). But since the result only has to be used inside the residue function, it is not necessary with an exact solution. It is enough to make a series ansatz around the pole p as $\psi \to \psi_p = \sum_{i=\min}^{\max} \psi_p^{(i)} z^i$ where min and max have to be chosen such that all powers that may contribute to the residue are included.

With the loop-by-loop version of Baikov representation briefly discussed above, it is possible to get quite a few Feynman integrals to a form that is univariate on the maximal cut, such that univariate intersection theory applies. This allows us to extract integral relations valid on the maximal cut, which corresponds to extracting the coefficients of master integrals in the highest sector. In ref. [52] we did this for quite a few Feynman integrals, illustrated in Fig. 1.

Yet to get the complete reduction of a Feynman integral, a multivariate intersection number is needed. One approach to the computation of multivariate intersection numbers [1, 56, 57], is given by the following set of equations:

$$_{\mathbf{n}}\langle\phi^{(\mathbf{n})}|\varphi^{(\mathbf{n})}\rangle = \sum_{p\in\mathcal{P}_n} \text{Res}_{z_n=p}\left(\psi_i^{(n)}\,_{\mathbf{n-1}}\langle e_i^{(\mathbf{n-1})}|\varphi^{(\mathbf{n})}\rangle\right) \tag{21}$$

$$\left(\delta_{ij}\partial_{z_n} + \hat{\mathbf{\Omega}}_{ji}^{(n)}\right)\psi_j^{(n)} = \hat{\phi}_i^{(n)} \tag{22}$$

$$\mathrm{d}_{z_n} v_i = \mathbf{\Omega}_{ij}^{(n)} v_j \qquad v_i = \int_{C_{n-1}} u\, e_i^{(\mathbf{n-1})} \tag{23}$$

$$\hat{\mathbf{\Omega}}_{ij}^{(n)} = \,_{\mathbf{n-1}}\langle(\partial_{z_n}+\hat{\omega}_n)e_i^{(\mathbf{n-1})}|h_k^{(\mathbf{n-1})}\rangle\left(\mathbf{C}_{(\mathbf{n-1})}^{-1}\right)_{kj} \tag{24}$$

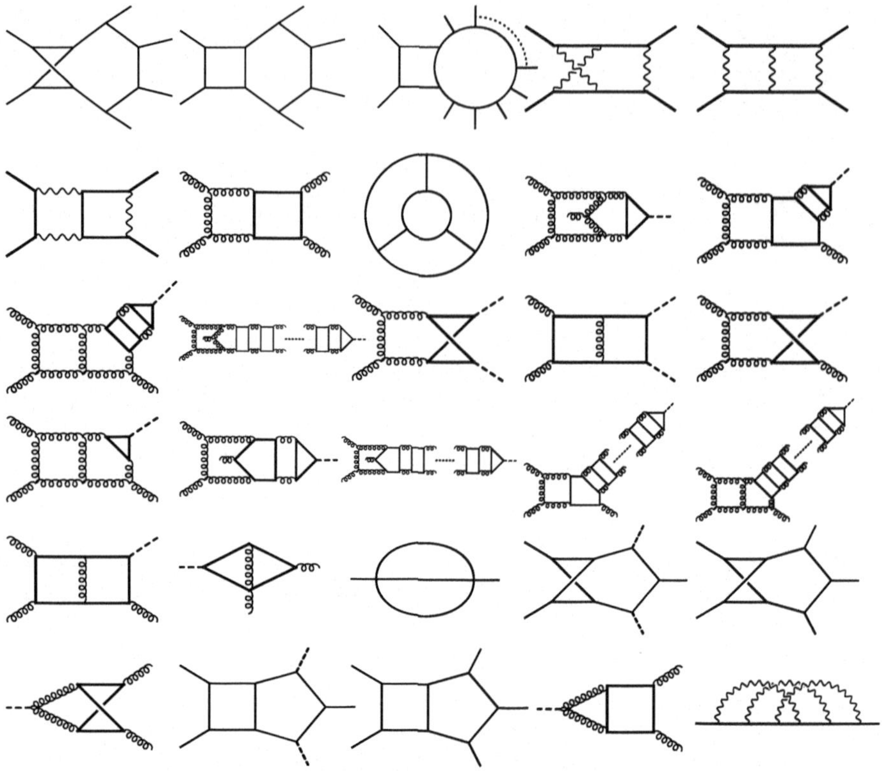

Fig. 1 Univariate examples from ref. [52] where they are reduced on the maximal cut. Please note the presence of L-loop and n-leg examples, something that would not be directly reducible with the public IBP-based codes

$$\phi_i^{(n)} = {}_{\mathbf{n-1}}\langle \phi^{(\mathbf{n})} | h_j^{(\mathbf{n-1})} \rangle \left(\mathbf{C}^{-1}_{(\mathbf{n-1})} \right)_{ji} \tag{25}$$

$$\mathbf{C}_{ij}^{(\mathbf{n-1})} = {}_{\mathbf{n-1}}\langle e_i^{(\mathbf{n-1})} | h_j^{(\mathbf{n-1})} \rangle \tag{26}$$

We will not go through in detail how to interpret or use these formulae, merely discuss a few of their features. ${}_{\mathbf{n}}\langle * | * \rangle$ denotes an n-variate intersection number. We see from Eq. (21) that the n-variate intersection number is given in terms of $n-1$-variate intersection numbers. This makes the formula recursive, and the recursion will end since we know how to compute the univariate intersection number. We also see that the intersection number is given as a sum of residues as in the univariate case, and that a ψ appears in the argument of the residue which is defined as a solution to a differential equation (22), a property also shared by the univariate case. The main difference from the univariate case is the presence of the connection-matrix Ω which replaces the ω of Eq. (20). Ω is defined through Eq. (23), and we

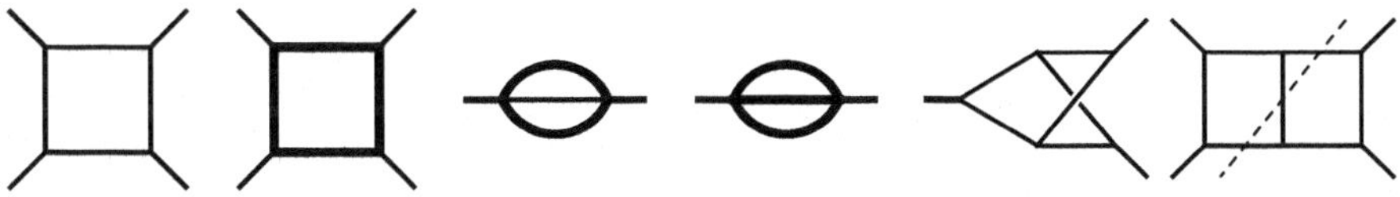

Fig. 2 Multivariate examples from ref. [1]

see that for $n = 1$ $v_i = u$ and $\Omega^{(1)} = \mathrm{d}\log(u)$ is a solution, but in general Ω is not given as a $\mathrm{d}\log$ of anything and must be computed using Eq. (24).

With the multivariate intersection number we are able to perform the full reduction of a number of Feynman integrals at one and two loops. See Fig. 2. In refs. [1, 57] we developed a number of different approaches to this complete reduction. The "straight decomposition" consist of the direct use of Eq. (19). The "bottom-up decomposition" consist of performing the reduction on a set of "spanning cuts" which together allows for the extraction of all the master integral coefficients. Finally there is the "top-down decomposition" which is the subject of the following section.

But before discussing the top-down approach, let us show an example of a reduction done in the "straight decomposition" approach. This will also be the example we will discuss in the context of top-down in the following section. The example is that of a one-loop massless box-integral reduced unto the set of master integrals consisting of a box and an s and a t-channel bubble. The dots on the LHS denote higher propagator powers.

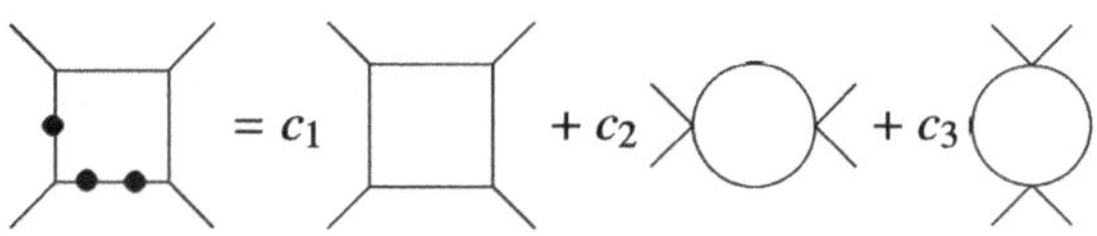

(27)

Performing the Baikov parametrization yields

$$u(\mathbf{x}) = \big((st - sx_4 - tx_3)^2 - 2tx_1(s(t+2x_3-x_2-x_4)+tx_3) + s^2x_2^2 + t^2x_1^2 - 2sx_2(t(s-x_3)+x_4(s+2t))\big)^{\frac{D-5}{2}} \tag{28}$$

with $u = \mathcal{B}^\gamma$ as given by Eq. (16). The integrals are then given as $I_i = \int_C u\hat{\phi}_i \mathrm{d}^4x$ with

$$\hat{\phi} = \frac{1}{x_1^3 x_2^2 x_3 x_4}, \quad \hat{\phi}_1 = \frac{1}{x_1x_2x_3x_4}, \quad \hat{\phi}_2 = \frac{1}{x_1x_3}, \quad \hat{\phi}_3 = \frac{1}{x_2x_4}. \tag{29}$$

From the u of Eq. (28) we may compute $\omega = \mathrm{d}\log(u)$, but then we encounter a problem: For the theory to be valid, it is a requirement that there are no poles of the ϕ_i that are not poles of ω. But the $x_i = 0$ poles of the ϕ_i are not present in ω so this assumption is violated! A solution to this is to regulate omega by performing

$$u \to u\, x_1^{\rho} x_2^{\rho} x_3^{\rho} x_4^{\rho} \tag{30}$$

where ρ is a regulator that can be put to zero at the end of the computation. This replacement corresponds to $\omega \to \omega + \sum_i \rho / x_i$ and therefore no illegal poles are left. With this regulator in place we may go ahead and apply Eq. (19). To do so requires the computation of 12 4-variate intersection numbers. That is between the four ϕ_i of Eq. (29) and the three members of a dual basis φ_i which can be chosen arbitrarily. We will not here write the individual intersection numbers, but combining them according to Eq. (19) and taking the $\rho \to 0$ limit, gives the result

$$c_1 = \frac{-(D-7)(D-6)(D-5)}{2s^2 t}, \quad c_2 = \frac{2(D-7)(D-5)(D-3)}{s^4 t},$$
$$c_3 = \frac{2(D-7)(D-5)(D-3)(2s+(D-8)t)}{(D-8)s^2 t^4} \tag{31}$$

for the three master integral coefficients, a result which is in agreement with that produced by standard IBP-codes such as FIRE [58].

4 The Top-Down Decomposition Approach

Let us try to redo the reduction of Eq. (27) with the top-down decomposition approach, which is best explained through an example. The philosophy is to try to emulate the integrand-level reduction as given by Eq. (1). Performing the maximal cut isolates the box-coefficient as in Eq. (3), since the bubble-terms vanish as the cut corresponds to taking a residue where none is present. That is

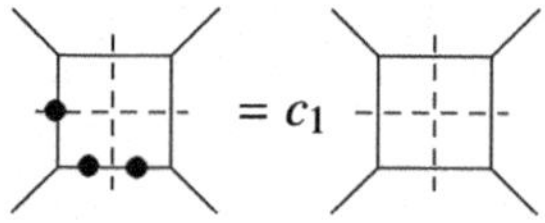

(32)

Using the u of Eq. (28), this gets evaluated[2] as

$$\oint \frac{u}{x_1^3 x_2^2 x_3 x_4} = c_1 \oint \frac{u}{x_1 x_2 x_3 x_4} \;\Leftrightarrow\; \tfrac{1}{2}\partial_{x_1}^2 \partial_{x_2} u\big|_{x_i \to 0} = c_1 u\big|_{x_i \to 0} \;\Leftrightarrow$$
$$c_1 = \frac{\tfrac{1}{2}\partial_{x_1}^2 \partial_{x_2} u}{u}\bigg|_{x_i \to 0} = \frac{-(D-7)(D-6)(D-5)}{2s^2 t} \tag{33}$$

in agreement with Eqs. (31).

[2]We note that this step has nothing to do with intersection theory, and was done for instance in ref. [45].

Next we do the double-cut corresponding to the s-channel bubble. We may move the now known box-coefficient to the LHS of the equation (as in Eq. (5)), giving

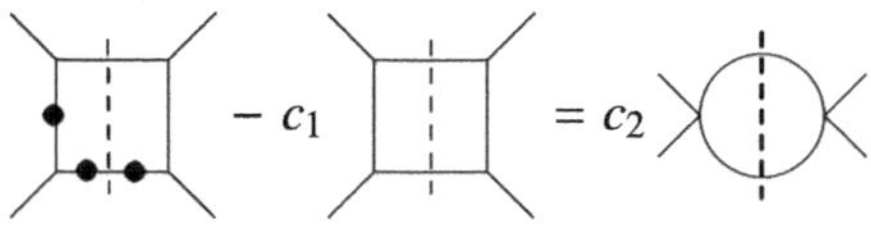

(34)

The LHS may be put together as one expression

$$\left.\left(\frac{\frac{1}{2}\partial_{x_1}^2 u}{x_2^2 x_4} - \frac{c_1 u}{x_2 x_4}\right)\right|_{x_{1,3}\to 0} = \Phi\, u_{\text{bub}} \tag{35}$$

where u_{bub} is u on the bubble-cut

$$u_{\text{bub}} = u|_{x_1,x_3\to 0} = \mathcal{B}_{\text{bub}}^{(D-5)/2} \tag{36}$$

$$\mathcal{B}_{\text{bub}} = \Big(st^2 + s(x_2-x_4)^2 - 2t(s(x_2+x_4)+2x_2x_4)\Big) \tag{37}$$

and $\Phi = \hat{\Phi}\, \mathrm{d}x_2 \wedge \mathrm{d}x_4$ is a rational two-form

$$\hat{\Phi} = \left.\frac{\frac{1}{2}\partial_{x_1}^2 u}{u x_2^2 x_4}\right|_{x_1,x_3\to 0} - \frac{c_1}{x_2 x_4} = \frac{\mathcal{P}(x_2,x_4)}{x_2^2 x_4 \mathcal{B}_{\text{bub}}^2} \tag{38}$$

where $\mathcal{P}$ is a polynomial in x_2 and x_4. We want to reduce this unto the bubble which on the cut has $\hat{\phi}_{\text{bub}} = 1$. But once again we encounter the problem that Φ has poles (in x_2 and x_4) which are not present in $\omega_{\text{bub}} = \mathrm{d}\log(u_{\text{bub}})$. As in the previous section we could fix this by introducing regulators $u_{\text{bub}} \to u_{\text{bub}} x_2^\rho x_4^\rho$ which would introduce the poles and make the procedure of Eq. (19) go through. But to do so would invalidate the point of performing the subtractions in the first place, since their entire purpose is to remove these poles as it happened in the integrand-level computation in Sect. 2. But of course there is no guarantee that these poles will cancel explicitly as we see. Rather what is guaranteed is that the resulting Φ is in the same equivalence class as a $\tilde{\Phi}$ without the poles, that is there exist a ξ such that

$$\tilde{\Phi} = \Phi - (\mathrm{d} + \omega_{\text{bub}})\xi = \frac{\tilde{\mathcal{P}}(x_2,x_4)}{\mathcal{B}_{\text{bub}}^2}\, \mathrm{d}x_2 \wedge \mathrm{d}x_4 \tag{39}$$

where $\tilde{\Phi}$ is guaranteed to describe the same integral as the original Φ. $\tilde{\mathcal{P}}$ is another polynomial which will be different from $\mathcal{P}$. Our approach will be to find such a ξ

explicitly. We make an ansatz

$$\xi = \frac{\sum_{i=-1,j=-1}^{2,2} \kappa_{1ij} x_2^i x_4^j \mathrm{d}x_2 + \sum_{i=-2,j=0}^{2,2} \kappa_{2ij} x_2^i x_4^j \mathrm{d}x_4}{\mathcal{B}_{\mathrm{bub}}} \tag{40}$$

We may then insert it in Eq. (39), and impose that $\hat{\tilde{\Phi}} \mathcal{B}_{\mathrm{bub}}^2 = \sum_{ij} q_{ij} x_2^i x_4^j$ has all q_{ij} with either $i < 0$ or $j < 0$ vanishing. Doing so will fix the κ-coefficients of the ansatz,[3] and result in a $\tilde{\Phi}$ without the left-over poles from the box-propagators.

With this in place we may now directly apply Eq. (19) to extract the bubble-coefficient without the need for any regulators:

$$c_2 = \frac{\langle \tilde{\Phi} | 1 \rangle}{\langle 1 | 1 \rangle} = \frac{2(D-7)(D-5)(D-3)}{s^4 t} \tag{41}$$

in agreement with Eqs. (31). The other bubble-coefficient c_3 may be extracted from the other bubble-cut in a similar fashion, resulting in all the master integral coefficients being known.

We see that the top-down approach to integral decomposition in this example required the calculation of 4 2-variable intersection numbers, as opposed to the 12 4-variate intersection numbers needed for the straight approach to the decomposition discussed in the previous section. Thus the top-down approach is an improvement in all respects. The trade-of is the need for solving the linear system of equations that impose the vanishing of the poles from $\tilde{\Phi}$, which is somewhat reminiscent of the linear systems that have to be solved in the traditional IBP-based approach to integral decomposition.

Let us finish the section by describing the top-down approach as an algorithm in the general case. Starting from the sectors with the most propagators we may compute

$$\tilde{\Phi} = \phi - \sum_{i \in \text{higher}} c_i \phi_i - (\mathrm{d} + \omega)\xi \tag{42}$$

where ω is computed from the u of that sector, and where ξ should be fitted such that all unwanted poles of $\tilde{\Phi}$ vanish. Then we may compute the coefficients of the masters in that sector with Eq. (19):

$$c_i = \langle \tilde{\Phi} | \varphi_j \rangle (\mathbf{C}^{-1})_{ji} \qquad \text{with} \qquad \mathbf{C}_{ij} = \langle \phi_i | \varphi_j \rangle \tag{43}$$

[3]We note that had we replaced the subtraction of the box-coefficient in Eq. (38) with a free subtraction term ($c_1 \to \kappa_1$), this κ-fitting step would give a solution only if $\kappa_1 = c_1$, so treating κ_1 on equal footing with the other κs, is another approach to fixing the box-coefficient.

This procedure should be repeated sector by sector until all master integral coefficients c_i are known.

5 Discussion

For now, the most efficient intersection-based approach to integral-decomposition, is the bottom-up approach described in detail in refs. [1, 57]. Despite the fact that the bottom-up approach computes higher sector coefficients several times, it beats the top-down approach due to the step involving the fitting of the coefficients of ξ that makes the higher sector poles vanish. This has to be done for each sector, and can be highly involved in low sectors of multi-loop amplitudes. If there were a way to identify a suitable ξ without this fitting-step the situation would be very different, but if this is possible is a good open question.

Yet the main bottleneck to intersection-based integral decomposition is not specific to the top-down approach. It involves the multivariate intersection number, which in our approach is computed using Eqs. (21–26). That approach has a number of subtle issues and open questions (partially discussed in refs. [1, 57]) and to find a different approach to computing the multivariate intersection number would be of both theoretical interest and practical use. It is known that much simpler formulae exist for the special case of all ϕ_i having $d\log$-form [55], and to find a formula of similar complexity for the multivariate intersection number valid for general ω and ϕ would be a huge improvement. (See ref. [59] for some work in that direction.)

As we saw above the top-down approach avoids the introduction of the regulators ρ discussed around Eq. (30). But another approach to this may be to define the cohomology-theory relative to the location of the unwanted poles. This is known as "relative cohomology" and is discussed mathematically in the univariate case in ref. [60]. To turn this into a form where it is directly applicable, and to generalize it to the multivariate case, would be a worthwhile research direction.

In these proceeding and in most of the work of refs. [1, 52, 57] we focused on general Feynman integrals. Yet it is very possible that simplifications occur for integrals with special "nice" properties that one might be interested in in other contexts as well. This refers mainly to integrals of uniform transcendentality [61, 62] (see refs. [63, 64] for developments in that direction), but also for instance finite integrals may be worth studying in the context of intersection theory. As a related question, also an investigation into connections between the intersection number and the pole- and branch cut-structure of the integral seems worthwhile. (See also ref. [65]).

At the moment the intersection number-based approach to integral decomposition is not a competitor in terms of speed or applicability to the highly optimized public [4–8] or private codes based on the IBP approach and Laporta's algorithm. This is due to the issues discussed above, particularly the intricateness of the multivariate intersection algorithm. The authors of refs. [1, 52, 57] have their private implementation of the multivariate algorithm as given by Eqs. ((21)–(26)), but that algorithm and the implementation still requires some amount of tweaking for individual cases. How satisfactory these issues will be solved in the future is of course unknown, but I believe that due to its mathematical simplicity—being merely projections in the vector space of Feynman integrals—the intersection-based algorithm has the potential to become a serious competitor to IBP-based approaches.

As we saw on the previous pages, it is possible to extend a lot of the structure of the OPP method to the integral level. This has been made possible thanks to the introduction of the intersection number as an effective inner product between Feynman integrals. If these developments are going to bring about progress comparable to the NLO revolution of the early 2000s is a question for the future.

Acknowledgments I would like to thank my collaborators on the work presented here, Federico Gasparotto, Stefano Laporta, Manoj K. Mandal, Pierpaolo Mastrolia, Luca Mattiazzi, and Sebastian Mizera. This project has received funding from the European Union's Horizon 2020 research and innovation program under the Marie Skłodowska-Curie grant agreement No. 847523 'INTERACTIONS'. The work of HF has been partially supported by a Carlsberg Foundation Reintegration Fellowship.

References

1. H. Frellesvig, F. Gasparotto, S. Laporta, M.K. Mandal, P. Mastrolia, L. Mattiazzi, S. Mizera, *Decomposition of Feynman Integrals by Multivariate Intersection Numbers*. arXiv:2008.04823
2. K.G. Chetyrkin, F.V. Tkachov, Integration by parts: the algorithm to calculate beta functions in 4 loops. Nucl. Phys. **B192**, 159–204 (1981)
3. S. Laporta, High precision calculation of multiloop Feynman integrals by difference equations. Int. J. Mod. Phys. **A15**, 5087–5159 (2000)
4. C. Anastasiou, A. Lazopoulos, Automatic integral reduction for higher order perturbative calculations. J. High Energy Phys. **07**, 046 (2004)
5. A.V. Smirnov, F.S. Chuharev, *FIRE6: Feynman Integral REduction with Modular Arithmetic*. arXiv:1901.07808
6. A. von Manteuffel, C. Studerus, *Reduze 2 - Distributed Feynman Integral Reduction*. arXiv:1201.4330
7. R.N. Lee, *Presenting LiteRed: A Tool for the Loop InTEgrals REDuction*. arXiv:1212.2685
8. J. Klappert, F. Lange, P. Maierhöfer, J. Usovitsch, *Integral Reduction with Kira 2.0 and Finite Field Methods*. arXiv:2008.06494
9. P. Mastrolia, S. Mizera, Feynman integrals and intersection theory. J. High Energy Phys. **02**, 139 (2019). arXiv:1810.03818
10. Z. Bern, L.J. Dixon, D.C. Dunbar, D.A. Kosower, Fusing gauge theory tree amplitudes into loop amplitudes. Nucl. Phys. **B435**, 59–101 (1995). hep-ph/9409265
11. Z. Bern, L.J. Dixon, D.C. Dunbar, D.A. Kosower, One loop n point gauge theory amplitudes, unitarity and collinear limits. Nucl. Phys. **B425**, 217–260 (1994). hep-ph/9403226
12. R. Britto, F. Cachazo, B. Feng, Generalized unitarity and one-loop amplitudes in N=4 super-Yang-Mills. Nucl. Phys. B **725**, 275–305 (2005). hep-th/0412103
13. G. Ossola, C.G. Papadopoulos, R. Pittau, Reducing full one-loop amplitudes to scalar integrals at the integrand level. Nucl. Phys. B **763**, 147–169 (2007). hep-ph/0609007
14. R. Ellis, W. Giele, Z. Kunszt, A numerical unitarity formalism for evaluating one-loop amplitudes. J. High Energy Phys. **03**, 003 (2008). arXiv:0708.2398
15. C. Anastasiou, R. Britto, B. Feng, Z. Kunszt, P. Mastrolia, D-dimensional unitarity cut method. Phys. Lett. B **645**, 213–216 (2007). hep-ph/0609191

16. D. Forde, Direct extraction of one-loop integral coefficients. Phys. Rev. D **75**, 125019 (2007). arXiv:0704.1835
17. E. Nigel Glover, C. Williams, One-loop gluonic amplitudes from single unitarity cuts. J. High Energy Phys. **12**, 067 (2008). arXiv:0810.2964
18. P. Mastrolia, Double-cut of scattering amplitudes and Stokes' theorem. Phys. Lett. **B678**, 246–249 (2009). arXiv:0905.2909
19. S. Badger, Direct extraction of one loop rational terms. J. High Energy Phys. **01**, 049 (2009). arXiv:0806.4600
20. R. Britto, B. Feng, Solving for tadpole coefficients in one-loop amplitudes. Phys. Lett. B **681**, 376–381 (2009). arXiv:0904.2766
21. R. Ellis, W.T. Giele, Z. Kunszt, K. Melnikov, Masses, fermions and generalized D-dimensional unitarity. Nucl. Phys. B **822**, 270–282 (2009). arXiv:0806.3467
22. P. Mastrolia, E. Mirabella, T. Peraro, Integrand reduction of one-loop scattering amplitudes through Laurent series expansion. J. High Energy Phys. **06**, 095 (2012). arXiv:1203.0291 [Erratum: J. High Energy Phys. 11, 128 (2012)]
23. P. Mastrolia, G. Ossola, On the integrand-reduction method for two-loop scattering amplitudes. J. High Energy Phys. **11**, 014 (2011). arXiv:1107.6041
24. P. Mastrolia, E. Mirabella, G. Ossola, T. Peraro, Scattering amplitudes from multivariate polynomial division. Phys. Lett. B **718**, 173–177 (2012). arXiv:1205.7087
25. D.A. Kosower, K.J. Larsen, Maximal unitarity at two loops. Phys. Rev. **D85**, 045017 (2012). arXiv:1108.1180
26. S. Badger, H. Frellesvig, Y. Zhang, Hepta-cuts of two-loop scattering amplitudes. J. High Energy Phys. **04**, 055 (2012). arXiv:1202.2019
27. S. Badger, H. Frellesvig, Y. Zhang, An integrand reconstruction method for three-loop amplitudes. J. High Energy Phys. **08**, 065 (2012). arXiv:1207.2976
28. S. Badger, H. Frellesvig, Y. Zhang, A two-loop five-gluon helicity amplitude in QCD. J. High Energy Phys. **12**, 045 (2013). arXiv:1310.1051
29. Y. Zhang, Integrand-level reduction of loop amplitudes by computational algebraic geometry methods. J. High Energy Phys. **09**, 042 (2012). arXiv:1205.5707
30. M. Søgaard, Global residues and two-loop hepta-cuts. J. High Energy Phys. **09**, 116 (2013). arXiv:1306.1496
31. M. Sogaard, Y. Zhang, Massive nonplanar two-loop maximal unitarity. J. High Energy Phys. **12**, 006 (2014). arXiv:1406.5044
32. M. Søgaard, Y. Zhang, Elliptic functions and maximal unitarity. Phys. Rev. D **91**(8), 081701 (2015). arXiv:1412.5577
33. P. Mastrolia, T. Peraro, A. Primo, Adaptive Integrand Decomposition in parallel and orthogonal space. J. High Energy Phys. **08**, 164 (2016). arXiv:1605.03157
34. S. Badger, G. Mogull, A. Ochirov, D. O'Connell, A complete two-loop, five-gluon helicity amplitude in Yang-Mills theory. J. High Energy Phys. **10**, 064 (2015). arXiv:1507.08797
35. S. Badger, G. Mogull, T. Peraro, Local integrands for two-loop all-plus Yang-Mills amplitudes. J. High Energy Phys. **08**, 063 (2016). arXiv:1606.02244
36. S. Badger, C. Brønnum-Hansen, H.B. Hartanto, T. Peraro, First look at two-loop five-gluon scattering in QCD. Phys. Rev. Lett. **120**(9), 092001 (2018). arXiv:1712.02229
37. S. Badger, C. Brønnum-Hansen, H.B. Hartanto, T. Peraro, Analytic helicity amplitudes for two-loop five-gluon scattering: the single-minus case. J. High Energy Phys. **01**, 186 (2019). arXiv:1811.11699
38. S. Badger, D. Chicherin, T. Gehrmann, G. Heinrich, J. Henn, T. Peraro, P. Wasser, Y. Zhang, S. Zoia, Analytic form of the full two-loop five-gluon all-plus helicity amplitude. Phys. Rev. Lett. **123**(7), 071601 (2019). arXiv:1905.03733
39. H. Ita, Two-loop integrand decomposition into master integrals and surface terms. Phys. Rev. D **94**(11), 116015 (2016). arXiv:1510.05626
40. S. Abreu, F. Febres Cordero, H. Ita, M. Jaquier, B. Page, M. Zeng, Two-loop four-gluon amplitudes from numerical unitarity. Phys. Rev. Lett. **119**(14), 142001 (2017). arXiv:1703.05273

41. S. Abreu, F. Febres Cordero, H. Ita, B. Page, M. Zeng, Planar two-loop five-gluon amplitudes from numerical unitarity. Phys. Rev. D **97**(11), 116014 (2018). arXiv:1712.03946
42. S. Abreu, F. Febres Cordero, H. Ita, B. Page, V. Sotnikov, Planar two-loop five-parton amplitudes from numerical unitarity. J. High Energy Phys. **11**, 116 (2018). arXiv:1809.09067
43. S. Abreu, J. Dormans, F. Febres Cordero, H. Ita, B. Page, V. Sotnikov, Analytic form of the planar two-loop five-parton scattering amplitudes in QCD. J. High Energy Phys. **05**, 084 (2019). arXiv:1904.00945
44. H.A. Frellesvig, Generalized unitarity cuts and integrand reduction at higher loop orders. Ph.D. thesis, Copenhagen U., 2014
45. P.A. Baikov, Explicit solutions of the multiloop integral recurrence relations and its application. Nucl. Instrum. Meth. **A389**, 347–349 (1997). hep-ph/9611449
46. R.N. Lee, Calculating multiloop integrals using dimensional recurrence relation and D-analyticity. Nucl. Phys. Proc. Suppl. **205–206**, 135–140 (2010). arXiv:1007.2256
47. A.G. Grozin, Integration by parts: an introduction. Int. J. Mod. Phys. **A26**, 2807–2854 (2011). arXiv:1104.3993
48. K.J. Larsen, Y. Zhang, Integration-by-parts reductions from unitarity cuts and algebraic geometry. Phys. Rev. **D93**(4), 041701 (2016). arXiv:1511.01071
49. J. Bosma, K.J. Larsen, Y. Zhang, *Differential Equations for Loop Integrals in Baikov Representation*. arXiv:1712.03760
50. H. Frellesvig, C.G. Papadopoulos, Cuts of Feynman integrals in Baikov representation. J. High Energy Phys. **04**, 083 (2017). arXiv:1701.07356
51. M. Harley, F. Moriello, R.M. Schabinger, Baikov-Lee representations of cut Feynman integrals. J. High Energy Phys. **06**, 049 (2017). arXiv:1705.03478
52. H. Frellesvig, F. Gasparotto, S. Laporta, M.K. Mandal, P. Mastrolia, L. Mattiazzi, S. Mizera, Decomposition of Feynman integrals on the maximal cut by intersection numbers. J. High Energy Phys. **05**, 153 (2019). arXiv:1901.11510
53. K. Cho, K. Matsumoto, Intersection theory for twisted cohomologies and twisted Riemann's period relations I. Nagoya Math. J. **139**, 67–86 (1995)
54. K. Matsumoto, Intersection numbers for logarithmic k-forms. Osaka J. Math. **35**(4), 873–893 (1998)
55. S. Mizera, Scattering amplitudes from intersection theory. Phys. Rev. Lett. **120**(14), 141602 (2018). arXiv:1711.00469
56. S. Mizera, Aspects of scattering amplitudes and moduli space localization. Ph.D. thesis, Princeton, Inst. Advanced Study, 2020. arXiv:1906.02099
57. H. Frellesvig, F. Gasparotto, M.K. Mandal, P. Mastrolia, L. Mattiazzi, S. Mizera, Vector space of Feynman integrals and multivariate intersection numbers. Phys. Rev. Lett. **123**(20), 201602 (2019). arXiv:1907.02000
58. A.V. Smirnov, FIRE5: a C++ implementation of Feynman integral reduction. Comput. Phys. Commun. **189**, 182–191 (2015). arXiv:1408.2372
59. S. Weinzierl, *On the Computation of Intersection Numbers for Twisted Cocycles*. arXiv:2002.01930
60. K. Matsumoto, *Relative Twisted Homology and Cohomology Groups Associated with Lauricella's F_D*. arXiv:1804.00366
61. N. Arkani-Hamed, J.L. Bourjaily, F. Cachazo, J. Trnka, Local integrals for planar scattering amplitudes. J. High Energy Phys. **06**, 125 (2012). arXiv:1012.6032
62. J.M. Henn, Multiloop integrals in dimensional regularization made simple. Phys. Rev. Lett. **110**, 251601 (2013). arXiv:1304.1806
63. S. Mizera, A. Pokraka, From infinity to four dimensions: higher residue pairings and Feynman integrals. J. High Energy Phys. **02**, 159 (2020). arXiv:1910.11852
64. J. Chen, X. Xu, L.L. Yang, *Constructing Canonical Feynman Integrals with Intersection Theory*. arXiv:2008.03045
65. S. Abreu, R. Britto, C. Duhr, E. Gardi, J. Matthew, From positive geometries to a coaction on hypergeometric functions. J. High Energy Phys. **02**, 122 (2020). arXiv:1910.08358

Hypergeometric Functions and Feynman Diagrams

Mikhail Kalmykov, Vladimir Bytev, Bernd A. Kniehl, Sven-Olaf Moch, Bennie F. L. Ward, and Scott A. Yost

Abstract The relationship between Feynman diagrams and hypergeometric functions is discussed. Special attention is devoted to existing techniques for the construction of the ε-expansion. As an example, we present a detailed discussion of the construction of the ε-expansion of the Appell function F_3 around rational values of parameters via an iterative solution of differential equations. As a by-product, we have found that the one-loop massless pentagon diagram in dimension $d = 3 - 2\epsilon$ is not expressible in terms of multiple polylogarithms. Another interesting example is the Puiseux-type solution involving a differential operator generated by a hypergeometric function of three variables. The holonomic properties of the F_N hypergeometric functions are briefly discussed.

1 Introduction

Recent interest in the analytical properties of Feynman diagrams has been motivated by processes at the LHC. The required precision demands the evaluation of a huge number of diagrams having many scales to a high order, so that a new

---

M. Kalmykov (✉)
JINR, Dubna, Russia

V. Bytev
BLTP JINR, Dubna, Moscow Region, Russia

B. A. Kniehl · S.-O. Moch
II. Institut fuer Theoretische Physik, Universitaet Hamburg, Hamburg, Germany
e-mail: kniehl@desy.de; sven-olaf.moch@desy.de

B. F. L. Ward
Department of Physics, Baylor University, Waco, TX, USA
e-mail: BFL_Ward@baylor.edu

S. A. Yost
Department of Physics, The Citadel, Charleston, SC, USA
e-mail: scott.yost@citadel.edu

J. Blümlein, C. Schneider (eds.), *Anti-Differentiation and the Calculation of Feynman Amplitudes*, Texts & Monographs in Symbolic Computation,
https://doi.org/10.1007/978-3-030-80219-6_9

branch of mathematics emerges, which we may call *the Mathematical Structure of Feynman Diagrams* [1, 2], which includes elements of algebraic geometry, algebraic topology, the analytical theory of differential equations, multiple hypergeometric functions, elements of number theory, modular functions and elliptic curves, multidimensional residues, and graph theory. This mathematical structure has been extensively developed, studied and applied. For a more detailed discussion of the oldest results and their relation to modern techniques, see Refs. [3, 4]). One of these approaches is based on the treatment of Feynman diagrams in terms of multiple hypergeometric functions [5]. For example, in the series of papers [6–8], the one-loop diagrams have been associated with the R-function (a particular case of the F_D-function [9–11]).

1.1 Mellin–Barnes Representation, Asymptotic Expansion, NDIM

A universal technique based on the Mellin–Barnes representation of Feynman diagrams has been applied to one-loop diagrams in Ref. [12, 13] and to two-loop propagator diagrams in Ref. [14–18].[1] The multiple Mellin–Barnes representation for a Feynman diagram in covariant gauge can be written in the form

$$\Phi(\mathbf{A}, \mathbf{B}; \mathbf{C}, \mathbf{D}; \mathbf{z}) = \int_{-i\infty}^{+i\infty} \phi(\mathbf{t}) d\mathbf{t} = \int_{-i\infty}^{+i\infty} \prod_{a,b,c,r} \frac{\Gamma(\sum_{i=1}^{m} A_{ai} t_i + B_a)}{\Gamma(\sum_{j=1}^{r} C_{bj} t_j + D_b)} dt_c z_k^{\sum_l \alpha_{kl} t_l} , \tag{1}$$

where z_k are ratios of Mandelstam variables and A, B, C, D are matrices and vectors depending linearly on the dimension of space-time n and powers of the propagators. Closing the contour of integration on the right (on the left), this integral can be presented around zero values of $\mathbf{z}$ in the form

$$\Phi(\mathbf{A}, \mathbf{B}; \mathbf{C}, \mathbf{D}; \mathbf{z}) = \sum_{\alpha} f_{\alpha} H(\mathbf{A}, \mathbf{B}; \mathbf{C}, \mathbf{D}; \mathbf{z}) \mathbf{z}^{\alpha} , \tag{2}$$

where the coefficients f_α are ratios of Γ-functions and the functions H are Horn-type hypergeometric functions [22] (see Sect. 2 for details). The analytic continuation of the hypergeometric functions $H(\mathbf{z})$ into another region of the variables $\mathbf{z}$ can be constructed via the integral representation (when available) [23, 24], $H(\mathbf{z}) \to H(1 - \mathbf{z})$. However, for more complicated cases of Horn-type hypergeometric functions, this type of analytic continuation is still under construction [25, 26].

[1] Several programs are available for the automatic generation of the Mellin–Barnes representation of Feynman diagrams [19–21].

A major set of mathematical results (see, for example, [27–29]) is devoted to the construction of the analytic continuation of a series around $z_j = 0$ to a series of the form $\frac{z_A}{z_B}$: $H(\mathbf{z}) \rightarrow H(\frac{z_A}{z_B})$, where the main physical application is the construction of an expansion about Landau singularities $L(\mathbf{z})$: $H(\mathbf{z}) \rightarrow H(L(\mathbf{z}))$. For example, the singular locus L of the Appell function $F_4(z_1, z_2)$ is $L = \{(z_1, z_2) \in \mathbb{C}^2 | z_1 z_2 R(z_1, z_2) = 0\} \cup L_\infty$ where $R(z_1, z_2) = (1-z_1-z_2)^2-4z_1z_2$, and the physically interesting case of an expansion around the singularities corresponds to an analytical continuation $F_4(z_1, z_2) \rightarrow F_4\left(\frac{R(z_1,z_2)}{z_1}, \frac{R(z_1,z_2)}{z_2}\right)$.

A similar problem, the construction of convergent series of multiple Mellin–Barnes integrals in different regions of parameters, has been analyzed in detail for the case of two variables [30–32]. However, to our knowledge, there are no systematic analyses of the relation between these series and the singularities of multiple Mellin–Barnes integrals.

It was understood long ago that there is a one-to-one correspondence between the construction of convergent series from Mellin–Barnes integrals and the asymptotic expansions; see Ref. [33] for example. The available software, e.g. Ref. [34], allows the construction of the analytical continuation of a Mellin–Barnes integral in the limit when some of the variables z go to 0, or ∞. These are quite useful in the evaluation of Feynman diagrams, but do not solve our problem. The current status of the asymptotic expansions is discussed in Ref. [35, 36].

Another technique for obtaining a hypergeometric representation is the so-called "Negative Dimensional Integration Method" (NDIM) [37–42]. However, it is easy to show [43] that all available results follow directly from the Mellin–Barnes integrals [12].

For some Feynman diagrams, the hypergeometric representation follows from a direct integration of the parametric representation, see Ref. [44–52].

We also mention that the "Symmetries of Feynman Integrals" method [53–55] can also be used to obtain the hypergeometric representation for some types of diagrams.

1.2 About GKZ and Feynman Diagrams

There are a number of different though entirely equivalent ways to describe hypergeometric functions:

- as a multiple series;
- as a solution of a system of differential equations (hypergeometric D-module);
- as an integral of the Euler type;
- as a Mellin–Barnes integral.

In a series of papers, Gel'fand et al. [56–58] (to mention only a few of their series of papers devoted to the systematic development of this approach) have developed a uniform approach to the description of hypergeometric functions.[2] The formal solution of the A-system is a so-called multiple Γ-series having the following form:

$$\sum_{(l_1,\cdots,l_N)\in\mathbf{L}} \frac{z_1^{l_1+\gamma_1}\cdots z_N^{l_N+\gamma_M}}{\Gamma(l_1+\gamma_1+1)\cdots\Gamma(l_N+\gamma_N+1)} ,$$

where Γ is the Euler Γ-function and the lattice $\mathbf{L}$ has rank d. When this formal series has a non-zero radius of convergence, it coincides (up to a factor) with a Horn-type hypergeometric series [58] (see Sect. 2). Any Horn-type hypergeometric function can be written in the form of a Γ-series by applying the reflection formula $\Gamma(a+n) = (-1)^n \frac{\Gamma(a)\Gamma(1-a)}{\Gamma(1-a-n)}$. Many examples of such a conversion—all Horn-hypergeometric functions of two variables—have been considered in Ref. [62].

The Mellin–Barnes representation was beyond Gelfand's consideration. It was worked out later by Fritz Beukers [63]; see also the recent paper [64]. Beukers analyzed the Mellin–Barnes integral

$$\int \Pi_{i=1}^N \Gamma(-\gamma_i - \mathbf{b}_i \mathbf{s}) v_i^{\gamma_i + \mathbf{b}_i \mathbf{s}} ds ,$$

and pointed out that, under the assumption that the Mellin–Barnes integral converges absolutely, it satisfies the set of A-hypergeometric equations. The domain of convergence for the A-hypergeometric series and the associated Mellin–Barnes integrals have been discussed recently in Ref. [65].

Following Beuker's results, we conclude that any Feynman diagram with a generic set of parameters (to guarantee convergence, we should treat the powers of propagators as non-integer parameters) could be treated as an A-function. However our analysis has shown that, typically, a real Feynman diagram corresponds to an A-function with reducible monodromy.

Let us explain our point of view. By studying Feynman diagrams having a onefold Mellin–Barnes representation [66], we have found that certain Feynman diagrams (E^q_{1220}, B^2_{1220}, V^q_{1220}, J^q_{1220} in the notation of Ref. [66]) with powers of propagator equal to one (the so-called master-integrals) have the following hypergeometric structure (we drop the normalization constant for simplicity):

$$\Phi(n, \mathbf{1}; z) = {}_3F_2(a_1, a_2, a_3; b_1, b_2; z) + z^\sigma {}_4F_3(1, c_1, c_2, c_3; p_1, p_2, p_3; z) , \quad (3)$$

where the dimension n of space-time [67] is not an integer and the difference between any two parameters of the hypergeometric function also are not integers. The holonomic rank of the hypergeometric function ${}_pF_{p-1}$ is equal to p, so that the Feynman diagram is a linear combination of two series having different holonomic

[2] The detailed discussion of A-functions and their properties is beyond our current consideration. There are many interesting papers on that subject, including (to mention only a few) Refs. [59–61].

rank. What could we say about the holonomic rank of a Feynman diagram Φ? To answer that question, let us find the differential equation for the Feynman diagram $\Phi(n, \mathbf{1}; z)$ starting from the representation Eq. (3). This could be done by the Holonomic Function Approach [68] or with the help of a programs developed by Frederic Chyzak [69] (it is a MAPLE package) or by Christoph Koutschan[3] [70] (it is MATHEMATICA package). We used a private realization of this approach based on ideas from the Gröbner basis technique. Finally, we obtained the result that the Feynman diagram Φ satisfies the homogeneous differential equation of the hypergeometric type of order 4 with a left-factorizable differential operator of order 1:

$$(\theta + A)\left[(\theta + B_1)(\theta + B_2)(\theta + B_3) + z\theta(\theta + C_1)(\theta + C_2)\right]\Phi(n, \mathbf{1}; z) = 0\,, \tag{4}$$

where none of the B_j and C_a are integers and $\theta = z\frac{d}{dz}$.

It follows from this differential equation that the holonomic rank of the Feynman diagram Φ is equal to 4, and factorization means that the space of solutions splits into a direct sum of two spaces of dimension one and three: $\Phi_{\text{dim}} = 4 = 1 \otimes 3$. As it follows from [71], the monodromy representation of Eq. (4) is reducible and there is a one-dimensional invariant subspace. Consequently, there are three non-trivial solutions (master-integrals) and the one-dimensional invariant subspace corresponds to an integral having a Puiseux-type solution (expressible in terms of Γ-functions).

We pointed out in Ref. [66] that a Feynman diagram can be classified by the dimension of its irreducible representation. This can be evaluated by the construction of differential equations or by using the dimension of the irreducible representation of the hypergeometric functions entering in the r.h.s. of Eq. (2). Indeed, in the example considered in Eq. (3), the dimension of the irreducible representation ${}_4F_3(1, \mathbf{c}; \mathbf{p}; z)$ is equal to 3 [71], so that the dimension of the irreducible space of the Feynman diagram Φ is equal to 3, see Eq. (4), and Φ is expressible via the sum of a series (see Eq. (3)) with an irreducible representation of dimension 3.

The results of the analysis performed in Ref. [66], are summarized in the following proposition:

Proposition *A Feynman diagram can be treated as a linear combination of Horn-type hypergeometric series where each term has equal* ***irreducible*** *holonomic rank.*

Examining this new "quantum number," irreducible holonomic rank, we discover, and can rigorously prove, an extra relation between master-integrals [72]. In many other examples we found a complete agreement between the results of differential reduction and and the results of a reduction based on the IBP relations [73, 74].

---

[3] See Koutschan's paper in the present volume.

The Feynman diagram J considered in Ref. [72] satisfies the differential equation

$$\left(\theta-\frac{n}{2}+I_1\right)(\theta-n+I_2)\left[\theta\left(\theta-n+\frac{1}{2}+I_3\right)+z\left(\theta-\frac{3n}{2}+I_4\right)\right]J=0\,, \quad (5)$$

where I_1, I_2, I_3, I_4 are integers, n is the dimension of space-time and $\theta = z\frac{d}{dz}$. The dimension of J is 4 and there are two one-dimensional invariant subspaces, corresponding to two first-order differential operators: $J_{\text{dim}} = 4 = 1 \otimes 1 \otimes 2$. Indeed, after integrating twice, we obtained

$$\left[\theta\left(\theta-n+\frac{1}{2}+I_3\right)+z\left(\theta-\frac{3n}{2}+I_4\right)\right]J = C_1 z^{n/2-I_1} + C_2 z^{n-I_2}\,.$$

Surprisingly, this simple relation has not been not reproduced (as of the end of 2016) by any of the powerful programs for the reduction of Feynman diagrams (see the discussion in Chapter 6 of Ref. [75]). In [76], it was shown that the extra relation [72] could be deduced from a diagram of more general topology by exploring a new relation derived by taking of the derivative with respect to the mass with a subsequent reduction with the help of IBP relations. However, it was not shown that the derivative with respect to mass can be deduced from derivatives with respect to momenta, so that the result of Ref. [76] could be considered as an alternative proof that, in the massive case, there may exist an extra relation between diagrams that does not follow from classical IBP relations.

Finally, we have obtained a very simple result [77]: Eq. (4) follows directly from the Mellin–Barnes representation for a Feynman diagram. (See Sect. 2 and Eq. (12) for details.) Based on this observation and on the results of our analysis performed in Ref. [66], and extending the idea of the algorithm of Ref. [78], we have constructed a simple and fast algorithm for the algebraic reduction of any Feynman diagram having a onefold Mellin–Barnes integral representation to a set of master-integrals without using the IBP relations. In particular, our approach and our program cover some types of Feynman diagrams with arbitrary powers of propagators considered in Refs. [79–81].

In a similar manner, one can consider the multiple Mellin–Barnes representation of a Feynman diagram [75, 82]. In contrast to the one variable case, the factorization of the partial differential operator is much more complicated. The dimension of the Pfaff system[4] related to the multiple Mellin–Barnes integral can be evaluated with the help of a prolongation procedure (see the discussion in Sect. 2). However, in this case, there may exist a Puiseux type solution even for a generic set of parameters (see for example Sect. 3.1.2).

---

[4]Rigorously speaking, this system of equations is correct when there is a contour in $\mathbf{C}^n$ that is not changed under translations by an arbitrary unit vector, see Ref. [83], so that we treat the powers of propagator as parameters.

Exploring the idea[5] presented in Ref. [71], one possibility is to construct an explicit solution of the invariant subspace (see [75] and Sect. 3.1.2) and find the dimension of the irreducible representation. Our results presented in [75] were confirmed by another technique in Ref. [84].

Let us illustrate the notion of irreducible holonomic rank (or an irreducible representation) in an application to Feynman diagrams. As follows from our analysis of sunset diagrams [75], the dimension of the irreducible representation of two-loop sunset with three different masses is equal to 4. There is only one hypergeometric function of three variables having holonomic rank 4, the F_D function. Then we expect that there is a linear combination of four two-loop sunsets and the product of one-loop tadpoles that are expressible in terms of a linear combination of the F_D functions.

Another approach to the construction of a GKZ representation of Feynman Diagrams was done recently in the series of papers in Refs. [85–87]. Based on the observation made in Ref. [88] about the direct relation between A-functions and Mellin transforms of rational functions, and exploring the Lee-Pomeransky representation [89], the authors studied a different aspect of the GKZ representation mainly considering the examples of massless or one-loop diagrams. Two non-trivial examples have been presented in Ref. [86]: the two-loop sunset with two different masses and one zero mass, which corresponds to a linear combination of two Appell functions F_4 (see Eq. (3.11) in [90])[6] and a two-loop propagator with three different masses related to the functions F_C of three variables [15].

A different idea on how to apply the GKZ technique to the analysis of Feynman Diagrams has been presented in [91] and has received further development in [92, 93].

1.3 One-Loop Feynman Diagrams

Let us give special attention to one-loop Feynman diagrams. In this case, two elegant approaches have been developed [94, 95] that allow us to obtain compact hypergeometric representations for the master-integrals. The authors of the first paper [94] explored the internal symmetries of the Feynman parametric representation to get a onefold integral representation for one-loop Feynman diagrams (see also [96, 97]). The second approach [95] is based on the solution of difference equations with respect to the dimension of space-time [98] for the one-loop integrals. In spite of

[5]The monodromy group is the group of linear transformations of solutions of a system of hypergeometric differential equations under rotations around its singular locus. In the case when the monodromy is reducible, there is a finite-dimensional subspace of holomorphic solutions of the hypergeometric system on which the monodromy acts trivially.

[6]It is interesting to note, that on-mass shell $z = 1$, this diagram has two Puiseux type solutions that do not have analytical continuations.

different ideas on the analysis of Feynman diagrams, both approaches, [94] and [95], produce the same results for one-loop propagator and vertex diagrams [99–101]. However, beyond these examples, the situation is less complete: it was shown in Ref. [95] that the off-shell one-loop massive box is expressible in terms of a linear combination of F_S Horn-type hypergeometric functions of three variables (see also discussions in Refs. [102–105]), or in terms of F_N Horn-type hypergeometric functions of three variables [106] (see Sect. 3.1.1).

Recently, it was observed [107–109] that massive conformal Feynman diagrams are invariant under a Yangian symmetry that allows to get the hypergeometric representation for the conformal Feynman diagrams.

1.4 Construction of ε-Expansion

For physical applications, the construction of the analytical coefficients of the Laurent expansions of hypergeometric functions around particular values of parameters (integer, half-integer, rational) is necessary. Since the analytic continuations of hypergeometric functions is still an unsolved problem, the results are written in some region of variables in each order of the ε-expansion in terms of special functions like classical or multiple polylogarithms, [110–116], and then these functions are analytically continued to another region. For this reason, the analytical properties of special functions were analyzed in detail [117–123]. Also, tools for the numerical evaluation of the corresponding functions are important ingredients [124–137].

Each of the hypergeometric function representations (series, integral, Mellin–Barnes, differential equation) can be used for the construction of the ε-expansion, and each of them has some technical advantages or disadvantages in comparison with the other ones. The pioneering ε-expansion of the hypergeometric function ${}_pF_{p-1}$ around $z = \pm 1$ was done by David Broadhurst [138, 139]. The expansion was based on the analysis of multiple series and it was interesting from a mathematical point of view [140] as well as for its application to quantum field theory [141]. The integral representation was mainly developed by Andrei Davydychev and Bas Tausk [142, 143], so that, finally, the all-order ε-expansion for the Gauss hypergeometric functions around a rational parameter, a case that covers an important class of diagrams, has been constructed [144] in terms of generalized log-sine [110] functions or in term of Nielsen polylogarithms [145, 146].

The integral representation was also the starting point for the construction of the ε-expansion of ${}_pF_{p-1}$ hypergeometric functions, [147, 148], and also the F_1 [40] and F_D functions around integer values of parameters [149–152].

Purely numerical approaches [153–155] can be applied for arbitrary values of the parameters. However, this technique typically does not produce a stable numerical result in regions around singularities of the hypergeometric functions.

A universal technique which does not depend on the order of the differential equation is based on the algebra of multiple sums [156–158]. For the hypergeometric functions[7] for which the nested-sum algorithms [156] are applicable, the results of the ε-expansion are automatically obtained in terms of multiple polylogarithms.

The nested-sum algorithms [156] have been implemented in a few packages [161, 162] and allow for the construction of the ε-expansion of hypergeometric functions ${}_pF_{p-1}$ and Appell functions F_1 and F_2 around integer values of parameters.[8] However, the nested-sum approach fails for the ε-expansion of hypergeometric functions around rational values of parameters and it is not applicable to some specific classes of hypergeometric functions (for example, the F_4 function, see [165]).

In the series of papers [166, 167], the generating function technique [168, 169] has been developed for the analytical evaluation of multiple sums. Indeed, the series generated by the ε-expansion of hypergeometric functions has the form $\sum_k c(k)z^k$, where the coefficients $c(k)$ include only products of the harmonic sums [170, 171] $\Pi_{a,b} S_a(k-1) S_b(2k-1)$ and $S_a(k) = \sum_{j=1}^{k} \frac{1}{j^a}$. The harmonic sums satisfy the recurrence relations

$$S_a(k) = S_a(k-1) + \frac{1}{k^a} \quad , \qquad S_a(2k+1) = S_a(2k-1) + \frac{1}{(2k+1)^a} + \frac{1}{(2k)^a} \, ,$$

so that the coefficients $c(k)$ satisfy the first order difference equation:[9]

$$P(k+1)c(k+1) = Q(k)c(k) + R(k) \, ,$$

where P and Q are polynomial functions that can be defined from the original series. This equation could be converted into a first order differential equation for the generating function $F(z) = \sum_k c(k)z^k$,

$$\frac{1}{z} P\left(z\frac{d}{dz}\right) F(z) - P(1)C(1)z = Q\left(z\frac{d}{dz}\right) F(z) + \sum_{k=1} R(k)z^k \, .$$

One of the remarkable properties of this technique, that the non-homogeneous part of the differential equation, the function $R(k)$, has one-unit less depth in contrast to the original sums, so that, step-by-step, all sums could be evaluated analytically. Based on this technique, all series arising from the ε-expansion of hypergeometric functions around half-integer values of parameters have been evaluated [166] up to weight 4. The limits considered were mainly motivated by physical reasons (at O(NNLO) only functions of weight 4 are generated) and, in this limit, only one new function [111] $H_{-1,0,0,1}(z)$ was necessary to introduce. These results [166] allow

---

[7] It was shown in [159, 160] that multiple Mellin–Barnes integrals related to Feynman diagrams could be evaluated analytically/numerically at each order in ε via multiple sums, without requiring a closed expression in terms of Horn-type hypergeometric functions.

[8] See also Refs. [163, 164] for an alternative realization.

[9] In general, it could be a more generic recurrence, $\sum_{1=0}^{k} p_{k+j}(k+j)c(k+j) = r(k)$.

us to construct the ε-expansion of the hypergeometric functions ${}_pF_{p-1}$ around half-integer values of parameters, see [172, 173].

Other results and theorems relevant for the evaluation of Feynman diagrams are related with the appearance of a factor $1/\sqrt{3}$ in the ε-expansion of some diagrams [174] expressible in terms of hypergeometric functions[10] were derived in Refs. [146, 178, 179].[11]

Let us consider typical problems arising in this program. We follow our analysis presented in Ref. [181], see also the closely related discussion in Ref. [182]. First, the construction of the difference equation for the coefficient functions $c(z)$ is not an easy task [183–185]. In the second step, the differential operator(s) coming from the difference equation $P\left(z\frac{d}{dz}\right) - zQ\left(z\frac{d}{dz}\right)$ should be factorized into a product of differential operators of the first order,

$$P\left(z\frac{d}{dz}\right) - zQ\left(z\frac{d}{dz}\right) = \Pi_{k=1}\left[p_k(z)\frac{d}{dz} - q_k(z)\right] ,$$

where $p_k(z)$ and $q_k(z)$ are rational functions. Unfortunately, the factorization of differential operators into irreducible factors is not unique [186]:

$$\left(\frac{d^2}{dx^2} - \frac{2}{x}\frac{d}{dx} + \frac{2}{x^2}\right) = \left(\frac{d}{dx} - \frac{1}{x}\right)\left(\frac{d}{dx} - \frac{1}{x}\right) = \left(\frac{d}{dx} - \frac{1}{x(1+ax)}\right)\left(\frac{d}{dx} - \frac{(1+2ax)}{x(1+ax)}\right) ,$$

where a is a constant.

However, the following theorem is valid (see [187]): Any two decompositions of a linear differential operator $L^{(p)}$ into a product (composition) of irreducible linear differential operators

$$L^{(p)} = L_1^{(a_1)} L_2^{(a_2)} \cdots L_m^{(a_m)} = P_1^{(r_1)} P_2^{(r_2)} \cdots P_k^{(r_k)}$$

have equal numbers of components $m = k$ and the factors L_j and P_a have the same order of differential operators: $L_a = P_j$ (up to commutation). In the application to the ε-expansion of hypergeometric functions this problem has been discussed in Ref. [188].

After factorization, the iterated integral over rational functions (which is not uniquely defined, as seen in the previous example) would be generated that in general is not expressible in terms of hyperlogarithms. Indeed, the solution of the differential equation

$$\left[R_1(z)\frac{d}{dz} + Q_1(z)\right]\left[R_2(z)\frac{d}{dz} + Q_2(z)\right] h(z) = F(z) .$$

[10] Recent results on the analytical evaluation of inverse binomial sums for particular values of the arguments have been presented in [175–177].

[11] The appearance of $1/\sqrt{3}$ in RG functions in seven loops was quite intriguing [122, 180].

has the form

$$h(z) = \int^{z} \frac{dt_3}{R_2(t_3)} \left[\exp^{-\int_0^{t_3} \frac{Q_2(t_4)}{R_2(t_4)} dt_4} \right] \int^{t_3} \frac{dt_1}{R_1(t_1)} \left[\exp^{-\int_0^{t_1} \frac{Q_1(t_2)}{R_1(t_2)} dt_2} \right] F(t_1) \, .$$

From this solution it follows [189] that the following conditions are enough to convert the iterated integral into hyperlogarithms: there are new variables ξ and x so that

$$\int^{z} \frac{Q_i(t)}{R_i(t)} dt = \ln \frac{M_i(\xi)}{N_i(\xi)} \Rightarrow \left. \frac{dt}{R_i(t)} \right|_{t=t(\xi)} \frac{N_i(\xi)}{M_i(\xi)} = dx \frac{K_i(x)}{L_i(x)} \, ,$$

where M_i, N_i, K_i, L_i are polynomial functions.

The last problem is related to the Abel-Ruffini theorem: the polynomial is factorizable into a product of its primitive roots, but there are not solutions in radicals for polynomial equations of degree five or more. The last problem got a very elegant solution by the introduction of cyclotomic polylogarithms [158], with the integration over irreducible cyclotomic polynomials $\Phi_n(x)$. The first two irreducible polynomials (see Eqs. (3.3)–(3.14) in [158]) are Φ_7 and Φ_9 (the polynomial of order 6). Two other polynomials of order 4, Φ_5 and Φ_{10}: $(x^4 \pm x^3 + x^2 \pm x + 1)$, have non-trivial primitive roots. But up to now, all these polynomials were not generated by Feynman diagrams. Surprisingly, by increasing the number of loops or number of scales, other mathematical structures are generated [190, 191]. Detailed analyses of properties of the new functions have been presented in Refs. [192, 193] and automated by Jakob Ablinger [194–196]. The problem of integration over algebraic functions (typically square roots of polynomials) was solved by the introduction of a new type of functions [197], intermediate between multiple and elliptic polylogarithms.

The series expansion is not very efficient for the construction of the ε-expansion, since the number of series increases with the order of the ε-expansion and increases the complexity of the individual sums. Let us recall that the Laurent expansion of a hypergeometric function contains a linear combination of multiple sums. From this point of view, the construction of the analytical coefficients of the ε-expansion of a hypergeometric function can be carried out independently of existing analytical results for each individual multiple sum. The "internal" symmetry of a Horn-type hypergeometric function is uniquely defined by the corresponding system of differential equations. While exploring this idea, a new algorithm was presented in Refs. [198, 199], based on factorization, looking for a linear parametrization and direct iterative solution of the differential equation for a hypergeometric function. This approach allows the construction of the analytical coefficients of the ε-expansion of a hypergeometric function, as well as obtaining analytical expressions for a large class of multiple series without referring to the algebra of nested sums.

Based on this approach, the all-order ε-expansion of the Gauss hypergeometric function around half-integer and rational values of parameters has been constructed [198, 200], so that the first 20 coefficients around half-integer values of parameters, the 12 coefficients for $q = 4$ and 10 coefficients for $q = 6$ have been generated already in 2012.[12] Another record is the generation of 24 coefficients for the Clausen hypergeometric function ${}_3F_2$ around integer values of parameters, relevant for the analysis[13] performed in [202]. To our knowledge, at the present moment, this remains the fastest and most universal algorithm.

Moreover, it was shown in Refs. [198, 199], that when the coefficients of the ε-expansion of a hypergeometric function are expressible in terms of multiple polylogarithms, there is a set of parameters (not uniquely defined) such that, at each order of ε, the coefficients of the ε-expansion include multiple polylogarithms of a single uniform weight. A few years later, this property was established not only for hypergeometric functions, but for Feynman Diagrams [203].

A multivariable generalization [181] of the algorithm of Refs. [198, 199] has been described. The main difference with respect to the case of one variable is the construction of a system of differential equations of triangular form to avoid the appearance of elliptic functions. As a demonstration of the validity of the algorithm, the first few coefficients of the ε-expansion of the Appell hypergeometric functions F_1, F_2, F_3 and F_D around integer values of parameters have been evaluated analytically [102].

The ε-expansion of the hypergeometric functions F_3 and F_D are not covered by the nested sums technique or its generalization. The differential equation technique can be applied to the construction of analytic coefficients of the ε-expansions of hypergeometric functions of several variables (which is equivalent to the multiple series of several variables) around any rational values of parameters via direct solution of the linear systems of differential equations.

The differential equation approach [198, 199] allows us to analyze arbitrary sets of parameters simultaneously and to construct the solution in terms of iterated integrals, but for any hypergeometric function the Pfaff system of differential equations should be constructed. That was the motivation for creation of the package(s) (the HYPERDIRE project) [82] for the manipulation of the parameters of Horn-type hypergeometric functions of several variables. For illustration, we describe in detail how it works in the application to the F_3 hypergeometric function in Sect. 3.2.

Recently, a new technique [204] for the construction of the ε-expansion of Feynman diagrams [205] as well as for hypergeometric functions has been presented [206]. It is based on the construction of a coaction[14] of certain hypergeometric functions. The structures of the ε-expansion of the Appell hypergeometric functions F_1, F_2, F_3 and F_4 as well as F_D (for the last function F_D see also the discussion in Ref. [208]) around integer values of parameters are in agreement with our analysis and partial results presented in Refs. [102, 181]. However, the structure

[12]The results have been written in terms of hyperlogarithms of primitive q-roots of unity.

[13]The results of [201] were relevant for the reduction of multiple zeta values to the minimal basis.

[14]An interesting construction of the coaction for the Feynman graph has been presented recently in Ref. [207].

of the ε-expansion around rational values of parameters has not been discussed in [206], nor in [208].

2 Horn-Type Hypergeometric Functions

2.1 Definition and System of Differential Equations

The study of solutions of linear partial differential equations (PDEs) of several variables in terms of multiple series, i.e. a multi-variable generalization of the Gauss hypergeometric function [209], began long ago [210].

Following the Horn definition [22], a multiple series is called a "Horn-type hypergeometric function," if, about the point $\mathbf{z} = \mathbf{0}$, there is a series representation

$$H(\mathbf{z}) = \sum_{\mathbf{m}} C(\mathbf{m})\mathbf{z}^{\mathbf{m}}, \tag{6}$$

where $\mathbf{z}^{\mathbf{m}} = z_1^{m_1} \cdots z_r^{m_r}$ for any integer multi-index $\mathbf{m} = (m_1, \cdots, m_r)$, and the ratio of two coefficients can be represented as a ratio of two polynomials:

$$\frac{C(\mathbf{m} + \mathbf{e}_j)}{C(\mathbf{m})} = \frac{P_j(\mathbf{m})}{Q_j(\mathbf{m})}, \tag{7}$$

where $\mathbf{e}_j$ denotes the unit vector with unity in its jth entry, $\mathbf{e}_j = (0, \cdots, 0, 1, 0, \cdots, 0)$. The coefficients $C(\mathbf{m})$ of such a series can be expressed as products or ratios of Gamma-functions (up to some factors irrelevant for our consideration) [211, 212]:

$$C(\mathbf{m}) = \frac{\prod_{j=1}^{p} \Gamma\left(\sum_{a=1}^{r} \mu_{ja} m_a + \gamma_j\right)}{\prod_{k=1}^{q} \Gamma\left(\sum_{b=1}^{r} \nu_{kb} m_b + \sigma_k\right)}, \tag{8}$$

where $\mu_{ja}, \nu_{kb}, \sigma_j, \gamma_j \in \mathbb{Z}$ and m_a are elements of $\mathbf{m}$.

The Horn-type hypergeometric function, Eq. (7), satisfies the following system of differential equations:

$$0 = L_j(\mathbf{z})H(\mathbf{z}) = \left[Q_j\left(\sum_{k=1}^{r} z_k \frac{\partial}{\partial z_k}\right)\frac{1}{z_j} - P_j\left(\sum_{k=1}^{r} z_k \frac{\partial}{\partial z_k}\right)\right] H(\mathbf{z}), \tag{9}$$

where $j = 1, \ldots, r$. Indeed,

$$Q_j \left(\sum_{k=1}^{r} z_k \frac{\partial}{\partial z_k} \right) \frac{1}{z_j} \sum_{\mathbf{m}} C(\mathbf{m}) \mathbf{z}^{\mathbf{m}} = \sum_{\mathbf{m}} Q_j(\mathbf{m}) C(\mathbf{m}+\mathbf{e}_j) \mathbf{z}^{\mathbf{m}}$$

$$= \sum_{\mathbf{m}} P_j(\mathbf{m}) C(\mathbf{m}) \mathbf{z}^{\mathbf{m}} = P_j \left(\sum_{k=1}^{r} z_k \frac{\partial}{\partial z_k} \right) \sum_{\mathbf{m}} C(\mathbf{m}) \mathbf{z}^{\mathbf{m}} .$$

The degrees of the polynomials P_i and Q_i are p_i and q_i, respectively. The largest of these, $r = \max\{p_i, q_j\}$, is called the order of the hypergeometric series. To close the system of differential equations, the *prolongation procedure* should be applied: by applying the differential operator ∂_i to L_j we can convert the system of linear PDEs with polynomial coefficients into Pfaff form (for simplicity, we assume that system is closed):

$$L_j H(\mathbf{z}) = 0 \Rightarrow \left\{ d\omega_i(\mathbf{z}) = \Omega_{ij}^k(\mathbf{z}) \omega_j(\mathbf{z}) dz_k \; , \quad d\left[d\omega_i(\mathbf{z})\right] = 0 \right\} . \tag{10}$$

Instead of a series representation, one can use a Mellin–Barnes integral representation (see the discussion in [77]). Indeed, the multiple Mellin–Barnes representation for a Feynman diagram could be written in the form in Eq. (1). Let us define the polynomials P_i and Q_i as

$$\frac{P_i(\mathbf{t})}{Q_i(\mathbf{t})} = \frac{\phi(\mathbf{t} + e_i)}{\phi(\mathbf{t})} . \tag{11}$$

The integral (1) then satisfies the system of linear differential equations (9)

$$Q_i(\mathbf{t})|_{t_j \to \theta_j} \frac{1}{z_i} \Phi(\mathbf{A}, \mathbf{B}; \mathbf{C}, \mathbf{D}; \mathbf{z}) = P_i(\mathbf{t})|_{t_j \to \theta_j} \Phi(\mathbf{A}, \mathbf{B}; \mathbf{C}, \mathbf{D}; \mathbf{z}) , \tag{12}$$

where $\theta_i = z_i \frac{d}{dz_i}$. Systems of equations such as Eq. (12) are left ideals in the Weyl algebra of linear differential operators with polynomial coefficients.

2.2 Contiguous Relations

Any Horn-type hypergeometric function is a function of two types of variables, *continuous* variables, $z_1, z_2, \cdots, z_r$ and *discrete* variables: $\{J_a\} := \{\gamma_k, \sigma_r\}$, where the latter can change by integer numbers and are often referred to as the *parameters* of the hypergeometric function.

For any Horn-hypergeometric function, there are linear differential operators changing the value of the discrete variables by one unit. Indeed, let us consider a multiple series defined by Eq. (6).

Two hypergeometric functions H with sets of parameters shifted by unity, $H(\boldsymbol{\gamma}+\mathbf{e_c};\boldsymbol{\sigma};\mathbf{z})$ and $H(\boldsymbol{\gamma};\boldsymbol{\sigma};\mathbf{z})$, are related by a linear differential operator:

$$H(\boldsymbol{\gamma}+\mathbf{e_c};\boldsymbol{\sigma};\mathbf{z}) = \left(\sum_{a=1}^{r} \mu_{ca} z_a \frac{\partial}{\partial z_a} + \gamma_c\right) H(\boldsymbol{\gamma};\boldsymbol{\sigma};\mathbf{z}) \,. \tag{13}$$

Similar relations also exist for the lower parameters:

$$H(\boldsymbol{\gamma};\boldsymbol{\sigma}-\mathbf{e}_c;\mathbf{z}) = \left(\sum_{b=1}^{r} \nu_{cb} z_b \frac{\partial}{\partial z_b} + \sigma_c - 1\right) H(\boldsymbol{\gamma};\boldsymbol{\sigma};\mathbf{z}) \,. \tag{14}$$

Let us rewrite these relations in a symbolic form:

$$R_K(\mathbf{z}) \frac{\partial}{\partial \mathbf{z}_K} H(\mathbf{J};\mathbf{z}) = H(\mathbf{J} \pm e_K;\mathbf{z}) \,, \tag{15}$$

where $R_K(\mathbf{z})$ are polynomial (rational) functions.

In Refs. [59, 78] it was shown that there is an algorithmic construction of inverse linear differential operators:

$$B_{L,N}(\mathbf{z}) \frac{\partial^L}{\partial \mathbf{z}_N} \left(R_K(\mathbf{z}) \frac{\partial}{\partial \mathbf{z}_K} \right) H(\mathbf{J};\mathbf{z}) \equiv B_{L,N}(\mathbf{z}) \frac{\partial^L}{\partial \mathbf{z}_N} H(\mathbf{J} \pm e_K;\mathbf{z}) = H(\mathbf{J};\mathbf{z}) \,. \tag{16}$$

Applying the direct or inverse differential operators to the hypergeometric function the values of the parameters can be changed by an arbitrary integer:

$$S(\mathbf{z}) H(\mathbf{J}+\mathbf{m};\mathbf{z}) = \sum_{j=0}^{r} S_j(\mathbf{z}) \frac{\partial^j}{\partial \mathbf{z}} H(\mathbf{J};\mathbf{z}) \,, \tag{17}$$

where $\mathbf{m}$ is a set of integers, S and S_j are polynomials and r is the holonomic rank (the number of linearly independent solutions) of the system of differential equations (9). At the end of the reduction, the differential operators acting on the function H can be replaced by a linear combination of the function evaluated with shifted parameters.

We note that special considerations are necessary when the system of differential operators, Eq. (9), has a Puiseux-type solution (see Sect. 3.1.2). In this case, the prolongation procedure gives rise to the Pfaffian form, but this set of differential equations is not enough to construct the inverse operators [213], so that new differential equations should be introduced. In the application to the Feynman diagrams, this problem is closely related with obtaining new relations between master integrals, see Ref. [72] for details. In the Sect. 3.1.2 we present an example of the Horn-type hypergeometric equation of second order of three variables having a Puiseux-type solution.

Another approach to the reduction of hypergeometric functions is based on the explicit algebraic solution of the contiguous relations, see the discussion in Ref. [214]. This technique is applicable in many particular cases, including ${}_2F_1$, ${}_3F_2$, and the Appell functions F_1, F_2, F_3, F_4 (see the references in Ref. [215]), and there is a general expectation that it could be solved for any Horn-type hypergeometric function. However, to our knowledge, nobody has analyzed the algebraic reduction in the application to general hypergeometric functions having a Puiseux-type solution.

The multiple Mellin–Barnes integral Φ defined by Eq. (1) satisfies similar differential contiguous relations:

$$\Phi(\mathbf{A}, \mathbf{B}+e_a; \mathbf{C}, \mathbf{D}; \mathbf{z}) = \left(\sum_{i=1}^{m} A_{ai}\theta_i + B_a \right) \Phi(\mathbf{A}, \mathbf{B}; \mathbf{C}, \mathbf{D}; \mathbf{z}) ,$$

$$\Phi(\mathbf{A}, \mathbf{B}; \mathbf{C}, \mathbf{D}-e_b; \mathbf{z}) = \left(\sum_{j=1}^{r} C_{bj}\theta_j + D_b \right) \Phi(\mathbf{A}, \mathbf{B}; \mathbf{C}, \mathbf{D}; \mathbf{z}) , \tag{18}$$

so that the original diagram may be explicitly reduced to a set of basis functions without examining the IBP relations [73, 74]. A non-trivial example of this type of reduction beyond IBP relations has been presented in Ref. [72] (see also the discussion in Chapter 6 of Ref. [75]).

3 Examples

3.1 Holonomic Rank & Puiseux-Type Solution

In addition to the examples presented previously in our series of publications, we present here a few new examples.

3.1.1 Evaluation of Holonomic Rank: The Hypergeometric Function F_N

The Lauricella–Saran hypergeometric function of three variables F_N is defined about the point $z_1 = z_2 = z_3 = 0$ by

$$F_N(a_1, a_2, a_3; b_1, b_2; c_1, c_2; z_1, z_2, z_3) = \sum_{m_1, m_2, m_3 = 0}^{\infty} \left[\Pi_{j=1}^{3} (a_j)_{m_j} \frac{z_j^{m_j}}{m_j!} \right] \frac{(b_1)_{m_1+m_3} (b_2)_{m_2}}{(c_1)_{m_1} (c_2)_{m_2+m_3}} . \tag{19}$$

This function is related to one-loop box diagrams in an arbitrary dimension considered by Andrei Davydychev [106].

Following the general algorithm [102], the following result is easily derived:

Theorem 1 *For generic values of the parameters, the holonomic rank of the function F_N is equal* 8.

In this way, for generic values of parameters, the result of differential reduction, Eq. (17), have the following form:

$$S(\mathbf{z}) F_N(\mathbf{J}+\mathbf{m};\mathbf{z}) = \left[S_0 + S_i \sum_{j=1}^{3} \theta_j + \sum_{\substack{i,j=1 \\ i<j}}^{3} S_{i,j} \theta_i \theta_j + S_{123}\, \theta_1 \theta_2 \theta_3 \right] F_N(\mathbf{J};\mathbf{z}) \,,$$

where $\theta_i = z_i \frac{d}{dz_i}$ and inverse differential operators can be easily constructed (Note that "easy" does not mean that these operators have a simple form, see Ref. [216]).

Theorem 2 *The system of differential equations defined by the series (19) is reducible when the one of the following combinations of parameters is an integer:*

$$\begin{aligned} &\{a_1, a_2, a_3, b_1, b_2, \} \in \mathbb{Z}\,, \\ &\{a_2 - c_1 - c_2 + b_1, \quad a_2 - c_2 + b_1, \quad b_2 - c_1 - c_2 + b_1, \quad b_2 - c_2 + b_1\} \in \mathbb{Z} \end{aligned} \tag{20}$$

When one or more conditions of the Theorem 2 are valid, the number of independent differential equations describing the function F_N reduces and some additional analysis is necessary (see for example Ref. [75]) to evaluate the value of irreducible holonomic rank.

3.1.2 Puiseux-Type Solution: Hypergeometric Function F_T

The Lauricella–Saran hypergeometric function of three variables F_T is defined about the point $z_1 = z_2 = z_3 = 0$ by

$$\begin{aligned} & F_T(a_1; a_2; b_1, b_2; c; z_1, z_2, z_3) \\ &= \sum_{m_1, m_2, m_3 = 0}^{\infty} \frac{(a_1)_{m_1} (a_2)_{m_2+m_3} (b_1)_{m_1+m_3} (b_2)_{m_2}}{(c)_{m_1+m_2+m_3}} \frac{z_1^{m_1} z_2^{m_2} z_3^{m_3}}{m_1! m_2! m_3!} \,. \end{aligned} \tag{21}$$

In this case, the differential operators, Eq. (9), are the following:

$$L_1 F_T: \quad \theta_1 \left(c-1+\sum_{j=1}^{3} \theta_j \right) F_T = z_1 \left(a_1+\theta_1\right) \left(b_1+\theta_1+\theta_3\right) F_T \,, \tag{22a}$$

$$L_2 F_T: \quad \theta_2 \left(c-1+\sum_{j=1}^{3} \theta_j \right) F_T = z_2 \left(a_2+\theta_2+\theta_3\right) \left(b_2+\theta_2\right) F_T \,, \tag{22b}$$

$$L_3 F_T: \quad \theta_3 \left(c-1+\sum_{j=1}^{3} \theta_j \right) F_T = z_3 \left(a_2+\theta_2+\theta_3\right) \left(b_1+\theta_1+\theta_3\right) F_T \,, \tag{22c}$$

where $F_T = F_T(a_1; a_2; b_1, b_2; c; z_1, z_2, z_3)$.

Let us introduce the function

$$\Phi_T = \frac{z_1^{1-c+a_2} z_2^{1-c+b_1}}{z_3^{1-c+b_1+a_2}} \,. \tag{23}$$

It is easy to check that

$$L_1 \Phi_T = L_2 \Phi_T = L_3 \Phi_T = 0 \,.$$

Theorem 3 *The system of differential equations defined by Eq. (22) has a Puiseux-type solution:*

$$\Phi_T = \frac{z_1^{1-c+a_2} z_2^{1-c+b_1}}{z_3^{1-c+b_1+a_2}} \,. \tag{24}$$

In particular, to construct the inverse contiguous relations for the function F_T, one extra differential equation should be added.

For completeness, we also note the following result:

Theorem 4 *The monodromy group of the system of differential equations defined by Eq. (22) is reducible when the one of the following combinations of parameters is an integer:*

$$\{a_1, a_2, b_1, b_2, \quad c-b_1-b_2, \quad c-a_1-a_2\} \in \mathbb{Z} \,.$$

A Puiseux-type solution for the hypergeometric differential equation of two variables was established by Erdelyi [217] still in the 50s and has been analyzed in detail in [218] in the framework of the GKZ approach.

3.2 Construction of the ε-Expansion via Differential Equations: The Appell Function F_3

To explain the technical details of our algorithm, let us analyze and construct the ε-expansion for the Appell hypergeometric function F_3. The preliminary results have been presented in [181, 188].

3.2.1 Notations

Let us consider the Appell hypergeometric function F_3 defined about $x = y = 0$ as

$$\omega_0 \equiv F_3(a_1, a_2, b_1, b_2, c; x, y) = \sum_{m=0}^{\infty}\sum_{n=0}^{\infty} \frac{(a_1)_m (a_2)_n (b_1)_m (b_2)_n}{(c)_{m+n}} \frac{x^m}{m!} \frac{y^n}{n!} , \quad (25)$$

It is symmetric with respect to simultaneous exchange

$$a_1 \Leftrightarrow a_2 , \quad b_1 \Leftrightarrow b_2 , \quad x \Leftrightarrow y . \quad (26)$$

so that $F_3(a_1, a_2, b_1, b_2, c; x, y) = F_3(a_2, a_1, b_2, b_1, c; y, x)$.

This function satisfies four differential equations (see Section 3.4 in [82]):

$$(1-x)\theta_{xx}\omega_0 = -\theta_{xy}\omega_0 + [(a_1+b_1)x-(c-1)]\,\theta_x\omega_0 + xa_1b_1\omega_0 , \quad (27a)$$

$$\begin{aligned}(xy-x-y)\theta_{xxy}\omega_0 &= [(1-y)(a_1+b_1)x - y(a_2+b_2+1-c)]\,\theta_{xy}\omega_0 \\ &\quad +(1-y)xa_1b_1\theta_y\omega_0 - ya_2b_2\theta_x\omega_0 , \end{aligned} \quad (27b)$$

and two other equations follow from Eqs. (27) and the symmetry relation (26). Equations (27) can be written in projective space with homogeneous coordinates on $\mathbf{P}_2(C)$ with $x = X/Z$ and $y = Y/Z$. We want also to remark that the system of differential equations for the Appell hypergeometric function $F_2(u, v)$ coincides with the system Eq. (27) by a redefinition $u = 1/x, v = 1/y$. This result follows also from analytic continuation of the Mellin–Barnes representation for Appell's functions F_2 and F_3.

3.2.2 Onefold Iterated Solution

Assuming the most general form of the parameters

$$a_j = \frac{p_j}{q} + a_1\varepsilon , \quad b_j = \frac{r_j}{q} + b_j\varepsilon , \quad c = 1 - \frac{p}{q} + c\varepsilon .$$

where p_i, r_i, p, q are integers, and writing the ε-expansion for the functions as

$$\omega_j = \sum_{k=0}^{\infty} \omega_j^{(k)} \varepsilon^k \,, \quad j = 0, 1, 2, 3, \tag{28}$$

we obtain the full system of differential equations (under the conditions $p_j r_j = 0 \,, j = 1, 2$):

$$\left[(1-x)\frac{d}{dx} - \frac{s_1}{q} - \frac{1}{x}\frac{p}{q}\right]\omega_1^{(r)} = -\frac{1}{x}\omega_3^{(r)} + \left[(a_1+b_1) - \frac{c}{x}\right]\omega_1^{(r-1)} + \left[\frac{a_1 r_1 + b_1 p_1}{q}\right]\omega_0^{(r-1)} + a_1 b_1 \omega_0^{(r-2)} \,, \tag{29a}$$

$$\left[(1-y)\frac{d}{dy} - \frac{s_2}{q} - \frac{1}{y}\frac{p}{q}\right]\omega_2^{(r)} = -\frac{1}{y}\omega_3^{(r)} + \left[(a_2+b_2) - \frac{c}{y}\right]\omega_2^{(r-1)} + \left[\frac{a_2 r_2 + b_2 p_2}{q}\right]\omega_0^{(r-1)} + a_2 b_2 s \omega_0^{(r-2)} \,, \tag{29b}$$

$$\begin{aligned} &\left[(xy-x-y)\frac{d}{dx} - (1-y)\frac{s_1}{q} + \frac{y}{x}\frac{(s_2+p)}{q}\right]\omega_3^{(r)} \\ &= \left[(1-y)(a_1+b_1) - \frac{y}{x}(a_2+b_2-c)\right]\omega_3^{(r-1)} \\ &+ (1-y)\left[\frac{a_1 r_1 + b_1 p_1}{q}\omega_2^{(r-1)} + a_1 b_1 \omega_2^{(r-2)}\right] \\ &- \frac{y}{x}\left[\frac{a_2 r_2 + b_2 p_2}{q}\omega_1^{(r-1)} + a_2 b_2 \omega_1^{(r-2)}\right] \,, \end{aligned} \tag{29c}$$

$$\begin{aligned} &\left[(xy-x-y)\frac{d}{dy} - (1-x)\frac{s_2}{q} + \frac{x}{y}\frac{(s_1+p)}{q}\right]\omega_3^{(r)} \\ &= \left[(1-x)(a_2+b_2) - \frac{x}{y}(a_1+b_1-c)\right]\omega_3^{(r-1)} \\ &+ (1-x)\left[\frac{a_2 r_2 + b_2 p_2}{q}\omega_1^{(r-1)} + a_2 b_2 \omega_1^{(r-2)}\right] \\ &- \frac{x}{y}\left[\frac{a_1 r_1 + b_1 p_1}{q}\omega_2^{(r-1)} + a_1 b_1 \omega_2^{(r-2)}\right] \,, \end{aligned} \tag{29d}$$

where we have introduced new notations:

$$s_1 = p_1 + r_1 \,, \quad s_2 = p_2 + r_2 \,. \tag{30}$$

We want to mention that this system of differential equations does not have the ε-form *a-la* Henn's form in [203, 219]. However, the system of equations (29) can be straightforwardly solved iteratively. Let us redefine functions $\omega_1, \omega_2, \omega_3$, as follows:

$$\omega_1^{(r)} = \theta_x \omega_0 = h_1(x)\phi_1^{(r)} , \quad \omega_2^{(r)} = \theta_y \omega_0 = h_2(y)\phi_2^{(r)} , \tag{31}$$

where $h_{1,2}(x)$ are new functions, defined as

$$h_1(x) = \sigma_1 \left[\frac{x^p}{(x-1)^{s_1+p}} \right]^{\frac{1}{q}} , \quad h_2(y) = \sigma_2 \left[\frac{y^p}{(y-1)^{s_2+p}} \right]^{\frac{1}{q}} , \tag{32}$$

and

$$\omega_3^{(r)} = \theta_x \theta_y \omega_0^{(r)} = H(x, y)\phi_3^{(r)} , \tag{33}$$

with

$$H(x, y) = \sigma_3 \left[\frac{x^{s_2+p} y^{s_1+p}}{(xy - x - y)^{s_1+s_2+p}} \right]^{\frac{1}{q}} , \tag{34}$$

where σ_j, $j = 1, 2, 3$ are some normalization constants.

Substituting it to the original system, Eq. (29), we have (for completeness we show all four equations explicitly):

$$\begin{aligned}(1-x)\frac{d}{dx}\phi_1^{(r)} &= -\frac{1}{x}\frac{H(x,y)}{h_1(x)}\phi_3^{(r)} \\ &+\left[(a_1+b_1)-\frac{c}{x}\right]\phi_1^{(r-1)} + \left[\frac{a_1 r_1 + b_1 p_1}{q}\right]\frac{1}{h_1(x)}\omega_0^{(r-1)} + \frac{1}{h_1(x)}a_1 b_1 \omega_0^{(r-2)} ,\end{aligned} \tag{35a}$$

$$\begin{aligned}(1-y)\frac{d}{dy}\phi_2^{(r)} &= -\frac{1}{y}\frac{H(x,y)}{h_2(y)}\phi_3^{(r)} \\ &+\left[(a_2+b_2)-\frac{c}{y}\right]\phi_2^{(r-1)} + \left[\frac{a_2 r_2 + b_2 p_2}{q}\right]\frac{1}{h_2(y)}\omega_0^{(r-1)} + \frac{1}{h_2(y)}a_2 b_2 \omega_0^{(r-2)} ,\end{aligned} \tag{35b}$$

$$\begin{aligned}(xy-x-y)\frac{d}{dx}\phi_3^{(r)} &= \left[(1-y)(a_1+b_1)-\frac{y}{x}(a_2+b_2-c)\right]\phi_3^{(r-1)} \\ &+\frac{a_1 r_1 + b_1 p_1}{q}(1-y)\frac{h_2(y)}{H(x,y)}\phi_2^{(r-1)} + a_1 b_1 (1-y)\frac{h_2(y)}{H(x,y)}\phi_2^{(r-2)} \\ &-\frac{a_2 r_2 + b_2 p_2}{q}\frac{y}{x}\frac{h_1(x)}{H(x,y)}\phi_1^{(r-1)} - a_2 b_2 \frac{y}{x}\frac{h_1(x)}{H(x,y)}\phi_1^{(r-2)} ,\end{aligned} \tag{36a}$$

$$
\begin{aligned}
(xy-x-y)\frac{d}{dy}\phi_3^{(r)} &= \left[(1-x)(a_2+b_2)-\frac{x}{y}(a_1+b_1-c)\right]\phi_3^{(r-1)} \\
&+\frac{a_2r_2+b_2p_2}{q}(1-x)\frac{h_1(x)}{H(x,y)}\phi_1^{(r-1)}+a_2b_2(1-x)\frac{h_1(x)}{H(x,y)}\phi_1^{(r-2)} \\
&-\frac{a_1r_1+b_1p_1}{q}\frac{x}{y}\frac{h_2(y)}{H(x,y)}\phi_2^{(r-1)}-a_1b_1\frac{x}{y}\frac{h_2(y)}{H(x,y)}\phi_2^{(r-2)}\,, \qquad (36b)
\end{aligned}
$$

This is a system of linear differential equations with algebraic coefficients. For completeness, it should be supplemented by the boundary condition. Boundary conditions are discussed below.

Remark Let us recall that the surface of singularities $L(x, y)$ of the F_3 hypergeometric function is defined by the system of equations (27) and has the form

$$
L : L_1 \cup L_2 \cup L_3 \cup L_4 \cup L_5 \equiv x \cup y \cup (1-x) \cup (1-y) \cup (x+y-xy)\,. \qquad (37)
$$

The extra functions, Eqs. (32) and (33), can be understood as the ratio of elements L_a of Eq. (37), and q-root is related with angle of rotations of curves L_j around zero.

Indeed, a multiple polylogarithm can be understood as a smooth map U from one region of singularities, where the solution of differential equation exists, to another one: $\mathrm{Li}_A\,(L_i) \xrightarrow{U} \mathrm{Li}_A\left(L_j\right)$. Such map is nothing but an analytic continuation that mixed the singularities of the differential system Eq. (27).

3.2.3 Boundary Conditions

The boundary conditions for the system of equations, Eqs. (35), (36), are defined by the series representation, so that

$$
\omega_0(z_1, z_2)|_{z_1=0} = {}_2F_1(a_2, b_2; c; z_2)\,, \quad \omega_0(z_1, z_2)|_{z_2=0} = {}_2F_1(a_1, b_1; c; z_1)\,. \qquad (38)
$$

Keeping in mind that

$$
\begin{aligned}
\omega_j(z_1, z_2) &= z_j \frac{a_j b_j}{c} F_3(a_1, a_2, b_1, b_2, 1+c; z_1, z_2)|_{\substack{a_j \to a_j+1 \\ b_j \to b_j+1}}\,, \quad j = 1, 2\,, \\
\omega_3(z_1, z_2) &= z_1 z_2 \frac{a_1 a_2 b_1 b_2}{c(1+c)} F_3(1+a_1, 1+a_2, 1+b_1, 1+b_2; 2+c; z_1, z_2)\,,
\end{aligned}
$$

we have

$$\omega_1(z_1, z_2)|_{z_1=0} = 0 \,, \quad \omega_1(z_1, z_2)|_{z_2=0} = z_1 \frac{a_1 b_1}{c} {}_2F_1(1+a_1, 1+b_1; 1+c; z_1) \,,$$
$$\omega_2(z_1, z_2)|_{z_2=0} = 0 \,, \quad \omega_2(z_1, z_2)|_{z_1=0} = z_2 \frac{a_2 b_2}{c} {}_2F_1(1+a_2, 1+b_2; 1+c; z_2) \,,$$
$$\omega_3(z_1, z_2)|_{z_1=0} = \omega_3(z_1, z_2)|_{z_2=0} = 0 \,. \quad (39)$$

The construction of the all-order ε-expansion of Gauss hypergeometric functions around rational values of parameters in terms of multiple polylogarithms has been constructed in Refs. [198, 200].

3.2.4 The Rational Parametrization: Towards Multiple Polylogarithms

It is well-known that onefold iterated integrals over algebraic functions are not, in general, expressible in terms of multiple polylogarithms but demand the introduction of a new class of functions [220–222].[15]

The iterative solution of the system Eq. (35), (36), have the following form. In the first two orders of ε-expansion, we have (these results follow from the series representation and a special choice of parameters):

$$\omega_0^{(0)} = 1 \,, \quad \phi_1^{(0)} = \phi_2^{(0)} = \phi_3^{(0)} = 0 \,, \quad (40a)$$
$$\omega_0^{(1)} = \phi_1^{(1)} = \phi_2^{(1)} = \phi_3^{(1)} = 0 \,, \quad (40b)$$

The second iteration produces:

$$\phi_3^{(2)}(x, y) = 0 \,, \quad (41a)$$
$$\phi_1^{(2)}(x, y) = a_1 b_1 \int_0^x \frac{dt}{(1-t)h_1(t)} \equiv a_1 b_1 R_1(x) \,, \quad (41b)$$
$$\phi_2^{(2)}(x, y) = a_2 b_2 \int_0^y \frac{dt}{(1-t)h_2(t)} \equiv a_2 b_2 R_2(y) \,, \quad (41c)$$
$$\omega_0^{(2)}(x, y) = \int_0^x \frac{dt_1}{t_1} \phi_1^{(2)} + {}_2F_1^{(2)}(x) + \int_0^y \frac{dt_1}{t_1} \phi_2^{(2)} + {}_2F_1^{(2)}(y) \quad (41d)$$
$$= \left[a_1 b_1 \int_0^x \frac{dt_1}{d_1} R_1(t) + a_2 b_2 \int_0^y \frac{dt_1}{t_1} R_2(t) \right] + {}_2F_1^{(2)}(x) + {}_2F_1^{(2)}(y) \,,$$

[15] Originally, such types of functions have been introduced in Ref. [197].

where ${}_2F_1^{(2)}(t)$ are the functions coming from boundary conditions. The finite part of the F_3 function is (in terms of R-functions):

$$\begin{aligned}
\phi_3^{(3)}(x,y) = & -\left(\frac{a_1 r_1 + b_1 p_1}{q}\right) a_2 b_2 h_2(y) R_2(y) \int_0^x \frac{dt}{H(t,y)\left(t + \frac{y}{1-y}\right)} \\
& + \left(\frac{a_2 r_2 + b_2 p_2}{q}\right) a_1 b_1 \int_0^x dt \frac{h_1(t) R_1(t)}{H(t,y)} \left(\frac{1}{t} - \frac{1}{t + \frac{y}{1-y}}\right) \\
& - \left(\frac{a_2 r_2 + b_2 p_2}{q}\right) a_1 b_1 h_1(x) R_1(x) \int_0^y dt \frac{dt}{H(x,t)\left(t + \frac{x}{1-x}\right)} \\
& + \left(\frac{a_1 r_1 + b_1 p_1}{q}\right) a_1 b_1 \int_0^y dt \frac{h_2(t) R_2(t)}{H(x,t)} \left(\frac{1}{t} - \frac{1}{t + \frac{x}{1-x}}\right) .
\end{aligned} \tag{42}$$

Up to some factor, the function $R(t)$ coincides with the Gauss hypergeometric function with a rational set of parameters:

$$R(z) \equiv \int_0^z \frac{dt}{(1-t)} \frac{1}{h(t)} \sim {}_2F_1\left(\frac{r_j}{q}, \frac{p_j}{q}, 1 - \frac{p}{q}; z\right), \quad j = 1, 2,$$

where $p_j r_j = 0$. The ε-expansion of the Gauss hypergeometric function around rational values of parameters has been analyzed in detail in Ref. [200]. It was shown that only the following cases are relevant:

- Integer set: $r_j = p_j = p = 0$,
- Zero-balance type: $r_j + p_j = -p, \quad r_j p_j = 0$.
- Binomial type: $p = p_j = 0, \quad r_j \neq 0$, (and symmetric one: $p = r_j = 0$, $p_j \neq 0$)
- Inverse binomial type: $r_j = p_j = 0, \quad p \neq 0$.
- Full type: $r_j = p_j = -p$.

For each particular set of parameters, the rational parametrization of Eqs. (35), (36) should be cross-checked. For illustration, let us consider a few examples.

3.2.5 The Rational Parametrization: Set 1

Let us consider the hypergeometric function

$$F_3\left(a_1\varepsilon, a_2\varepsilon, b_1\varepsilon, b_2\varepsilon, 1 - \frac{p}{q} + c\varepsilon; x, y\right),$$

so that

$$r_1 = p_1 = p_2 = s_2 = 0 \,, \quad p \neq 0 \,.$$

For this set of parameters, we have $s_1 = 0$, $s_2 = 0$, $p \neq 0$, and

$$h(x) = \left(\frac{x}{x-1} \right)^{\frac{p}{q}} \,, \quad h(y) = \left(\frac{y}{y-1} \right)^{\frac{p}{q}} \,, \quad H(x,y) = \left[\frac{xy}{((x-1)(y-1)-1)} \right]^{\frac{p}{q}} \,.$$

There is a parametrization that converts the functions h_1 and h_2 into rational functions P_1 and P_2:

$$\frac{x}{x-1} = P_1^q(\xi_1, \xi_2) \,, \quad \frac{y}{y-1} = P_2^q(\xi_1, \xi_2) \,,$$

where the functions P_1 and P_2 have the form

$$P_m(x,y) = \Pi_{i,j,k,l} \frac{(x-a_i)(y-b_j)}{(x-c_k)(y-d_l)} \,, \quad m = 1, 2 \,, \tag{43}$$

for a set of algebraic numbers $\{a_i, b_j, c_k, d_l\}$. In terms of new variables, the function H has the form

$$H(x,y) = \frac{xy}{x+y-xy} = \frac{1}{\frac{1}{x} + \frac{1}{y} - 1} = \frac{1}{\frac{1}{P_1^q} + \frac{1}{P_2^q} + 1} = P_3^q \,,$$

where P_3 is again a rational function of two variables of the type (43).

After a redefinition $\frac{1}{P_i^q} = Q_i^q$, $\quad i = 1, 2, 3$ we obtain the result that a statement about existence of a rational parametrization for the functions $h(x)$, $h(y)$ and $H(x,y)$ is equivalent to the existence of three rational functions of two variables satisfying the equation

$$Q_1^q + Q_2^q + Q_3^q = 1 \,. \tag{44}$$

To our knowledge, for $q > 2$, a solution exists only in terms of elliptic functions. However, it may happen that such a parametrization exists for $q = 2$, but we are not able to find it.

This problem is closely related to the solution of a functional equation. For example, for the equation

$$f^n + g^n = 1 \,,$$

the solution can be characterized as follows: [223, 224]:

- For $n = 2$, all solutions are the form

$$f = \frac{2\beta(z)}{1+\beta^2(z)} , \quad g = \frac{1-\beta^2(z)}{1+\beta^2(z)} ,$$

  where $\beta(z)$ is an arbitrary function.
- For $n = 3$, one solution is given by

$$f = \frac{1}{2\wp}\left(1+\frac{1}{\sqrt{3}}\wp'\right) , \quad g = -\frac{1}{2\wp}\left(1-\frac{1}{\sqrt{3}}\wp'\right) , \tag{45}$$

  where $\wp$ is the Weierstrass $\wp$-function satisfy $\left(\wp'\right)^2 = 4\wp^3 - 1$. For $n = 3$, the original equation is of genus 1, so that uniformization theorem assures the existence of an elliptic solution.

One of the most natural sets of variables for the set of parameters under consideration is the following: $P_1 = \xi_1$, $P_2 = \xi_2$, so that

$$H = \frac{1}{\xi_1^p \xi_2^p}\left(\xi_1^q\xi_2^q - \xi_1^q - \xi_2^q\right)^{\frac{p}{q}} . \tag{46}$$

3.2.6 The Rational Parametrization: Set 2

In a similar manner, let us analyze another set of parameters:

$$F_3\left(-\frac{p}{q}+a_1\varepsilon, a_2\varepsilon, b_1\varepsilon, -\frac{p}{q}+b_2\varepsilon, 1-\frac{p}{q}+c\varepsilon; x, y\right) .$$

In this case, $s_1 = s_2 = -p$, and the functions h have the form $h(x) = x^{\frac{p}{q}}$, $h(y) = y^{\frac{p}{q}}$. Applying the same trick with the introduction of new functions P_1 and P_2, we would find that the existence of a rational parametrization corresponds in the present case to the validity of Eq. (44). In particular, by introducing new variables $x^{\frac{1}{q}} = \xi_1, y^{\frac{1}{q}} = \xi_2$, we obtain $H = \left(\xi_1^q\xi_2^q - \xi_1^q - \xi_2^q\right)^{\frac{p}{q}}$.

3.2.7 The Rational Parametrization: Set 3

Let us analyze the following set of parameters: $p = 0$, $s_1, s_2 \neq 0$, corresponding to

$$F_3\left(\frac{p_1}{q}+a_1\varepsilon, \frac{p_2}{q}+b_1\varepsilon, b_2\varepsilon, 1+c\varepsilon; x, y\right) .$$

In this case,

$$h_1(x) = (1-x)^{\frac{p_1}{q}}\ , \quad h_2(y) = (1-y)^{\frac{p_2}{q}}\ , \quad H(x,y) = \left(\frac{x^{s_2} y^{s_1}}{(xy-x-y)^{s_1+s_2}}\right)^{\frac{1}{q}} .$$

For simplicity, we set $s_1 = -s_2 = s$, and put $1-x = P_1^q$ and $1-y = P_2^q$, so that $H = \left(\frac{1-P_2^q}{1-P_1^q}\right)^{\frac{s}{q}} \equiv P_3$. In particular, for $P_1 = \xi_1$, $P_2 = \xi_2$, $H = \left(\frac{1-\xi_2^q}{1-\xi_1^q}\right)^{\frac{s}{q}}$.

3.2.8 The Rational Parametrization: Set 4

Let us analyze the following set of parameters: $s_1 = -p, s_2 = 0$, so that hypergeometric function is

$$F_3\left(-\frac{p_1}{q} + a_1\varepsilon, a_2\varepsilon, b_1\varepsilon, b_2\varepsilon, 1 - \frac{p}{q} + c\varepsilon; x, y\right) .$$

In this case,

$$h_1(x) = x^{\frac{p}{q}}\ , \quad h_2(y) = \left(\frac{y}{y-1}\right)^{\frac{p}{q}}\ , \quad H(x,y) = x^{\frac{p}{q}} \equiv h(x)\ . \tag{47}$$

Let us take a new set of variables $(x, y) \to (\xi_1, \xi_2)$:

$$\xi_1 = x^{\frac{1}{q}}\ , \quad \xi_2 = \left(\frac{y}{y-1}\right)^{\frac{1}{q}}\ , \tag{48}$$

In terms of a new variables we have:

$$H(x,y) \equiv h_1(x) = \xi_1^p\ , \quad h_2(y) = \xi_2^p\ , \quad x + y - xy = \frac{\xi_1^q - \xi_2^q}{1 - \xi_2^q} . \tag{49}$$

Thus, a rational parametrization exists.

3.2.9 The Rational Parametrization: Set 5

Consider the set of parameters defined by $s_2 = p = 0, s_1 \neq 0$, that corresponds to the case $F_3\left(-\frac{p_1}{q} + a_1\varepsilon, a_2\varepsilon, b_1\varepsilon, b_2\varepsilon, 1 + c\varepsilon; x, y\right)$. In this case,

$$h_1(x) = (1-x)^{\frac{p_1}{q}}\ , \quad h_2(y) = 1\ , \quad H(x,y) = \left(\frac{y}{xy-x-y}\right)^{\frac{p_1}{q}}\ , \tag{50}$$

Let us suggest that

$$1 - x = P^q \,, \quad H(x,y) = \frac{y}{x+y-xy} = Q^q \,,$$

where P and Q are rational functions. Then, $y = \frac{(1-P^q)Q^q}{1-P^q Q^q}$. After the redefinition, $PQ = R$;, we get $y = \frac{(1-P^q)}{(1-R^q)}\frac{R^q}{P^q}$. The simplest version of P and R are polynomials: $P = \xi_1$, and $R = \xi_2$. In this parametrization,

$$h(x) = \xi_1^{p_1} \,, \quad H(x,y) = \left(\frac{\xi_2}{\xi_1}\right)^{p_1} \,, \quad y = \frac{1-\xi_1^q}{1-\xi_2^q}\left(\frac{\xi_2}{\xi_1}\right)^q \,.$$

In this case, the rational parametrization exists.

3.2.10 Explicit Construction of Expansion: Integer Values of Parameters

Let us consider the construction of the ε-expansion around integer values of parameters. If we put

$$\omega_0 = F_3(a_1\varepsilon, b_1\varepsilon, a_2\varepsilon, b_2\varepsilon, 1 + c\varepsilon; x, y),$$

then the system of differential equations can be presented in the form

$$\frac{\partial}{\partial x}\omega_1 = \left[\frac{1}{x-1} - \frac{1}{x}\right]\omega_3 - \left[\frac{c}{x} + \frac{(a_1+b_1-c)}{x-1}\right]\varepsilon\omega_1 - a_1 b_1 \frac{1}{x-1}\varepsilon^2\omega_0 \,, \tag{51}$$

$$\frac{\partial}{\partial y}\omega_2 = \left[\frac{1}{y-1} - \frac{1}{y}\right]\omega_3 - \left[\frac{c}{y} + \frac{(a_2+b_2-c)}{y-1}\right]\varepsilon\omega_2 - a_2 b_2 \frac{1}{y-1}\varepsilon^2\omega_0 \,, \tag{52}$$

$$\frac{\partial}{\partial x}\omega_3 = \left[\frac{(a_2+b_2-c)}{x} - \frac{(a_1+b_1+a_2+b_2-c)}{x+\frac{y}{1-y}}\right]\varepsilon\omega_3 - \frac{a_1 b_1}{x+\frac{y}{1-y}}\varepsilon^2\omega_2 + \left[\frac{1}{x} - \frac{1}{x+\frac{y}{1-y}}\right]a_2 b_2\varepsilon^2\omega_1 \,, \tag{53}$$

$$\frac{\partial}{\partial y}\omega_3 = \left[\frac{(a_1+b_1-c)}{y} - \frac{(a_1+b_1+a_2+b_2-c)}{y+\frac{x}{1-x}}\right]\varepsilon\omega_3 - \frac{a_2 b_2}{y+\frac{x}{1-x}}\varepsilon^2\omega_1 + \left[\frac{1}{y} - \frac{1}{y+\frac{x}{1-x}}\right]a_1 b_1\varepsilon^2\omega_2 \,. \tag{54}$$

This system can be straightforwardly integrated in terms of multiple polylogarithms, defined via a onefold iterated integral G, where

$$G(z; a_k, \mathbf{a}) = \int_0^z \frac{dt}{t - a_k} G(t; \mathbf{a}) \ . \tag{55}$$

In addition, the ε-expansion of a Gauss hypergeometric function around integer values of parameters is needed. It has the following form (see Eq. (34) in [173]):

$$\begin{aligned} {}_2F_1(a\varepsilon, b\varepsilon; 1+c\varepsilon; z) &= 1+ab\varepsilon^2 \mathrm{Li}_2(z) \\ &+ab\varepsilon^3 \left[(a+b-c)\mathrm{S}_{1,2}(z) - c\mathrm{Li}_3(z)\right] + O(\varepsilon^4) \ . \end{aligned} \tag{56}$$

The first iteration gives rise to

$$\omega_0^{(0)} = 1 \ , \quad \omega_1^{(0)} = \omega_2^{(0)} = \omega_3^{(0)} = 0 \ , \quad \omega_0^{(1)} = \omega_1^{(1)} = \omega_2^{(1)} = \omega_3^{(1)} = 0 \ .$$

The results of the second iteration are the following:

$$\begin{aligned} &\omega_3^{(2)} = 0 \ , \quad \omega_1^{(2)} = -a_1 b_1 \ln(1-x) \ , \quad \omega_2^{(2)} = -a_2 b_2 \ln(1-y) \ , \\ &\omega_0^{(2)} = a_1 b_1 \mathrm{Li}_2(x) + a_2 b_2 \mathrm{Li}_2(y) \ , \end{aligned} \tag{57}$$

where the classical polylogarithms $\mathrm{Li}_n(z)$ are defined as

$$\mathrm{Li}_1(z) = -\ln(1-z) \ , \quad \mathrm{Li}_{n+1}(z) = \int_0^z \frac{dt}{t} \mathrm{Li}_n(t) \ , \quad n \geq 1. \tag{58}$$

After the third iteration, we have

$$\omega_3^{(3)} = 0 \tag{59}$$

$$\omega_1^{(3)} = \frac{1}{2} a_1 b_1 (a_1 + b_1 - c) \ln^2(1-x) - a_1 b_1 c \mathrm{Li}_2(x) \ , \tag{60}$$

$$\omega_2^{(3)} = \frac{1}{2} a_2 b_2 (a_2 + b_2 - c) \ln^2(1-y) - a_2 b_2 c \mathrm{Li}_2(y) \ , \tag{61}$$

$$\begin{aligned} \omega_0^{(3)} &= -a_1 b_1 c \mathrm{Li}_3(x) - a_2 b_2 c \mathrm{Li}_3(y) \\ &+a_1 b_1 (a_1 + b_1 - c) \mathrm{S}_{1,2}(x) + a_2 b_2 (a_2 + b_2 - c) \mathrm{S}_{1,2}(y) \ , \end{aligned} \tag{62}$$

where $\mathrm{S}_{a,b}(z)$ are the Nielsen polylogarithms:

$$z \frac{d}{dz} \mathrm{S}_{a,b}(z) = \mathrm{S}_{a-1,b}(z) \ , \quad \mathrm{S}_{1,b}(z) = \frac{(-1)^b}{b!} \int_0^1 \frac{dx}{x} \ln^b(1 - zx) \ . \tag{63}$$

The result of the next iteration, $\omega_3^{(4)}(x, y)$, can be expressed in several equivalent forms:

$$\frac{\omega_3^{(4)}(x, y)}{a_1 a_2 b_1 b_2} = \ln(1-y) G_1\left(x; -\frac{y}{1-y}\right) - G_2(x; 1) + G_{1,1}\left(x; -\frac{y}{1-y}, 1\right),$$

$$\frac{\omega_3^{(4)}(x, y)}{a_1 a_2 b_1 b_2} = \ln(1-x) G_1\left(y; -\frac{x}{1-x}\right) - G_2(y; 1) + G_{1,1}\left(y; -\frac{x}{1-x}, 1\right).$$

Keeping in mind that

$$G_{1,1}\left(x; -\frac{y}{1-y}, 1\right) = \int_0^x \frac{dt}{t+\frac{y}{1-y}} \ln(1-t) \tag{64}$$
$$= -\ln(1-y)\ln(x+y-xy) + \ln(1-y)\ln y - \text{Li}_2\left(x+y-xy\right) + \text{Li}_2\left(y\right),$$

the result can be written in a very simple form,

$$\frac{\omega_3^{(4)}(x, y)}{a_1 a_2 b_1 b_2} = \text{Li}_2\left(x\right) + \text{Li}_2\left(y\right) - \text{Li}_2\left(x+y-xy\right). \tag{65}$$

Taking into account that $\frac{\omega_3^{(4)}(x,y)}{a_1 a_2 b_1 b_2} = \frac{1}{2} xy F_3(1, 1, 1, 1; 3; x, y)$, we obtain the well-known result [225]

$$\frac{1}{2} xy F_3(1, 1, 1, 1; 3; x, y) = \text{Li}_2\left(x\right) + \text{Li}_2\left(y\right) - \text{Li}_2\left(x+y-xy\right).$$

There is also another form [226] for this integral,

$$\frac{1}{2} xy F_3(1, 1, 1, 1; 3; x, y) = \text{Li}_2\left(\frac{x}{x+y-xy}\right) - \text{Li}_2\left(\frac{x-xy}{x+y-xy}\right)$$
$$- \ln(1-y)\ln\left(\frac{y}{x+y-xy}\right).$$

One form can be converted to the other using the well-known dilogarithm identity

$$\text{Li}_2\left(\frac{1}{z}\right) = -\text{Li}_2\left(z\right) - \frac{1}{2}\ln^2(-z) - \zeta_2,$$

together with the attendant functional relations [110, 227].

The following expressions result from direct iterations in terms of G-functions:

$$\frac{\omega_1^{(4)}(x,y)}{a_1b_1} = a_2b_2\left\{ \ln(1-y)\left[G_{1,1}\left(x;1,-\frac{y}{1-y}\right) - G_2\left(x;-\frac{y}{1-y}\right)\right]\right.$$
$$\left. -G_{1,2}(x;1,1) + G_3(x;1) + G_{1,1,1}\left(x;1,-\frac{y}{1-y},1\right) - G_{2,1}\left(x;-\frac{y}{1-y},1\right)\right\}$$
$$+a_2b_2G_1(x;1)G_2(y;1) - c\Delta_1G_{2,1}(x;1,1) - \Delta_1^2G_{1,1,1}(x;1,1,1) - c^2G_3(x;1)$$
$$+(a_1b_1-c\Delta_1)G_{1,2}(x;1,1)\,, \tag{66}$$

where

$$\Delta_j = a_j + b_j - c\,.$$

The G-functions can be converted into classical or Nielsen polylogarithms with the help of the following relations:

$$G_{1,1}\left(x;1,-\frac{y}{1-y}\right) = G_1(x;1)G_1\left(x;-\frac{y}{1-y}\right) - G_{1,1}\left(x;-\frac{y}{1-y},1\right)\,, \tag{67a}$$

$$G_1(x;1)\,G_{1,1}\left(x;-\frac{y}{1-y},1\right) = G_{1,1,1}\left(x;1,-\frac{y}{1-y},1\right) \tag{67b}$$
$$+2G_{1,1,1}\left(x;-\frac{y}{1-y},1,1\right)\,,$$

$$G_{1,1,1}\left(x;-\frac{y}{1-y},1,1\right) = \mathrm{S}_{1,2}(x+y-xy) - \mathrm{S}_{1,2}(y) \tag{67c}$$
$$+\frac{1}{2}\ln^2(1-y)\left[\ln(x+y-xy)-\ln y\right] + \ln(1-y)\left[\mathrm{Li}_2(x+y-xy)-\mathrm{Li}_2(y)\right]\,,$$

$$G_{2,1}\left(x;-\frac{y}{1-y},1\right) + \ln(1-y)G_2\left(x;-\frac{y}{1-y}\right) = G_{1,2}\left(1;1-\frac{1}{y},\frac{1}{x}\right) + G_3(x;1)$$
$$= \int_0^x \frac{du}{u}\left[\mathrm{Li}_2(y) - \mathrm{Li}_2(u+y-uy)\right]\,. \tag{67d}$$

In a similar manner,

$$\frac{\omega_2^{(4)}(x,y)}{a_2b_2} = a_1b_1\left\{ \ln(1-x)\left[G_{1,1}\left(y;1,-\frac{x}{1-x}\right) - G_2\left(y;-\frac{x}{1-x}\right)\right]\right.$$
$$\left. -G_{1,2}(y;1,1) + G_3(y;1) + G_{1,1,1}\left(y;1,-\frac{x}{1-x},1\right) - G_{2,1}\left(y;-\frac{x}{1-x},1\right)\right\}$$
$$-c\Delta_2G_{2,1}(y;1,1) - \Delta_2^2G_{1,1,1}(y;1,1,1) - c^2G_3(y;1) - c\Delta_2G_{1,2}(y;1,1)$$
$$+a_1b_1G_1(y;1)G_2(x;1) + a_2b_2G_{1,2}(y;1,1)\,, \tag{68}$$

and the last term, expressible in terms of functions of order 3, is $\omega_3^{(5)}$.

3.2.11 Construction of ε-Expansion via Integral Representation

An alternative approach to construction of the higher order ε-expansion of generalized hypergeometric functions is based on their integral representation. We collect here some representations for the Appell hypergeometric function F_3 extracted from Refs. [10, 11].

For our purposes, the most useful expression is the following: (see Eq. (16) in Section 9.4. of [11]), [225, 226]:

$$\frac{\Gamma(c_1)\Gamma(c_2)}{\Gamma(c_1+c_2)} F_3(a_1, b_1, a_2, b_2, c_1+c_2; x, y)$$
$$= \int_0^1 du u^{c_1-1}(1-u)^{c_2-1} {}_2F_1(a_1, b_1; c_1; ux) {}_2F_1(a_2, b_2; c_2; (1-u)y) \,. \quad (69)$$

Indeed, expanding one of the hypergeometric functions as a power series leads to

$$\int_0^1 u^{c_1-1}(1-u)^{j+c_2-1} {}_2F_1(a_1, b_1; c_1; ux) \sum_{j=0}^{\infty} \frac{(a_2)_j (b_2)_j y^j}{(c_2)_j j!} \,.$$

The order of summation and integration can be interchanged in the domain of convergence of the series, and after integration we obtain Eq. (69).

The twofold integral representation [10],

$$\frac{\Gamma(b_1)\Gamma(b_2)\Gamma(c-b_1-b_2)}{\Gamma(c)} F_3(a_1, a_2; b_1, b_2; c; x, y) = \quad (70)$$
$$\int\int_{0\le u, 0\le v, u+v\le 1} du dv u^{b_1-1} v^{b_2-1} (1-u-v)^{c-b_1-b_2-1} (1-ux)^{-a_1} (1-vy)^{-a_2} \,,$$

can be reduced to the following integral:

$$\frac{\Gamma(b_1)\Gamma(b_2)\Gamma(c-b_1-b_2)}{\Gamma(c)} F_3(a_1, a_2; b_1, b_2; c; x, y) \quad (71)$$
$$= \int_0^1 du \int_0^{1-u} dv u^{b_1-1} v^{b_2-1} (1-u-v)^{c-b_1-b_2-1} (1-ux)^{-a_1} (1-vy)^{-a_2}$$
$$= \frac{\Gamma(b_2)\Gamma(c-b_1-b_2)}{\Gamma(c-b_1)}$$
$$\times \int_0^1 du u^{b_1-1} (1-ux)^{-a_1} (1-u)^{c-b_1-1} {}_2F_1\left(\left. \begin{matrix} a_2, b_2 \\ c-b_1 \end{matrix} \right| (1-u)y \right) .$$

For the particular values of parameters ($c_1 = a_1, c_2 = a_2$), the integral Eq. (69) can be reduced to the Appell function F_1:

$$F_3(a_1, a_2, b_1, b_2; a_1 + a_2; x, y) = \frac{1}{(1-y)^{b_2}} F_1 \left(a_1, b_1, b_2, a_1 + a_2; x, -\frac{y}{1-y} \right) .$$
$$= \frac{1}{(1-x)^{b_1}} F_1 \left(a_2, b_1, b_2, a_1 + a_2; -\frac{x}{1-x}, y \right) .$$

Using the onefold integral representation for the Appell function F_1, it is possible to prove the following relations:

$$F_3(a, c-a, b, c-b, c, x, y) = (1-y)^{c+a-b} {}_2F_1 \left(\begin{matrix} a, b \\ c \end{matrix} \middle| x + y - xy \right) . \quad (72)$$

Using Eqs. (69) and (71), the onefold integral representation can be written for the coefficients of the ε-expansion of the hypergeometric function F_3 via the ε-expansion of the Gauss hypergeometric function, constructed in Refs. [198, 200].

The coefficients of the ε-expansion of the Gauss hypergeometric function can be expressed in terms of multiple polylogarithms of a q-root of unity with arguments $\left(\frac{z}{z-1}\right)^{\frac{1}{q}}$, $z^{\frac{1}{q}}$ or $(1-z)^{\frac{1}{q}}$ (see also [157]), so that the problem of finding a rational parametrization reduces to the problem of finding a rational parametrization of the integral kernel of Eqs. (69) and (71) in terms of variables generated by the ε-expansion of the Gauss hypergeometric function.

The construction of the higher-order ε-expansion of the Gauss hypergeometric function around rational values of parameters [198, 200], plays an crucial role in construction of the higher-order ε-expansion of many (but not all) Horn-type hypergeometric functions.

3.2.12 Relationship to Feynman Diagrams

Let us consider the one-loop pentagon with vanishing external legs. The higher-order ε-expansion for this diagram has been constructed [165] in terms of iterated onefold integrals over algebraic functions. In Ref. [228], the hypergeometric representation for the one-loop pentagon with vanishing external momenta has been constructed as a sum of Appell hypergeometric functions F_3.

In [181], where our differential equation approach is presented, it was pointed out that the one-loop pentagon can be expressed in terms of multiple polylogarithms. Ref. [154] verified the numerical agreement between the results of Refs. [165, 228, 229] constructed the iterative solution of the differential equation [230].

Let us recall the results of Ref. [228]. The one-loop massless pentagon is expressible in terms of the Appell function F_3 with the following set of parameters:

$$\Phi_5^{(d)} \sim F_3\left(1, 1, \frac{7-d}{2}, 1, \frac{10-d}{2}; x, y\right), \tag{73}$$

where d is dimension of space-time. Another representation presented in [228] has the structure

$$H_5^{(d)} \sim F_3\left(\frac{1}{2}, 1, 1, \frac{d-2}{2}, \frac{d+1}{2}; \frac{y}{x+y-xy}, \frac{1}{x}\right). \tag{74}$$

Let us consider the case of $d = 4 - 2\varepsilon$. The first representation, Eq. (73), is

$$\Phi_5^{(4-2\varepsilon)} \sim F_3\left(1, 1, \frac{3}{2} - \varepsilon, 1, 4 - \varepsilon; x, y\right).$$

This case,

$$\{p_1 = p_2 = r_2 = p = 0\}, \{r_1 = 1, q = 2\} \Longrightarrow s_1 \neq 0; \quad s_2 = 0, \quad p = 0,$$

corresponds to our **set 5**, so that the ε-**expansion is expressible in terms of multiple polylogarithms**, defined by Eq. (55).

For the other representation, Eq. (74),

$$H_5^{(4-2\varepsilon)} \sim F_3\left(\frac{1}{2}, 1, 1, \varepsilon, \frac{5}{2} - \varepsilon; \frac{y}{x+y-xy}, \frac{1}{x}\right),$$

so that it is reducible to the following set of parameters:

$$\{p_2 = r_1 = r_2 = 0\}, \quad \{p_1 = 1, q = 2, \ p = 1\} \Longrightarrow s_1 = 1; \quad s_2 = 0, \quad p = -1.$$

This is our **set 4**, so that the ε-**expansion is expressible in terms of multiple polylogarithms**, defined by Eq. (55).

The ε-expansion of the one-loop pentagon about $d = 3 - 2\varepsilon$ could be treated in a similar manner. In this case, the first representation corresponds to **set 1**

$$\Phi^{(3-2\varepsilon)} \sim F_3\left(1, 1, 2+\varepsilon, 1, \frac{7}{2} + \varepsilon; x, y\right) \Longrightarrow p_1 = p_2 = r_1 = r_2 = 0, \quad q = 2, p = 1$$

and there is no rational parametrization, so that the result of the ε-**expansion is expressible in terms of a onefold iterated integral over algebraic functions.**

The other representation, Eq. (74), also cannot be expressed in terms of multiple polylogarithms:

$$H_5^{(3-2\varepsilon)} \sim F_3\left(\frac{1}{2}, 1, 1, \frac{1}{2} - \varepsilon, 2 - \varepsilon; \frac{y}{x + y - xy}, \frac{1}{x}\right) \Longrightarrow \begin{array}{l} p_2 = r_1 = p = 0\,, \\ q = 2 \quad p_1 = r_2 = 1\,, \\ s_1 = 1; \quad s_2 = 1\,. \end{array}$$

This corresponds to **set 3**, and the **ε-expansion is expressible in terms of onefold iterated integral over algebraic functions**.

In this way, the question of the all-order ε-expansion of a one-loop Feynman diagram in terms of multiple polylogarithms is reduced to the question of the existence of a rational parametrization for the (ratio) of singularities.

Remark The dependence of the coefficients of the ε-expansion (multiple polylogarithms or elliptic function) on the dimension of space-time is not new. In particular, it is well known that the two-loop sunset in $3 - 2\varepsilon$ dimension is expressible in terms of polylogarithms [231, 232] and demands introduction of new functions in $4 - 2\varepsilon$ dimension [233–239].

4 Conclusion

The deep relationship between Feynman diagrams and hypergeometric functions has been reviewed, and we have tried to enumerate all approaches and recent results on that subject. Special attention was devoted to the discussion of different algorithms for constructing the analytical coefficients of the ε-expansion of multiple hypergeometric functions. We have restricted ourselves to multiple polylogarithms and functions related to integration over rational functions (the next step after multiple polylogarithms). The values of parameters related to elliptic polylogarithms was beyond our consideration.

We have presented our technique for the construction of coefficients of the higher order ε-expansion of multiple Horn-type hypergeometric functions, developed by the authors[16] during the period 2006–2013. One of the main results of interest was the observation [102, 145, 146, 166, 198–200] that for each Horn-type hypergeometric function, a set of parameters can be found so that the coefficients of the ε-expansion include only functions of weight one (so-called "pure functions," in a modern terminology). As was understood in 2013 by Johannes Henn [203, 219], this property is valid not only for hypergeometric functions but also for generic Feynman diagrams.

[16]Unfortunately, the further prolongation of this project has not been supported by DFG, so that many interesting results remain unpublished.

Our approach is based on the systematic analysis of the system of hypergeometric differential equations (linear differential operators of hypergeometric type with polynomial coefficients) and does not demand the existence of an integral representation, which is presently unknown for a large class of multiple Horn-type hypergeometric functions (they could be deduced, but are not presently available in the mathematical literature).

Our approach is based on the factorization of the system of differential equations into a product of differential operators, together with finding a rational parametrization and constructing iterative solutions. To construct such a system, an auxiliary manipulation with parameters (shifting by integer values) is required, which can be done with the help of the HYPERDIRE set of programs [82]. This technique is applicable not only to hypergeometric functions defined by series but also to multiple Mellin–Barnes integrals [77]—one of the representations of Feynman diagrams in a covariant gauge. We expect that the present technique is directly applicable (with some technical modifications) to the construction of the ε-expansion of hypergeometric functions beyond multiple polylogarithms, specially, that the two-loop sunset has a simple hypergeometric representation [234].

There are two of our considerations that have not been solved algorithmically: (a) the factorization of linear partial differential operators into irreducible factors is not unique, as has been illustrated by Landau [186] (see also Ref. [187]): $(\partial_x + 1)(\partial_x + 1)(\partial_x + x\partial_y) = (\partial_x^2 + x\partial_{xy} + \partial_x + (2+x)\partial_y)(\partial_x + 1)$; (b) The choice of parametrization is still an open problem, but there is essential progress in this direction [240, 241].

Our example has shown that such a parametrization is defined by the locus of singularities of a system of differential equations, so that the problem of finding a rational parametrization is reduced to the parametrization of solutions of the Diophantine equation for the singular locus of a Feynman diagram and/or hypergeometric function. It is well known that in the case of a positive solution of this problem (which has no complete algorithmic solution), the corresponding system of partial differential equations of a few variables takes the simplest structure. At the same time, there is a relationship between the type of solution of the Diophantine equation for the singular locus and the structure of the coefficients of the ε-expansion: a linear solution allows us to write the results of the ε-expansion in terms of multiple polylogarithms. An algebraic solution gives rise to functions different from multiple polylogarithms and elliptic functions, etc. It is natural to expect that, in the case of an elliptic solution of the Diophantine equation for the singular locus, the results for the ε-expansion are related to the elliptic generalization of multiple polylogarithms.

Another quite interesting and still algorithmically open problem is the transformation of multiple Horn-type hypergeometric functions with reducible monodromy to hypergeometric functions with irreducible monodromy. In the application to Feynman diagrams, such transformations correspond to functional relations, studied recently by Oleg Tarasov [242–244], and by Andrei Davydychev [106].

Further analysis of the symmetries of the hypergeometric differential equations related to the Mellin–Barnes representation of Feynman diagrams (for simplicity we will call it the hypergeometric representation) has revealed their deep connection to the holonomic properties of Feynman diagrams. In particular, a simple and fast algorithm was constructed for the reduction of any Feynman diagram having a onefold Mellin–Barnes integral representation to a set of master integrals [77].

The importance of considering the dimension of the irreducible representation instead of generic holonomic rank has been pointed out in the application to Feynman diagrams [66]. In the framework of this approach, the set of irreducible, non-trivial master-integrals corresponds to the set of irreducible (with respect to analytical continuation of the variables, masses, and external momenta) solutions of the corresponding system of hypergeometric differential equations, whereas diagrams expressible in terms of Gamma-functions correspond to Puiseux-type solutions (monomials with respect to Mandelstam variables) of the original system of hypergeometric equations.

Acknowledgments MYK is very indebted to Johannes Blümlein, Carsten Schneider and Peter Marquard for the invitation and for creating such a stimulating atmosphere in the workshop "Anti-Differentiation and the Calculation of Feynman Amplitudes" at DESY Zeuthen. BFLW thanks Carsten Schneider for kind the hospitality at RISC. MYK thanks the Hamburg University for hospitality while working on the manuscript. SAY acknowledges support from The Citadel Foundation.

References

1. R. Hwa, V. Teplitz, *Homology and Feynman Integrals* (Benjamin, New York, 1966)
2. F. Pham, *Introduction a L'etude Topologuqie des Singularities de Landau* (Gauthier-Villars Editeur, Paris, 1967)
3. V.A. Golubeva, Some problems in the analytical theory of Feynman integrals. Russ. Math. Surv. **31** 139 (1976)
4. M.Yu. Kalmykov, B.A. Kniehl, B.F.L. Ward, S.A. Yost, Hypergeometric functions, their epsilon expansions and Feynman diagrams. arXiv:0810.3238 [hep-th]
5. D.S. Kershaw, Feynman amplitudes as power series. Phys. Rev. D **8**, 2708 (1973)
6. D. Kreimer, One loop integrals revisited. 1. The Two point functions. Z. Phys. C **54**, 667 (1992)
7. D. Kreimer, One loop integrals revisited. 2. The Three point functions. Int. J. Mod. Phys. A **8**, 1797 (1993)
8. L. Brucher, J. Franzkowski, D. Kreimer, Loop integrals, R functions and their analytic continuation. Mod. Phys. Lett. A **9**, 2335 (1994)
9. A. Erdelyi, *Higher Transcendental Functions* (McGraw-Hill, New York, 1953)
10. L.J. Slater, *Generalized Hypergeometric Functions* (Cambridge University Press, Cambridge 1966)
11. H.M. Srivastava, P.M. Karlsson, *Multiple Gaussian Hypergeometric Series* (Ellis Horwood Ltd., Chichester; Halsted Press, New York, 1985)
12. E.E. Boos, A.I. Davydychev, A method of evaluating massive Feynman integrals. Theor. Math. Phys. **89**, 1052 (1991)
13. A.I. Davydychev, General results for massive N point Feynman diagrams with different masses. J. Math. Phys. **33**, 358 (1992)
14. D.J. Broadhurst, J. Fleischer, O.V. Tarasov, Two loop two point functions with masses: asymptotic expansions and Taylor series, in any dimension. Z. Phys. C **60**, 287 (1993)

15. F.A. Berends, M. Buza, M. Böhm, R. Scharf, Closed expressions for specific massive multiloop selfenergy integrals. Z. Phys. C **63**, 227 (1994)
16. S. Bauberger, F.A. Berends, M. Böhm, M. Buza, Analytical and numerical methods for massive two loop selfenergy diagrams. Nucl. Phys. B **434**, 383 (1995)
17. A.I. Davydychev, A.G. Grozin, Effect of $m(c)$ on b quark chromomagnetic interaction and on-shell two loop integrals with two masses. Phys. Rev. D **59**, 054023 (1999)
18. I. Bierenbaum, S. Weinzierl, The Massless two loop two point function. Eur. Phys. J. C **32**, 67 (2003)
19. J. Gluza, K. Kajda, R. Riemann, AMBRE: a Mathematica package for the construction of Mellin-Barnes representations for Feynman integrals. Comput. Phys. Commun. **177**, 879 (2007)
20. A.V. Smirnov, V.A. Smirnov, On the resolution of singularities of multiple Mellin-Barnes integrals. Eur. Phys. J. C **62**, 445 (2009)
21. M. Prausa, Mellin-Barnes meets Methods of Brackets: a novel approach to Mellin-Barnes representation of Feynman integrals. Eur. Phys. J. C **77**, 594 (2017)
22. J. Horn, Hypergeometriche Funktionen zweier Veränder lichen. Math. Ann. **105**, 381 (1931)
23. A.C.T. Wu, Generalized Euler-Pochhammer integral representation for single-loop Feynman amplitudes. Phys. Rev. D **9**, 370 (1974)
24. K. Mano, Comment on generalized Fuler-Pochhammer integral representation for single-loop Feynman amplitudes. Phys. Rev. D **11**, 452 (1975)
25. B. Ananthanarayan, S. Friot, S. Ghosh, New series representations for the two-loop massive sunset diagram. Eur. Phys. J. C **80**, 606 (2020)
26. B. Ananthanarayan, S. Friot, S. Ghosh, A. Hurier, New analytic continuations for the Appell F_4 series from quadratic transformations of the Gauss ${}_2F_1$ function. arXiv:2005.07170 [hep-th]
27. S.I. Bezrodnykh, Analytic continuation of the Appell function F_1 and integration of the associated system of equations in the logarithmic case. Comput. Math. Math. Phys. **57**, 559 (2017)
28. S.I. Bezrodnykh, Analytic continuation of the Lauricella function $F_D^{(N)}$ with arbitrary number of variables. Integral Transforms Spec. Funct. **29**, 21 (2018)
29. S.I. Bezrodnykh, Analytic continuation of the Horn hypergeometric series with an arbitrary number of variables. Integral Transforns Spec. Funct. bf 31 788 (2020)
30. O.N. Zhdanov, A.K. Tsikh, Investigation of multiple Mellin-Barnes integrals by means of multidimensional residue. Sib. Math. J. **39**, 245 (1998)
31. M. Passare, A.K. Tsikh, A.A. Cheshel, Multiple Mellin-Barnes integrals as periods of Calabi-Yau manifolds with several moduli. Theor. Math. Phys. **109**, 1544 (1997)
32. S. Friot, D. Greynat, On convergent series representations of Mellin-Barnes integrals. J. Math. Phys. **53**, 023508 (2012)
33. S. Friot, D. Greynat, E. De Rafael, Asymptotics of Feynman diagrams and the Mellin-Barnes representation. Phys. Lett. B **628**, 73 (2005)
34. M. Czakon, Automatized analytic continuation of Mellin-Barnes integrals. Comput. Phys. Commun. **175**, 559 (2006)
35. T.Y. Semenova, A.V. Smirnov, V.A. Smirnov, On the status of expansion by regions. Eur. Phys. J. C **79**, 136 (2019)
36. N.D. Lenshina, A.A. Radionov, F.V. Tkachov, Finite Z-less integral expressions for β-functions in the MS4 scheme. arXiv:2005.03291 [hep-th]
37. I.G. Halliday, R.M. Ricotta, Negative dimensional integrals. 1. Feynman graphs. Phys. Lett. B **193**, 241 (1987)
38. D.J. Broadhurst, Two loop negative dimensional integration. Phys. Lett. B **197**, 179 (1987)
39. C. Anastasiou, E.W.N. Glover, C. Oleari, Application of the negative dimension approach to massless scalar box integrals. Nucl. Phys. B **565**, 445 (2000)
40. C. Anastasiou, E.W.N. Glover, C. Oleari, Scalar one loop integrals using the negative dimension approach. Nucl. Phys. B **572**, 307 (2000)

41. A.T. Suzuki, E.S. Santos, A.G.M. Schmidt, Massless and massive one loop three point functions in negative dimensional approach. Eur. Phys. J. C **26**, 125 (2002)
42. A.T. Suzuki, E.S. Santos, A.G.M. Schmidt, General massive one loop off-shell three point functions. J. Phys. A **36**, 4465 (2003)
43. A.T. Suzuki, E.S. Santos, A.G.M. Schmidt, One loop N-point equivalence among negative dimensional, Mellin-Barnes and Feynman parametrization approaches to Feynman integrals. J. Phys. A **36**, 11859 (2003)
44. G. Somogyi, Angular integrals in d dimensions. J. Math. Phys. **52**, 083501 (2011)
45. A.G. Grozin, A.V. Kotikov, HQET heavy-heavy vertex diagram with two velocities. arXiv:1106.3912 [hep-ph]
46. S. Abreu, R. Britto, H. Grönqvist, Cuts and coproducts of massive triangle diagrams. J. High Energy Phys. **1507**, 111 (2015)
47. J. Ablinger, A. Behring, J. Blümlein, A. De Freitas, A. von Manteuffel, C. Schneider, Calculating three loop ladder and v-topologies for massive operator matrix elements by computer algebra. Comput. Phys. Commun. **202**, 33 (2016)
48. T.F. Feng, C.H. Chang, J.B. Chen, Z.H. Gu, H.B. Zhang, Evaluating Feynman integrals by the hypergeometry. Nucl. Phys. B **927**, 516 (2018)
49. T.F. Feng, C.H. Chang, J.B. Chen, Z.H. Gu, H.B. Zhang, The system of partial differential equations for the C_0 function. Nucl. Phys. B **940**, 130 (2019)
50. X.Y. Yang, H.N. Li, The hypergeometric system for one-loop triangle integral. Int. J. Mod. Phys. A **34**, 1950232 (2020)
51. Z.H. Gu, H.B. Zhang, T.F. Feng, Hypergeometric expression for a three-loop vacuum integral. Int. J. Mod. Phys. A **35**, 2050089 (2020)
52. A.G. Grozin, HQET vertex diagram: ε expansion. Phys. Rev. D **102**, 054022 (2020)
53. P. Burda, B. Kol, R. Shir, Vacuum seagull: evaluating a three-loop Feynman diagram with three mass scales. Phys. Rev. D **96**, 125013 (2017)
54. B. Kol, R. Shir, The propagator seagull: general evaluation of a two loop diagram. J. High Energy Phys. **1903**, 083 (2019)
55. B. Kol, A. Schiller, R. Shir, Numerator seagull and extended symmetries of Feynman integrals. arXiv:2009.04947 [hep-th]
56. I.M. Gel'fand, M.M. Kapranov, A.V. Zelevinsky, Hypergeometric functions and toric varieties. Funck. Anal. i Priloz. **23**, 12 (1989)
57. I.M. Gel'fand, M.M. Kapranov, A.V. Zelevinsky, Generalized Euler integrals and *A*-hypergeometric functions, Adv. Math. **84**, 255 (1990)
58. I.M. Gel'fand, M.I. Graev, V.S. Retakh, General hypergeometric systems of equations and series of hypergeometric type. Russ. Math. Surv. **47**, 1 (1992)
59. M. Saito, B. Sturmfels, N. Takayama, *Gröbner Deformations of Hypergeometric Differential Equations* (Springer, Berlin, 2000)
60. F. Beukers, Notes on A-hypergeometric functions. https://webspace.science.uu.nl/~beuke106/AHGcourse.pdf
61. T. Reichelt, M. Schulze, C. Sevenheck, U. Walther, Algebraic aspects of hypergeometric differential equations. arXiv:2004.07262 [math.AG]
62. E. Bod, Algebraic A-hypergeometric functions and their monodromy. Dissertation, 2013. http://dspace.library.uu.nl/handle/1874/275540
63. F. Beukers, Monodromy of A-hypergeometric functions. arXiv:1101.0493 [math.AG]
64. S-J. Matsubara-Heo, On Mellin-Barnes integral representations for GKZ hypergeometric functions. arXiv:1802.04939 [math.CA]
65. L. Nilsson, M. Passare, A.K. Tsikh, Domains of convergence for A-hypergeometric series and integrals J. Sib. Fed. Univ. Math. Phys. **12**, 509 (2019)
66. V.V. Bytev, M.Yu. Kalmykov, B.A. Kniehl, Differential reduction of generalized hypergeometric functions from Feynman diagrams: one-variable case. Nucl. Phys. B **836**, 129 (2010)
67. G. 't Hooft, M.J.G. Veltman, Regularization and renormalization of Gauge fields. Nucl. Phys. B **44**, 189 (1972)

68. D. Zeilberger, A holonomic systems approach to special functions identities. J. Comp. Appl. Math. **32**, 321 (1990)
69. https://specfun.inria.fr/chyzak//mgfun.html
70. https://www3.risc.jku.at/research/combinat/software/ergosum/RISC/HolonomicFunctions.html
71. F. Beukers, G. Heckman, Monodromy for the hypergeometric function ${}_nF_{n-1}$. Invent. Math. **95**, 325 (1989)
72. M.Yu. Kalmykov, B.A. Kniehl, Counting master integrals: integration by parts versus differential reduction. Phys. Lett. B **702**, 268 (2011)
73. F.V. Tkachov, A theorem on analytical calculability of four loop renormalization group Phys. Lett. B **100**, 65 (1981)
74. K.G. Chetyrkin, F.V. Tkachov, Integration by parts: the algorithm to calculate beta functions in 4 loops. Nucl. Phys. B **192**, 159 (1981)
75. M.Yu. Kalmykov, B.A. Kniehl, Counting the number of master integrals for sunrise diagrams via the Mellin-Barnes representation. J. High Energy Phys. **1707**, 031 (2017)
76. V.A. Smirnov, V.A. Smirnov, FIRE4, LiteRed and accompanying tools to solve integration by parts relations. Comput. Phys. Commun. **184**, 2820 (2013)
77. M.Yu. Kalmykov, B.A. Kniehl, Mellin-Barnes representations of Feynman diagrams, linear systems of differential equations, and polynomial solutions. Phys. Lett. B **714**, 103 (2012)
78. N. Takayama, Gröbner basis and the problem of contiguous relations. Jpn. J. Appl. Math. **6**, 147 (1989)
79. H. Frellesvig, F. Gasparotto, S. Laporta, M.K. Mandal, P. Mastrolia, L. Mattiazzi, S. Mizera, Decomposition of Feynman integrals on the maximal cut by intersection numbers. J. High Energy Phys. **1905**, 153 (2019)
80. H. Frellesvig, F. Gasparotto, S. Laporta, M.K. Mandal, P. Mastrolia, L. Mattiazzi, S. Mizera, Vector space of Feynman integrals and multivariate intersection numbers. Phys. Rev. Lett. **123**, 201602 (2019)
81. H. Frellesvig, F. Gasparotto, S. Laporta, M.K. Mandal, P. Mastrolia, L. Mattiazzi, S. Mizera, Decomposition of Feynman integrals by multivariate intersection numbers. [arXiv:2008.04823 [hep-th]]
82. V.V. Bytev, M.Yu. Kalmykov, B.A. Kniehl, HYPERDIRE, HYPERgeometric functions DIfferential REduction: MATHEMATICA-based packages for differential reduction of generalized hypergeometric functions ${}_pF_{p-1}$, F_1,F_2,F_3,F_4. Comput. Phys. Commun. **184**, 2332 (2013)
83. T.M. Sadykov, On a multidimensional system of differential hypergeometric equations. Sibirsk. Mat. Zh. **39**, 1141 (1998)
84. T. Bitoun, C. Bogner, R.P. Klausen, E. Panzer, Feynman integral relations from parametric annihilators. Lett. Math. Phys. **109**, 497 (2019)
85. L. de la Cruz, Feynman integrals as A-hypergeometric functions. J. High Energy Phys. **1912**, 123 (2019)
86. R.P. Klausen, Hypergeometric series representations of Feynman Integrals by GKZ hypergeometric systems. J. High Energy Phys. **2004**, 121 (2020)
87. T.F. Feng, C.H. Chang, J.B. Chen, H.B. Zhang, GKZ-hypergeometric systems for Feynman integrals. Nucl. Phys. B **953**, 114952 (2020)
88. L. Nilsson, M. Passare, Mellin transforms of multivariate rational functions. J. Geom Anal.**23**, 24 (2013)
89. R.N. Lee, A.A. Pomeransky, Critical points and number of master integrals. J. High Energy Phys. **1311**, 165 (2013)
90. F. Jegerlehner, M.Yu. Kalmykov, O(alpha alpha(s)) correction to the pole mass of the t quark within the standard model. Nucl. Phys. B **676**, 365 (2004)
91. P. Vanhove, Feynman integrals, toric geometry and mirror symmetry. arXiv:1807.11466.
92. A. Klemm, C. Nega, R. Safari, The l-loop Banana Amplitude from GKZ Systems and relative Calabi-Yau Periods. J. High Energy Phys. **2004**, 088 (2020)

93. K. Bönisch, F. Fischbach, A. Klemm, C. Nega, R. Safari, Analytic structure of all loop banana amplitudes. arXiv:2008.10574 [hep-th].
94. A.I. Davydychev, R. Delbourgo, A geometrical angle on Feynman integrals. J. Math. Phys. **39**, 4299 (1998)
95. J. Fleischer, F. Jegerlehner, O.V. Tarasov, A New hypergeometric representation of one loop scalar integrals in d dimensions. Nucl. Phys. B **672**, 303 (2003)
96. S. Bloch, D. Kreimer, Feynman amplitudes and Landau singularities for 1-loop graphs. Commun. Num. Theor. Phys. **4**, 709 (2010)
97. J.L. Bourjaily, E. Gardi, A.J. McLeod, C. Vergu, All-mass n-gon integrals in n dimensions. J. High Energy Phys. **2008**, 029 (2020)
98. O.V. Tarasov, Connection between Feynman integrals having different values of the space-time dimension. Phys. Rev. D **54**, 6479 (1996)
99. A.I. Davydychev, Geometrical methods in loop calculations and the three-point function. Nucl. Instrum. Meth. A **559**, 293 (2006)
100. O.V. Tarasov, New relationships between Feynman integrals. Phys. Lett. B **670**, 67 (2008)
101. A.I. Davydychev, Geometrical splitting and reduction of Feynman diagrams. J. Phys. Conf. Ser. **762**, 012068 (2016)
102. V.V. Bytev, M.Yu. Kalmykov, S.O. Moch, HYPERgeometric functions DIfferential REduction (HYPERDIRE): MATHEMATICA based packages for differential reduction of generalized hypergeometric functions: F_D and F_S Horn-type hypergeometric functions of three variables. Comput. Phys. Commun. **185**, 3041 (2014)
103. J. Blümlein, K.H. Phan, T. Riemann, Scalar one-loop vertex integrals as meromorphic functions of space-time dimension d. Acta Phys. Polon. B **48**, 2313 (2017)
104. K.H. Phan, T. Riemann, Scalar 1-loop Feynman integrals as meromorphic functions in space-time dimension d. Phys. Lett. B **791**, 257 (2019)
105. K.H. Phan, Scalar 1-loop Feynman integrals as meromorphic functions in space-time dimension d, II: special kinematics. Eur. Phys. J. C **80**, 414 (2020)
106. A.I. Davydychev, Four-point function in general kinematics through geometrical splitting and reduction. J. Phys. Conf. Ser. **1085**, 052016 (2018)
107. F. Loebbert, D. Müller, H. Münkler, Yangian bootstrap for conformal Feynman integrals Phys. Rev. D **101**, 066006 (2020)
108. F. Loebbert, J. Miczajka, D. Müller, H. Münkler, Massive conformal symmetry and integrability for Feynman integrals. Phys. Rev. Lett. **125**, 091602 (2020)
109. F. Loebbert, D. Müller, J. Miczajka, H. Münkler, Yangian bootstrap for massive Feynman integrals. arXiv:2010.08552 [hep-th]
110. L. Lewin, *Polylogarithms and Associated Functions* (North Holland, New York, 1981)
111. E. Remiddi, J.A.M. Vermaseren, Harmonic polylogarithms. Int. J. Mod. Phys. A **15**, 725 (2000)
112. T. Gehrmann, E. Remiddi, T. Gehrmann, E. Remiddi, Two loop master integrals for gamma* —> 3 jets: the planar topologies. Nucl. Phys. B **601**, 248 (2001)
113. J.M. Borwein, D.M. Bradley, D.J. Broadhurst, P. Lisonek, Special values of multiple polylogarithms. Trans. Am. Math. Soc. **353**, 907 (2001)
114. A.B. Goncharov, Polylogarithms in arithmetic and geometry, in *Proceedings of the International Congress of Mathematicians, Zurich* (1994)
115. A.B. Goncharov, The double logarithm and Manin's complex for modular curves. Math. Res. Lett. **4**, 617 (1997)
116. A.B. Goncharov, Multiple polylogarithms, cyclotomy and modular complexes. Math. Res. Lett. **5**, 497 (1998)
117. A.B. Goncharov, Multiple polylogarithms and mixed Tate motives. math/0103059 [math.AG] (2001)
118. A.B. Goncharov, Galois symmetries of fundamental groupoids and noncommutative geometry.
119. A.B. Goncharov, M. Spradlin, C. Vergu, A. Volovich, Classical polylogarithms for amplitudes and Wilson loops. Phys. Rev. Lett. **105**, 151605 (2010)

120. C. Duhr, H. Gangl, J.R. Rhodes, From polygons and symbols to polylogarithmic functions. J. High Energy Phys. **1210**, 075 (2012)
121. C. Duhr, Hopf algebras, coproducts and symbols: an application to Higgs boson amplitudes. J. High Energy Phys. **1208**, 043 (2012)
122. E. Panzer, Feynman integrals and hyperlogarithms. arXiv:1506.07243 [math-ph]
123. C. Duhr, F. Dulat, PolyLogTools-polylogs for the masses. J. High Energy Phys. **1908**, 135 (2019)
124. T. Gehrmann, E. Remiddi, Numerical evaluation of harmonic polylogarithms. Comput. Phys. Commun. **141**, 296 (2001)
125. T. Gehrmann, E. Remiddi, Numerical evaluation of two-dimensional harmonic polylogarithms. Comput. Phys. Commun. **144**, 200 (2002)
126. J. Vollinga, S. Weinzierl, Numerical evaluation of multiple polylogarithms. Comput. Phys. Commun. **167**, 177 (2005)
127. M.Yu. Kalmykov, A. Sheplyakov, lsjk - a C++ library for arbitrary-precision numeric evaluation of the generalized log-sine functions. Comput. Phys. Commun. **172**, 45 (2005)
128. D. Maitre, HPL, a mathematica implementation of the harmonic polylogarithms. Comput. Phys. Commun. **174**, 222 (2006)
129. R. Bonciani, G. Degrassi, A. Vicini, On the generalized harmonic polylogarithms of one complex variable. Comput. Phys. Commun. **182**, 1253 (2011)
130. D. Maitre, Extension of HPL to complex arguments. Comput. Phys. Commun. **183**, 846 (2012)
131. S. Buehler, C. Duhr, CHAPLIN - complex harmonic polylogarithms in fortran. Comput. Phys. Commun. **185**, 2703 (2014)
132. H. Frellesvig, D. Tommasini, C. Wever, On the reduction of generalized polylogarithms to Li_n and $\mathrm{Li}_{2,2}$ and on the evaluation thereof. J. High Energy Phys. **1603**, 189 (2016)
133. H. Frellesvig, Generalized polylogarithms in Maple. arXiv:1806.02883 [hep-th]
134. A. Ablinger, J. Blümlein, M. Round, C. Schneider, Numerical implementation of harmonic polylogarithms to weight w = 8. Comput. Phys. Commun. **240**, 189 (2019)
135. L. Naterop, A. Signer, Y. Ulrich, handyG-Rapid numerical evaluation of generalised polylogarithms in Fortran. Comput. Phys. Commun. **253**, 107165 (2020)
136. C. Duhr, L. Tancredi, Algorithms and tools for iterated Eisenstein integrals. J. High Energy Phys. **2002**, 105 (2020)
137. M. Walden, S. Weinzierl, Numerical evaluation of iterated integrals related to elliptic Feynman integrals. arXiv:2010.05271 [hep-ph]
138. D.J. Broadhurst, On the enumeration of irreducible k fold Euler sums and their roles in knot theory and field theory. hep-th/9604128
139. D.J. Broadhurst, Massive three - loop Feynman diagrams reducible to SC* primitives of algebras of the sixth root of unity. Eur. Phys. J. C **8**, 311 (1999)
140. J.M. Borwein, D.M. Bradley, D.J. Broadhurst, Evaluations of K fold Euler/Zagier sums: a compendium of results for arbitrary k. [hep-th/9611004]
141. D.J. Broadhurst, J.A. Gracey, D. Kreimer, Beyond the triangle and uniqueness relations: Nonzeta counterterms at large N from positive knots. Z. Phys. C **75**, 559 (1997)
142. A.I. Davydychev, J.B. Tausk, Two loop selfenergy diagrams with different masses and the momentum expansion. Nucl. Phys. B **397**, 123 (1993)
143. A.I. Davydychev, J.B. Tausk, A Magic connection between massive and massless diagrams. Phys. Rev. D **53**, 7381 (1996)
144. A.I. Davydychev, Explicit results for all orders of the epsilon expansion of certain massive and massless diagrams. Phys. Rev. D **61**, 087701 (2000)
145. A.I. Davydychev, M.Yu. Kalmykov, Some remarks on the epsilon expansion of dimensionally regulated Feynman diagrams. Nucl. Phys. Proc. Suppl. **89**, 283 (2000)
146. A.I. Davydychev, M.Yu. Kalmykov, New results for the epsilon expansion of certain one, two and three loop Feynman diagrams. Nucl. Phys. B **605**, 266 (2001)
147. T. Huber, D. Maitre, HypExp: a mathematica package for expanding hypergeometric functions around integer-valued parameters. Comput. Phys. Commun. **175**, 122 (2006)

148. T. Huber, D. Maitre, HypExp 2, expanding hypergeometric functions about half-integer parameters. Comput. Phys. Commun. **178**, 755 (2008)
149. C. Bogner, F. Brown, Symbolic integration and multiple polylogarithms. PoS LL **2012**, 053 (2012)
150. C. Bogner, F. Brown, Feynman integrals and iterated integrals on moduli spaces of curves of genus zero. Commun. Num. Theor. Phys. **09**, 189 (2015)
151. E. Panzer, Algorithms for the symbolic integration of hyperlogarithms with applications to Feynman integrals. Comput. Phys. Commun. **188**, 148 (2015)
152. C. Bogner, MPL-A program for computations with iterated integrals on moduli spaces of curves of genus zero. Comput. Phys. Commun. **203**, 339 (2016)
153. Z.W. Huang, J. Liu, NumExp: numerical epsilon expansion of hypergeometric functions. Comput. Phys. Commun. **184**, 1973 (2013)
154. D. Greynat, J. Sesma, A new approach to the epsilon expansion of generalized hypergeometric functions. Comput. Phys. Commun. **185**, 472 (2014)
155. D. Greynat, J. Sesma, G. Vulvert, Derivatives of the Pochhammer and reciprocal Pochhammer symbols and their use in epsilon-expansions of Appell and Kampe de Feriet functions. J. Math. Phys. **55**, 043501 (2014)
156. S. Moch, P. Uwer, S. Weinzierl, Nested sums, expansion of transcendental functions and multi-scale multi-loop integrals. J. Math. Phys. **43**, 3363 (2002)
157. S. Weinzierl, Expansion around half-integer values, binomial sums and inverse binomial sums. J. Math. Phys. **45**, 2656 (2004)
158. J. Ablinger, J. Blümlein, C. Schneider, Harmonic sums and polylogarithms generated by cyclotomic polynomials. J. Math. Phys. **52**, 102301 (2011)
159. V.A. Smirnov, Analytical result for dimensionally regularized massless on shell double box. Phys. Lett. B **460**, 397 (1999)
160. J.B. Tausk, Nonplanar massless two loop Feynman diagrams with four on-shell legs. Phys. Lett. B **469**, 225 (1999)
161. S. Weinzierl, Symbolic expansion of transcendental functions. Comput. Phys. Commun. **145**, 357 (2002)
162. S. Moch, P. Uwer, XSummer: transcendental functions and symbolic summation in Form. Comput. Phys. Commun. **174**, 759 (2006)
163. C. Anzai, Y. Sumino, Algorithms to evaluate multiple sums for loop computations. J. Math. Phys. **54**, 033514 (2013)
164. A.J. McLeod, H. Munch, G. Papathanasiou, M. von Hippel, A novel algorithm for nested summation and hypergeometric expansions. J. High Energy Phys. **2011**, 122 (2020)
165. V. Del Duca, C. Duhr, E.W.N. Glover, V.A. Smirnov, The One-loop pentagon to higher orders in epsilon. J. High Energy Phys. **1001**, 042 (2010)
166. A.I. Davydychev, M.Yu. Kalmykov, Massive Feynman diagrams and inverse binomial sums. Nucl. Phys. B **699**, 3 (2004)
167. M.Yu. Kalmykov, Series and epsilon-expansion of the hypergeometric functions. Nucl. Phys. Proc. Suppl. **135**, 280 (2004)
168. H.S. Wilf, *Generatingfunctionology* (Academic, Boston, 1994)
169. P. Flajolet, R. Sedgewick, *Analytic Combinatorics* (Cambridge University Press, Cambridge, 2009)
170. J.A.M. Vermaseren, Int. J. Mod. Phys. A **14**, 2037 (1999). [hep-ph/9806280]
171. J. Blümlein, S. Kurth, Phys. Rev. D **60**, 014018 (1999). [hep-ph/9810241]
172. F. Jegerlehner, M.Yu. Kalmykov, O. Veretin, MS-bar versus pole masses of gauge bosons. 2. Two loop electroweak fermion corrections. Nucl. Phys. B **658**, 49 (2003)
173. M.Yu. Kalmykov, Gauss hypergeometric function: reduction, epsilon-expansion for integer/half-integer parameters and Feynman diagrams. J. High Energy Phys. **0604**, 056 (2006)
174. J. Fleischer, M.Yu. Kalmykov, Single mass scale diagrams: construction of a basis for the epsilon expansion. Phys. Lett. B **470**, 168 (1999)

175. J. Ablinger, Proving two conjectural series for $\zeta(7)$ and discovering more series for $\zeta(7)$. arXiv:1908.06631 [math.CO]
176. W. Wang, C. Xu, Alternating multiple zeta values, and explicit formulas of some Euler-Apery-type series. arXiv:1909.02943 [math.NT]
177. J. Braun, D. Romberger, H.J. Bentz, On four families of power series involving harmonic numbers and central binomial coefficients. arXiv:2006.13115 [math.NT]
178. M.Yu. Kalmykov, B.F.L. Ward, S.A. Yost, Multiple (inverse) binomial sums of arbitrary weight and depth and the all-order epsilon-expansion of generalized hypergeometric functions with one half-integer value of parameter. J. High Energy Phys. **0710**, 048 (2007)
179. M.Yu. Kalmykov, B.A. Kniehl, Sixth root of unity' and Feynman diagrams: Hypergeometric function approach point of view. Nucl. Phys. Proc. Suppl. **205–206**, 129 (2010)
180. O. Schnetz, Numbers and functions in quantum field theory. Phys. Rev. D **97**, 085018 (2018)
181. V.V. Bytev, M.Yu. Kalmykov, B.A. Kniehl, When epsilon-expansion of hypergeometric functions is expressible in terms of multiple polylogarithms: the two-variables examples. PoS LL **2012**, 029 (2012)
182. J. Ablinger, J. Blümlein, P. Marquard, N. Rana, C. Schneider, Automated solution of first order factorizable systems of differential equations in one variable. Nucl. Phys. B **939**, 253 (2019)
183. C. Schneider, Symbolic summation in difference fields. Technical Report 01-17, RISC-Linz, J. Kepler University, Nov 2001
184. C. Schneider, A new Sigma approach to multi-summation. Adv. Appl. Math. **34**, 740 (2005)
185. C. Schneider, A difference ring theory for symbolic summation. J. Symb. Comput. **72**, 82 (2016)
186. E. Landau, J. für Math. **124**, 115–120 (1901)
187. F. Schwarz, *Loewy Decomposition of Linear Differential Equations* (Springer, Berlin, 2012)
188. S.A. Yost, V.V. Bytev, M.Yu. Kalmykov, B.A. Kniehl, B.F.L. Ward, The epsilon expansion of Feynman diagrams via hypergeometric functions and differential reduction. arXiv:1110.0210 [math-ph]
189. M.Yu. Kalmykov, B.A. Kniehl, All-order epsilon-expansions of hypergeometric functions of one variable. Phys. Part. Nucl. **41**, 942 (2010)
190. F. Brown, O. Schnetz, A K3 in ϕ^4. Duke Math. J. **161**, 1817 (2012)
191. F. Brown, O. Schnetz, Modular forms in quantum field theory. Commun. Num. Theor Phys. **07**, 293 (2013)
192. J. Ablinger, J. Blümlein, C. Schneider, Analytic and algorithmic aspects of generalized harmonic sums and polylogarithms. J. Math. Phys. **54**, 082301 (2013)
193. J. Ablinger, J. Blümlein, C.G. Raab, C. Schneider, Iterated binomial sums and their associated iterated integrals. J. Math. Phys. **55**, 112301 (2014)
194. J. Ablinger, A computer algebra toolbox for harmonic sums related to particle physics. arXiv:1011.1176 [math-ph]
195. J. Ablinger, Computer algebra algorithms for special functions in particle physics. arXiv:1305.0687 [math-ph]
196. J. Ablinger, The package HarmonicSums: computer algebra and analytic aspects of nested sums. PoS LL **2014**, 019 (2014)
197. U. Aglietti, R. Bonciani, Master integrals with 2 and 3 massive propagators for the 2 loop electroweak form-factor - planar case. Nucl. Phys. B **698**, 277 (2004)
198. M.Yu. Kalmykov, B.F.L. Ward, S. Yost, All order epsilon-expansion of Gauss hypergeometric functions with integer and half/integer values of parameters. J. High Energy Phys. **0702**, 040 (2007)
199. M.Yu. Kalmykov, B.F.L. Ward, S. Yost, On the all-order epsilon-expansion of generalized hypergeometric functions with integer values of parameters. J. High Energy Phys. **0711**, 009 (2007)
200. M.Yu. Kalmykov, B.A. Kniehl, Towards all-order Laurent expansion of generalized hypergeometric functions around rational values of parameters. Nucl. Phys. B **809**, 365 (2009)

201. J. Blumlein, D.J. Broadhurst, J.A.M. Vermaseren, The multiple zeta value data mine. Comput. Phys. Commun. **181**, 582 (2010)
202. R.H. Boels, On the field theory expansion of superstring five point amplitudes. Nucl. Phys. B **876**, 215 (2013)
203. J.M. Henn, Multiloop integrals in dimensional regularization made simple. Phys. Rev. Lett. **110**, 251601 (2013)
204. S. Abreu, R. Britto, C. Duhr, E. Gardi, Algebraic structure of cut Feynman integrals and the diagrammatic coaction. Phys. Rev. Lett. **119**, 051601 (2017)
205. S. Abreu, R. Britto, C. Duhr, E. Gardi, Diagrammatic Hopf algebra of cut Feynman integrals: the one-loop case. J. High Energy Phys. **1712**, 090 (2017)
206. S. Abreu, R. Britto, C. Duhr, E. Gardi, From positive geometries to a coaction on hypergeometric functions. J. High Energy Phys. **2002**, 122 (2020)
207. D. Kreimer, Outer Space as a combinatorial backbone for Cutkosky rules and coactions. arXiv:2010.11781 [hep-th]
208. F. Brown, C. Dupont, Lauricella hypergeometric functions, unipotent fundamental groups of the punctured Riemann sphere, and their motivic coactions. arXiv:1907.06603 [math.AG]
209. C.F. Gauss, *Gesammelte Werke*, vol. 3 (Teubner, Leipzig, 1823), pp. 1866–1929
210. G. Lauricella, Sulle funzioni ipergeometriche a piu variabili. Rend. Circ. Math. Palermo **7**, 111–158 (1893)
211. O. Ore, J. Math. Pure Appl. **9**, 311 (1930)
212. M. Sato, Singular orbits of a prehomogeneous vector space and hypergeometric functions. Nagoya Math. J. **120**, 1 (1990)
213. V.V. Bytev, B.A. Kniehl, HYPERDIRE HYPERgeometric functions DIfferential REduction: mathematica-based packages for the differential reduction of generalized hypergeometric functions: Horn-type hypergeometric functions of two variables. Comput. Phys. Commun. **189**, 128 (2015)
214. M.J. Schlosser, Multiple hypergeometric series: Appell series and beyond. [arXiv:1305.1966 [math.CA]]
215. M.Yu. Kalmykov, V.V. Bytev, B.A. Kniehl, B.F.L. Ward, S.A. Yost, Feynman diagrams, differential reduction, and hypergeometric functions. PoS ACAT **08**, 125 (2008)
216. V.V. Bytev, B.A. Kniehl, HYPERDIRE - HYPERgeometric functions DIfferential REduction: Mathematica-based packages for the differential reduction of generalized hypergeometric functions: Lauricella function F_c of three variables. Comput. Phys. Commun. **206**, 78 (2016)
217. A. Erdelyi, Hypergeometric functions of two variables. Acta Math. **83**, 131 (1950)
218. A. Dickenstein, L. Matusevich, T.M. Sadykov, Bivariate hypergeometric D-modules. Adv. Math. **196**, 78 (2005)
219. J.M. Henn, Lectures on differential equations for Feynman integrals. J. Phys. A **48**, 153001 (2015)
220. J.M. Henn, V.A. Smirnov, Analytic results for two-loop master integrals for Bhabha scattering I. J. High Energy Phys. **1311**, 041 (2013)
221. S. Caron-Huot, J.M. Henn, Iterative structure of finite loop integrals. J. High Energy Phys. **1406**, 114 (2014)
222. F. Brown, C. Duhr, A double integral of d-log forms which is not polylogarithmic: arXiv:2006.09413 [hep-th].
223. F. Gross, On the equation $f^n + g^n = 1$. Bull. Am. Math. Soc. **72**, 86 (1966)
224. I.N. Baker, On a class of merpmorphic functions. Proc. Am. Math. Soc. **17**, 819 (1966)
225. A.M. Sanchis-Lozano, Simple connections between generalized hypergeomeric series and dilogarithms. J. Comput. Appl. Math. **85**, 325 (1997)
226. Yu.A. Brychkov, N. Saad, On some formulas for the Appell function $F_3(a, a', b, b', c; w, z)$. Integral Transform. Spec. Funct. **26**, 910 (2015)
227. A.N. Kirillov, Dilogarithm identities. Prog. Theor. Phys. Suppl. **118**, 61 (1995)
228. B.A. Kniehl, O.V. Tarasov, Analytic result for the one-loop scalar pentagon integral with massless propagators. Nucl. Phys. B **833**, 298 (2010)

229. M.G. Kozlov, R.N. Lee, One-loop pentagon integral in d dimensions from differential equations in ϵ-form. J. High Energy Phys. **1602**, 021 (2016)
230. A.V. Kotikov, Differential equations method: new technique for massive Feynman diagrams calculation. Phys. Lett. B **254**, 158 (1991)
231. A.K. Rajantie, Feynman diagrams to three loops in three-dimensional field theory. Nucl. Phys. B **480**, 729 (1996). Erratum: [Nucl. Phys. B **513**, 761 (1998)]
232. R.N. Lee, A.A. Pomeransky, Differential equations, recurrence relations, and quadratic constraints for L-loop two-point massive tadpoles and propagators. J. High Energy Phys. **1908**, 027 (2019)
233. S. Laporta, E. Remiddi, Analytic treatment of the two loop equal mass sunrise graph. Nucl. Phys. B **704**, 349 (2005)
234. O.V. Tarasov, Hypergeometric representation of the two-loop equal mass sunrise diagram. Phys. Lett. B **638**, 195 (2006)
235. S. Bloch, P. Vanhove, The elliptic dilogarithm for the sunset graph. J. Number Theor. **148**, 328 (2015)
236. L. Adams, C. Bogner, S. Weinzierl, The iterated structure of the all-order result for the two-loop sunrise integral. J. Math. Phys. **57**, 032304 (2016)
237. S. Bloch, M. Kerr, P. Vanhove, Local mirror symmetry and the sunset Feynman integral. Adv. Theor. Math. Phys. **21**, 1373 (2017)
238. J. Broedel, C. Duhr, F. Dulat, L. Tancredi, Elliptic polylogarithms and iterated integrals on elliptic curves II: an application to the sunrise integral. Phys. Rev. D **97**, 116009 (2018)
239. C. Bogner, S. Müller-Stach, S. Weinzierl, The unequal mass sunrise integral expressed through iterated integrals on $\overline{\mathcal{M}}_{1,3}$. Nucl. Phys. B **954**, 114991 (2020)
240. M. Besier, D. Van Straten, S. Weinzierl, Rationalizing roots: an algorithmic approach. Commun. Num. Theor. Phys. **13**, 253 (2019)
241. M. Besier, D. Van Straten, S.S. Weinzierl, RationalizeRoots: software package for the rationalization of square roots. Comput. Phys. Commun. **253**, 107197 (2020)
242. O.V. Tarasov, New relationships between Feynman integrals. Phys. Lett. B **670**, 67 (2008).
243. O.V. Tarasov, Derivation of functional equations for Feynman integrals from algebraic relations. J. High Energy Phys. **1711**, 038 (2017)
244. O.V. Tarasov, Functional reduction of Feynman integrals. J. High Energy Phys. **1902**, 173 (2019)

Differential Equations and Feynman Integrals

Anatoly V. Kotikov

Abstract The role of differential equations in the process of calculating Feynman integrals is reviewed. An example of a diagram is given for which the method of differential equations was introduced, the properties of the inverse-mass-expansion coefficients are shown, and modern methods based on differential equations are discussed.

1 Introduction

The calculation of Feynman integrals (FIs) provides basic information both for the matrix elements of the experimentally studied processes and for the characteristics of the physical models themselves, i.e. their renormalization, critical behavior, etc. When studying renormalization and the critical behavior, it is usually sufficient to restrict oneself to the limit of massless particles at which the corresponding two-point FIs are fairly simple. However, starting at the 2 or 3 loop level, there is a need to use modern methods such as integration by parts (IBP) [1] and the Gegenbauer polynomial method [2].[1]

Calculating FIs having massive propagators is a much more complex problem. Simple results, in the form of a product of Γ-functions exist for simple tadpoles only, see Eq. (10) below. A massive one-loop diagram is already given by a one-fold integral, see Eq. (15) below.

[1]See also Ref. [3] and the reviews [4] and [5]. Note that multipoint massless FIs are as complex as massive 2-point FIs. For the relationship between 2-point massive FIs and 3-point massless FIs, cf. [6].

A. V. Kotikov (✉)
Bogoliubov Laboratory of Theoretical Physics, Joint Institute for Nuclear Research, Dubna, Russia
e-mail: kotikov@theor.jinr.ru

J. Blümlein, C. Schneider (eds.), *Anti-Differentiation and the Calculation of Feynman Amplitudes*, Texts & Monographs in Symbolic Computation,
https://doi.org/10.1007/978-3-030-80219-6_10

It turns out, however, that massive FIs satisfy IBP procedures [1], which lead to relations between the FIs equivalent to the original ones, but with different powers of the propagators, including powers equal to zero. Diagrams containing propagators with degrees equal to zero are equivalent to simpler diagrams obtained by canceling these propagators and reducing the points they join to one point.

Such relations can be understood in two ways. First, considering them algebraically, one can understand them as connections between diagrams that are not independent and can be reduced to a certain set of independent diagrams, which are called master integrals (or masters) [7].

Second, propagators with powers greater than one can be considered as derivatives, with respect to the corresponding mass or external momentum, from the propagator with a degree of one. Thus, the relations between the master integrals can be considered as differential equations (DEs) for these masters. An example is given in Sect. 2, containing inhomogeneous terms, including only simpler diagrams, which are obtained from the original diagrams by reducing some propagator. For these simpler diagrams one can obtain similar DEs by applying the IBP procedure, see Appendix. They contain inhomogeneous terms, including only even simpler diagrams, which are obtained from simple diagrams by propagator reduction. By repeating the original procedure several times, it is usually possible to obtain DEs containing inhomogeneous terms, including only tadpoles, which in turn are easily computable exactly. Note, however, that starting from the 2-loop level, obtaining results for massive tadpoles requires the use of modern methods of FI calculation, cf. [8] and references and discussions therein. Sometimes it is convenient to stop the considered procedure on one-loop massive FI and to perform the integration after introducing Feynman parameters, cf. [9]. More complicated diagrams can be obtained from these tadpoles by solving successively obtained DEs with certain boundary conditions. For dimensionally regularized massive FIs a good boundary condition is obtained in the limit of large masses, $m \to \infty$, at which these diagrams usually vanish.

The paper is organized as follows. In Sect. 2 we will consider a two-loop FI, the calculation of which leads to the use of differential equations. The calculation of massive diagrams is given in Sect. 3. Here rules are given for their efficient calculation, examples of two- and three-point diagrams are considered. The recurrence relations for the coefficients of decomposition in the inverse mass are considered. In Sect. 4 a short review of modern computing technology is given. Appendix contains the derivation of the DEs for massive diagrams from the inhomogeneous term of the DE for the diagram considered in Sect. 2.

2 History

As mentioned in the introduction, integral representations for one-loop FIs (obtained, for example, using the Feynman parameter method [9]) are hypergeometric functions[2] and, thus, can be represented as solutions of some DEs. The importance of DEs for FIs was recognized long ago, see, for example, [11, 12]. However, in my opinion, the practical application awaited the emergence the of IBP procedure [1] for FIs and is based on the use of IBP relations, see Eqs. (16) and (17) below.

IBP-based DEs appeared in the nineties in several papers, studying FIs: for massive two-point functions in [13, 14], for massive three-point functions in [15], and for four-point in [16]. Also n-point functions were considered in [17, 18]. A short overview was given in Ref. [19] dedicated to the 70th anniversary of Academician O.S. Parasyuk, the co-author of the BPHZ renormalization procedure [20], cf. e.g. [21]. The results for massless diagrams can be sometimes obtained more easily in x-space, cf. [22–24]. It is convenient to compute the so-called dual diagrams in x-space, cf. [23, 25]. A dual diagram is obtained from the initial one by replacement of all momenta p by x with the rules of correspondence between the graph and the integral. Massive two-point and three-point diagrams were studied in x-space in Refs. [26, 27], respectively.

In Refs. [13, 26] we studied a preprint of the excellent yet unpublished work [7] on the calculation of two-loop massive FIs. Despite its excellent results, the paper itself turned out to be quite difficult to understand.

I therefore decided to reproduce these results using the IBP relations, which proved to be very successful for calculating the correction to the longitudinal structure function of the deep-inelastic scattering (DIS) [25, 28]. Indeed, the method developed [23, 25] for calculating massless FIs containing the (traceless) product of impulses in the numerators of propagators was based on the application of IBPs to such diagrams. This method, extended to 3-, 4- and 5-loop diagrams and built into computer algebra programs, is the basis of the modern calculations, starting with the excellent work in which NNLO corrections for anomalous dimensions of Wilson operators were obtained, see e.g. [29], and references and discussions therein. A similar method has also been developed [30] to calculate massive corrections in the DIS process, cf. [31, 32] and the review [33] and details given therein.

[2]Investigations of hypergeometric functions related to the calculation of FI are recently presented [10] as a contribution to this volume.

The first example which was studied in Refs. [13, 26] was the diagram

$$I_1(q^2, m^2) = $$ 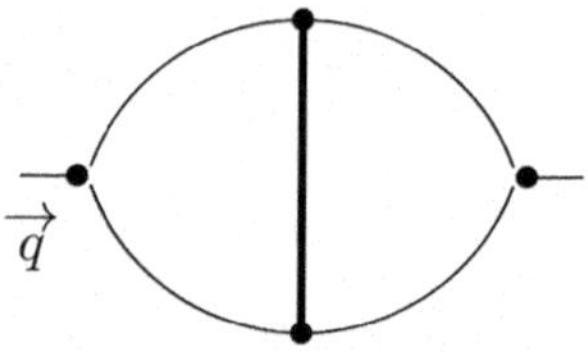 , (1)

having the vertical massive propagator, see Eq. (9) for the definitions. The diagram has left-right and top-bottom symmetries.

Applying IBP relations (16) to the left triangle of the diagram $I_1(q^2, m^2)$ in succession with vertical and lateral distinguished lines, we get

$$(d-4)I_1(q^2, m^2) = 2\Bigg[$$ 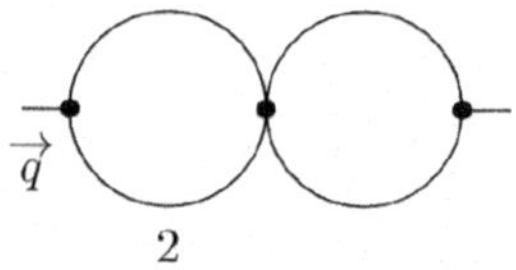 $-$ 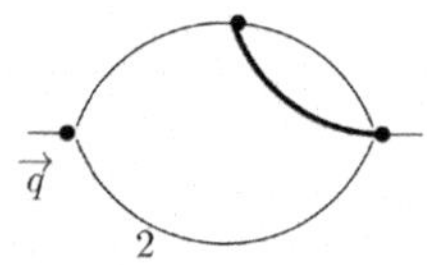

$-m^2$ 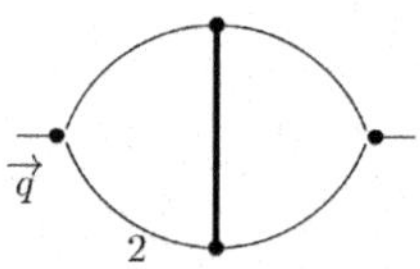 $\Bigg] - 2m^2$ 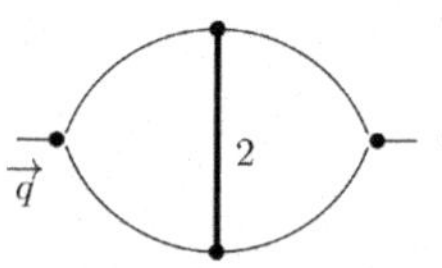 , (2)

$$(d-4)I_1(q^2, m^2) = $$ 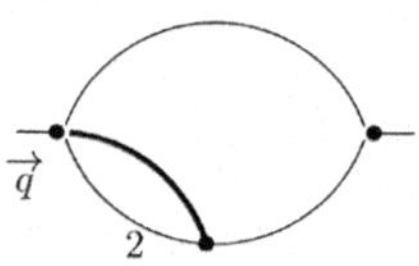 $-q^2$ 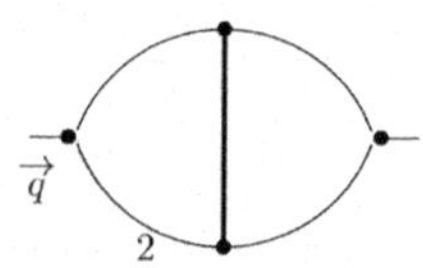

$-m^2$ 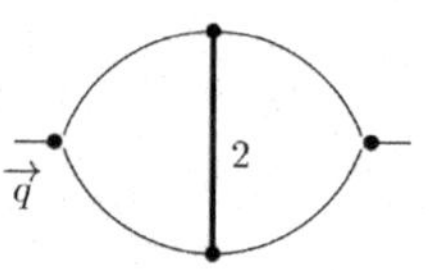 . (3)

Taking the combination of these equations: Eq. (2)–$2(m^2/q^2)\times$ Eq. (3), we have

$$(d-4)\left(1-\frac{2m^2}{q^2}\right) I_1(q^2, m^2) = 2J_1(q^2, m^2) - 2m^2\left(1-\frac{m^2}{q^2}\right)$$ 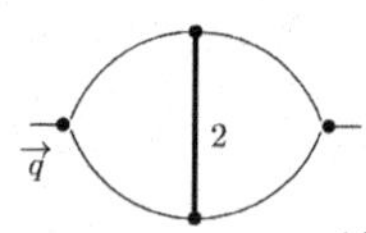 , (4)

where

$$J_1(q^2, m^2) = $$ 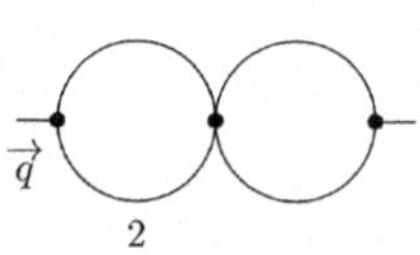 $-$ 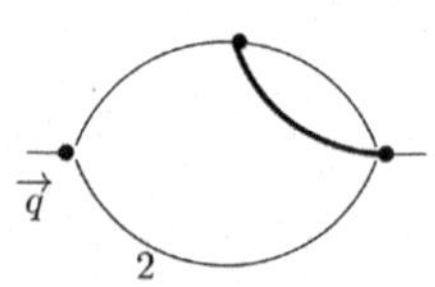 $-\frac{m^2}{q^2}$ 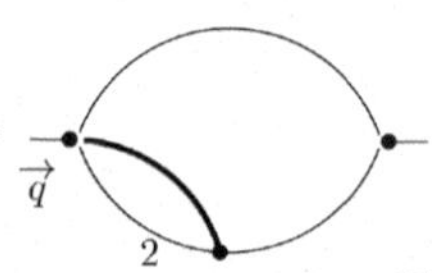 . (5)

Because

$$\frac{1}{(q^2+m^2)^2} = -\frac{d}{dm^2}\frac{1}{(q^2+m^2)^2}, \tag{6}$$

Eq. (5) can be rewritten in the form

$$\left[(d-4)\left(1-\frac{2m^2}{q^2}\right) - 2m^2\left(1-\frac{m^2}{q^2}\right)\frac{d}{dm^2}\right] I_1(q^2,m^2) = 2\,J_1(q^2,m^2)\,, \tag{7}$$

i.e. the first order DE[3] for the original diagram with the inhomogeneous term $J_1(q^2, m^2)$ containing only simpler diagrams, i.e. those obtained from the original expression by canceling one of the propagators, see Eq. (5).

The first diagram in the inhomogeneous term $J_1(q^2, m^2)$ is independent of the mass and can therefore be easily calculated as a product of the Γ-functions, see Eq. (11) below,

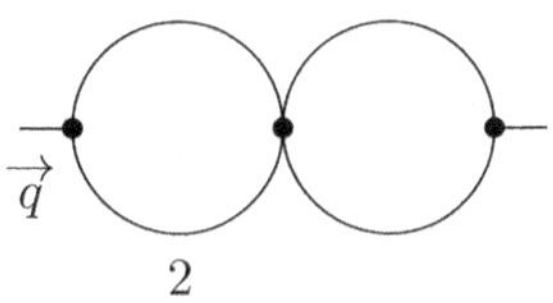

$$= L_{2,1}(q^2)L_{1,1}(q^2) = \frac{1}{(4\pi)^d}\frac{A(2,1)A(1,1)}{q^{2(5-d)}}, \tag{8}$$

where $A(\alpha_1, \alpha_2)$ is given in Eq. (13) below.

Using IBP relations, for the remaining two diagrams in the inhomogeneous $J_1(q^2, m^2)$ term diagrams, one can obtain similar equations with inhomogeneous terms containing only even simpler diagrams, i.e. those obtained from the original by canceling two propagators. These results are given in Appendix.

3 Calculation of Massive Feynman Integrals

Let us briefly consider the rules for calculating diagrams having massive propagators.

1. The massless propagator and the propagator with mass m will be represented as

$$\frac{1}{q^{2\alpha}} =$$ 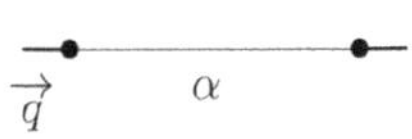 $$, \quad \frac{1}{(q^2+m^2)^\alpha} =$$ 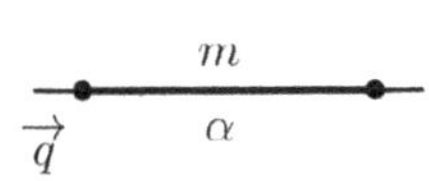 $$, \tag{9}$$

[3]Hereafter we consider only first order DEs. The consideration of the higher order DEs can be found in Section 7 of the review [33]. See also the recent papers [34].

where the symbol m will be omitted in the single-mass case (as in the case of $I_1(q^2, m^2)$ in Eq. (1)).

2. The massive one-loop tadpole $T_{\alpha_1,\alpha_2}(m^2)$ and the massless loop $L_{\alpha_1,\alpha_2}(q^2)$ can be calculated exactly as combinations of the Γ-functions:

$$T_{\alpha_1,\alpha_2}(m^2) = \int \frac{Dk}{k^{2\alpha_1}(k^2+m^2)^{\alpha_2}} =$$

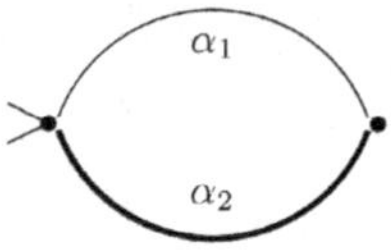

$$= \frac{R(\alpha_1,\alpha_2)}{m^{2(\alpha_1+\alpha_2-d/2)}} \tag{10}$$

$$L_{\alpha_1,\alpha_2}(q^2) = \int \frac{Dk}{(q-k)^{2\alpha_1}k^{2\alpha_2}} =$$

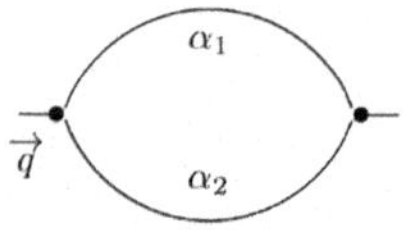

$$= \frac{A(\alpha_1,\alpha_2)}{q^{2(\alpha_1+\alpha_2-d/2)}}\,, \tag{11}$$

where

$$A(\alpha_1,\alpha_2) = \frac{a(\alpha_1)a(\alpha_2)}{a(\alpha_1+\alpha_2-d/2)}, \quad a(\alpha) = \frac{\Gamma(\tilde{\alpha})}{\Gamma(\alpha)}, \quad \tilde{\alpha} = \frac{d}{2} - \alpha\,, \tag{12}$$

$$R(\alpha_1,\alpha_2) = \frac{\Gamma(d/2-\alpha_1)\Gamma(\alpha_1+\alpha_2-d/2)}{\Gamma(d/2)\Gamma(\alpha_2)} \tag{13}$$

and

$$Dk = \frac{d^dk}{\pi^{d/2}} = (4\pi)^{d/2}\, D_Ek, \quad D_Ek = \frac{d^dk}{(2\pi)^d}\,. \tag{14}$$

Here D_Ek is the usual Euclidean measure in $d = 4 - 2\varepsilon$ dimensions.

3. A simple loop of two massive propagators with masses m_1 and m_2 can be represented as hypergeometric function, which can be calculated in a general form, for example, by the Feynman-parameter method, see [9]. It is very convenient, using this approach to represent the loop as an integral of a propagator with the "effective mass" μ [13, 35–40]:

$$\int \frac{Dk}{[(q-k)^2+m_1^2]^{\alpha_1}[k^2+m_2^2]^{\alpha_2}} = \frac{\Gamma(\alpha_1+\alpha_2-d/2)}{\Gamma(\alpha_1)\Gamma(\alpha_2)}$$
$$\times \int_0^1 \frac{ds\, s^{\alpha_1-1}\,(1-s)^{\alpha_2-1}}{[s(1-s)q^2+m_1^2 s+m_2^2(1-s)]^{\alpha_1+\alpha_2-d/2}} = \frac{\Gamma(\alpha_1+\alpha_2-d/2)}{\Gamma(\alpha_1)\Gamma(\alpha_2)}$$
$$\times \int_0^1 \frac{ds}{s^{1-\tilde{\alpha}_2}\,(1-s)^{1-\tilde{\alpha}_1}}\, \frac{1}{[q^2+\mu^2]^{\alpha_1+\alpha_2-d/2}}, \quad \left(\mu^2 = \frac{m_1^2}{1-s} + \frac{m_2^2}{s}\right).$$

It is useful to rewrite the equation graphically as

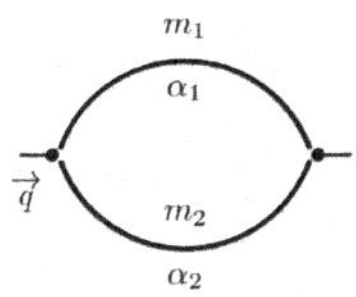

$$= \frac{\Gamma(\alpha_1+\alpha_2-d/2)}{\Gamma(\alpha_1)\Gamma(\alpha_2)} \int_0^1 \frac{ds}{s^{1-\tilde{\alpha}_2}(1-s)^{1-\tilde{\alpha}_1}}$$

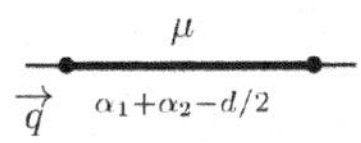

(15)

The rule is very convenient in the cases $m_2 = 0$ and $m_1 = m_2$, where the variable μ is equal to $\mu^2 = m_1^2/s$ and $\mu^2 = m_1^2/s(1-s)$, respectively. Such simple forms of μ provide the possibility to use directly an inverse-mass expansion without applying the Mellin-Barnes representation, which is essentially a more complicated procedure.

4. For any triangle with indices α_i $(i = 1, 2, 3)$ and masses m_i there is the following relation, which is based on integration by parts procedure [1, 13, 15]

$(d - 2\alpha_1 - \alpha_2 - \alpha_3)$

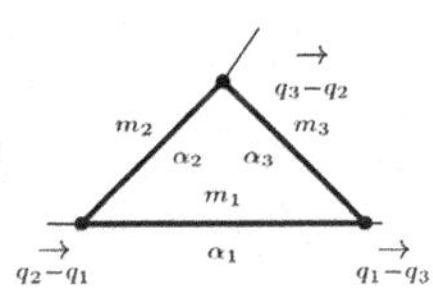

$= \alpha_2 \Big[$

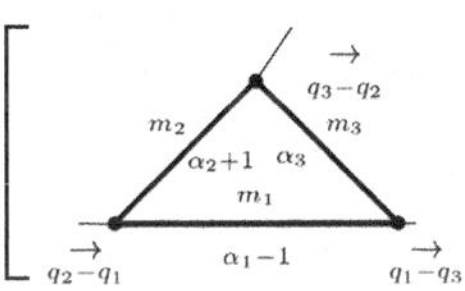

$-\Big[(q_2-q_1)^2 + m_1^2 + m_2^2\Big] \times$

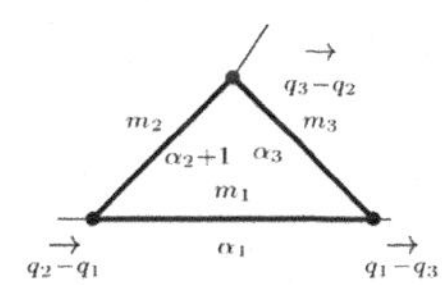

$\Big]$

$+\alpha_3\Big[\alpha_2 \leftrightarrow \alpha_3, m_2 \leftrightarrow m_3\Big] - 2m_1^2\alpha_1 \times$

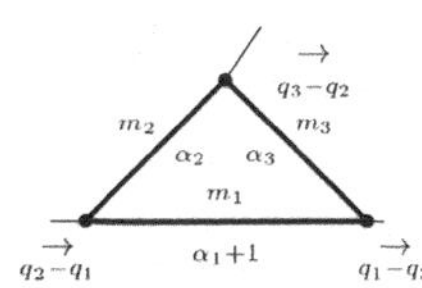

. (16)

Eq. (16) can been obtained by introducing the factor $(\partial/\partial k_\mu)\,(k-q_1)^\mu$ to the subintegral expression of the triangle, shown below as [...], and using the integration by parts procedure as follows:

$$d\int Dk\,[...] = \int Dk\left(\frac{\partial}{\partial k_\mu}(k-q_1)^\mu\right)[...] = \int Dk\,\frac{\partial}{\partial k_\mu}\left((k-q_1)^\mu\,[...]\right)$$
$$-\int Dk\,(k-q_1)^\mu\,\frac{\partial}{\partial k_\mu}\left([...]\right) \quad (17)$$

The first term in the r.h.s. becomes to be zero because it can be represented as a surface integral on the infinite surface. Evaluating the second term in the r.h.s. we reproduce Eq. (16). Note that Eq. (17) can also be applied to the n-point subgraph, see, for example, [17].

As it is possible to see from Eqs. (16) and (17) the line with the index α_1 is distinguished. The contributions of the other lines are the same. So, we will denote below the line with the index α_1 as a "distinguished line". It is clear that a various choices of the distinguished line produce different types of the IBP relations.

3.1 Basic Massive Two-Loop Integrals

Below we will concentrate mostly on two-loop two-point and three-point diagrams, which can be taken from the diagrams shown in Fig. 1. We will call them as:

$$\begin{aligned}
\hat{I}_j &= \hat{I}(q, m_j = m \neq 0, m_p = 0, p \neq j), \\
\hat{I}_{ij} &= \hat{I}(q, m_i = m_j = m \neq 0, m_p = 0, p \neq i \neq j), \\
\hat{I}_{ijs} &= \hat{I}(q, m_i = m_j = m_s = m \neq 0, m_p = 0, p \neq i \neq j \neq s), \\
\hat{I}_{ijst} &= \hat{I}(q, m_i = m_j = m_s = m_t = m \neq 0, M_p = 0, p \neq i \neq j \neq s \neq t),
\end{aligned} \tag{18}$$

$$\begin{aligned}
\hat{P}_j &= \hat{P}(q, m_j = m \neq 0, m_p = 0, p \neq j), \\
\hat{P}_{ij} &= \hat{P}(q, m_i = m_j = m \neq 0, m_p = 0, p \neq i \neq j), \\
\hat{P}_{ijs} &= \hat{P}(q, m_i = m_j = m_s = m \neq 0, m_p = 0, p \neq i \neq j \neq s), \\
\hat{P}_{ijst} &= \hat{P}(q, m_i = m_j = m_s = m_t = m \neq 0, m_p = 0, p \neq i \neq j \neq s \neq t).
\end{aligned} \tag{19}$$

Now we repeat once again the procedure of the DE method. Application of the IBP procedure [1] to loop internal momenta leads to relations between various FIs and, therefore, to the necessity of calculating only some of them, which in a sense are independent. These independent diagrams (which were chosen completely arbitrarily, of course) are called master integrals [7].

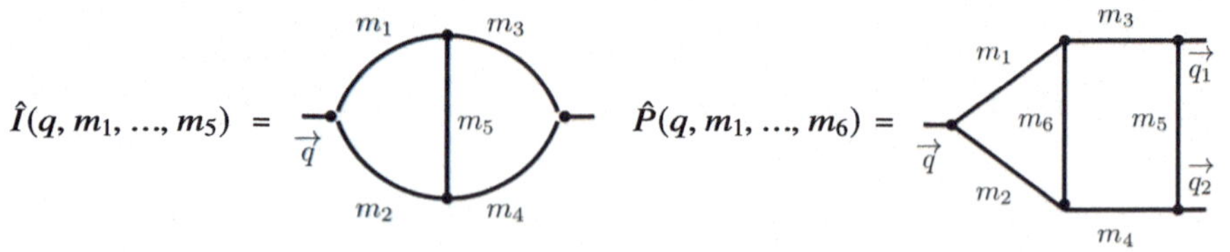

Fig. 1 Two-loop two-point diagram $\hat{I}(q, m_1, ..., m_5)$ and three-point diagram $\hat{P}(q, m_1, ..., m_6)$ with $q_1^2 = q_2^2 = 0$

Applying the IBP procedure [1] to the master-integrals themselves leads to DEs [13, 26] for them with the inhomogeneous terms containing less complex diagrams. Applying the IBP procedure to diagrams in inhomogeneous terms leads to new DEs for them with new inhomogeneous terms containing even more less complex diagrams ($\equiv$ less^2 complex ones). By repeating the procedure several times, in the last step we can obtain inhomogeneous terms containing mainly tadpoles, which can be easily calculated in-turn.

By solving the corresponding DEs in this last step, the diagrams for the inhomogeneous terms of the DEs in the previous step can be reproduced. Repeating the procedure several times, we can get the results for the original Feynman diagram.

Thus, the DE method procedure is well defined, but it requires a lot of manual work and a lot of time. So, the calculations [36] of each of the diagrams P_6 and P_{126} took about a month of work (of course, along with checking the results). It would be nice, however, to transfer some of the work to the computer. The first attempt is based on the properties of the inverse mass expansion coefficients of the master integrals. It is presented in the next section. A more modern and efficient technique is discussed in Sect. 5.

4 Evaluation of Series

Calculations of the two-point diagrams shown in Fig. 1, which do not contain elliptic structures, see Fig. 2 in Ref. [37],[4] as well as calculations of some three-point diagrams shown in Fig. 1, see also Fig. 3 in Ref. [37], lead to results with interesting properties of their inverse mass expansion coefficients.

4.1 *Properties of Series*

The inverse-mass expansion of two-loop two-point and three-point diagrams[5] with one nonzero mass (massless and massive propagators are shown by dashed and solid

[4] In fact, the results for these two-point diagrams were found in the late eighties and early nineties, and were planned to be published in a long paper summarizing the results obtained in Refs. [13, 15]. However, this paper has not been published. These results, after verification, were published in Ref. [37].

[5] The diagrams are complicated two-loop FIs that do not have cuts of three massive particles. Thus, their results should be expressible as combinations of polylogarithms. Note that we consider only three-point diagrams with independent upward momenta q_1 and q_2, which satisfy the conditions $q_1^2 q_2^2 = 0$ and $(q_1+q_2)^{=} q^2 \neq 0$, where q is a downward momentum.

lines, respectively), can be considered as

$$\mathrm{FI} = \frac{\hat{N}}{q^{2\alpha}} \sum_{n=1} C_n\, (\eta x)^n \biggl\{ F_0(n) + \biggl[\ln x\, F_{1,1}(n) + \frac{1}{\varepsilon}\, F_{1,2}(n) \biggr] \tag{20}$$
$$+ \biggl[\ln^2 x\, F_{2,1}(n) + \frac{1}{\varepsilon}\, \ln x\, F_{2,2}(n) + \frac{1}{\varepsilon^2}\, F_{2,3}(n) + \zeta(2)\, F_{2,4}(n) \biggr] + \cdots \biggr\},$$

where $x = q^2/m^2$, $\eta = 1$ or -1 and $\alpha = 1$ and 2 for two-point and three-point cases, respectively. The normalization factor is $\hat{N} = (\overline{\mu}^2/m^2)^{2\varepsilon}$, where the mass scale $\overline{\mu} = 4\pi e^{-\gamma_E}\mu$ is the standard one of the $\overline{MS}$-scheme and γ_E is Euler's constant. Moreover,

$$C_n = \frac{(n!)^2}{(2n)!} \equiv \hat{C}_n \tag{21}$$

for diagrams with two-massive-particle-cuts ($2m$-cuts). For the diagrams with one-massive-particle-cuts (m-cuts) one has $C_n = 1$.

For the m-cut case, the coefficients $F_{N,k}(n)$ should have the form

$$F_{N,k}(n) \sim \frac{S_{\pm a,\ldots}}{n^b}, \quad \frac{\zeta(\pm a)}{n^b}, \tag{22}$$

where $S_{\pm a,\ldots} \equiv S_{\pm a,\ldots}(j-1)$ are nested sums [41]: [6]

$$S_{\pm a}(j) = \sum_{m=1}^{j} \frac{(-1)^m}{m^a}, \quad S_{\pm a,\pm b,\ldots}(j) = \sum_{m=1}^{j} \frac{(-1)^m}{m^a} S_{\pm b,\ldots}(m), \tag{23}$$

and $\zeta(\pm a) = S_{\pm a}(\infty)$ and $\zeta(\pm a, , \pm b, \ldots) = S_{\pm a,\pm b,\ldots}(\infty)$ are the Euler-Zagier constants.

For $2m$-cut case, the coefficients $F_{N,k}(n)$ can be more complicated

$$F_{N,k}(n) \sim \frac{S_{\pm a,\ldots}}{n^b}, \frac{V_{a,\ldots}}{n^b}, \frac{W_{a,\ldots}}{n^b}, \tag{24}$$

where $W_{\pm a,\ldots} \equiv W_{\pm a,\ldots}(j-1)$ and $V_{\pm a,\ldots} \equiv V_{\pm a,\ldots}(j-1)$ with [37]

$$W_a(j) = \sum_{m=1}^{j} \frac{\hat{C}_m^{-1}}{m^a}, \quad W_{a,b,c,\cdots}(j) = \sum_{m=1}^{j} \frac{\hat{C}_m^{-1}}{m^a} S_{b,c,\cdots}(m), \tag{25}$$

$$V_a(j) = \sum_{m=1}^{j} \frac{\hat{C}_m}{m^a}, \quad V_{a,b,c,\cdots}(j) = \sum_{m=1}^{j} \frac{\hat{C}_m}{m^a} S_{b,c,\cdots}(m), \tag{26}$$

[6] In our previous papers [23, 25, 36, 37] the nested sums $K_{a,b,\ldots}(j) = \sum_{m=1}^{j} \frac{(-1)^{m+1}}{m^a} S_{b,\ldots}(m) = -S_{-a,b,\ldots}(j)$ have been used together with their analytic continuations [25, 42].

The terms $\sim V_{a,...}$ and $\sim W_{a,...}$ can appear only in the case of the $2m$-cut. The origin of the appearance of these terms is the product of series (20) with the different coefficients $C_n = 1$ and $C_n = \hat{C}_n$.

4.2 Two-Point Examples

As an example, consider the two-loop two-point diagrams $\hat{I}_5$ and $\hat{I}_{12}$ studied in [37]

$\hat{I}_5 =$ 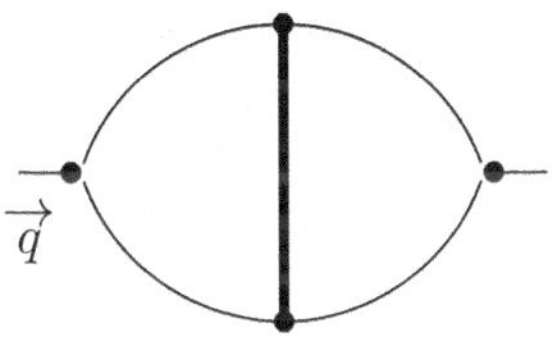 , $\hat{I}_{12} =$ 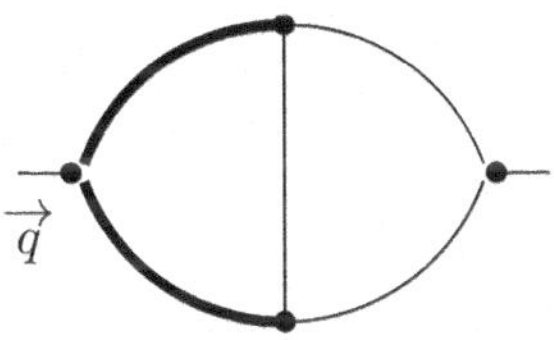 (27)

where $\hat{I}_5$ coincides with $I_1(q^2, m^2)$ considered in Sect. 2.

Their results are

$$\hat{I}_5 = \frac{\hat{N}}{q^2} \sum_{n=1} \frac{x^n}{n} \left\{ \ln^2 x - \frac{2}{n} \ln x + 2\zeta(2) + 4S_{-2} + 2\frac{2}{n^2} + \frac{2(-)^n}{n^2} \right\}, \quad (28)$$

$$\hat{I}_{12} = -\frac{\hat{N}}{q^2} \sum_{n=1} \frac{(-x)^n}{n^2} \left\{ \frac{1}{n} + \hat{C}_n \left(-2\ln x - 3W_1 + \frac{2}{n} \right) \right\}. \quad (29)$$

From (28) one can see that the corresponding functions $F_{N,k}(n)$ have the form

$$F_{N,k}(n) \sim \frac{1}{n^{3-N}}, \quad (N \geq 2), \quad (30)$$

if we introduce the following complexity of the sums ($\overline{\Phi} = (S, V, W)$)

$$\overline{\Phi}_{\pm a} \sim \overline{\Phi}_{\pm a_1, \pm a_2} \sim \overline{\Phi}_{\pm a_1, \pm a_2, \cdots, \pm a_m} \sim \zeta_a \sim \frac{1}{n^a}, \quad (\sum_{i=1}^{m} a_i = a). \quad (31)$$

The number $3 - N$ determines the level of transcendentality (or complexity, or weight) of the coefficients $F_{N,k}(n)$. The property greatly reduces the number of the possible elements in $F_{N,k}(n)$. The level of transcendentality decreases if we consider the singular parts of diagrams and/or coefficients in front of ζ-functions and of logarithm powers. Thus, finding the parts we can predict, the rest is obtained using the ansatz based on the results known already, but contains elements with a higher level of transcendentality.

Other two-loop two-point integrals in [37] have a similar form. They were exactly calculated by DE method [13, 26]. Their representations in the form of Nielsen polylogarithms [43] can be found also in Ref. [37].

4.3 Three-Point Examples

Now we consider the two-loop three-point diagrams, $\hat{P}_5$ and $\hat{P}_{12}$:

$\hat{P}_5 =$ 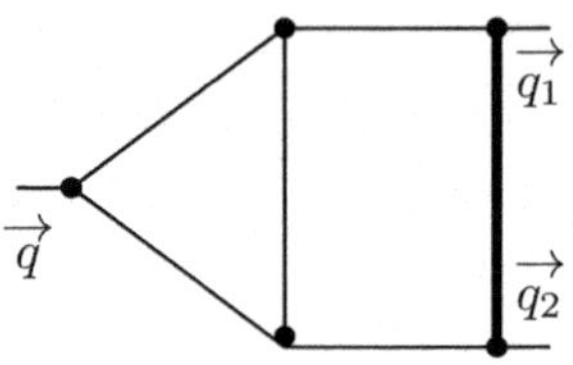 , $\hat{P}_{12} =$ 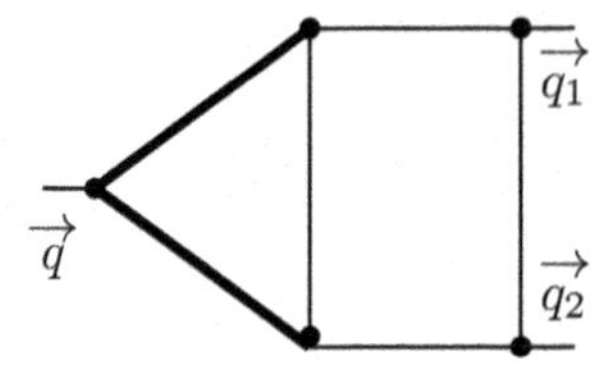 .

Their results are (see [37]):

$$\hat{P}_5 = \frac{\hat{N}}{(q^2)^2} \sum_{n=1} \frac{x^n}{n} \Bigg\{ -6\zeta_3 + 2S_1\zeta_2 + 6S_3 - 2S_1S_2 + 4\frac{S_2}{n} - \frac{S_1^2}{n} + 2\frac{S_1}{n^2} + \left(-4S_2 + S_1^2 - 2\frac{S_1}{n} \right) \ln x + S_1 \ln^2 x \Bigg\}, \tag{32}$$

$$\hat{P}_{12} = \frac{\hat{N}}{(q^2)^2} \sum_{n=1} \frac{(-x)^n}{n^2} \hat{C}_n \Bigg\{ \frac{2}{\varepsilon^2} + \frac{2}{\varepsilon}\left(S_1 - 3W_1 + \frac{1}{n} - \ln x \right) - 6W_2 - 18W_{1,1} - 13S_2 + S_1^2 - 6S_1W_1 + 2\frac{S_1}{n} + \frac{2}{n^2} - 2\left(S_1 + \frac{1}{n} \right) \ln x + \ln^2 x \Bigg\}. \tag{33}$$

Now the coefficients $F_{N,k}(n)$ have the form

$$F_{N,k}(n) \sim \frac{1}{n^{4-N}}, \quad (N \geq 3), \tag{34}$$

The diagram P_5 (and also P_1, P_3, P_6 and P_{126} in [37]) was calculated exactly by the differential equation method [13, 26].[7] To find the results for P_{12} (and also all others in [37]) we have used the knowledge of the several n terms in the inverse-mass expansion (20) (usually less than $n = 100$) and the following arguments:

- If a two-loop two-point diagram with a "similar topology" (for example, I_{12} for P_{12}, etc.) was already calculated, we should consider a similar set of basic

[7] The evaluation of the inverse mass expansion coefficients is demonstrated in Ref. [38].

elements for corresponding $F_{N,k}(n)$ of two-loop three-point diagrams but with a higher level of complexity.

- Let the diagram under consideration contain singularities and/or powers of logarithms. Since the coefficients are very simple before the leading singularity, or the largest degree of the logarithm, or the largest ζ-function, they can often be predicted directly from the first few terms of the expansion.
  Moreover, often we can calculate the singular part using a different technique (see [37] for extraction of $\sim W_1(n)$ part). Then we should expand the singular parts, find the main elements and try to use them (with the corresponding increase in the level of complexity) in order to predict the regular part of the diagram. If we need to find ε-suppressed terms, we should increase the level of complexity of the corresponding basic elements.

Later, using the ansatz for $F_{N,k}(n)$ and several terms (usually less than 100) in the above expression, which can be exactly calculated, we obtain a system of algebraic equations for the parameters of the ansatz. Solving the system, we can obtain the analytical results for FIs without exact calculations. To check the results, we only need to calculate a few more terms in the above inverse-mass expansion (20) and compare them with the predictions of our ansatz with the fixed coefficients indicated above.

Thus, the considered arguments give a possibility to find results for many complicated two-loop three-point diagrams without direct calculations. Several process options have been successfully used to calculate Feynman diagrams for many processes (see [36–40, 44]).

Note that properties similar to (30) and (34) but $b = 0$ in (22) were found for the eigenvalues of anomalous dimensions [45] and coefficient functions [46], as well as in the next-to-leading corrections [47] to the BFKL equation [48] for $N = 4$ the Super Yang-Mills (SYM) model. Such a strong restriction made it possible to obtain anomalous dimensions in the first three orders of the perturbation theory directly from the corresponding results for QCD (the "most complicated" parts are the same in $N = 4$ SYM and QCD) [49, 50], as well as in the 4th, 5th, 6th and 7th orders (see [51], [52], [53] and [54], respectively) in the algebraic Bethe ansatz [55].

Note that the series (28), (29) and (32) can be expressed as a combination of the Nielsen [43] and Remiddi-Vermaseren [56] polylogarithms with weight $4 - N$ (see [36, 37]). More complicated cases were examined in [57].

4.4 *Properties of Massive Diagrams*

Coefficients of the inverse-mass-series expansions of the two-point and three-point FIs have the structure (30) and (34) with the rule (31). Note that these conditions greatly reduce the number of possible harmonic sums. In turn, the restriction is associated with a DE specific form for the considered FIs. The DEs can be formally represented as [58, 59] (see the example $I_1(q^2, m^2)$ considered in Sect. 2)

$$\left((x+a)\frac{d}{dx} - \overline{k}(x)\varepsilon\right) \text{FI} = \text{less complicated diagrams}(\equiv \text{FI}_1), \qquad (35)$$

with some number a and some function $\overline{k}(x)$. This form is generated by the IBP procedure for diagrams including an inner n-leg one-loop subgraph, which in turn contains the product $k^{\mu_1}...k^{\mu_m}$ of its internal momenta k with $m = n - 3$.

Indeed, for ordinary degrees $\alpha_i = 1 + a_i\varepsilon$ with arbitrary a_i of subgraph propagators, the IBP relation (16) gives the coefficient $d - 2\alpha_1 - \sum_{i=2}^{p} \alpha_i + m \sim \varepsilon$ for $m = n - 3$. Important examples of applying the rule are the diagrams $\hat{I}_5$, $\hat{I}_{12}$ and $\hat{P}_5$, $\hat{P}_{12}$ (for the case $n = 2$ and $n = 3$) and also the diagrams in Ref. [60] (for the case $n = 3$ and $n = 4$). However, we note that the results for the non-planar diagrams (see Fig. 3 of [37]) obey Eq. (34) but their subgraphs do not comply with the above rule. The disagreements may be related to the on-shell vertex of the subgraph, but this requires additional research.

Taking the set of less complicated Feynman integrals FI_1 as diagrams having internal n-leg subgraphs, we get their result structure similar to the one given above (34), but with a lower level of complexity.

So, the integrals FI_1 should obey to the following equation (see $J_2^{(1)}(q^2, m^2)$ in Appendix)

$$\left((x+a_1)\frac{d}{dx} - \overline{k}_1(x)\varepsilon\right) \text{FI}_1 = \text{less}^2 \text{ complicated diagrams}(\equiv \text{FI}_2). \qquad (36)$$

Thus, we will have the a set of equations for all Feynman integrals FI_n as

$$\left((x+a_n)\frac{d}{dx} - \overline{k}_n(x)\varepsilon\right) \text{FI}_n = \text{less}^{n+1} \text{ complicated diagrams}(\equiv \text{FI}_{\text{n}+1}), \qquad (37)$$

with the last integral $\text{FI}_{\text{n}+1}$ containing only tadpoles. Note that for the case $n = 2$ the diagrams corresponding to the example $I_1(q^2, m^2)$, satisfy the system of equations, formally represented as Eq. (37).

5 Modern Technique of Massive Diagrams

In the last decade, several popular applications of DEs have emerged, allowing the use of computer resources and thus to obtain results for very complicated FIs.

In my opinion, the most successfully used approach is the so-called canonical form representation [61] of DEs (and its generalizations in Refs. [62, 63]), the method [64] of simplified DEs, and the ability to use the effective mass (see Eq.

(15)), as well as their combinations. DEs are also effectively used in calculating FIs with an elliptical structure (see [65]).

5.1 Canonical Form of Differential Equations

In our notation (see Eqs. (35)–(37)), the canonical form [61], which was introduced by Johannes Henn in 2013 and is widely popular now (there is a huge number of publications, which simply cannot be listed here), represents a homogeneous matrix equation of the form (see also the review [66])

$$\frac{d}{dx}\widehat{FI} - \varepsilon\widehat{K}(x)\widehat{FI} = 0, \tag{38}$$

for the vector

$$\widehat{FI} = \begin{pmatrix} \mathrm{FI} \\ \mathrm{FI_1}/\varepsilon \\ \dots \\ \mathrm{FI_n}/\varepsilon^{\mathrm{n}} \end{pmatrix},$$

where the matrix $\widehat{K}$ contains the functions $\overline{k}_j/(x + a_j)$ as its elements. The form (38) is called as the "canonic basic".

Note that obtaining it is far from trivial (see, for example, Appendix for $\mathrm{FI_{n=2}}$ diagrams). Moreover, it is not always achievable (see [62, 63]), where FIs were considered that are not reducible to (38)), and to obtain it is sometimes associated with a non-trivial analysis (see Refs. [67] and [68] containing methods and criteria to obtain the equation, respectively). However, the form of (38) is very convenient as it can be easily diagonalized. Note that formally for real calculations of $\mathrm{FI_n}$ it is convenient to replace

$$\mathrm{FI}_n = \widetilde{\mathrm{FI}}_n\overline{\mathrm{FI}}_n,$$

where the term $\overline{\mathrm{FI}}_n$ obeys the corresponding homogeneous equation

$$\left((x + a_n)\frac{d}{dx} - \overline{k}_n(x)\varepsilon\right)\overline{\mathrm{FI}}_n = 0, \tag{39}$$

The replacement simplifies the above Eq. (37) to the following form

$$(x + a_n)\frac{d}{dx}\widetilde{\mathrm{FI}}_n = \widetilde{\mathrm{FI}}_{n+1}\frac{\overline{\mathrm{FI}}_{n+1}}{\overline{\mathrm{FI}}_n}, \tag{40}$$

having the solution

$$\widetilde{\mathrm{FI}}_n(x) = \int_0^x \frac{dx_1}{x_1 + a_n} \widetilde{\mathrm{FI}}_{n+1}(x_1) \frac{\overline{\mathrm{FI}}_{n+1}(x_1)}{\overline{\mathrm{FI}}_n(x_1)} \tag{41}$$

Usually there are some cancellations in the ratio $\overline{\mathrm{FI}}_{n+1}/\overline{\mathrm{FI}}_n$ and sometimes it is equal to 1. In the last case, Eq. (41) coincides with the definition of Goncharov polylogarithms [69] (see also the review [70] and the references therein).

Sometimes the integrand in (41) can have a quadratic form in the denominator, for example, $x_1^2 \pm x_1 + 1$ (th signs $\pm$ can change, including when passing from the Euclidean metric to the Minkowski metric). Such forms appeared in two-point FIs, $\hat{I}_{14}$, $\hat{I}_{15}$ and $\hat{I}_{123}$ and can be represented as Nielsen three-logarithm with complicated argument, i.e. $\mathrm{Li}_3(-y^3)$, where $y = (\sqrt{x+4} - x)/(\sqrt{x+4} + x)$ is so-called conformal variable, as well as in the transform in [71] of $H(-r, ...)$ functions, introduced in [72], to the Remiddi-Vermaseren polylogarithms [56] of variable $\sim y$ where one integral representation contains the factor $x_1^2 \pm x_1 + 1$ in the denominator and is thus left in this form. Terms of this kind have appeared recently in [73] also and could be shown to be mapped into cyclotomic harmonic polylogarithms [74] in Ref. [75]. We note that such terms appear also in contributions of the massive form factors at 3-loop order [76]. Already before, the study of such and related integral representations lead to the discovery of cyclotomic polylogarithms, see [74] and Ref. [33] for a review.

5.2 *Other Approaches*

Here we will consider other methods that can be connected both with each other and with the canonical form and its generalizations, Unfortunately, we cannot pretend here to be complete in listing all the approaches.[8]

1. The *simplified DE approach* [64] is based on the violation of momentum conservation by the parameter x, with some propagator. Using the IBP relations, we can obtain set of equations which depends on x. We can solve it with the boundary conditions at $x = 0$ and take the limit $x \to 1$. The equations in this approach are usually representable in canonical form, which leads to very important results (see [78]).
2. *Series expansions in singular and regular fixed points* [79] (see also Ref. [80] and discussion therein) for DE systems, which generate Eq. (38), for example, as

$$\varepsilon \widehat{K}(x) \to \widehat{K}_1(x) + \varepsilon \widehat{K}_2(x) \,. \tag{42}$$

[8] A short review of many approaches has recently been presented as an introduction to this volume [77].

The results are obtained in the form of Goncharov polylogarithms [69] and, in some complicated cases, numerically.

3. *Symmetries of FIs* is a general method introduced in [81] which associates with any given Feynman diagram a system of partial DEs. The method uses the same variations which are used in the DE method [13] and the IBP technique [1], but distinguishes itself by associating with any diagram a natural Lie group which acts on the diagram's parameter space. This approach was further developed and numerous diagrams have been analyzed within it (see the recent paper [82] and discussions and references therein).
4. Using *the effective mass* (15) reduces the number of loops in the considered diagram. In the cases under consideration, two-loop diagrams were reduced to one-loop ones. Then, one-loop diagrams were easily calculated using the DE method, and the required two-loop diagrams were presented as integrals of the obtained one-loop results (see Ref. [39]).

5.3 Elliptic Structure

Recently, the scientific community has centered its attention to the study of FIs whose geometric properties are defined by elliptic curves. We already have a lot of progress in understanding simplest functions beyond usual polylogarithms, the so-called elliptic polylogarithms (see the recent papers [65, 83–85] and references and discussions therein). Unfortunately, this topic is beyond the scope of this consideration (discussions about elliptic polylogarithms can be found in Ref. [65], which is a contribution to this Volume), but we would like to point out only some of the integral representations that can be used in conjunction with elliptic polylogarithms or even instead of elliptic polylogarithms.

The effective mass form (15) turned out to be convenient for integrals containing an elliptic structure, since it allows one to represent the final result (see Ref. [39]) as an integral containing an elliptic kernel (i.e., a root of a polynomial of the 3rd or 4th degree) and a remainder represented in the form of an ordinary (Goncharov) polylogarithms. This approach can be an alternative to the introduction of elliptic polylogarithms, which have a very complex structure (see, for example, the recent paper [86], where the study of sunsets in special kinematics was carried out both in the form of elliptic polylogarithms (following Ref. [87]), as well as in the form of integral representations containing an elliptic kernel and ordinary polylogarithms. Notice, that such analysis has been done in all orders of the dimensional regulator following the corresponding results in Ref. [88]).

At the end of the section, we would like to note a recent paper [89], where the results for the most complex two-point single-mass diagrams containing an elliptical structure were obtained in the following form: using the effective mass representation, the original FIs were presented as integrals of one-loop diagrams dependent on the ratio μ/m. These one-loop diagrams were considered in a generalized canonical form (42). The authors of Ref. [89] have obtained a very convenient representations for extremely complicated FIs.

6 Conclusions

In this short review we examined the applicability of DEs for calculating FIs. We have considered an example $I_1(q^2, m^2)$, which led to the DE method sometime ago. The consistent application of the IBP relations to $I_1(q^2, m^2)$, and then to the diagrams of the inhomogeneous terms that arise each time, made it possible to obtain a DE hierarchy for increasingly simple diagrams obtained at each step by reducing one propagator. As noted in Sect. 3.1, the DE method is well defined but requires a lot of manual work and a lot of time.

Next, we showed an effective method restoring the exact result for two-point and three-point two-loop diagrams in terms of inverse-mass-expansion coefficients, which have a beautiful structure and can be predicted using the corresponding coefficients at the poles or at transcendental constants such as Euler's ζ-functions. These predictions were verified by analytical calculations of the first few terms using computer programs. Thus, this method is, apparently, the first, where computer programs were used for FI calculations using differential equations.

We have also given a brief overview of modern popular techniques such as the '*canonical form* of DEs [61], the *simplified DE approach* [64] and the method of the *effective mass*', see, for example, Ref. [40]. Section 5.2 lists other popular approaches as well.

The canonical form [61], and its generalizations [62, 63], are probably the most commonly used approaches, at least as a part of the calculations.

The effective mass method [40] allows one to actually work with diagrams that have fewer loops than the original ones. The results for the original diagrams are obtained in the form of integral representations, where the integrand expressions are determined by calculating the diagrams with fewer loops. So, in Ref. [39] the two-loop diagrams with an elliptic structure were considered. The corresponding one-loop diagrams depending on the effective mass *has no elliptic structure*. Thus, the results of the original diagrams were presented in the form of integral representations containing an elliptic kernel, i.e., a root of a polynomial of the 3rd or 4th degree, and ordinary polylogarithms. This representations can be used instead of elliptic polylogarithms, and even more complex objects than elliptic polylogarithms, see [89] and discussions therein.

Following the discussion in Sect. 5.3, the combined application of the *effective-mass* approach and generalizations of the canonical form for effective-mass-dependent diagrams can yield results for very complicated FIs. Such an analysis has already been carried out in the recent article [89] and, in our opinion, similar calculations can be performed in the near future for many complicated FIs.

Acknowledgments The author thanks Johannes Blümlein for invitation to present a contribution to the Proceedings of International Conference "Antidifferentiation and the Calculation of Feynman Amplitudes" (4–9 October 2020, Zeuthen, Germany) and Andrey Pikelner for help with Axodraw2 [90]. The author thanks Kay Schönwald also for a strong improvement in article style.

Appendix: Massive Part of $J_1(q^2, m^2)$ in Eq. (5)

In this appendix we consider the following diagrams

$I_2^{(\alpha)}(q^2, m^2) =$ 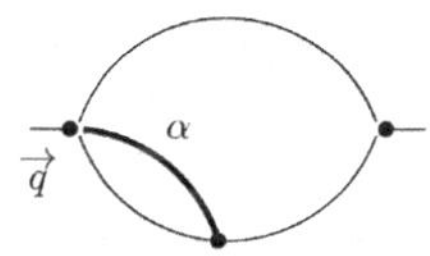

, $S^{(\beta,\alpha)}(q^2, m^2) =$ 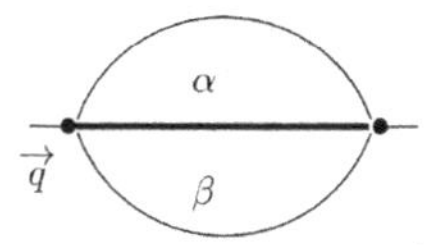

(43)

The IBP relations for the internal loop of the diagram produce two equations:

$$(d-1-2\alpha)\, I_2^{(\alpha)}(q^2, m^2) = \alpha\, J_2^{(\alpha+1)}(q^2, m^2) - m^2\, \alpha\, I_2^{(\alpha+1)}(q^2, m^2)\,, \quad (44)$$

$$(d-3)\, I_2^{(1)}(q^2, m^2) = T_{0,2}(m^2=0)\, L_{1,1}(q^2) \;-\; S^{(2,1)}(q^2, m^2)$$

$-m^2$ 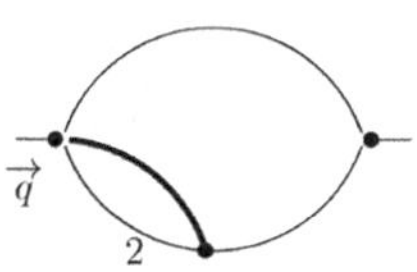

$- 2m^2\, I_2^{(2)}(q^2, m^2)\,,$ (45)

where

$$J_2^{(\alpha)}(q^2, m^2) = T_{0,\alpha}(m^2)\, L_{1,1}(q^2) \;-\; S^{(1,2)}(q^2, m^2)\,. \quad (46)$$

We note that $T_{0,2}(m^2 = 0) = 0$ in dimensional regularization and

$$T_{0,2}(m^2) L_{1,1}(q^2) = \frac{1}{(4\pi)^d} \frac{R(0,2) A(1,1)}{m^{2(2-d/2)} q^{2(2-d/2)}}\,, \quad (47)$$

where $R(\alpha_1, \alpha_2)$ and $A(\alpha_1, \alpha_2)$ are given in Eqs. (13) and (12), respectively.

The IBP relations for internal triangles of the diagram $I_2^{(1)}(q^2, m^2)$ produce two additional equations:

$$(d-4)\, I_2^{(1)}(q^2, m^2) = S^{(2,1)}(q^2, m^2) - J_2^{(2)}(q^2, m^2) - m^2\, I_2^{(2)}(q^2, m^2)$$

$-q^2$ 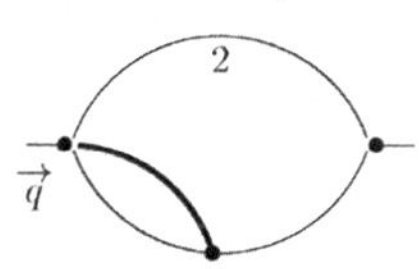

, (48)

$(d-4)$ 

$= S^{(2,1)}(q^2,m^2) - T_2(m^2=0)\, L_{1,1}(q^2)$

$+m^2$ 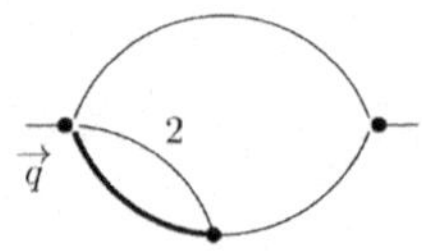

$-\, q^2$ 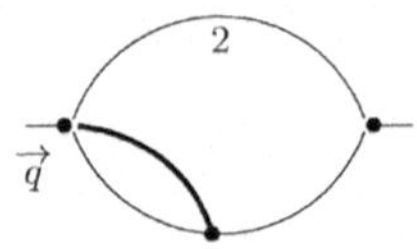

. (49)

Using Eqs. (45) and (49) in the combination: $2\times(45) + (49)$, we have

$(3d-10)\, I_2^{(1)}(q^2,m^2) = -4m^2\, I_2^{(2)}(q^2,m^2) - m^2$ 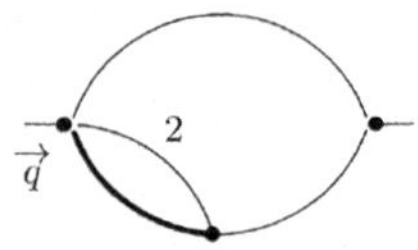

$-q^2$ 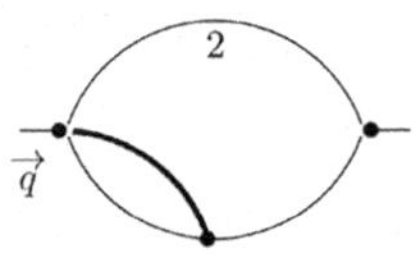

. (50)

So, we have for the mass-dependent part of $J_1(q^2,m^2)$, see Eq. (5),

m^2 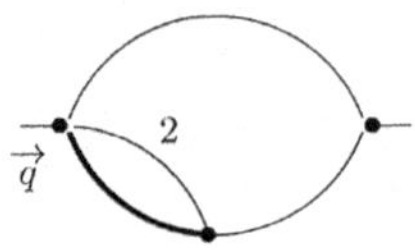

$+\, q^2$ 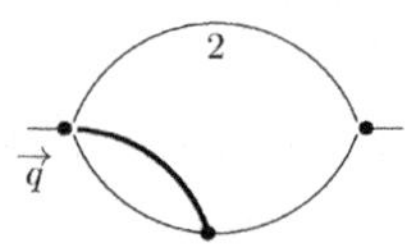

$$= \left[3d - 10 - 4m^2 \frac{d}{dm^2}\right] I_2^{(1)}(q^2,m^2)\,, \tag{51}$$

i.e. the mass-dependent combination is expressed through the diagram $I_2^{(1)}(q^2,m^2)$ and its derivative.

Using Eq. (44) one obtains

$$\left[d - 2 - \alpha - m^2 \frac{d}{dm^2}\right] I_2^{(\alpha)}(q^2,m^2) = \alpha\, J_2^{(\alpha+1)}(q^2,m^2)\,, \tag{52}$$

i.e. diagram $I_2^{(\alpha)}(q^2,m^2)$ obeys the differential equation with the inhomogeneous term $J_2^{(\alpha+1)}(q^2,m^2)$ having a very simple form: it contains only one-loop diagrams. We see that the last term in $J_2^{(\alpha)}(q^2,m^2)$, see Eq. (46), is expressed through the massive one loop integral $M_{\alpha_1,\alpha_2}(q^2, m^2)$:

$$M_{\alpha_1,\alpha_2}(q^2, m^2) = \int \frac{Dk}{(q-k)^{2\alpha_1}(k^2+m^2)^{\alpha_2}} =$$ 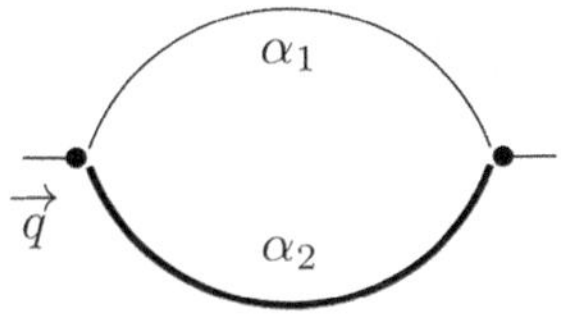

. (53)

Indeed,

$$S^{(1,\alpha)}(q^2, m^2) = A(1,1)\, M_{2-d/2,\alpha}(q^2, m^2)\,. \tag{54}$$

The one-loop diagram $M_{2-d/2,\alpha}(q^2, m^2)$ can be evaluated by one of some effective methods, for example, by Feynman parameters.

We would like to note that $I_2^{(1)}(q^2, m^2)$ satisfies Eq. (52) with $\alpha = 1$ that is not of the type of (35). But the integral $I_2^{(2)}(q^2, m^2)$ satisfies Eq. (52) with $\alpha = 2$ and is of the type of (35). So, it is convenient to rewrite (51) with $I_2^{(2)}(q^2, m^2)$ in its r.h.s.:

m^2 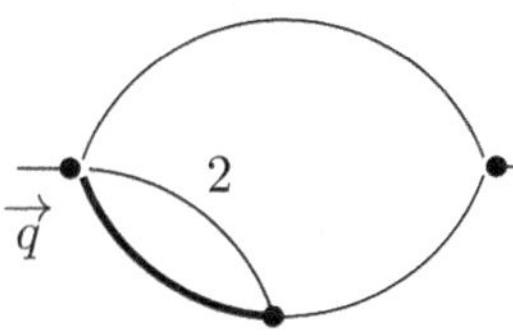

$+q^2$ 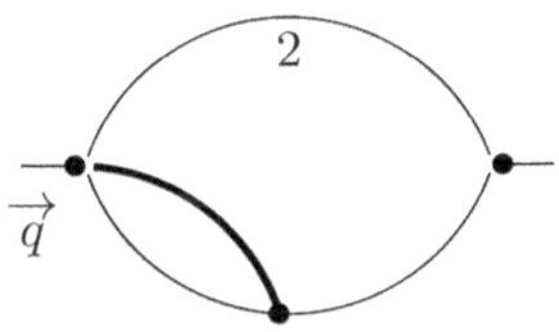

$$= \frac{3d-10}{d-3}\, J_2^{(2)}(q^2, m^2) - \frac{d-2}{d-3}\, m^2\, I_2^{(2)}(q^2, m^2)\,. \tag{55}$$

Now we should compare the IBP-based equations for $J_2^{(2)}(q^2, m^2)$ and $J_2^{(3)}(q^2, m^2)$ obtained in the right-hand sides of (55) and (52), respectively, with Eq. (35). Since $J_2^{(3)}(q^2, m^2) = -(d/dm^2)\, J_2^{(2)}(q^2, m^2)$, consider only $J_2^{(2)}(q^2, m^2)$.

So, we should prepare the IBP-based equations for the massive one-loop diagrams $M_{\varepsilon,2}(q^2, m^2)$ and $M_{\varepsilon,3}(q^2, m^2)$. Applying the IBP procedure with massive distinguished line to $M_{\varepsilon,2}(q^2, m^2)$, we have

$$(-3\varepsilon)M_{\varepsilon,2}(q^2, m^2) = \varepsilon\big[M_{1+\varepsilon,1}(q^2, m^2) - (q^2+m^2)\, M_{1+\varepsilon,2}(q^2, m^2)\big] \\ -4m^2\, M_{\varepsilon,3}(q^2, m^2)\,. \tag{56}$$

The corresponding applications of the IBP procedure with massless distinguished line to $M_{1+\varepsilon,1}(q^2, m^2)$ and $M_{1+\varepsilon,2}(q^2, m^2)$ leads to the following results:

$$(1-4\varepsilon)M_{1+\varepsilon,1}(q^2, m^2) = M_{\varepsilon,2}(q^2, m^2) - (q^2+m^2)\, M_{1+\varepsilon,2}(q^2, m^2)\,, \tag{57}$$

$$-4\varepsilon M_{1+\varepsilon,2}(q^2, m^2) = 2\, M_{\varepsilon,3}(q^2, m^2) - 2\,(q^2+m^2)\, M_{1+\varepsilon,3}(q^2, m^2). \tag{58}$$

The last equation has the following form

$$\left[-4\varepsilon - (q^2+m^2)\,\frac{d}{dm^2}\right] M_{1+\varepsilon,2}(q^2,m^2) = -\,\frac{d}{dm^2}\,M_{\varepsilon,2}(q^2,m^2). \tag{59}$$

Putting (57) to (56), we have after little algebra

$$\begin{aligned} -4\varepsilon(1-3\varepsilon)M_{\varepsilon,2}(q^2,m^2) &= -2\varepsilon(1-4\varepsilon)\,(q^2+m^2)\,M_{1+\varepsilon,2}(q^2,m^2)] \\ &\quad -4(1-4\varepsilon)\,m^2\,M_{\varepsilon,3}(q^2,m^2)\,, \end{aligned} \tag{60}$$

which transforms to

$$-\left[4\varepsilon(1-3\varepsilon)+2(1-4\varepsilon)\frac{d}{dm^2}\right]M_{\varepsilon,2}(q^2,m^2) = -2\varepsilon(1-4\varepsilon)(q^2+m^2)M_{1+\varepsilon,2}(q^2,m^2) \tag{61}$$

So, Eqs. (59) and (61) can be considered as a system of equations having a form similar to Eq. (35).

References

1. K.G. Chetyrkin, F.V. Tkachov, Nucl. Phys. B **192**, 159 (1981); F.V. Tkachov, Phys. Lett. B **100**, 65 (1981); A.N. Vasiliev, Y.M. Pismak, Yu.R. Khonkonen, Theor. Math. Phys. **46**, 104 (1981)
2. K.G. Chetyrkin et al., Nucl. Phys. B **174**, 345 (1980); A.V. Kotikov, Phys. Lett. B **375**, 240 (1996)
3. A.V. Kotikov, S. Teber, Phys. Rev. D **89**(6), 065038 (2014)
4. A. Grozin, *Lectures on QED and QCD* (2005), pp. 1–156 [hep-ph/0508242]; A.G. Grozin, Int. J. Mod. Phys. A **27**, 1230018 (2012)
5. S. Teber, A.V. Kotikov, Theor. Math. Phys. **190**(3), 446 (2017); A.V. Kotikov, S. Teber, Phys. Part. Nucl. **50**(1), 1 (2019)
6. A.I. Davydychev, J.B. Tausk, Phys. Rev. D **53**, 7381 (1996)
7. D.J. Broadhurst, Z. Phys. C **47**, 115 (1990)
8. B.A. Kniehl, A.F. Pikelner, O.L. Veretin, J. High Energy Phys. **1708**, 024 (2017); A. Pikelner, Comput. Phys. Commun. **224**, 282 (2018)
9. L.H. Ryder, *Quantum Field Theory* (Cambridge University Press, Cambridge, 1996)
10. M. Kalmykov, V. Bytev, B.A. Kniehl, S.O. Moch, B.F.L. Ward, S.A. Yost, Hypergeometric Functions and Feynman Diagrams. arXiv:2012.14492 [hep-th]
11. T. Regge, *Algebraic Topology Methods in the Theory of Feynman Rela- tivistic Amplitudes, Battelle Rencontres: 1967 Lectures in Mathematics and Physics*, ed. by C.M. DeWitt, J.A. Wheeler (W.A. Benjamin, New York, 1968), pp. 433–458
12. V.A. Golubeva, Some problems in the analytical theory of Feynman integrals. Russ. Math. Surv. **31**, 139 (1976)
13. A.V. Kotikov, Phys. Lett. B **254**, 158 (1991)
14. E. Remiddi, Nuovo Cim. A **110**, 1435 (1997)
15. A.V. Kotikov, Phys. Lett. B **259**, 314 (1991)
16. T. Gehrmann, E. Remiddi, Nucl. Phys. B **580**, 485 (2000)
17. A.V. Kotikov, Phys. Lett. B **267**, 123 (1991); Mod. Phys. Lett. A **6**, 3133 (1991)

18. Z. Bern, L.J. Dixon, D.C. Dunbar, D.A. Kosower, Nucl. Phys. B **425**, 217 (1994)
19. A.V. Kotikov, Ukr. Fiz. Zh. (Russ. Ed.) **37**, 303 (1992)
20. N.N. Bogoliubov, O.S. Parasiuk, Acta Math. **97**, 227 (1957); K. Hepp, Commun. Math. Phys. **2**, 301 (1966); W. Zimmermann, Commun. Math. Phys. **15**, 208 (1969) [Lect. Notes Phys. **558**, 217 (2000)]
21. N.N. Bogolyubov, D.V. Shirkov, Intersci. Monogr. Phys. Astron. **3**, 1 (1959)
22. D.I. Kazakov, Phys. Lett. **133B**, 406 (1983); Theor. Math. Phys. **58**, 223 (1984); D.I. Kazakov, Theor. Math. Phys. **62**, 84 (1985); N.I. Usyukina, Theor. Math. Phys. **54**, 78 (1983); V.V. Belokurov, N.I. Usyukina, J. Phys. A **16**, 2811 (1983); Theor. Math. Phys. **79**, 385 (1989)
23. D.I. Kazakov, A.V. Kotikov, Theor. Math. Phys. **73**, 1264 (1988); A.V. Kotikov, Theor. Math. Phys. **78**, 134 (1989)
24. J.A. Gracey, Conformal methods for massless Feynman integrals and large N_f *methods*, in *Computer Algebra in Quantum Field Theory*, ed. by C. Schneider, J. Blümlein (Springer, Wien, 2013), pp. 97–118. arXiv:1301.7583 [hep-th]
25. D.I. Kazakov, A.V. Kotikov, Nucl. Phys. B **307**, 721 (1988) [Nucl. Phys. B **345**, 299 (1990)]
26. A.V. Kotikov, Mod. Phys. Lett. A **6**, 677 (1991)
27. A.V. Kotikov, Int. J. Mod. Phys. A **7**, 1977 (1992)
28. D.I. Kazakov, A.V. Kotikov, Phys. Lett. B **291**, 171 (1992); Yad. Fiz. **46**, 1767 (1987); D.I. Kazakov, A.V. Kotikov, G. Parente, O.A. Sampayo, J. Sanchez Guillen, Phys. Rev. Lett. **65**, 1535 (1990)
29. F. Herzog, S. Moch, B. Ruijl, T. Ueda, J.A.M. Vermaseren, A. Vogt, Phys. Lett. B **790**, 436 (2019)
30. J. Blümlein, S. Klein, B. Tödtli, Phys. Rev. D **80**, 094010 (2009)
31. A. Behring, J. Blümlein, A. De Freitas, A. von Manteuffel, K. Schönwald, C. Schneider, The Polarized Transition Matrix Element $A_{gq}(N)$ of the Variable Flavor Number Scheme at $O(\alpha_s^3)$. arXiv:2101.05733 [hep-ph]
32. J. Ablinger, J. Blümlein, A. De Freitas, A. Goedicke, M. Saragnese, C. Schneider, K. Schönwald, Nucl. Phys. B **955**, 115059 (2020)
33. J. Blümlein, C. Schneider, Int. J. Mod. Phys. A **33**(17), 1830015 (2018)
34. A. Klemm, C. Nega, R. Safari, J. High Energy Phys. **2004**, 088 (2020); K. Bönisch, F. Fischbach, A. Klemm, C. Nega, R. Safari, Analytic Structure of all Loop Banana Amplitudes. arXiv:2008.10574 [hep-th]
35. J. Fleischer et al., Phys. Lett. B **462**, 169 (1999)
36. J. Fleischer et al., Phys. Lett. B **417**, 163 (1998)
37. J. Fleischer et al., Nucl. Phys. B **547**, 343 (1999); Acta Phys. Polon. B **29**, 2611 (1998)
38. A.V. Kotikov, Particles **3**(2), 394 (2020)
39. B.A. Kniehl et al., Nucl. Phys. B **738**, 306 (2006); Nucl. Phys. B **948**, 114780 (2019)
40. B.A. Kniehl, A.V. Kotikov, Phys. Lett. B **638**, 531 (2006); Phys. Lett. B **712**, 233 (2012)
41. J.A.M. Vermaseren, Int. J. Mod. Phys. A **14**, 2037 (1999); J. Blümlein, S. Kurth, Phys. Rev. D **60**, 014018 (1999)
42. A.V. Kotikov, Phys. Atom. Nucl. **57**, 133 (1994); A.V. Kotikov, V.N. Velizhanin, Analytic continuation of the Mellin moments of deep inelastic structure functions. hep-ph/0501274
43. A. Devoto, D.W. Duke, Riv. Nuovo Cim. **7N6**, 1 (1984)
44. B.A. Kniehl et al., Phys. Rev. Lett. **97**, 042001 (2006); Phys. Rev. D **79**, 114032 (2009); Phys. Rev. Lett. **101**, 193401 (2008); Phys. Rev. A **80**, 052501 (2009)
45. A.V. Kotikov, L.N. Lipatov, Nucl. Phys. B **661**, 19 (2003); *Proceedings of the XXXV Winter School*. Repino, S'Peterburg, 2001 (hep-ph/0112346)
46. L. Bianchi et al., Phys. Lett. B **725**, 394 (2013)
47. A.V. Kotikov, L.N. Lipatov, Nucl. Phys. **B582**, 19 (2000)
48. L.N. Lipatov, Sov. J. Nucl. Phys. **23**, 338 (1976); V.S. Fadin et al., Phys. Lett. B **60**, 50 (1975); E.A. Kuraev et al., Sov. Phys. JETP **44**, 443 (1976); Sov. Phys. JETP **45**, 199 (1977); I.I. Balitsky, L.N. Lipatov, Sov. J. Nucl. Phys. **28**, 822 (1978); JETP Lett. **30**, 355 (1979)
49. A.V. Kotikov et al., Phys. Lett. B **557**, 114 (2003)
50. A.V. Kotikov et al., Phys. Lett. B **595**, 521 (2004)

51. A.V. Kotikov et al., J. Stat. Mech. **0710**, P10003 (2007); Z. Bajnok, R.A. Janik, T. Lukowski, Nucl. Phys. B **816**, 376 (2009)
52. T. Lukowski et al., Nucl. Phys. B **831**, 105 (2010)
53. C. Marboe et al., J. High Energy Phys. **1507**, 084 (2015)
54. C. Marboe, V. Velizhanin, J. High Energy Phys. **1611**, 013 (2016)
55. M. Staudacher, J. High Energy Phys. **0505**, 054 (2005); N. Beisert, M. Staudacher, Nucl. Phys. B **727**, 1 (2005)
56. E. Remiddi, J.A.M. Vermaseren, Int. J. Mod. Phys. A **15**, 725 (2000)
57. A.I. Davydychev, M.Y. Kalmykov, Nucl. Phys. B **699**, 3 (2004)
58. A.V. Kotikov, The property of maximal transcendentality in the $N = 4$ supersymmetric Yang-Mills, in *Subtleties in Quantum Field Theory*, ed. by D. Diakonov (2010), pp. 150–174 [arXiv:1005.5029 [hep-th]]; Phys. Part. Nucl. **44**, 374 (2013); A.V. Kotikov, A.I. Onishchenko, DGLAP and BFKL equations in $\mathcal{N} = 4$ SYM: from weak to strong coupling. arXiv:1908.05113 [hep-th]
59. A.V. Kotikov, Theor. Math. Phys. **176**, 913 (2013); Theor. Math. Phys. **190**(3), 391 (2017)
60. T. Gehrmann et al., J. High Energy Phys. **1203**, 101 (2012)
61. J.M. Henn, Phys. Rev. Lett. **110**, 251601 (2013); J. Phys. A **48**, 153001 (2015)
62. L. Adams, S. Weinzierl, Phys. Lett. B **781**, 270 (2018)
63. L. Adams, C. Bogner, A. Schweitzer, S. Weinzierl, J. Math. Phys. **57**(12), 122302 (2016)
64. C.G. Papadopoulos, J. High Energy Phys. **1407**, 088 (2014)
65. S. Weinzierl, Simple differential equations for Feynman integrals associated to elliptic curves. arXiv:1912.02578 [hep-ph]; arXiv:2012.08429 [hep-th]
66. M. Argeri, P. Mastrolia, Int. J. Mod. Phys. A **22**, 4375 (2007)
67. R.N. Lee, J. High Energy Phys. **1504**, 108 (2015); J. High Energy Phys. **1810**, 176 (2018); R.N. Lee, A.I. Onishchenko, J. High Energy Phys. **1912**, 084 (2019)
68. R.N. Lee, A.A. Pomeransky, Normalized Fuchsian form on Riemann sphere and differential equations for multiloop integrals. arXiv:1707.07856 [hep-th]
69. A.B. Goncharov, Multiple polylogarithms and mixed Tate motives. math/0103059 [math.AG]
70. C. Duhr, Mathematical aspects of scattering amplitudes. arXiv:1411.7538 [hep-ph]
71. A. Kotikov et al., Nucl. Phys. B **788**, 47 (2008)
72. U. Aglietti, R. Bonciani, Nucl. Phys. B **668**, 3 (2003); U. Aglietti, R. Bonciani, G. Degrassi, A. Vicini, Phys. Lett. B **595**, 432 (2004); Phys. Lett. B **600**, 57 (2004); J. High Energy Phys. **0701**, 021 (2007)
73. R.N. Lee, M.D. Schwartz, X. Zhang, The Compton Scattering Total Cross Section at Next-to-Leading Order (2021). [arXiv:2102.06718 [hep-ph]]
74. J. Ablinger, J. Blümlein, C. Schneider, J. Math. Phys. **52**, 102301 (2011)
75. J. Ablinger, J. Blümlein, C. Schneider, Iterated Integrals Over Letters Induced by Quadratic Forms (2021) [arXiv:2103.08330 [hep-th]]
76. J. Henn, A.V. Smirnov, V.A. Smirnov, M. Steinhauser, R.N. Lee, J. High Energy Phys. **1703**, 139 (2017); J. Ablinger, J. Blümlein, P. Marquard, N. Rana, C. Schneider, Phys. Lett. B **782**, 528 (2018); J. Ablinger, J. Blümlein, P. Marquard, N. Rana, C. Schneider, Nucl. Phys. B **939**, 253 (2019)
77. J. Blümlein, Analytic Integration Methods in Quantum Field Theory: An Introduction (2021). arXiv:2103.10652 [hep-th]
78. D.D. Canko, C.G. Papadopoulos, N. Syrrakos, J. High Energy Phys. **2101**, 199 (2021); D.D. Canko, N. Syrrakos, Resummation Methods for Master Integrals (2020). arXiv:2010.06947 [hep-ph]; N. Syrrakos, Pentagon Integrals to Arbitrary Order in the Dimensional regulator (2020). arXiv:2012.10635 [hep-ph]
79. R.N. Lee, A.V. Smirnov, V.A. Smirnov, J. High Energy Phys. **1803**, 008 (2018)
80. R.N. Lee, Libra: A Package for Transformation of Differential Systems for Multiloop Integrals (2020). arXiv:2012.00279 [hep-ph]
81. B. Kol, Symmetries of Feynman Integrals and the Integration by Parts Method (2015). arXiv:1507.01359 [hep-th]

82. B. Kol, A. Schiller, R. Shir, Numerator Seagull and extended symmetries of Feynman integrals. J. High Energy Phys. **2101**, 165 (2021)
83. P. Vanhove, Feynman Integrals, Toric Geometry and Mirror Symmetry (2018). arXiv:1807.11466 [hep-th]
84. J. Broedel, C. Duhr, F. Dulat, B. Penante, L. Tancredi, From Modular Forms to Differential Equations for Feynman Integrals (2018). arXiv:1807.00842 [hep-th]; J. Broedel, A. Kaderli, J. Phys. A **53**(24), 245201 (2020)
85. M. Walden, S. Weinzierl, Numerical Evaluation of Iterated Integrals Related to Elliptic Feynman Integrals (2020). arXiv:2010.05271 [hep-ph]; L. Adams, E. Chaubey, S. Weinzierl, PoS LL **2018**, 069 (2018). [arXiv:1807.03599 [hep-ph]]
86. J. Campert, F. Moriello, A. Kotikov, Sunrise Integral with Two Internal Masses and Pseudo-Threshold Kinematics in Terms of Elliptic Polylogarithms (2020). arXiv:2011.01904 [hep-ph]
87. J. Broedel, C. Duhr, F. Dulat, L. Tancredi, J. High Energy Phys. **1805**, 093 (2018)
88. M.Y. Kalmykov, B.A. Kniehl, Nucl. Phys. B **809**, 365 (2009)
89. M.A. Bezuglov, A.I. Onishchenko, O.L. Veretin, Nucl. Phys. B **963**, 115302 (2021)
90. J.C. Collins, J.A.M. Vermaseren, Axodraw Version 2 (2016). [arXiv:1606.01177 [cs.OH]]

Holonomic Anti-Differentiation and Feynman Amplitudes

Christoph Koutschan

Abstract Computer algebra methods within the scope of the holonomic systems approach provide a versatile toolbox to integration problems in the context of Feynman diagrams. This is demonstrated with the aid of several benchmark problems, ranging from hypergeometric series evaluations to Bessel integrals of sunrise diagrams.

1 Introduction

Recent interest in the mathematical structure of Feynman diagrams has been inspired by the persistent accuracy of high-energy experiments at LHC. Still in the 1970s it was pointed out that Feynman diagrams can be understood as a special class of functions satisfying some system of differential equations. Later, it was shown in [1] within analytical regularization, that any regularized Feynman integral satisfies some holonomic system of linear differential equations. In dimension regularization, this statement was later presented by a few groups [2–4].

It was a popular idea to explore the holonomic systems approach, as originally formulated by Zeilberger [5], for the reduction of Feynman diagrams to the set of so-called master integrals [6–8]. Unfortunately, this idea was not followed up, due to the complexity of the problem. Nevertheless, we claim that the holonomic approach can be quite useful for solving other problems, related to Feynman diagrams. One goal of this paper is to substantiate this claim with the aid of a well-chosen set of problems, which we are going to tackle with the HolonomicFunctions package [9, 10].

This work was initiated at the WPC workshop "Anti-Differentiation and the Calculation of Feynman Amplitudes", that took place in October 2020 at DESY

C. Koutschan (✉)
Johann Radon Institute for Computational and Applied Mathematics (RICAM), Austrian Academy of Sciences, Linz, Austria
e-mail: christoph.koutschan@ricam.oeaw.ac.at

J. Blümlein, C. Schneider (eds.), *Anti-Differentiation and the Calculation of Feynman Amplitudes*, Texts & Monographs in Symbolic Computation,
https://doi.org/10.1007/978-3-030-80219-6_11

Zeuthen. The material presented here reflects the outcome of several discussions during this meeting. Specifically, we have to give Mikhail Kalmykov credit for compiling the collection of challenge problems.

2 The Holonomic Systems Approach

Before we start, we give a very brief introduction to the main mathematical tool that is used in this paper, that is the holonomic systems approach [5]. For more details and background on the methods employed here, we refer to the survey articles [11, 12].

In order to write mixed difference-differential equations in a concise way, we employ the following usual operator notation: let D_x denote the partial derivative operator with respect to x (x is then called a continuous variable) and S_n the forward shift operator with respect to n (n is then called a discrete variable); they act on a function f by

$$D_x f = \frac{\partial f}{\partial x} \quad \text{and} \quad S_n f = f\big|_{n \to n+1}.$$

They allow us to write linear homogeneous difference-differential equations in terms of operators, e.g.,

$$\frac{\partial}{\partial x} f(k, n+1, x, y) + n \frac{\partial}{\partial y} f(k, n, x, y) + x f(k+1, n, x, y) - f(k, n, x, y) = 0$$

turns into

$$\big(D_x S_n + n D_y + x S_k - 1\big) f(k, n, x, y) = 0;$$

in other words, such equations are represented by polynomials in the operator symbols D_x, S_n, etc., with coefficients in some field $\mathbb{K}$ which is typically some rational function field in the variables x, n, etc., and possibly in some additional parameters. Note that in general the polynomial ring $\mathbb{K}\langle D_x, S_n, \ldots \rangle$ is not commutative (this fact is indicated by the angle brackets): the coefficients from the field $\mathbb{K}$ do not commute with the polynomial variables D_x, S_n, etc. For instance, multiplication with $a(x, n) \in \mathbb{K}$ is subject to the rules

$$D_x \cdot a(x, n) = a(x, n) \cdot D_x + \frac{\partial}{\partial x} a(x, n) \quad \text{and} \quad S_n \cdot a(x, n) = a(x, n+1) \cdot S_n.$$

Such non-commutative rings of operators are called Ore algebras, and we typically denote them by $\mathbb{O}$; concise definitions and specifications of the properties of such algebras can be found, for instance, in [9].

We define the annihilator (with respect to some Ore algebra $\mathbb{O}$) of a function f by:

$$\operatorname{ann}_{\mathbb{O}}(f) := \{P \in \mathbb{O} \mid P(f) = 0\}.$$

It can easily be seen that $\operatorname{ann}_{\mathbb{O}}(f)$ is a left ideal in $\mathbb{O}$. Every left ideal $I \subseteq \operatorname{ann}_{\mathbb{O}}(f)$ is called an annihilating ideal for f.

Definition 1 Let $\mathbb{O} = \mathbb{K}\langle \dots \rangle$ be an Ore algebra. A function f is called ∂-finite w.r.t. $\mathbb{O}$ if $\mathbb{O}/\operatorname{ann}_{\mathbb{O}}(f)$ is a finite-dimensional $\mathbb{K}$-vector space. The dimension of this vector space is called the (holonomic) rank of f w.r.t. $\mathbb{O}$.

In the holonomic systems approach, the representing data structures of functions are (generators of) annihilating ideals (plus initial values). When working with (left) ideals, we use (left) Gröbner bases [13, 14] which are an important tool for executing certain operations (e.g., the ideal membership test) in an algorithmic way.

Without proof we state the following theorem about closure properties of ∂-finite functions; its proof can be found in [9, Chap. 2.3]. We remark that all of them are algorithmically executable, and the algorithms work with the above mentioned data structure.

Theorem 1 *Let $\mathbb{O}$ be an Ore algebra and let f and g be ∂-finite w.r.t. $\mathbb{O}$ of rank r and s, respectively. Then*

(i) $f + g$ *is ∂-finite of rank* $\leqslant r + s$.

(ii) $f \cdot g$ *is ∂-finite of rank* $\leqslant rs$.

(iii) f^2 *is ∂-finite of rank* $\leqslant r(r+1)/2$.

(iv) Pf *is ∂-finite of rank* $\leqslant r$ *for any* $P \in \mathbb{O}$.

(v) $f|_{x \to A(x,y,\dots)}$ *is ∂-finite of rank* $\leqslant rd$ *if* $x, y, \dots$ *are continuous variables and if the algebraic function* A *satisfies a polynomial equation of degree* d.

(vi) $f|_{n \to A(n,k,\dots)}$ *is ∂-finite of rank* $\leqslant r$ *if* A *is an integer-linear expression in the discrete variables* $n, k, \dots$.

Note that in most examples the bounds on the rank are sharp.

If we want to consider integration and summation problems, then the function in question needs to be holonomic, a concept that is closely related to ∂-finiteness. The precise definition is a bit technical and therefore skipped here; the interested reader can find it, e.g., in [5, 9, 15]. All functions that appear in this paper are both ∂-finite and holonomic. The following theorem establishes the closure of holonomic functions with respect to sums and integrals; for its proof, we once again refer to [5, 9].

Theorem 2 *Let the function f be holonomic w.r.t. D_x (resp. S_n). Then also $\int_a^b f \,\mathrm{d}x$ (resp. $\sum_{n=a}^{b} f$) is holonomic.*

If a function is ∂-finite and holonomic then Chyzak's algorithm [16] can be used to compute an annihilating ideal for the integral (resp. sum), or a heuristic approach proposed in [17]. In either case, the treatment of integrals and summations is based on the method of creative telescoping [18]. For example, for a parametrized integral of the form $\int_{a(t)}^{b(t)} f(x,t)\,\mathrm{d}x$, one has to determine a pair (P, Q), called the telescoper and the certificate, with the properties that $P + D_x Q \in \operatorname{ann}(f)$ and that the operator P is free of x and D_x. Then, integrating the equation $(P + D_x Q)(f) = 0$ and using the fundamental theorem of calculus, yields a linear differential equation for the integral.

In our calculations we will use the software package HolonomicFunctions [10], implemented in Mathematica by the author, where all the above mentioned algorithms are available.

3 Particular Values of Hypergeometric Functions

In [19] the authors present nice evaluations of hypergeometric functions at particular values. Let ε be an arbitrary parameter, then the following holds:

$$
{}_2F_1\left(\left.\begin{matrix}2\varepsilon, 3\varepsilon\\ \frac{1}{2}+2\varepsilon\end{matrix}\,\right|\frac{1}{4}\right) = \frac{\Gamma(1+\varepsilon)\,\Gamma(1+4\varepsilon)}{\Gamma(1+2\varepsilon)\,\Gamma(1+3\varepsilon)} \tag{1}
$$

see also [20] for similar hypergeometric evaluations at $\frac{1}{4}$. In this section we are demonstrating the usage of computer algebra for proving identities like (1). Other software packages, specialized to the treatment of hypergeometric series, include HYP [21] and HYPERDIRE [22].

3.1 Evaluation of a ${}_2F_1$

We write down the definition of the ${}_2F_1$ hypergeometric function as an infinite sum

$$
{}_2F_1\left(\left.\begin{matrix}2\varepsilon, 3\varepsilon\\ \frac{1}{2}+2\varepsilon\end{matrix}\,\right|\frac{1}{4}\right) = \sum_{k=0}^{\infty} \frac{(2\varepsilon)_k\,(3\varepsilon)_k}{(\frac{1}{2}+2\varepsilon)_k\,k!}4^{-k}
$$

and denote the expression inside the sum by $f_{k,\varepsilon}$. By viewing k and ε as discrete variables, one can immediately construct two difference equations, one in k and one in ε, and both of first order, which are satisfied by $f_{k,\varepsilon}$:

$$
2(k+1)(4\varepsilon+2k+1)\,f_{k+1,\varepsilon} = (2\varepsilon+k)(3\varepsilon+k)\,f_{k,\varepsilon},
$$

$$
\begin{aligned}
&6\varepsilon^2(2\varepsilon+1)(3\varepsilon+1)(3\varepsilon+2)(4\varepsilon+2k+1)(4\varepsilon+2k+3)\,f_{k,\varepsilon+1} = \\
&\quad (4\varepsilon+1)(4\varepsilon+3)(2\varepsilon+k)(2\varepsilon+k+1)(3\varepsilon+k)(3\varepsilon+k+1)(3\varepsilon+k+2)\,f_{k,\varepsilon}.
\end{aligned}
$$

Applying the creative telescoping algorithm to these recurrence equations yields a telescoper

$$P = 3(3\varepsilon + 1)(3\varepsilon + 2)S_\varepsilon - 4(4\varepsilon + 1)(4\varepsilon + 3)$$

and a certificate Q that is given by the following rational function:

$$\frac{(4\varepsilon + 1)(4\varepsilon + 3)k\big(74\varepsilon^3 + 3(19k + 18)\varepsilon^2 + (12k^2 + 27k + 10)\varepsilon + k(k + 1)(k + 2)\big)}{3\varepsilon^2(2\varepsilon + 1)(4\varepsilon + 2k + 1)}.$$

They satisfy the telescopic relation $(P + (S_k - 1)Q)(f_{k,\varepsilon}) = 0$, a fact that can be easily verified by applying the operator $P + (S_k - 1)Q$ to $f_{k,\varepsilon}$ and by subsequent simplification (which is straightforward, but tedious by hand). Summing this relation for k from 0 to ∞ yields

$$\sum_{k=0}^{\infty} P(f_{k,\varepsilon}) + \underbrace{\lim_{k\to\infty} Qf_{k,\varepsilon} - Qf_{k,\varepsilon}\big|_{k=0}}_{=0},$$

which reveals that the ${}_2F_1$ function from Eq. (1), let us denote it by $F(\varepsilon)$, satisfies the following recurrence equation:

$$3(1 + 3\varepsilon)(2 + 3\varepsilon)F(\varepsilon + 1) = 4(1 + 4\varepsilon)(3 + 4\varepsilon)F(\varepsilon).$$

Plugging in the right-hand side of (1) into the above recurrence, and simplifying it, reveals that also the closed form, the quotient of Gamma functions, satisfies the same recurrence. By comparing a single initial value ($\varepsilon = 0$), we establish the identity: for $\varepsilon = 0$ the infinite sum reduces to a finite one since only the first summand (which equals 1) survives, thanks to the definition of the Pochhammer symbol. Similarly, all Gamma functions on the right-hand side evaluate to 1 when ε is sent to 0.

3.2 *Evaluations of ${}_3F_2$ Hypergeometric Functions*

In an analogous fashion, one can prove identities like

$$\frac{1}{(1-\varepsilon)(1+2\varepsilon)}\, {}_3F_2\left(\begin{matrix}1, 1+\varepsilon, 1+2\varepsilon\\ \frac{3}{2}+\varepsilon, 2-\varepsilon\end{matrix}\,\middle|\,\frac{1}{4}\right) = \frac{1}{3\varepsilon^2}\left(\frac{\Gamma(1+2\varepsilon)\,\Gamma(1-\varepsilon)}{\Gamma(1+\varepsilon)} - 1\right). \tag{2}$$

This one is a consequence of Eq. (1), but can also be proven directly with the holonomic approach. The summand here (after expanding the definition of ${}_3F_2$) is

$$f_{k,\varepsilon} = \frac{(\varepsilon+1)_k\,(2\varepsilon+1)_k}{(1-\varepsilon)(2\varepsilon+1)\,4^k\,(2-\varepsilon)_k\,\left(\varepsilon+\frac{3}{2}\right)_k}.$$

Again, creative telescoping yields

$$P = (\varepsilon+1)^2 S_\varepsilon + 2\varepsilon(2\varepsilon+1),$$
$$Q = \frac{(\varepsilon-k-1)(10\varepsilon^2+5\varepsilon k+9\varepsilon+k^2+3k+2)}{3\varepsilon}$$

with $(P+(S_k-1)Q)(f_{k,\varepsilon}) = 0$. In contrast to the previous example, one gets an inhomogeneous contribution

$$Qf_{k,\varepsilon}\big|_{k=0} = \frac{(\varepsilon-1)(10\varepsilon^2+9\varepsilon+2)}{3\varepsilon}\cdot\frac{1}{(1-\varepsilon)(1+2\varepsilon)} = -\frac{2+5\varepsilon}{3\varepsilon},$$

which gives rise to the inhomogeneous recurrence equation

$$(\varepsilon+1)^2F(\varepsilon+1) + 2\varepsilon(2\varepsilon+1)F(\varepsilon) = -\frac{5\varepsilon+2}{3\varepsilon}.$$

It is easy to check that also the right-hand side of Eq. (2) satisfies this recurrence.

It is well known, that the two-loop massless propagator diagram possesses a large class of symmetries under exchange of indices (see [23–25]). Recently, the following relations between Clausen's hypergeometric function of arguments $z = \pm 1$ was proven (see Eq. (5) in [26]):

$$\begin{aligned}
&{}_3F_2\left(\left.\begin{matrix}1, B, 2A\\ 1+B, 2-A\end{matrix}\right|-1\right) + \frac{B}{1+A-B}\cdot {}_3F_2\left(\left.\begin{matrix}1, 2A, 1+A-B\\ 2-A, 2+A-B\end{matrix}\right|-1\right)\\
&\quad= B\frac{\Gamma(2-A)\,\Gamma(B+A-1)\,\Gamma(B-A)\,\Gamma(1+A-B)}{\Gamma(2A)\,\Gamma(1+B-2A)}\\
&\qquad - \frac{1-A}{B+A-1}\cdot {}_3F_2\left(\left.\begin{matrix}1, B, 2A\\ 1+B, A+B\end{matrix}\right|1\right),
\end{aligned}$$

where A and B are arbitrary numbers. Also such type of identities can be treated, by applying creative telescoping to the expression

$$\frac{(-1)^k(2A)_k(B)_k}{(2-A)_k(B+1)_k} + \frac{(-1)^k B\,(2A)_k(A-B+1)_k}{(A-B+1)\,(2-A)_k(A-B+2)_k}$$

in order to obtain a set of recurrences in A and B for the left-hand side (of holonomic rank 3), and then by analogously computing an annihilator for the right-hand side, which turns out to consist of exactly the same recurrences.

3.3 *Finding More ${}_2F_1$ Identities*

Not only can we apply the holonomic systems approach to prove identities like (1) or (2) or to evaluate the hypergeometric functions appearing there (i.e., without knowing the right-hand sides), but the holonomic approach also allows one to find, almost automatically, many more, similar identities. We exemplify this with Eq. (1), i.e., we seek identities of the form

$$ {}_2F_1\left(\begin{matrix} a+2\varepsilon, b+3\varepsilon \\ c+2\varepsilon \end{matrix} \middle| x\right) = H(a,b,c,\varepsilon,x), \tag{3}$$

where H stands for some hypergeometric expression with respect to ε: the shift-quotient $H(a,b,c,\varepsilon+1,x)/H(a,b,c,\varepsilon,x)$ should be a rational function, when regarded as a function in ε. In practice, it will be the case the H is hypergeometric-hyperexponential in all parameters a, b, c, ε, x, which means that it can be expressed in closed form in terms of powers, Gamma functions, and the like.

In the algebraic language, the problem is to identify conditions on the parameters a, b, c, x such that the telescoper of the summand

$$ s_{k,\varepsilon} = s_{k,\varepsilon}(a,b,c,x) = \frac{(a+2\varepsilon)_k\,(b+3\varepsilon)_k}{(c+2\varepsilon)_k}\,\frac{x^k}{k!} $$

is a first-order operator in S_ε. Following the approach proposed in [17], one can construct an ansatz for the telescopic operator $P + (S_k - 1)Q$ using the following specification:

$$ P = P(\varepsilon, S_\varepsilon) = p_1 S_\varepsilon + p_0, \quad Q = Q(k,\varepsilon) = \frac{1}{c+2\varepsilon+k}\cdot\sum_{i=0}^{4} q_i k^i $$

Note that Q need not depend on S_k or S_ε, because the input $s_{k,\varepsilon}$ is hypergeometric. All unknowns here, namely the seven symbols $p_0, p_1, q_0, \ldots, q_4$, are assumed to be rational functions in $\mathbb{Q}(a,b,c,\varepsilon)$ and should not depend on k. Applying the telescopic operator to the summand $s_{k,\varepsilon}$, and by subsequently dividing by $s_{k,\varepsilon}$ yields

$$ p_1 \cdot \frac{s_{k,\varepsilon+1}}{s_{k,\varepsilon}} + p_0 + Q(k+1,\varepsilon)\cdot\frac{s_{k+1,\varepsilon}}{s_{k,\varepsilon}} - Q(k,\varepsilon) = 0 $$

where

$$\frac{s_{k,\varepsilon+1}}{s_{k,\varepsilon}} = \frac{(a+2\varepsilon+k)_2\,(b+3\varepsilon+k)_3\,(c+2\varepsilon)_2}{(a+2\varepsilon)_2\,(b+3\varepsilon)_3\,(c+2\varepsilon+k)_2},$$

$$\frac{s_{k+1,\varepsilon}}{s_{k,\varepsilon}} = \frac{x(a+2\varepsilon+k)(b+3\varepsilon+k)}{(k+1)(c+2\varepsilon+k)}.$$

By clearing denominators, i.e., multiplying by $(c+2\varepsilon+k)(c+2\varepsilon+k+1)$, this identity of rational functions is turned into a polynomial equation of degree 6 in k. Coefficient comparison with respect to the variable k ensures that the parameters of the ansatz will not depend on k, as required, and will lead to a linear system of seven equations for the seven unknowns $p_0, p_1, q_0, \ldots, q_4$.

Since we are seeking a nontrivial solution of this system, we are interested in the cases where the system matrix is singular. We note that this matrix, although with 7×7 being small in dimension, has a nontrivial size in terms of byte count (totaling to about 1 MB), due to the appearance of the parameters a, b, c, ε, x. The determinant of the matrix is given (in fully factored form) by

$$\begin{aligned} &x(c-a)(a-c-1)(a+2\varepsilon-1)(a+2\varepsilon)^6(a+2\varepsilon+1)^6(b+3\varepsilon-1)(b+3\varepsilon)^6 \\ &\times (b+3\varepsilon+1)^6(b+3\varepsilon+2)^6(c+2\varepsilon)(c+2\varepsilon+1)(b-c+\varepsilon-1)(b-c+\varepsilon) \\ &\times \Big((x+2)(4x-1)\varepsilon^2 + (4ax^2+2ax+8bx-2b-7cx+c+2x^2+9x-2)\varepsilon \\ &\qquad + a(a+1)x^2 + x(2a+b+2)(b-c+1) + c(c-b-1)\Big). \end{aligned}$$

The first three factors of the determinant correspond to trivial or well-known evaluations of the hypergeometric function:

$$\begin{aligned} x=0: &\quad {}_2F_1\!\left(\begin{matrix} a+2\varepsilon, b+3\varepsilon \\ c+2\varepsilon \end{matrix} \middle| 0\right) = 1, \\ c=a: &\quad {}_2F_1\!\left(\begin{matrix} a+2\varepsilon, b+3\varepsilon \\ a+2\varepsilon \end{matrix} \middle| x\right) = \frac{1}{(1-x)^{b+3\varepsilon}}, \\ c=a-1: &\quad {}_2F_1\!\left(\begin{matrix} a+2\varepsilon, b+3\varepsilon \\ a+2\varepsilon-1 \end{matrix} \middle| x\right) = \frac{a(1-x)+bx+\varepsilon(x+2)+x-1}{(a+2\varepsilon-1)(1-x)^{b+3\varepsilon+1}}. \end{aligned}$$

All remaining factors that are linear in ε do not give useful conditions: since the ε appears with a constant coefficient and since the parameters a, b, c, x are not supposed to depend on ε, these factors can never become 0. The only interesting factor is the last one, a quadratic polynomial in ε, which is zero if and only if all its three coefficients vanish. This yields three nonlinear polynomial equations in the parameters a, b, c, x. A (lexicographic) Gröbner basis of the ideal generated by these polynomials is given by the following six polynomials:

$$
\begin{aligned}
&(x+2)(4x-1),\\
&(x+2)(2a-2c+1),\\
&a^2-2ac+a+c^2-c+x,\\
&12a+8bx-2b-12cx-9c+8x+4,\\
&12ab-18ac+12a-12bc+6b+18c^2-21c+8x+4,\\
&12b^2-36bc+24b+27c^2-36c+5x+10.
\end{aligned}
$$

Thanks to their triangular shape, they allow us to determine the complete set of solutions (a, b, c, x) to our polynomial equations, parametrized by a:

$$
\begin{aligned}
&\left(a, \frac{3(a-1)}{2}, a-1, -2\right), &&\left(a, \frac{3a+2}{2}, a+2, -2\right),\\
&\left(a, \frac{3a}{2}, \frac{2a+1}{2}, \frac{1}{4}\right), &&\left(a, \frac{3a-1}{2}, \frac{2a+1}{2}, \frac{1}{4}\right).
\end{aligned}
$$

Clearly, the first two families of solutions are not interesting, since the corresponding hypergeometric series are not convergent. In contrast, the two families in the second row do give us valid identities:

$$
{}_2F_1\left(\begin{matrix} a+2\varepsilon, \frac{3}{2}a+3\varepsilon \\ \frac{1}{2}(2a+1)+2\varepsilon \end{matrix} \,\middle|\, \frac{1}{4}\right) = \frac{\Gamma\left(\frac{a}{2}+\varepsilon+1\right)\Gamma(2a+4\varepsilon+1)}{\Gamma(a+2\varepsilon+1)\,\Gamma\left(\frac{3a}{2}+3\varepsilon+1\right)}, \tag{4}
$$

$$
{}_2F_1\left(\begin{matrix} a+2\varepsilon, \frac{1}{2}(3a-1)+3\varepsilon \\ \frac{1}{2}(2a+1)+2\varepsilon \end{matrix} \,\middle|\, \frac{1}{4}\right) = \left(\frac{4}{3}\right)^{\frac{3a}{2}+3\varepsilon} \cdot \frac{\Gamma\left(\frac{a}{2}+\varepsilon+1\right)\Gamma\left(a+2\varepsilon+\frac{1}{2}\right)}{\Gamma\left(\frac{a}{2}+\varepsilon+\frac{1}{2}\right)\Gamma(a+2\varepsilon+1)}. \tag{5}
$$

Note that Eq. (4) is a generalization of the original identity (1) we started with, which is recovered for $a = 0$.

Also, we should remark that this approach is not restricted to the special form where we have 2ε and 3ε in the top parameters of the ${}_2F_1$, and 2ε in the bottom parameter, but also to other situation where the epsilon coefficients $2, 2, 3$ are replaced by other integers. In this fashion one could potentially find many more similar identities.

However, we do not claim that the two identities stated above are necessarily new. There is a vast literature on special evaluations of hypergeometric functions, and it is likely that they already appear somewhere. For example, large classes of such identities were presented in [27] and [28], and in particular the latter seems to take a similar approach as the one discussed here. We nevertheless would like to point out that, although the holonomic systems approach may not be the most efficient way for *finding* new identities, it is definitely a very useful tool for *proving* them.

4 Holonomic Integration

Some of the multiloop Feynman diagrams contain the one-loop diagram or the product of one-loop diagrams insertions [29–33]. In particular, the L-loop bubble type diagram, can be understood as the integration of $L - 1$ propagators with an external massive line. The diagrams of that type have been studied from mathematical [34, 35] as well as from practical evaluation point of view [36, 37]. In particular, it was pointed out, that both types of diagrams are expressible in terms of F_C hypergeometric functions [38].

In this section, we will be interested in the integral

$$I(a) = \int_0^\infty \frac{t^{\alpha-1}}{(t+a)^j} f(t)\,\mathrm{d}t = \int_0^\infty F(t,a)\,\mathrm{d}t. \tag{6}$$

where $j \in \mathbb{Z}$ and α is a parameter, and where the unspecified function $f(t)$ satisfies the following linear non-homogeneous differential equation with polynomial coefficients:

$$(t+1)(t+9)f''(t) + (b_2t^2 + b_1t + b_0)f'(t) + c_1(t+3)f(t) = c_2t. \tag{7}$$

where b_0, b_1, b_2, c_1, c_2 are parameters (or numerical constants). Note that such type of equation appears originally in the paper by Broadhurst-Fleischer-Tarasov [39] in the context of analytical evaluation of two-loop sunset diagrams with equal masses.

Of course, one natural question that one could ask in this context, is whether the solutions of Eq. (7) can be expressed in closed form, e.g., in terms of known special functions or as hypergeometric series. However, here we want to focus on the integral (6) and ask the question: does this integral satisfy a similarly nice relation as the original function $f(t)$, that is to say: a linear differential equation, and if so, how can we find it?

From the theory of holonomy it follows immediately that this is the case: the property of $f(t)$ being holonomic transfers to the whole integral, because the kernel is just a simple combination of power functions (and hence holonomic), and because holonomicity is preserved under definite integration.

In order to perform actual calculation, we shall first devise a holonomic description for the function f, by artificially viewing it as a bivariate function $f(t,a)$. In other words, we want to give generators of a holonomic ideal in the operator algebra $\mathbb{O} = \mathbb{K}(t,a)\langle D_t, D_a\rangle$ where $\mathbb{K} = \mathbb{Q}(\alpha, b_0, b_1, b_2, c_1, c_2, j)$. The first generator is readily obtained Eq. (7), which one has to homogenize in order to get an annihilating operator. In terms of operators, this corresponds to left-multiplying the operator given by the left-hand side of (7) by an annihilating operator of its right-hand side:

$$\begin{aligned}(tD_t - 1) \cdot \big((t+1)(t+9)D_t^2 + (b_2t^2 + b_1t + b_0)D_t + c_1(t+3)\big)\\ = (t^3 + 10t^2 + 9t)D_t^3 + (b_0t + b_1t^2 + b_2t^3 + t^2 - 9)D_t^2\\ + (-b_0 + b_2t^2 + c_1t^2 + 3c_1t)D_t - 3c_1.\end{aligned}$$

The second generator is just D_a since $f(t, a)$ actually does not depend on a. From the noncommutative version of Buchberger's product criterion it follows that these two operators form a Gröbner basis, and by the definition of f, it is clear that they generate $\operatorname{ann}_{\mathbb{O}}(f)$, the annihilator of f with respect to $\mathbb{O}$.

Simple transformations convert the annihilating operators for f into operators that annihilate the whole integrand of (6), let us denote this integrand by $F(t, a)$. Algorithmically we can do it by exploiting the closure property that the product of two holonomic functions is again holonomic, but in such simple instances, one could even do it by hand. In any case, the result is as follows:

$$\begin{aligned}&(a+t)D_a + j,\\ &t(1+t)(9+t)D_t^3 + \big(-9 + b_0t + t^2 + b_1t^2 + b_2t^3 - 3t(9+10t+t^2)R\big)D_t^2\\ &+ \big(b_2t^2 + c_1t^2 + 3c_1t - b_0 - 2(-9 + b_0t + (1+b_1)t^2 + b_2t^3)R\\ &\quad + 3t(9+10t+t^2)R^2 - 3t(9+10t+t^2)R'\big)D_t\\ &- 3c_1 - (-9 + b_0t + (1+b_1)t^2 + b_2t^3)R' - (9t + 10t^2 + t^3)R''\\ &+ (b_0 - t(b_2t + c_1(3+t)) + 3t(9+10t+t^2)R'R\\ &+ (-9 + b_0t + (1+b_1)t^2 + b_2t^3)R^2 - t(9+10t+t^2)R^3\end{aligned}$$

where $R = R(t, a) = \frac{(a+t)(\alpha-1)-jt}{t(a+t)}$ and where R' refers to the differentiation with respect to t. In expanded form, this annihilator covers about a page.

Equipped with this holonomic description of the integrand, we can now employ the creative telescoping algorithm as implemented in the HolonomicFunctions package [10], in order to obtain two operators, namely a telescoper $P = P(a, D_a)$ and a certificate $Q = Q(t, a, D_t, D_a)$ with the property that $P + D_tQ$ is an element in $\operatorname{ann}_{\mathbb{O}}(F)$. To keep the exposition concise, we first look at the special case $\alpha = 1$. Then these two operators are given as follows:

$$\begin{aligned}P = -(a-9)(a-1)D_a^3 + (a^2b_2 - ab_1 - 2aj + b_0 + 10j)D_a^2\\ + (2ab_2j + ac_1 - b_1j - 3c_1 - j^2 + j)D_a + j(b_2j - b_2 + c_1),\end{aligned}$$

and

$$Q = \frac{(t+1)(t+9)(a+jt)}{(j-1)t} D_t^2 + \Big(a^2(b_0+t(b_1+b_2t)) + a(j(t(b_0+t(b_1+b_2t+2)) + 20) + 18) + t(b_0+t(b_1+b_2t))) + jt(t(b_0+t(b_1+b_2t-1)-10) + 3j(t+1)(t+9)-9)\Big) \Big/ \Big((j-1)t(a+t)\Big) D_t$$
$$+ \Big(a^3c_1(t+3) + a^2(b_0j + jt(b_1+b_2t) + c_1(j+2)t(t+3)) + a\big(j^2(t(2b_0+t(2b_1+2b_2t-1))+9) + j(t+3)\big(2c_1t^2+t-3\big) + c_1t^2(t+3)\big) + jt\big(t(t(-b_1-b_2t+c_1(t+3)+1)-b_0) + 2j(t(b_0+t(b_1+b_2t-2)-15)-9) + 3j^2(t+1)(t+9)-9\big)\Big) \Big/ \Big((j-1)t(a+t)^2\Big).$$

By denoting the result of applying the operator Q to the integrand F by $g(t, a)$, we express the above property as the equation

$$P\big(F(t,a)\big) = -\frac{\mathrm{d}}{\mathrm{d}t} g(t,a).$$

Integrating both side of this equation (almost) yields the desired relation for the integral:

$$P\big(I(a)\big) = g(0,a) - \lim_{t\to\infty} g(t,a) = g(0,a)$$

(the latter simplification under appropriate convergence assumptions on the given integral). Since the right-hand side of this (potentially) inhomogeneous differential equation is not given explicitly, but in terms of the unspecified function $f(t)$, it may be desirable to convert it to a holonomic description, i.e., into a homogeneous linear differential equation.

For this purpose, one shall derive a linear differential equation for $g(0, a)$ which, thanks to holonomic closure properties, is possible even without knowing its explicit closed form. The procedure consists of two steps: (1) derive an annihilator for $Q(F)$, which is possible by the closure under operator application (the command `DFiniteOreAction` yields an output of several pages), and (2) by applying the closure property "algebraic substitution" (the corresponding command is called `DFiniteSubstitute`). As a result, one receives the following operator that annihilates $g(0, a)$:

$$a^3D_a^3 + (3a^2j + 5a^2)D_a^2 + (3aj^2 + 7aj + 2a)D_a + (j^3 + 2j^2 - j - 2).$$

Multiplying this operator from the left to the telescoper P yields an order-6 annihilating operator for the integral $I(a)$ (not printed here for space reasons).

When we want to deal with the case of general α, then the approach is completely analogous, with the difference that all expressions get more unhandy, and that the telescoper for general α is an operator of order 6.

In this way, starting from the linear differential equation (7) for the 2-loop sunset, we have obtained the differential equation for the *3-loop* bubble diagram with two masses, studied recently in [40, 41].

5 Sunrise in Terms of Bessel-K Functions

In this section, we are studying a family of integrals that correspond to sunrise Feynman diagrams. Within dimensional regularization [42] in the momentum space it is defined as

$$J^{(L)}(\overrightarrow{M_j^2}; \overrightarrow{\alpha_j}; p^2) = \int \prod_{j=1}^{L} \frac{\mathrm{d}^n k_j}{\left(k_j^2 + M_j^2\right)^{\alpha_j}} \cdot \frac{1}{\left((p-k_1-\ldots-k_L)^2 + M_{L+1}^2\right)^{\alpha_{L+1}}},$$

where α_j are positive integers and M_j^2 and p^2 are some (in general, complex) parameters and n is an (in general, non-integer) parameter of dimensional regularization. Using the coordinate representation for the Feynman propagator and performing an integration over the angle,

$$\int \exp(ipx)\,\mathrm{d}^n\hat{x} = 2\pi^{\frac{n}{2}} \left(\frac{2}{qx}\right)^{\frac{n}{2}-1} J_{\frac{n}{2}-1}(qx),$$

where $q^2 = -p^2$ and where $J_\nu(z)$ denotes the Bessel function of the first kind, it is easy to get a one-fold integral representation for this type diagram [43]:

$$\begin{aligned}
J^{(L)}(\overrightarrow{M_j^2}; \overrightarrow{\alpha_j}; p^2) &= \int \mathrm{d}^n k \exp(iqx) \prod_{i=1}^{L+1} \int \frac{\mathrm{d}^n k_i \exp(ik_i x)}{\left(k^2 + M_i^2\right)^{\alpha}} \\
&= \int k^{n-1}\,\mathrm{d}k \int \mathrm{d}\hat{k} \exp(iqx) \prod_{i=1}^{L+1} \int \frac{\mathrm{d}^n k_i \exp(ik_i x)}{\left(k^2 + M_i^2\right)^{\alpha}} \\
&= 2\pi^{\frac{n}{2}} \int x^{n-1} \left(\frac{2}{qx}\right)^{\frac{n}{2}-1} J_{\frac{n}{2}-1}(qx) \prod_{i=1}^{L+1} \frac{2\pi^{\frac{n}{2}}}{\Gamma(\alpha_i)} \left(\frac{2M_i}{x}\right)^{\frac{n}{2}-\alpha_i} K_{\frac{n}{2}-\alpha}(M_i x)\,\mathrm{d}x \\
&= \frac{\left(\pi^{\frac{n}{2}}\right)^{L+2}}{2^{\alpha-\frac{n}{2}(L+2)-L-1}} \left(\frac{1}{q}\right)^{\frac{n}{2}-1} \int_0^\infty t^{\alpha-\frac{n}{2}L} J_{\frac{n}{2}-1}(tq) \prod_{j=1}^{L+1} \left(K_{\frac{n}{2}-\alpha_j}(M_j t) \frac{M_j^{\frac{n}{2}-\alpha_j}}{\Gamma(\alpha_j)}\right) \mathrm{d}t
\end{aligned} \quad (8)$$

where $\alpha = \sum_{k=1}^{L+1} \alpha_k$, and n is the dimension of space-time, where $q^2 = -p^2$, and $K_\nu(z)$ denotes the modified Bessel function of the second kind. This integral has been studied in [44–47].

In the rest of this section, we will focus on the integral representation (8). Note that the integrand contains a product of $L + 2$ Bessel functions (both J and K counted together). Since a Bessel function has holonomic rank 2, it follows by Theorem 1(ii) that the expression in the integral has holonomic rank at most 2^{L+2}. Unfortunately, it turns out that the bound in this instance is tight, i.e., the holonomic rank of the integrand is exactly 2^{L+2}. Since all creative telescoping algorithms are very sensitive concerning the holonomic rank of the input annihilator, this class of integrals is going to pose challenges for our package.

For computing telescopers of holonomic integrals, we have several algorithms at hand: we mention just Chyzak's algorithm [16] and a heuristic ansatz proposed by the author [17]. The advantage of the former algorithm is that it is theoretically sound and is guaranteed to terminate and to return the minimal-order telescoper, while the latter uses several heuristics to shape the ansatz, which may result in a non-minimal telescoper and in some instances this "algorithm" even fails to terminate. The disadvantage of Chyzak's algorithm is that it is very sensitive to the holonomic rank of the input due to the uncoupling step. The other approach [17] was designed specifically to address this issue and to circumvent the costly uncoupling step. Since our examples have relatively large holonomic rank, we will use the heuristic approach, and hence the reported telescopers need not necessarily be minimal. On the other side, the usage of this algorithm enables us to get some results at all: for example, in the most simple example (first line of Table 1), we obtain a result after about 2 seconds, while the algorithm [16] was aborted after 1000 seconds without yielding any result.

As a toy example, we start with the case $L = 1$. Hence, in this case the integral depends on the six parameters $q, n, \alpha_1, \alpha_2, M_1, M_2$. Nevertheless, the HolonomicFunctions program is able to compute a telescoper within a few seconds. This telescoper is a third-order operator in D_q, but is still too long to be printed here. Instead, we display the specialized version with $M_1 = M_2 = 1$:

$$\begin{aligned}
&-8q^3(q^2+4)D_q^3 + 4q^2\big(5nq^2 - 16\alpha_1 - 16\alpha_2 + 12n - 8\alpha_1 q^2 - 8\alpha_2 q^2 - 12q^2 - 24\big)D_q^2 \\
&-2q\big(16\alpha_1^2 + 16\alpha_2^2 + 48\alpha_1 + 32\alpha_1\alpha_2 + 48\alpha_2 + 7n^2q^2 + 12n^2 - 32\alpha_1 n - 32\alpha_2 n - 24n \\
&+ q^2(-24\alpha_1 n - 24\alpha_2 n - 30n + 16\alpha_1^2 + 16\alpha_2^2 + 48\alpha_1 + 48\alpha_2 + 48\alpha_1\alpha_2 + 28)\big)D_q \\
&- 32\alpha_1^2 - 32\alpha_2^2 + 32\alpha_1 - 64\alpha_1\alpha_2 + 32\alpha_2 + 3n^3q^2 + 4n^3 - 16\alpha_1 n^2 - 16\alpha_2 n^2 \\
&- 16\alpha_1 n^2 q^2 - 16\alpha_2 n^2 q^2 - 14n^2q^2 + 16\alpha_1^2 n + 16\alpha_2^2 n + 16\alpha_1 n + 32\alpha_1\alpha_2 n + 16\alpha_2 n \\
&+ 16\alpha_1^2 nq^2 + 16\alpha_2^2 nq^2 + 48\alpha_1 nq^2 + 48\alpha_2 nq^2 + 80\alpha_1\alpha_2 nq^2 + 20nq^2 - 16n - 8q^2 \\
&- 32\alpha_1^2 q^2 - 32\alpha_2^2 q^2 - 64\alpha_1\alpha_2^2 q^2 - 32\alpha_1 q^2 - 32\alpha_2 q^2 - 64\alpha_1^2\alpha_2 q^2 - 96\alpha_1\alpha_2 q^2.
\end{aligned}$$

Table 1 Some benchmark computations: the first column gives the specification of an instance of integral (8) (parameters that are not mentioned are kept symbolic), "Rank" refers to the holonomic rank of the integrand, "Order" to the order (degree w.r.t. D_q) of the telescoper, "Time" to the computation time (in seconds), and "Size" to the byte size of the telescoper (not the certificate), using Mathematica's `ByteCount`. One computation didn't finish within 36 hours, one computation crashed (ERR); nevertheless, the order of the telescoper could be extracted from the log files

Case	Rank	Order	Time	Size
$L = 1, M_j = 1$	8	3	2 s	15 KB
$L = 1$	8	3	6 s	149 KB
$L = 2, M_j = 1, \alpha_j = j$	16	4	34 s	36 KB
$L = 2, M_j = 1$	16	7	797 s	6.8 MB
$L = 2, M_1 = q, M_2 = M_3 = 1$	16	7	8003 s	4.6 MB
$L = 2, M_1 = M_2 = M_3$	16	7	ERR	–
$L = 2, \alpha_1 = 5, \alpha_2 = 11, \alpha_3 = 14$	16	7	>36 h	–
$L = 3, M_j = 1, \alpha_j = j$	32	11	83006 s	1.4 MB
$L = 3, M_1 = q, M_2 = M_3 = M_4 = 1, \alpha_j = j$	32	11	763 s	1.0 MB

We have also looked at some "less trivial" cases of the integral (8). As the above output suggests, it will not be reasonable to print the resulting operators, but instead we tabularize our findings (see Table 1), together with some information on timings and sizes of the outputs. These data allow us to acquire an intuition on how the difficulty of the computation depends on the holonomic rank of the input, the number of parameters, etc.

Table 1 gives an impression that the computation of integrals related to sunrise Feynman diagrams is challenging but not completely hopeless for the holonomic systems approach. We plan to explore further the applicability of this approach to Feynman integrals in a forthcoming publication.

Concluding, we have shortly discussed a set of problems related to the evaluation of Feynman diagrams, where the holonomic systems approach, implemented in the package HolonomicFunctions, could be quite useful. We are looking forward to many other exciting collaborations between computer algebra and particle physics.

Acknowledgments We would like to thank Johannes Blümlein and Carsten Schneider for organizing the WPC workshop on "Anti-Differentiation and the Calculation of Feynman Amplitudes" at DESY Zeuthen, despite the difficult circumstances due to the corona pandemic. We are extremely grateful to Mikhail Kalmykov for providing us with this list of challenge problems and for many references. The discussions with him during the workshop enlightened the physics background of these computational tasks.

References

1. M. Kashiwara, T. Kawai, Commun. Math. Phys. **54**, 121 (1977). https://doi.org/10.1007/BF01614133

2. F.V. Tkachov, Nucl. Instrum. Meth. **A389**, 309 (1997). https://doi.org/10.1016/S0168-9002(97)00110-1
3. A.V. Smirnov, A.V. Petukhov, Lett. Math. Phys. **97**, 37 (2011). https://doi.org/10.1007/s11005-010-0450-0
4. T. Bitoun, C. Bogner, R.P. Klausen, E. Panzer, Lett. Math. Phys. **109**(3), 497 (2019). https://doi.org/10.1007/s11005-018-1114-8
5. D. Zeilberger, J. Comput. Appl. Math. **32**(3), 321 (1990)
6. O.V. Tarasov, Nucl. Instrum. Meth. **A534**, 293 (2004). https://doi.org/10.1016/j.nima.2004.07.104
7. V.P. Gerdt, Nucl. Phys. Proc. Suppl. **135**, 232 (2004). https://doi.org/10.1016/j.nuclphysbps.2004.09.011
8. A.V. Smirnov, JHEP **04**, 026 (2006). https://doi.org/10.1088/1126-6708/2006/04/026
9. C. Koutschan, Advanced applications of the holonomic systems approach. Ph.D. thesis, RISC, Johannes Kepler University, Linz, Austria (2009)
10. C. Koutschan, HolonomicFunctions (user's guide). Tech. Rep. 10-01, RISC Report Series, Linz, Austria (2010). http://www.risc.jku.at/research/combinat/software/HolonomicFunctions/
11. C. Koutschan, in *Computer Algebra in Quantum Field Theory: Integration, Summation and Special Functions*, ed. by C. Schneider, J. Blümlein. Texts & Monographs in Symbolic Computation (Springer, Wien, 2013), pp. 171–194
12. F. Chyzak, The ABC of creative telescoping. Habilitation a diriger des recherches (HDR) (2014). http://specfun.inria.fr/chyzak/Chyzak-2014-ABC.pdf
13. B. Buchberger, Ein Algorithmus zum Auffinden der Basiselemente des Restklassenrings nach einem nulldimensionalen Polynomideal. Ph.D. thesis, University of Innsbruck, Austria (1965)
14. A. Kandri-Rody, V. Weispfenning, J. Symb. Comput. **9**(1), 1 (1990)
15. S.C. Coutinho, *A primer of algebraic D-modules*, *London Mathematical Society Student Texts*, vol. 33 (Cambridge University Press, 1995)
16. F. Chyzak, Discrete Math. **217**(1-3), 115 (2000). https://doi.org/10.1016/S0012-365X(99)00259-9
17. C. Koutschan, Math. Comp. Sci. **4**(2-3), 259 (2010). https://doi.org/10.1007/s11786-010-0055-0
18. D. Zeilberger, J. Symb. Comput. **11**, 195 (1991)
19. A.I. Davydychev, M.Y. Kalmykov, Nucl. Phys. B **605**, 266 (2001)
20. P.W. Karlsson, J. Math. Anal. Appl. **196**(1), 172 (1995)
21. C. Krattenthaler, J. Symb. Comput. **20**, 737 (1995)
22. V.V. Bytev, M.Y. Kalmykov, S.O. Moch, Comput. Phys. Commun. **185**, 3041 (2014)
23. D.T. Barfoot, D.J. Broadhurst, Z. Phys. **C41**, 81 (1988). https://doi.org/10.1007/BF01412581
24. A.G. Grozin, Int. J. Mod. Phys. **A27**, 1230018 (2012). https://doi.org/10.1142/S0217751X12300189
25. A.V. Kotikov, S. Teber, Phys. Part. Nucl. **50**(1), 1 (2019). https://doi.org/10.1134/S1063779619010039
26. A.V. Kotikov, S. Teber, Teor. Mat. Fiz. **194**(2), 331 (2018). https://doi.org/10.4213/tmf9340
27. A. Ebisu, Special values of the hypergeometric series. Tech. Rep., arXiv:1308.5588 (2013)
28. F. Beukers, J. Forsgård, Γ-evaluations of hypergeometric series. Tech. Rep., arXiv:2004.08117 (2020)
29. T. Regge, E.R. Speer, M.J. Westwater, Fortsch. Phys. **20**, 365 (1972). https://doi.org/10.1002/prop.19720200603
30. S. Friot, D. Greynat, E. De Rafael, Phys. Lett. **B628**, 73 (2005). https://doi.org/10.1016/j.physletb.2005.08.126
31. J.P. Aguilar, D. Greynat, E. De Rafael, Phys. Rev. **D77**, 093010 (2008). https://doi.org/10.1103/PhysRevD.77.093010
32. S. Bekavac, A.G. Grozin, D. Seidel, V.A. Smirnov, Nucl. Phys. **B819**, 183 (2009). https://doi.org/10.1016/j.nuclphysb.2009.04.015
33. R. Lee, P. Marquard, A.V. Smirnov, V.A. Smirnov, M. Steinhauser, JHEP **03**, 162 (2013). https://doi.org/10.1007/JHEP03(2013)162

34. G. Ponzano, T. Regge, E.R. Speer, M.J. Westwater, Commun. Math. Phys. **15**, 83 (1969). https://doi.org/10.1007/BF01645374
35. P. Aluffi, M. Marcolli, Commun. Num. Theor. Phys. **3**, 1 (2009). https://doi.org/10.4310/CNTP.2009.v3.n1.a1
36. D.J. Broadhurst, Z. Phys. **C47**, 115 (1990). https://doi.org/10.1007/BF01551921
37. D.J. Broadhurst, Eur. Phys. J. **C8**, 311 (1999). https://doi.org/10.1007/s100529900935
38. F.A. Berends, M. Buza, M. Bohm, R. Scharf, Z. Phys. **C63**, 227 (1994). https://doi.org/10.1007/BF01411014
39. D.J. Broadhurst, J. Fleischer, O.V. Tarasov, Z. Phys. C **60**, 287 (1993)
40. J. Ablinger, J. Blümlein, A. De Freitas, M. van Hoeij, E. Imamoglu, C.G. Raab, C.S. Radu, C. Schneider, J. Math. Phys. **59**(6), 062305 (2018). https://doi.org/10.1063/1.4986417
41. S. Abreu, M. Becchetti, C. Duhr, R. Marzucca, JHEP **02**, 050 (2020). https://doi.org/10.1007/JHEP02(2020)050
42. G. 't Hooft, M. Veltman, Nucl. Phys. B **44**(1), 189 (1972)
43. E. Mendels, Il Nuovo Cimento A **45**, 87 (1978). https://doi.org/10.1007/BF02729917
44. D.H. Bailey, J.M. Borwein, D. Broadhurst, M.L. Glasser, J. Phys. **A41**, 205203 (2008). https://doi.org/10.1088/1751-8113/41/20/205203
45. J.M. Borwein, B. Salvy, Exp. Math. **17**, 223 (2011). https://doi.org/10.1080/10586458.2008.10129032
46. D. Broadhurst, D.P. Roberts, in *PoS LL2018*, 053 (2018). https://doi.org/10.22323/1.303.0053
47. R.N. Lee, A.A. Pomeransky, JHEP **08**, 027 (2019). https://doi.org/10.1007/JHEP08(2019)027

Outer Space as a Combinatorial Backbone for Cutkosky Rules and Coactions

Dirk Kreimer

Abstract We consider a coaction which exists for any bridge-free graph. It is based on the cubical chain complex associated to any such graph by considering two boundary operations: shrinking edges or removing them. Only if the number of spanning trees of a graph G equals its number of internal edges we find that the graphical coaction Δ^G constructed here agrees with the coaction Δ_{Inc} proposed by Britto and collaborators. The graphs for which this is the case are one-loop graphs or their duals, multi-edge banana graphs. They provide the only examples discussed by Britto and collaborators so far. We call such graphs simple graphs. The Dunce's cap graph is the first non-simple graph. The number of its spanning trees (five) exceeds the number of its edges (four). We compare the two coactions which indeed do not agree and discuss this result. We also point out that for kinematic renormalization schemes the coaction Δ^G simplifies.

1 Introduction

The notion of a coaction has gained prominence recently in the context of amplitude computations in high energy physics [1].

This is motivated by the appearance of multiple polylogarithms and their elliptic cousins in such computations [2–4]. For such functions the existence of such a coaction is known. Indeed Francis Brown gave a masterful account of is appearance and conceptual role [5, 6] in particular also with regards to the small graphs principle, see for example [7, 8].

D. Kreimer (✉)
Department of Mathematics, Humboldt University, Berlin, Germany

Department of Physics, Humboldt University, Berlin, Germany
e-mail: kreimer@math.hu-berlin.de

J. Blümlein, C. Schneider (eds.), *Anti-Differentiation and the Calculation of Feynman Amplitudes*, Texts & Monographs in Symbolic Computation,
https://doi.org/10.1007/978-3-030-80219-6_12

For physicists it is second nature to regard any Feynman integral computation as a manipulation on Feynman graphs.

One hence wishes to identify coactions in combinatorial manipulations of Feynman graphs which are in accordance with their appearance in the study of such polylogarithms.

A possible approach is based on reverse engineering by pulling back the coaction structure of the functions into which Feynman graphs evaluate to the graphs themselves. This is an approach successfully employed by Britto et al. [1] and they conjecture a graphical coaction which by construction is correct for the known graphs amenable to computation.

These are the simple graphs alluded to in the abstract above plus a few non-simple graphs with kinematics chosen such that a large number of terms in their conjectured graphical coactions is bound to vanish. As a consequence they evaluate to multiple polylogarithms (MPLs) and one is again in safe terrain and can again pull back to the coaction on MPLs.

Here we introduce a coaction which exists independent of any physics consideration as a purely mathematical construct.

Its existence follows from known studies in graph complexes and graph homology [9–11]. We derive it in all detail.

We then compare the constructions of Britto et al. and ours and show that the two constructions agree on simple graphs and on graphs which evaluate to mere MPLs.

Next we discuss differences for generic kinematics for non simple graphs and argue why the suggestion of Britto et al. [1] for a graphical coaction is bound to fail.

We also discuss simplifications apparent in kinematical renormalization schemes and relate the Steinmann relations [12] to the structure of the cubical chain complex.

2 Incidence Hopf Algebras for (Lower) Triangular Matrices

Let us first define incidence Hopf algebras following Schmitt [13, 14]. Apart from changes in notation this material is similar to the presentation in Appendix C.3 of Britto et al. [1].

We start from a (partially) ordered set P with partial order $\leq$.

For $x, y \in P$, $x \leq y$, consider the interval

$$[x, y] = \{z \in P | x \leq z \leq y\}.$$

Let P_2 be the $\mathbb{Q}$-algebra generated by such intervals [1, 13] through multiplication as a free product by disjoint union of intervals.

It gives rise to an incidence bialgebra I_P upon setting

$$\Delta([x, y]) = \sum_{x \leq z \leq y} [x, z] \otimes [z, y],$$

for the co-product Δ and

$$\hat{\mathbb{I}}([x,y]) = \delta_{x,y}, \tag{1}$$

for the co-unit $\hat{\mathbb{I}}$, where

$$\delta_{x,y} = 1,\ x = y,\ \delta_{x,y} = 0, \text{ else.}$$

Note that $\Delta([x,x]) = [x,x] \otimes [x,x]$ is group-like.

Following Schmitt we can turn this bialgebra I_P into a Hopf algebra $I_{\tilde{P}}$ by augmenting I_P by multiplicative inverses $[x,x]^{-1}$ for group-like $[x,x]$, for all $x \in P$.

The antipode $S : I_{\tilde{P}} \to I_{\tilde{P}}$, $S \circ S = \text{id}$ is defined by $S([x,x]) = [x,x]^{-1}$ and

$$S([x,y]) = \sum_{x=z_0 \le z_1 \le \ldots \le z_k = y} (-1)^k \frac{1}{[x,x]} \prod_{i=1}^{k} \frac{[z_{i-1}, z_i]}{[z_i, z_i]}.$$

2.1 Example: Lower Triangular Matrices

As an example consider lower triangular $n \times n$ matrices M, $M_{i,j} = 0,\ j \gneq i$. P is provided by the first n integers and the intervals $[ji]$, $1 \le j \le i \le n$, are represented as $M_{i,j}$.

As $M_{i,j} \in H_{GF}$, a Hopf algebra [15], the algebra structure of P_2 agrees with the algebra structure of H_{GF} and is a free commutative algebra. We have $\Delta(M_{i,j} M_{l,s}) = \Delta(M_{i,j})\Delta(M_{l,s})$.

$I_P \equiv I_M$ gets a different bialgebra structure though. Instead of using the coproduct Δ_{GF} of H_{GF} the coproduct is

$$\Delta M_{i,j} = \sum_{k=j}^{i} M_{k,j} \otimes M_{i,k}.$$

Coassociativity of this map is obvious.

$$(\Delta \otimes \text{id})\Delta(M_{j,k}) = \sum_{h,i} M_{h,k} \otimes M_{i,h} \otimes M_{j,i},$$

$$(\text{id} \otimes \Delta)\Delta(M_{j,k}) = \sum_{h,i} M_{i,k} \otimes M_{h,i} \otimes M_{j,h},$$

and the two expressions on the right obviously agree.

Consider the $\mathbb{Q}$-vectorspace V_1 generated by elements $M_{i,1}$, $i \gneq 1$. Let $\rho_\Delta : V_1 \to V_1 \otimes I_{\tilde{P}}$,

$$\rho_\Delta(M_{i,1}) = \sum_{k=2}^{i} M_{k,1} \otimes M_{i,k},$$

be the restriction of Δ to V_1.

Then coassociativity of Δ delivers

$$(\mathrm{id} \otimes \Delta)\rho_\Delta = (\rho_\Delta \otimes \mathrm{id})\rho_\Delta,$$

and we also have by Eq.(1) that $(\mathrm{id} \otimes \hat{\mathbb{I}})\rho_\Delta = \mathrm{id}$. We conclude

Proposition 2.1 *ρ_Δ is a coaction on V_1.*

Note that we get such a Hopf algebra and coaction for any chosen lower triangular matrix M. We write Δ^M whenever necessary.

It is useful to define matrices $M_{\mathbb{I}}$

$$(M_{\mathbb{I}})_{i,j} = M_{i,j}/M_{i,i},$$

which has unit entries along the diagonal and the diagonal matrix M_{D}

$$(M_{\mathsf{D}})_{i,j} = 0,\ i \neq j,\ (M_{\mathsf{D}})_{i,i} = M_{i,i}.$$

Then,

$$(M_{\mathsf{D}}) \times M_{\mathbb{I}} = M. \tag{2}$$

For lower triangular matrices there are two maps which are natural to consider: shifting to the row above or to the column to the right.

So consider the map

$$m_r : M_{i,j} \to M_{i-1,j},$$

where we set $M_{0,j} = 0$ and the map

$$m_c : M_{i,j} \to M_{i,j+1},$$

where we set $M_{n,n+1} = 0$.

Proposition 2.2 *We have*

$$(\mathrm{id} \otimes m_r) \circ \Delta = \Delta \circ m_r, \tag{3}$$

and

$$(m_c \otimes \mathrm{id}) \circ \Delta = \Delta \circ m_c. \tag{4}$$

Proof m_r: Δ maps entries from the j-th row to entries in the j-th row on the rhs of the tensorproduct, and m_r shifts $j \to j-1$ on both sides of the equation.
m_c: Δ maps entries from the k-th column to entries in the k-th column on the lhs of the tensorproduct, and m_c shifts $k \to k+1$ on both sides of the equation. □

Example 2.3 Let us consider an example for all the above:

$$M = \begin{pmatrix} 1 & 0 & 0 & 0 \\ a & b & 0 & 0 \\ c & d & e & 0 \\ f & g & h & j \end{pmatrix}. \tag{5}$$

Then,

$$M_{\mathrm{I}} = \begin{pmatrix} 1 & 0 & 0 & 0 \\ a/b & 1 & 0 & 0 \\ c/e & d/e & 1 & 0 \\ f/j & g/j & h/j & 1 \end{pmatrix}, \tag{6}$$

and

$$M_{\mathsf{D}} = \begin{pmatrix} 1 & 0 & 0 & 0 \\ 0 & b & 0 & 0 \\ 0 & 0 & e & 0 \\ 0 & 0 & 0 & j \end{pmatrix}. \tag{7}$$

We have

$$\rho_\Delta(f) = f \otimes j + c \otimes h + a \otimes g,$$

when we regard f as an element of the $\mathbb{Q}$-vectorspace V_1 spanned by a, c, f. Also $\Delta(g) = g \otimes j + d \otimes h + b \otimes g$, where $g \in I_M \setminus V_1$, and I_M is the $\mathbb{Q}$-vectorspace spanned by the entries of M.

Furthermore

$$\Delta(m_r(g)) = \Delta(d) = d \otimes e + b \otimes d = (\mathrm{id} \otimes m_r)(g \otimes j + d \otimes h + b \otimes g),$$

as $m_r(j) = 0$, and

$$\Delta(m_c(g)) = \Delta(h) = h \otimes j + e \otimes h = (m_c \otimes \mathrm{id})(g \otimes j + d \otimes h + b \otimes g),$$

as $m_c(b) = 0$.

Let us introduce some more terminology. For any lower triangular $n \times n$ matrix M let us call the entries $M_{i,j}$ the Galois correspondents of M, $M_{i,j} \in \mathsf{Gal}(M)$. We regard Δ as a map

$$\mathsf{Gal}(M) \to \mathsf{Gal}(M) \otimes \mathsf{Gal}(M).$$

For several say k such matrices M_i, $1 \leq i \leq k$, each of them giving rise to a coproduct and coaction $\Delta^i \equiv \Delta^{M_i}$ we associate the set $\mathsf{Gal}_k := \cup_{i=1}^k \mathsf{Gal}(M_i)$. The union is not a disjoint union as a single Galois correspondent can be contained in various sets $\mathsf{Gal}(M_i)$ of such correspondents simultaneously.

We then define for all $x \in \mathsf{Gal}_k$,

$$\Delta(x) = \sum_{j=1}^{k} \Delta^j(x),$$

where we set $\Delta^j(x) = 0\ \forall x \notin \mathsf{Gal}(M_j)$.

In fact there is a matrix M which we can assign to Gal_k. Of particular interest to us is the case where the entries in the upper left and lower right corner are all equal: $(M_i)_{1,1} = (M_j)_{1,1}$ and $(M_i)_{n,n} = (M_j)_{n,n}$, $\forall i, j$.

The generic construction is an obvious iteration of the following example on two matrices.

Example 2.4 Assume M_1, M_2 are lower $k \times k$ square matrices.

So the M_i^B below are lower $(k-2) \times (k-2)$ square matrices, while the M_i^C are $(k-2) \times 1$ column matrices, the M_i^R are $1 \times (k-2)$ row matrices, $i \in \{1, 2\}$.

$$M_1 = \left(\begin{array}{c|ccc|c} 1 & 0 & \sim & 0 & 0 \\ \hline . & . & . & . & 0 \\ M_1^C & . & M_1^B & . & \wr \\ . & . & . & . & 0 \\ \hline g & . & M_1^R & . & c \end{array}\right),$$

$$M_2 = \left(\begin{array}{c|ccc|c} 1 & 0 & \sim & 0 & 0 \\ \hline . & . & . & . & 0 \\ M_2^C & . & M_2^B & . & \wr \\ . & . & . & . & 0 \\ \hline h & . & M_2^R & . & c \end{array}\right),$$

Then,

$$M = \left(\begin{array}{c|ccc|ccc|c} 1 & 0 & \sim & 0 & 0 & \sim & 0 & 0 \\ \hline \cdot & \cdot & \cdot & \cdot & \sim & \sim & \sim & 0 \\ M_1^C & \cdot & M_1^B & \cdot & \sim & 0 & \sim & \wr \\ \cdot & \cdot & \cdot & \cdot & \sim & \sim & \sim & 0 \\ \hline \cdot & \sim & \sim & \sim & \cdot & \cdot & \cdot & 0 \\ M_2^C & \sim & 0 & \sim & \cdot & M_2^B & \cdot & \wr \\ \cdot & \sim & \sim & \sim & \cdot & \cdot & \cdot & 0 \\ \hline g+h & \cdot & M_1^R & \cdot & \cdot & M_2^R & \cdot & c \end{array}\right),$$

And indeed immediately checks

$$\Delta^M = \Delta^{M_1} + \Delta^{M_2}.$$

Furthermore note that for any entry M_{ij} in a $n \times n$ matrix M there exists a lower triangular $(i - j + 1) \times (i - j + 1)$ matrix M^{ij}, with

$$\Delta^M(x) = \Delta^{M^{ij}}(x),\ \forall x \in \mathsf{Gal}(M^{ij}).$$

Here the lowest leftmost entry of the matrix M^{ij} is $M_{i,j}$.

3 Lower Triangular Matrices from the Cubical Chain Complex

Lower triangular matrices derived from the cubical chain complex played a prominent role already in [16]. We refine their construction here to derive a graphical coaction.

A most prominent role in the study of Feynman graphs G is played by their $|G|$ independent loops. They provide the basis for the subsequent loop integrations of Feynman integrals.

Assume given a bridgefree graph G together with a spanning tree T for it constituting a pair $(G, T) \equiv G_T \in H_{GF}$.

We can route all external momentum flow through edges $e \in E_T$ of the spanning tree. The remaining $|G| = e_G - e_T$ edges e_i generate a basis L_{G_T} for the cycle space of G,

$$L_{G_T} = \{\cup_{i=1}^{|G|} l_i\},$$

where each l_i is a cycle of edges given by a pair (e_i, p_i) where $e_i \notin E_T$ and $p_i \subseteq E_T$ is the unique path of edges connecting source and target of e_i.

The Feynman integral is independent of the choice of T as long as T is a spanning tree of G, $T \in \mathcal{T}(G)$.

Example 3.1 Here is the wheel with three spokes w_3:

$w_3 =$

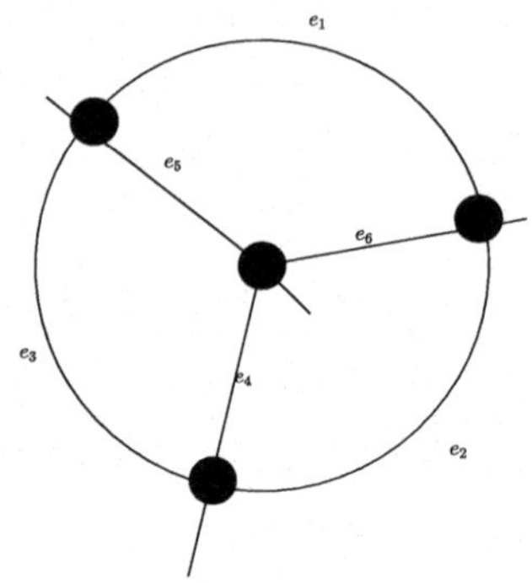

Its cycles are

$$\{(e_1, e_2, e_3), (e_1, e_5, e_6), (e_2, e_4, e_6), (e_3, e_5, e_4), (e_2, e_3, e_5, e_6), (e_6, e_4, e_3, e_1), (e_4, e_5, e_1, e_2)\}.$$

It has $16 = |\mathcal{T}(w_3)|$ spanning trees T on three edges. Choose $T = (e_1, e_6, e_4)$. e_2, e_3, e_5 generate a basis for the cycles $l_1 := (e_2, p_2)$, $l_2 := (e_3, p_3)$ and $l_3 := (e_5, p_5)$, and $p_2 = (e_4, e_6)$, $p_3 = (e_4, e_6, e_1)$ and $p_5 = (e_1, e_6)$.

This setup suggests to study Culler–Vogtmann *Outer Space* [9]. It assigns a k-dimensional cell $C(G)$ to any graph G on $k + 1$ edges and the Feynman integral becomes an integral over the volume over this cell.

This is evident in parametric space where we can identify the edge length A_e of an edge e with the parametric variable.

The renormalized Feynman form (see [17] for notation and for other than log-divergent singularities in renormalization)

$$\mathsf{Int}_R(G)(q, p) = \sum_F (-1)^{|F|} \frac{\ln\left(\frac{\Phi(G/F)\psi(F)+\Phi_0(F)\psi(G/F)}{\Phi_0(G/F)\psi(F)+\Phi_0(F)\psi(G/F)}\right)}{\psi_G^2} \Omega_G$$

as provided by the Symanzik polynomials Φ, ψ gives the volume form for $C(G)$. The sum is over the forests of G as demanded by renormalization and we get

$$\Phi_R(G)(q) = \int_{C(G)} \mathsf{Int}_R(G).$$

Here, $q \in \mathbf{Q}_G$ is a vector spanned by Lorentz invariants in $\mathbf{Q}_G$ as provided by the external momentum vectors of G and $p \in \mathbb{P}_G$ is a point in the projective space spanned by positive real edge variables.

A sum over all vacuum Feynman graphs then integrates Outer Space (OS). A sum over all bridgefree graphs with n loops and a given number s of external legs sums the corresponding classes of graphs $X_{n,s}$ with s fixed marked points and n loops [11].

The codimension k boundaries of cells $C(G)$ are cells themselves which are assigned to reduced graphs where k edges of T and therefore of G shrink.

All codimension one boundaries appear for example as

$$C(G) \to C(G/e),\ e \in E_G.$$

The lowest dimensional boundary is of codimension e_T and the graph assigned to it is the rose G/E_T.

When studying $C(G)$ and the cells apparent as codimension k boundaries a prominent role its played by the barycenters of all these cells.

Example 3.2 Consider the simplest cell $C(b)$, a one-dimensional line for the bubble graph b on two edges with different masses (green or red lines) with two 1-edge spanning trees indicated by double edges.

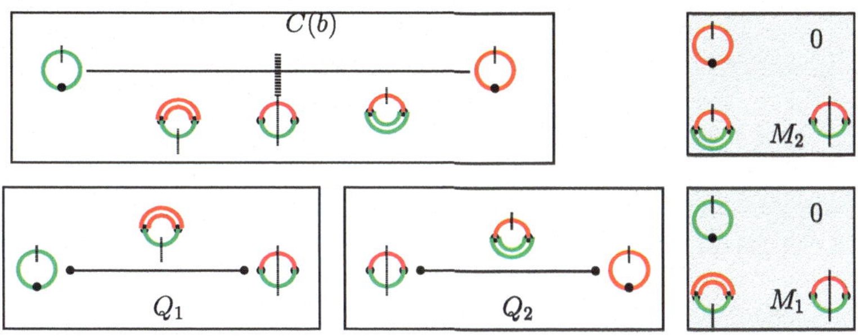

The codimension one ends are zero-dimensional cells to which a tadpole graph is associated. At the barycentric middle of the cell we have the graph with its two vertices as spanning forest, and the two internal edges on-shell. The barycentric middle is determined by

$$A_r m_r = A_g m_g$$

as a point $p \in \mathbb{P}_b \equiv \mathbb{P}^1$ which determines the ratio $p_b = A_r/A_g = m_g/m_r$, and there is $q_b \in \mathbf{Q}_b \sim \mathbb{R}$ (generated by q^2 with $q_b : q^2 = (m_1+m_2)^2$), so that (q_b, p_b) determines a threshold divisor in $\mathsf{Int}_R(b)(q, p)$.

The cell $C(b)$ gives by the two spanning trees rise to two one-dimensional unit cubes $Q_1 \sim [0, 1]$, $Q_2 \sim [0, 1]$. Approaching zero, the graph shrinks to a tadpole, approaching one it approaches the leading singularity

$$\mathrm{Int}_R(b)(q, p),$$

corresponding to the barycenter and the evaluation of $\Phi_R(b)$ at $q^2 = (m_1 + m_2)^2$. As each cube is one-dimensional it gives rise to a single lower triangular Hodge matrix as indicated.

For each cell $C(G)$ with associated (reduced) graph g the barycenter corresponds to the leading singular graph g_F, where $F = \cup_{v \in V_g}$, so that all internal edges of g are on the mass-shell.

Such barycenters define paths from the barycenter of $C(G)$ through barycenters of lower and lower dimensional hypersurfaces until we reach the barycenter of codimension e_T cells.

The collection of all these paths defines the *spine of Outer Space* as a deformation retract of OS [9].

The barycenters then provide the coordinates in parametric space of the threshold divisors which generate monodromy of Feynman amplitudes.

Physical thresholds are determined by solving a variational problem [16] determining the minimal kinematical configuration so as to make the discriminant of the second Symanzik polynomial Φ vanish for the associated leading singular graph.

The spine determines a set of $|\mathcal{T}(G)|$ e_T-dimensional cubes and with it $|\mathcal{T}(G)| \times e_T!$ paths from the midpoint of $C(G)$ to the rose G/E_T. Each such path defines a lower triangular matrix corresponding to the $e_T!$ simplices into which a cube decomposes.

We will thus now turn to the cubical chain complex for Feynman graphs [15, 16] where a pair G_T of a graph G and a spanning tree T for it gives rise to a e_T-cube.

Any e_T-cube gives rise to a natural cell decomposition into $e_T!$ simplices and therefore generates $e_T!$ lower triangular matrices corresponding to the $e_T!$ possible orderings of the edges. Figure 1 is instructive.

Note that the cell $C(t)$ as well as the cubes Q_i are two dimensional and in fact the cell $C(t)$ can be dissected in $|\mathcal{T}(t)| = 3$ open cubes so that $C(t) = \overline{\amalg_i Q_i}$ is the completion of their union. This is in fact typical for one-loop graphs:

$$C(G) = \bar{X},\ X = \amalg_{j=1}^{|\mathcal{T}(G)|} Q_j,\ 1 \leq j \leq |\mathcal{T}(G)|$$

and $\bar{X} \setminus X$ is the spine of G. This is very different for generic graphs where $\dim(C(G)) = \dim(Q_j) = -1 + |G|$ [15].

Let us study one cube say for the spanning tree on blue and red edges, so the cube Q_1 containing  .

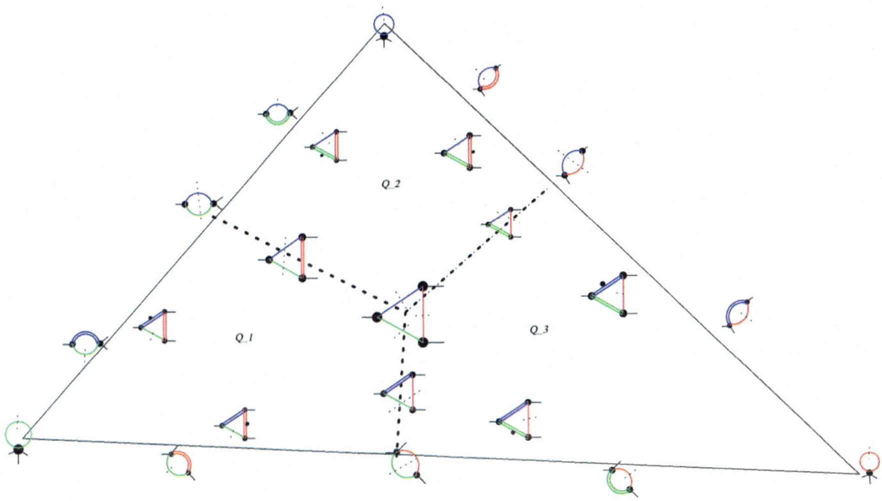

Fig. 1 A two dimensional cell $C(t)$ (a 2-simplex, so itself a triangle) in outer space for the triangle graph t on three different masses (indicated by colored edges). On-shell edges are thin and marked by a hashed line, off-shell edges are double lines, a dot orders the two-edge spanning trees with the dotted edge the longer one. For this simple graph the spine gives a simplicial decomposition of $C(t)$ into three open 2-cubes Q_1, Q_2, Q_3

It provides nine cells: a two-dimensional square, four one-dimensional edges, and four zero-dimensional corners.

$Q^{(1)} =$

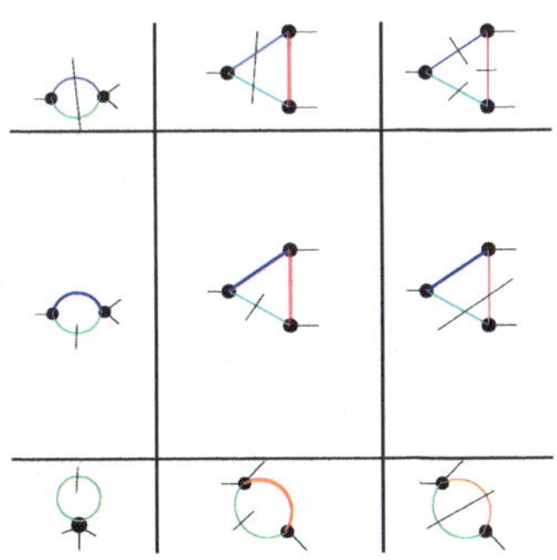

The three graphs in the anti-diagonal from the lowest left corner (the origin of the cube) to the upper right corner appear in both Hodge matrices $M^{(1)}$ and $M^{(2)}$ defined below and associated to this cube.

In such n-cubes, the $n!$ paths from the rose (the origin of the cube) to the leading singular graph (diagonally opposed for the main diagonal) share their origin and their endpoint and do not intersect otherwise. This reflects the Steinmann relations [12] as threshold divisors do not overlap between graphs and sectors assigned to different paths. An example:

Example 3.3 We list all six matrices $M^{(i)}$, $1 \leq i \leq 6$ for the triangle graph.

$M^{(1)} =$ 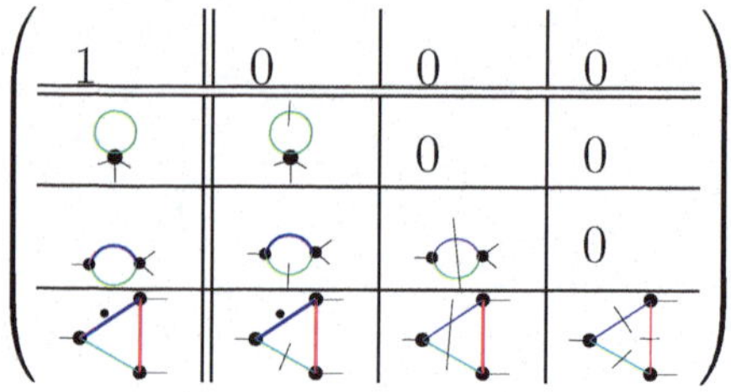

,

for sectors

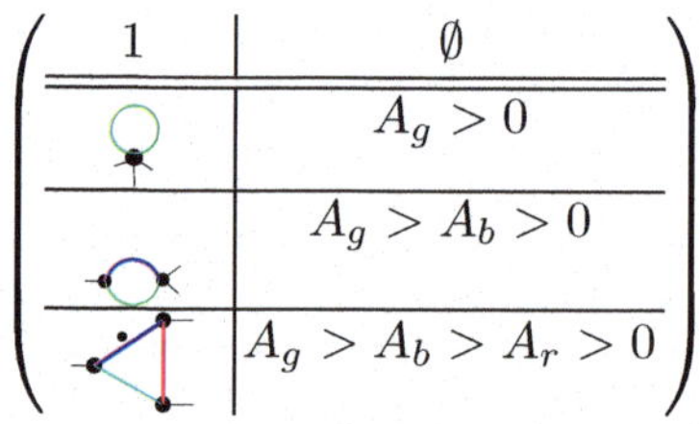

,

$M^{(2)} =$ 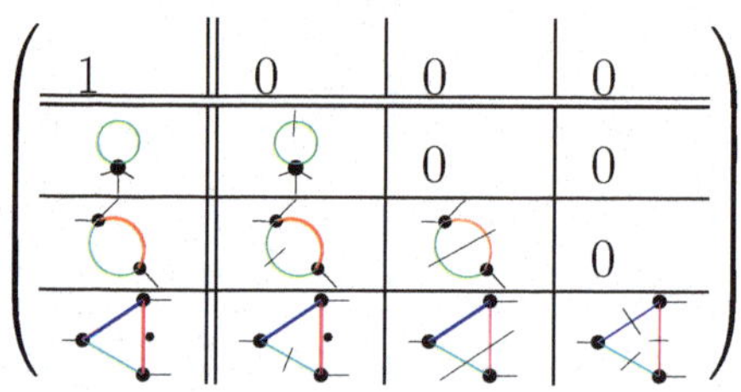

,

for sectors

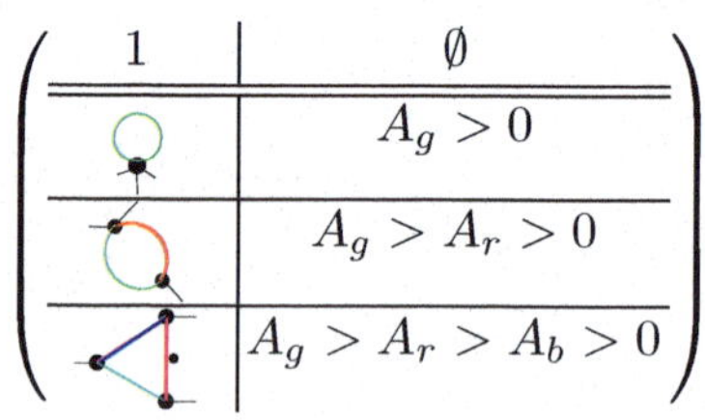

.

As $A_r > A_b$ and $A_b > A_r$ do not coexist, both sectors give monodromy in different regions of $\mathbf{Q}_t$.

$M^{(3)} =$ 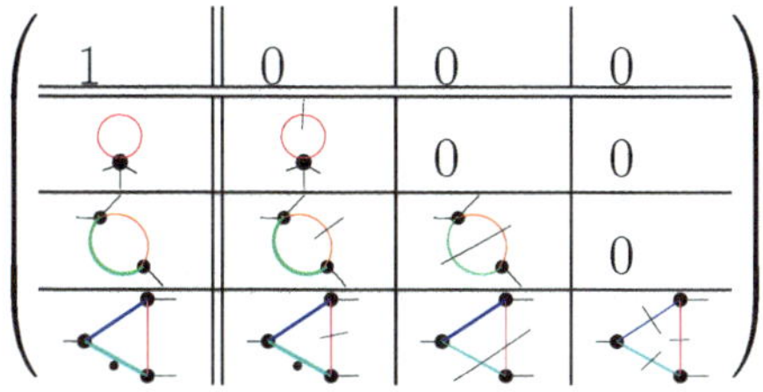

,

for sectors

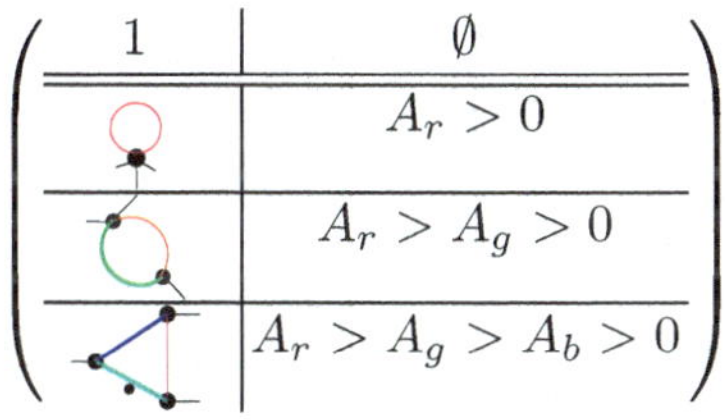

,

$M^{(4)} =$ 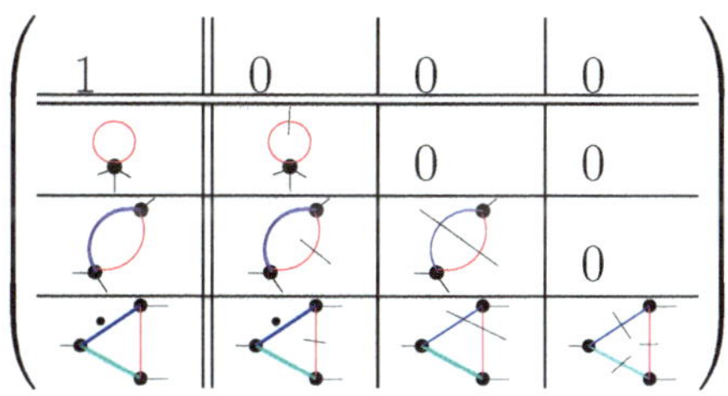

,

for sectors

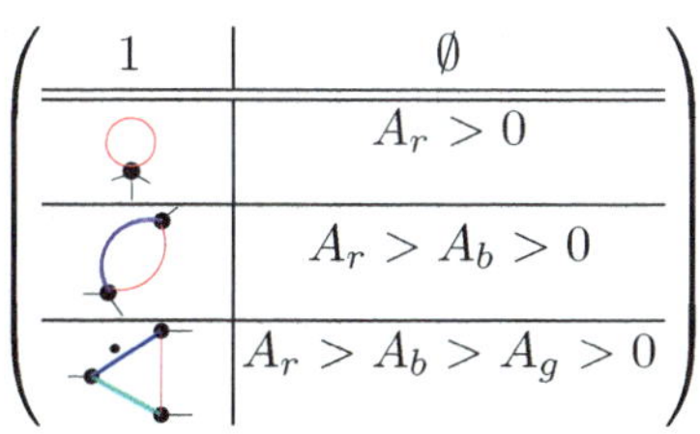

,

$M^{(5)} =$ 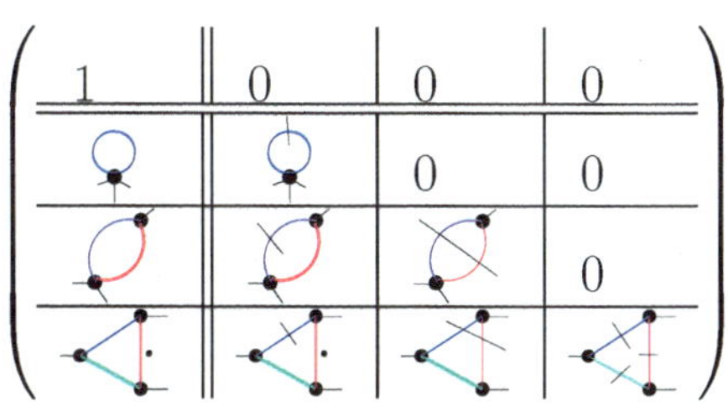

,

for sectors

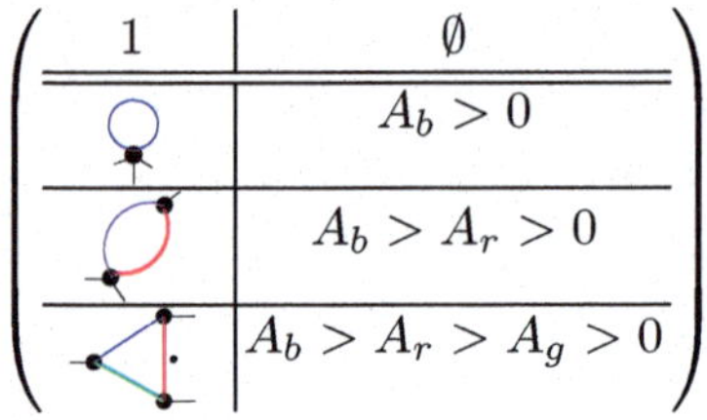

,

$M^{(6)} =$ 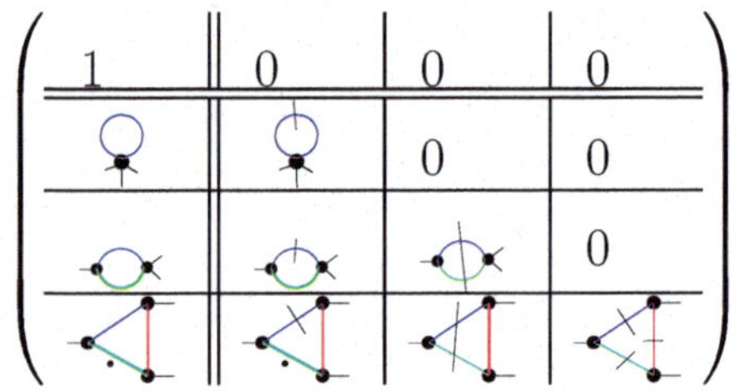

,

for sectors

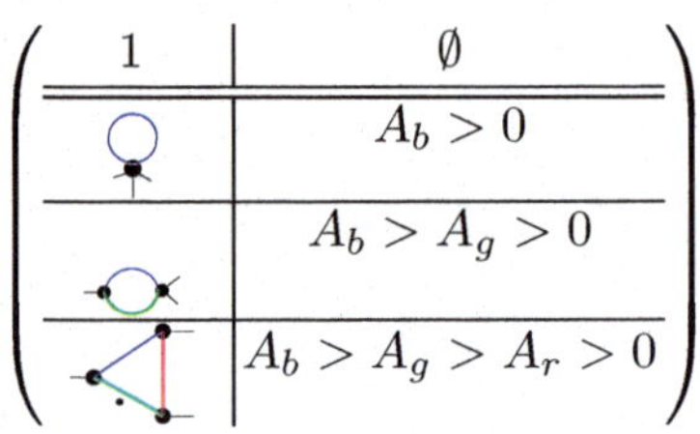

.

Note that each such Hodge matrix is well-defined in its sector.

We define

$$\Delta^M = \sum_{j=1}^{6} \Delta^{M^{(j)}},$$

where we find M as

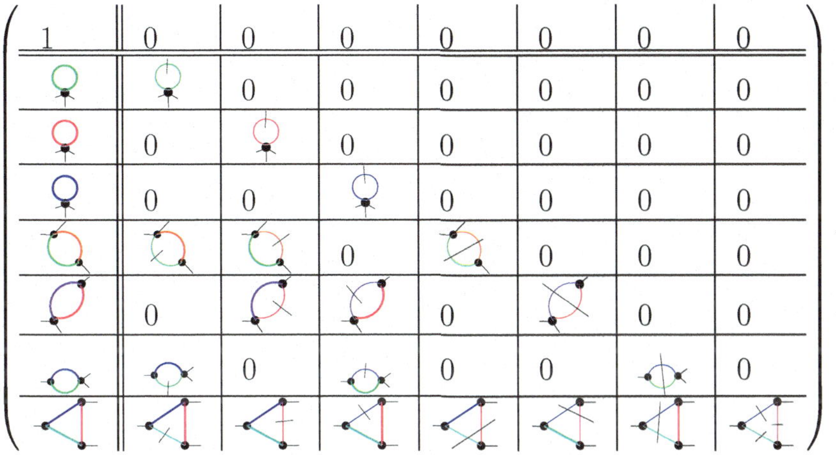

Let us work out a few coactions.

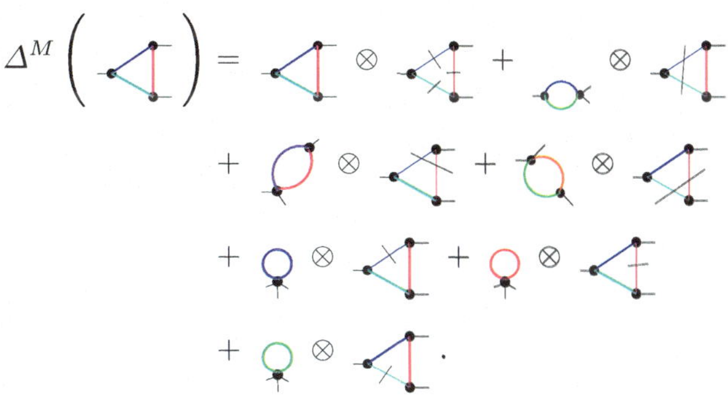

Note that $\Delta^M(M_{i,i}) = M_{i,i} \otimes M_{i,i}$ is group-like, so for example

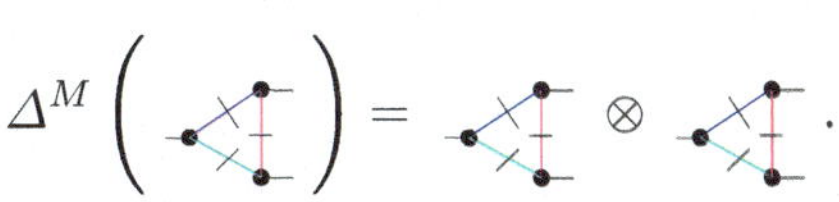

Furthermore the tadpole graphs fulfill

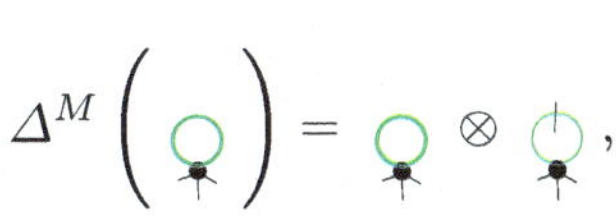

and similar for blue, red.

This construction is generic and constructs a graphical coaction for any graph with any number of loops and legs.

For example a n-loop vertex graph G in Φ^4 theory has $2n$ edges and all spanning trees $T \in \mathcal{T}(G)$ have n edges.

We get $|\mathcal{T}(G)| \times n!$ Hodge matrices $M(G, T_{\mathfrak{O}})$, where the number of spanning trees

$$|\mathcal{T}(G)| = \psi(1, \ldots, 1),$$

is given through the first Symanzik polynomial evaluated at unit arguments.

The required graphical coaction Δ^G then comes as above by summing the individual coactions $\Delta^{M(G,T_{\mathfrak{O}})}$ for $M(G, T_{\mathfrak{O}})$ which corresponds to a construction of a matrix $M = M^G$ from all the matrices $M(G, T_{\mathfrak{O}})$.

The situation simplifies if we use kinematical renormalization schemes which set tadpole graphs to zero and use (see [15])

$$\Phi_R(G) = \sum_{T \in \mathcal{T}(G)} \boldsymbol{\Phi}_R((G, T)). \tag{8}$$

Here on the rhs we use Feynman rules $\boldsymbol{\Phi}_R((G, T))$ integrating the space-like parts of loop momenta after the energy components $k_{i,0}$ have been integrated out as residue integrals. These residue integrals generated the sum over spanning trees on the right [15].

This allows to erase the leftmost column and uppermost row in the matrices $M^{(i)}$ and we get six 3×3 matrices $N^{(i)}$ which we can combine to a matrix N as follows:

$N =$

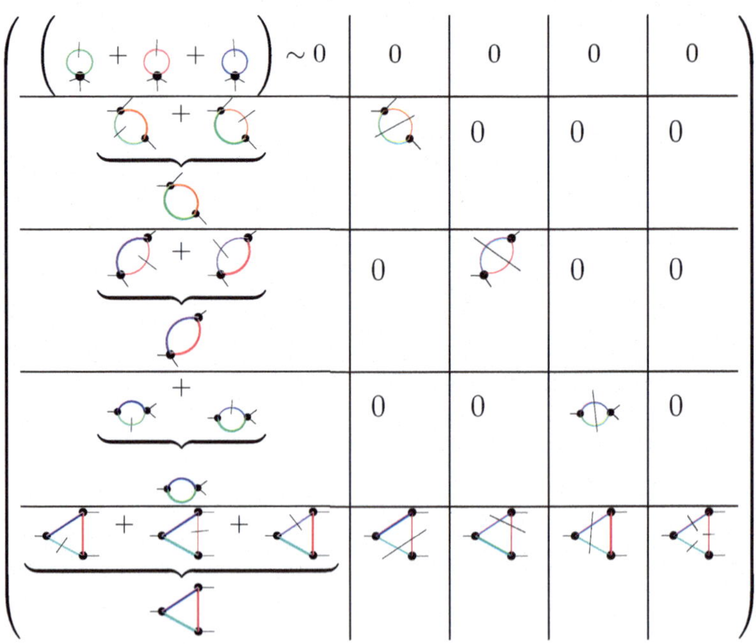

.

Here we could write

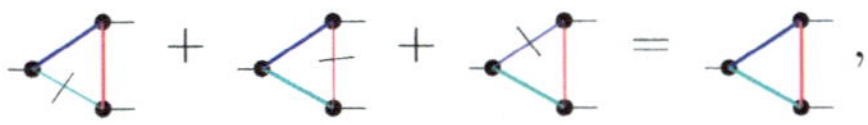

as the sum over residues when doing the $dk_{i,0}$ (contour) integrations for any graph G pairs off with the spanning trees of G by Eq. (8). Here an entry $(G, F) \in H_{GF}$ in the matrix is shorthand for $\Phi_R((G, F))$.

Note that the corresponding coaction Δ^N is utterly based on Cutkosky graphs:

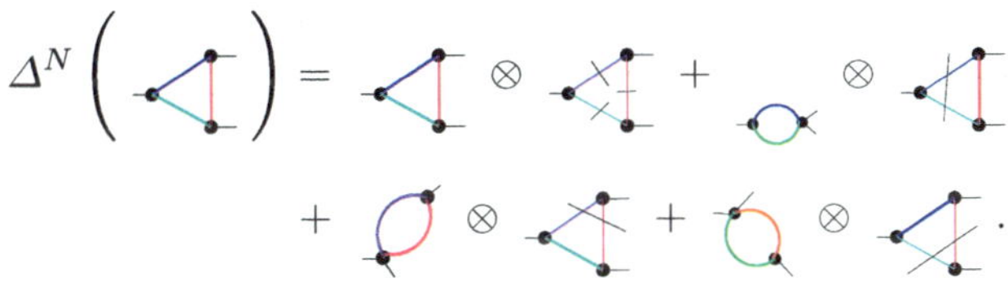

Also,

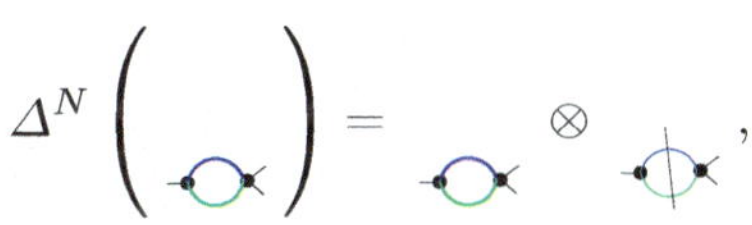

and so on. This is particularly useful when using kinematical renormalization schemes where indeed any tadpole vanishes.

There is much more information in our matrices (where we understand that entries are evaluated by renormalized Feynman rules)

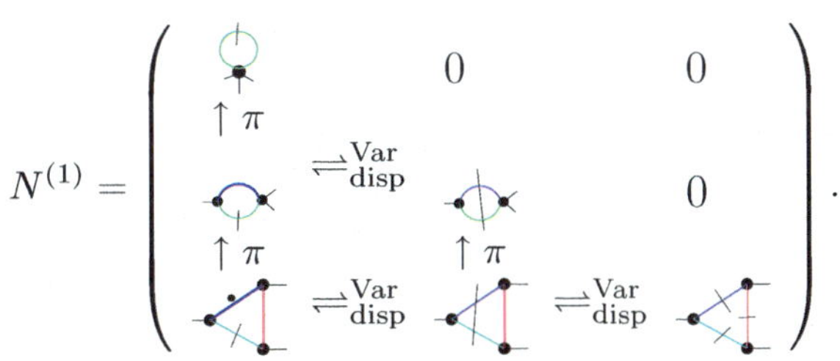

Some properties:

- The boundary d of the cubical chain complex [18] and its action on a graph $G_F \in H_{GF}$ is realized on $\Phi_R(G_F)$ as indicated for $N^{(1)}$ above.

$$d = d_0 + d_1,\ d \circ d = d_0 \circ d_0 = d_1 \circ d_1 = 0,$$

  goes to the right: $\mathrm{Var}(\Phi_R(G_F)) = \Phi_R(d_0(G_F))$, and up: $\pi \circ \Phi_R(G_F) = \Phi_R(d_1(G_F))$, corresponding to m_r and m_c in the coaction.

- Any variation induces a transition in the columns $C_i \to C_{i+1}$ by putting an edge e with quadric $Q(e)$ on the mass-shell. Therefore Hodge matrices.
- This determines a point in a fiber determined by the zero locus $Q(e) = 0$ of the quadric $Q(e)$ assigned to edge e. π is the corresponding projection onto a base space provided by the reduced graph. It also determines a sequence of iterated integrals associated to the order $\mathfrak{o}$ in either parametric or quadric Feynman rules. The next Sec.(3.1) gives an example.
- Any row R_{i+1} is a fibration over R_i by a one-dimensional fiber. For example the z-integral in Eq. (12) is an integral over such a one-dimensional fiber.
- Boundaries of the dispersion integral are provided by the leading singularities stored in M_{D}.

3.1 The Triangle Graph

Consider the one-loop triangle with vertices $\{A, B, C\}$ and edges

$$\{(A, B), (B, C), (C, A)\},$$

and quadrics (in this example we use both p, q to indicate 4-momenta as we are not invoking parametric variables) :

$$P_{AB} = k_0^2 - k_1^2 - k_2^2 - k_3^2 - M_1,$$

$$P_{BC} = (k_0 + q_0)^2 - k_1^2 - k_2^2 - k_3^2 - M_2,$$

$$P_{CA} = (k_0 - p_0)^2 - (k_1)^2 - (k_2)^2 - (k_3 - p_3)^2 - M_3.$$

Here, we Lorentz transformed into the rest frame of the external Lorentz 4-vector $q = (q_0, 0, 0, 0)^T$, and oriented the space like part of $p = (p_0, \mathbf{p})^T$ in the 3-direction: $\mathbf{p} = (0, 0, p_3)^T$.

Using $q_0 = \sqrt{q^2}$, $q_0 p_0 = q_\mu p^\mu \equiv q.p$, $\mathbf{p} \cdot \mathbf{p} = \frac{q.p^2 - p.pq.q}{q^2}$, we can express everything in covariant form whenever we want to.

We consider first the two quadrics P_{AB}, P_{BC} which intersect in $\mathbb{C}^4$.

The real locus we want to integrate is $\mathbb{R}^4$, and we split this as $\mathbb{R} \times \mathbb{R}^3$, and the latter three dimensional real space we consider in spherical variables as $\mathbb{R}_+ \times S^1 \times [-1, 1]$, by going to coordinates $k_1 = \sqrt{s} \sin\phi \sin\theta$,$k_2 = \sqrt{s} \cos\phi \sin\theta$, $k_3 = \sqrt{s} \cos\theta$, $s = k_1^2 + k_2^2 + k_3^2$, $z = \cos\theta$.

We have

$$P_{AB} = k_0^2 - s - M_1,$$

$$P_{BC} = (k_0 + q_0)^2 - s - M_2.$$

So we learn say $s = k_0^2 - M_1$ from the first and

$$k_0 = k_r := \frac{M_2 - M_1 - q_0^2}{2q_0}$$

from the second, so we set

$$s_r := \frac{M_2^2 + M_1^2 + (q_0^2)^2 - 2(M_1 M_2 + q_0^2 M_1 + q_0^2 M_2)}{4q_0^2}.$$

The integral over the real locus transforms to

$$\int_{\mathbb{R}^4} d^4k \rightarrow \frac{1}{2}\int_{\mathbb{R}}\int_{\mathbb{R}_+} \sqrt{s}\delta_+(P_{AB})\delta_+(P_{BC})dk_0 ds \times \int_0^{2\pi}\int_{-1}^{1} d\phi\delta_+(P_{CA})dz.$$

We consider k_0, s to be base space coordinates, while P_{CA} also depends on the fibre coordinate $z = \cos\theta$. Nothing depends on ϕ (for the one-loop box it would).

Integrating in the base and integrating also ϕ trivially in the fibre gives

$$\frac{1}{2}\frac{\sqrt{s_r}}{2q_0}2\pi \int_{-1}^{1} \delta_+(P_{CA}(s = s_r, k_0 = k_r))dz.$$

For P_{CA} we have

$$P_{CA} = (k_r - p_0)^2 - s_r - \mathbf{p}\cdot\mathbf{p} - 2|\mathbf{p}|\sqrt{s_r}z - M_3 =: \alpha + \beta z. \tag{9}$$

Integrating the fibre gives a very simple expression (the Jacobian of the δ-function is $1/(2\sqrt{s_r}|\mathbf{p}|)$, and we are left with the Omnès factor[1]

$$\frac{\pi}{4|\mathbf{p}|q_0} = \frac{\pi}{2\sqrt{\lambda(q^2, p^2, (q+p)^2)}} = \Phi_R\left(\quad \right). \tag{10}$$

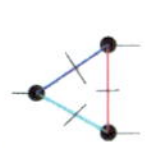

This contributes as long as the fibre variable

$$z = \frac{(k_r - p_0)^2 - s_r - \mathbf{p}\cdot\mathbf{p} - M_3}{2|\mathbf{p}|\sqrt{s_r}} \tag{11}$$

lies in the range $(-1, 1)$. This is just the condition that the three quadrics intersect.

[1] For any 4-vector r we have $r^2 = r_0^2 - \mathbf{r}\cdot\mathbf{r}$. Let q be a time-like 4-vector, p an arbitrary 4-vector. Then, $(q\cdot p^2 - q^2 p^2)/q^2 = \lambda(q^2, p^2, (q+p)^2)/4q^2$ and in the rest frame of q, $(q\cdot p^2 - q^2p^2)/q^2 = \mathbf{p}\cdot\mathbf{p}$ where $\lambda(a, b, c) = a^2 + b^2 + c^2 - 2(ab + bc + ca)$, as always.

An anomalous threshold below the normal threshold appears when $(m_1 - m_2)^2 < q^2 < (m_1 + m_2)^2$.

On the other hand, when we leave the propagator P_{CA} uncut, we have the integral

$$\frac{1}{2}\frac{\sqrt{s_r}}{2q_0}2\pi \int_{-1}^{1} \frac{1}{P_{CA\,(s=s_r,k_0=k_r)}} dz.$$

This delivers a result as foreseen by S-Matrix theory [19, 20].

The two δ_+-functions constrain the k_0- and t-variables, so that the remaining integrals are over the compact domain S^2. Here the fiber is provided by the one-dimensional z-integral and the compactum C_{G/E_F} is the two-dimensional S^2 while for C_G it is the one-dimensional S^1.

As the integrand does not depend on ϕ, this gives a result of the form

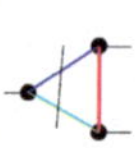

$$\Phi_R\left(\quad \right) = 2\pi C \underbrace{\int_{-1}^{1} \frac{1}{\alpha + \beta z} dz}_{=:J_{CA}} = 2\pi \frac{C}{\beta} \ln \frac{\alpha + \beta}{\alpha - \beta} \qquad (12)$$

$$= \frac{1}{2} \underbrace{\mathrm{Var}(\Phi_R(b_2))}_{\Phi_R\left(\quad \right)} \times J_{CA},$$

where $C = \sqrt{s_r}/2q_0$ is intimately related to $\mathrm{Var}(\Phi_R(b_2))$ for b_2 the reduced triangle graph (the bubble), and the factor $1/2$ here is $\mathrm{Vol}(S^1)/\mathrm{Vol}(S^2)$.

Here, α and β are given through (see Eq. (9)) $l_1 \equiv \mathbf{p}^2 = \lambda(q^2, p^2, (p+q)^2)/4q^2$ and $l_2 := s_r = \lambda(q^2, M_1, M_2)/4q^2$ as

$$\alpha = (k_r - p_0)^2 - l_2 - l_1 - M_3,\ \beta = 2\sqrt{l_1 l_2}.$$

Note that

$$\frac{C}{\beta} = \frac{1}{\sqrt{\lambda(q^2, p^2, (q+p)^2)}} = \frac{1}{2q_0|\mathbf{p}|},$$

in Eq. (12) is proportional to the Omnès factor Eq. (10).

In summary, there is a Landau singularity in the reduced graph in which we shrink P_{CA}. It is located at

$$q_0^2 = s_{normal} = (\sqrt{M_1} + \sqrt{M_2})^2 = s_{\left(\quad \right)}.$$

It corresponds to the threshold divisor defined by the intersection $(P_{AB} = 0) \cap (P_{BC} = 0)$ at the point

$$\left(q^2 = (\sqrt{M_1} + \sqrt{M_2})^2, \quad \underbrace{\frac{A_1}{A_2} = \frac{\sqrt{M_2}}{\sqrt{M_1}}}_{\text{barycenter}\left(\quad \right)} \right)$$

This is not a Landau singularity when we unshrink P_{CA} though. A (leading) Landau singularity appears in the triangle when we also intersect the previous divisor with the locus $(P_{CA} = 0)$.

It has a location which can be computed from the parametric approach. One finds

$$\begin{aligned} q_0^2 &= s_{anom} = (\sqrt{M_1} + \sqrt{M_2})^2 + \\ &+ \frac{4M_3(\sqrt{\lambda_2}\sqrt{M_1} - \sqrt{\lambda_1}\sqrt{M_2})^2 - \big(\sqrt{\lambda_1}(p^2 - M_2 - M_3)}{4M_3\sqrt{\lambda_1}\sqrt{\lambda_2}} \\ &+ \frac{\sqrt{\lambda_2}((p+q)^2 - M_1 - M_3)\big)^2}{4M_3\sqrt{\lambda_1}\sqrt{\lambda_2}} \\ &= s_{\left(\quad \right)}, \end{aligned}$$

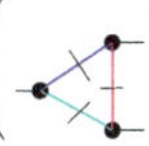

with $\lambda_1 = \lambda(p^2, M_2, M_3)$ and $\lambda_2 = \lambda((p+q)^2, M_1, M_3)$.

Equation (12) above is the promised result: the leading singularity of the reduced graph t/P_{CA} and the non-leading singularity of t have the same location and both involve $\mathrm{Var}(\Phi_R(b_2))$ and the non-leading singularity of t factorizes into the (fibre) amplitude $J_{CA} \times \mathrm{Var}(\Phi_R(b_2))$.

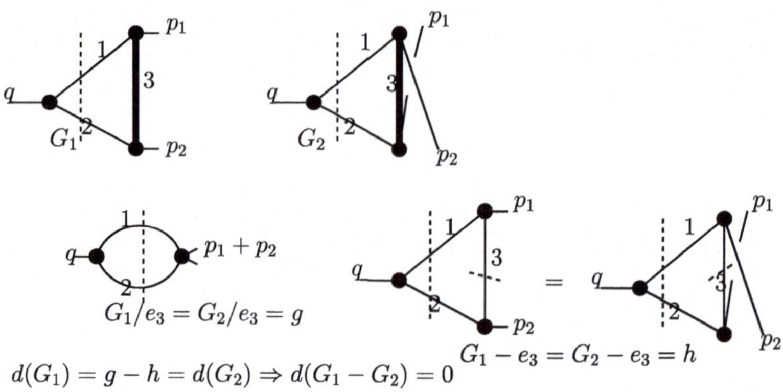

Fig. 2 The two Cutkosky triangle graphs G_1, G_2 are distinguished by a permutation of external edges p_1, p_2. Edges e_1, e_2 are on-shell, e_3 is off-shell and hence in the forest. Shrinking or removing it delivers in both cases the same reduced (g) or leading (h) graph. As a result we get a cycle $d(G_1 - G_2) = 0$. Obviously there is no X such that $dX = G_1 - G_2$

This gives rise to a cycle which is a generator in the cohomology of the cubical chain complex as Fig. 2 demonstrates [15]. As for dispersion, we get a result effectively mapping $C_3 \to C_2 \to C_1$:

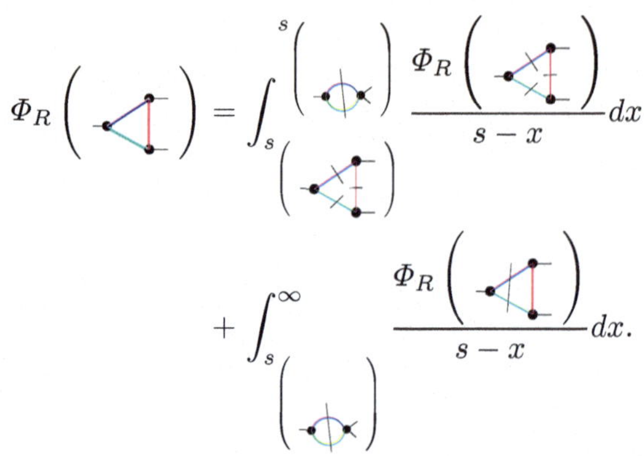

The situation is very similar for the Dunce's cap graph dc. Again we have spanning trees of length two and monodromy generated from partitioning its three vertices in all possible ways by cuts.

Look first at a single term for a chosen ordered spanning tree (Fig. 3):

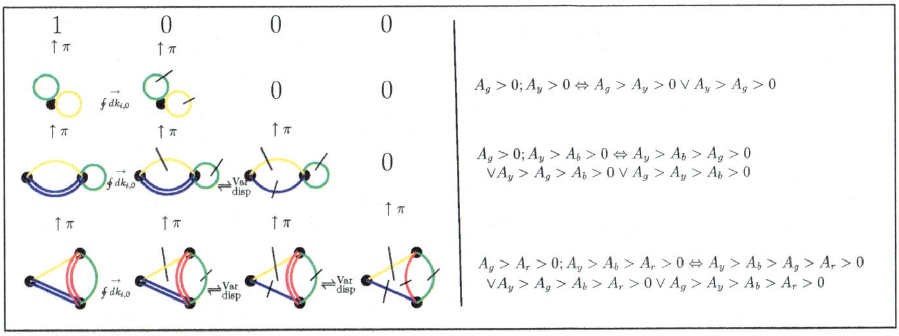

Fig. 3 Dispersion in the Dunce's cap. Here the order is blue before red, so red shrinks first and blue is cut first. Note that due to the presence of more than one loop, choosing a spanning tree (blue, red: the thick double edges) and an order does not single out a single sector as it would in the one-loop case. Here we get three sectors. See the right column. Summing over trees and orders correctly delivers all 24 sectors from the ten ordered spanning trees partitioning them as $24 = 3 + 2 + 3 + 2 + 3 + 2 + 3 + 2 + 2 + 2$ as we see below

3.2 *Summing Up*

We use Eq. (8) where Φ_R has integrated out all energy integrals $\oint \prod_{i=1}^{|G|} dk_{i,0}$ by contour integrations closing in the upper halfplane.

This leads to a graphical coaction:

Theorem 3.4

$$\Delta^G = \sum_{(T,\mathfrak{o})\sim G} \Delta^{G_{T_\mathfrak{o}}}(g),$$

defines a graphical coaction for all $g \in \mathsf{Gal}(M^G)$. For kinematical renormalization schemes it can be written as a coaction on Cutkosky graphs.

Corollary 3.5 *Assume the number of spanning trees equals the number of edges of a graph, $|\mathcal{T}(G)| = e_G$ which is true for one-loop graphs and their duals, banana graphs. We call them simple graphs (in blunt ignorance of the analytical complexity of an n-edge banana graph, $n \geq 3$). Then*

$$\sum_{(T,\mathfrak{o})\sim G} \Delta^{G_{T_\mathfrak{o}}}(G) = \Delta_{\mathsf{Inc}}(G),$$

where Δ_{Inc} is the incidence Hopf algebra and coaction used by Britto et al. [1].

Here,

$$\Delta_{\mathsf{Inc}}(G) = \Delta_{\mathsf{Inc}}(G, \emptyset) = \sum_{\substack{\emptyset \leq X \leq E_G \\ X \neq \emptyset}} (G_X, \emptyset) \otimes (G, X),$$

in their notation. G_X has all edges e contracted, $e \notin X$. In fact one-loop graphs evaluate to dilogs [2] and hence provide the first examples to pull back the coaction from such functions to graphs.

There are non simple graphs in dedicated kinematics (massless internal edges, light-like external momenta) where Δ_{Inc} agrees with Δ^G as well, but not in a generic situation:

Corollary 3.6 *The first non simple graph is the Dunce's cap graph dc with two loops and three vertices. It has four edges and five spanning trees. $\Delta_{\mathsf{Inc}}(dc) \neq \Delta^{dc}(dc)$.*

Similar for all other non simple graphs in generic kinematics.

3.3 *The Dunce's Cap*

Example 3.7 Let us work out the Dunce's cap. We start with Fig. 4.

There are $6 = \binom{4}{2}$ choices for two out of four edges. One of these does not form a possible basis for two loops in the graph, the other five choices determine the five spanning trees of the graph as in Fig. 5. Correspondingly the co-dimension two edge BC is not part of the cell of the Dunce's cap, nor are the four corners. We give one cube as an example in Fig. 6, see also Fig. 7.

For example for the spanning tree T_3 with order blue before red so that we shrink red first we find the matrix $M^{(3)}$ given in Fig. 8. Applying Δ^M is in Fig. 9. If we change the order to red before blue we get a different matrix. Finally the case of a spanning tree on the yellow and blue edge, with order yellow before blue (Figs. 10 and 11). Next we can get rid of dangling tadpole graphs using for example $M_{\mathsf{D}}^{(3)}$ in $M^{(3)}$ using the matrix of Fig. 12. And also use the matrix Fig. 13.

We construct $\tilde{M}^{(3)} = M_{\mathsf{D'}}^{(3)} \times (M_{\mathsf{D}}^{(3)})^{-1} \times M^{(3)}$. We now sum over orders and spanning trees for all $\tilde{M}^{(i)}$, and use hence kinematic renormalization schemes for which we have

$\Phi_R \Bigg($  $\Bigg) = \Phi_R \Bigg($  $\Bigg) = 0.$ (13)

This then allows to eliminate the leftmost column and topmost row from the coaction matrices and allows to sum over spanning trees so that we can formulate the coaction on Cutkosky graphs.

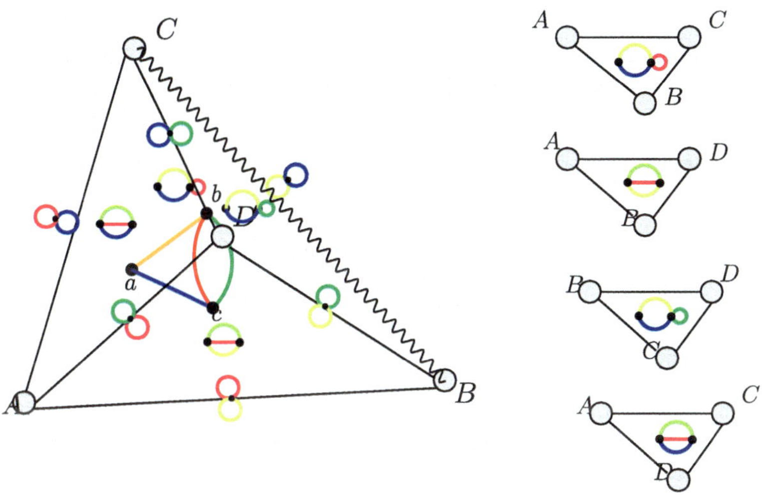

Fig. 4 The Dunce's cap dc and its cell $C(dc)$, a tetrahedron. We also indicate the four triangular cells which are its co-dimension one hypersurfaces. It is a graph on four edges, its cell in OS is thus the three-dimensional tetrahedron $C(dc)$. Its spine gives rise to five two-dimensional cubes Q_i which can not provide a triangulation of $C(dc)$. Instead $C(dc)$ gives rise to a fibration of the cubes Q_i. The spine is a union of ten paths. Six of them give rise to two sectors, and four of them to three sectors, adding up to the 24 sectors in $C(dc)$. Renormalization makes the extra sector in the latter four paths well-defined

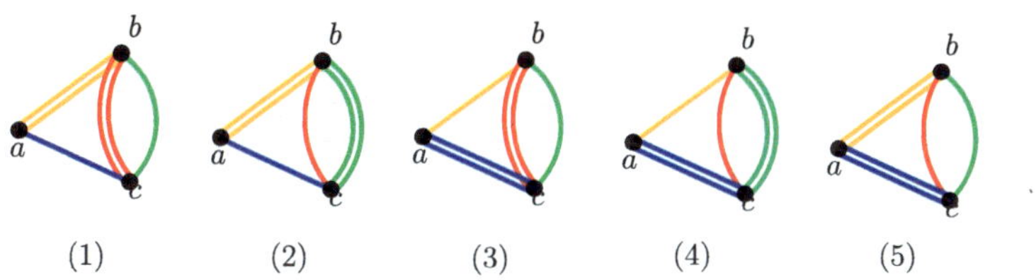

Fig. 5 The five spanning trees of the Dunce's cap dc. They give rise to five cubes Q_i and ten matrices $M(dc, T_\mathfrak{o})$. Spanning trees are on two edges so we get two possible orders and hence ten matrices. Integrating the energy variables indeed generates residues $\sum_T \mathbf{\Phi}_R(G_T)$ for those trees

Again we find a matrix $N = N(dc)$ which defines a coaction which only involves Cutkosky graphs as in Fig. 14.

Remark 3.8 Deformed coactions. Pulling back the known coaction of (elliptic) polylogs to a graphical coaction Britto et.al. find the need to deform their coaction in a systematic way using the parity of the number of edges. We can incorporate this in Δ^G in a similar fashion but attempt at an approach using the $\mathbb{Z}_2$ grading of graph homology in future work (Fig. 15).

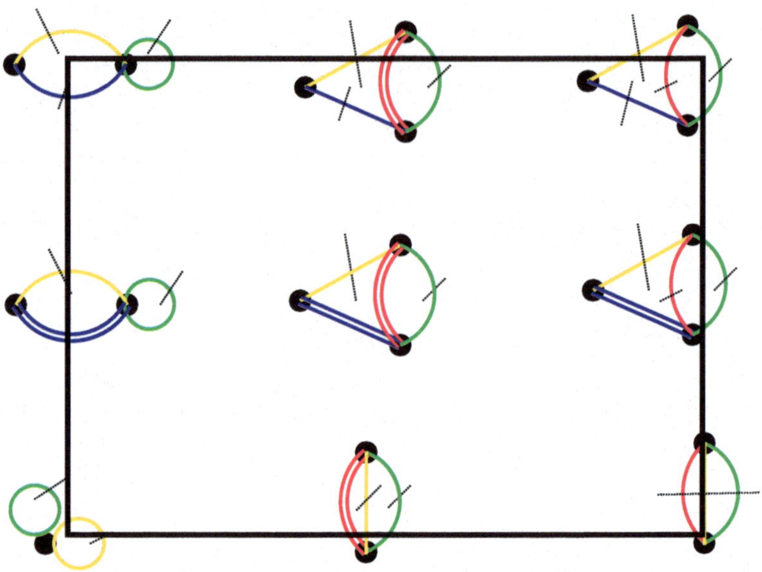

Fig. 6 A cube Q for the graph dc. It gives rise to two matrices $M^{(j)}$. Note that it contains the six entries of $M^{(3)}$. Note that in all nine entries of the cube graphs are evaluated at $A_g m_g = A_y m_y$ and the cube describes a codimension 1 surface of $\mathbb{P}^3 = \mathbb{P}_{dc}$. The one-dimensional fibre which has the cube as base is given by the variable A_y/A_g

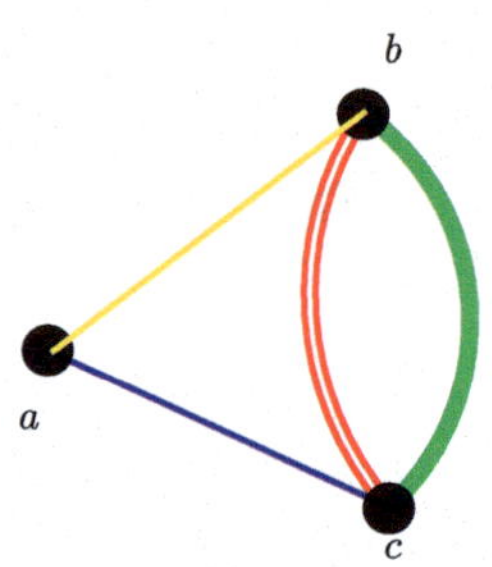

Fig. 7 This is illegal. The green and red edge do not form a spanning tree. Correspondingly there is neither matrix nor residue assigned to this configuration and hence for the nonsimple Dunce's cap graph the coaction Δ_{Inc} of [1] (which includes this graph) deviates from the structure of a cubical complex

Remark 3.9 First entry condition. Steinman relations. Note that to any entry $M_{j,2}$ belongs a 2-partition $V_G = V_G^{(1)} \amalg V_G^{(2)}$. This defines a variable $s = (\sum_{v \in V_G^{(1)}} q(v))^2$. The Matrix M then describes the monodromy of functions $\Phi_R(M_{j,1})$ in the leftmost column through the entries $M_{j,2}$ in the next column when varying this variable s. $M_{j,2}$ are by construction the first entries which have a non-trivial cut each originating from a distinct non-overlapping sector.

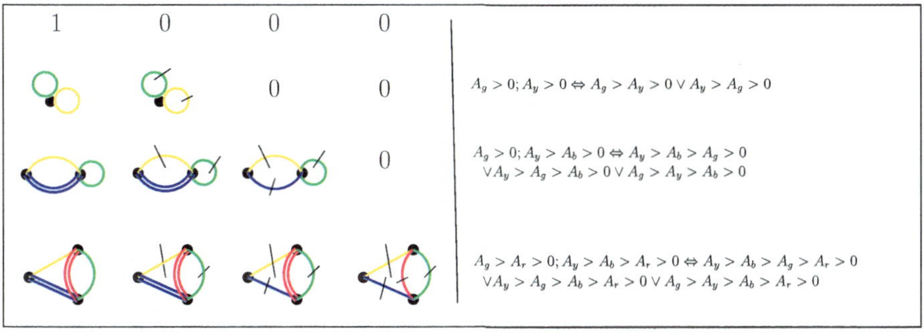

Fig. 8 The matrix $M^{(3)}$ which we had before. We have obviously four such matrices giving three sectors each

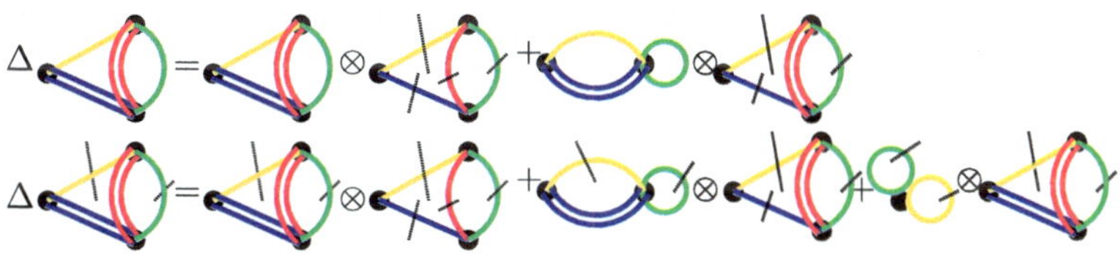

Fig. 9 In the upper row Δ^M acts as a coaction, in the lower as a coproduct

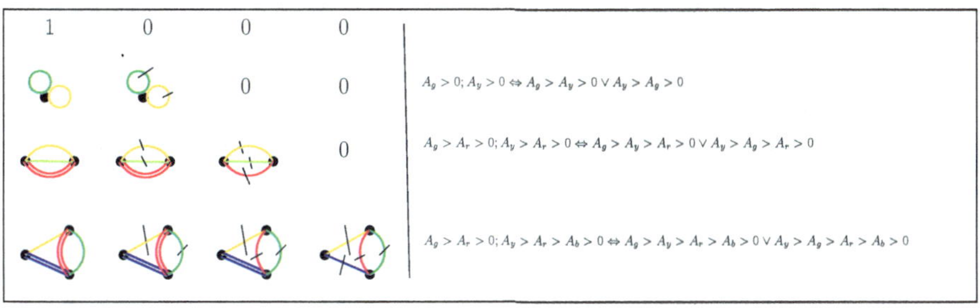

Fig. 10 The matrix $M^{(2)}$. We also give the sectors to which its entries contribute of which there are two and we have four such matrices

One interpretation of the Steinmann relation is that two different 2-partitions which define two different variables s, t indeed do not interfere. The monodromy in a chosen variable s is solely determined by subdividing the associated 2-partition further.

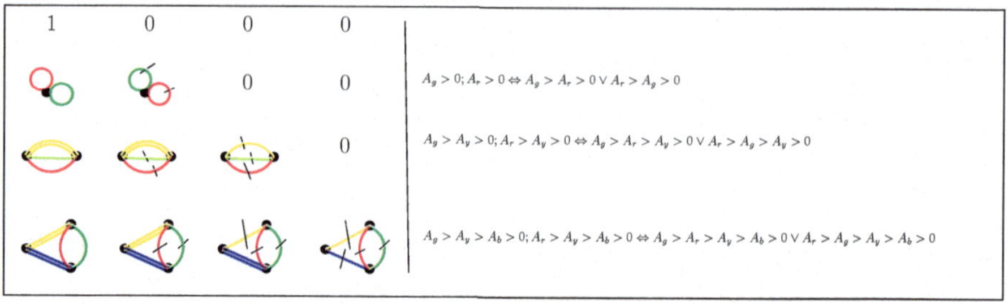

Fig. 11 The matrix $M^{(9)}$. We also give the sectors to which its entries contribute of which there are two and we have two such matrices from the two possible orders

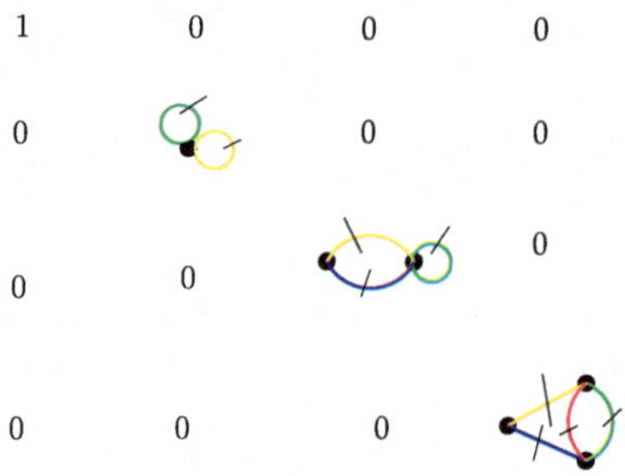

Fig. 12 The matrix $M_{\mathrm{D}}^{(3)}$. Multiplying from the left with its inverse unifies the diagonal and eliminates all tadpoles due to Eq. (13)

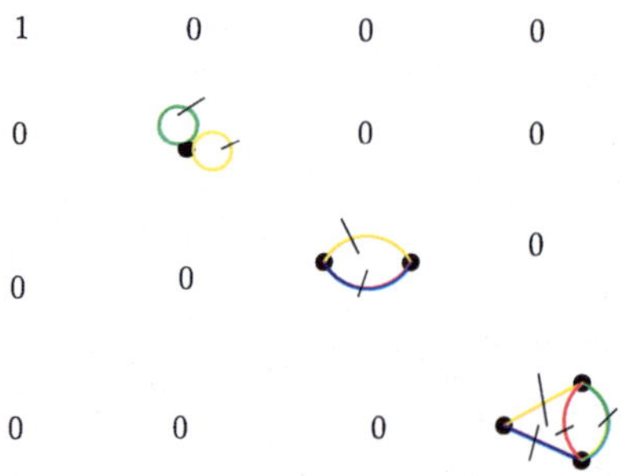

Fig. 13 The matrix $M_{\mathrm{D'}}^{(3)}$. Multiplying it from the right reinserts all diagonal entries apart from tadpoles

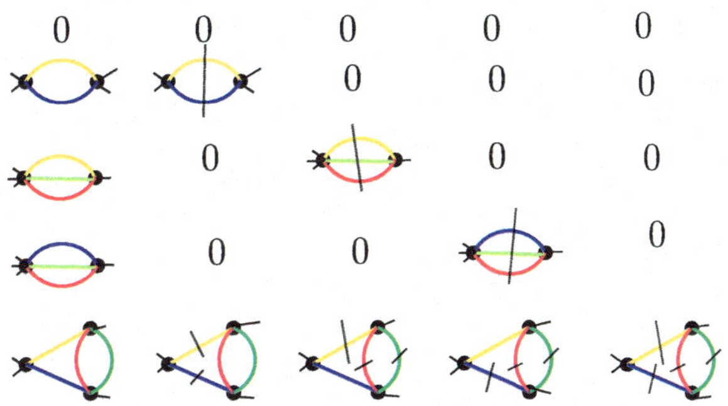

Fig. 14 The matrix $N(dc)$. The second entry in the lowest row is a shorthand given in Fig. 15

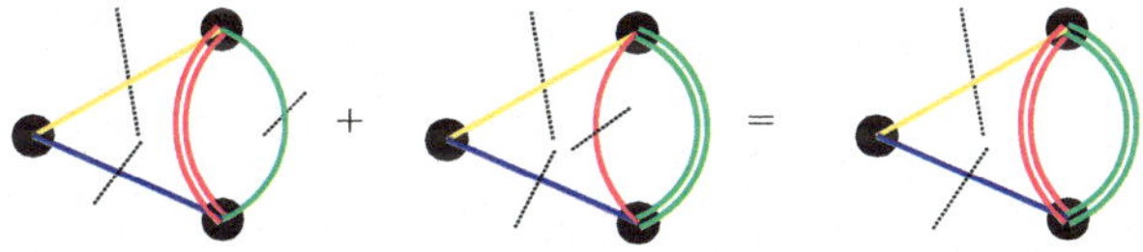

Fig. 15 Integrating the subloop sums two residues by putting either the red or green edge on shell. We can combine this into one entry in the matrix $N(dc)$ thanks to the fact that tadpoles vanish

4 Conclusions

- The cell decomposition of OS together with the corresponding spine provide a cubical complex for Feynman graphs organized by spanning trees.
- Boundaries correspond to either reduced or Cutkosky-cut graphs.
- Each cube has an accompanying simplex decomposition giving Hodge matrices according to a chosen order of edges in a spanning tree.
- Each Hodge matrix defines a coaction.
- Summing over trees and orders defines a coproduct and graphical coaction Δ^G for any Feynman graph G.
- Only in simple cases it agrees with Δ_{Inc}. For generic kinematics Δ_{Inc} is maximally wrong.
- The use of dimensional regularization is neither necessary nor sufficient to find a valid graphical coaction.
- Task: interprete Δ^G in terms of Brown's approach [5, 6] in particular on the possibly not so mysterious rhs (the 'de Rham side').
- Question: What is Brown's small graphs principle making out of the simplifications in kinematic renormalization?
- This so far is a story on principal sheets and variations in the real domain. For a complete understanding in algebraic geometry one must make room for complex variations of masses and kinematics. Need to take into account finer structure of OS. Whilst here we worked with the spine of OS, one needs to consider markings and bordification of OS itself.

Acknowledgments It is a pleasure to thank Marko Berghoff, Spencer Bloch, Karen Vogtmann and Karen Yeats for helpful advice. I thank Spencer in particular for numerous discussions on the matrices investigated here. I am grateful to Johannes Blümlein and all the organizers for their efforts.

Appendix 1: The Cubical Chain Complex

We assume the reader is familiar with the notion of a graph and of spanning trees and forests. See [15] where these notions are reviewed. We follow the notation there. In particular $|G|$ is the number of independent cycles of G, e_G the number of internal edges and v_G the number of vertices of G. For a pair of a graph and a spanning forest we write (G, F) or G_F. If a spanning forest has k connected components we call it a k-forest. A spanning tree T is a 1-forest. Its number of edges hence e_T. $\mathcal{T}(G)$ is the set of spanning trees of G.

Pairs (G, F) are elements of a Hopf algebra H_{GF} based on the core Hopf algebra H_{core} of bridgefree graphs [15].

As an algebra H_{GF} is the free commutative $\mathbf{Q}$-algebra generated by such pairs. Product is disjoint union and the empty graph and empty tree provide the unit.

A k-cube is a k-dimensional cube $[0,1]^k \subsetneq \mathbb{R}^k$.

Consider G_T. We define a cube complex for e_T-cubes Cub_G^T assigned to G. There are $e_T!$ orderings $\mathfrak{o} = \mathfrak{o}(T)$ which we can assign to the internal edges of T.

We define a boundary for any elements $G_F \equiv (G, F)$ of H_{GF}. For this consider such an ordering

$$\mathfrak{o} : E_F \to [1, \dots, e_F]$$

of the e_F edges of F. There might be other labels assigned to the edges of G and we assume that removing an edge or shrinking an edge will not alter the labels of the remaining edges. In fact the whole Hopf algebra structure of H_{core} and H_{GF} is preserved for arbitrarily labeled graphs [21].

The (cubical) boundary map d is defined by $d := d_0 + d_1$ where

$$d_0(G_F^{\mathfrak{o}(F)}) := \sum_{j=1}^{e_F} (-1)^j (G_{F\setminus e_j}^{\mathfrak{o}(F\setminus e_j)}), \quad d_1(G_F^{\mathfrak{o}(F)}) := \sum_{j=1}^{e_F} (G/e_j{}_{F/e_j}^{\mathfrak{o}(F/e_j)}). \tag{14}$$

We understand that all edges $e_k, k \gneq j$ on the right are relabeled by $e_k \to e_{k-1}$ which defines the corresponding $\mathfrak{o}(T/e_j)$ or $\mathfrak{o}(T\setminus e_j)$. Similar if T is replaced by F.

From [18] we know that d is a boundary:

Theorem 4.1 *[18]*

$$d \circ d = 0,\ d_0 \circ d_0 = 0,\ d_1 \circ d_1 = 0.$$

Starting from G_T for any chosen $T \in \mathcal{T}(G)$ each chosen order $\mathfrak{o}$ defines one of $e_T!$ simplices of a e_T-cube Cub_G^T. We write $T_\mathfrak{o}$ for a spanning tree T with a chosen order $\mathfrak{o}$ of its edges. It identifies one such simplex.

Such simplices will each provide one of the lower triangular matrices defining our coactions. If $spt(G)$ is the number of spanning trees of a graph G, we get $spt(G) \times e_T!$ such matrices where we use that $e_T = e_G - |G|$ is the same for all spanning trees T of G, as there are $e_T!$ different matrices for each of the $spt(G)$ different e_T-cubes Cub_G^T.

Appendix 2: The Lower Triangular Matrices $M(G, T_\mathfrak{o})$

Consider a pair $(G, T_\mathfrak{o})$ where G is a bridgeless Feynman graph and $T_\mathfrak{o}$ a spanning tree T of G with an ordering $\mathfrak{o}$ of its edges $e \in E_T$. There are $e_T!$ such orderings where e_T is the number of edges of T.[2]

To such a pair we associate a $(e_T + 1) \times (e_T + 1)$ lower triangular square matrix

$$M = M(G, T_\mathfrak{o})$$

with $M_{ij} \in H_{GF}$.

More precisely, $M_{ij} \in \mathsf{Gal}(M) \equiv \mathsf{Gal}(G, T_\mathfrak{o})$, where $\mathsf{Gal}(G, T_\mathfrak{o}) \subsetneq H_{GF}$ is the set of Galois correspondents of $(G, T_\mathfrak{o})$, i.e. the graphs which can be obtained from G by removing or shrinking edges of T in accordance with $\mathfrak{o}$.

As stated above for a pair (G, T) there are $e_T!$ such matrices $M(G, T_\mathfrak{o})$ generated by the corresponding e_T-cube of the cubical chain complex associated to any pair $(G; T) \in G_F$ [15].

M is defined through its entries $M_{ij} \in \mathsf{Gal}(G, T_\mathfrak{o})$, $j \leq i$,

$$M_{ij} := (G/E_j, (T/E_j \backslash E^i).$$

Here E_j is the set given by the first $(e_T - j + 1)$-entries of the set

$$\{\emptyset, e_{e_T}, e_{j-1}, \ldots, e_1\}$$

and E^i by the first i entries of $\{\emptyset, e_1, \ldots, e_{e_T}\}$. We shrink edges in reverse order and remove them in order.

Define the map $\Delta^M \equiv \Delta^{(G,T_\mathfrak{o})} : \mathsf{Gal}(G, T_\mathfrak{o}) \to \mathsf{Gal}(G, T_\mathfrak{o}) \otimes \mathsf{Gal}(G, T_\mathfrak{o})$,

$$\Delta^{(G,T_\mathfrak{o})} (M)_{jk} = \sum_{i=1}^{e_T+1} (M)_{ik} \otimes (M)_{ji}\,, \tag{15}$$

[2] In the parametric representation $\mathfrak{o}$ orders them by length.

as before. We often omit the superscript $\{\}^{(G,T_{\mathfrak{D}})}$ when not necessary.

Let us now define

$$V_1 =: \mathsf{Gal}(G, T_{\mathfrak{D}})_/ \subsetneq \mathsf{Gal}(G, T_{\mathfrak{D}}),$$

as the $\mathbb{Q}$-span of elements $M_{j,1}$, $j \geq 2$.

Then we can regard the coproduct Δ^M as a coaction

$$\rho_{\Delta^M} : \mathsf{Gal}(G, T_{\mathfrak{D}})_/ \to \mathsf{Gal}(G, T_{\mathfrak{D}})_/ \otimes \mathsf{Gal}(G, T_{\mathfrak{D}}).$$

Soon we will evaluate entries in M_{I} by Feynman rules.

$$(M_{\mathrm{I}})ij \to \Phi_R(M_{ij})/\Phi_R(M_{jj}).$$

This normalization $M \to M_{\mathsf{D}} \times M_{\mathrm{I}}$ to the leading singularities is common [1].

Appendix 3: Summing Orders and Trees

Let us first consider the sectors we are integrating over. A graph G provides $e_G!$ sectors. We partition them as follows. We have $e_G = e_T + |G|$. Then

$$\frac{e_G!}{e_T! \times |G|!} \geq spt(G),$$

with equality only for $|G| = 1$ and the dual of one-loop graphs ('bananas') and $spt(G)$ is the number of spanning trees of G (see also [15]). We note that $e_T! \times |G|!$ is the number of sectors

$$a_{e_i} \geq a_{e_f} \Leftrightarrow e_i \in E_G \setminus E_T \wedge e_f \in E_T,$$

where each edge not in the spanning tree is larger than each edge in the spanning tree. This allows to shrink all e_T edges in the spanning tree in any order in accordance with the spine being a deformation retract in the Culler–Vogtmann Outer Space [9].

The difference

$$e_G! - spt(G) \times e_T! \times |G|!$$

are the sectors where at least one loop shrinks. Any spanning tree T defines a basis of $|G|$ loops l_i, $1 \leq i \leq |G|$, provided by a path p_i in T connecting the two ends of an edge $e_i \in E_G \setminus E_T$. We say that e_i generates l_i.

For any given T the sectors where a loop shrinks fulfill two conditions

(i) for any l_i, $a_{e_i} \geq a_e$, $\forall e \in E_{p_i}$,
(ii) it is not a sector for which $a_{e_i} \geq a_{e_f} \Leftrightarrow e_i \in E_G \setminus E_T \wedge e_f \in E_T$, holds.

The latter condition (ii) ensures that when shrinking e_T edges at least one edge in $E_G \setminus E_T$ and hence a loop shrinks. The former condition (i) ensures that each loop l_i retracts to its generator e_i.

Example 4.2 As an example we consider the Dunce's cap and the wheel with three spokes graph.

The Dunce's cap: Each spanning tree T gives rise to $2! \times 2!$ sectors $e_T! \times |G|!$. There are five spanning trees, so this covers 20 sectors where no loop shrinks. There are four edges in the Dunce's cap so we get 4! sectors. For the four missing sectors four spanning trees provide one each.

The wheel with three spoke graph:

$e_T! \times |G|! = 3! \times 3! = 36$ and there are 16 spanning trees giving us 576 sectors. The 16 spanning trees correspond to 16 choices of three edges while there are $20 = \binom{6}{3}$ such choices altogether. There are $6! = 720$ sectors. The missing $144 = (20 - 16) \times 3! \times 3!$ sectors come from the four triangle subgraphs providing $4 \times 3! \times 3!$ sectors.

This ends our example.

As a result if we let $n(T_{\mathfrak{o}})$ be the number of sectors provided by an ordered spanning tree we have

Lemma 4.3

$$e_G! = \sum_{T \in \mathcal{T}} \sum_{\mathfrak{o}} n(T_{\mathfrak{o}}).$$

It thus makes sense to assign a union of sectors $\mathsf{sec}_{T_{\mathfrak{o}}} = \amalg_{j=1}^{n(T_{\mathfrak{o}})} \mathsf{sec}_j$ to each ordered spanning tree $T_{\mathfrak{o}}$. Here $\mathsf{sec}_j \in \mathcal{SEC}_T^{\mathfrak{o}}$, the set of sectors compatible with T and its order of edges $\mathfrak{o}$.

We have a coaction $\rho_{\Delta^{T^{\mathfrak{o}}}}$ and coproduct $\Delta^{T^{\mathfrak{o}}}$ for each ordered spanning tree $T^{\mathfrak{o}}$ with a corresponding set $\mathsf{Gal}(G, T_{\mathfrak{o}})$ for each.

We define

$$\mathsf{Gal}(G) := \amalg_T \amalg_{\mathfrak{o}} \mathsf{Gal}(G, T_{\mathfrak{o}}).$$

This gives rise to a corresponding matrix M_G formed from $M(G, T_{\mathfrak{o}})$ and corresponding coproduct and coaction Δ^G.

References

1. S. Abreu, R. Britto, C. Duhr, E. Gardi, Diagrammatic Hopf algebra of cut Feynman integrals: the one-loop case. J. High Energy Phys. **1712**, 090 (2017). https://doi.org/10.1007/JHEP12(2017)026. Cite as: arXiv:1704.07931 [hep-th]
2. S. Bloch, D. Kreimer, Feynman amplitudes and Landau singularities for one-loop graphs. Commun. Number Theory Phys. **4**(4), 709–753 (2010)
3. S. Bloch, M. Kerr, P. Vanhove, A Feynman integral via higher normal functions. Compos. Math. **151**, 2329–2375 (2015). https://doi.org/10.1112/S0010437X15007472. arXiv:1406.2664 [hep-th]
4. C. Bogner, S. Müller-Stach, S. Weinzierl, The unequal mass sunrise integral expressed through iterated integrals on $\overline{\mathcal{M}}_{1,3}$. Nucl. Phys. B **954**, 114991 (2020). e-Print: 1907.01251 [hep-th]
5. F. Brown, Feynman amplitudes, coaction principle, and cosmic Galois group. Commun. Number Theory Phys. **11**(3), 453–556 (2017) https://doi.org/10.4310/CNTP.2017.v11.n3.a1
6. F. Brown, Notes on motivic periods. Commun. Number Theory Phys. **11**(3), 557–655 (2017). https://doi.org/10.4310/CNTP.2017.v11.n3.a2
7. F. Brown, Coaction structure for Feynman amplitudes and a small graphs principle (2014). http://www.ihes.fr/~brown/OxfordCoaction.pdf
8. M. Tapuskovic, Motivic Galois coaction and one-loop Feynman graphs (2019). arXiv:1911.01540 [math.AG]
9. M. Culler, K. Vogtmann, Moduli of graphs and automorphisms of free groups. Invent. Math. **84**(1), 91–119 (1986)
10. J. Conant, K. Vogtmann, On a theorem of Kontsevich. Algebr. Geom. Topol. **3**(2), 1167–1224 (2003)
11. J. Conant, A. Hatcher, M. Kassabov, K. Vogtmann, Assembling homology classes in automorphism groups of free groups. Comment. Math. Helv. **91**(4), 751–806 (2016)
12. O. Steinmann, Über den Zusammenhang zwischen den Wightmanfunktionen und den retardierten Kommutatoren. Helv. Phys. Acta **33**, 257 (1960); O. Steinmann, Wightman-Funktionen und retardierten Kommutatoren. II. Helv. Phys. Acta **33**, 347 (1960)
13. W.R. Schmitt, Incidence Hopf algebras. J. Pure Appl. Algebra **96**, 299–330 (1994)
14. W.R. Schmitt, Antipodes and incidence coalgebras. J. Combin. Theory Ser. A **46**, 264–290 (1987)
15. M. Berghoff, D. Kreimer, Graph Complexes and Feynman Rules (2020). arXiv:2008.09540 [hep-th]
16. K. Bloch, Cutkosky Rules and Outer Space (2018). arXiv:1512.01705
17. F. Brown, D. Kreimer, Angles, scales and parametric renormalization. Lett. Math. Phys. **103**, 933-1007 (2013). e-Print: 1112.1180 [hep-th]
18. A. Hatcher, K. Vogtmann, Rational homology of $\mathrm{Aut}(F_n)$. Math. Res. Lett. **5**, 759–780 (1998)
19. M.J.W. Bloxham, D.I. Olive, J.C. Polkinghorne, S-matrix singularity structure in the physical region. III. General discussion of simple Landau singularities. J. Math. Phys. **10**, 553 (1969). https://doi.org/10.1063/1.1664876
20. R. Eden, P. Landshoff, D. Olive, J. Polkinghorne, *The analytic S-matrix* (University Press, Cambridge, 1966)
21. V. Turaev, Loops on Surfaces, Feynman Diagrams, and Trees arXiv:hep-th/0403266

Integration-by-Parts: A Survey

Peter Marquard

Abstract We present an overview of the field of Integration-By-Parts with special emphasis on Laporta's algorithm. We give an overview of the problems associated with Laporta's algorithm and try to illustrate possible ways out.

1 Introduction

The concept of Integration-By-Parts as a tool to facilitate multi-loop calculations was first introduced in [1].

Let us consider a family of Feynman integrals with L loops with corresponding loop momenta k_i and $E+1$ legs of the form

$$\mathcal{J}(a_1, \ldots, a_N, -b_{N+1}, \ldots, -b_I) = \int \left(\prod_{i=1}^{L} d^d k_i\right) \prod_{j=1}^{N} \frac{1}{(P_j^2)^{a_j}} \prod_{j=N+1}^{I} (P_j^2)^{b_j}, \tag{1}$$

where

$$P_j^2 = \left(\sum_{m=1}^{L} A_{jm} k_m + \sum_{m=1}^{E} B_{jm} q_m\right)^2 - m_j^2 . \tag{2}$$

Here, q_i are the external momenta fulfilling momentum conservation

$$0 = \sum_{i=0}^{E+1} q_i, \tag{3}$$

P. Marquard (✉)
DESY, Zeuthen, Germany
e-mail: peter.marquard@desy.de

J. Blümlein, C. Schneider (eds.), *Anti-Differentiation and the Calculation of Feynman Amplitudes*, Texts & Monographs in Symbolic Computation,
https://doi.org/10.1007/978-3-030-80219-6_13

such that we only have E independent momenta and m_i are the masses on the internal lines. In total we then have

$$\mathcal{I} = L(L+1)/2 + LE \tag{4}$$

invariants (scalar products) involving the loop momenta. In order to have well-defined integrals we work in dimensional regularization with $d = 4 - 2\epsilon$.

The Feynman integrals in this family are not independent, but are related by the Integration-By-Parts (IBP) identities and it follows

$$0 = \int \left(\prod_{i=1}^{L} d^d k_i \right) \frac{\partial}{\partial k^\mu} p^\mu \prod_{j=1}^{N} \frac{1}{(P_j^2)^{a_j}} \prod_{j=N+1}^{\mathcal{I}} (P_j^2)^{b_j} \tag{5}$$

with $k \in \{k_i\}$, $p \in \{k_i, q_i\}$. This relation is an immediate consequence of the fact that

$$0 = \int d^d k \frac{\partial}{\partial k^\mu} f(k) \tag{6}$$

in dimensional regularization for a suitable function $f(k)$.

Defining raising and lowering operators $\mathbf{i}^+, \mathbf{i}^-$ with properties

$$\mathbf{i}^+ \mathcal{J}(n_1, \ldots, n_\mathcal{I}) = n_i \mathcal{J}(n_1, , \ldots, n_i + 1, \ldots, n_\mathcal{I}) \tag{7}$$

$$\mathbf{i}^- \mathcal{J}(n_1, \ldots, n_\mathcal{I}) = \mathcal{J}(n_1, , \ldots, n_i - 1, \ldots, n_\mathcal{I}) \tag{8}$$

then the general form of an IBP relation is

$$\begin{aligned} O_{\text{IBP}}(k, p)\mathcal{J}(n_1, \ldots, n_\mathcal{I}) = &\left\{ d\delta_{kp} + \sum C_{ij}\, \mathbf{i}^+ \mathbf{j}^- + \sum D_k(s_{ij}, m_i^2)\, \mathbf{k}^+ \right\} \\ &\times \mathcal{J}(n_1, \ldots, n_\mathcal{I}) \,. \end{aligned} \tag{9}$$

This operator notation has turned out to be particularly useful to find explicit recursions and discuss the mathematical structure of the integration-by-parts relations. The Lie-algebraic structure formed by the IBP operators has been discussed in [2].

Using IBP identities all integrals belonging to the same family can represented as a linear combination of a small set of basis integrals, so-called master integrals M_i

$$\mathcal{J}(c_1, \ldots, c_\mathcal{I}) = \sum_k C_k(d, s_{ij}, m_i) M_k \,, \tag{10}$$

where the coefficients C_k, in general, are rational functions of the space-time dimension d, the kinematic invariants $s_{ij} = (q_i + q_j)^2$, and masses m_i. Thus, Feynman integrals are part of a linear space. Representing an integral by a linear combination of master integrals has been coined *integral reduction* in the literature. How to obtain this *reduction* is the main topic of this article.

IBP reduction has evolved into one of the cornerstone of most multi-loop calculations. We can not give a exhaustive list of reference here but refer the reader to [3], where many examples for state of the art multi-loops calculations are given.

There are two major ways to make use of the IBP identities: find explicit rules that reduce the powers of the propagators or use Laporta's algorithm [4]. Finding explicit reduction rules can be found manually on a case-by-case basis and automated by using `LITERED` [5]. For two very versatile classes of integrals that appear in many applications, massive tadpoles and massless propagators, the corresponding reduction rules have been implemented in publicly available computer programs, e.g. `MATAD` [6], `FMFT` [7], `TARCER` [8], `MINCER` [9, 10], and `FORCER` [11] which have been used in many physics applications, see also [12]. For a possible solution to the reduction problem that uses a different approach see [13]. Laporta's algorithm has also been implemented in public `REDUCE` [14, 15], `FIRE` [16–19], `KIRA` [20, 21]) as well as many private codes and we will focus our attention on this approach in the remainder of this article.

Before we come to the main part of this contribution we like to mention, that integration-by-parts is also a crucial ingredient for the calculation of master integrals using differential equations, cf. [22].

2 Laporta's Algorithm

Laporta's algorithm works, in short, by explicitly generating all Integration-By-Parts identities for a well-chosen set of seed integrals and solving each of the equations in the resulting system of linear equations for the, according to a chosen order, most complicated integral present. This guarantees that we express integrals by simpler ones and finally by only master integrals.

The problem is thus given by solving the linear system of equations

$$\begin{pmatrix} * & * & * & * & \cdots\cdots & * & * \\ * & * & * & * & \cdots\cdots & * & * \\ * & * & * & * & \cdots\cdots & * & * \\ \vdots & \vdots & \vdots & \vdots & \cdots\cdots & \vdots & \vdots \\ \vdots & \vdots & \vdots & \vdots & \cdots\cdots & \vdots & \vdots \\ \vdots & \vdots & \vdots & \vdots & \cdots\cdots & \vdots & \vdots \\ \vdots & \vdots & \vdots & \vdots & \cdots\cdots & \vdots & \vdots \\ * & * & * & * & \cdots\cdots & * & * \\ * & * & * & * & \cdots\cdots & * & * \\ * & * & * & * & \cdots\cdots & * & * \end{pmatrix} \begin{pmatrix} J_1 \\ J_2 \\ J_3 \\ J_4 \\ \vdots \\ \vdots \\ J_{N-1} \\ J_N \end{pmatrix} = \begin{pmatrix} 0 \\ 0 \\ 0 \\ \vdots \\ \vdots \\ \vdots \\ \vdots \\ 0 \\ 0 \\ 0 \end{pmatrix} \tag{11}$$

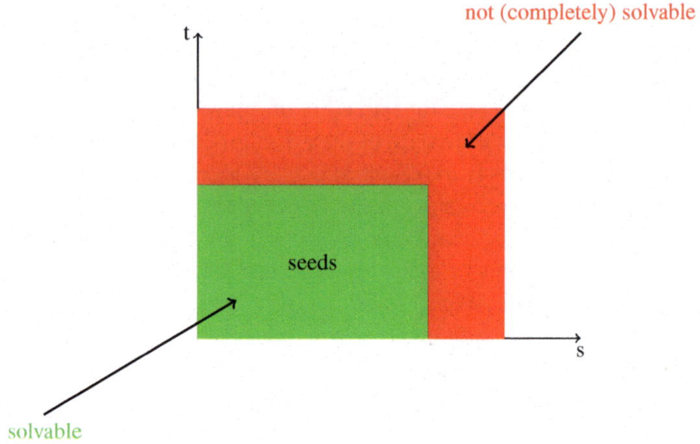

Fig. 1 Integral content of the linear system of equations. s the sum of propagator powers and t the sum of numerator powers

Fig. 2 Example for a highly symmetric case. Removing one of the lines in the diagram on the left-hand side yields one of the diagrams on the right-hand side, which are all identical

or in short

$$\mathcal{M}\mathbf{J} = \mathbf{0}, \tag{12}$$

where $\mathbf{J}$ is the vector of integrals and the coefficient matrix $\mathcal{M}$, which, in general, is an $M \times N$ matrix with $M > N$, but rank $(\mathcal{M}) < N$.

As is depicted in Fig. 1 the integrals contained in the system of equations are, in general, all seed integrals and also integrals where the sum of propagator and/or numerator powers is raised by one unit with respect to the seed integrals. Experience shows that in most cases a complete reduction for the seed integrals can be found but only a partial reduction for the additional integrals.

One important aspect when dealing with Integration-By-Parts identities is, that it does not know anything about symmetries of the Feynman integrals. Consider the example in Fig. 2: Removing any of the lines of the diagram on the left-hand side leads to the same diagram on the right-hand although all of them are given in a different representation in terms of the corresponding propagators. Thus, the diagrams on the right-hand side have to be mapped to each other to avoid redundant calculation of identical sectors.

Using a Gauss-Jordan elimination the system of equations can be transformed to reduced row echelon form. For a dense system of linear equations Gauss elimination scales like $O(N^3)$ with the number of equations or variables, iff we assume constant time for the necessary integer or algebraic arithmetic. Solving the system of equations can become very costly when one considers multi-loop calculations where the number of equations can exceed 10^9. On the other hand of the spectrum in multi-leg calculations many kinematic invariants appear which leads to very complex rational arithmetic.

If we order the integrals in a suitable way the structure of the solved system is given by

$$
\left(\begin{array}{ccc||ccccc||ccc}
0 & \cdots & 0 & 0 & 0 & \cdots & \cdots & 0 & 0 & \cdots & 0 \\
\vdots & & \vdots & \vdots & \vdots & \cdots & \cdots & \vdots & \vdots & & \vdots \\
0 & \cdots & 0 & 0 & 0 & \cdots & \cdots & 0 & 0 & \cdots & 0 \\
\hline\hline
* & \cdots & * & 0 & 0 & \cdots & \cdots & 0 & * & \cdots & * \\
\vdots & & \vdots & \vdots & \vdots & \cdots & \cdots & \vdots & \vdots & & \vdots \\
* & \cdots & * & 0 & 0 & \cdots & \cdots & 0 & * & \cdots & * \\
\hline\hline
0 & \cdots & 0 & 1 & 0 & \cdots & \cdots & 0 & * & \cdots & * \\
0 & \cdots & 0 & 0 & 1 & 0 & \cdots & 0 & * & \cdots & * \\
0 & \cdots & 0 & 0 & 0 & \ddots & \ddots & \vdots & \vdots & & \vdots \\
\vdots & & \vdots & \vdots & \vdots & \ddots & 1 & 0 & * & \cdots & * \\
0 & \cdots & 0 & 0 & 0 & \cdots & 0 & 1 & * & \cdots & *
\end{array}\right)
\left(\begin{array}{c}
J_N \\ \vdots \\ J_{K+1} \\ \hline J_K \\ \vdots \\ J_{M+1} \\ \hline J_M \\ \vdots \\ J_1
\end{array}\right)
=
\left(\begin{array}{c}
0 \\ 0 \\ 0 \\ \vdots \\ \vdots \\ \vdots \\ \vdots \\ \vdots \\ 0 \\ 0 \\ 0
\end{array}\right)
\tag{13}
$$

where the double lines separate parts of the solution that we will discuss in the following. $J_1, \ldots, J_M$ denote the master integrals, $J_{M+1}, \ldots, J_K$ the integrals for which we have obtained a complete reduction to master integrals and finally $J_{K+1}, \ldots, J_N$ the set of integrals present in the system of equations, for which only a partial solution but no complete reduction to master integrals could be achieved. The top row of the matrix indicates that a considerable part of the equations only contains redundant information.

The depicted structure of the solution immediately shows some of the shortcomings of Laparta's algorithm. We will discuss these and possible solutions next.

Redundancy

The identically vanishing first rows of (13) show that we are considering a significant amount of equations that are redundant and do not contain any new information. Since these equations eventually vanish from the system of equations they can be considered a nuisances only and do not pose any significant problems. These equations can also rather easily be eliminated by preconditioning the system of equations with a run, where all variables are replaced by numbers, see [23] for an implementation.

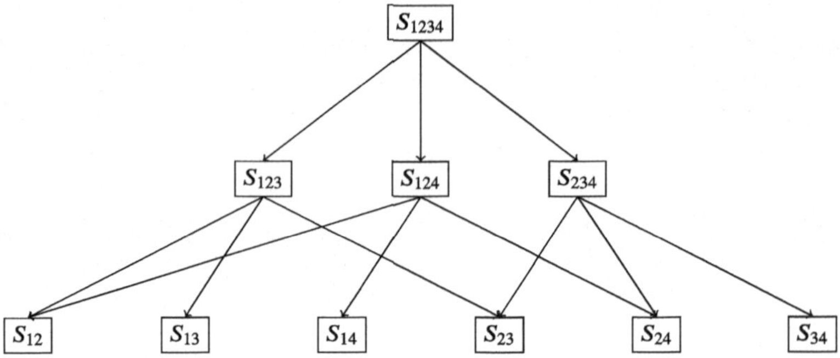

Fig. 3 Hierarchical structure of the Integration-By-Parts identities

Fringe Integrals

As discussed before, the system of equations generated by the chosen seeds integrals contains, in general, integrals that are not part of the set of seed integrals. The presence of these fringe integrals leads to corresponding parts of the system of equation that are dense and in addition contain complicated coefficients. If possible, considering these integrals should be avoided.

The appearance of these fringe integrals can be partially be avoided by using syzygies [24, 25]. Syzygies come into play when we try to choose the operator that generates the Integration-by-Parts identities from the beginning in a way that no integrals with additional propagator or numerator powers are generated. This can help to reduce the number of integrals for which no complete solution can be found.

Parallelization

The structure of the Integration-By-Parts identities immediately generate the structure illustrated in Fig. 3. If we define a sector by the denominators that are present, we find that the system of equations for a given sector only depends on integrals in the sector under considerations and on integrals from simpler sectors, i.e. with a smaller number of denominators. In the given example, this means that in order to solve, e.g., sector S_{124} we only need to consider this sector and in addition the sectors S_{12}, S_{14}, S_{24}. This also means that all sectors in the same row in Fig. 3 can be solved in parallel. In addition to allowing for parallelization this also serves to keep the systems of equations to be solved small.

Finite Fields

Probably the most promising approach to improve and extend the reach of Integration-By-Parts methods employing Laporta's algorithm is based on finite field methods. As it turns out, for complicated problems the rational arithmetics needed to handle and simplify the coefficients of the matrix in every step of the Gauss-Jordan elimination very quickly becomes the bottleneck of the operation. Especially, in intermediate stages of the Gauss-Jordan elimination the problem suffers from a large swell of the expression sizes. This can be overcome by performing the reduction for specific integer values of all the appearing variables and reconstructing the rational functions only in the end. Note, that in the end, many integrals that appear in the system of equations are not needed for the physics problem at hand and thus, only a small fraction of the total number of integrals has to be reconstructed. In addition, since arbitrary precision rational numbers are not very suitable for fast operations also the use of finite fields is advisable. These ideas have successfully been implemented in `FIRE`, `KIRA`, and in private codes. For the rational reconstruction from data given over finite fields `FireFly` [26] is available. It has also been suggested to perform the whole calculation over finite fields and only perform the rational reconstruction at the very end for the final result. A framework to facilitate this in available in `FiniteFlow` [27].

3 Conclusions

We presented an overview of the field of integral reduction using integration-by-parts techniques with particular emphasis on Laporta's algorithm. We discussed the problems related to the special structure of integration-by-parts identities and illustrated possible ways to overcome them. Especially finite field methods are very promising to push the machinery to a higher level.

References

1. K. Chetyrkin, F. Tkachov, Nucl. Phys. B **192**, 159 (1981). https://doi.org/10.1016/0550-3213(81)90199-1
2. R.N. Lee, JHEP **07**, 031 (2008). https://doi.org/10.1088/1126-6708/2008/07/031
3. J. Blümlein, *Analytic integration methods in quantum field theory: an Introduction*, [arXiv:2103.10652 [hep-th]], contribution to this volume
4. S. Laporta, Int. J. Mod. Phys. A **15**, 5087 (2000). https://doi.org/10.1016/S0217-751X(00)00215-7
5. R.N. Lee, *Presenting LiteRed: a tool for the Loop InTEgrals REDuction*, arXiv:1212.2685 [hep-ph]
6. M. Steinhauser, Comput. Phys. Commun. **134**, 335 (2001). https://doi.org/10.1016/S0010-4655(00)00204-6
7. A. Pikelner, Comput. Phys. Commun. **224**, 282 (2018). https://doi.org/10.1016/j.cpc.2017.11.017
8. R. Mertig, R. Scharf, Comput. Phys. Commun. **111**, 265 (1998). https://doi.org/10.1016/S0010-4655(98)00042-3
9. S.A. Larin, F.V. Tkachov, J.A.M. Vermaseren, *The FORM version of MINCER*, NIKHEF-H-91-18 (1991)
10. S.G. Gorishnii, S.A. Larin, L.R. Surguladze, F.V. Tkachov, Comput. Phys. Commun. **55**, 381 (1989). https://doi.org/10.1016/0010-4655(89)90134-3
11. B. Ruijl, T. Ueda, J.A.M. Vermaseren, Comput. Phys. Commun. **253**, 107198 (2020). https://doi.org/10.1016/j.cpc.2020.107198
12. J. Vermaseren, *Some steps towards improving IBP calculations and related topics*, contribution to this volume

13. H. Frellesvig, *Top-down Decomposition: A Cut-based Approach to Integral Reductions*, contribution tro this volume
14. A. von Manteuffel, C. Studerus, *Reduze 2 - Distributed Feynman Integral Reduction*, arXiv:1201.4330 [hep-ph]
15. C. Studerus, Comput. Phys. Commun. **181**, 1293 (2010). https://doi.org/10.1016/j.cpc.2010.03.012
16. A.V. Smirnov, F.S. Chuharev, Comput. Phys. Commun. **247**, 106877 (2020). https://doi.org/10.1016/j.cpc.2019.106877
17. A.V. Smirnov, Comput. Phys. Commun. **189**, 182 (2015). https://doi.org/10.1016/j.cpc.2014.11.024
18. A.V. Smirnov, V.A. Smirnov, Comput. Phys. Commun. **184**, 2820 (2013). https://doi.org/10.1016/j.cpc.2013.06.016
19. A.V. Smirnov, JHEP **10**, 107 (2008). https://doi.org/10.1088/1126-6708/2008/10/107
20. P. Maierhöfer, J. Usovitsch, P. Uwer, Comput. Phys. Commun. **230**, 99 (2018). https://doi.org/10.1016/j.cpc.2018.04.012
21. J. Klappert, F. Lange, Comput. Phys. Commun. **247**, 106951 (2020). https://doi.org/10.1016/j.cpc.2019.106951
22. A.V. Kotikov, *Differential equations and Feynman integrals*, arXiv:2102.07424 [hep-ph], contribution to this volume
23. P. Kant, Comput. Phys. Commun. **185**, 1473 (2014). https://doi.org/10.1016/j.cpc.2014.01.017
24. J. Gluza, K. Kajda, D.A. Kosower, Phys. Rev. D **83**, 045012 (2011). https://doi.org/10.1103/PhysRevD.83.045012
25. B. Agarwal, S.P. Jones, A. von Manteuffel, *Two-loop helicity amplitudes for $gg \to ZZ$ with full top-quark mass effects*, arXiv:2011.15113 [hep-ph]
26. J. Klappert, F. Lange, P. Maierhöfer, J. Usovitsch, *Integral Reduction with Kira 2.0 and Finite Field Methods*, arXiv:2008.06494 [hep-ph]
27. T. Peraro, JHEP **07**, 031 (2019). https://doi.org/10.1007/JHEP07(2019)031

Calculating Four-Loop Corrections in QCD

Sven-Olaf Moch and Vitaly Magerya

Abstract We review the current status of perturbative corrections in QCD at four loops for scattering processes with space- and time-like kinematics at colliders, with specific focus on deep-inelastic scattering and electron-positron annihilation. The calculations build on the parametric reduction of loop and phase space integrals up to four-loop order using computer algebra programs such as `Form`, designed for large scale computations.

1 Introduction

Perturbation theory forms the backbone of theory predictions for scattering processes at high energy colliders. Given the size of the coupling constant α_s in the theory of strong interactions, Quantum Chromodynamics (QCD), this requires the computation of quantum corrections at higher orders. For collisions involving hadrons, either in the initial state or identified in the final state, the theory description is based on QCD factorization, which allows for the separation of long- and short-distance physics. Within this framework, quantum corrections to the hard scattering cross section driven by short-distance physics are calculated typically at the next-to-next-to-leading order (NNLO) in order to reach an accuracy of the order of a few percent from the truncation of the perturbative expansion. The long-distance physics part of the interaction is encoded in hadronic matrix elements which are inaccessible to perturbation theory. Based on the description of a hadron as an incoherent combination of parton states, it is possible, though, to compute matrix elements

S.-O. Moch (✉)
II. Institut für Theoretische Physik, Universität Hamburg, Hamburg, Germany
e-mail: sven-olaf.moch@desy.de

V. Magerya
Institut für Theoretische Physik, Campus Süd, Karlsruher Institut für Technologie (KIT), Karlsruhe, Germany
e-mail: vitalii.maheria@kit.edu

J. Blümlein, C. Schneider (eds.), *Anti-Differentiation and the Calculation of Feynman Amplitudes*, Texts & Monographs in Symbolic Computation,
https://doi.org/10.1007/978-3-030-80219-6_14

of partonic operators and, in particular, their scale dependence in perturbative QCD. The gained knowledge, encoded in splitting functions, serves as input to the parton evolution equations derived from the renormalization group and forms an essential ingredient in the determination of the non-perturbative parton distribution functions (PDFs) or parton fragmentation functions (FFs) from fits to experimental data. The current description of QCD evolution equations for PDFs and FFs is complete at NNLO. This requires the splitting functions in space-like and time-like kinematics at the three-loop level [1–6] as well as the coefficient functions for the hard scattering at two-loop order entering, e.g., in DIS structure functions [7–10] or in fragmentation functions in e^+e^- annihilation [3, 11–14].

With the increasing precision of the experimental data collected at the Large Hadron Collider (LHC) for Standard Model (SM) processes used to extract fundamental theory parameters such as the strong coupling α_s or the PDFs [15], the step towards the next-to-next-to-next-to-leading order (N^{3}LO) becomes necessary. This is particularly crucial in preparation for the physics program at a future Electron-Ion Collider (EIC) [16, 17], where PDFs as well as parton FFs are expected to be accessible with high precision, but also in view of the ongoing future circular collider (FCC) studies [18]. The push beyond the state-of-the-art requires the calculation of four-loop corrections, building on known results for the renormalization of QCD at four-loop [19, 20] and even five-loop order [21–23].

The simplest cross section computations at the four-loop level involve semi-inclusive (single-scale) observables, such as DIS structure functions and e^+e^- fragmentation functions and the current status of their calculation will be discussed and reviewed in detail below.

2 Space-Like Kinematics

The scattering reaction for unpolarized DIS reads

$$l(k) + \text{nucl}(p) \rightarrow l'(k') + X \,, \tag{1}$$

where l and l' denote the scattered lepton and 'nucl' a nucleon with respective momenta k, k' and p. X summarizes the remaining hadronic final states. The inclusive DIS cross section factorizes as $d\sigma \sim L^{\mu\nu} W_{\mu\nu}$ in terms of leptonic and hadronic tensors $L_{\mu\nu}$ and $W_{\mu\nu}$. The latter one encodes the strong interaction dynamics and can be expanded to define the unpolarized structure functions $F_{2,3,L}$,

$$\begin{aligned} W_{\mu\nu}(p,q) &= \frac{1}{4\pi} \int \mathrm{d}^4 z \, \mathrm{e}^{\mathrm{i}q\cdot z} \, \langle \text{nucl}, p | J^\dagger_\mu(z) J_\nu(0) | \text{nucl}, \text{p} \rangle \\ &= \frac{e_{\mu\nu}}{2x_B} \, F_L(x_B, Q^2) + \frac{d_{\mu\nu}}{2x_B} \, F_2(x_B, Q^2) + \mathrm{i} \frac{\epsilon_{\mu\nu pq}}{p \cdot q} \, F_3(x_B, Q^2) \,. \end{aligned} \tag{2}$$

Here J_μ represents an electro-magnetic or weak current. The momentum q is transferred by the gauge-boson with space-like kinematics, $Q^2 \equiv -q^2 > 0$, and the Bjorken variable is defined as

$$x_B = \frac{Q^2}{2p \cdot q}, \tag{3}$$

with $0 < x_B \leq 1$. The symmetric tensors $e_{\mu\nu}$ and $d_{\mu\nu}$ multiplying the structure functions $F_{2,L}$ are dependent on p and q, while the totally antisymmetric one $\epsilon_{\mu\nu\alpha\beta}$ in front of the structure function F_3 arises from the vector/axial-vector interference, see [24, 25] for definitions.

QCD factorization allows for the decomposition of the DIS structure functions in terms of (space-like) coefficient functions $C_{a,\mathrm{f}}$ and PDFs ϕ_f,

$$F_a(x_B, Q^2) = \sum_{\mathrm{f=q,\bar{q},g}} \int_{x_B}^1 \frac{dz}{z}\, \phi_\mathrm{f}\left(\frac{x_B}{z}, \mu^2\right) C_{a,\mathrm{f}}\left(z, \alpha_\mathrm{s}(\mu^2), \frac{\mu^2}{Q^2}\right) + O\left(\frac{1}{Q^2}\right), \tag{4}$$

up to higher-twist corrections $O(1/Q^2)$. The coefficient functions can be computed in perturbation theory via expansions in the strong coupling $a_\mathrm{s} \equiv \alpha_\mathrm{s}/(4\pi)$ as

$$C_{a,\mathrm{f}}(x, \alpha_\mathrm{s}) = \delta(1-x) + a_\mathrm{s}\, c^{(1)}_{a,\mathrm{f}}(x) + a_\mathrm{s}^2\, c^{(2)}_{a,\mathrm{f}}(x) + a_\mathrm{s}^3\, c^{(3)}_{a,\mathrm{f}}(x) + a_\mathrm{s}^4\, c^{(4)}_{a,\mathrm{f}}(x) + \dots, \tag{5}$$

and are completely known up to N^{3}LO [24, 25], i.e. all terms $c^{(3)}_{a,\mathrm{f}}$. At four-loop order a low number of fixed Mellin moments, defined as

$$c(N) = \int_0^1 dx\, x^{N-1}\, c(x), \tag{6}$$

are available [26] as well as the complete soft corrections in the limit $x \to 1$ using threshold resummation and QCD factorization in d-dimensions [27].

The scale dependence of the PDFs is governed by the well-known evolution equations

$$\frac{d}{d\ln\mu^2}\phi_\mathrm{f} = \sum_{\mathrm{f'=q,\bar{q},g}} P_{\mathrm{ff'}}(\alpha_s(\mu^2)) \otimes \phi_{\mathrm{f'}}(x, \mu^2). \tag{7}$$

For QCD with n_f quark flavors and with $'\otimes'$ denoting the standard convolution these are commonly expressed in terms of $2n_f - 1$ scalar equations in the flavor non-singlet case, and a coupled set of 2×2 matrix equations in the flavor singlet case. The evolution kernels, i.e. the space-like splitting functions $P_{\mathrm{ff'}}$ are calculable in perturbative QCD as well,

$$P_{\mathrm{ff'}}(x, \alpha_s) = a_s P^{(0)}_{\mathrm{ff'}}(x) + a_s^2 P^{(1)}_{\mathrm{ff'}}(x) + a_s^3 P^{(2)}_{\mathrm{ff'}}(x) + a_s^4 P^{(3)}_{\mathrm{ff'}}(x) + a_s^5 P^{(4)}_{\mathrm{ff'}}(x) + \dots \,. \tag{8}$$

The NNLO results $P^{(2)}_{\mathrm{ff'}}$ are all known [1, 2]. At N^{3}LO, i.e., at four loops, the non-singlet quark-quark splitting functions have been computed in the large-N_c limit and number of Mellin moments for the remaining color coefficients are known [28, 29] for a general $SU(N_c)$ gauge theory. In the flavor-singlet sector the leading large-n_f terms and those proportional to quartic color Casimirs are known [29, 30]. Beyond this order, even some low-N Mellin moments of the five-loop contributions to the non-singlet quark-quark splitting function $P^{(4)}_{\mathrm{ns}}$ have been determined [31].

In the following, we give a brief overview of the computational set-up and work-flow underlying the computations at four loops and beyond.

2.1 Computational Work-Flow

Using the operator product expansion in DIS one can relate the product of currents J_μ in the hadronic tensor in Eq. (2) to Mellin moments of the structure functions $F_{2,3,L}$, see, e.g. [32]. The latter, parameterizing the (semi)-inclusive cross section, are then obtained with the help of the optical theorem from the imaginary part of the forward Compton amplitude for the gauge boson-nucleon scattering. Thus, the computation of QCD corrections in DIS starts from the forward Compton amplitude of the corresponding gauge boson-parton scattering process, using the kinematics of Eq. (3), which gives access to both, the coefficient functions in Eq. (5) and the splitting functions in Eq. (8). In case, one is only interested in the latter, the direct computation of operator matrix elements, evaluated in parton two-point functions, proves more efficient and allows for the determination of the anomalous dimensions $\gamma(N)$, i.e., the Mellin transforms of the splitting functions, cf. Eq. (6).

The required Feynman diagrams up to four loops can generated using the diagram generator `Qgraf` [33] and the group theory factors for a general color $SU(N_c)$ gauge theory can be obtained with algorithms described in [34]. The loop integrals are considered in dimensional regularization [35, 36], $d = 4 - 2\varepsilon$, which is the standard framework in perturbative QCD at higher orders and the integral reductions are performed by means of integration-by-parts identities (IBP) [37, 38]. The solution of the IBP reductions are encoded in the program `Forcer` [39], which performs a parametric reduction of four-loop massless propagator diagrams to master integrals. The latter are shown in Fig. 1 and their analytic expressions as a Laurent series in ε have been computed in [41, 42]. The symbolic manipulations employ the computer algebra system `Form` [43–45] and its multi-threaded version `TForm` [46] in order to handle both, the run times and the size of the intermediate expressions occurring in the reduction of diagrams with high Mellin moments N.

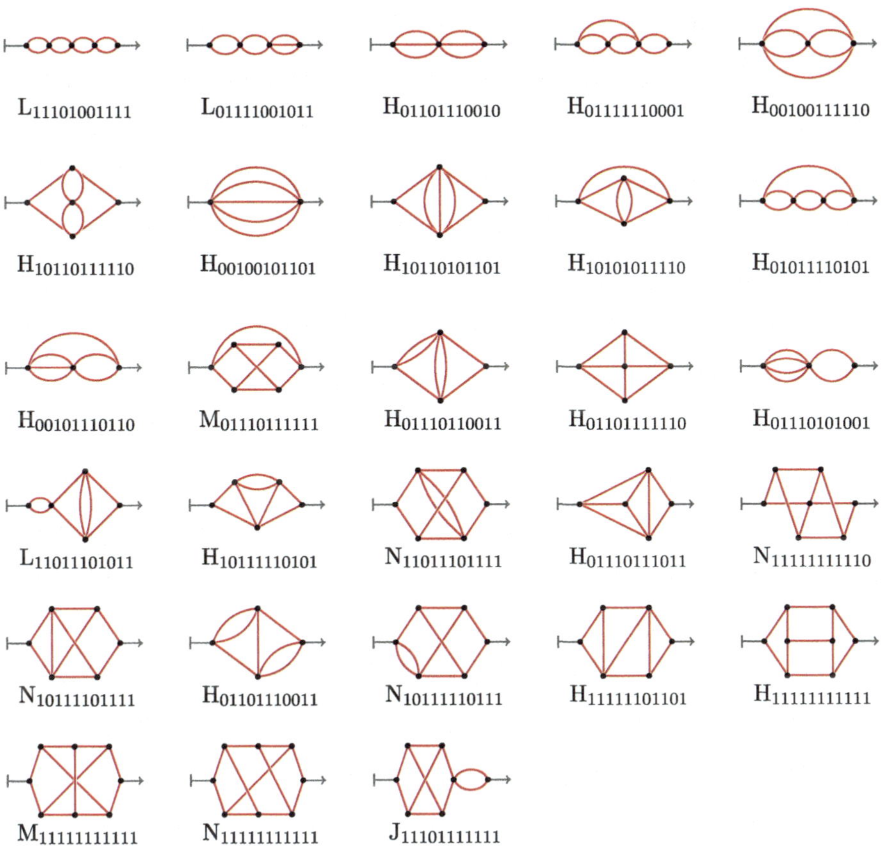

Fig. 1 Master integrals for four-loop propagators (figure from [40])

The approach delivers results for fixed Mellin moments of the anomalous dimensions and DIS coefficient functions. When enough fixed Mellin moments are available, one can follow the approach of [47], and attempt the reconstruction of an analytic expression as a function of N in terms of harmonic sums [48, 49]. In the planar limit, i.e., for large N_c, the exact four-loop results for moments up to $N = 20$ are sufficient to determine the analytic expressions of the non-singlet quark-quark anomalous dimensions $\gamma_{\rm ns}^{(3)}(N)$ as a function of N by `LLL`-based techniques [50–53] and solving systems of Diophantine equations, cf. [28] for details.

The bottleneck of the approach via fixed Mellin moments is caused by the high powers of propagators, which need to undergo the parametric reduction with the program `Forcer` [39]. This leads to large intermediate expressions of the order of TByte and to long run times of the computer algebra system `Form`. For example, the computation of the Mellin moment $N = 10$ of the quark coefficient function in the projection on F_L in Eq. (4) at four loops requires the evaluation $O(3200)$ diagrams with a total of $O(800000)$ hrs CPU time, i.e. almost 100 years altogether.

Fortunately, the multi-threaded version `TForm` delivers an average speed-up factor of $O(10)$ and with a cluster of sufficiently many servers, the problem is doable within half a year of "wall time".

Extensions to five-loop low-N Mellin moments of the non-singlet anomalous dimension $\gamma_{\rm ns}^{(4)}(N)$, i.e. the Mellin transform of $P_{\rm ns}^{(4)}$, as achieved in [31], require the computation of five-loop self-energy integrals, which can be accomplished with an implementation [54] of the local R* operation [55–57]. This allows for the reduction to four-loop integrals, that can be evaluated again by the FORCER program [39]. However, the size of intermediate expressions and the run times of `Form` become prohibitively large beyond the fixed values $N = 2$ and $N = 3$.

3 Time-Like Kinematics

Semi-inclusive e^+e^- annihilation via a virtual photon or Z-boson with time-like momentum q proceeds as

$$e^- + e^+ \;\to\; \gamma/Z(q) \;\to\; h(p) + X \;, \tag{9}$$

where $h(p)$ stands for a specific species of identified hadrons in the final state. The time-like kinematics are characterized by the momentum transfer $Q^2 \equiv q^2 > 0$ and the Feynman variable is

$$x_F \;=\; \frac{2\,p \cdot q}{Q^2} \,, \tag{10}$$

with $0 < x_F \leq 1$. In the center-of-mass frame x_F is the fraction of the beam energy carried by the hadron h. The space- and time-like processes (1) and (9) are related by crossing which implies a mapping $x_B \to x_F$ for the kinematics in Eqs. (3) and (10) and the use of analytic continuation.

In perturbative QCD, the total (angle-integrated) fragmentation function

$$\frac{1}{\sigma_{\rm tot}}\,\frac{{\rm d}\sigma^{\,h}}{{\rm d}x} \;=\; F^h(x, Q^2) \;, \tag{11}$$

as well as the transverse (F_T^h), longitudinal (F_L^h) and asymmetric (F_A^h) ones parameterizing the double-differential cross section $d\sigma^{\,h}/dx\,d\cos\theta_h$ [58], are given by

$$F_a^h(x, Q^2) \;=\; \sum_{\rm f=q,\,\bar{q},\,g} \int_x^1 \frac{{\rm d}z}{z}\; C^T_{a,\rm f}\Big(z, \alpha_{\rm s}(Q^2)\Big)\; D_{\rm f}^h\Big(\frac{x}{z}, Q^2\Big) \;+\; O\Big(\frac{1}{Q}\Big) \,. \tag{12}$$

in terms of the parton fragmentation functions (FFs) D^h_{f} and the (time-like) coefficient functions $C^T_{a,\mathrm{f}}$,

$$C^T_{a,\mathrm{f}}(x, \alpha_{\mathrm{s}}) \;=\; \sigma_{\mathrm{ew}}\big(\delta(1-x) \,+\, a_{\mathrm{s}}\, c^{T,(1)}_{a,\mathrm{f}}(x) \,+\, a_{\mathrm{s}}^2\, c^{T,(2)}_{a,\mathrm{f}}(x) \,+\, a_{\mathrm{s}}^3\, c^{T,(3)}_{a,\mathrm{f}}(x) \,+\, \ldots\big)\,, \tag{13}$$

where σ_{ew} denotes the electroweak prefactors [58] and the second-order coefficient functions are known [3, 11–14], while the three-loop corrections $c^{T,(3)}_{a,\mathrm{f}}(x)$ have not been derived so far.

The parton FFs D^h_{f} obey evolution equations analogous to the PDFs in Eq. (7),

$$\frac{\mathrm{d}}{\mathrm{d}\ln\mu^2}\, D^h_{\mathrm{f}} \;=\; \sum_{\mathrm{f}'=\mathrm{q},\,\bar{\mathrm{q}},\,\mathrm{g}} P^T_{\mathrm{f}'\mathrm{f}}(\alpha_s(\mu^2)) \,\otimes\, D^h_{\mathrm{f}'}(x, \mu^2)\,, \tag{14}$$

with time-like splitting functions, but with 2×2 matrix $P^T_{\mathrm{f}'\mathrm{f}}$ in the flavor singlet case transposed compared to PDFs in Eq. (7). In perturbative QCD the time-like splitting functions can be expanded as

$$P^T_{\mathrm{f}'\mathrm{f}}(x, \alpha_{\mathrm{s}}) \;=\; a_{\mathrm{s}} P^{T,(0)}_{\mathrm{f}'\mathrm{f}}(x) + a_{\mathrm{s}}^2 P^{T,(1)}_{\mathrm{f}'\mathrm{f}}(x) + a_{\mathrm{s}}^3 P^{T,(2)}_{\mathrm{f}'\mathrm{f}}(x) + a_{\mathrm{s}}^4 P^{T,(3)}_{\mathrm{f}'\mathrm{f}}(x) + \ldots\;, \tag{15}$$

where the NNLO results $P^{T,(2)}_{\mathrm{f}'\mathrm{f}}$ are all known [3–6], while at $\mathrm{N}^3\mathrm{LO}$ only the non-singlet quark-quark splitting functions are available in the large-N_c limit [28]. These results are all based on analytic continuation from space- to time-like kinematics and on exploiting reciprocity relations for collinear splitting functions in QCD. In the sequel, we discuss the computational work-flow for the direct calculation of QCD corrections to semi-inclusive e^+e^- annihilation (9).

3.1 *Inclusive Cross-Sections*

A practical indirect way of calculating total cross-sections is the optical theorem. Through it, $O(\alpha_s^3)$ corrections for e^+e^- annihilation in Eq. (9) can be expressed in terms of the four-loop propagator diagrams. In the massless case all the 22 master integrals for these propagators shown in Fig. 1 have been calculated in [41, 42].

The direct way on the other hand requires the calculation of all squared amplitudes with 2, 3, 4, and 5 particles in the final state (with 3, 2, 1, and 0 loops respectively), e.g.

$$\sigma \sim \sum_n \int \mathrm{dPS}_n\, |\langle p_1, \ldots, p_n | iT | q\rangle|^2 = \int \mathrm{dPS}_3 \left| \text{[diagram]} + \text{[diagram]} + \ldots \right|^2 + \ldots, \tag{16}$$

and integration of those over the respective phase space,

$$\mathrm{dPS}_n \equiv \left(\prod_{i=1}^{n} \frac{\mathrm{d}^d p_i}{(2\pi)^{d-1}} \delta^+(p_i^2) \right) (2\pi)^d \, \delta^d(q - \sum_{j=1}^{n} p_j) \, . \tag{17}$$

Performing the phase-space integration analytically quickly turns out to be the bottleneck: the parameterization of 4- and 5- particle phase spaces necessarily requires the introduction of square roots into the integrand, preventing an analytic solution, see, e.g., the "tripole parameterization" of the 4-particle phase space in [59]. Instead one should consider both loop and phase-space integration appearing in the squared amplitude together, as a single "cut" diagram:

(18)

Then, by applying the idea of "reverse unitarity" [60]: that is, replacing on-shell conditions for final state particles in Eq. (17) by denominators,

$$\delta^+(p^2) = \frac{1}{2\pi i} \left(\frac{1}{p^2 - i0} - \frac{1}{p^2 + i0} \right) , \tag{19}$$

one can treat each outgoing line as a "cut propagator", and thus construct IBP relations for cut diagrams.

In this way the direct calculation is reduced to calculating the master integrals for 2-, 3-, 4-, and 5-particle cuts of four-loop propagators. For this task conventional IBP software can be largely reused with two modifications: first, any cut propagator raised to a non-negative power can be set to zero (because $x\,\delta(x) = 0$), and second, when symmetries between diagrams are constructed, cut propagators should not be symmetrized with the regular ones.

The full set of (massless) master integrals for 5-particle cuts of four-loop propagators has been first calculated in [61], for 4- and 3-particle cuts in [40], and for 2-particle cuts in [62–64]. As an example, the set of master integrals for 4-particle cuts is shown in Fig. 2. Because these integrals are single-scale, it is convenient to calculate them by solving dimensional recurrence relations (DRR) [65, 66], which relate the values of these integrals at different values of the space-time dimension d:

$$I_i(d+2) = M_{ii} I_i(d) + \sum_{j \neq i} M_{ij} I_j(d) \, . \tag{20}$$

As explained in [40], there is at most a single master integral per sector, and thus the matrix M_{ij} is triangular. With the help of the "dimensional recurrence and analyticity" method of [67], an ansatz for the full solution can be constructed, leaving only a number of constants undetermined. Once enough extra information is gathered to fix these constants (i.e. values of the leading pole coefficients, or

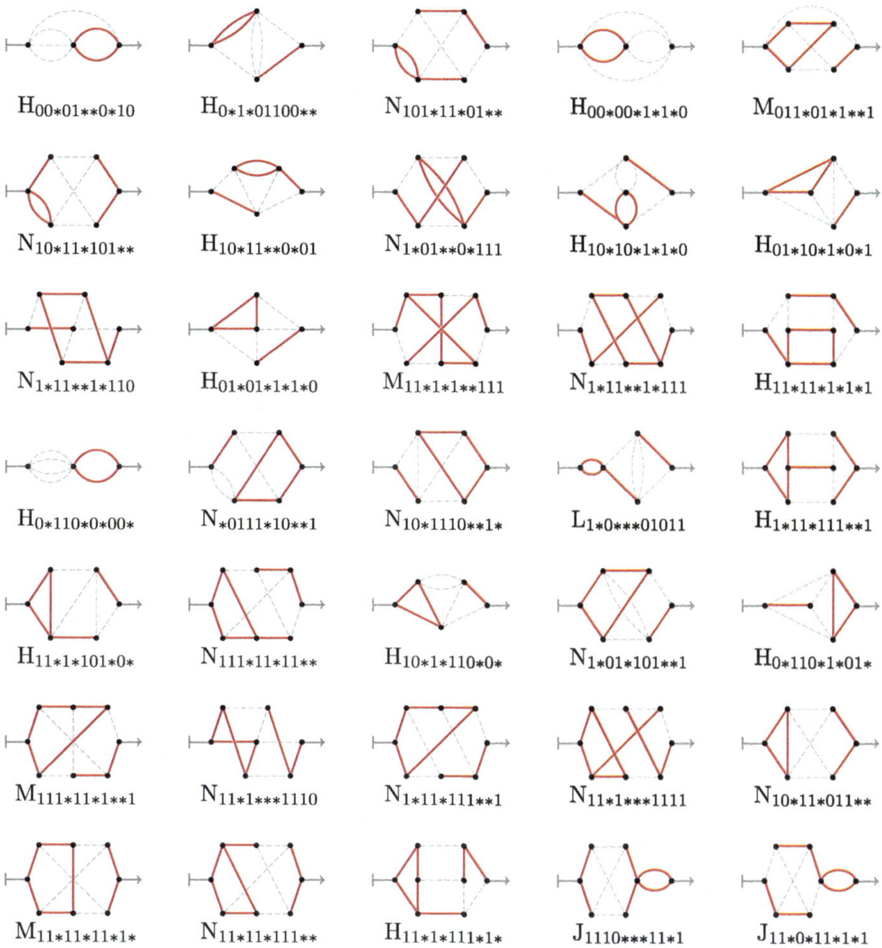

Fig. 2 Master integrals for all four-particle cuts of four-loop propagators (figure from [40])

several terms of the ε-expansion computed by alternative means), `DREAM` [68] can be used to evaluate $I_i(4-2\varepsilon)$ as a series in ε with arbitrary precision (thousands of digits), and these numerical values can then be turned into analytic expressions in terms of multiple zeta values [69] with the help of an integer relation reconstruction algorithm like `PSQL` [70].

The optical theorem (or rather Cutkosky rules [71, 72]), being the alternative way of computing fully inclusive quantities, provides an essential cross-check on these master integrals: the imaginary part of each four-loop propagator diagram must be equal to a combination of its cuts. For example:

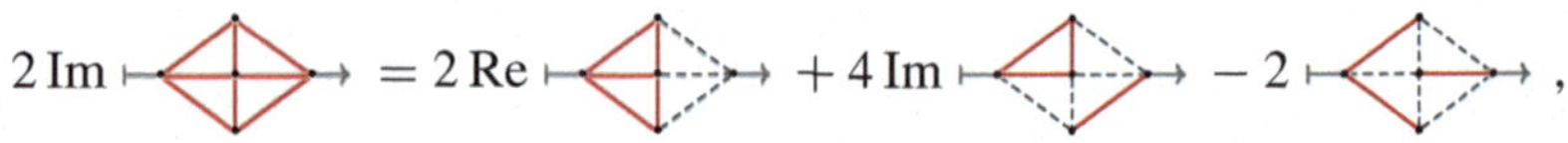

(21)

where the dashed lines indicate the propagators cut according to Eq. (19), see also Fig. 2.

3.2 *Semi-Inclusive Cross-Sections*

Integrals for semi-inclusive cross-sections differ from the inclusive case by the presence of the scaling parameter x (i.e., x_F in Eq. (10)) in the integration measure,

$$\mathrm{dPS}_n(x) \equiv \mathrm{dPS}_n\, \delta(x - 2q \cdot p_1/q^2), \tag{22}$$

so that the semi-inclusive cut diagrams now have the form of

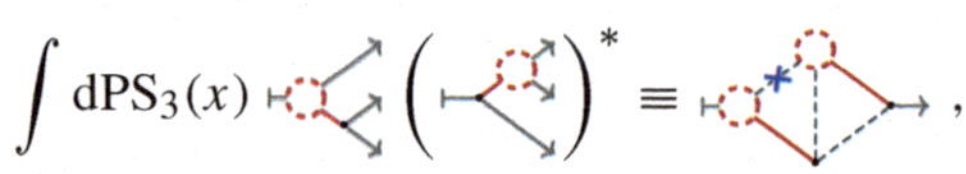

(23)

where the crossed dashed line corresponds to the constraint Eq. (22). The inclusion of x still allows for an IBP reduction if one applies Eq. (19) to transform this additional δ-function into a cut propagator, this time a massive one. This complicates calculations:

- first, by introducing linear dependencies between denominators of a given diagram; these need to be split through partial fractioning, with the end result that a single Feynman diagram can now contribute terms to several different meta topologies;
- second, by the increased number of master integrals: there are 693 semi-inclusive master integrals (298 for 5-particle cuts, 277 for 4-particle cuts, 96 for 3-particle cuts, and 22 for 2-particle cuts) vs. 115 for the inclusive case;
- third, by the increase in the size of IBP expressions, and the increased computational requirements of the reduction;
- and finally by the fact that one can not easily solve DRR for these integrals: the method used for the inclusive case largely relied on numerical evaluation, and having a free parameter x makes that impractical (if not impossible).

The master integrals for semi-inclusive cuts for three-loop propagators were completed in [73, 74] and for four-loop propagators in [75].

A convenient way to calculate the values of these master integrals is the method of differential equations [76, 77]: the integrands of cut master integrals can be differentiated with respect to x, and the obtained expressions can then be reduced back to the same integrals via IBP relations, resulting in systems of differential equations of the form

$$\frac{\partial}{\partial x} I_i(d,x) = M_{ij}(d,x) I_j(d,x)\,. \tag{24}$$

To solve these equations, one follows the observation from [78]: if there is a basis transformation

$$J_i(d,x) = T_{ij}(d,x) I_j(d,x)\,, \tag{25}$$

such that once substituted into Eq. (24) factorizes the dependence of M on $d = 4-2\varepsilon$, transforming the equation into an ε-form,

$$\frac{\partial}{\partial x} J_i(d,x) = \varepsilon S_{ij}(x) J_j(d,x)\,, \tag{26}$$

then the solution can easily found as a series in ε,

$$I_i(4-2\varepsilon, x) \equiv \sum_k \varepsilon^k I_i^k(x), \qquad I_i^{(k)}(x) = \int \mathrm{d}x\; S_{ij}(x) I_j^{(k-1)}(x) + C_i^{(k)}\,. \tag{27}$$

Only two issues remain: how to find the transformation matrix T_{ij} from Eq. (25), and how to fix the integration constants $C_i^{(k)}$.

A general algorithm of constructing ε-form transformations directly from the matrix M_{ij} was presented in [79] and improved upon in [18, 80]. We rely upon the public implementation of this algorithm, `Fuchsia` [81, 82], to find T_{ij}, specifically on the new version available at [83].

To fix the integration constants observe that if one integrates a semi-inclusive integral over all x, the result should be a fully inclusive integral. So by writing down equations of the form

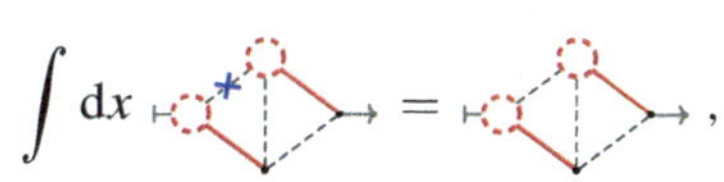
(28)

for each master integral, and inserting the series' in ε for both the semi-inclusive and the (known) inclusive integrals, all $C_i^{(k)}$ can be recovered. The only complication here is that the solution for $I_i(d,x)$ may contain terms $\sim 1/x$, which would make the integral on the left-hand side divergent if taken order-by-order in the series. This can be side-stepped by multiplying the integrand on the left-hand side by x^m with high enough m, and inserting a denominator of the form $\left(2q\cdot p_1/q^2\right)^m$ into the diagram on the right-hand side.

At this stage, it remains to apply the IBP reductions for the semi-inclusive case to the Feynman diagrams of the individual parton processes contributing to the semi-inclusive e^+e^- annihilation in Eq. (9). This will check the NNLO results for the time-like splitting functions $P^{T,(2)}_{\rm f'f}(x)$ in Eq. (15) by a direct computation and determine the hitherto unknown three-loop corrections $c^{T,(3)}_{a,\rm f}(x)$ in Eq. (13).

4 Conclusions

The push towards N^{3}LO accuracy in QCD for DIS structure functions or fragmentation functions in e^+e^- annihilation requires calculations at four-loop order. The efforts are realized with a largely automated work-flow for the generation of all Feynman diagrams, the parametric IBP reduction to master integrals of loop and phase space integrals, for the latter after a mapping with "reverse unitarity" to loop integrals with cuts, and the computation of the master integrals with various algorithms, such as DRR or differential equations. The complexity of the computations, i.e., the size of the expressions, the run times for IBP reductions and the algorithms for the solution of master integrals poses challenges to currently available computer algebra programs and requires continuous improvements. We have presented a brief review of the current status, listing available results as well as indicating the needs for future improvements.

Acknowledgments S.M. is support by Deutsche Forschungsgemeinschaft (DFG) through the Research Unit FOR 2926, "Next Generation pQCD for Hadron Structure: Preparing for the EIC", project MO 1801/5-1.

References

1. S. Moch, J.A.M. Vermaseren, A. Vogt, Nucl. Phys. B **688**, 101 (2004). arXiv:hep-ph/0403192
2. A. Vogt, S. Moch, J.A.M. Vermaseren, Nucl. Phys. B **691**, 129 (2004). arXiv:hep-ph/0404111
3. A. Mitov, S. Moch, A. Vogt, Phys. Lett. B **638**, 61 (2006). arXiv:hep-ph/0604053
4. S. Moch, A. Vogt, Phys. Lett. B **659**, 290 (2008). arXiv:0709.3899
5. A.A. Almasy, S. Moch, A. Vogt, Nucl. Phys. B **854**, 133 (2012). arXiv:1107.2263
6. H. Chen, T.-Z. Yang, H.X. Zhu, Y.J. Zhu, Chin. Phys. C **45**, 043101 (2021). arXiv:2006.10534
7. W.L. van Neerven, E.B. Zijlstra, Phys. Lett. B **272**, 127 (1991)
8. E.B. Zijlstra, W.L. van Neerven, Phys. Lett. B **273**, 476 (1991)
9. E.B. Zijlstra, W.L. van Neerven, Nucl. Phys. B **383**, 525 (1992)
10. S. Moch, J.A.M. Vermaseren, Nucl. Phys. B **573**, 853 (2000). arXiv:hep-ph/9912355
11. P.J. Rijken, W.L. van Neerven, Phys. Lett. B **386**, 422 (1996). arXiv:hep-ph/9604436
12. P.J. Rijken, W.L. van Neerven, Nucl. Phys. B **487**, 233 (1997). arXiv:hep-ph/9609377
13. P.J. Rijken, W.L. van Neerven, Phys. Lett. B **392**, 207 (1997). arXiv:hep-ph/9609379
14. A. Mitov, S.-O. Moch, Nucl. Phys. B **751**, 18 (2006). arXiv:hep-ph/0604160
15. A. Accardi et al., Eur. Phys. J. C **76**, 471 (2016). arXiv:1603.08906
16. D. Boer et al., (2011). arXiv:1108.1713
17. A. Accardi et al., Eur. Phys. J. A **52**, 268 (2016). arXiv:1212.1701
18. A. Blondel et al., Standard model theory for the FCC-ee Tera-Z stage, in *Mini Workshop on Precision EW and QCD Calculations for the FCC Studies : Methods and Techniques*, CERN Yellow Reports: Monographs Vol. 3/2019, Geneva, 2018, CERN. arXiv:1809.01830

19. T. van Ritbergen, J.A.M. Vermaseren, S.A. Larin, Phys. Lett. B **400**, 379 (1997). arXiv:hep-ph/9701390
20. M. Czakon, Nucl. Phys. B **710**, 485 (2005). arXiv:hep-ph/0411261
21. P.A. Baikov, K.G. Chetyrkin, J.H. Kühn, Phys. Rev. Lett. **118**, 082002 (2017). arXiv:1606.08659
22. F. Herzog, B. Ruijl, T. Ueda, J.A.M. Vermaseren, A. Vogt, JHEP **02**, 090 (2017). arXiv:1701.01404
23. T. Luthe, A. Maier, P. Marquard, Y. Schroder, JHEP **10**, 166 (2017). arXiv:1709.07718
24. J.A.M. Vermaseren, A. Vogt, S. Moch, Nucl. Phys. B **724**, 3 (2005). arXiv:hep-ph/0504242
25. S. Moch, J.A.M. Vermaseren, A. Vogt, Nucl. Phys. B **813**, 220 (2009). arXiv:0812.4168
26. B. Ruijl, T. Ueda, J.A.M. Vermaseren, J. Davies, A. Vogt, PoS **LL2016**, 071 (2016). arXiv:1605.08408
27. G. Das, S.-O. Moch, A. Vogt, JHEP **03**, 116 (2020). arXiv:1912.12920
28. S. Moch, B. Ruijl, T. Ueda, J.A.M. Vermaseren, A. Vogt, JHEP **10**, 041 (2017). arXiv:1707.08315
29. S. Moch, B. Ruijl, T. Ueda, J.A.M. Vermaseren, A. Vogt, Phys. Lett. B **782**, 627 (2018). arXiv:1805.09638
30. J. Davies, A. Vogt, B. Ruijl, T. Ueda, J.A.M. Vermaseren, Nucl. Phys. B **915**, 335 (2017). arXiv:1610.07477
31. F. Herzog et al., Phys. Lett. B **790**, 436 (2019). arXiv:1812.11818
32. A.J. Buras, Rev. Mod. Phys. **52**, 199 (1980)
33. P. Nogueira, J. Comput. Phys. **105**, 279 (1993)
34. T. van Ritbergen, A.N. Schellekens, J.A.M. Vermaseren, Int. J. Mod. Phys. A **14**, 41 (1999). arXiv:hep-ph/9802376
35. G. 't Hooft, M.J.G. Veltman, Nucl. Phys. B **44**, 189 (1972)
36. C.G. Bollini, J.J. Giambiagi, Nuovo Cim. B **12**, 20 (1972)
37. F.V. Tkachov, Phys. Lett. B **100**, 65 (1981)
38. K.G. Chetyrkin, F.V. Tkachov, Nucl. Phys. B **192**, 159 (1981)
39. B. Ruijl, T. Ueda, J.A.M. Vermaseren, Comput. Phys. Commun. **253**, 107198 (2020). arXiv:1704.06650
40. V. Magerya, A. Pikelner, JHEP **12**, 026 (2019). arXiv:1910.07522
41. P.A. Baikov, K.G. Chetyrkin, Nucl. Phys. B **837**, 186 (2010). arXiv:1004.1153
42. R.N. Lee, A.V. Smirnov, V.A. Smirnov, Nucl. Phys. B **856**, 95 (2012). arXiv:1108.0732
43. J.A.M. Vermaseren, (2000). arXiv:math-ph/0010025
44. J. Kuipers, T. Ueda, J.A.M. Vermaseren, J. Vollinga, Comput. Phys. Commun. **184**, 1453 (2013). arXiv:1203.6543
45. B. Ruijl, T. Ueda, J. Vermaseren, (2017). arXiv:1707.06453
46. M. Tentyukov, J.A.M. Vermaseren, Comput. Phys. Commun. **181**, 1419 (2010). arXiv:hep-ph/0702279
47. V.N. Velizhanin, Nucl. Phys. B **864**, 113 (2012). arXiv:1203.1022
48. J.A.M. Vermaseren, Int. J. Mod. Phys. A **14**, 2037 (1999). arXiv:hep-ph/9806280
49. J. Blümlein, S. Kurth, Phys. Rev. D **60**, 014018 (1999). arXiv:hep-ph/9810241
50. A.K. Lenstra, H.W. Lenstra, L. Lovász, Mathematische Annalen **261**, 515 (1982)
51. K. Matthews, (unpublished), summarized in [52]; see pp. 16/17
52. J.H. Silverman, Des. Codes Cryptography **20**, 5 (2000)
53. http://www.numbertheory.org/calc/krm_calc.html
54. F. Herzog, B. Ruijl, JHEP **05**, 037 (2017). arXiv:1703.03776
55. K.G. Chetyrkin, F.V. Tkachov, Phys. Lett. B **114**, 340 (1982)
56. K.G. Chetyrkin, V.A. Smirnov, Phys. Lett. B **144**, 419 (1984)
57. K.G. Chetyrkin, (2017). arXiv:1701.08627
58. P. Nason, B.R. Webber, Nucl. Phys. B **421**, 473 (1994) [Erratum: Nucl. Phys.B 480, 755 (1996)]
59. A. Gehrmann-De Ridder, T. Gehrmann, G. Heinrich, Nucl. Phys. B **682**, 265 (2004). arXiv:hep-ph/0311276

60. C. Anastasiou, K. Melnikov, Nucl. Phys. B **646**, 220 (2002). arXiv:hep-ph/0207004
61. O. Gituliar, V. Magerya, A. Pikelner, JHEP **06**, 099 (2018). arXiv:1803.09084
62. G. Heinrich, T. Huber, D. Maitre, Phys. Lett. B **662**, 344 (2008). arXiv:0711.3590
63. G. Heinrich, T. Huber, D.A. Kosower, V.A. Smirnov, Phys. Lett. B **678**, 359 (2009). arXiv:0902.3512
64. R.N. Lee, A.V. Smirnov, V.A. Smirnov, JHEP **04**, 020 (2010). arXiv:1001.2887
65. O.V. Tarasov, Phys. Rev. D **54**, 6479 (1996). arXiv:hep-th/9606018
66. O.V. Tarasov, Nucl. Phys. B Proc. Suppl. **89**, 237 (2000). arXiv:hep-ph/0102271
67. R.N. Lee, Nucl. Phys. B **830**, 474 (2010). arXiv:0911.0252
68. R.N. Lee, K.T. Mingulov, (2017). arXiv:1712.05173
69. J. Blümlein, D.J. Broadhurst, J.A.M. Vermaseren, Comput. Phys. Commun. **181**, 582 (2010). arXiv:0907.2557
70. H.R.P. Ferguson, D.H. Bailey, S. Arno, Math. Comput. **68**, 351 (1999)
71. R.E. Cutkosky, J. Math. Phys. **1**, 429 (1960)
72. G. 't Hooft, M.J.G. Veltman, NATO Sci. Ser. B **4**, 177 (1974)
73. O. Gituliar, JHEP **02**, 017 (2016). arXiv:1512.02045
74. O. Gituliar, S. Moch, Acta Phys. Polon. B **46**, 1279 (2015). arXiv:1505.02901
75. V. Magerya, (2021). PhD thesis (Universität Hamburg)
76. A.V. Kotikov, Phys. Lett. B **254**, 158 (1991)
77. A.V. Kotikov, Phys. Lett. B **267**, 123 (1991) [Erratum: Phys. Lett. B **295**, 409–409 (1992)]
78. J.M. Henn, Phys. Rev. Lett. **110**, 251601 (2013). arXiv:1304.1806
79. R.N. Lee, JHEP **04**, 108 (2015). arXiv:1411.0911
80. R.N. Lee, A.A. Pomeransky, (2017). arXiv:1707.07856
81. O. Gituliar, V. Magerya, Comput. Phys. Commun. **219**, 329 (2017). arXiv:1701.04269
82. O. Gituliar, V. Magerya, PoS **LL2016**, 030 (2016). arXiv:1607.00759
83. https://github.com/magv/fuchsia.cpp

Contiguous Relations and Creative Telescoping

Peter Paule

Abstract This article presents an algorithmic theory of contiguous relations. Contiguous relations, first studied by Gauß, are a fundamental concept within the theory of hypergeometric series. In contrast to Takayama's approach, which for elimination uses non-commutative Gröbner bases, our framework is based on parameterized telescoping and can be viewed as an extension of Zeilberger's creative telescoping paradigm based on Gosper's algorithm. The wide range of applications include elementary algorithmic explanations of the existence of classical formulas for non-terminating hypergeometric series such as Gauß, Pfaff-Saalschütz, or Dixon summation. The method can be used to derive new theorems, like a non-terminating extension of a classical recurrence established by Wilson between terminating ${}_4F_3$-series. Moreover, our setting helps to explain the non-minimal order phenomenon of Zeilberger's algorithm.

1 Preamble

A first version of this article [20] has been produced about 20 years ago. At the occasion of the Wolfgang Pauli Centre Workshop "Antidifferentiation and the Calculation of Feynman Amplitudes" at DESY (Deutsches Elektronen-Synchrotron, Zeuthen, October 4–9, 2020) the organizers Johannes Blümlein and Carsten Schneider asked (and encouraged!) me to speak about this work. Without their initiative this updated version of [20] would have never appeared.

P. Paule (✉)
Johannes Kepler University Linz, Research Institute for Symbolic Computation (RISC), Linz, Austria
e-mail: Peter.Paule@risc.jku.at

J. Blümlein, C. Schneider (eds.), *Anti-Differentiation and the Calculation of Feynman Amplitudes*, Texts & Monographs in Symbolic Computation,
https://doi.org/10.1007/978-3-030-80219-6_15

2 Introduction

Contiguous relations are a fundamental concept within the theory of hypergeometric series and orthogonal polynomials; see, for instance, [1]. An example in connection with the thematical scope of this volume is [13], which is contained in this book and whose subsection 2.2 is devoted to contiguous relations for multivariate hypergeometric functions.

As often, the story begins with Gauß [10]. Let

$$ {}_2F_1\left(\begin{matrix} a, b \\ c \end{matrix}; z\right) = \sum_{k=0}^{\infty} \frac{(a)_k (b)_k}{(c)_k k!} z^k, $$

where $(x)_k$ is the shifted factorial

$$ (x)_k = x(x+1)\cdots(x+k-1) \text{ if } k \geq 1 \text{ and } (x)_0 = 1. $$

Gauß defined two such ${}_2F_1$ series as *contiguous*, if two of the parameters are pairwise equal, and if the third pair differs by 1. In particular, Gauß showed that a ${}_2F_1$ series and any two others contiguous to it are linearly related. For instance,

$$ (a-c)\,{}_2F_1\left(\begin{matrix} a-1, b \\ c \end{matrix}; z\right) + (c - 2a - (b-a)z)\,{}_2F_1\left(\begin{matrix} a, b \\ c \end{matrix}; z\right) $$
$$ + a(1-z)\,{}_2F_1\left(\begin{matrix} a+1, b \\ c \end{matrix}; z\right) = 0, \tag{1} $$

is the first entry [10, 7.1] in Gauß' list of fifteen ($= 6 \cdot 5/2$) fundamental contiguous relations. Moreover, in Section 11 of [10] Gauß describes how to obtain relations between

$$ {}_2F_1\left(\begin{matrix} a, b \\ c \end{matrix}; z\right), {}_2F_1\left(\begin{matrix} a+\lambda, b+\mu \\ c+\nu \end{matrix}; z\right) \text{ and } {}_2F_1\left(\begin{matrix} a+\lambda', b+\mu' \\ c+\nu' \end{matrix}; z\right), $$

where the $\lambda, \lambda', \mu, \mu', \nu, \nu'$ are integers taken from $\{-1, 0, 1\}$. This gives in total 325 ($= 26 \cdot 25/2$) relations.[1]

Today Gauß' notion of contiguous relations is used in a more general sense. Namely, two ${}_pF_q$ series, i.e.,

$$ {}_pF_q\left(\begin{matrix} a_1, \ldots, a_p \\ b_1, \ldots, b_q \end{matrix}; z\right) = \sum_{k=0}^{\infty} \frac{(a_1)_k \cdots (a_p)_k}{(b_1)_k \cdots (b_q)_k} \frac{z^k}{k!}, $$

[1] The number of these relations can be reduced further by taking symmetries (e.g., swapping a and b) into account.

are said to be *contiguous* if their parameters differ by integers. A first systematic textbook treatment of contiguous relations was presented by E. Rainville [25]. For an excellent up-to-date account the reader is referred to the book by G.E. Andrews et al. [1].

In particular, this book has played an influential role for the work presented in this article. Section 3.12 of [1] is devoted to summation, in particular, to a comparison of the classic method of contiguous relations to the W–Z method, more precisely, to Zeilberger's algorithm [41] also called *creative telescoping*. It is pointed out that "the W–Z method is an effective algorithm for discovering useful instances of contiguous relations." Moreover, it is explained that a specific use of contiguous relations, called *Pfaff's method*, can serve as a valuable alternative. Namely, "There is a somewhat different summation method due to Pfaff. This method is less algorithmic than the W–Z method. However, it spreads out the algebraic complications to systems of recurrences. Consequently, it may provide new summations in addition to the one we wish to prove and it may allow the required algebra to be considerably simpler than that required by the W–Z method. Pfaff's method rather resembles the W–Z method; however, it allows the various additional parameters in the summation to play an important role" [1, p. 171].

In order to illustrate this point, various examples are given. We consider one of these, namely Bailey's summation of a balanced ${}_4F_3$ series,

$$S(n) = S(n,a,b) = {}_4F_3\left(\begin{matrix} a/2, (a+1)/2, b+n, -n \\ b/2, (b+1)/2, a+1 \end{matrix}; 1\right) = \frac{(b-a)_n}{(b)_n}; \quad n \geq 0; \tag{2}$$

see [1, (3.11.7)]. By $F(n,k)$ we denote the kth summand of this ${}_4F_3$ series.

As a note, when dealing with such hypergeometric series we always assume that the summand terms are well-defined; this convention is made explicit also in (16). In view of the fact that the given sum is terminating at n, in this particular case, $a+1+\ell \neq 0$, $b/2+\ell \neq 0$, and $(b+1)/2+\ell \neq 0$ for $0 \leq \ell \leq n$.

Following the presentation in [1], we first prove (2) by applying Zeilberger's algorithm[2] which computes the telescoping summand recurrence

$$\begin{aligned}(n+1)(-n-b+a)\,F(n,k) + (-a^2+ab-a+2nb+3b+2+4n+2n^2)\,F(n+1,k)\\ -(b+n+1)(a+n+2)\,F(n+2,k) = \Delta_k\,G(n,k-1),\end{aligned} \tag{3}$$

where the (forward) difference operator is defined as usual as

$$\Delta_k\,f(k) = f(k+1) - f(k), \tag{4}$$

[2]We are using the Mathematica package "fastZeil" presented in [22]; it is freely available as described at https://combinatorics.risc.jku.at/software.

and where

$$G(n,k) = -\frac{(a+1+2k)(a+2k)(b+n+k)(n+1)}{(b+n)(n-k+1)}F(n,k). \tag{5}$$

Summing (3) over k from 0 to n implies a recurrence for the sum $S(n)$,

$$(n+1)(-n-b+a)\,S(n)+(-a^2+ab-a+2nb+3b+2+4n+2n^2)\,S(n+1)$$
$$-(b+n+1)(a+n+2)\,S(n+2)=0. \tag{6}$$

Since $S(0) = 1$, $S(1) = (b-a)/b$, and since $(b-a)_n/(b)_n$ satisfies (6), the summation (2) is proven. We note that owing to the constraint on b, the denominator on the right-hand side of (5) is non-zero for $0 \le k \le n$.

Notice that Zeilberger's algorithm has not produced a minimal recurrence for $S(n)$. This is in contrast to the closed form representation $(b-a)_n/(b)_n$ which corresponds to a recurrence of order 1. The reason for this phenomenon will be explained below.

Again following [1], Pfaff's method works as follows. By subtracting the summands term by term one finds that

$$S(n,a,b) - S(n-1,a,b) = \frac{a(1-b-2n)}{b(b+1)}T(n-1,a+2,b+2), \tag{7}$$

where

$$T(n,a,b) = {}_4F_3\left(\begin{matrix} a/2, (a+1)/2, b+n-1, -n \\ b/2, (b+1)/2, a \end{matrix}; 1\right).$$

Then inspection of $T(n,a,b)$ for some concrete values of n leads to the conjecture

$$T(n,a,b) = \frac{(b-a)_n}{(b+2n-1)(b)_{n-1}}. \tag{8}$$

Next one repeats this step by subtracting $S(n-1,a,b)$ from $T(n,a,b)$, which yields,

$$T(n,a,b) - S(n-1,a,b) = -\frac{(a+n)(b+n-1)}{b(b+1)}T(n-1,a+2,b+2). \tag{9}$$

The proof is then completed by observing that (7) and (9), together with the initial values $S(0,a,b) = T(0,a,b) = 1$, completely define $S(n,a,b)$ and $T(n,a,b)$, and by verifying that the right sides of (2) and (8) satisfy the same recurrences and initial values.

Summarizing, as pointed out in [1, Sect. 3.12], both proofs rely on contiguous relations. The Zeilberger output recurrence (6) is of the form

$$c_0 \cdot {}_4F_3\left(\begin{matrix} a_1, a_2, a_3, a_4 \\ b_1, b_2, b_3 \end{matrix}; 1\right) + c_1 \cdot {}_4F_3\left(\begin{matrix} a_1, a_2, a_3+1, a_4-1 \\ b_1, b_2, b_3 \end{matrix}; 1\right)$$
$$+ c_2 \cdot {}_4F_3\left(\begin{matrix} a_1, a_2, a_3+2, a_4-2 \\ b_1, b_2, b_3 \end{matrix}; 1\right) = 0, \tag{10}$$

where the ${}_4F_3$ parameters a_i and b_j are taken as in the ${}_4F_3$ series in (2), and where the c_l are the coefficients in the recurrence (6). The existence of this relation is predicted by Theorem 1C in Sect. 10. To compute the relation as in (3), the rational function version (116) applies.

The proof by Pfaff's method relies on *two* contiguous relations. Namely, if the ${}_4F_3$ parameters a_i and b_j are taken again as in the ${}_4F_3$ series in (2), then

$$c_0 \cdot {}_4F_3\left(\begin{matrix} a_1, a_2, a_3, a_4 \\ b_1, b_2, b_3 \end{matrix}; 1\right) + c_1 \cdot {}_4F_3\left(\begin{matrix} a_1, a_2, a_3-1, a_4+1 \\ b_1, b_2, b_3 \end{matrix}; 1\right)$$
$$+ c_2 \cdot {}_4F_3\left(\begin{matrix} a_1+1, a_2+1, a_3, a_4+1 \\ b_1+1, b_2+1, b_3+1 \end{matrix}; 1\right) = 0 \tag{11}$$

with $c_0 = 1$, $c_1 = -1$ and $c_2 = -a(1-b-2n)/(b(b+1))$ corresponds to (7), and

$$c_0 \cdot {}_4F_3\left(\begin{matrix} a_1, a_2, a_3-1, a_4 \\ b_1, b_2, b_3-1 \end{matrix}; 1\right) + c_1 \cdot {}_4F_3\left(\begin{matrix} a_1, a_2, a_3-1, a_4+1 \\ b_1, b_2, b_3 \end{matrix}; 1\right)$$
$$+ c_2 \cdot {}_4F_3\left(\begin{matrix} a_1+1, a_2+1, a_3, a_4+1 \\ b_1+1, b_2+1, b_3+1 \end{matrix}; 1\right) = 0 \tag{12}$$

with $c_0 = 1$, $c_1 = -1$ and $c_2 = (a+n)(b+n-1)/(b(b+1))$ corresponds to (9). The existence of both of these contiguous relations is implied by Theorem 1B in Sect. 9.

We note that proving (2) by Pfaff's method leads in a direct manner to the discovery and the proof of a 'companion summation', namely (8). Another difference between the methods is that, when executing the 'Pfaff proof', the relations (11) and (12) in [1] have been derived 'by hand', whereas (10) was delivered automatically by Zeilberger's algorithm. This latter aspect of 'hand-computation' will be removed by the main theorems in this article. In particular, we will see that:

Any contiguous relation between terminating and most of the contiguous relations between non-terminating hypergeometric series can be found automatically by the computer.

In particular, this means that in its essence Pfaff's method is as algorithmic as the W–Z method. For example, if we do not know the coefficients in (11) and (12), we simply compute them by the algorithm described in Sect. 3.

It is important to note that conceptually even more is true. As explained in Sect. 4, contiguous relations can be found automatically by the same mechanism which is applied to find Zeilberger recurrences, namely, creative telescoping [41].

Zeilberger's algorithm [39] is based on the observation that a straightforward extension of Gosper's algorithm [11] for *indefinite* hypergeometric summation (hypergeometric telescoping) can be used for automatic *definite* hypergeometric summation (creative telescoping). For the sake of completeness of the presentation, the essence of creative telescoping, the *parameterized Gosper algorithm*, is briefly sketched in Sect. 3; in Sect. 7 we show how the RISC package `fastZeil`, which implements this extended Gosper algorithm, can be brought into action. For further information on Zeilberger's algorithm, creative telescoping, and the W–Z method, the reader is referred to the book [23] by M. Petkovšek, H.S. Wilf, and D. Zeilberger.

Before listing the contents of this article, we illustrate this new application of creative telescoping by having another look at Bailey's summation (2).

First of all we note that running Zeilberger's algorithm with order 1 results in an empty output. This proves that there does not exist a contiguous relation of the form

$$c_0\, S(n,a,b) + c_1\, S(n-1,a,b) = 0.$$

To overcome this issue, in the spirit of creative telescoping, we introduce a further shift—but not with respect to n as we would do when using Zeilberger's algorithm! Instead we shift one of the parameters, say a, and take as a new ansatz,

$$c_0\, S(n,a,b) + c_1\, S(n-1,a,b) + c_2\, S(n-1,a-1,b) = 0. \tag{13}$$

Then we apply the package described in Sect. 7 which computes indeed a relation of type (13) with the coefficients

$$\begin{aligned} c_0 &= -(a+n)(n+b-1)(2n+b-2), \\ c_1 &= (n+b-a-1)\left((a^2+a(2n-1)+n(2n+b-2)\right), \quad \text{and} \\ c_2 &= a(a-b)(1+a-b). \end{aligned} \tag{14}$$

This provides a new proof of Bailey's summation (2), since (13) with the values c_l from (14) together with the initial value $S(0,a,b)=1$ completely define $S(n,a,b)$, and it is easy to verify that $(b-a)_n/(b)_n$ satisfies the same recurrence and initial value.

As described in Sect. 4, all such contiguous relations can be computed automatically by creative telescoping via *telescoping contiguous relations*; see Theorems 1 and 2. For instance, in order to obtain (13), the corresponding telescoping contiguous relation for all $k \geq 0$ is computed as

$$c_0 \frac{(a_1)_k(a_2)_k(a_3)_k(a_4)_k}{(b_1)_k(b_2)_k(b_3)_k k!} + c_1 \frac{(a_1)_k(a_2)_k(a_3-1)_k(a_4+1)_k}{(b_1)_k(b_2)_k(b_3)_k k!}$$
$$+ c_2 \frac{(a_1)_k(a_2-1)_k(a_3-1)_k(a_4+1)_k}{(b_1)_k(b_2)_k(b_3-1)_k k!} = \Delta_k\, C(k) \frac{(a_1)_k(a_2)_k(a_3)_k(a_4)_k}{(b_1)_k(b_2)_k(b_3)_k k!} \tag{15}$$

with c_l as in (14),

$$a_1 = \frac{a}{2},\ a_2 = \frac{a+1}{2},\ a_3 = b+n,\ a_4 = -n,\ b_1 = \frac{b}{2},\ b_2 = \frac{b+1}{2},\ b_3 = a+1,$$

and

$$C(x) = \frac{x(x+a)(2x+b-2)(2x+b-1)(b+n-1)(a+2n-1)}{n(2x+a-1)(x+b+n-1)}.$$

Summing both sides of (15) over k from 0 to n results in (13). We want to stress that the existence of (15), and thus that of (13), is predicted by Theorem 1C in Sect. 10.

There is quite some literature where contiguous relations are used. Most often this usage is more or less of implicit nature, for instance, as part of a method or a derivation. Much less literature can be found where general aspects of how to compute contiguous relations are treated; but there is still some. In addition to the books [1] and [25], there are articles such as [26, 27, 35, 36], or [8]; the latter devoted to q-series summation. In particular we want to stress the pioneering work of Takayama [35], where for elimination in difference-differential operator rings, non-commutative Gröbner bases methods are introduced. This theme reoccurs in the context of Zeilberger's holonomic systems approach to special functions identities [40]; see, for instance, the work of Chyzak [6] and of Koutschan [16], and also the references given there.

Despite all this work, we feel that our viewpoint and methods described in this article have particular advantages. Based on difference equations our approach is elementary and connects directly to Zeilberger's extension of Gosper's algorithm. An independent development in this direction is [3]. This article mentions connections to Karr-Schneider summation theory which also applies here: Basically all what we describe can be algorithmically realized using Schneider's Sigma package [30]. Further references to Schneider's work are given in Sect. 3.1.

Nevertheless, there are other aspects like summation theory for non-terminating hypergeometric series. In our setting, the existence of fundamental summation theorems like the Gauß ${}_2F_1$ or the Pfaff-Saalschütz ${}_3F_2$ formulas find natural explanations; see the Sects. 8.2 and 10.2. This, in particular, includes contiguous relations between non-terminating hypergeometric series. Another spin-off concerns the fact that our setting in many cases admits explanations of the phenomenon why Zeilberger's algorithm does not always deliver the minimal recurrence and why 'creative symmetrizing' sometimes can help; see the Sects. 11.2 and 11.3. For related work see [5]. Finally, all what we say in this article, including the explanations for non-minimality of Zeilberger orders, carries over to the case of q-hypergeometric series and q-contiguous relations.

The organization of the rest of this paper is as follows. In Sect. 3 a brief description of the parameterized Gosper algorithm is given. This section is kept as short as possible since this algorithm is essentially the same as when used as the computational engine in Zeilberger's algorithm for proving definite hypergeometric summation identities.

Section 4 presents Theorem 1 which states the existence of telescoping contiguous relations for terms which are summands of hypergeometric ${}_pF_q$-series with argument $z = 1$ or when $p \neq q + 1$. In Sect. 5 a detailed Proof of Theorem 1 is given.

The examples presented in Sect. 6 should give a first impression of the variety of potential applications for telescoping contiguous relations and the methods described. In this section the examples are enriched with some details for proper illustration of the method, a theme which is continued further in Sect. 7. There we give a brief description of the computer algebra package we have used for our algorithmic applications.

The Sects. 8, 9, and 10 describe three somehow disjoint refinements of Theorem 1 which imply the existence of contiguous relations for hypergeometric ${}_{q+1}F_q$-series with argument $z = 1$. Illustrating examples concern formulas such as the non-terminating versions of Gauß' ${}_2F_1$ and the Pfaff-Saalschütz ${}_3F_2$ summations, but include also the examples from the Introduction as Bailey's ${}_4F_3$-series summation.

Finally, Sect. 11 presents further, more involved applications. Using parameterized telescoping, we derive a generalization of a result which arose in the classical work by James Wilson on hypergeometric recurrences and contiguous relations. In addition, we discuss non-minimality of Zeilberger recurrences from telescoping contiguous relations point of view. In particular, we explain why 'creative symmetrizing' in some instances successfully reduces the order. This discussion includes a new (algorithmic) proof of the non-terminating version of Dixon's well-poised ${}_3F_2$-series. The concluding Sect. 12 points to the fact that all what has been said in this article carries over to q-hypergeometric series and to q-contiguous relations.

3 The Parameterized Gosper Algorithm

Zeilberger, [39] and [41], was the first who discovered that Gosper's algorithm [11] finds a straightforward extension that can be used for creative telescoping. In other words, Zeilberger observed that Gosper's algorithm for indefinite hypergeometric summation can be used to solve also definite hypergeometric summation problems. On this basis, Wilf and Zeilberger developed a rich theoretical framework which, for instance, includes also W–Z pairs and companion identities; see [23].

We present the essence of creative telescoping in the form of an input/output description of the corresponding parameterized extension of Gosper's algorithm. To this end we need to introduce a few definitions.

Throughout this article, p and q denote fixed non-negative integers; Δ_k is the difference operator defined in (4). The parameters a_i, b_j, and the argument z range over the complex numbers; for z we assume $z \neq 0$, unless explicitly mentioned otherwise.

Remark As in the computer algebra examples presented, for the purpose of symbolic computation the a_i, b_j, and z usually are taken as indeterminates; i.e., instead of $\mathbb{K} = \mathbb{C}$, one takes

$$\mathbb{K} = \mathbb{C}(a_1, \ldots, a_p, b_1, \ldots, b_q, z);$$

or, even more precisely,

$$\mathbb{K} = \mathbb{F}(a_1, \ldots, a_p, b_1, \ldots, b_q, z),$$

where the field $\mathbb{F}$ is a computable algebraic extension of $\mathbb{Q}$ depending on extra parameters involved.

However, when seeing $\mathbb{K}$ in this article, the reader should feel free to interpret it as $\mathbb{K} = \mathbb{C}$.

In contrast to complex variables like a_i, b_j, or z, the variable x will always denote an indeterminate. As usual, with $\mathbb{K}$ as the coefficient domain, $\mathbb{K}[x]$ is the ring of polynomials in x; $\mathbb{K}(x)$ is its quotient field, the rational functions in x.

Throughout, $\mathbb{N} := \mathbb{Z}_{\geq 0}$ is the set of non-negative integers. The variables n and k always denote non-negative integers; i.e., $n, k \in \mathbb{N}$.

Following [12], for $k \in \mathbb{N}$ we will use the notation

$$_pF_q\left(\begin{matrix} a_1, \ldots, a_p \\ b_1, \ldots, b_q \end{matrix}; z\right)_k := \frac{(a_1)_k \cdots (a_p)_k}{(b_1)_k \cdots (b_q)_k} \frac{z^k}{k!}. \tag{16}$$

When dealing with such a term we always assume it is well-defined; this means, $b_j + \ell \neq 0$ for $0 \leq \ell \leq k-1$ and all j. The analogous convention applies to hypergeometric series,

$$_pF_q\left(\begin{matrix} a_1, \ldots, a_p \\ b_1, \ldots, b_q \end{matrix}; z\right) = \sum_{k=0}^{\infty} {_pF_q}\left(\begin{matrix} a_1, \ldots, a_p \\ b_1, \ldots, b_q \end{matrix}; z\right)_k.$$

The notation $_pF_q(a_1, \ldots, a_p; b_1, \ldots, b_q; z)$ and $_pF_q(a_1, \ldots, a_p; b_1, \ldots, b_q; z)_k$ will be used within text lines.

Definition 1 A sequence $t(k)$ over $\mathbb{K}$ is called a *hypergeometric term* if there exists a rational function $\rho \in \mathbb{K}(x)$ such that $t(k+1) = \rho(k) \cdot t(k)$ for all sufficiently large k.

Definition 2 Two hypergeometric terms $s(k)$ and $t(k)$ over $\mathbb{K}$ are *similar* if there exists a rational function $\rho \in \mathbb{K}(x)$ such that $s(k) = \rho(k)t(k)$ for all sufficiently large k.

3.1 The Parameterized Gosper Algorithm

Input. Hypergeometric terms $t(k), t_0(k), \ldots, t_d(k)$ over $\mathbb{K}$ where each $t_l(k)$, $0 \leq l \leq d$, is similar to $t(k)$.

Remark We want to comment on the way how to actually input a hypergeometric term. According to Definition 1, this, for instance, can be done by specifying a homogeneous first-order recurrence with polynomial coefficients plus an initial value. Alternatively, one can give an expression in closed form, for instance, in the form of a hypergeometric term as in (16).

Output. All hypergeometric terms $g(k)$ over $\mathbb{K}$ and all tuples $(c_0, \ldots, c_d) \in \mathbb{K}^{d+1}$ such that for all sufficiently large k,

$$c_0\, t_0(k) + \cdots + c_d\, t_d(k) = g(k+1) - g(k) \quad (= \Delta_k g(k)). \tag{17}$$

One can show that each such $g(k)$ must be of the form

$$g(k) = r(k)\, t(k) \tag{18}$$

where $r \in \mathbb{K}(x)$ is a rational function which is computed by the algorithm.

Note. Gosper's algorithm is the special case $d = 0$ with $t(k) = t_0(k)$. Let $t(k) = f(n,k)$ be a term which is hypergeometric with respect to k and n (plus mild side conditions), then Zeilberger's algorithm is the special case with $t_l(k) = f(n+l,k)$ for $0 \leq l \leq d$, and $\mathbb{K}$, for instance, chosen as $\mathbb{K} = \mathbb{C}(n)$. For more detailed descriptions of these algorithms see, for instance, [23].

We also note that the zero term $g(k) = 0$ together with $(c_0, \ldots, c_d) = (0, \ldots, 0)$ always form a solution to (17). All solutions $(c_0, \ldots, c_d, g(k))$ form a vector space over $\mathbb{K}$, hence the output of the parameterized Gosper algorithm can be given in terms of a basis.

Independent Verification

It is important to note that running the algorithm delivers all the information necessary to prove the correctness of the telescoping recurrence (17) *independently* from the steps of the algorithm. Namely, suppose we want to verify (17) for certain c_l and $g(k)$, where $g(k)$ is given as in (18) by the rational function $r \in \mathbb{K}(x)$.

Since all terms $t_l(k)$ are similar to $t(k)$, the left hand side of (17) can be written as a rational function multiple of $t(k)$. Due to

$$g(k+1) - g(k) = \left(r(k+1)\frac{t(k+1)}{t(k)} - r(k) \right) t(k)$$

we can divide both sides of (17) by $t(k)$, and checking (17) then reduces to checking the resulting equality of rational functions. Wilf and Zeilberger [37] call $r(x)$ the *certificate*.

Remark As already mentioned, the parameterized Gosper algorithm (creative telescoping) is the driving engine of Zeilberger's algorithm. It is described in detail in a slightly different form in [23].—It is interesting to note that the parameterized Gosper algorithm, in much more general form, has been used extensively by M. Karr [14, 15] in his difference field approach to symbolic summation. However, Karr has never linked it to *definite* summation. In the framework of difference fields and, more recently, difference rings, this step has been carried out, accompanied by other substantial theoretical and algorithmical enhancements, by Carsten Schneider in [29–34]; see also the references given there.

4 Telescoping Contiguous Relations for $z \neq 1$ or $p \neq q+1$

This section contains the first main theorem of the paper, Theorem 1. It states the existence of telescoping contiguous relations with respect to non-negative integer shifts if $z \neq 1$ or $p \neq q+1$ (or both).

Despite the existence of telescoping contiguous relations apriori is independent from the question whether they involve summands of convergent series, for applications such as taking the infinite sum over such summands,

$${}_pF_q\left(\begin{matrix} a_1,\ldots,a_p \\ b_1,\ldots,b_q \end{matrix};z\right) = \sum_{k=0}^{\infty} {}_pF_q\left(\begin{matrix} a_1,\ldots,a_p \\ b_1,\ldots,b_q \end{matrix};z\right)_k,$$

we need to consider the conditions for the convergence of such series in case they are non-terminating; i.e., where none of the a_i is zero or a negative integer.

According to [1, Th. 2.1.1] such series converge absolutely for all z if $p \leq q$ and for $|z| < 1$ if $p = q+1$; if $p > q+1$ they diverge for all $z \neq 0$. The remaining case, "The case $|z| = 1$ when $p = q+1$ is of great interest"; see [1, p. 62]. According to [1, Th. 2.1.2], the ${}_{q+1}F_q$-series with $|z| = 1$ converges absolutely if

$$\mathrm{Re}\Big(\sum_{j=1}^{q} b_j - \sum_{i=1}^{q+1} a_j\Big) > 0. \tag{19}$$

For $|z| = 1$ and $z \neq 1$ it converges conditionally if

$$0 \geq \mathrm{Re}\Big(\sum_{j=1}^{q} b_j - \sum_{i=1}^{q+1} a_j\Big) > -1, \tag{20}$$

and it diverges if this real part is less or equal -1.

We want to note explicitly that if $z = 1$ and $p = q + 1$, the criteria for the existence of telescoping contiguous relations, given by the Theorems 1A, 1B, and 1C, come in the form of refinements of (19).

Theorem 1 *Suppose $z \neq 1$ or $p \neq q+1$. Let $d = \max\{p, q+1\}$. For $0 \leq l \leq d$ let $(\alpha_1^{(l)}, \ldots, \alpha_p^{(l)}, \beta_1^{(l)}, \ldots, \beta_q^{(l)})$ be pairwise different tuples with non-negative integer entries. Then there exist $c_0, \ldots, c_d$ in $\mathbb{K}$, not all 0, and a polynomial $C(x) \in \mathbb{K}[x]$ such that for all $k \geq 0$,*

$$\sum_{l=0}^{d} c_l \cdot {}_pF_q\left(\begin{matrix} a_1 + \alpha_1^{(l)}, \ldots, a_p + \alpha_p^{(l)} \\ b_1 - \beta_1^{(l)}, \ldots, b_q - \beta_q^{(l)} \end{matrix}; z\right)_k = \Delta_k\, C(k)\, {}_pF_q\left(\begin{matrix} a_1, \ldots, a_p \\ b_1, \ldots, b_q \end{matrix}; z\right)_k . \tag{21}$$

Moreover, $C(0) = 0$, and if $C(x) \neq 0$, for the polynomial degree of $C(x)$ one has

$$\deg C(x) \leq q + 1 - d + \max_{0 \leq l \leq d} \{\alpha_1^{(l)} + \cdots + \alpha_p^{(l)} + \beta_1^{(l)} + \cdots + \beta_q^{(l)}\}; \tag{22}$$

in addition, if $p \leq q + 1$,

$$\lim_{k \to \infty} C(k)\, {}_pF_q\left(\begin{matrix} a_1, \ldots, a_p \\ b_1, \ldots, b_q \end{matrix}; z\right)_k = 0. \tag{23}$$

If $p > q$ the limit (23) is valid if one of the a_i is a non-positive integer.

Proof Since $C(x)$ is a polynomial, the limit (23) is immediate from classical asymptotics as Theorem 2.2.1 in [1]. The rest of Theorem 1 is proven in Sect. 5. □

Remark According to the convergence criteria stated above, when summing (21) over k from 0 to ∞, imposing $|z| < 1$ or $p \leq q$ guarantees the absolute convergence of ${}_pF_q(a_1, \ldots, a_p; b_1, \ldots, b_q; z)$ and of the series

$${}_pF_q\left(\begin{matrix} a_1 + \alpha_1^{(l)}, \ldots, a_p + \alpha_p^{(l)} \\ b_1 - \beta_1^{(l)}, \ldots, b_q - \beta_q^{(l)} \end{matrix}; 1\right), \quad l \in \{0, \ldots, d\}.$$

If $p = q + 1$ and $z \neq 1$ in such applications of Theorem 1, one needs to check whether the criteria for absolute or conditional convergence are satisfied. An example for $z = -1$ and $p = 2 = q + 1$ is provided by Kummer's summation formula (152). There we also show that, alternatively, Kummer's identity can be derived as a limiting case of Dixon summation (150) which can be obtained from Theorem 1A with $p = 3 = q + 1$ and argument $z = 1$.

Remark In Theorem 1 one can allow arbitrary integer parameters instead of restricting to non-negative integers. More precisely, for arbitrary parameters $\alpha_i^{(l)}$ and $\beta_j^{(l)}$ this gives a relation,

$$\sum_{l=0}^{d} c_l \cdot {}_pF_q\left(\begin{matrix} a_1+\alpha_1^{(l)},\ldots,a_p+\alpha_p^{(l)} \\ b_1-\beta_1^{(l)},\ldots,b_q-\beta_q^{(l)} \end{matrix};z\right)_k = \Delta_k\, R(k)\, {}_pF_q\left(\begin{matrix} a_1,\ldots,a_p \\ b_1,\ldots,b_q \end{matrix};z\right)_k, \tag{24}$$

with a *rational* function $R(x) \in \mathbb{K}(x)$ instead of a *polynomial* $C(x) \in \mathbb{K}[x]$. For this extension, in view of (30) and (31), it is important to notice that because of possible poles of the $R(x)$, not all integer choices of $\alpha_i^{(l)}$ and $\beta_j^{(l)}$ are admissible.

Definition 3 The relations (21) and (24) are called *telescoping contiguous relations.*

The next corollary shows that the restriction to non-negative integer shifts in Theorem 1 is not an essential one.

Corollary 1 *Any telescoping contiguous relation of the form* (21) *and, if poles of* $R(x)$ *cause no problem, in the version of* (24) *can be computed by the parameterized Gosper algorithm.*

Proof It suffices to prove the statement with respect to the form (24). For the parameterized Gosper algorithm described in Sect. 4, take as input

$$t_l(k) = {}_pF_q\left(\begin{matrix} a_1+\alpha_1^{(l)},\ldots,a_p+\alpha_p^{(l)} \\ b_1-\beta_1^{(l)},\ldots,b_q-\beta_q^{(l)} \end{matrix};z\right)_k$$

and

$$t(k) = {}_pF_q\left(\begin{matrix} a_1,\ldots,a_p \\ b_1,\ldots,b_q \end{matrix};z\right)_k$$

with $\mathbb{K} = \mathbb{C}(a_1,\ldots,a_p,b_1,\ldots,b_q,z)$ as the field of constants. Note that $t(k)$ and the $t_l(k)$ are hypergeometric terms; in addition, all terms $t_l(k)$ are similar to $t(k)$ as required. The parameterized Gosper algorithm finds *all* $(c_0,\ldots,c_d) \in \mathbb{K}^{d+1}$ and $R(x) \in \mathbb{K}(x)$ such that for $g(k) = R(k)t(k)$ the tuple $(c_0,\ldots,c_d,g(k))$ satisfies (17). Therefore the solution $(c_0,\ldots,c_d,R(k)t(k))$ as in (24) will be found by the algorithm. □

Before proving Theorem 1 in Sect. 5, we present some immediate consequences. Further applications are given in Sect. 6.

4.1 Telescoping Contiguous Relations for $z \neq 1$ and $(p, q) = (1, 0)$

Suppose $z \neq 1$ and $(p, q) = (1, 0)$. In this case, $d = \max\{p, q+1\} = 1$. According to Theorem 1 there exist c_0 and c_1, not all 0, and a polynomial $C(x)$ with $C(0) = 0$ such that for all $k \geq 0$,

$$c_0\, t_0(k) + c_1\, t_1(k) = \Delta_k\, C(k)\, t(k), \tag{25}$$

where

$$t(k) = t_0(k) = {}_1F_0\left(\begin{matrix} a \\ - \end{matrix}; z\right)_k \quad \text{and} \quad t_1(k) = {}_1F_0\left(\begin{matrix} a+\alpha \\ - \end{matrix}; z\right)_k$$

with $\alpha \in \mathbb{Z}_{>0}$ According to (22),

$$\deg C(x) \leq q + 1 - d + \alpha = \alpha.$$

For fixed α, the c_j and $C(x)$ can be computed automatically as described in Sect. 7. For example, for $\alpha = 1$ one obtains (25) with $c_0 = a$, $c_1 = a(z-1)$, and $C(x) = x$. For $|z| < 1$ one can sum the resulting telescoping relation over k from 0 to infinity, which using (23) gives,

$$a \cdot {}_1F_0\left(\begin{matrix} a \\ - \end{matrix}; z\right) - a(1-z) \cdot {}_1F_0\left(\begin{matrix} a+1 \\ - \end{matrix}; z\right) = 0; \tag{26}$$

this is in accordance with the binomial expansion

$${}_1F_0\left(\begin{matrix} a \\ - \end{matrix}; z\right) = (1-z)^a.$$

4.2 Telescoping Contiguous Relations for $z \neq 1$ and $(p, q) = (2, 1)$

Suppose $z \neq 1$ and $(p, q) = (2, 1)$. In this case, $d = \max\{p, q+1\} = 2$. According to Theorem 1 there exist c_0, c_1 and c_2, not all 0, and a polynomial $C(x)$ with $C(0) = 0$ such that for all $k \geq 0$,

$$c_0\, t_0(k) + c_1\, t_1(k) + c_2\, t_2(k) = \Delta_k\, C(k)\, t(k), \tag{27}$$

where $t(k) = t_0(k)$ and

$$t_0(k) = {}_2F_1\left(\begin{matrix} a, b \\ c \end{matrix}; z\right)_k, t_1(k) = {}_1F_0\left(\begin{matrix} a+\alpha_1, b+\beta_1 \\ c-\gamma_1 \end{matrix}; z\right)_k, t_2(k) = {}_2F_1\left(\begin{matrix} a+\alpha_2, b+\beta_2 \\ c-\gamma_2 \end{matrix}; z\right)_k.$$

Here the $(\alpha_l, \beta_l, \gamma_l)$ for $l = 1$ and $l = 2$ are different triples of non-negative integers, each with entries not all 0. For $|z| < 1$ one can sum the telescoping relation (27) over k from 0 to infinity, which using (23) gives,

$$c_0 \cdot {}_2F_1\left(\begin{matrix} a, b \\ c \end{matrix}; z\right) + c_1 \cdot {}_1F_0\left(\begin{matrix} a+\alpha_1, b+\beta_1 \\ c-\gamma_1 \end{matrix}; z\right) + c_2 \cdot {}_2F_1\left(\begin{matrix} a+\alpha_2, b+\beta_2 \\ c-\gamma_2 \end{matrix}; z\right) = 0. \tag{28}$$

For fixed α_l,β_l, and γ_l, the c_j and $C(x)$ can be computed automatically as described in Sect. 7; concrete examples are given there.

We want to conclude this section with the remark that Theorem 1 together with our implementation of parameterized telescoping presented in Sect. 7 (or any other implementation meeting the specification given in Sect. 3) settles the existence and the computation of the general three term contiguous ${}_2F_1$-relations treated by Gauß in [10].

5 Proof of Theorem 1

In this section we prove Theorem 1. To this end we make use of several elementary facts which are presented in the form of lemmas.

5.1 Preparatory Lemmas

Definition 4 Let x be an indeterminate, $c \in \mathbb{C} \setminus \mathbb{Z}_{\leq 0}$, and $m \in \mathbb{N}$. Define $\mu_0(c; x) := 1$, and

$$\mu_m(c; x) := \left(1 + \frac{x}{c}\right)\left(1 + \frac{x}{c+1}\right)\cdots\left(1 + \frac{x}{c+m-1}\right), \quad m \geq 1.$$

For the degree of the polynomial $\mu_m(c; x) \in \mathbb{K}[x]$ we have

$$\deg \mu_m(c; x) = m. \tag{29}$$

Lemma 1 *Let* $(\alpha_1, \ldots, \alpha_p, \beta_1, \ldots, \beta_q) \in \mathbb{N}^{p+q}$ *such that for all* $i \in \{1, \ldots, p\}$ *and* $j \in \{1, \ldots, q\}$,

$$a_i + \alpha_i \notin \{1, \ldots, \alpha_i\} \text{ and } b_j - \beta_j \notin \mathbb{Z}_{\leq 0}.$$

Then for all $k \geq 0$,

$$\begin{aligned} &{}_pF_q \left(\begin{matrix} a_1 + \alpha_1, \ldots, a_p + \alpha_p \\ b_1 - \beta_1, \ldots, b_q - \beta_q \end{matrix} ; z \right)_k \\ &= \prod_{i=1}^{p} \mu_{\alpha_i}(a_i; k) \cdot \prod_{j=1}^{q} \mu_{\beta_j}(b_j - \beta_j; k) \cdot {}_pF_q \left(\begin{matrix} a_1, \ldots, a_p \\ b_1, \ldots, b_q \end{matrix} ; z \right)_k . \end{aligned}$$

Proof For $k \geq 0$,

$$(a+1)_k = \left(1 + \frac{k}{a}\right)(a)_k \tag{30}$$

and

$$\frac{1}{(b-1)_k} = \left(1 + \frac{k}{b-1}\right)\frac{1}{(b)_k}. \tag{31}$$

The lemma is proven by iterated application of (30) and (31). □

For

$$t(k) = \frac{(a_1)_k \cdots (a_p)_k}{(b_1)_k \cdots (b_q)_k} \frac{z^k}{k!} \tag{32}$$

let us consider all pairs $P(x) \in \mathbb{K}[x]$ and $R(x) \in \mathbb{K}(x)$ such that for all $k \geq 0$,

$$P(k)\, t(k) = \Delta_k R(k)\, t(k). \tag{33}$$

It turns out that if (33) holds, then $R(x)$ has to be a polynomial, too. More precisely, all such pairs can be characterized as follows. We note that the essence of this characterization is closely related to what is called the Gosper, resp. Gosper-Petkovšek, form and to the author's concept of greatest factorial factorization; see [19].

Lemma 2 *Suppose* $P(x) \in \mathbb{K}[x]$ *and* $R(x) \in \mathbb{K}(x)$ *satisfy the relation*

$$P(k)\, {}_pF_q \left(\begin{matrix} a_1, \ldots, a_p \\ b_1, \ldots, b_q \end{matrix} ; z \right)_k = \Delta_k\, R(k)\, {}_pF_q \left(\begin{matrix} a_1, \ldots, a_p \\ b_1, \ldots, b_q \end{matrix} ; z \right)_k , \quad k \geq 0. \tag{34}$$

Then there exists a polynomial $P_1(x) \in \mathbb{K}[x]$ *such that*

$$P(x) = z \prod_{i=1}^{p}(x + a_i) \cdot P_1(x+1) - x \prod_{j=1}^{q}(x + b_j - 1) \cdot P_1(x) \tag{35}$$

and

$$R(x) = x \prod_{j=1}^{q}(x + b_j - 1) \cdot P_1(x). \tag{36}$$

Vice versa, if $P(x)$ *and* $R(x)$ *are of the form* (35) *and* (36) *with* $P_1(x)$ *being an arbitrary polynomial in* $\mathbb{K}[x]$*, then relation* (34) *is satisfied.*

Proof First we prove the statement assuming that

$$a_i \notin \mathbb{Z} \text{ and } a_i - b_j \notin \mathbb{N}, \quad i \in \{1, \ldots, p\}, j \in \{1, \ldots, q\}.$$

Then, since (34) is a finite sum, analytic continuation proves the statement without these restrictions.

With $t(k)$ as in (32), relation (34) turns into $P(k)t(k) = R(k+1)t(k+1) - R(k)t(k)$. Dividing out $t(k)$ results in an equality between rational functions; in other words, relation (34) is equivalent to the relation

$$\frac{P(x) + R(x)}{R(x+1)} = \frac{(x + a_1) \cdots (x + a_p)}{(x + b_1) \cdots (x + b_q)} \frac{z}{x+1}. \tag{37}$$

Suppose $P(x)$ and $R(x)$ are of the form (35) and (36) with $P_1(x)$ being an arbitrary polynomial in $\mathbb{K}[x]$. It is easily verified that then relation (37) is satisfied. Therefore it remains to prove the other direction of the lemma.

To this end, suppose that (34) holds for some $P(x) \in \mathbb{K}[x]$ and $R(x) = R_1(x)/R_2(x) \in \mathbb{K}(x)$ with R_1 and R_2 being coprime polynomials in $\mathbb{K}[x]$. So we can rewrite relation (37) as

$$(x+1) \prod_{j=1}^{q}(x + b_j) \cdot \overline{P}(x)\, R_2(x+1) = z \prod_{i=1}^{p}(x + a_i) \cdot R_1(x+1)\, R_2(x) \tag{38}$$

where

$$\overline{P}(x) = R_1(x) + P(x)R_2(x). \tag{39}$$

In the following we will use that $\gcd(R_1(x), R_2(x)) = 1$, $\gcd\left(\overline{P}(x), R_2(x)\right) = 1$, and the fact that if $h(x) \in \mathbb{K}[x]$ is irreducible then $\gcd(h(x), h(x+l)) = 1$ for all non-zero integers l.

Suppose that $x+b \mid R_2(x)$ where $b = 1$ or $b = b_j$ for some $j \in \{1, \dots, q\}$. Then $x + b + 1$ divides $R_2(x + 1)$, and (38) implies that $x + b + 1 \mid R_2(x)$. By iterating this observation we obtain that $x + b + l \mid R_2(x)$ for all $l \in \mathbb{N}$, a contradiction to $R_2(x)$ being a non-zero polynomial. Consequently $(x + 1) \prod_{j=1}^{q}(x + b_j)$ must divide $R_1(x + 1)$; in other words, there exists a polynomial $P_1(x) \in \mathbb{K}[x]$ such that

$$R_1(x) = x \prod_{j=1}^{q}(x + b_j - 1) \cdot P_1(x). \tag{40}$$

By an analogous reasoning one can show that $\prod_{i=1}^{p}(x + a_i)$ must divide $\overline{P}(x)$; this means, there exists a polynomial $Q(x) \in \mathbb{K}[x]$ such that

$$\overline{P}(x) = \prod_{i=1}^{p}(x + a_i) \cdot Q(x). \tag{41}$$

By (40) and (41), Eq. (38) reduces to

$$Q(x)\, R_2(x + 1) = z\, P_1(x + 1)\, R_2(x). \tag{42}$$

Without loss of generality we may assume that the leading coefficient of $R_2(x)$ is equal to 1. The next proof step will show that $R_2(x)$ must have degree 0 which implies that

$$R_2(x) = 1 \qquad \text{and} \qquad Q(x) = z\, P_1(x + 1). \tag{43}$$

For proving this, suppose that an arbitrary irreducible polynomial $h(x)$ divides $R_2(x)$. Then $h(x + 1) \mid R_2(x + 1)$, and (42) implies that $h(x + 1) \mid R_2(x)$. Iterating this observation we obtain that $h(x + l) \mid R_2(x)$ for all $l \in \mathbb{N}$. Therefore $R_2(x)$ can only have irreducible factors which are constants, and (43) is proved.

Finally, Eq. (39) together with (40) and (41) imply (35). Since $R_2(x) = 1$ we have $R(x) = R_1(x)$, and Eq. (36) is nothing but relation (40). This completes the proof of the lemma. □

We are interested in polynomial solutions $P(x)$ and $R(x)$ to (34) which are minimal with respect to their degree in x; see also Lemma 3.

Corollary 2 *The minimal non-trivial choice for $P(x) \in \mathbb{K}[x]$ and $R(x) \in \mathbb{K}(x)$ such that* (34) *holds is the following:*

$$P(x) = z \prod_{i=1}^{p}(x + a_i) - x \prod_{j=1}^{q}(x + b_j - 1) \tag{44}$$

and

$$R(x) = x \prod_{j=1}^{q}(x + b_j - 1). \tag{45}$$

Proof Immediate from Lemma 2 choosing $P_1(x) = 1$. □

The minimally chosen polynomials $P(x)$ and $Q(x)$ from Corollary 2 will play a fundamental role which gives rise to the following definition.

Definition 5 To any hypergeometric term ${}_pF_q(a_1, \ldots, a_p; b_1, \ldots, b_q; z)_k$ we associate the polynomials

$${}_pP_q(x) := z \prod_{i=1}^{p}(x + a_i) - x \prod_{j=1}^{q}(x + b_j - 1) \tag{46}$$

and

$${}_pR_q(x) := x \prod_{j=1}^{q}(x + b_j - 1). \tag{47}$$

For the proof of Theorem 1 we need a bit more than the minimal non-trivial choice specified in Corollary 2; we also need the cases where $P_1(x) = x^n$.

Definition 6 For any non-negative integer n,

$${}_pP_q^{(n)}(x) := z\,(x + 1)^n \prod_{i=1}^{p}(x + a_i) - x^{n+1} \prod_{j=1}^{q}(x + b_j - 1) \in \mathbb{K}[x]. \tag{48}$$

Notice that ${}_pP_q^{(0)}(x) = {}_pP_q(x)$.

Lemma 3 *Suppose $z \neq 1$ or $p \neq q + 1$. Then*

$$\deg {}_pP_q^{(n)}(x) = \max\{n + p, n + 1 + q\} = n + \max\{p, q + 1\} = n + \deg {}_pP_q(x), \tag{49}$$

and for the leading coefficient,

$$\operatorname{lcf} {}_pP_q^{(n)}(x) = \begin{cases} -1 \;, \textit{if} \;\; p \leq q \\ z - 1 \;, \textit{if} \;\; p = q + 1 \\ z \;, \textit{if} \;\; p > q + 1 \end{cases}. \tag{50}$$

Proof Immediate by inspection. □

Lemma 4 ("Reduction Lemma") *Suppose $z \neq 1$ or $p \neq q+1$. Fix $n \in \mathbb{N}$. Let*

$$d = \deg {}_pP_q(x) \;\; and \;\; c_n = \operatorname{lcf} {}_pP_q^{(n)}(x)$$

be the degree and the leading coefficient, respectively, of the polynomial ${}_pP_q^{(n)}(x)$. Then there exists a polynomial ${}_pQ_q^{(n)}[x] \in \mathbb{K}[x]$ with $\deg {}_pQ_q^{(n)}(x) \leq n+d-1$ such that for all $k \geq 0$,

$$c_n \cdot k^{n+d} \; {}_pF_q\left(\begin{matrix} a_1, \ldots, a_p \\ b_1, \ldots, b_q \end{matrix}; z\right)_k = \left({}_pQ_q^{(n)}(k) + \Delta_k\, k^n \; {}_pR_q(k)\right) {}_pF_q\left(\begin{matrix} a_1, \ldots, a_p \\ b_1, \ldots, b_q \end{matrix}; z\right)_k . \tag{51}$$

Proof Let $P_1(x) = x^n$. Then according to Lemma 2 we have for all $k \geq 0$,

$${}_pP_q^{(n)}(k) \; {}_pF_q\left(\begin{matrix} a_1, \ldots, a_p \\ b_1, \ldots, b_q \end{matrix}; z\right)_k = \Delta_k\, k^n \, {}_pR_q(k) \; {}_pF_q\left(\begin{matrix} a_1, \ldots, a_p \\ b_1, \ldots, b_q \end{matrix}; z\right)_k . \tag{52}$$

By Lemma 3 we have $\deg {}_pP_q^{(n)}(x) = n+d$; i.e., choosing $c_n := \operatorname{lcf} {}_pP_q^{(n)}(x) \in \mathbb{K}$ we can define

$${}_pQ_q^{(n)}(x) := c_n \cdot x^{n+d} - {}_pP_q^{(n)}(x) \in \mathbb{K}[x]$$

where $\deg {}_pQ_q^{(n)}(x) \leq n+d-1$. Hence Lemma 4 follows from (52) after replacing ${}_pP_q^{(n)}(k)$ by $c_n \cdot k^{n+d} - {}_pQ_q^{(n)}(k)$. □

The Reduction Lemma implies the following result in a straightforward manner.

Corollary 3 *Suppose $z \neq 1$ or $p \neq q+1$. Let $d = \deg {}_pP_q(x)$. For any fixed $n \in \mathbb{N}$ there exist polynomials $u_n(x)$ and $v_n(x)$ in $\mathbb{K}[x]$ with*

$$\deg u_n(x) \leq d-1 \;\; and \;\; \deg v_n(x) = n \tag{53}$$

such that for all $k \geq 0$,

$$k^{n+d} \; {}_pF_q\left(\begin{matrix} a_1, \ldots, a_p \\ b_1, \ldots, b_q \end{matrix}; z\right)_k = u_n(k) \; {}_pF_q\left(\begin{matrix} a_1, \ldots, a_p \\ b_1, \ldots, b_q \end{matrix}; z\right)_k + \Delta_k\, v_n(k) \; {}_pR_q(k) \; {}_pF_q\left(\begin{matrix} a_1, \ldots, a_p \\ b_1, \ldots, b_q \end{matrix}; z\right)_k . \tag{54}$$

Proof The proof proceeds by induction on n. For $n = 0$ we invoke Lemma 4 with $n = 0$; i.e., we can choose $u_0(x) = 1/c \cdot {}_pQ_q^{(0)}(x)$ with $\deg u_0(x) \leq d-1$ and $v_0(x) = 1/c$ where $c = \operatorname{lcf} {}_pP_q^{(0)}(x)$.

For proving the induction step, let $t(k)$ be as in (32). According to Lemma 4 we have with $c = \operatorname{lcf}\, {}_pP_q^{(n+1)}(x)$ that for all $k \geq 0$,

$$k^{n+1+d}\, t(k) = \frac{1}{c} \cdot {}_pQ_q^{(n+1)}(k)\, t(k) + \Delta_k \frac{1}{c} \cdot k^{n+1}\, {}_pR_q(k)\, t(k). \tag{55}$$

Since $\deg {}_pQ_q^{(n+1)}(x) \leq n + d$, the polynomial ${}_pQ_q^{(n+1)}(x)$ can be written in the form

$${}_pQ_q^{(n+1)}(x) = \sum_{j=0}^{n+d} Q_j x^j \text{ with } Q_j \in \mathbb{K}.$$

For all m with $0 \leq m \leq n$ we apply the induction hypothesis to $Q_{m+d}\, k^{m+d}\, t(k)$, and thus by (54) we obtain polynomials $u(x)$ and $v(x)$ in $\mathbb{K}[x]$ with $\deg u(x) \leq d-1$ and $\deg v(x) \leq n$ such that for all $k \geq 0$,

$${}_pQ_q^{(n+1)}(k)\, t(k) = \sum_{j=0}^{d-1} Q_j k^j\, t(k) + u(k)\, t(k) + \Delta_k\, v(k)\, {}_pR_q(k)\, t(k). \tag{56}$$

Finally combining (56) with (55) we obtain the polynomials

$$u_{n+1}(x) = \frac{1}{c}\Big(u(x) + \sum_{j=0}^{d-1} Q_j x^j\Big) \text{ and } v_{n+1}(x) = \frac{1}{c}\left(x^{n+1} + v(x)\right),$$

which satisfy (53) and (54) for $n + 1$ instead of n. This completes the proof of Corollary 3. □

We shall utilize Corollary 3 in the following form. We note explicitly that if $0 \in \mathbb{K}[x]$ is the zero polynomial, we use the convention $\deg 0 = -1$.

Corollary 4 *Suppose $z \neq 1$ or $p \neq q+1$. Let $d = \deg {}_pP_q(x)$. For any $M \in \mathbb{K}[x]$ with $\deg M(x) = m$ there exist polynomials $U(x)$ and $V(x)$ in $\mathbb{K}[x]$ with*

$$\deg U(x) \leq d - 1 \quad \textit{and} \quad \deg V(x) = \max\{m - d, -1\} \tag{57}$$

such that for all $k \geq 0$,

$$M(k)\, {}_pF_q\left(\begin{matrix} a_1, \ldots, a_p \\ b_1, \ldots, b_q \end{matrix}; z\right)_k = U(k)\, {}_pF_q\left(\begin{matrix} a_1, \ldots, a_p \\ b_1, \ldots, b_q \end{matrix}; z\right)_k + \Delta_k\, V(k)\, {}_pR_q(k)\, {}_pF_q\left(\begin{matrix} a_1, \ldots, a_p \\ b_1, \ldots, b_q \end{matrix}; z\right)_k. \tag{58}$$

Proof Without loss of generality we can rewrite $M(x)$ into the form

$$M(x) = \sum_{j=0}^{d-1} M_j x^j + \sum_{n=0}^{m-d} M_{n+d} x^{n+d}$$

with coefficients M_i in $\mathbb{K}$.

If $m < d$ the second sum is zero. This means, in this case we can choose $U(x) = M(x)$ and $V(x) = 0$, and both, (57) and (58), are satisfied. In particular we have that $\deg V(x) = \deg 0 = \max\{m - d, -1\} = -1$.

In order to prove the corollary also for $m \geq d$, let $t(k)$ be as in (32). By invoking Corollary 3 we obtain that for all $k \geq 0$,

$$M(k)\,t(k) = \sum_{j=0}^{d-1} M_j k^j\, t(k)$$

$$+ \sum_{n=0}^{m-d} M_{n+d}\, u_n(k)\, t(k) + \Delta_k \sum_{n=0}^{m-d} M_{n+d}\, v_n(k)\, {}_pR_q(k)\, t(k).$$

But this proves Corollary 4 since we can choose,

$$U(x) = \sum_{j=0}^{d-1} M_j x^j + \sum_{n=0}^{m-d} M_{n+d}\, u_n(x) \text{ and } V(x) = \sum_{n=0}^{m-d} M_{n+d}\, v_n(x),$$

and it is easily verified that both, (57) and (58), are satisfied. □

5.2 *Proof of Theorem 1*

With the results of the preceding subsection we are ready to prove Theorem 1.

Let $t(k)$ be as in (32). First we prove the statement of Theorem 1 assuming that the condition to apply Lemma 1 holds; namely, for $i \in \{1, \ldots, p\}, j \in \{1, \ldots, q\}, l \in \{0, \ldots, d\}$,

$$a_i + \alpha_i^{(l)} \notin \{1, \ldots, \alpha_i^{(l)}\} \text{ and } b_j - \beta_j^{(l)} \notin \mathbb{Z}_{\leq 0}.$$

Then, analytic continuation proves the statement without these restrictions an the a_i.[3]

[3]The conditions on the $b_j - \beta_j^{(l)}$ remain valid as being those for ${}_pF_q$ bottom parameters.

According to Lemma 1 the left hand side of (21), with unspecified $c_l \in \mathbb{K}$ which will be specialized further in a later step, can be rewritten as

$$\Big(\sum_{l=0}^{d} c_l\, M_l(k)\Big)\, t(k)$$

where

$$M_l(x) = \prod_{i=1}^{p} \mu_{\alpha_i^{(l)}}(a_i; x) \prod_{j=1}^{q} \mu_{\beta_j^{(l)}}(b_j - \beta_j^{(l)}; x) \in \mathbb{K}[x],$$

with $\mu_m(c, x)$ as in Definition 4 of Sect. 5.1.

According to Corollary 4 there exist polynomials $U_l(x)$ and $V_l(x)$ in $\mathbb{K}[x]$ with

$$\deg U_l(x) \leq d - 1 \quad \text{and} \quad \deg V_l(x) = \max\{\deg M_l(x) - d, -1\} \tag{59}$$

such that for all $k \geq 0$,

$$\Big(\sum_{l=0}^{d} c_l\, M_l(k)\Big)\, t(k) = \Big(\sum_{l=0}^{d} c_l\, U_l(k)\Big)\, t(k) + \Delta_k \Big(\sum_{l=0}^{d} c_l\, V_l(k)\Big)\, {}_pR_q(k)\, t(k). \tag{60}$$

If we can choose $c_l \in \mathbb{K}$, not all zero, such that

$$\sum_{l=0}^{d} c_l\, U_l(x) = 0, \tag{61}$$

then Theorem 1 is proven. Namely, using these specific solutions $c_0, \ldots, c_d$, not all 0, we can set

$$C(x) := \Big(\sum_{l=0}^{d} c_l\, V_l(x)\Big)\, {}_pR_q(x). \tag{62}$$

And, choosing the c_l and $C(x)$ this way, (60) is nothing but (21); moreover, we have $C(0) = 0$ owing to ${}_pR_q(0) = 0$. In addition, if $C \neq 0$ the degree estimate (22) holds which can be seen as follows. From (62) and (59) we have that

$$\deg C(x) \leq \deg {}_pR_q(x) - d + \max_{0 \leq l \leq d}\{\deg M_l(x)\},$$

and (22) is implied by $\deg {}_pR_q(x) = q + 1$ together with

$$\deg M_l(x) = \alpha_1^{(l)} + \cdots + \alpha_p^{(l)} + \beta_1^{(l)} + \cdots + \beta_q^{(l)},$$

according to (29).

Finally we show that a non-trivial choice of c_l satisfying (61) indeed exists. To this end we define for $0 \leq m \leq d - 1$ and $0 \leq l \leq d$,

$$u_{m,l} := \text{ coefficient of } x^m \text{ in } U_l(x).$$

This gives rise to a $d \times (d+1)$ matrix U via

$$U := (u_{m,l}) = \begin{pmatrix} u_{0,0} & u_{0,1} & \cdots & u_{0,d} \\ \vdots & \vdots & \ddots & \vdots \\ u_{d-1,0} & u_{d-1,1} & \cdots & u_{d-1,d} \end{pmatrix}. \tag{63}$$

Now finding all c_l satisfying (61) is equivalent to finding all solutions $(c_0, \ldots, c_d) \in \mathbb{K}^{d+1}$ to the homogeneous nullspace problem

$$\begin{pmatrix} u_{0,0} & u_{0,1} & \cdots & u_{0,d} \\ \vdots & \vdots & \ddots & \vdots \\ u_{d-1,0} & u_{d-1,1} & \cdots & u_{d-1,d} \end{pmatrix} \begin{pmatrix} c_0 \\ c_1 \\ \vdots \\ c_d \end{pmatrix} = \begin{pmatrix} 0 \\ \vdots \\ 0 \end{pmatrix}.$$

Since we have $d+1$ unknowns and d equations, there exists a solution $(c_0, \ldots, c_d) \in \mathbb{K}^{d+1}$ where the c_l are not all 0. This completes the proof of Theorem 1. □

Remark In symbolic summation various articles describe algorithms that split the summand into a summable and a non-summable part. Then computing a recurrence only for the non-summable part often yields a speed-up. For the hypergeometric case this has been considered, e.g., in [4]. The approach presented in this subsection is different: the main goal is to find optimal estimates on the shift-set and to guarantee that the certificate $C(x)$ is a polynomial. As pointed out by the anonymous referee, the approach of Sect. 5.2 might also yield a refined method to compute the parameterized telescoping solution, and that it would be interesting to check if the underlying system to be solved is simpler than the system one has to solve in the standard parameterized Gosper method.

5.3 Connection to Differential Equations

This section is not necessary for understanding the flow of the arguments. Nevertheless, we feel that we should at least mention how things are related to the classical hypergeometric differential equations.

We begin by recalling a fact which is straight-forward. In this section we suppose that $p \leq q+1$.

Lemma 5 *Let D_x be the differential operator with respect to x. For $n, k \in \mathbb{N}$ such that $n \geq k$,*

$$(xD_x)^n \,{}_pF_q\left(\begin{matrix}a_1,\ldots,a_p\\ b_1,\ldots,b_q\end{matrix};x\right)_k = k^n \,{}_pF_q\left(\begin{matrix}a_1,\ldots,a_p\\ b_1,\ldots,b_q\end{matrix};x\right)_k. \tag{64}$$

Next, for the choice $P(x) = {}_pP_q(x)$ and $R(x) = {}_pR_q(x)$, we sum both sides of (34) over all k from 0 to ∞ to obtain,

$$\begin{aligned}\sum_{k=0}^{\infty} {}_pP_q(k)\,{}_pF_q\left(\begin{matrix}a_1,\ldots,a_p\\ b_1,\ldots,b_q\end{matrix};z\right)_k &= -{}_pR_q(0) + \lim_{k\to\infty} {}_pR_q(k)\,{}_pF_q\left(\begin{matrix}a_1,\ldots,a_p\\ b_1,\ldots,b_q\end{matrix};z\right)_k\\ &= 0,\end{aligned} \tag{65}$$

where the last equality is owing to (23) and (36). Setting $\theta = zD_z$ and using Lemma 5, the left hand side of (65) turns into

$$\Big(z\prod_{i=1}^{p}(\theta + a_i) - \theta\prod_{j=1}^{q}(\theta + b_j - 1)\Big)\,{}_pF_q\left(\begin{matrix}a_1,\ldots,a_p\\ b_1,\ldots,b_q\end{matrix};z\right) \tag{66}$$

which is the non-trivial side of the classic homogeneous differential equation for ${}_pF_q$; see, for instance, [25, §47, eq. (2)].

Thus we can summarize as follows: *Relation* (34) *with the minimal choice $P(x) = {}_pP_q(x)$ and $R(x) = {}_pR_q(x)$ can be considered as a finite, telescoping version of the homogeneous differential equation for ${}_pF_q$ series.*

6 Applications of Theorem 1

Example We begin by considering one of Gauß' fifteen classical contiguous relations [10, 7.2],

$$(b-a)\,{}_2F_1\left(\begin{matrix}a,b\\ c\end{matrix};z\right) + a\,{}_2F_1\left(\begin{matrix}a+1,b\\ c\end{matrix};z\right) - b\,{}_2F_1\left(\begin{matrix}a,b+1\\ c\end{matrix};z\right) = 0. \tag{67}$$

We have $p = 2$, $q = 1$, and thus $d = \deg {}_2P_1(x) = 2$; in addition, $c \notin \mathbb{Z}_{\leq 0}$, and $|z| < 1$ as a condition for convergence.

Following the proof of Theorem 1 in Sect. 5.2, let us determine complex numbers c_0, c_1, and c_2, and a polynomial $C(x) \in \mathbb{C}[x]$ with $C(0) = 0$ and

$$\deg C(x) \leq q + 1 - d + \max\{0, 1\} = 1 + 1 - 2 + \max\{0, 1\} = 1,$$

such that for all $k \geq 0$,

$$c_0 \cdot {}_2F_1\left(\begin{matrix} a, b \\ c \end{matrix}; z\right)_k + c_1 \cdot {}_2F_1\left(\begin{matrix} a+1, b \\ c \end{matrix}; z\right)_k + c_2 \cdot {}_2F_1\left(\begin{matrix} a, b+1 \\ c \end{matrix}; z\right)_k$$
$$= \Delta_k\, C(k)\, {}_2F_1\left(\begin{matrix} a, b \\ c \end{matrix}; z\right)_k. \tag{68}$$

As in the proof of Lemma 2, w.l.o.g. we may assume that $a \notin \{0\}$ and $b \notin \{0\}$. Let

$$t(k) = \frac{(a)_k (b)_k}{(c)_k} \frac{z^k}{k!}. \tag{69}$$

According to Lemma 1 the left hand side of (68) can be written as

$$\Big(\sum_{l=0}^{2} c_l\, M_l(k)\Big)\, t(k)$$

where

$$M_0(x) = 1, \quad M_1(x) = 1 + \frac{x}{a}, \quad \text{and} \quad M_2(x) = 1 + \frac{x}{b}.$$

Hence, according to Corollary 4, to establish the relation (60) we can choose

$$U_l(x) = M_l(x) \quad \text{and} \quad V_l(x) = 0, \quad 0 \leq l \leq 2. \tag{70}$$

Then (62) implies that $C(x) = 0$; i.e., $C(0) = 0$ and $\deg C(x) = -1 \leq 1$.

Finally we have to choose $c_l \in \mathbb{C}$, not all 0, such that $\sum_{l=0}^{2} c_l\, U_l(x) = 0$. Because of (70) we obtain according to (63),

$$U = \begin{pmatrix} 1 & 1 & 1 \\ 0 & 1/a & 1/b \end{pmatrix}.$$

It is easily verified that

$$(c_0, c_1, c_2) = (b - a, a, -b)$$

generates the one-dimensional nullspace of

$$U = \begin{pmatrix} 1 & 1 & 1 \\ 0 & 1/a & 1/b \end{pmatrix} \begin{pmatrix} c_0 \\ c_1 \\ c_2 \end{pmatrix} = \begin{pmatrix} 0 \\ 0 \end{pmatrix}.$$

Consequently we obtain as the desired telescoping contiguous relation

$$(b-a)\,{}_2F_1\left(\begin{matrix} a, b \\ c \end{matrix}; z\right)_k + a\,{}_2F_1\left(\begin{matrix} a+1, b \\ c \end{matrix}; z\right)_k - b\,{}_2F_1\left(\begin{matrix} a, b+1 \\ c \end{matrix}; z\right)_k = 0. \tag{71}$$

Note that the right hand side is 0 which is due to $C(x) = 0$; this means, in this case the contiguous relation (67) is already true when restricted to the kth summand. Of course, as any (telescoping) contiguous relation, the equality (71) can be verified independently from its derivation. Namely, after dividing both sides by $t(k)$, relation (71) reduces to

$$b - a + a\,\frac{a+k-1}{a} - b\,\frac{b+k-1}{b} = 0. \tag{72}$$

Example As a second example we again consider relation (1),

$$(a+1-c)\,{}_2F_1\left(\begin{matrix} a, b \\ c \end{matrix}; z\right) + ((a+1-b)z - 2(a+1) + c)\,{}_2F_1\left(\begin{matrix} a+1, b \\ c \end{matrix}; z\right)$$
$$+ (1-z)(a+1)\,{}_2F_1\left(\begin{matrix} a+2, b \\ c \end{matrix}; z\right) = 0, \tag{73}$$

which is the first of Gauß' fifteen fundamental contiguous relations with a replaced by $a+1$. As explained in [1, (2.5.19)], this relation gives rise to a set of orthogonal polynomials.—As in the previous example we have $p = 2$, $q = 1$, and thus $d = \deg {}_2P_1(x) = 2$; in addition, $c \notin \mathbb{Z}_{\leq 0}$, and $|z| < 1$ as a condition for convergence.

Again by following the proof of Theorem 1 in Sect. 5.2, we determine complex numbers c_0, c_1, and c_2 and a polynomial $C(x) \in \mathbb{C}[x]$ with $C(0) = 0$ and

$$\deg C(x) \leq 1 + 1 - 2 + \max\{0, 1, 2\} = 2, \tag{74}$$

and such that for all $k \geq 0$,

$$c_0 \cdot {}_2F_1\left(\begin{matrix} a, b \\ c \end{matrix}; z\right)_k + c_1 \cdot {}_2F_1\left(\begin{matrix} a+1, b \\ c \end{matrix}; z\right)_k + c_2 \cdot {}_2F_1\left(\begin{matrix} a+2, b \\ c \end{matrix}; z\right)_k$$
$$= \Delta_k\, C(k)\,{}_2F_1\left(\begin{matrix} a, b \\ c \end{matrix}; z\right)_k. \tag{75}$$

As in the proof of Lemma 2, w.l.o.g. we may assume that $a \notin \{0, -1\}$. Let $t(k)$ be as in (69).

According to Lemma 1 the left hand side of (75) can be written as

$$\Big(\sum_{l=0}^{2} c_l\, M_l(k)\Big)\, t(k)$$

where

$$M_0(x) = 1, \quad M_1(x) = 1 + \frac{x}{a}, \quad \text{and}$$
$$M_2(x) = \left(1 + \frac{x}{a}\right)\left(1 + \frac{x}{a+1}\right) = 1 + \frac{2a+1}{a(a+1)}x + \frac{1}{a(a+1)}x^2.$$

From Lemma 4 (with $n = 0$) we obtain that for all $k \geq 0$,

$$(z-1) \cdot k^2\, t(k) = {}_2Q_1^{(0)}(k)\, t(k) + \Delta_k\, {}_2R_1(k)\, t(k) \tag{76}$$

where

$${}_2Q_1^{(0)}(x) = \operatorname{lcf} {}_2P_1(x) \cdot x^2 - {}_2P_1(x) = -\left((a+b)z - c + 1\right)x - abz,$$

and

$${}_2R_1(x) = x(x + c - 1).$$

Utilizing (76) we obtain

$$\begin{aligned}
&\Big(\sum_{l=0}^{2} c_l\, M_l(k)\Big)\, t(k)\\
&= \left(c_0\, M_0(k) + c_1\, M_1(k) + c_2\left(1 + \frac{2a+1}{a(a+1)}k\right)\right) t(k) + c_2 \frac{k^2}{a(a+1)} t(k)\\
&= \left(c_0\, U_0(k) + c_1\, U_1(k) + c_2\, U_2(k)\right) t(k) + \Delta_k\, C(k)\, t(k)
\end{aligned}$$

where

$$U_0(x) = 1 \quad (= M_0(x)), \quad U_1(x) = 1 + \frac{x}{a}, \quad (= M_1(x)), \quad \text{and}$$
$$U_2(x) = \frac{(a-b+1)z - a - 1}{(a+1)(z-1)} + \frac{(a-b+1)z - 2a - 2 + c}{a(a+1)(z-1)}x. \tag{77}$$

and

$$C(x) = \frac{c_2}{a(a+1)(z-1)}\, {}_2R_1(x). \tag{78}$$

Now (78) implies that $C(0) = 0$ and $\deg C(x) = 2$ in accordance with (74).

Finally we have to choose $c_l \in \mathbb{C}$, not all 0, such that $\sum_{l=0}^{2} c_l\, U_l(x) = 0$. Because of (77) we obtain according to (63),

$$U = \begin{pmatrix} 1 & 1 & \frac{(a-b+1)z-a-1}{(a+1)(z-1)} \\ 0 & \frac{1}{a} & \frac{(a-b+1)z-2a-2+c}{a(a+1)(z-1)} \end{pmatrix}.$$

It is easily verified that

$$(c_0, c_1, c_2) = (a(a-c+1), a((a-b+1)z-2a-2+c), a(a+1)(1-z)) \tag{79}$$

generates the one-dimensional nullspace of

$$U = \begin{pmatrix} 1 & 1 & \frac{(a-b+1)z-a-1}{(a+1)(z-1)} \\ 0 & \frac{1}{a} & \frac{(a-b+1)z-2a-2+c}{a(a+1)(z-1)} \end{pmatrix} \begin{pmatrix} c_0 \\ c_1 \\ c_2 \end{pmatrix} = \begin{pmatrix} 0 \\ 0 \end{pmatrix}.$$

Consequently, by choosing the c_l as in (79), and $C(x)$ with substituting $c_2 = a(a+1)(1-z)$ into (78), we obtain as the desired telescoping contiguous relation (75) from which (73) is obtained as usual by summation over all $k \geq 0$.

In practice, as stated and proven in Corollary 1, the coefficients c_l and the polynomial $C(x)$ are computed by the Parameterized Gosper Algorithm; see Sect. 7 and the examples presented subsequently.

7 A Package for Computing Telescoping Contiguous Relations

In Sect. 4 we began to explain that telescoping and classical contiguous relations can be computed automatically—up to restrictions imposed by computational complexity—by creative telescoping. Each computer algebra package that implements Zeilberger's algorithm is in its essence based on the parameterized Gosper algorithm which executes creative telescoping. Consequently each of these packages can be easily adapted to contiguous relations computations.

Already at the time of the prototype version [20] of this article, Axel Riese has carried out such an adaption within the Paule–Schorn [22] package `fastZeil`,

written in the Mathematica system and available from the Web at
https://combinatorics.risc.jku.at/software
To use this package, follow the installation instructions, open a Mathematica session, and read in the package as follows:

```
In[1]:= << RISC`fastZeil`
```

```
Fast Zeilberger Package version 3.61
written by Peter Paule, Markus Schorn, and Axel Riese
© RISC-JKU
```

For better readability, we write the rising factorials in 'pretty print' format:

In[2]:= $(\mathrm{x\_})_{\mathrm{k\_}}$:= **Pochhammer[x, k]**

In[3]:= $\{(\mathrm{x})_0, (\mathrm{x})_1, (\mathrm{x})_2, (\mathrm{x})_5\}$

Out[3]= $\{1, \mathrm{x}, \mathrm{x}(1+\mathrm{x}), \mathrm{x}(1+\mathrm{x})(2+\mathrm{x})(3+\mathrm{x})(4+\mathrm{x})\}$

7.1 *Computer Discovery and Proof of (67)*

To do the example (67), resp. (68), we consider the problem to compute c_0, c_1, c_2, not all 0, and a polynomial $C(x)$ such that for all $k \geq 0$,

$$c_0\, t_0(k) + c_1\, t_1(k) + c_2\, t_2(k) = \Delta_k\, C(k)\, t(k), \tag{80}$$

where $t(k) = t_0(k)$, and

$$t_0(k) = {}_2F_1\left(\begin{matrix} a, b \\ c \end{matrix}; z\right)_k, \; t_1(k) = {}_2F_1\left(\begin{matrix} a+1, b \\ c \end{matrix}; z\right)_k, \; t_2(k) = {}_2F_1\left(\begin{matrix} a, b+1 \\ c \end{matrix}; z\right)_k.$$

To invoke the package, we need to make explicit use of the similarity[4] between the $t_j(k)$:

In[4]:= **t[a_, b_, c_, k_] :=** $\frac{(\mathbf{a})_{\mathbf{k}}(\mathbf{b})_{\mathbf{k}}}{(\mathbf{c})_{\mathbf{k}}\mathbf{k}!}\mathbf{z}^{\mathbf{k}}$

In[5]:= **ra = t[a + 1, b, c, k]/t[a, b, c, k] //FullSimplify**

Out[5]= $\frac{\mathrm{a}+\mathrm{k}}{\mathrm{a}}$

In[6]:= **rb = t[a, b + 1, c, k]/t[a, b, c, k] //FullSimplify**

Out[6]= $\frac{\mathrm{b}+\mathrm{k}}{\mathrm{b}}$

[4] Recall Definition 2.

Hence we have,

$$t_1(k) = \text{ra} \cdot t(k) = \frac{a+k}{a} \cdot t(k) \text{ and } t_2(k) = \text{rb} \cdot t(k) = \frac{b+k}{b} \cdot t(k).$$

To solve the problem related to (80), we call parameterized telescoping as follows:

```
In[7]:= Gosper[t[a, b, c, k], {k, n1, n2}, Parameterized → {1, ra, rb}]
If '-n1+n2' is a natural number, then: :

Out[7]= {Sum [(a − b)F0(k) − aF1(k) + bF2(k), {k, n1, n2}] = 0}
```

This output has to be interpreted as follows: in the setting

$$F_0(k) = t_0(k), F_1(k) = t_1(k), \text{ and } F_2(k) = t_2(k),$$

one has for all $n_j \in \mathbb{N}$ such that $n_1 \leq n_2$:

$$(a-b)\sum_{k=n_1}^{n_2} t(k) - a\sum_{k=n_1}^{n_2} t_1(k) + b\sum_{k=n_1}^{n_2} t_2(k) = 0. \tag{81}$$

For $n = n_1 = n_2$ this is (71).

Remark As already mentioned, despite being a trivial relation on the summand level, relation (71), resp. (81), cannot be handled with the standard Zeilberger algorithm owing to the fact that we have shifts in *two* parameters: $a \to a+1$ and $b \to b+1$.

7.2 Computer Discovery and Proof of (73)

To do the example (73), resp. (75), we consider the problem to compute c_0, c_1, c_2, not all 0, and a polynomial $C(x)$ such that for all $k \geq 0$,

$$c_0\, t_0(k) + c_1\, t_1(k) + c_2\, t_2(k) = \Delta_k\, C(k)\, t(k), \tag{82}$$

where $t(k) = t_0(k)$, and

$$t_0(k) = {}_2F_1\left(\begin{matrix} a, b \\ c \end{matrix}; z\right)_k, t_1(k) = {}_2F_1\left(\begin{matrix} a+1, b \\ c \end{matrix}; z\right)_k, t_2(k) = {}_2F_1\left(\begin{matrix} a+2, b \\ c \end{matrix}; z\right)_k.$$

To invoke the package, we need again use similarity between the $t_j(k)$:

In[8]:= $\mathbf{t[a\_, b\_, c\_, k\_] := \dfrac{(a)_k (b)_k}{(c)_k k!} z^k}$

```
In[9]:= ra1 = t[a + 1, b, c, k]/t[a, b, c, k]  //FullSimplify
```

Out[9]= $\frac{a + k}{a}$

```
In[10]:= ra2 = (t[a + 2, b, c, k]/t[a + 1, b, c, k]   //FullSimplify) *
              (t[a + 1, b, c, k]/t[a, b, c, k]   //FullSimplify)//Factor
```

Out[10]= $\frac{(a + k)(1 + a + k)}{a(1 + a)}$

Hence we have,

$$t_1(k) = \mathrm{ra1}\cdot t(k) = \frac{a+k}{a}\cdot t(k) \text{ and } t_2(k) = \mathrm{ra2}\cdot t(k) = \frac{(a+k)(1+a+k)}{a(1+a)}\cdot t(k).$$

To solve the problem related to (82), we again call parameterized telescoping:

```
In[11]:= Gosper[t[a, b, c, k], {k, 0, n}, Parameterized → {1, ra1, ra2}]
```

If 'n' is a natural number, then: :

Out[11]= $\{\mathrm{Sum}[-a(a-c+1)F_0(k) - a(az-2a-bz+c+z-2)F_1(k) + a(a+1)(z-1)F_2(k), \{k, 0, n\}] = \frac{(a+n)(b+n)z^{n+1}(a)_n(b)_n}{n!(c)_n}\}$

This output has to be interpreted as follows: taking

$$F_0(k) = t_0(k),\ F_1(k) = t_1(k), \text{ and } F_2(k) = t_2(k),$$

one has for all $n \in \mathbb{N}$:

$$a(a+1-c)\sum_{k=0}^{n} t(k) + a((a+1-b)z - 2(a+1) + c)\sum_{k=0}^{n} t_1(k) + a(a+1)(1-z)\sum_{k=0}^{n} t_2(k) = -(n+1)(c+n)\cdot t(n+1). \tag{83}$$

For $n \to \infty$ this relation becomes Gauß' relation (73) since the right-hand side turns to zero owing to the limit property (23).

Finally, we note that subtracting from (83) the case $n - 1$ results in

$$a(a+1-c)t(n) + a((a+1-b)z - 2(a+1) + c)t_1(n) + a(a+1)(1-z)t_2(n) = -\Delta_n n(c+n-1)t(n),$$

which confirms the choice of the c_i as in (79). With these c_i we obtained the desired telescoping contiguous relation of the form (75) with $C(x) = -x(c+x-1)$.

8 Telescoping Contiguous Relations for $z = 1$: Case A

An important class of contiguous relations concerns the case $z = 1$ and $p = q+1 \geq 1$; i.e., involving summands of the form

$$ {}_{q+1}F_q\left(\begin{matrix} a_1, \ldots, a_{q+1} \\ b_1, \ldots, b_q \end{matrix}; 1\right)_k . \tag{84}$$

To establish versions of Theorem 1 for this situation, we need to refine further.

Definition 7 (Case-A Condition) We say that the complex parameters in (84) satisfy the Case-A condition, if

$$ \sum_{j=1}^{q} b_j - \sum_{i=1}^{q+1} a_i - q \notin \mathbb{Z}_{\geq 0}. \tag{85}$$

This section gives a Case-A version of Theorem 1; in the Sects. 9 and 10 corresponding theorems, Theorems 1B and 1C, for other parameter conditions, Case-B and Case-C, respectively, are presented.

Theorem 1A *Suppose $z = 1$ and $p = q+1$. Let the complex parameters a_i and b_j satisfy the Case-A condition* (85). *For $0 \leq l \leq q$ let $(\alpha_1^{(l)}, \ldots, \alpha_{q+1}^{(l)}, \beta_1^{(l)}, \ldots, \beta_q^{(l)})$ be pairwise different tuples with non-negative integer entries.*

Then there exist $c_0, \ldots, c_q$ in $\mathbb{K}$, not all 0, and a polynomial $C(x) \in \mathbb{K}[x]$ such that for all $k \geq 0$,

$$ \sum_{l=0}^{q} c_l \cdot {}_{q+1}F_q\left(\begin{matrix} a_1+\alpha_1^{(l)}, \ldots, a_{q+1}+\alpha_{q+1}^{(l)} \\ b_1-\beta_1^{(l)}, \ldots, b_q-\beta_q^{(l)} \end{matrix}; 1\right)_k = \Delta_k\, C(k)\, {}_{q+1}F_q\left(\begin{matrix} a_1, \ldots, a_{q+1} \\ b_1, \ldots, b_q \end{matrix}; 1\right)_k . \tag{86}$$

Moreover, $C(0) = 0$, and if $C(x) \neq 0$, for the polynomial degree of $C(x)$ one has

$$ \deg C(x) \leq 1+M \quad \text{where} \quad M := \max_{0 \leq l \leq q} \{\alpha_1^{(l)}+\cdots+\alpha_{q+1}^{(l)}+\beta_1^{(l)}+\cdots+\beta_q^{(l)}\}; \tag{87}$$

in addition,

$$ \mathrm{Re}\Big(\sum_{j=1}^{q} b_j - \sum_{i=1}^{q+1} a_i\Big) > M \Rightarrow \lim_{k\to\infty} C(k)\, {}_pF_q\left(\begin{matrix} a_1, \ldots, a_{q+1} \\ b_1, \ldots, b_q \end{matrix}; 1\right)_k = 0. \tag{88}$$

Remark According to [1, Thm. 2.1.2], the condition on the left-hand side of (88) is exactly the condition needed for the absolute convergence of all series

$$ {}_{q+1}F_q\left(\begin{matrix} a_1+\alpha_1^{(l)},\dots,a_{q+1}+\alpha_{q+1}^{(l)} \\ b_1-\beta_1^{(l)},\dots,b_q-\beta_q^{(l)} \end{matrix};1\right),\quad l\in\{0,\dots,q\}. $$

Proof Observe that when $z=1$,

$$ {}_{q+1}P_q(x)=x^q\Big(\sum_{i=1}^{q+1}a_i-\sum_{j=1}^{q}b_j+q\Big)+O(x^{q-1}), $$

hence $\deg {}_{q+1}P_q(x)=q$ and $\deg {}_{q+1}P_q^{(n)}(x)=n+q$. Using these degree estimates the statement is proven analogously to the proof of Theorem 1. The limit (88) follows from (87) by using,

$$ {}_{q+1}F_q\left(\begin{matrix} a_1,\dots,a_{q+1} \\ b_1,\dots,b_q \end{matrix};1\right)\sim\frac{\Gamma(b_1)\dots\Gamma(b_q)}{\Gamma(a_1)\dots\Gamma(a_{q+1})}k^{-1+\sum_i a_i-\sum_j b_j},\quad k\to\infty; \tag{89} $$

see [1, proof of Thm. 2.1.2] for this asymptotic estimate. □

Remark As in Theorem 1 one can allow arbitrary integer parameters instead of restricting to non-negative integers. More precisely, for arbitrary parameters $\alpha_i^{(l)}$ and $\beta_j^{(l)}$ this gives a relation,

$$ \sum_{l=0}^{q}c_l\cdot{}_{q+1}F_q\left(\begin{matrix} a_1+\alpha_1^{(l)},\dots,a_{q+1}+\alpha_{q+1}^{(l)} \\ b_1+\beta_1^{(l)},\dots,b_q+\beta_q^{(l)} \end{matrix};1\right)_k=\Delta_k\,R(k)\,{}_{q+1}F_q\left(\begin{matrix} a_1,\dots,a_{q+1} \\ b_1,\dots,b_q \end{matrix};1\right)_k, \tag{90} $$

with a *rational* function $R(x)\in\mathbb{K}(x)$ instead of a *polynomial* $C(x)\in\mathbb{K}[x]$. Again, as in (24), it is important to notice that because of possible poles of the $R(x)$, not all integer choices of $\alpha_i^{(l)}$ and $\beta_j^{(l)}$ are admissible.

Definition 8 Also the relations (86) and (90) are called *telescoping contiguous relations.*

Corollary 5 *Any telescoping contiguous relation of the form* (86) *and, if poles of* $R(x)$ *cause no problem, in the version of* (90) *can be computed by the parameterized Gosper algorithm.*

Proof Analogous to that for Corollary 1. □

We present some illustrating applications.

8.1 Telescoping Contiguous Relations for $z = 1$ and $(p, q) = (1, 0)$

Suppose $z = 1$ and $(p, q) = (1, 0)$. According to Theorem 1 there exist $c_0 \neq 0$ and a polynomial $C(x)$ with $C(0) = 0$ such that for all $k \geq 0$,

$$c_0 \cdot {}_1F_0\left(\begin{matrix} a+\alpha \\ - \end{matrix}; 1\right)_k = \Delta_k\, C(k)\, {}_1F_0\left(\begin{matrix} a \\ - \end{matrix}; 1\right)_k . \tag{91}$$

For all $\alpha \in \mathbb{Z}_{\geq 0}$ this is true if the case-A condition,

$$\sum_{j=1}^{q} b_j - \sum_{i=1}^{q+1} a_i - q = -a \notin \mathbb{Z}_{\geq 0}, \tag{92}$$

holds. It turns out that $C(x) = x/(a + \alpha)$, and summing (91) over k from 0 to n produces a telescoping hypergeometric sum,

$$\sum_{k=0}^{n} {}_1F_0\left(\begin{matrix} a+\alpha \\ - \end{matrix}; 1\right)_k = \sum_{k=0}^{n} (-1)^k \binom{-(a+\alpha)}{k} = \frac{n+1}{a+\alpha} \frac{(a)_{n+1}}{(n+1)!}. \tag{93}$$

Independently from Theorem 1A, this is obtained —including a confirmation of the Case-A condition—by our implementation of parameterized telescoping:

In[12]:= $\text{Gosper}\left[\frac{(\mathbf{a}+\alpha)_\mathbf{k}}{\mathbf{k}!}, \{\mathbf{k}, 0, \mathbf{n}\}\right]$

If 'n' is a natural number and $a + \alpha \neq 0$, then: :

Out[12]= $\text{Sum}\left[\frac{(\mathrm{a}+\alpha)_\mathrm{k}}{\mathrm{k}!}, \{\mathrm{k}, 0, \mathrm{n}\}\right] = \frac{(\mathrm{a}+\alpha+\mathrm{n})(\mathrm{a}+\alpha)_\mathrm{n}}{\mathrm{n}!(\mathrm{a}+\alpha)}$

Finally, we remark that applying the limit formula (88) the relation (93) turns into

$$\sum_{k=0}^{\infty} \frac{(a+\alpha)_k}{k!} = \sum_{k=0}^{\infty} (-1)^k \binom{-(a+\alpha)}{k} = 0,$$

matching $(1-1)^{-(a+\alpha)} = 0$.

8.2 Computer Proof of Gauß' ${}_2F_1$ Summation

Most of the classical non-terminating ${}_pF_q$ series summation formulas can be proved using contiguous relations. With the means of telescoping contiguous relations the essential part of these proofs can be done automatically by the computer.

For example, let us take Gauß' ${}_2F_1$ summation theorem [1, Thm. 2.2.2]:

$$\sum_{k=0}^{\infty} \frac{(a)_k (b)_k}{(c)_k k!} = {}_2F_1\left(\begin{matrix} a, b \\ c \end{matrix}; 1\right) = \frac{\Gamma(c)\Gamma(c-a-b)}{\Gamma(c-a)\Gamma(c-b)}, \quad \mathrm{Re}(c-a-b) > 0. \tag{94}$$

The condition on the real part is to guarantee absolute convergence; see [1, Thm. 2.1.2]. We will follow a variant of the proof idea presented in [1, 2.2]. Its key ingredient is the contiguous relation

$${}_2F_1\left(\begin{matrix} a, b \\ c \end{matrix}; 1\right) = \frac{(c-1)(c-a-b-1)}{(c-a-1)(c-b-1)} {}_2F_1\left(\begin{matrix} a, b \\ c-1 \end{matrix}; 1\right). \tag{95}$$

Once this relation is found, the rest of the proof of (94) follows by unfolding (95),

$$\begin{aligned} {}_2F_1\left(\begin{matrix} a, b \\ c \end{matrix}; 1\right) &= \frac{(c-1)(c-a-b-1)}{(c-a-1)(c-b-1)} \frac{(c-2)(c-a-b-2)}{(c-a-2)(c-b-2)} {}_2F_1\left(\begin{matrix} a, b \\ c-2 \end{matrix}; 1\right) = \dots \\ &= \frac{\Gamma(c)\Gamma(c-a-b)}{\Gamma(c-a)\Gamma(c-b)} \cdot \frac{\Gamma(c-a-n)\Gamma(c-b-n)}{\Gamma(c-n)\Gamma(c-a-b-n)} {}_2F_1\left(\begin{matrix} a, b \\ c-n \end{matrix}; 1\right), \end{aligned}$$

and by observing that

$$\lim_{n\to\infty} \frac{\Gamma(c-a-n)\Gamma(c-b-n)}{\Gamma(c-n)\Gamma(c-a-b-n)} {}_2F_1\left(\begin{matrix} a, b \\ c-n \end{matrix}; 1\right) = 1.$$

Using the parameterized Gosper algorithm, relation (95) is found automatically as follows. For the ansatz

$$c_0 \cdot {}_2F_1\left(\begin{matrix} a, b \\ c \end{matrix}; 1\right)_k + c_1 \cdot {}_2F_1\left(\begin{matrix} a, b \\ c-1 \end{matrix}; 1\right)_k = \Delta_k C(k)\, {}_2F_1\left(\begin{matrix} a, b \\ c \end{matrix}; 1\right)_k \tag{96}$$

the algorithm computes

$$c_0 = (c-a-1)(c-b-1),\ c_1 = -(c-1)(c-a-b-1), \quad \text{and } C(x) = x(x+c-1). \tag{97}$$

Consequently, $\deg C(x) \le 1+M$ where $M=1$. To ensure convergence of all series involved, one has to require also $\mathrm{Re}(c-a-b) > 1 = M$. This allows to apply (88) after summing (96) over k from 0 to ∞, which gives the desired (95).

9 Telescoping Contiguous Relations for $z = 1$: Case B

The next refinement of Theorem 1 concerns the following violation of the Case-A condition

Definition 9 (Case-B Condition) The parameters $a_1, \ldots, a_{q+1}$ and $b_1, \ldots, b_q$ satisfy the Case-B condition, if

$$\sum_{j=1}^{q} b_j - \sum_{i=1}^{q+1} a_i - q \in \mathbb{Z}_{\geq 1}. \tag{98}$$

Hence the remaining violation of the Case-A condition is when $\sum_{j=1}^{q} b_j - \sum_{i=1}^{q+1} a_i = q$; this is Case-C which is treated in Sect. 10.

Theorem 1B *Suppose $z = 1$ and $p = q + 1 \geq 2$. Let the complex parameters a_i and b_j satisfy the Case-B condition* (98) *and let $d = q - 1$ or $d = q$. For $0 \leq l \leq d$ let $(\alpha_1^{(l)}, \ldots, \alpha_{q+1}^{(l)}, \beta_1^{(l)}, \ldots, \beta_q^{(l)})$ be pairwise different tuples with non-negative integer entries. Suppose,*

$$M := \max_{0 \leq l \leq d} \{\alpha_1^{(l)} + \cdots + \alpha_{q+1}^{(l)} + \beta_1^{(l)} + \cdots + \beta_q^{(l)}\} < \sum_{j=1}^{q} b_j - \sum_{i=1}^{q+1} a_i. \tag{99}$$

Then for $d = q - 1$ or $d = q$ there exist $c_0, \ldots, c_d$ in $\mathbb{K}$, not all 0, *and a polynomial $C(x) \in \mathbb{K}[x]$ such that for all $k \geq 0$,*

$$\sum_{l=0}^{d} c_l \cdot {}_{q+1}F_q\left(\begin{matrix} a_1 + \alpha_1^{(l)}, \ldots, a_{q+1} + \alpha_{q+1}^{(l)} \\ b_1 - \beta_1^{(l)}, \ldots, b_q - \beta_q^{(l)} \end{matrix}; 1\right)_k = \Delta_k\, C(k)\, {}_{q+1}F_q\left(\begin{matrix} a_1, \ldots, a_{q+1} \\ b_1, \ldots, b_q \end{matrix}; 1\right)_k. \tag{100}$$

Moreover, $C(0) = 0$, and

$$\deg C(x) \leq 1 + \sum_{j=1}^{q} b_j - \sum_{i=1}^{q+1} a_i. \tag{101}$$

In addition, owing to (89), *if strict inequality in* (101) *holds,*

$$\lim_{k \to \infty} C(k)\, {}_pF_q\left(\begin{matrix} a_1, \ldots, a_{q+1} \\ b_1, \ldots, b_q \end{matrix}; 1\right)_k = 0; \tag{102}$$

otherwise,

$$\lim_{k\to\infty} C(k)_{q+1}F_q\left(\begin{matrix} a_1,\dots,a_{q+1} \\ b_1,\dots,b_q \end{matrix}; 1\right)_k = \textit{leading coefficient of } C(x)\cdot\frac{\Gamma(b_1)\dots\Gamma(b_q)}{\Gamma(a_1)\dots\Gamma(a_{q+1})}. \tag{103}$$

Remark According to [1, Thm. 2.1.2], the condition (99) is exactly the condition needed for the absolute convergence of all series

$${}_{q+1}F_q\left(\begin{matrix} a_1+\alpha_1^{(l)},\dots,a_{q+1}+\alpha_{q+1}^{(l)} \\ b_1-\beta_1^{(l)},\dots,b_q-\beta_q^{(l)} \end{matrix}; 1\right), \quad l\in\{0,\dots,q-1\}.$$

Proof Let $k_0 := \sum_{j=1}^{q} b_j - \sum_{i=1}^{q+1} a_i$, and suppose that $k_0 - q = \alpha \in \mathbb{Z}_{\geq 1}$. Observe that

$$\deg{}_{q+1}P_q(x) = q,\dots,\deg{}_{q+1}P_q^{(\alpha-1)}(x) = q+\alpha-1, \deg{}_{q+1}P_q^{(\alpha)}(x) = q+\alpha-1, \tag{104}$$

but

$$\deg{}_{q+1}P_q^{(\alpha+1)}(x) = q+\alpha+1.$$

Owing to (104), one can find a polynomial $P_1(x)$ with $\deg P_1(x) = \alpha$ such that

$$P(x) = \prod_{i=1}^{q+1}(x+a_i)\cdot P_1(x+1) - x\prod_{j=1}^{q}(x+b_j-1)\cdot P_1(x) = c\cdot x^{q-1} + O(x^{q-2}), \tag{105}$$

where c is some constant in $\mathbb{K}$. Analogously to the proof of Theorem 1, for

$$t_l(k) := {}_{q+1}F_q\left(\begin{matrix} a_1+\alpha_1^{(l)},\dots,a_{q+1}+\alpha_{q+1}^{(l)} \\ b_1-\beta_1^{(l)},\dots,b_q-\beta_q^{(l)} \end{matrix}; 1\right)_k \text{ and } t(k) := {}_{q+1}F_q\left(\begin{matrix} a_1,\dots,a_{q+1} \\ b_1,\dots,b_q \end{matrix}; 1\right)$$

one carries out the following transformation into a telescoping form:

$$\begin{aligned}\sum_{l=0}^{d} c_l t_l(k) &= \sum_{l=0}^{d} c_l \prod_{i=1}^{q+1} \mu_{\alpha_i^{(l)}}(a_i;k) \prod_{j=1}^{q} \mu_{\beta_j^{(l)}}(b_j-\beta_j^{(l)};k) t(k) \\ &= \sum_{l=0}^{d} c_l M_l(k) t(k) \\ &= \Big(\sum_{l=0}^{d} c_l\, U_l(k)\Big)\, t(k) + \Delta_k\Big(\sum_{l=0}^{d} c_l\, V_l(k)\Big)\,{}_pR_q(k)\, t(k).\end{aligned}$$

For each $l \in \{0, \dots, d\}$: if the constant c in (105) is non-zero, then $\deg U_l(x) \le q-2$; otherwise, $\deg U_l(x) \le q-1$. In the first case one has $d := q-1$; otherwise $d := q$. For both cases, with the same argument as in Theorem 1, there is a choice of the c_l, not all zero, such that $\sum_{l=0}^{d} c_l\, U_l(x) = 0$. Notice that for the transformation into a telescoping form we used the condition (99) for the estimate,

$$\deg M_l(x) = \alpha_1^{(l)} + \cdots + \alpha_{q+1}^{(l)} + \beta_1^{(l)} + \cdots + \beta_q^{(l)} \le M < k_0 = q + \alpha,$$

together with (104) and (105). To prove the remaining statements, one again uses the arguments as in the proof of Theorem 1. □

Remark Again one can allow arbitrary integer parameters instead of restricting to non-negative integers. More precisely, for arbitrary parameters $\alpha_i^{(l)}$ and $\beta_j^{(l)}$ this gives a relation,

$$\sum_{l=0}^{d} c_l \cdot {}_{q+1}F_q\left(\begin{matrix} a_1+\alpha_1^{(l)}, \dots, a_{q+1}+\alpha_{q+1}^{(l)} \\ b_1+\beta_1^{(l)}, \dots, b_q+\beta_q^{(l)} \end{matrix}; 1\right)_k = \Delta_k\, R(k)\, {}_{q+1}F_q\left(\begin{matrix} a_1, \dots, a_{q+1} \\ b_1, \dots, b_q \end{matrix}; 1\right)_k, \tag{106}$$

with a *rational* function $R(x) \in \mathbb{K}(x)$ instead of a *polynomial* $C(x) \in \mathbb{K}[x]$. Again, as in (24), it is important to notice that because of possible poles of the $R(x)$, not all integer choices of $\alpha_i^{(l)}$ and $\beta_j^{(l)}$ are admissible.

For the sake of better visibility, we state the criterion for the choice of d, which emerged from the proof, as a particular corollary.

Corollary 6 *Let c be the constant as in* (105). *Then a criterion for the choice of d in Theorem 1B is this:*

$$d = \begin{cases} q-1, & \text{if } c \neq 0 \\ q, & \text{if } c = 0 \end{cases}. \tag{107}$$

Definition 10 Also the relations (100) and (106) are called *telescoping contiguous relations.*

Corollary 7 *Any telescoping contiguous relation of the form* (86) *and, if poles of $R(x)$ cause no problem, in the version of* (90) *can be computed by the parameterized Gosper algorithm.*

Proof Analogous to that for Corollary 1. □

An example is provided by the existence of the contiguous relation (11), representing the Pfaff relation (7), which is predicted by Theorem 1B with $d = 2 = q-1$ and with $M = 4$ for the maximum of the shift sums; moreover, connecting to the proof of Theorem 1B, $k_0 = 5$. With the same data, Theorem 1B also implies the existence of the contiguous relation (12), representing the Pfaff relation (9).

The next subsections show two more illustrating examples.

9.1 Telescoping Contiguous Relations for $z = 1$ and $(p, q) = (2, 1)$

The minimal choice for Theorem 1B is $q = 1$; i.e., assuming that the constant c in Corollary 6 is non-zero, we seek for a relation,

$$c_0 \cdot {}_2F_1\left(\begin{matrix} a_1+\alpha_1, a_2+\alpha_2 \\ b_1-\beta_1 \end{matrix}; 1\right)_k = \Delta_k\, C(k)\, {}_2F_1\left(\begin{matrix} a_1, a_2 \\ b_1 \end{matrix}; 1\right)_k, \quad k \geq 0.$$

The Case-B condition is $b_1 - a_1 - a_2 - 1 := \alpha \in \mathbb{Z}_{\geq 1}$; hence $\alpha := 1$ is again a minimal choice. Finally, we need condition (99) to be satisfied. This means, $M := \alpha_1 + \alpha_2 + \beta_1 < b_1 - a_1 - a_2 = 2$, and $(\alpha_1, \alpha_2, \beta_1) = (0, 0, 0)$ is a minimal choice. Then Theorem 1B tells us that the sum

$$\sum_{k=0}^{n} \frac{(a_1)_k (a_2)_k}{(a_1+a_2+2)_k k!}$$

has a closed form by telescoping. In view of (99), other admissible choices are $(\alpha_1, \alpha_2, \beta_1) = (1, 0, 0)$ or $(\alpha_1, \alpha_2, \beta_1) = (0, 0, 0)$, which says that the sums

$$\sum_{k=0}^{n} \frac{(a_1+1)_k (a_2)_k}{(a_1+a_2+2)_k k!} \quad \text{and} \quad \sum_{k=0}^{n} \frac{(a_1)_k (a_2)_k}{(a_1+a_2+1)_k k!}$$

are also telescoping.[5] This can be confirmed by running Gosper's algorithm, for example:

In[13]:= $\text{Gosper}\left[\dfrac{(\mathbf{a_1})_\mathbf{k}(\mathbf{a_2})_\mathbf{k}}{(\mathbf{a_1}+\mathbf{a_2}+\mathbf{1})_\mathbf{k}\mathbf{k!}}, \{\mathbf{k}, \mathbf{0}, \mathbf{n}\}\right]$

If 'n' is a natural number and $a_1 a_2 \neq 0$,then: :

Out[13]= $\text{Sum}\left[\dfrac{(a_1)_k(a_2)_k}{(a_1+a_2+1)_k k!}, \{k, 0, n\}\right] = \dfrac{(a_1+n)(a_2+n)}{a_1 a_2} \dfrac{(a_1)_n (a_2)_n}{(a_1+a_2+1)_n n!}$

The algorithm computes $C(x) = x(x + a_1 + a_2)/(a_1 a_2)$, which means that we have equality in the bound estimate (101), and the limit (103) of Theorem 1B implies,

$$\sum_{k=0}^{\infty} \frac{(a_1)_k (a_2)_k}{(a_1+a_2+1)_k k!} = \frac{1}{a_1 a_2} \frac{\Gamma(a_1+a_2+1)}{\Gamma(a_1)\Gamma(a_2)} = \frac{\Gamma(a_1+a_2+1)}{\Gamma(a_1+1)\Gamma(a_2+1)}.$$

Notice that this is a telescoping special case of Gauß' summation formula (94).

[5]Notice that the right-hand sum is obtained by replacing a_1 with $a_1 - 1$ in the left sum.

10 Telescoping Contiguous Relations for $z = 1$: Case C

It remains to consider the final possibility for a violation of the Case-A condition, namely, when $\sum_{j=1}^{q} b_j - \sum_{i=1}^{q+1} a_i = q$.

Definition 11 (Case-C Condition) The parameters $a_1, \ldots, a_{q+1}$ and $b_1, \ldots, b_q$ satisfy the Case-C condition, if

$$\sum_{j=1}^{q} b_j - \sum_{i=1}^{q+1} a_i = q. \tag{108}$$

Theorem 1C *Suppose $z = 1$ and $p = q + 1 \geq 2$. Let the complex parameters a_i and b_j satisfy the Case-C condition* (108) *and let*

$$d := \begin{cases} q-1, & \text{if } \deg_{q+1} P_q(x) = q-1 \\ q, & \text{if } \deg_{q+1} P_q(x) < q-1 \end{cases}. \tag{109}$$

For $0 \leq l \leq d$ let $(\alpha_1^{(l)}, \ldots, \alpha_{q+1}^{(l)}, \beta_1^{(l)}, \ldots, \beta_q^{(l)})$ be pairwise different tuples with non-negative integer entries. Suppose,

$$M := \max_{0 \leq l \leq d} \{\alpha_1^{(l)} + \cdots + \alpha_{q+1}^{(l)} + \beta_1^{(l)} + \cdots + \beta_q^{(l)}\} < q. \tag{110}$$

Then for d there exist $c_0, \ldots, c_d$ in $\mathbb{K}$, not all 0, and a polynomial $C(x) \in \mathbb{K}[x]$ such that for all $k \geq 0$,

$$\sum_{l=0}^{d} c_l \cdot {}_{q+1}F_q\left(\begin{matrix} a_1 + \alpha_1^{(l)}, \ldots, a_{q+1} + \alpha_{q+1}^{(l)} \\ b_1 - \beta_1^{(l)}, \ldots, b_q - \beta_q^{(l)} \end{matrix}; 1\right)_k = \Delta_k\, C(k)\, {}_{q+1}F_q\left(\begin{matrix} a_1, \ldots, a_{q+1} \\ b_1, \ldots, b_q \end{matrix}; 1\right)_k. \tag{111}$$

Moreover, $C(0) = 0$, and

$$\deg C(x) \leq 1 + q. \tag{112}$$

In addition, owing to (89)*, if strict inequality in* (112) *holds,*

$$\lim_{k \to \infty} C(k)\, {}_pF_q\left(\begin{matrix} a_1, \ldots, a_{q+1} \\ b_1, \ldots, b_q \end{matrix}; 1\right)_k = 0; \tag{113}$$

otherwise,

$$\lim_{k\to\infty} C(k)_{q+1}F_q\left(\begin{matrix} a_1,\ldots,a_{q+1} \\ b_1,\ldots,b_q \end{matrix};1\right)_k = \textit{leading coefficient of } C(x)\cdot\frac{\Gamma(b_1)\ldots\Gamma(b_q)}{\Gamma(a_1)\ldots\Gamma(a_{q+1})}. \tag{114}$$

Remark According to [1, Thm. 2.1.2], the condition (99) is exactly the condition needed for the absolute convergence of all series

$${}_{q+1}F_q\left(\begin{matrix} a_1+\alpha_1^{(l)},\ldots,a_{q+1}+\alpha_{q+1}^{(l)} \\ b_1-\beta_1^{(l)},\ldots,b_q-\beta_q^{(l)} \end{matrix};1\right),\quad l\in\{0,\ldots,q-1\}.$$

Proof Observe that

$$\deg {}_{q+1}P_q(x)\le q-1,\ \text{ but }\ \deg {}_{q+1}P_q^{(n)}(x)=q+n,\ \ n\ge 1; \tag{115}$$

hence the degree $q-1$ provides a natural bound. Analogously to the proof of Theorem 1, for

$$t_l(k) := {}_{q+1}F_q\left(\begin{matrix} a_1+\alpha_1^{(l)},\ldots,a_{q+1}+\alpha_{q+1}^{(l)} \\ b_1-\beta_1^{(l)},\ldots,b_q-\beta_q^{(l)} \end{matrix};1\right)_k \text{ and } t(k) := {}_{q+1}F_q\left(\begin{matrix} a_1,\ldots,a_{q+1} \\ b_1,\ldots,b_q \end{matrix};1\right)$$

one carries out the following transformation into a telescoping form:

$$\begin{aligned}\sum_{l=0}^{q-1} c_l t_l(k) &= \sum_{l=0}^{q-1} c_l \prod_{i=1}^{q+1}\mu_{\alpha_i^{(l)}}(a_i;k)\prod_{j=1}^{q}\mu_{\beta_j^{(l)}}(b_j-\beta_j^{(l)};k)t(k)\\ &= \sum_{l=0}^{q-1} c_l M_l(k)t(k)\\ &= \Big(\sum_{l=0}^{q-1} c_l\,U_l(k)\Big)\,t(k)+\Delta_k\Big(\sum_{l=0}^{q-1} c_l\,V_l(k)\Big)\,{}_pR_q(k)\,t(k).\end{aligned}$$

For each $l\in\{0,\ldots,d\}$: if ${}_{q+1}P_q(x)=q-1$ then $\deg U_l(x)\le q-2$; otherwise, $\deg U_l(x)\le q-1$. In the first case one has $d:=q-1$; otherwise $d:=q$. For both cases, with the same argument as in Theorem 1, there is a choice of the c_l, not all zero, such that $\sum_{l=0}^{d} c_l\,U_l(x)=0$. Notice that for the transformation into a telescoping form we used the condition (110) for the estimate,

$$\deg M_l(x)=\alpha_1^{(l)}+\cdots+\alpha_{q+1}^{(l)}+\beta_1^{(l)}+\cdots+\beta_q^{(l)}\le M<q.$$

To prove the remaining statements, one again uses the arguments as in the proof of Theorem 1. □

Remark Again one can allow arbitrary integer parameters instead of restricting to non-negative integers. More precisely, for arbitrary parameters $\alpha_i^{(l)}$ and $\beta_j^{(l)}$ this gives a relation,

$$\sum_{l=0}^{q-1} c_l \cdot {}_{q+1}F_q\left(\begin{matrix} a_1+\alpha_1^{(l)}, \ldots, a_{q+1}+\alpha_{q+1}^{(l)} \\ b_1+\beta_1^{(l)}, \ldots, b_q+\beta_q^{(l)} \end{matrix}; 1\right)_k = \Delta_k\, R(k)\, {}_{q+1}F_q\left(\begin{matrix} a_1, \ldots, a_{q+1} \\ b_1, \ldots, b_q \end{matrix}; 1\right)_k, \tag{116}$$

with a *rational* function $R(x) \in \mathbb{K}(x)$ instead of a *polynomial* $C(x) \in \mathbb{K}[x]$. Again, as in (24), it is important to notice that because of possible poles of the $R(x)$, not all integer choices of $\alpha_i^{(l)}$ and $\beta_j^{(l)}$ are admissible.

Definition 12 Also the relations (111) and (116) are called *telescoping contiguous relations.*

Corollary 8 *Any telescoping contiguous relation of the form* (111) *and, if poles of* $R(x)$ *cause no problem, in the version of* (116) *can be computed by the parameterized Gosper algorithm.*

Proof Analogous to that for Corollary 1. □

An example is provided by the existence of the contiguous relation (10), representing the Zeilberger output recurrence (6), which is predicted by Theorem 1C with $d = \deg {}_4P_3(x) = 2 = q - 1$ and with $M = 2$ for the maximum of the shift sums. With the same data, Theorem 1C implies the contiguous relation (15) representing the new mixed relation (13) for Bailey's summation. To compute the relation as in (15), one uses the rational function variation (116) of Theorem 1C.

The next subsections show two more illustrating examples.

10.1 Telescoping Contiguous Relations for $z = 1$ and $(p, q) = (2, 1)$

The minimal choice for Theorem 1C is $q = 1$. Let us seek for a relation,

$$c_0 \cdot {}_2F_1\left(\begin{matrix} a_1+\alpha_1, a_2+\alpha_2 \\ b_1-\beta_1 \end{matrix}; 1\right)_k = \Delta_k\, C(k)\, {}_2F_1\left(\begin{matrix} a_1, a_2 \\ b_1 \end{matrix}; 1\right)_k, \quad k \geq 0.$$

To guarantee existence, according to Theorem 1C we consider,

$$\begin{aligned}\deg {}_2P_1(x) &= \deg\Big((x+a_1)(x+a_2) - x(x+b_1-1)\Big)\\ &= \deg\Big((a_1+a_2-b_1+1)x + a_1a_2\Big) = \deg a_1a_2 = 0,\end{aligned}$$

invoking the Case-C condition, $b_1 - a_1 - a_2 = 1$. Hence, if $a_1 a_2 \neq 0$ then $\deg_2 P_1(x) = 0 = q - 1$, and we can call Theorem 1C with $d = q - 1 = 0$. Moreover, we need $M := \alpha_1 + \alpha_2 + \beta_1 < q = 1$, thus $(\alpha_1, \alpha_2, \beta_1) = (0, 0, 0)$ is the only choice, and Theorem 1C tells us that the sum,

$$\sum_{k=0}^{n} \frac{(a_1)_k (a_2)_k}{(a_1 + a_2 + 1)_k k!} = \frac{(a_1 + 1)_n (a_2 + 1)_n}{(a_1 + a_2 + 1)_n n!}, \tag{117}$$

has a closed form by telescoping; for the evaluation see `Out[13]`. In other words, this minimal case coincides with that of Theorem 1B presented in Sect. 9.1.

10.2 Telescoping Contiguous Relations for $z = 1$ and $(p, q) = (3, 2)$

We proceed with $q = 2$ as the next to minimal choice for Theorem 1C. Let us seek for a relation,

$$c_0 \cdot {}_3F_2\left(\begin{matrix} a_1 + \alpha_1^{(0)}, a_2 + \alpha_2^{(0)}, a_3 + \alpha_3^{(0)} \\ b_1 - \beta_1^{(0)}, b_2 - \beta_1^{(0)} \end{matrix}; 1\right)_k + c_1 \cdot {}_3F_2\left(\begin{matrix} a_1 + \alpha_1^{(1)}, a_2 + \alpha_2^{(1)}, a_3 + \alpha_3^{(1)} \\ b_1 - \beta_1^{(1)}, b_2 - \beta_1^{(1)} \end{matrix}; 1\right)_k$$

$$= \Delta_k\, C(k)\, {}_3F_2\left(\begin{matrix} a_1, a_2, a_3 \\ b_1, b_2 \end{matrix}; 1\right)_k, \quad k \geq 0.$$

To guarantee existence, according to Theorem 1C, we consider,

$$\deg_3 P_2(x) = \deg\Big((x + a_1)(x + a_2)(x + a_3) - x(x + b_1 - 1)(x + b_2 - 1)\Big)$$

$$= \deg\Big((a_1 + a_2 + a_3 - b_1 - b_2 + 2)x^2 + (a_1 a_2 + a_1 a_3 + a_2 a_3 - (b_1 - 1)(b_2 - 1))x + a_1 a_2 a_3\Big)$$

$$= \deg\Big((a_1 a_2 + a_1 a_3 + a_2 a_3 - (b_1 - 1)(b_2 - 1))x + a_1 a_2 a_3\Big)$$

invoking the Case-C condition, $b_1 + b_2 - a_1 - a_2 - a_3 = 2$. Hence we assume that

$$a_1 a_2 + a_1 a_3 + a_2 a_3 - (b_1 - 1)(b_2 - 1) \neq 0,$$

because then $\deg_3 P_2(x) = 1$, and we can call Theorem 1C with $d = q - 1 = 1$. Moreover, we need to have,

$$M := \max_{0 \leq l \leq 1} \{\alpha_1^{(l)} + \alpha_2^{(l)} + \alpha_3^{(l)} + \beta_1^{(l)} + \beta_2^{(l)}\} < q = 2.$$

Thus, $(\beta_2^{(0)}, \alpha_3^{(1)}) = (1, 1)$ and all other parameters equal to zero, is an admissible choice. For this choice, Theorem 1C tells us that a non-trivial telescoping relation of the form

$$c_0 \cdot {}_3F_2\left(\begin{matrix} a_1, a_2, a_3 \\ b_1, b_2 - 1 \end{matrix}; 1\right)_k + c_1 \cdot {}_3F_2\left(\begin{matrix} a_1, a_2, a_3 + 1 \\ b_1, b_2 \end{matrix}; 1\right)_k$$
$$= \Delta_k\, C(k)\, {}_3F_2\left(\begin{matrix} a_1, a_2, a_3 \\ b_1, b_2 \end{matrix}; 1\right)_k, \quad k \geq 0,$$

exists provided that $b_1 + b_2 - a_1 - a_2 - a_3 = 2$. As described in Sect. 7, the RISC package `fastZeil` by applying parameterized telescoping computes:

$$c_0 = -a_3(1+a_3-b_1)(1+a_1+a_2+a_3-b_1),\, c_1 = a_3(1+a_1+a_3-b_1)(1+a_2+a_3-b_1)$$

and

$$C(x) = x(x + b_1 - 1)(x + a_1 + a_2 + a_3 - b_1 + 1);$$

for the computation b_2 is replaced by $a_1 + a_2 + a_3 - b_1 + 2$. Applying (114) one obtains in the limit $k \to \infty$ the relation:

$$c_0 \cdot {}_3F_2\left(\begin{matrix} a_1, a_2, a_3 \\ b_1, a_1 + a_2 + a_3 - b_1 + 1 \end{matrix}; 1\right) + c_1 \cdot {}_3F_2\left(\begin{matrix} a_1, a_2, a_3 + 1 \\ b_1, a_1 + a_2 + a_3 - b_1 + 2 \end{matrix}; 1\right)$$
$$= \frac{\Gamma(b_1)\Gamma(a_1 + a_2 + a_3 - b_1 + 2)}{\Gamma(a_1)\Gamma(a_2)\Gamma(a_3)}. \tag{118}$$

Another special case is obtained by replacing a_3 with $-n \in \mathbb{Z}_{\leq 0}$ and then taking the limit $k \to \infty$,

$$S(n) = \frac{(b_1 - a_1 + n - 1)(b_1 - a_2 + n - 1)}{(b_1 + n - 1)(b_1 - a_1 - a_2 + n - 1)} S(n-1), \quad n \geq 1. \tag{119}$$

where

$$S(n) := {}_3F_2\left(\begin{matrix} a_1, a_2, -n \\ b_1, a_1 + a_2 - b_1 - (n+1) \end{matrix}; 1\right).$$

Unfolding the relation (119) gives the celebrated Pfaff-Saalschütz formula [1, Thm. 2.2.6],

$$S(n) = \frac{(b_1 - a_1)_n (b_1 - a_2)_n}{(b_1)_n (b_1 - a_1 - a_2)_n}. \tag{120}$$

Finally, setting $b_1 = -n$ in $S(n)$ gives the telescoping special case (117) of Gauß' summation formula.

11 Further Applications

The examples presented in this section should deepen the impression of a wide spectrum of potential applications of telescoping contiguous relations and the methods described. Using parameterized telescoping, we derive a generalization, Theorem 2, of a result which arose in the classical work by James Wilson on hypergeometric recurrences and contiguous relations. Two further subsections discuss non-minimality of Zeilberger recurrences from telescoping contiguous relations point of view. In particular, we explain why 'creative symmetrizing' in some instances successfully reduces the order. This discussion includes a new (algorithmic) proof of the non-terminating version of Dixon's well-poised ${}_3F_2$-series.

11.1 Generalizing a Theorem by James A. Wilson

As mentioned, in [1] various approaches for deriving contiguous relations are described, for instance, by integration or by using Wilson's method [38]. We choose an example that is given in [1, (3.7.5)] for explaining Wilson's technique, namely

$$fg\,{}_4F_3\left(\begin{matrix}a, b, c, d\\ e, f, g\end{matrix}; 1\right) - (f-a)(g-a)\,{}_4F_3\left(\begin{matrix}a, b+1, c+1, d+1\\ e+1, f+1, g+1\end{matrix}; 1\right)$$
$$+\frac{a(e-b)(e-c)(e-d)}{e(e+1)}\,{}_4F_3\left(\begin{matrix}a+1, b+1, c+1, d+1\\ e+2, f+1, g+1\end{matrix}; 1\right) = 0, \tag{121}$$

where *one of the upper parameters is a negative integer*, and where

$$a+b+c+d+1 = e+f+g. \tag{122}$$

Throughout this section we assume (122) to hold. To connect to classical terminology we remark that Wilson's contiguous relation is between *balanced* ${}_4F_3$-series; i.e., as in (122) the sum of the top parameters plus 1 equals the sum of the bottom parameters. More generally, if the parameters of a ${}_{q+1}F_q$-series satisfy the relation

$$\sum_{j=1}^{q} b_j - \sum_{i=1}^{q+1} a_i = m \tag{123}$$

for $m \in \mathbb{Z}$, the series is called m-balanced.

To fit (121) into our framework we set

$$a_1 = a, a_2 = b, a_3 = c, a_4 = d, b_1 = e + 2, b_2 = f + 1, b_3 = g + 1,$$

which translates (121) into the telescoped version,

$$c_0\, {}_4F_3\left(\begin{matrix} a_1, a_2, a_3, a_4 \\ b_1 - 2, b_2 - 1, b_3 - 1 \end{matrix}; 1\right)_k + c_1\, {}_4F_3\left(\begin{matrix} a_1, a_2 + 1, a_3 + 1, a_4 + 1 \\ b_1 - 1, b_2, b_3 \end{matrix}; 1\right)_k$$
$$+ c_2\, {}_4F_3\left(\begin{matrix} a_1 + 1, a_2 + 1, a_3 + 1, a_4 + 1 \\ b_1, b_2, b_3 \end{matrix}; 1\right)_k = \Delta_k C(k)\, {}_4F_3\left(\begin{matrix} a_1, a_2, a_3, a_4 \\ b_1, b_2, b_3 \end{matrix}; 1\right)_k, \quad k \geq 0, \tag{124}$$

with

$$b_1+b_2+b_3-(a_1+a_2+a_3+a_4)-q = e+2+f+1+g+1-(a+b+c+d)-3 = 2.$$

Consequently, this turns the series into ones which are 5-balanced and the Case-B condition (98) is satisfied. Moreover, in view of

$$M := \max_{0 \leq l \leq q-1} \{\alpha_1^{(l)} + \cdots + \alpha_{q+1}^{(l)} + \beta_1^{(l)} + \cdots + \beta_q^{(l)}\} = 4 < \sum_{j=1}^{q} b_j - \sum_{i=1}^{q+1} a_i = 5,$$

the existence of (124) is guaranteed by Theorem 1B with $d = q - 1 = 2$, and where the coefficients c_j and the polynomial $C(x)$ can be computed by parameterized telescoping.

We remark explicitly that to this end, instead of renaming the variables, one can work directly in the original setting (121). More precisely, using our package we compute coefficients c_0, c_1, c_2, not all 0, and a polynomial $C(x)$ such that for all $k \geq 0$,

$$c_0\, t_0(k) + c_1\, t_1(k) + c_2\, t_2(k) = \Delta_k\, C(k)\, t(k), \tag{125}$$

where

$$t(k) = {}_4F_3\left(\begin{matrix} a, b, c, d \\ e + 2, f + 1, g + 1 \end{matrix}; 1\right)_k,$$

$$t_0(k) = {}_4F_3\left(\begin{matrix} a, b, c, d \\ e, f, g \end{matrix}; 1\right)_k,$$

$$t_1(k) = {}_4F_3\left(\begin{matrix} a, b+1, c+1, d+1 \\ e+1, f+1, g+1 \end{matrix}; 1\right)_k, \quad \text{and}$$

$$t_2(k) = {}_4F_3\left(\begin{matrix} a+1, b+1, c+1, d+1 \\ e+2, f+1, g+1 \end{matrix}; 1\right)_k.$$

For running the program we supply the rational functions r_l that are induced by the hypergeometric similarity relations

$$t_l(k) = r_l(k)\, t(k);$$

recall Definition 2. More concretely:

In[14]:= $\mathbf{t[k\_] := \dfrac{(a)_k (b)_k (c)_k (d)_k}{(e+2)_k (f+1)_k (g+1)_k k!}}$

In[15]:= $\mathbf{r0 = \dfrac{(a)_k (b)_k (c)_k (d)_k}{k!(e)_k (f)_k (g)_k} \dfrac{1}{t[k]}}$ **//FullSimplify**

Out[15]= $\dfrac{(k+e)(k+e+1)(k+f)(k+g)}{e(e+1)fg}$

In[16]:= $\mathbf{r1 = \dfrac{(a)_k (b+1)_k (c+1)_k (d+1)_k}{k!(e+1)_k (f+1)_k (g+1)_k} \dfrac{1}{t[k]}}$ **//FullSimplify**

Out[16]= $\dfrac{(k+b)(k+c)(k+d)(k+e+1)}{bcd(e+1)}$

In[17]:= $\mathbf{r2 = \dfrac{(a+1)_k (b+1)_k (c+1)_k (d+1)_k}{k!(e+2)_k (f+1)_k (g+1)_k} \dfrac{1}{t[k]}}$ **//FullSimplify**

Out[17]= $\dfrac{(k+a)(k+b)(k+c)(k+d)}{abcd}$

In[18]:= **RatFuMults = {r0, r1, r2} /. g → a + b + c + d − e − f + 1**

After these preparations we are ready to compute the desired telescoping relation:

In[19]:= **Gosper[t[k] /.g → a + b + c + d − e − f + 1, {k, 0, n − 1}, Parameterized → RatFuMults]**

If ‘n’ is a natural number, then: :

Out[19]= Sum[bcde(e+1)f(a+b+c+d−e−f+1)t$_0$(k)+bcde(e+1)(a−f)(b+c+d−e−f+1)t$_1$[k] − abcd(b − e)(c − e)(d − e)t$_2$[k], {k, 0, n − 1}]
$= -n(n+e+1)(n+f)(n+a+b+c+d-e-f+1)(bcd-bce-bde-ben-cde-cen-den-en^2) \dfrac{(a)_n(b)_n(c)_n(d)_n}{n!(e+2)_n(f+1)_n(a+b+c+d-e-f+2)_n}$

In other words, the telescoping relation (125) is constituted by

$$c_0 = bcde(e+1)fg, \quad c_1 = -bcde(e+1)(f-a)(g-a),$$
$$c_2 = -abcd(b-e)(c-e)(d-e),$$

and the polynomial

$$C(x) = -x(x+e+1)(x+f)(x+g)\left(bcd - bce - bde - cde - (b+c+d)ex - ex^2\right).$$

We remark that after executing the call `In[19]` for parameterized telescoping, the program allows to retrieve the polynomial $C(x)$ explicitly with

```
In[20]:= show[R]

Out[20]= k(1+e+k)(1+a+b+c+d−e−f+k)(f+k)(−bcd+bce+bde+cde+(b+c+d)ek+ek^2)
```

Obviously, if one of the entries a, b, c, d is a negative integer, relation (125) in the limit $k \to \infty$ implies Wilson's relation (121). But applying the limit property (103), the telescoping relation (125) as a "bonus" implies a generalization of Wilson's (121), *which does not require that one of the upper parameters is a negative integer*:

Theorem 2 *If* $a+b+c+d+1 = e+f+g$ *then*

$$fg\,{}_4F_3\left(\begin{matrix} a, b, c, d \\ e, f, g \end{matrix}; 1\right) - (f-a)(g-a)\,{}_4F_3\left(\begin{matrix} a, b+1, c+1, d+1 \\ e+1, f+1, g+1 \end{matrix}; 1\right)$$
$$+ \frac{a(e-b)(e-c)(e-d)}{e(e+1)}\,{}_4F_3\left(\begin{matrix} a+1, b+1, c+1, d+1 \\ e+2, f+1, g+1 \end{matrix}; 1\right)$$
$$= \frac{\Gamma(e+1)\Gamma(f+1)\Gamma(g+1)}{\Gamma(a)\Gamma(b+1)\Gamma(c+1)\Gamma(d+1)}. \tag{126}$$

11.2 Non-minimality of Zeilberger Recurrences

Bailey's summation (2) already has shown that Zeilberger's algorithm does not always deliver a recurrence of minimal order for the sum in question. Another such example is the summation

$$S_d(n) = \sum_{k=0}^{n} (-1)^k \binom{n}{k} \binom{d\,k}{n} = (-d)^n, \quad n \geq 0, \tag{127}$$

where d is any positive integer.

We remark that this evaluation is an immediate consequence of the following elementary fact which is implied by the binomial theorem. For any choice of complex numbers a_i,

$$\sum_{k=0}^{n}(-1)^k\binom{n}{k}(a_0 + a_1\,k + \cdots + a_n\,k^n) = (-1)^n n!\,a_n;$$

see, for instance, [12, Ch. 5].

However, in [22] it has been pointed out that the Zeilberger recurrence for $S_d(n)$, $d \geq 2$, is of order $d-1$. For instance, for $d = 3$ by running Zeilberger's algorithm one obtains

$$2(2n+3)\,S_3(n+2) + 3(5n+7)\,S_3(n+1) + 9(n+1)\,S_3(n) = 0$$

as the output recurrence for the sum $S_3(n)$.

In order to consider the problem from contiguous relations point of view, we translate $S_3(n)$ for $n = 3m$ into hypergeometric notation. One can easily verify that

$$S_3(3m) = (-1)^m\binom{3m}{m}\,T(m) \qquad (m \geq 0) \tag{128}$$

where

$$T(m) = \sum_{k=0}^{2m}\frac{(-2m)_k(m+1/3)_k(m+2/3)_k}{(1/3)_k(2/3)_k k!} = {}_3F_2\left(\begin{matrix}-2m,\, m+1/3,\, m+2/3\\ 1/3,\, 2/3\end{matrix};1\right). \tag{129}$$

According to (127) we have for $m \geq 0$,

$$T(m) = (-1)^m\binom{3m}{m}^{-1}(-3)^{3m}. \tag{130}$$

The fact that the Zeilberger recurrence for $T(m)$ is of order 2 tells us that there is no contiguous relation with $c_l \in \mathbb{C}(m)$ of the form

$$c_0\cdot{}_3F_2\left(\begin{matrix}-2m,\, m+1/3,\, m+2/3\\ 1/3,\, 2/3\end{matrix};1\right) + c_1\cdot{}_3F_2\left(\begin{matrix}-2m-2,\, m+4/3,\, m+5/3\\ 1/3,\, 2/3\end{matrix};1\right) = 0,$$

where in the second ${}_3F_2$-series m is replaced by $m+1$.

However, one can try another ansatz for a (telescoping) contiguous relation, for instance,

$$c_0\,{}_3F_2\left(\begin{matrix}-2m, m+1/3, m+2/3\\1/3, 5/3\end{matrix};1\right)_k + c_1\,{}_3F_2\left(\begin{matrix}-2m, m+1/3, m+2/3\\1/3, 2/3\end{matrix};1\right)_k$$
$$+ c_2\,{}_3F_2\left(\begin{matrix}-2m-2, m+4/3, m+5/3\\1/3, 2/3\end{matrix};1\right)_k = \Delta_k\, C(k)\,{}_3F_2\left(\begin{matrix}-2m-2, m+1/3, m+2/3\\1/3, 5/3\end{matrix};1\right)_k. \tag{131}$$

Summing (131) over k from 0 to $2m+2$ would then give the recurrence

$$c_0\, P(m) + c_1\, T(m) + c_2\, T(m+1) = 0, \quad m \geq 0, \tag{132}$$

where

$$P(m) = {}_3F_2\left(\begin{matrix}-2m, m+1/3, m+2/3\\1/3, 5/3\end{matrix};1\right) \tag{133}$$

is a *balanced* series which owing to the Pfaff-Saalschütz formula (120) evaluates to

$$P(m) = 0 \text{ for } m \geq 1. \tag{134}$$

Running the parameterized Gosper algorithm shows that a formula of type (131) indeed exists. Our package computes that (131), and thus (132), holds for

$$\begin{aligned}
c_0 &= 9m(3m-1)(21m^2+27m+8),\\
c_1 &= 18(m+1)(2m+1)(9m+1),\\
c_2 &= -(3m+1)(3m+2)(9m+1), \quad \text{and}\\
C(x) &= \frac{x(9x^2-4)\tilde{C}(x)}{2(m+1)(3m+2)(3m+4)}
\end{aligned}$$

where

$$\tilde{C}(x) = 3x(162m^3+405m^2+261m+40)-(3m+1)(189m^3+549m^2+555m+184).$$

Summarizing, in contrast to Zeilberger's algorithm the contiguous relations approach allows additional integer shifts in other parameters. So in the present example this enables one to invoke the Pfaff-Saalschütz evaluation (134) to zero, which finally has led to the desired order 1 recurrence for $T(m)$, namely

$$\frac{T(m+1)}{T(m)} = 18\frac{(m+1)(2m+1)}{(3m+1)(3m+2)},$$

which together with $T(0) = 1$ proves (130). *In other words, why Zeilberger's algorithm sometimes misses to compute the minimal recurrence simply is explained by the fact that this algorithm searches only within a restricted subclass of contiguous relations.*

Remark We want to note explicitly that this example is remarkable also with regard to the existence of (131). Renaming the variables as follows,

$$a_1 = -2m - 2, a_2 = m + 1/3, a_3 = m + 2/3, b_1 = 1/3, b_2 = 5/3,$$

turns (131) into

$$c_0 \cdot {}_3F_2\left(\begin{matrix} a_1+2, a_2, a_3 \\ b_1, b_2 \end{matrix}; 1\right)_k + c_1 \cdot {}_3F_2\left(\begin{matrix} a_1+2, a_2, a_3 \\ b_1, b_2-1 \end{matrix}; 1\right)_k$$

$$+ \; c_2 \cdot {}_3F_2\left(\begin{matrix} a_1, a_2+1, a_3+1 \\ b_1, b_2-1 \end{matrix}; 1\right)_k = \Delta_k \, C(k)\, {}_3F_2\left(\begin{matrix} a_1, a_2, a_3 \\ b_1, b_2 \end{matrix}; 1\right)_k. \quad (135)$$

In this case,

$$b_1+b_2-(a_1+a_2+a_3)-q = 1/3+5/3-(-2m-2+m+1/3+m+2/3)-2 = 1,$$

which matches the Case-B condition (98). But

$$M := \max_{0\le l\le 2} \{\alpha_1^{(l)} + \alpha_2^{(l)} + \alpha_3^{(l)} + \beta_1^{(l)} + \beta_2^{(l)}\} = 3 < \sum_{j=1}^{2} b_j - \sum_{i=1}^{3} a_i = 3$$

violates the requirement (99), and thus the existence of (131) cannot be derived from the generic form of Theorem 1B with $d = q = 2$. Nevertheless, the following refinement for $q = 2$ applies.

Corollary 1B *Suppose $z = 1$ and $p = q + 1 = 3$. Let the complex parameters a_i and b_j satisfy the Case-B condition* (98). *For* $0 \le l \le 2$ *let* $(\alpha_1^{(l)}, \alpha_2^{(l)}, \alpha_3^{(l)}, \beta_1^{(l)}, \beta_2^{(l)})$ *be pairwise different tuples with non-negative integer entries such that*

$$M := \max_{0\le l\le 2} \{\alpha_1^{(l)} + \alpha_2^{(l)} + \alpha_3^{(l)} + \beta_1^{(l)} + \beta_2^{(l)}\} = 3. \quad (136)$$

Then there exist c_0, c_1, c_2 *in* $\mathbb{K}$, *not all* 0, *and a polynomial* $C(x) \in \mathbb{K}[x]$ *such that for all* $k \ge 0$,

$$\sum_{l=0}^{2} c_l \cdot {}_3F_2\left(\begin{matrix} a_1+\alpha_1^{(l)}, a_2+\alpha_2^{(l)}, a_3+\alpha_3^{(l)} \\ b_1-\beta_1^{(l)}, b_2-\beta_2^{(l)} \end{matrix}; 1\right)_k = \Delta_k \, C(k)\, {}_3F_2\left(\begin{matrix} a_1, a_2, a_3 \\ b_1, b_2 \end{matrix}; 1\right)_k. \quad (137)$$

Proof To prove the statement one modifies the proof of Theorem 1; we restrict to presenting a sketch. In the case of Corollary 1B we have $k_0 := \sum_{j=1}^{q} b_j - \sum_{i=1}^{q+1} a_i = 3$; this means, $k_0 - q = \alpha$ with $q = 2$ and $\alpha = 1$. Observe that

$$\deg {}_3P_2(x) = 2, \deg {}_3P_2^{(1)}(x) = 2, \text{ but } \deg {}_3P_2^{(2)}(x) = 4. \tag{138}$$

Owing to (138), one can find a polynomial $P_1(x)$ with $\deg P_1(x) = 1$, or $\deg P_1(x) = 0$, such that

$$P(x) = \prod_{i=1}^{3}(x + a_i) \cdot P_1(x+1) - x \prod_{j=1}^{2}(x + b_j - 1) \cdot P_1(x) = p_1 x + p_0, \tag{139}$$

where $p_0, p_1 \in \mathbb{K}$ are not both zero. Suppose $p_1 \neq 0$. Then for $j = 1, 2$ there are $\gamma_j \in \mathbb{K}$ and polynomials $C_j(x) \in \mathbb{K}[x]$ such that

$$k^j\, t(k) = \gamma_j\, t(k) + \Delta_k C_j(x) t(k), \tag{140}$$

where $t(k) = {}_3F_2(a_1, a_2, a_3; b_1, b_2; 1)_k$. Again with the notation used in the proof of Theorem 1B, the left-hand side of (137) turns into,

$$\sum_{l=0}^{2} c_l t_k(k) = \sum_{l=0}^{2} c_l M_l(k) t(k),$$

with polynomials $M_l(x)$ of the form,

$$M_l(x) = \gamma_{l,0} + \gamma_{l,1}x + \gamma_{l,2}x^2 + \gamma_{l,3}x^3, \quad l = 0, 1, 2.$$

Finally, owing to (140),

$$\sum_{l=0}^{2} c_l M_l(k) t(k) = \Big(\sum_{l=0}^{2} c_l\, U_l(k)\Big)\, t(k) + \Delta_k \Big(\sum_{l=0}^{2} c_l\, V_l(k)\Big)\, {}_pR_q(k)\, t(k),$$

with $U_l(x)$ of the form $U_l(x) = u_{l,0} + u_{l,3}x^3$. Hence there exist $c_0, c_1, c_2 \in \mathbb{K}$, not all zero, such that $\sum_{l=0}^{2} c_l\, U_l(x) = 0$. □

11.3 *Creative Symmetrizing Revisited*

The discussion in Sect. 11.2 has shed new light on the fact that Zeilberger's algorithm does not always deliver a minimal recurrence for a given sum. In several such instances, by the method of 'creative symmetrizing', introduced in [18], it is possible to transform the original sum in such a way that for the transformed version Zeilberger's recurrence is minimal. In this section we shall see that the contiguous relations point of view can help to understand why creative symmetrizing can help.

As an illustrating example we consider a sum which Helmut Prodinger [24] has brought to our attention; namely for $n \geq 1$ let

$$S(n) = \sum_{k=1}^{n} (-1)^k \binom{n}{k}^2 \binom{n}{k-1}. \tag{141}$$

Carlitz [2], using Pfaff-Saalschütz summation, gave the evaluation

$$S(2m) = (-1)^m \frac{(3m)!}{(m!)^2\,(m-1)!\,(2m+1)} \tag{142}$$

for $m \geq 1$. However, he did not mention what happens if n is odd.

Before applying contiguous relations we explain what creative symmetrizing is about. Let $A(m) = S(2m)$ for $m \geq 1$. Again Zeilberger's algorithm does not deliver the first order recurrence corresponding to (142); rather than this it outputs,

$$\begin{aligned}
&-18(2m+1)(3m+1)(3m+2)(4m+7)(6m+5)(6m+7)\,A(m)\\
&-12(2m+3)(4m+5)(36m^4+180m^3+341m^2+290m+90)\,A(m+1)\\
&-2(m+1)(m+2)(2m+3)(2m+5)^2(4m+3)\,A(m+2) = 0,
\end{aligned}$$

which together with the corresponding certificate, which is too huge to be displayed here, is sufficient to prove (142).

However, as observed by Axel Riese [28] creative symmetrizing reduces the order to the minimal one. Namely, consider

$$\begin{aligned}
2A(m) &= \sum_{k=1}^{2m} (-1)^k \binom{2m}{k}^2 \binom{2m}{k-1} + \sum_{k=1}^{2m} (-1)^{2m+1-k} \binom{2m}{2m+1-k}^2 \binom{2m}{2m-k}\\
&= \sum_{k=1}^{2m} (-1)^k \binom{2m}{k}^2 \binom{2m}{k-1} \left(1 - \binom{2m}{k-1}\binom{2m}{k}^{-1}\right)\\
&= \sum_{k=1}^{2m} (-1)^k \frac{2m-2k+1}{2m-k+1} \binom{2m}{k}^2 \binom{2m}{k-1}.
\end{aligned}$$

This way we obtain an equivalent but transformed sum presentation $a(m)$ of $A(m)$, where

$$a(m) = \sum_{k=1}^{2m} (-1)^k \frac{2m-2k+1}{2(2m-k+1)} \binom{2m}{k}^2 \binom{2m}{k-1}. \tag{143}$$

Now, when we take the summand of $a(m)$ as input for Zeilberger's algorithm, it outputs as recurrence for the sum $a(m)$,

$$3(2m+1)(3m+1)(3m+2)\,a(m)+m(m+1)(2m+3)\,a(m+1)=0 \tag{144}$$

which, in view of $a(1)=A(1)=S(2)=-2$ immediately implies (142).

We note that in the odd case, i.e., if $n=2m+1$, Zeilberger's algorithm again gives a second order recurrence; but creative symmetrizing also helps here. Namely, analogous to above, for $B(m)=S(2m+1)$, $m\geq 0$, one obtains an equivalent but transformed sum presentation $b(m)$, where

$$b(m)=\sum_{k=1}^{2m+1}(-1)^k\frac{m+1}{2(2m-k+2)}\binom{2m+1}{k}^2\binom{2m+1}{k-1}. \tag{145}$$

For this rearrangement, Zeilberger's algorithm again outputs the minimal recurrence, namely

$$-3(3m+4)(3m+5)\,b(m)-(m+2)^2\,b(m+1)=0. \tag{146}$$

Consequently, since $b(0)=B(0)=S(1)=-1$, we obtain for $m\geq 0$,

$$S(2m+1)=b(m)=(-1)^{m+1}\frac{(3m+2)!}{2\,(m+1)!^2\,m!} \tag{147}$$

as a closed form evaluation for the odd case.

Summarizing, we have seen that creative symmetrizing, i.e., rearranging the summation by combining the first and the last summand, the second and the term before the last one, a.s.o., resulted in an order reduction of Zeilberger's output recurrence.

In the remaining part of this section we show that an explanation of this phenomenon is provided by the contiguous relations point of view.

To this end, let us consider the odd case (the even case can be treated analogously) and rewrite $B(m)=S(2m+1)$ into hypergeometric notation, i.e.,

$$B(m)=-(2m+1)^2\,{}_3F_2\left(\begin{matrix}-(2m+1),-2m,-2m\\2,2\end{matrix};1\right). \tag{148}$$

The ${}_3F_2$ series is nearly-poised, this means, the second top and the first bottom parameter add up to the same number as the third top and the second bottom parameter; in the given example this is $-2m+2$. The series would be well-poised, if the remaining top parameter increased by 1 would be the same number. As we will explain below, well-poised series behaves "more nicely" with respect to (telescoping) contiguous relations.

First, we point out that creative symmetrizing converts the sum representation (145) into a (terminating) well-poised series; namely,

$$b(m) = -(m+1)(2m+1)\,{}_3F_2\left(\begin{matrix}-2m, -(2m+1), -(2m+1)\\ 2, 2\end{matrix}; 1\right). \qquad (149)$$

In fact, this well-poised ${}_3F_2$ is the special case $a = -2m, b = c = -(2m+1)$ of Dixon's summation formula [1],

$${}_3F_2\left(\begin{matrix}a, b, c\\ a+1-b, a+1-c\end{matrix}; 1\right) = \frac{\Gamma(1+\frac{a}{2})\Gamma(1+\frac{a}{2}-b-c)\Gamma(1+a-b)\Gamma(1+a-c)}{\Gamma(1+a)\Gamma(1+a-b-c)\Gamma(1+\frac{a}{2}-b)\Gamma(1+\frac{a}{2}-c)}, \qquad (150)$$

where $\mathrm{Re}(1+\frac{1}{2}-b-c) > 0$. The substitution $a = -2m, b = c = -(2m+1)$ gives a hypergeometric term on the right-hand side of (150), hence $b(m)$ satisfies an order 1 recurrence.

Second, we explain why well-poised series behave better with respect to (telescoping) contiguous relations than nearly-poised series. Namely, Theorem 1A with $d = q = 2$ and parameterized telescoping gives,

$$c_0 \cdot {}_3F_2\left(\begin{matrix}a, b, c\\ a+1-b, a+1-c\end{matrix}; 1\right)_k + c_1 \cdot {}_3F_2\left(\begin{matrix}a+1, b, c\\ a+2-b, a+2-c\end{matrix}; 1\right)_k$$
$$+ c_2 \cdot {}_3F_2\left(\begin{matrix}a+2, b, c\\ a+3-b, a+3-c\end{matrix}; 1\right)_k = \Delta_k R(k)\,{}_3F_2\left(\begin{matrix}a, b, c\\ a+1-b, a+1-c\end{matrix}; 1\right)_k, \qquad (151)$$

where

$$c_0 = -a(1+a-b)(2+a-b)(2+a-2b-2c)(1+a-c)(2+a-c),$$
$$c_1 = 0,$$
$$c_2 = a(1+a)(2+a-2b)(2+a-2c)(1+a-b-c)(2+a-b-c),$$

and

$$R(k) = p(k)\frac{(-2-a+b)(-1-a+b)(-2-a+c)(-1-a+c)k}{(1+a-b+k)(1+a-c+k)}$$

with

$$p(k) = -2 - a + 3a^2 + 2a^3 + 4b + ab - 2a^2b - 2b^2 + 4c + ac - 2a^2c - 6bc - abc$$
$$+ 2b^2c - 2c^2 + 2bc^2 + 3ak + 3a^2k - 2abk - 2ack + ak^2.$$

Remark Notice that in view of the pattern of the shifts of the bottom parameters in the ${}_3F_2$-series, we applied Theorem 1A in the version of (106) which gives a *rational* function $R(x) \in \mathbb{K}(x)$ instead of a *polynomial* $C(x) \in \mathbb{K}[x]$.

Inspection of the coefficients c_j reveals the crucial feature of the well-poised property: it puts c_1 to zero!

Besides its relevance for our example, the fact that $c_1 = 0$ allows a proof of Dixon's identity along the same lines as our proof of Gauß' summation formula (94). Using the abbreviation,

$$F(a,b,c) := {}_3F_2\left(\begin{matrix} a, b, c \\ a+1-b, a+1-c \end{matrix}; 1\right),$$

in the limit $k \to \infty$ relation (151) turns into

$$\begin{aligned} F(a,b,c) &= \frac{(1+a)(2+a-2b)(2+a-2c)(1+a-b-c)(2+a-b-c)}{(1+a-b)(2+a-b)(2+a-2b-2c)(1+a-c)(2+a-c)} F(a+2,b,c) \\ &= \frac{(a)_{2n}(\frac{a}{2}-b+1)_n(\frac{a}{2}-c+1)_n(a-b-c+1)_{2n}}{(\frac{a}{2})_n(\frac{a}{2}-b-c+1)_n(a-b+1)_{2n}(a-c+1)_{2n}} F(a+2n,b,c). \end{aligned}$$

Finally, applying

$$\Gamma(x) = \lim_{n\to\infty} \frac{n!n^{x-1}}{(x)_n} \quad \text{and} \quad \lim_{n\to\infty} F(a+2n,b,c) = 1$$

proves (150).

Remark Connecting to the remarks given after the statement of Theorem 1 in Sect. 4, we note that Dixon's identity (150) in the limit $c \to \infty$ gives

$${}_2F_1\left(\begin{matrix} a, b \\ a+1-b \end{matrix}; -1\right) = \frac{\Gamma(1+\frac{a}{2})\Gamma(1+a-b)}{\Gamma(1+a)\Gamma(1+\frac{a}{2}-b)}, \quad \mathrm{Re}(b) < 1, \tag{152}$$

which is Kummer's summation theorem [1, Cor. 3.1.2], Alternatively, one can compute the telescoping relation of the form,

$$\begin{aligned} &c_0 \cdot {}_2F_1\left(\begin{matrix} a, b \\ a+1-b \end{matrix}; -1\right)_k + c_1 \cdot {}_2F_1\left(\begin{matrix} a+1, b \\ a+2-b \end{matrix}; -1\right)_k \\ &+ c_2 \cdot {}_2F_1\left(\begin{matrix} a+2, b \\ a+3-b \end{matrix}; -1\right)_k = \Delta_k R(k) {}_2F_1\left(\begin{matrix} a, b \\ a+1-b \end{matrix}; -1\right)_k, \end{aligned} \tag{153}$$

where $c_0 = -(a+1-b)(a+2-b)$, $c_1 = 0$, $c_2 = (a+1)(a+2-2b)$ and $R(x) = (a+1-b)(a+2-b)(b-1)x/(a(a+1-b+x))$. As with (151), this relation is a recurrence with shifts in a only, hence it can be computed already with Zeilberger's algorithm. However, we want to emphasize that its existence is predicted by Theorem 1 applied with the condition $p = 2 = q+1$ and $z = -1$.

Finally note that Kummer's summation follows by taking the limit $k \to \infty$ in (153), and by iterating the resulting relation as we did to obtain the Dixon sum. To arrive at (152), one has to apply the binomial theorem [1, (2.1.6)] in the form,

$$ {}_1F_0\left(\begin{matrix} b \\ - \end{matrix}; -1\right) = 2^{-b}. $$

12 Conclusion: q-Case

There are many variations like Corollary 1B of the method presented. Such variations depend on the particular application, needless to say. But even when facing a problem not generically covered by one of the theorems in this article, using a computer algebra implementation of parameterized telescoping could lead to the desired (telescoping) contiguous relation. Still this algorithmic possibility does not make tables of such relations obsolete. An excellent reference in this regard is [17], a huge collection of hypergeometric series summation and transformation identities including contiguous relations; most importantly, the table look-up is greatly supported by coming in the form of a Mathematica package.

Another aspect is that all what has been said in this article carries over to q-hypergeometric series and to q-contiguous relations. We are planning to treat the q-case in a subsequent paper.

As a kind of a "preview": already at the time of [20], Axel Riese has implemented a q-version of the algorithm for computing (telescoping) contiguous relations described in Sect. 4. This extension of his Mathematica package `qZeil` [21] allows to derive automatically (telescoping) q-contiguous relations, for example, those of Heine [9, Exercise 1.9]. Also in the scope are q-functional relations like

$$ F(a, b; t) = \frac{1 - atq}{1 - t} + \frac{(1 - aq)(b - atq)}{(1 - bq)(1 - t)} tqF(aq, bq; tq) $$

where

$$ F(a, b; t) = 1 + \sum_{n=1}^{\infty} \frac{(1 - aq)(1 - aq^2) \cdots (1 - aq^n)}{(1 - bq)(1 - bq^2) \cdots (1 - bq^n)} t^n; $$

see, for instance, the book by N.J. Fine [7, (4.1)].

Acknowledgments Part of this work has been carried out at the Department of Mathematics of the Pennsylvania State University. I wish to thank George Andrews for his overwhelming hospitality and for providing a wonderful working environment. For valuable feed-back on an early version of this article I want to acknowledge my thanks to George Andrews, Richard Askey, Frédéric Chyzak, Tom Koornwinder, Christian Krattenthaler, Bruno Salvy, Raimundas Vidunas, Herb Wilf, and Doron Zeilberger. Special thanks go to the anonymous referee who read the manuscript extremely carefully and whose suggestions for improvement were highly appreciated.

Partially supported by SFB-grant F50-06 of the Austrian Science Fund FWF, and by a visiting researcher grant of the Department of Mathematics of the Pennsylvania State University.

References

1. G.E. Andrews, R. Askey, R. Roy, *Special Functions*. Encyclopedia of Mathematics and Its Applications, vol. 71, ed. by G.-C. Rota (Cambridge University Press, Cambridge, 1999)
2. L. Carlitz, Some binomial sums. Fibonacci Quart. **14**, 249–253 (1976)
3. W.Y.C. Chen, Q.-H. Hou, Y.-P. Mu, The extended Zeilberger algorithm with parameters. J. Symb. Comp. **47**, 643–654 (2012)
4. S. Chen, H. Huang, M. Kauers, Z. Li, An improved Abramov-Petkovšek reduction and creative telescoping for hypergeometric terms, in *The 2015 International Symposium on Symbolic and Algebraic Computation* (ACM, New York, 2015), pp. 117–124
5. F. Chyzak, *About the non-minimality of the outputs of Zeilberger's algorithm*, RISC Technical Report 00-12, Research Institute for Symbolic Computation (RISC), Johannes Kepler University Linz (2000), p. 20. Available at https://www3.risc.jku.at/publications
6. F. Chyzak, *The ABC of Creative Telescoping: Algorithms, Bounds, Complexity*, Memoir of accreditation to supervise research (HDR), Université d'Orsay, (2014), p. 64. Available at https://specfun.inria.fr/chyzak/publications
7. N.J. Fine, *Basic Hypergeometric Series and Applications*. Mathematical Surveys and Monographs, vol. 27 (American Mathematical Society, Providence, 1989)
8. F. Gao, V.J.W. Guo, Contiguous relations and summation and transformation formulae for basic hypergeometric series. J. Diff. Equ. Appl. **19**, 2029–2041 (2013)
9. G. Gasper, M. Rahman, *Basic Hypergeometric Series* (Cambridge University Press, Cambridge, 1990)
10. C.F. Gauss, *Disquisitiones generales circa seriem infinitam* $1 + \frac{\alpha\beta}{1\gamma}x + \dots$, pars prior, Commentationes societatis regiae scientarum Gottingensis recentiores 2 (classis mathematicae) (1813), pp. 3–46. https://gdz.sub.uni-goettingen.de/id/PPN235999628
11. R.W. Gosper, Jr., Decision procedure for indefinite hypergeometric summation. Proc. Natl. Acad. Sci. U.S.A. **75**, 40–42 (1978)
12. R.L. Graham, D.E. Knuth, O. Patashnik, *Concrete Mathematics*, 2nd edn. (Addison–Wesley, Reading, 1994)
13. M. Kalmykov, V. Bytev, B. Kniehl, S.-O. Moch, B. Ward, S. Yost, *Hypergeometric Functions and Feynman Diagrams*, in *Anti-Differentiation and the Calculation of Feynman Amplitudes*, ed. by C. Schneider, J. Blümlein. Texts & Monographs in Symbolic Computation (Springer, Berlin, 2021)
14. M. Karr, Summation in finite terms. J. ACM **28**, 305–350 (1981)
15. M. Karr, Theory of summation in finite terms. J. Symbolic Comput. **1**, 303–315 (1985)
16. C. Koutschan, Creative telescoping for holonomic functions, in *Computer Algebra in Quantum Field Theory: Integration, Summation and Special Functions*, ed. by C. Schneider, J. Blümlein. Texts & Monographs in Symbolic Computation (Springer, Berlin, 2013), pp. 171–194
17. C. Krattenthaler, HYP and QHYP. J. Symbolic Comput. **20**, 737–744 (1995)
18. P. Paule, Short and easy computer proofs of the Rogers-Ramanujan identities and of identities of similar type. Electron. J. Combin. **1**, R10, 1–9 (1994)
19. P. Paule, Greatest factorial factorization and symbolic summation. J. Symbolic Comput. **20**, 235–268 (1995)
20. P. Paule, *Contiguous Relations and Creative Telescoping*, Technical Report, RISC (2001)

21. P. Paule, A. Riese, A Mathematica q-analogue of Zeilberger's algorithm based on an algebraically motivated approach to q-hypergeometric telescoping, in Fields Institute Communications, vol. 14 (American Mathematical Society, Providence, 1997), pp. 179–210
22. P. Paule, M. Schorn, A Mathematica version of Zeilberger's algorithm for proving binomial coefficient identities. J. Symbolic Comput. **20**, 673–698 (1995)
23. M. Petkovšek, H.S. Wilf, and D. Zeilberger, $A=B$ (Peters, Wellesley, 1996)
24. H. Prodinger, *Email from January 12th, 2001.*
25. E.D. Rainville, *Special Functions* (MacMillan, New York, 1960)
26. M.A. Rakha, A.K. Ibrahim, On the contiguous relations of hypergeometric series. J. Comp. Appl. Math. **192**, 396–410 (2006)
27. M.A. Rakha, A.K. Ibrahim, A.K. Rathie, On the computations of contiguous relations for ${}_2F_1$ *hypergeometric series*. Commun. Korean Math. Soc. **24**, 291–302 (2009)
28. A. Riese, *Carlitz.nb* Mathematica Notebook (2001)
29. C. Schneider, Symbolic summation in difference fields, Ph.D. Thesis, RISC, J. Kepler University, Linz, 2001
30. C. Schneider, Symbolic summation assists combinatorics. Sém. Lothar. Combin. **56**, 1–36 (2007)
31. C. Schneider, Parameterized telescoping proves algebraic independence of sums. Ann. Comb. **14**, 533–552 (2010)
32. C. Schneider, Fast algorithms for refined parameterized telescoping in difference fields, in *Computer Algebra and Polynomials*. Lecture Notes in Computer Science, vol. 8942 (Springer, Berlin, 2015), pp. 157–191
33. C. Schneider, A difference ring theory for symbolic summation. J. Symb. Comput. **72**, 82–127 (2016)
34. C. Schneider, Summation theory ii: characterizations of $R\Pi\Sigma$-extensions and algorithmic aspects. J. Symb. Comput. **80**, 616–664 (2017)
35. N. Takayama, Gröbner bases and the problem of contiguous relations. J. Appl. Math. **6**, 147–169 (1989)
36. R. Vidunas, Contiguous relations of hypergeometric series. J. Comp. Appl. Math. **153**, 507–519 (2003)
37. H.S. Wilf, D. Zeilberger, Rational functions certify combinatorial identities. J. Amer. Math. Soc. **3**, 147–158 (1990)
38. J.A. Wilson, Hypergeometric series, recurrence relations and some new orthogonal polynomials, Ph.D. Thesis, University of Wisconsin, Madison, 1978
39. D. Zeilberger, A fast algorithm for proving terminating hypergeometric identities. Discrete Math. **80**, 207–211 (1990)
40. D. Zeilberger, A holonomic systems approach to special functions identities. J. Comp. Appl. Math. **32**, 321–368 (1990)
41. D. Zeilberger, The method of creative telescoping. J. Symbolic Comput. **11**, 195–204 (1991)

Nested Integrals and Rationalizing Transformations

Clemens G. Raab

Abstract A brief overview of some computer algebra methods for computations with nested integrals is given. The focus is on nested integrals over integrands involving square roots. Rewrite rules for conversion to and from associated nested sums are discussed. We also include a short discussion comparing the holonomic systems approach and the differential field approach. For simplification to rational integrands, we give a comprehensive list of univariate rationalizing transformations, including transformations tuned to map the interval [0, 1] bijectively to itself.

1 Introduction

By nested integrals, we mean multiple integrals whose integrand is a product of individual integrands depending only on one integration variable each and where each integration variable occurs as an integration bound for the next inner integral. Commonly, they use the origin as their base point and have the form

$$\int_0^x f_1(t_1) \int_0^{t_1} f_2(t_2) \dots \int_0^{t_{k-1}} f_k(t_k)\, dt_k \dots dt_1, \tag{1}$$

where each f_i is allowed to depend only on t_i and on external parameters, but not on any other integration variable t_j. Note that here and in all that follows, the possible dependence on external parameters is not denoted explicitly. Also conventions with base points other than the origin are possible, e.g.

$$\int_x^1 f_1(t_1) \int_{t_1}^1 f_2(t_2) \dots \int_{t_{k-1}}^1 f_k(t_k)\, dt_k \dots dt_1. \tag{2}$$

C. G. Raab (✉)
Inst. f. Algebra, Johannes Kepler Universität Linz, Linz, Austria
e-mail: clemensr@algebra.uni-linz.ac.at

J. Blümlein, C. Schneider (eds.), *Anti-Differentiation and the Calculation of Feynman Amplitudes*, Texts & Monographs in Symbolic Computation,
https://doi.org/10.1007/978-3-030-80219-6_16

Choosing the integrands from certain classes of functions, the nested integrals give rise to various classes of functions. Already Kummer [30] considered nested integrals over rational functions and determined that it is sufficient to consider integrands that are reciprocals of linear polynomials in the integration variable. These integrals are often referred to as *hyperlogarithms*. They occurred also in the works of Poincaré, Lappo-Danilevski, and many others, see e.g. [27, 31, 35, 48]. In the context of QFT, special choices of rational integrands have been used frequently, giving *harmonic polylogarithms* [41], *cyclotomic harmonic polylogarithms* [10], and *generalized harmonic polylogarithms* [11]. Generalizing the notation of Ref. [41], integrands for generalized harmonic polylogarithms $\mathrm{H}_{a_1,\ldots,a_k}(x)$ are given by

$$f_a(x) := \frac{c_a}{x-a} \tag{3}$$

with $c_a := \operatorname{sgn}(-a+0)$ such that $f_a(x) > 0$ for $x > 0$ close to zero, which is the base point of these integrals following (1). For integrals of the form (2), the choice $c_a := \operatorname{sgn}(1-a-0)$ is preferred in order to have $f_a(x) > 0$ for $x < 1$ close to one.

Going beyond rational integrands, one can also allow more general functions by requiring only the square of the integrand to be a rational function. A minimal set of integrands needed to express these integrals (and hence also those over any algebraic functions expressible by non-nested square roots) was determined in Ref. [8] based on work by Hermite [28], which consists of integrands (3) and the following ones.

$$f_{\{a_1,\ldots,a_k\}}(x) := f_{a_1}(x)^{1/2} \ldots f_{a_k}(x)^{1/2} \qquad \text{for } k \geq 2 \tag{4}$$

$$f_{(a,\{a_1,\ldots,a_k\})}(x) := f_a(x) f_{\{a_1,\ldots,a_k\}}(x) \qquad \text{for } k \geq 1, a \notin \{a_1,\ldots,a_k\} \tag{5}$$

$$f_{(\{a_1,\ldots,a_k\},j)}(x) := x^j f_{\{a_1,\ldots,a_k\}}(x) \qquad \text{for } j \in \{1,\ldots,k-2\} \tag{6}$$

In the context of QFT, certain explicit non-rational integrands of this type occurred already in Ref. [13], for instance, and continue to arise more often in computations in recent years, see e.g. [9, 16–18].

More generally, one can use integrands that are *hyperexponential* functions, i.e. their logarithmic derivative is a rational function. Rational functions and square roots of rational functions mentioned so far are hyperexponential too. Nested integrals over hyperexponential functions give rise to the *d'Alembertian functions* [12].

In the context of QFT, also nested integrals with other types of integrands beyond hyperexponential functions arise as well. For instance, integrands appear that are solutions of second-order differential equations represented in terms of complete elliptic integrals or ${}_2F_1$-functions, see e.g. [7].

It is well known that, for reasonably regular integrands, nested integrals satisfy the shuffle relations [40]. These relations apply, for instance, if integrands have at most simple poles. In general, however, additional terms arise in the shuffle relations, which are given by nested integrals of lower depth. An algebraic theory covering that is worked out by the author and Georg Regensburger in Ref. [39].

For a given class of integrands, if one chooses a minimal set of integrands among them such that still all nested integrals of that type can be expressed, then one can deduce that nested integrals over that set of integrands do not satisfy any additional algebraic relation beyond the shuffle relations. This in turn allows to compute canonical forms of quantities that are expressed polynomially in terms of nested integrals over the given class of integrands. For rational integrands, for example, the set of all integrands of the form (3), for $a \in \mathbb{C}$, has this property. Hence, over the rational functions, all algebraic relations of generalized harmonic polylogarithms are given by the shuffle relations, cf. also Ref. [25]. The analogous statement for nested integrals over functions whose square is rational holds when one selects the set of integrands given by Eqs. (3) through (6) and branch cuts are ignored. A minimal set of hyperexponential integrands needed to express all d'Alembertian functions as well as a corresponding canonical form was worked out by the author [37].

For the rest of this chapter, we focus on integrands whose square is rational. In Sect. 2, by briefly looking at how nested integrals may arise from nested sums, we highlight computational methods introduced in Ref. [8] relying on special identities that have been constructed to be used as rewrite rules. These rewrite rules allow to do certain computations with nested integrals and nested sums more directly than via e.g. constructing and solving differential equations as it is done in more general methods. We briefly compare two computer algebra approaches to the construction of such differential equations. Section 3 is devoted to expressing nested integrals involving square roots in terms of hyperlogarithms by suitable change of variables.

Many of the formulae and algorithms discussed below are implemented in the package `HarmonicSums` [1–3].

2 Obtaining Nested Integrals from Nested Sums

In analogy to nested integrals, nested sums are multiple sums of a summand that is the product of summands that depend only on one summation variable and where each summation variable occurs as one bound of the summation range of the next inner sum. For example, they may take the general form

$$\sum_{i_1=1}^{n} f_1(i_1) \sum_{i_2=1}^{i_1} f_2(i_2) \cdots \sum_{i_k=1}^{i_{k-1}} f_k(i_k).$$

Passing from sequences indexed by a discrete variable n to functions depending on a continuous variable x can work in two main ways. On the one hand, we can view the sequence given as the sequence of Taylor coefficients of a function, the *generating function* of the sequence. On the other hand, we can aim at an integral representation of the sequence, i.e. a definite integral over an integrand depending on a parameter n that reproduces the entries of the sequence. Often, such integral representations are based on integral transforms. Below, we utilize the Mellin transform, modified to take the following form.

$$\mathbf{M}(f(x))(n) = \int_0^1 x^n f(x)\,dx \tag{7}$$

If the summands $f_j(n)$ are *P-finite* (also called *P-recursive*) sequences, i.e. sequences that satisfy a linear recurrence with polynomial coefficients, then so are the corresponding nested sums. Closely related is the concept of *holonomic* sequences, which is equivalent in case of univariate sequences. In particular, for summands satisfying first-order recurrences the nested sums are *d'Alembertian* sequences [12], like the harmonic sums [47], generalized harmonic sums (S-sums) [33], or nested (inverse) binomial sums [8], for example. This allows general strategies and algorithms for P-finite/holonomic sequences to be applied to nested sums as well.

In the following, however, our focus lies on approaches and algorithms that exploit the nested structure of the sums. Similarly, also for integrals, there are specialized methods that are able to exploit the structure of nested integrals, in addition to the general algorithms that do not. Such dedicated approaches not only reduce the computational burden, but also enable general theoretical statements to be proven that would be very hard to obtain otherwise, see also [8].

2.1 Generating Functions

Generating functions are defined by infinite sums of the form

$$F(x) = \sum_{n=0}^{\infty} f_n x^n. \tag{8}$$

It is well known that any linear recurrence with polynomial coefficients for the sequence (f_n) can be translated into a linear differential equation with polynomial coefficients for the function $F(x)$. This provides a general strategy to compute the function $F(x)$ by constructing and solving a differential equation, starting form a recurrence for the sequence (f_n), independent of the explicit form of that sequence. In practice, however, constructing and solving differential equations can be avoided altogether in many cases by exploiting the syntactic presentation of the sequence.

This can be done by utilizing general properties of generating functions that can be interpreted as rewrite rules. The aim of such rewrite rules is to express $F(x)$ in terms of other generating functions of sequences that are simpler. For instance, a well-known identity for generating functions is

$$\sum_{n=1}^{\infty} x^n \frac{f_n}{n} = \int_0^x \frac{1}{t} \sum_{n=1}^{\infty} t^n f_n \, dt, \tag{9}$$

which holds for arbitrary sequences (f_n). It reduces computing the generating function of a sequence $(\frac{f_n}{n})$ to computing the generating function of (f_n). In general, one is interested in identities that allow to simplify generating functions of the forms $\sum_{n=0}^{\infty} x^n g_n f_n$ and $\sum_{n=0}^{\infty} x^n g_n \sum_{i=0}^{n} f_i$ for concrete g but arbitrary f. In Eq. (9), we have g such that $g_0 = 0$ and $g_n = \frac{1}{n}$ for $n \geq 1$.

Equations (7.1) through (7.11) in Ref. [8] give rewrite rules for evaluating generating functions involving sums. Among them, the rules

$$\sum_{n=1}^{\infty} x^n \binom{2n}{n} \sum_{i=1}^{n} f_i = \frac{1}{4\sqrt{\frac{1}{4} - x}} \int_0^x \frac{1}{t\sqrt{\frac{1}{4} - t}} \sum_{n=1}^{\infty} t^n n \binom{2n}{n} f_n \, dt \tag{10}$$

$$\sum_{n=1}^{\infty} \frac{x^n}{n\binom{2n}{n}} \sum_{i=1}^{n} f_i = \frac{\sqrt{x}}{\sqrt{4-x}} \int_0^x \frac{1}{\sqrt{t}\sqrt{4-t}} \sum_{n=0}^{\infty} \frac{t^n}{\binom{2n}{n}} f_{n+1} \, dt \tag{11}$$

$$= \sum_{n=1}^{\infty} \frac{x^n}{n\binom{2n}{n}} f_n + \frac{\sqrt{x}}{\sqrt{4-x}} \int_0^x \frac{1}{\sqrt{t}\sqrt{4-t}} \sum_{n=1}^{\infty} \frac{t^n}{\binom{2n}{n}} f_n \, dt \tag{12}$$

as well as

$$\sum_{n=1}^{\infty} \frac{x^n}{(2n+1)\binom{2n}{n}} \sum_{i=1}^{n} f_i = \frac{2}{\sqrt{x}\sqrt{4-x}} \int_0^x \frac{1}{\sqrt{t}\sqrt{4-t}} \sum_{n=1}^{\infty} \frac{t^n}{\binom{2n}{n}} f_n \, dt \tag{13}$$

involve also the central binomial coefficient. We illustrate the use of such rewrite rules by the following small example.

Example 1 Consider the generating function given by

$$\sum_{n=1}^{\infty} x^n \frac{1}{n^2\binom{2n}{n}} \sum_{i=1}^{n} \frac{1}{i}.$$

In order to apply one of the rules involving the binomial coefficient, we first need to use Eq. (9). Applying that rule to $f_n = \frac{1}{n\binom{2n}{n}} \sum_{i=1}^{n} \frac{1}{i}$, we obtain

$$\sum_{n=1}^{\infty} x^n \frac{1}{n^2\binom{2n}{n}} \sum_{i=1}^{n} \frac{1}{i} = \int_0^x \frac{1}{t} \sum_{n=1}^{\infty} t^n \frac{1}{n\binom{2n}{n}} \sum_{i=1}^{n} \frac{1}{i} \, dt.$$

Proceeding with the new generating function inside the integrand, we apply Eq. (12) to $f_n = \frac{1}{n}$ in order to obtain

$$\sum_{n=1}^{\infty} x^n \frac{1}{n\binom{2n}{n}} \sum_{i=1}^{n} \frac{1}{i} = \sum_{n=1}^{\infty} \frac{x^n}{n^2\binom{2n}{n}} + \frac{\sqrt{x}}{\sqrt{4-x}} \int_0^x \frac{\sum_{n=1}^{\infty} \frac{t^n}{n\binom{2n}{n}}}{\sqrt{t}\sqrt{4-t}} dt.$$

To treat the first term on the right hand side, we apply Eq. (9) with $f_n = \frac{1}{n\binom{2n}{n}}$ yielding

$$\sum_{n=1}^{\infty} \frac{x^n}{n^2\binom{2n}{n}} = \int_0^x \frac{1}{t} \sum_{n=1}^{\infty} \frac{t^n}{n\binom{2n}{n}} dt.$$

It remains to evaluate the last generating function by instantiating Eq. (11) with the Kronecker delta $f_n = \delta_{1,n}$.

$$\sum_{n=1}^{\infty} \frac{x^n}{n\binom{2n}{n}} \sum_{i=1}^{n} \delta_{1,i} = \frac{\sqrt{x}}{\sqrt{4-x}} \int_0^x \frac{\sum_{n=0}^{\infty} \frac{t^n}{\binom{2n}{n}} \delta_{1,n+1}\, dt}{\sqrt{t}\sqrt{4-t}} = \frac{\sqrt{x}}{\sqrt{4-x}} \mathrm{H}_{\{0,4\}}(x)$$

Recall that the integrand $f_{\{0,4\}}$ is defined by Eqs. (3) and (4) so that one could also write $\mathrm{H}_{\{0,4\}}(x) = \arccos(1-\frac{x}{2})$ explicitly. Altogether, we obtained the generating function as a sum of two nested integrals performing hardly any computation and without constructing any differential equation.

$$\sum_{n=1}^{\infty} x^n \frac{1}{n^2\binom{2n}{n}} \sum_{i=1}^{n} \frac{1}{i} = \mathrm{H}_{0,\{0,4\},\{0,4\}}(x) + \mathrm{H}_{\{0,4\},4,\{0,4\}}(x)$$

2.2 Mellin Representations

An integral representation of a given nested sum in terms of the Mellin transform (7) usually does not take the form of just one term $\mathbf{M}(f(x))\,(n)$. In general, Mellin representations of nested sums take the form

$$c_0 + \sum_{j=1}^{k} c_j^n\, \mathbf{M}\big(f_j(x)\big)\,(n)\,, \tag{14}$$

for some $k \geq 1$ with $c_0, \ldots, c_k$ and $f_1(x), \ldots, f_k(x)$ being independent of n. Often, the integral (7) defining the Mellin transform needs to be regularized accordingly due to singularities of $f_j(x)$. Such Mellin representations can be used to compute asymptotic expansions of complicated expressions involving nested sums as described by Eqs. (2.15) and (2.16) from Ref. [8], like it was done e.g. in Ref. [9].

In Ref. [4], a general method for computing Mellin representations of P-finite sequences based on constructing and solving differential equations is presented. Specialized on sequences that are given as nested sums, a refined version [6] of that method was given later, which exploits the nested structure of the input but still relies on constructing and solving differential equations.

Instead, one can use basic identities of the Mellin transform to exploit the structure of the nested sum to be represented. Several basic identities that allow to build Mellin representations of sequences from Mellin representations of simpler sequences are collected in Sect. 2 of Ref. [8], for example. Among them,

$$\sum_{i=1}^{n} c^i\, \mathbf{M}(f(x))\,(i) = c^n\, \mathbf{M}\left(\frac{x}{x-\frac{1}{c}} f(x)\right)(n) - \mathbf{M}\left(\frac{x}{x-\frac{1}{c}} f(x)\right)(0)\,, \tag{15}$$

which reduces the Mellin representation of a sum to the Mellin representation of its summand, as well as

$$\mathbf{M}(f(x))\,(n) \cdot \mathbf{M}(g(x))\,(n) = \mathbf{M}\left(\int_x^1 \frac{f(\frac{x}{t})g(t)}{t}\,dt\right)(n)\,, \tag{16}$$

which allows to compute the Mellin representation of a product from the Mellin representations of the factors by evaluating Mellin convolution integrals.

2.2.1 Computer algebra approaches to parameter integrals

We see that performing a Mellin convolution amounts to computing a definite integral depending on a parameter. An overview of this topic and related algorithms is given in Ref. [38], for example. Here, we just give a short explanation. There are essentially two main approaches in computer algebra for treating rather general parameter integrals, which we briefly compare below without going into details. The key concept is that of *creative telescoping*, which, for a given integrand $f(t)$ depending on additional parameters, aims to construct a linear operator L such that L commutes with $\frac{d}{dt}$ and an explicit antiderivative $g(t)$ of $L(f(t))$ can be found:

$$L(f(t)) = \frac{d}{dt} g(t). \tag{17}$$

Then, by the properties imposed on L, integrating from a to b yields an implicit equation $L\left(\int_a^b f(t)\,dt\right) = g(b) - g(a)$ for the parameter integral, which typically is a differential or recurrence equation depending on how L acts on the parameters in the integrand. If a or b depends on additional parameters acted on by L, then additional terms arise in the equation from the difference $L\left(\int_a^b f(t)\,dt\right) - \int_a^b L(f(t))\,dt$. To obtain an evaluation of the parameter integral, the implicit equation still has to be solved by other means. For computing telescoping relations (17), on the one hand, many algorithms based on *holonomic systems* have been developed over the past 30 years, see e.g. [20, 22–24, 46, 49]. On the other hand, integration algorithms based on *differential fields* have been developed for more than 50 years, many of which are suitable for creative telescoping, see e.g. [19, 32, 34, 36, 42, 45].

One main difference between these two approaches lies in the way functions are represented. Therefore, there is a fundamental difference in what kind of antiderivatives $g(t)$ can be found by the algorithms to fulfill Eq. (17). In the holonomic systems approach, g is restricted to the form $g(t) = Q(f(t))$ for some linear (typically differential/recurrence) operator Q acting on the integrand. Hence, essentially no additional functions can appear in g that did not already appear in the integrand f. In contrast, with algorithms using differential fields, antiderivatives $g(t)$ can be found that involve certain new functions that do not already occur in the integrand $f(t)$. As a result, potentially simpler operators L may allow the antidifferentiation in Eq. (17) to be carried out by the algorithm in question, yielding a differential or recurrence equation of smaller order for the parameter integral. Another difference lies in the type of integrands that can be handled. In the holonomic systems approach, most general algorithms work with *D-finite functions*, i.e. integrands that satisfy linear homogeneous differential equations with polynomial coefficients. Algorithms using differential fields can deal with Liouvillian integrands [45] and also a large class of non-Liouvillian functions, see e.g. [19, 36]. The two classes of integrands accessible by algorithms of the two approaches are very large, each covering a majority of common special functions. Despite the fact that many functions (e.g. all d'Alembertian functions) are both D-finite and Liouvillian, there are many Liouvillian functions that are not D-finite, and vice versa. In particular, arbitrary quotients and compositions of Liouvillian functions are Liouvillian again, whereas the same is not true for D-finite functions. The holonomic systems approach is not specific to integration and many algorithms using this approach can be used for summation as well. Inspired by integration algorithms using differential fields, an algorithmic analog has been introduced for summation by Karr [29] using *difference fields*, which was developed further by Schneider, see [43] and references therein.

In fact, there is also a third approach to compute parameter integrals, which is more specialized. It relies on collections of identities that either evaluate the given parameter integral or relate it to other integrals. Such identities can be used as rewrite rules for evaluating or simplifying parameter integrals of specific form. Coming back to Mellin convolutions (16), many such identities are provided in Sect. 4 of Ref. [8] specially for rewriting integrals $\int_x^1 t^{-1} g(\frac{x}{t}) h(t) f(t)\,dt$, with certain concrete choices of g and h but arbitrary f, in terms of similar integrals which only involve the derivative f' instead of f. If these rewrite rules are applied to f being a nested integral, then in the resulting integral the function f' will only involve a nested integral with lower depth. Iterating this reduction, a Mellin convolution involving a nested integral can be performed, provided sufficiently many rewrite rules are available. This is analogous to rewrite rules like Eqs. (10) through (13) mentioned above, which remove one summation step each. Due to the page limit, we refer to Ref. [8] for examples of such rules and how they can be combined to obtain general patterns for Mellin representations of nested sums with certain prefactors.

There are methods to design new rewrite rules of this type for Mellin convolutions with other choices of g and h or even for other types of parameter integrals, e.g. integrals arising from integral transforms. For example, Sect. 7 of Ref. [8] also lists some rewrite rules for the integral transform $\int_0^1 \frac{1}{1-tx} f(t)\,dt$, that allow to convert Mellin representations of sequences into their generating function. Specialized identities can also be derived for doing other tasks by rewriting, e.g. evaluating Mellin transforms in terms of nested sums.

3 Rationalizing Transformations

In this section, we discuss simplification of square roots in the integrands by suitable changes of variables. That is, if $\sqrt{f(x)}$ appears with $f(x)$ a rational function, we want to apply a transformation $x = g(y)$ such that $\sqrt{f(g(y))}$ can be simplified to a rational function. To preserve the nested structure of the integrals properly, the same change of variables has to be applied to all integrands of a nested integral. For instance, a nested integral of the form (1) would become the nested integral

$$\int_{g^{-1}(0)}^{g^{-1}(x)} f_1(g(u_1))g'(u_1) \int_{g^{-1}(0)}^{u_1} f_2(g(u_2))g'(u_2) \ldots \\ \ldots \int_{g^{-1}(0)}^{u_{k-1}} f_k(g(u_k))g'(u_k)\,du_k \ldots du_1. \qquad (18)$$

Since this transformation should not introduce any new functions in those integrands that are already rational (nor via $g'(y)$), $g(y)$ is required to be a rational function.

Example 2 Formula (5.30) in [8] gives an integral representation of the nested sum

$$\sum_{i=1}^{n} \frac{\binom{2i}{i}}{i^2} \sum_{j=1}^{i} \frac{(-1)^j}{j^2}.$$

Among others, it involves the nested integral

$$\int_x^1 \frac{1}{t_1} \int_{t_1}^1 \frac{1}{\sqrt{t_2}\sqrt{1+t_2}} \int_{t_2}^1 \frac{1}{\sqrt{t_3}\sqrt{1+t_3}} \int_{t_3}^1 \frac{1}{\sqrt{t_4}\sqrt{1-t_4}}\,dt_4\,dt_3\,dt_2\,dt_1$$

of the form (2). In order to obtain integrands that are rational functions, based on Eqs. (71) and (72) below for $a_1 = -1$ and $a_2 = 1$, we can use the change of variables

$$x = \frac{2y^2}{y^4+1} \quad \text{respectively} \quad y = \frac{\sqrt{1+x}-\sqrt{1-x}}{\sqrt{2}\sqrt{x}},$$

which maps the interval [0, 1] bijectively to itself to preserve the form (2). Then, by Eqs. (75) and (76), we have $\sqrt{x}\sqrt{1\pm x} = \frac{\sqrt{2}y(1\pm y^2)}{y^4+1}$ and the differentials occurring in the above nested integral are transformed as follows.

$$\frac{dx}{x} = \left(\frac{2}{y} - \frac{4y^3}{y^4+1}\right)dy \qquad \frac{dx}{\sqrt{x}\sqrt{1\pm x}} = \frac{2\sqrt{2}(1\mp y^2)}{y^4+1}dy$$

To obtain properly nested integrals again, the change of variable has to be applied uniformly to all levels within the nested integral, which in the present case yields

$$32\sqrt{2}\int_y^1 \left(\frac{1}{u_1} - \frac{2u_1^3}{u_1^4+1}\right)\int_{u_1}^1 \frac{1-u_2^2}{u_2^4+1}\int_{u_2}^1 \frac{1-u_3^2}{u_3^4+1}\int_{u_3}^1 \frac{1+u_4^2}{u_4^4+1}du_4\,du_3\,du_2\,du_1$$

having rational integrands. Altogether, the nested integral above can hence be written in terms of cyclotomic harmonic polylogarithms [10] evaluated at $\frac{\sqrt{1+x}-\sqrt{1-x}}{\sqrt{2}\sqrt{x}}$.

Below, we let C be any field of characteristic zero, i.e. a field extension of the rational numbers possibly containing indeterminates. Its algebraic closure is denoted by $\overline{C}$. Elements of C will be considered as constants. In full generality, the problem of finding a rationalizing transformation for square roots of univariate rational functions can be stated as follows.

Problem 1 Given a set of nonzero rational functions $F \subset C(x)$, find, if possible, a non-constant rational function $g \in C(y)$ such that, upon substituting $g(y)$ for x, every element of F can be written in the form $c \cdot f(y)^2$ for some $c \in C$ and $f \in C(y)$.

Remark 1 Not for every set of radicands $F \subset C(x)$ a rationalizing transformation $g \in C(y)$ exists, however. If F is such that there are $f_1, \dots, f_n \in F$ and $r \in C(x)$ such that $p := f_1 \cdot \ldots \cdot f_n \cdot r^2$ is a squarefree polynomial (i.e. not divisible by the square of any non-constant polynomial) of degree ≥ 3, then a rationalizing transformation cannot exist for F. This is because, for any rationalizing transformation $g \in C(y)$, there would exist nonzero $c \in C$ and $f \in C(y)$ such that $p(g(y)) = c \cdot f(y)^2$. However, the irreducible algebraic curve defined by $p(X) = c \cdot Y^2$ has a rational parameterization $X = g(y)$ and $Y = f(y)$ with $f, g \in \overline{C}(y)$ if and only if the curve has genus 0 (see e.g. [44]), which happens only for $\deg_x(p) \leq 2$ since p is squarefree.

In the past, many ad-hoc transformations have been used in practice for explicitly converting integrands with square roots into rational integrands, see e.g. [5], or for evaluating certain quantities in terms of polylogarithms at arguments involving square roots, see e.g. [21, 26]. In the rest of this section, we will give an exhaustive overview of explicit formulae for rationalizing transformations solving Problem 1. In general, also a multivariate analog of Problem 1 can be considered, for which a general method to construct rationalizing transformations was given recently [14] that works in many cases and has been implemented [15].

Remark 2 In all that follows, we exclude the case when all elements of F are squares in $C(x)$, which trivially admits the rationalizing transformation $g(y) = y$. Without loss of generality, we can assume that the elements of F are monic squarefree polynomials in $C[x]$, since multiplying an element of F by a nonzero square in $C(x)$ or by a nonzero constant from C does not change the possible transformations g. Furthermore, we can assume without loss of generality that no polynomial in F divides another, since distinct $f_1, f_2 \in F$ with $f_1 | f_2$ allow to replace f_2 by the quotient f_2/f_1 (or to remove f_1 from F, if f_1 is constant) without changing g.

For simplicity, one is interested in rationalizing transformations g of low degree $\max(\deg(\text{num}(g)), \deg(\text{den}(g)))$. As soon as one rationalizing transformation g is known for a given set F, infinitely many rationalizing transformations can be obtained by composition $g(h(y))$ with any non-constant $h \in C(y)$. If $h \in C(y)$ has degree 1, i.e. it is of the form $\frac{ay+b}{cy+d}$, then composition $g(h(y))$ does not change the degree of g, otherwise the degree is increased.

Next, we give explicit formulae for rationalizing transformations for radicands which have coefficients in arbitrary field extensions of $\mathbb{Q}$. The transformations are chosen such that they map $y = 0$ to $x = g(0) = 0$. After that, we will give dedicated formulae for the fields $C = \mathbb{R}$ and $C = \mathbb{C}$, where we impose in addition that the transformations map the interval $[0, 1]$ bijectively to itself. All rationalizing transformations given below have been constructed by the author on various occasions distributed over the past few years, using the computer algebra systems MATHEMATICA, SINGULAR, and MAPLE, with verification of their properties in MATHEMATICA.

3.1 General Transformations Mapping 0 to 0

Let C be a field of characteristic zero. Any $F \subset C(x)$, for which a rationalizing transformation exists, see Remark 1, can be reduced, as described in Remark 2, to one of the following four essentially different cases (or to the trivial case $F = \{\}$).

Note that all rationalizing transformations $g \in C(y)$ given below are of lowest possible degree $\max(\deg(\text{num}(g)), \deg(\text{den}(g)))$. Therefore, in each of the cases, any other rationalizing transformation can be obtained via composition $g(h(y))$ with $h \in \overline{C}(y)$. Moreover, the coefficients in the given transformations are rational expressions in the coefficients of the radicand polynomials from F. More precisely, if the field C is the smallest field extension of $\mathbb{Q}$ that contains the coefficients of these polynomials, then the coefficients of g lie in the same field C and do not involve any new algebraic or transcendental numbers.

Expressions for an inverse of g are not unique and may require new algebraic numbers in general. All inverses below are expressed as elements of $\overline{C}(x)[r_1, \ldots, r_k]$, where $r_1, \ldots, r_k$ are square roots of the polynomials in F. In addition to directly satisfying $g(g^{-1}(x)) = x$, the formulae for $g^{-1}(x)$ below are selected to yield the unique Puiseux series in $\overline{C}((x^{1/2}))$ that also satisfies $g^{-1}(g(y)) = y$.

We start with the simplest case of one linear polynomial

$$F = \{x - a\}, \tag{19}$$

where $a \in C$. Then, for instance, we can use one of the following two rationalizing transformations of degree 2, depending on whether a is zero or not.

$$g(y) = y^2 \tag{20}$$

$$g(y) = -4ay(y + 1) \tag{21}$$

An inverse transformation is straightforwardly obtained as

$$g^{-1}(x) = \sqrt{x} \quad \text{respectively} \quad g^{-1}(x) = \frac{\sqrt{x - a} - \sqrt{-a}}{2\sqrt{-a}}. \tag{22}$$

Next, we consider one quadratic radicand polynomial

$$F = \{x^2 + c_1 x + c_0\}, \tag{23}$$

where $c_0, c_1 \in C$ are such that $c_1^2 \neq 4c_0$, i.e. the polynomial is not a square. Depending on whether c_0 is zero or not, we can use one of the following two rationalizing transformations of degree 2, for example.

$$g(y) = \frac{c_1 y^2}{4(y + 1)} \tag{24}$$

$$g(y) = \frac{4c_0 y}{(c_1^2 - 4c_0)y^2 - 2c_1 y + 1} \tag{25}$$

Also in this case, the respective algebraic inverses are straightforwardly obtained.

$$g^{-1}(x) = \frac{2}{c_1}\left(x + \sqrt{x^2 + c_1 x}\right) \tag{26}$$

$$g^{-1}(x) = \frac{c_1 x + 2c_0 - 2\sqrt{c_0}\sqrt{x^2 + c_1 x + c_0}}{(c_1^2 - 4c_0)x} \tag{27}$$

By Remark 1, F cannot contain a squarefree polynomial of degree ≥ 3. So, we continue with the case of two linear polynomials

$$F = \{x - a_1, x - a_2\}, \tag{28}$$

where $a_1, a_2 \in C$ are distinct. Again, we distinguish the two cases whether 0 is a root of one of these polynomials or not. Assuming $a_1 \neq 0$, without loss of generality, we can use one of the following two rationalizing transformations of degree 4, depending on whether a_2 is zero or not.

$$g(y) = \frac{4a_1 y^2}{(y^2 + 1)^2} \tag{29}$$

$$g(y) = \frac{4a_1 a_2 y(y - a_1)(y - a_2)}{(y^2 - a_1 a_2)^2} \tag{30}$$

Writing the inverse transformations in terms of square roots of polynomials in F, we obtain the following expressions for the inverse of (29) and (30), respectively.

$$g^{-1}(x) = \frac{\sqrt{a_1}\sqrt{x}}{\sqrt{-a_1 x}} \left(\sqrt{-a_1} - \sqrt{x - a_1}\right) \tag{31}$$

$$g^{-1}(x) = \frac{\left(a_1 + \sqrt{-a_1}\sqrt{x - a_1}\right)\left(a_2 + \sqrt{-a_2}\sqrt{x - a_2}\right)}{x} \tag{32}$$

Finally, we conclude with the case of two quadratic polynomials, which by Remark 1 need to have a common root. This is equivalent to the more symmetric case of three quadratic polynomials that pairwise have exactly one common root, i.e.

$$F = \{(x - a_1)(x - a_2), (x - a_1)(x - a_3), (x - a_2)(x - a_3)\}, \tag{33}$$

where $a_1, a_2, a_3 \in C$ are pairwise distinct. Assuming $0 \notin \{a_1, a_2\}$, without loss of generality, we can use one of the following two rationalizing transformations of degree 4, for instance, depending on whether a_3 is zero or not.

$$g(y) = \frac{4a_1 a_2 y^2}{(a_1 - a_2)^2 y^4 + 2(a_1 + a_2)y^2 + 1} \tag{34}$$

$$g(y) = -\frac{4a_1 a_2 a_3 y(y - a_1)(y - a_2)(y - a_3)}{(s_1^2 - 4s_2)y^4 + 8s_3 y^3 - 2s_1 s_3 y^2 + s_3^2} \tag{35}$$

For shorter notation in Eq. (35), we used the elementary symmetric polynomials

$$s_1 = a_1 + a_2 + a_3, \quad s_2 = a_1 a_2 + a_1 a_3 + a_2 a_3, \quad \text{and} \quad s_3 = a_1 a_2 a_3. \tag{36}$$

Using square roots of only two of the three polynomials in F, we can write the respective inverses of (34) and (35) as follows. For shorter notation, we abbreviate the square roots $r_1 := \sqrt{(x-a_1)(x-a_2)}$ and $r_2 := \sqrt{(x-a_1)(x-a_3)}$ in Eq. (38).

$$g^{-1}(x) = \frac{\sqrt{a_1a_2}}{(a_1-a_2)x}\left(\frac{\sqrt{x(x-a_1)}}{\sqrt{-a_1}} - \frac{\sqrt{x(x-a_2)}}{\sqrt{-a_2}}\right) \tag{37}$$

$$\begin{aligned} g^{-1}(x) = \frac{s_3}{(s_1^2-4s_2)x+4s_3}\Bigg(& s_1 - 2x - \frac{s_1-2a_3}{\sqrt{a_1a_2}}r_1 \\ & - \frac{s_1-2a_2}{\sqrt{a_1a_3}}r_2 + \frac{a_1\cdot(s_1-2a_1)}{\sqrt{a_1a_2}\sqrt{a_1a_3}(x-a_1)}r_1r_2\Bigg) \end{aligned} \tag{38}$$

3.2 *Real-Valued Square Roots on the Interval* [0, 1]

Here, we consider $C = \mathbb{R}$ and assume that all radicands $f \in F$ are such that $f(x) \geq 0$ for all $x \in [0, 1]$, so that all the square roots are real-valued on $[0, 1]$. In addition, we require rationalizing transformations which not only map 0 to 0 like the ones above but which also map the interval $[0, 1]$ bijectively to itself and hence are monotonically increasing on that interval. Such bijections of $[0, 1]$ allow to preserve this common integration range of integrals and avoid non-real integration bounds. Indeed, as the exhaustive collection below shows, this is possible for all cases discussed above, where rationalizing transformations exist at all, and the degrees of the transformations remain the same. This is achieved by carefully constructed Möbius transformations of y in the general formulae given in Sect. 3.1.

For each rationalizing transformation $g(y)$, we give an explicit formula for the inverse $g^{-1}(x)$ so that, for all $x \in [0, 1]$, the unique $y \in [0, 1]$ with $x = g(y)$ is given by $y = g^{-1}(x)$. As above, it will be a common property of all rationalizing transformations given below that their inverse is given in terms of the square roots of the original radicand polynomials. In order to avoid additional case distinctions, we use radicand polynomials like $a^{-1}(a - x)$ instead of $x - a$ and $a - x$, for instance. It is straightforward to adapt the formulae to other normalizations of radicands if necessary. Outside the interval $[0, 1]$, unless stated otherwise, the explicit expressions given below for $g(y)$ and $g^{-1}(x)$ still satisfy $g(g^{-1}(x)) = x$, but $g^{-1}(g(y)) = y$ need not hold due to the multivalued nature of the inverse of g.

Once a rationalizing transformation $g(y)$ is known for a given set F that maps $[0, 1]$ to itself, infinitely many such rationalizing transformations of the same degree can be obtained by composition $g(h(y))$ with

$$h(y) = \frac{y}{(1-\lambda)y+\lambda}, \tag{39}$$

where $\lambda > 0$ is arbitrary. In fact, all Möbius transformations that map $[0, 1]$ to itself in a bijective and monotonically increasing way are given by Eq. (39) for some

$\lambda > 0$. The special value $\lambda = 1$ yields the identity map and replacing λ by $1/\lambda$ gives the inverse transformation of (39).

In contrast to Sect. 3.1 above, however, it is no longer possible in all cases to give transformations of the same degree which have coefficients that are rational expressions in the coefficients of the radicand polynomials from F. More precisely, the coefficients of the transformations given below involve taking square roots, but still they are real numbers. Hence, if we take C as the smallest field extension of $\mathbb{Q}$ that contains the coefficients of the polynomials in F, then the coefficients of the transformations may lie in an algebraic extension of C. The rationalizing transformations below are chosen so that the degree of this field extension over C is minimal.

3.2.1 One Square Root

Starting with the simplest case of just a linear radicand, the radicand polynomial (up to a positive constant factor) has to have the form

$$x \quad \text{or} \quad a^{-1}(a-x), \tag{40}$$

for some $a < 0$ or $a \geq 1$, in order to be non-negative for all $x \in [0, 1]$. In the former case, the rationalizing transformation (20) already has the property of mapping the interval $[0, 1]$ bijectively to itself, so we trivially have

$$g(y) = y^2 \quad \text{and} \quad g^{-1}(x) = \sqrt{x} \tag{41}$$

on $[0, 1]$. In the latter case, however, we need to modify the transformation (21). Introducing $\alpha := \sqrt{1 - a^{-1}} \geq 0$ for any $a < 0$ or $a \geq 1$, we have that both

$$g(y) = y\cdot\big((1-2a+2a\alpha)y + 2a\cdot(1-\alpha)\big) \quad \text{and} \tag{42}$$

$$g^{-1}(x) = a\cdot(1+\alpha)\left(1 - \sqrt{a^{-1}(a-x)}\right) \tag{43}$$

map the interval $[0, 1]$ bijectively to itself in a monotonically increasing way. The Möbius transformation used to obtain (42) from (21) is $h(y) = -\frac{y}{2a(1+\alpha)}$. From Eq. (41) resp. (43), we easily obtain also the explicit expressions of the square root

$$\sqrt{x} = y \quad \text{respectively} \quad \sqrt{a^{-1}(a-x)} = (\alpha - 1)y + 1 \tag{44}$$

as rational functions in y whenever $y = g^{-1}(x)$. For later reference, we note that the inverses $g^{-1}(x)$ given by Eqs. (41) and (43) can be rewritten in the form

$$g^{-1}(x) = \frac{x}{\sqrt{x}} \quad \text{respectively} \quad g^{-1}(x) = \frac{(1+\alpha)x}{1+\sqrt{a^{-1}(a-x)}}. \tag{45}$$

Next, we consider radicands that are quadratic polynomials as in Eq. (23). First, we assume that $x = 0$ is a root of the radicand, which then, being squarefree and nonnegative on the interval $[0, 1]$, necessarily (up to a positive constant factor) equals

$$a^{-1}x(a-x), \tag{46}$$

for some $a < 0$ or $a \geq 1$. For $a < 0$ or $a > 1$, the transformations

$$g(y) = \frac{ay^2}{y^2+a-1} \quad \text{and} \tag{47}$$

$$g^{-1}(x) = a\alpha\frac{\sqrt{a^{-1}x(a-x)}}{a-x} \tag{48}$$

map the interval $[0, 1]$ bijectively to itself in a monotonically increasing way, where the inverse transformation (48) involves $\alpha := \sqrt{1-a^{-1}} > 0$. In fact, Eq. (47) can be obtained from Eq. (24) with $c_1 = -a$ by substituting $h(y) = -\frac{2y}{y\pm\sqrt{1-a}}$ for y, which has complex coefficients if $a > 1$ even though the transformations (24) and (47) have real coefficients. Note that Eqs. (47) and (48) are not valid if $a = 1$. For the special value $a = 1$, the following rationalizing transformation maps the interval $[0, 1]$ bijectively to itself in a monotonically increasing way, for example.

$$g(y) = \frac{y^2}{2y^2-2y+1} \tag{49}$$

$$g^{-1}(x) = \frac{x-\sqrt{x(1-x)}}{2x-1} \tag{50}$$

Observe that the singularity at $x = \frac{1}{2}$ in Eq. (50) is removable. The transformation (49) can be obtained from Eq. (24) with $c_1 = -1$ by replacing y with $h(y) = -\frac{2y}{(1-i)y+i}$ or its complex conjugate. By Eq. (48) respectively (50), the square root is easily expressed as rational function in $y = g^{-1}(x)$ by

$$\sqrt{a^{-1}x(a-x)} = \frac{a\alpha y}{y^2+a-1} \quad \text{respectively} \quad \sqrt{x(1-x)} = \frac{y(1-y)}{2y^2-2y+1}. \tag{51}$$

Furthermore, the inverses given by Eqs. (48) and (50) can also be written as

$$g^{-1}(x) = \frac{\alpha x}{\sqrt{a^{-1}x(a-x)}} \quad \text{respectively} \quad g^{-1}(x) = \frac{x}{x+\sqrt{x(1-x)}}. \tag{52}$$

If $x = 0$ is not a root of the quadratic radicand, then (up to a positive constant factor) the quadratic polynomial has the form

$$c_0^{-1}(x^2 + c_1 x + c_0), \tag{53}$$

where $c_0, c_1 \in \mathbb{R}$, $c_0 \neq 0$, are such that the polynomial is not a square and is nonnegative for all $x \in [0, 1]$. These conditions are equivalent to requiring $c_0, c_1 \in \mathbb{R}$ to be such that the conditions $c_0 \neq 0$, $c_1^2 \neq 4c_0$, and $c_0(c_0 + c_1 + 1) \geq 0$ hold and at least one of the inequalities $-1 < 2c_0 + c_1 < 0$ and $c_1^2 > 4c_0$ does not hold. In this case, we set $\alpha := \sqrt{1 + c_0^{-1} \cdot (c_1 + 1)} \geq 0$ and the following rationalizing transformation maps the interval $[0, 1]$ bijectively to itself in a monotonically increasing way.

$$g(y) = \frac{y(c_1 y + 2c_0 \cdot (1 + \alpha))}{1 + 2c_0 + c_1 + 2c_0\alpha - y^2} \tag{54}$$

$$g^{-1}(x) = c_0 \cdot (1 + \alpha) \frac{\sqrt{c_0^{-1}(x^2 + c_1 x + c_0)} - 1}{x + c_1} \tag{55}$$

In Eq. (55), the singularity at $x = -c_1$ is removable. Note that these formulae work regardless whether the quadratic polynomial (53) has two real roots (i.e. $c_1^2 > 4c_0$) or two conjugate complex roots (i.e. $c_1^2 < 4c_0$). We can obtain Eq. (54) also from Eq. (25) via the Möbius transformation $h(y) = \frac{y}{c_1 y + 2c_0(1+\alpha)}$. By virtue of Eq. (55), we have the following expression of the square root as rational function in terms of $y = g^{-1}(x)$.

$$\sqrt{c_0^{-1}(x^2 + c_1 x + c_0)} = \frac{1 + 2c_0 + c_1 + 2c_0\alpha + c_1(1 + \alpha)y + y^2}{1 + 2c_0 + c_1 + 2c_0\alpha - y^2} \tag{56}$$

In fact, the inverse (55) can also be written in different form.

$$g^{-1}(x) = \frac{(1 + \alpha)x}{1 + \sqrt{c_0^{-1}(x^2 + c_1 x + c_0)}} \tag{57}$$

3.2.2 Two Square Roots

For two real-valued square roots of linear polynomials on the interval $[0, 1]$, up to a positive constant factor, the two radicand polynomials are given by the set

$$F = \{x, a^{-1}(a - x)\} \tag{58}$$

for some $a < 0$ or $a \geq 1$, if one of the square roots vanishes at $x = 0$. The generic case when neither of the two square roots vanishes at $x = 0$ gives rise to radicands (63) below. First, for the radicands (58), we introduce $\alpha := \sqrt{1-a^{-1}} \geq 0$ to express the rationalizing transformation

$$g(y) = \frac{4y^2}{((1-\alpha)y^2+1+\alpha)^2} \tag{59}$$

$$g^{-1}(x) = a\cdot(1+\alpha)\frac{1-\sqrt{a^{-1}(a-x)}}{\sqrt{x}} \tag{60}$$

mapping the interval [0, 1] bijectively to itself in a monotonically increasing way. The limit of formula (60) at $x = 0$ is finite and zero. Although, with $a_1 = a$, both transformations (29) and (59) have real coefficients, the Möbius transformation $h(y) = \frac{y}{\sqrt{2a-1+2a\alpha}}$, which changes the former into the latter, has complex coefficients if $a < 0$. Based on Eqs. (59) and (60), the rational function representations

$$\sqrt{x} = \frac{2y}{(1-\alpha)y^2+1+\alpha} \quad \text{and} \quad \sqrt{a^{-1}(a-x)} = \frac{1-2a(1+\alpha)+y^2}{1-2a(1+\alpha)-y^2} \tag{61}$$

of the square roots in terms of $y = g^{-1}(x)$ hold. Also in this case, the inverse (60) can be rewritten to obtain the following expression.

$$g^{-1}(x) = \frac{(1+\alpha)\sqrt{x}}{1+\sqrt{a^{-1}(a-x)}} \tag{62}$$

If none of the two square roots vanishes at $x = 0$, the radicand polynomials (up to a positive constant factor) are given by

$$F = \{a_1^{-1}(a_1-x), a_2^{-1}(a_2-x)\}, \tag{63}$$

with distinct $a_1, a_2 \in \mathbb{R}$ such that $a_i < 0$ or $a_i \geq 1$ for each i. With

$$\alpha := \left(1+\sqrt{1-a_1^{-1}}\right)\left(1+\sqrt{1-a_2^{-1}}\right) > 1, \tag{64}$$

a rationalizing transformation that maps the interval [0, 1] bijectively to itself in a monotonically increasing way can be obtained from Eq. (30) by replacing y with $\frac{y}{\alpha}$.

$$g(y) = \frac{4a_1a_2\alpha y(y-a_1\alpha)(y-a_2\alpha)}{(y^2-a_1a_2\alpha^2)^2} \tag{65}$$

For $x \in [0, 1]$, the inverse can be given as

$$g^{-1}(x) = a_1a_2\alpha\frac{\left(1-\sqrt{a_1^{-1}(a_1-x)}\right)\left(1-\sqrt{a_2^{-1}(a_2-x)}\right)}{x}, \tag{66}$$

where the singularity at $x = 0$ is removable. With $y = g^{-1}(x)$ as in Eq. (66), the two square roots become rational functions in y as follows.

$$\sqrt{a_1^{-1}(a_1 - x)} = \frac{y^2 - 2a_2\alpha y + a_1 a_2 \alpha^2}{-y^2 + a_1 a_2 \alpha^2} \tag{67}$$

$$\sqrt{a_2^{-1}(a_2 - x)} = \frac{y^2 - 2a_1\alpha y + a_1 a_2 \alpha^2}{-y^2 + a_1 a_2 \alpha^2} \tag{68}$$

Equivalently, the inverse (66) can be written as

$$g^{-1}(x) = \frac{\alpha x}{\left(1 + \sqrt{a_1^{-1}(a_1 - x)}\right)\left(1 + \sqrt{a_2^{-1}(a_2 - x)}\right)}. \tag{69}$$

3.2.3 Three Square Roots

Dealing with real-valued square roots of three quadratic polynomials, which pairwise have exactly one common root, we first consider the case when $x = 0$ is among their roots. Up to a positive constant factor, the three radicand polynomials being nonnegative on the whole interval $[0, 1]$ necessarily are of the form

$$F = \{a_1^{-1}x(a_1 - x), a_2^{-1}x(a_2 - x), a_1^{-1}a_2^{-1}(a_1 - x)(a_2 - x)\} \tag{70}$$

with distinct $a_1, a_2 \in \mathbb{R}$ such that $a_i < 0$ or $a_i \geq 1$ for each i. Then, a rationalizing transformation and its inverse, both mapping $[0, 1]$ bijectively to itself in a monotonically increasing way, can be given as follows.

$$g(y) = \frac{4s_2 y^2}{(-s_1 + 2s_2(1 - \alpha))y^4 + 2s_1 y^2 - s_1 + 2s_2(1 + \alpha)} \tag{71}$$

$$g^{-1}(x) = a_1 a_2 \left(\sqrt{1 - a_1^{-1}} + \sqrt{1 - a_2^{-1}}\right) \frac{\sqrt{a_1^{-1}x(a_1 - x)} - \sqrt{a_2^{-1}x(a_2 - x)}}{(a_1 - a_2)x} \tag{72}$$

For shorter notation in Eq. (71) and also below, we use

$$\alpha := \sqrt{(1 - a_1^{-1})(1 - a_2^{-1})} \geq 0 \tag{73}$$

as well as the elementary symmetric polynomials

$$s_1 = a_1 + a_2 \quad \text{and} \quad s_2 = a_1 a_2 \tag{74}$$

as abbreviations. Note that the limit of the formula (72) at $x = 0$ is finite and zero. The transformation (71) can also be obtained from Eq. (34) via the Möbius transformation $h(y) = \frac{y}{\sqrt{-a_1-a_2+2a_1a_2(1+\alpha)}}$, which has complex coefficients if $a_1a_2 < 0$ even though Eqs. (34) and (71) have real coefficients. With $y = g^{-1}(x)$ given by Eq. (72), for $x \in [0, 1]$, the three square roots of the polynomials in F can be written as the rational functions

$$\sqrt{a_1^{-1}x(a_1-x)} = \frac{2s_2y\left((\beta_1-\beta_2)y^2+\beta_1+\beta_2\right)}{(-s_1+2s_2(1-\alpha))y^4+2s_1y^2-s_1+2s_2(1+\alpha)} \tag{75}$$

$$\sqrt{a_2^{-1}x(a_2-x)} = \frac{2s_2y\left((\beta_2-\beta_1)y^2+\beta_1+\beta_2\right)}{(-s_1+2s_2(1-\alpha))y^4+2s_1y^2-s_1+2s_2(1+\alpha)} \tag{76}$$

$$\sqrt{\frac{(a_1-x)(a_2-x)}{a_1a_2}} = \frac{(s_1-2s_2(1-\alpha))y^4-s_1+2s_2(1+\alpha)}{(-s_1+2s_2(1-\alpha))y^4+2s_1y^2-s_1+2s_2(1+\alpha)} \tag{77}$$

in y using also $\beta_i := \sqrt{1-a_i^{-1}} \geq 0$ with $i = 1, 2$ for shorter notation. Moreover, the inverse (72) can be written more symmetrically as

$$g^{-1}(x) = \left(\sqrt{1-a_1^{-1}}+\sqrt{1-a_2^{-1}}\right)\frac{x}{\sqrt{a_1^{-1}x(a_1-x)}+\sqrt{a_2^{-1}x(a_2-x)}}. \tag{78}$$

While Eqs. (75) and (76) also hold for x outside the interval $[0, 1]$, Eq. (77) does not hold in the same generality. This restriction can be lifted by replacing the square roots of the polynomials from F by pairwise products of $\sqrt{x}$, $\sqrt{a_1^{-1}(a_1-x)}$, and $\sqrt{a_2^{-1}(a_2-x)}$ in Eqs. (72) through (78). This modification was used in Example 2.

If $x = 0$ is not among the roots of the three quadratic polynomials, then (up to a positive constant factor) they are given by

$$F = \left\{\frac{(a_1-x)(a_2-x)}{a_1a_2}, \frac{(a_1-x)(a_3-x)}{a_1a_3}, \frac{(a_2-x)(a_3-x)}{a_2a_3}\right\}, \tag{79}$$

with pairwise distinct $a_1, a_2, a_3 \in \mathbb{R}$ satisfying $a_i < 0$ or $a_i \geq 1$ for each i, since the radicand polynomials are nonnegative for all $x \in [0, 1]$. To express the following more compactly, we use the elementary symmetric polynomials (36) and we let

$$\alpha := \sqrt{(1-a_1^{-1})(1-a_2^{-1})}+\sqrt{(1-a_1^{-1})(1-a_3^{-1})}+\sqrt{(1-a_2^{-1})(1-a_3^{-1})} > 0. \tag{80}$$

Then, replacing y with $\frac{y}{1+\alpha}$ in Eq. (35), we obtain the rationalizing transformation

$$g(y) = -\frac{4s_3y(y-a_1(1+\alpha))(y-a_2(1+\alpha))(y-a_3(1+\alpha))}{(s_1^2-4s_2)y^4+8s_3(1+\alpha)y^3-2s_1s_3(1+\alpha)^2y^2+s_3^2(1+\alpha)^4}, \tag{81}$$

which maps the interval [0, 1] bijectively to itself in a monotonically increasing way. Its inverse on the interval [0, 1] is given by

$$g^{-1}(x) = -\frac{s_3(1+\alpha)}{(s_1^2-4s_2)x+4s_3}\left(2x - s_1 + (s_1-2a_3)\sqrt{\frac{(a_1-x)(a_2-x)}{a_1a_2}}\right.$$
$$\left.+(s_1-2a_2)\sqrt{\frac{(a_1-x)(a_3-x)}{a_1a_3}} + (s_1-2a_1)\sqrt{\frac{(a_2-x)(a_3-x)}{a_2a_3}}\right) \tag{82}$$

and, if $s_1^2 \neq 4s_2$, has a singularity at $x = -\frac{4s_3}{s_1^2-4s_2} \neq 0$, which is removable whenever it lies in the interior of the interval [0, 1], or at least has a finite limit if $x = 1$. On the interval [0, 1], the three square roots can be expressed as the rational functions

$$\sqrt{\frac{(a_i-x)(a_j-x)}{a_ia_j}} = \frac{p_i(y)p_j(y)}{q(y)} \tag{83}$$

in y, for $i, j \in \{1, 2, 3\}$, with $y = g^{-1}(x)$ given by Eq. (82), where for shorter notation the following abbreviations were used.

$$p_i(y) := (s_1-2a_i)y^2 - 2\frac{s_3}{a_i}(1+\alpha)y + s_3(1+\alpha)^2 \tag{84}$$

$$q(y) := (s_1^2-4s_2)y^4 + 8s_3(1+\alpha)y^3 - 2s_1s_3(1+\alpha)^2y^2 + s_3^2(1+\alpha)^4 \tag{85}$$

Alternatively, Eq. (82) can be written more compactly as

$$g^{-1}(x) = \frac{(1+\alpha)x}{1+\sqrt{\frac{(a_1-x)(a_2-x)}{a_1a_2}}+\sqrt{\frac{(a_1-x)(a_3-x)}{a_1a_3}}+\sqrt{\frac{(a_2-x)(a_3-x)}{a_2a_3}}}. \tag{86}$$

With the expressions (82) and (86) for $g^{-1}(x)$ on the interval [0, 1], we do not have $g(g^{-1}(x)) = x$ for x outside the interval [0, 1] in general (unless one of the a_i is the sum of the other two). To satisfy $g(g^{-1}(x)) = x$ as well as Eq. (83) in full generality, one can replace the square roots of the polynomials from F by the pairwise products of the square roots $\sqrt{a_i^{-1}(a_i-x)}$, $i = 1, 2, 3$, in the formulae (82), (83), and (86).

Remark 3 Note the similarity of the formulae (45), (52), (57), (62), (69), (78), and (86). Despite the large variety of rationalizing transformations $g(y)$ given on [0, 1], in each case, the inverse can be expressed in the form $g^{-1}(x) = \frac{w(1)x}{w(x)}$, where $w(x)$ is some simple expression in terms of the respective square roots.

3.3 *Complex-Valued Square Roots on the Interval* [0, 1]

In the following, we no longer require the square roots to take real values on the interval [0, 1] like before in Sect. 3.2. Moreover, we consider $C = \mathbb{C}$, i.e. we also treat radicands with complex coefficients. Still, we aim at rationalizing transformations that, when considered on the interval [0, 1], give monotonically increasing bijections of [0, 1] to itself. This imposes some restrictions on the rationalizing transformation g and hence also on the radicands in F, which we explain now.

If, for a given set F of radicands, there is a rationalizing transformation $g \in \mathbb{C}(y)$ that maps the interval [0, 1] to itself, then g necessarily can also be written with real coefficients, since it is real-valued on [0, 1] and any rational function $g \in \mathbb{C}(y)$ can be written as $g = g_1 + ig_2$ for some $g_1, g_2 \in \mathbb{R}(y)$. Consequently, if some $f \in \mathbb{C}(x)$ becomes a square $f(g(y))$ by the change of variable $x = g(y)$, then also its complex conjugate $\overline{f}$ becomes a square in $\mathbb{C}(y)$ by the same transformation g. Altogether, we have that a set $F \subset \mathbb{C}(x)$ does not admit a rationalizing transformation that maps [0, 1] to itself, if the set $F \cup \overline{F}$ does not. This restricts most of the general cases listed in Sect. 3.1 to special choices of the coefficients of radicands.

For instances that can be reduced (cf. Remark 2) to radicands of real-valued square roots on [0, 1], we refer to the transformations in Sect. 3.2. As it turns out, the remaining cases can also be treated by some of the formulae given in that section. The simplest case not reducible to real-valued square roots is given by one square root whose only singularity lies outside the real line. The set $F \cup \overline{F}$ of radicands can be reduced to the form (63) with $a_2 = \overline{a_1} \neq a_1$. Similarly, for a square root with two singularities, where exactly one of them is on the real line, $F \cup \overline{F}$ can be reduced to the form (70) or (79) with one a_i being the complex conjugate of one of the others. All other cases of complex-valued roots that admit a rationalizing transformation mapping [0, 1] to itself can also be reduced to one of these cases detailed below.

3.3.1 Two Square Roots

We consider F as in Eq. (63), where $a_1, a_2 \in \mathbb{C} \setminus \mathbb{R}$ are such that $a_2 = \overline{a_1}$. With α as in Eq. (64), we have that the rationalizing transformation given by Eq. (65) again maps the interval [0, 1] bijectively to itself in a monotonically increasing way. Also the formulae (66) through (68) still hold.

Alternatively, we can express all of these formulae also with real coefficients using $\mathrm{Re}(a_1)$, $\mathrm{Im}(a_1)$, $|a_1|^2$, and $|\frac{a_1-1}{a_1}|$. More explicitly, $\alpha > 1$ can be written as

$$\alpha = 1 + \left|\frac{a_1 - 1}{a_1}\right| + \sqrt{2\left(1 - \frac{\mathrm{Re}(a_1)}{|a_1|^2} + \left|\frac{a_1 - 1}{a_1}\right|\right)} \tag{87}$$

and the rationalizing transformation (65) and its inverse (66) read as follows.

$$g(y) = \frac{4|a_1|^2\alpha y(y^2 - 2\,\mathrm{Re}(a_1)\alpha y + |a_1|^2\alpha^2)}{(y^2 - |a_1|^2\alpha^2)^2} \tag{88}$$

$$g^{-1}(x) = 4|a_1|^2\alpha \frac{\left(1 - \sqrt{a_1^{-1}(a_1 - x)}\right)\left(1 - \sqrt{\overline{a_1}^{-1}(\overline{a_1} - x)}\right)}{x} \tag{89}$$

Moreover, the two square roots can be written explicitly as $r_1(y) \pm i r_2(y)$ in terms of $y = g^{-1}(x)$ with $r_1, r_2 \in \mathbb{R}(y)$. Then, $r_1(y)$ and $r_2(y)$ give the real and imaginary parts as long as y is real, which happens whenever x is real.

$$\sqrt{a_1^{-1}(a_1 - x)} = \frac{y^2 - 2\,\mathrm{Re}(a_1)\alpha y + |a_1|^2\alpha^2}{-y^2 + |a_1|^2\alpha^2} + i\frac{2\,\mathrm{Im}(a_1)\alpha y}{-y^2 + |a_1|^2\alpha^2} \tag{90}$$

$$\sqrt{\overline{a_1}^{-1}(\overline{a_1} - x)} = \frac{y^2 - 2\,\mathrm{Re}(a_1)\alpha y + |a_1|^2\alpha^2}{-y^2 + |a_1|^2\alpha^2} - i\frac{2\,\mathrm{Im}(a_1)\alpha y}{-y^2 + |a_1|^2\alpha^2} \tag{91}$$

3.3.2 Three Square Roots

First, we treat the set of radicands F as in Eq. (70) with $a_1, a_2 \in \mathbb{C} \setminus \mathbb{R}$ such that $a_2 = \overline{a_1}$. With α as in Eq. (73), the rationalizing transformation given by Eqs. (71) and (74) again maps the interval $[0, 1]$ bijectively to itself in a monotonically increasing way. Its inverse is given by Eq. (72) and also Eqs. (75) through (77) remain valid to the extent mentioned there. While the formula (72) for $y = g^{-1}(x)$ gives a real value whenever x is real, it does not make Eq. (77) true for general x. In the following, we consider the inverse

$$g^{-1}(x) = a_1 a_2 \left(\sqrt{1 - a_1^{-1}} + \sqrt{1 - a_2^{-1}}\right) \frac{\sqrt{a_1^{-1}(a_1 - x)} - \sqrt{a_2^{-1}(a_2 - x)}}{(a_1 - a_2)\sqrt{x}} \tag{92}$$

instead, which gives the same values if $x > 0$, but does not give a real value if $x < 0$. In return, the right hand sides of Eqs. (75) through (77) give correct expressions for the pairwise products of $\sqrt{x}$, $\sqrt{a_1^{-1}(a_1 - x)}$, and $\sqrt{a_2^{-1}(a_2 - x)}$ for general x.

We can write these formulae with real coefficients using $\mathrm{Re}(a_1)$, $\mathrm{Im}(a_1)$, $|a_1|^2$, and $|\frac{a_1 - 1}{a_1}|$. With

$$\alpha = \left|\frac{a_1 - 1}{a_1}\right| > 0, \tag{93}$$

we introduce the abbreviations

$$\beta := \sqrt{\frac{1}{2}\left(1 - \frac{\operatorname{Re}(a_1)}{|a_1|^2} + \alpha\right)} \tag{94}$$

$$q(y) := (|a_1|^2 - \operatorname{Re}(a_1) - |a_1|^2\alpha)y^4 + 2\operatorname{Re}(a_1)y^2 + |a_1|^2 - \operatorname{Re}(a_1) + |a_1|^2\alpha \tag{95}$$

for shorter notation. Then, we have that

$$g(y) = \frac{2|a_1|^2y^2}{q(y)} \tag{96}$$

$$g^{-1}(x) = -i\frac{|a_1|^2\beta}{\operatorname{Im}(a_1)} \cdot \frac{\sqrt{a_1^{-1}(a_1 - x)} - \sqrt{\overline{a_1^{-1}}(\overline{a_1} - x)}}{\sqrt{x}} \tag{97}$$

and with $y = g^{-1}(x)$ we have the following rational expressions in y for the pairwise products of square roots, which reveal the real and imaginary part as long as y is real.

$$\sqrt{x}\sqrt{\frac{a_1 - x}{a_1}} = \frac{(|a_1|^2 - \operatorname{Re}(a_1) + |a_1|^2\alpha)y}{\beta q(y)} + i\frac{\operatorname{Im}(a_1)y^3}{\beta q(y)} \tag{98}$$

$$\sqrt{x}\sqrt{\frac{\overline{a_1} - x}{\overline{a_1}}} = \frac{(|a_1|^2 - \operatorname{Re}(a_1) + |a_1|^2\alpha)y}{\beta q(y)} - i\frac{\operatorname{Im}(a_1)y^3}{\beta q(y)} \tag{99}$$

$$\sqrt{\frac{a_1 - x}{a_1}}\sqrt{\frac{\overline{a_1} - x}{\overline{a_1}}} = \frac{-(|a_1|^2 - \operatorname{Re}(a_1) - |a_1|^2\alpha)y^4 + |a_1|^2 - \operatorname{Re}(a_1) + |a_1|^2\alpha}{q(y)} \tag{100}$$

Finally, we turn to radicands F as in Eq. (79) with $a_1 \in \mathbb{R}$ such that $a_1 < 0$ or $a_1 \geq 1$ and $a_2, a_3 \in \mathbb{C} \setminus \mathbb{R}$ such that $a_3 = \overline{a_2}$. With α as in Eq. (80), the rationalizing transformation given by Eqs. (81) and (36) is a monotonically increasing bijection of the interval $[0, 1]$ to itself and its inverse on $[0, 1]$ can be given by Eq. (82) resp. (86). Furthermore, also the expressions (83) for the three square roots remain valid for $y = g^{-1}(x)$ with $x \in [0, 1]$. To obtain formulae that hold for general x, we instead use the inverse

$$\begin{aligned} g^{-1}(x) = -\frac{s_3(1+\alpha)}{(s_1^2 - 4s_2)x + 4s_3}&\left(2x - s_1 + (s_1 - 2a_3)\sqrt{\frac{a_1 - x}{a_1}}\sqrt{\frac{a_2 - x}{a_2}}\right. \\ &\left. +(s_1 - 2a_2)\sqrt{\frac{a_1 - x}{a_1}}\sqrt{\frac{a_3 - x}{a_3}} + (s_1 - 2a_1)\sqrt{\frac{a_2 - x}{a_2}}\sqrt{\frac{a_3 - x}{a_3}}\right) \end{aligned} \tag{101}$$

below. In terms of the real quantities a_1, $\mathrm{Re}(a_2)$, $\mathrm{Im}(a_2)$, $|a_2|^2$, and $|\frac{a_2-1}{a_2}|$, we can write

$$\alpha = \left|\frac{a_2-1}{a_2}\right| + \sqrt{\frac{a_1-1}{a_1}}\sqrt{2\left(1-\frac{\mathrm{Re}(a_2)}{|a_2|^2}+\left|\frac{a_2-1}{a_2}\right|\right)} \tag{102}$$

and

$$q(y) = (a_1^2 - 4a_1\,\mathrm{Re}(a_2) - 4\,\mathrm{Im}(a_2)^2)y^4 + 8a_1|a_2|^2(1+\alpha)y^3 \\ - 2a_1|a_2|^2(a_1+2\,\mathrm{Re}(a_2))(1+\alpha)^2y^2 + a_1^2|a_2|^4(1+\alpha)^4 \tag{103}$$

for Eqs. (80) and (84). Then, we have

$$g(y) = -4a_1|a_2|^2\frac{y(y-a_1(1+\alpha))(y^2-2\,\mathrm{Re}(a_2)(1+\alpha)y+|a_2|^2(1+\alpha)^2)}{q(y)} \tag{104}$$

and

$$g^{-1}(x) = -\frac{a_1|a_2|^2(1+\alpha)}{(a_1^2-4a_1\,\mathrm{Re}(a_2)-4\,\mathrm{Im}(a_2)^2)x+4a_1|a_2|^2}\Bigg(2x-a_1-2\,\mathrm{Re}(a_2) \\ +a_1\sqrt{\frac{a_1-x}{a_1}}\left(\sqrt{\frac{a_2-x}{a_2}}+\sqrt{\frac{\overline{a_2}-x}{\overline{a_2}}}\right)+(2\,\mathrm{Re}(a_2)-a_1)\sqrt{\frac{a_2-x}{a_2}}\sqrt{\frac{\overline{a_2}-x}{\overline{a_2}}} \\ +2i\,\mathrm{Im}(a_2)\sqrt{\frac{a_1-x}{a_1}}\left(\sqrt{\frac{a_2-x}{a_2}}-\sqrt{\frac{\overline{a_2}-x}{\overline{a_2}}}\right)\Bigg). \tag{105}$$

In terms of $y = g^{-1}(x)$, the pairwise products of square roots can be expressed as

$$\sqrt{\frac{a_1-x}{a_1}}\sqrt{\frac{a_2-x}{a_2}} = \frac{a_1(y^2-2\,\mathrm{Re}(a_2)(1+\alpha)y+|a_2|^2(1+\alpha)^2)p(y)}{q(y)} \\ - i\frac{2\,\mathrm{Im}(a_2)y(y-a_1(1+\alpha))p(y)}{q(y)} \tag{106}$$

and

$$\sqrt{\frac{a_1-x}{a_1}}\sqrt{\frac{\overline{a_2}-x}{\overline{a_2}}} = \frac{a_1(y^2-2\,\mathrm{Re}(a_2)(1+\alpha)y+|a_2|^2(1+\alpha)^2)p(y)}{q(y)} \\ + i\frac{2\,\mathrm{Im}(a_2)y(y-a_1(1+\alpha))p(y)}{q(y)} \tag{107}$$

as well as

$$\sqrt{\frac{a_2 - x}{a_2}}\sqrt{\frac{\overline{a_2} - x}{\overline{a_2}}} = \frac{r(y)}{q(y)}, \tag{108}$$

where for shorter notation, in addition to Eq. (103), we also use the abbreviations

$$p(y) := (2\,\mathrm{Re}(a_2) - a_1)y^2 - 2|a_2|^2(1+\alpha)y + a_1|a_2|^2(1+\alpha)^2 \tag{109}$$

and

$$r(y) := (a_1^2 + 4\,\mathrm{Im}(a_2)^2)y^4 - 4a_1(a_1\,\mathrm{Re}(a_2) + 2\,\mathrm{Im}(a_2)^2)(1+\alpha)y^3$$
$$+ 6a_1^2|a_2|^2(1+\alpha)^2y^2 - 4a_1^2\,\mathrm{Re}(a_2)|a_2|^2(1+\alpha)^3y + a_1^2|a_2|^4(1+\alpha)^4. \tag{110}$$

Also, Eqs. (106) through (108) exhibit the real and imaginary parts whenever the quantity $y = g^{-1}(x)$ is real.

Acknowledgments The author wants to thank Jakob Ablinger for useful discussions. The author is supported by the Austrian Science Fund (FWF) grant P 31952.

References

1. J. Ablinger, A computer algebra toolbox for harmonic sums related to particle physics. Diploma Thesis, Johannes Kepler Univ. Linz, 2009 [arXiv:1011.1176]
2. J. Ablinger, Computer algebra algorithms for special functions in particle physics. PhD Thesis, Johannes Kepler Univ. Linz, 2012 [arXiv:1305.0687]
3. J. Ablinger, The package HarmonicSums: computer algebra and analytic aspects of nested sums. PoS **LL2014**, 019 (2014) [arXiv:1407.6180]
4. J. Ablinger, Inverse Mellin transform of holonomic sequences. PoS **LL2016**, 067 (2016) [arXiv:1606.02845]
5. J. Ablinger, Discovering and proving infinite binomial sums identities. Exp. Math. **26**, 62–71 (2017) [arXiv:1507.01703]
6. J. Ablinger, An improved method to compute the inverse Mellin transform of holonomic sequences. PoS **LL2018**, 063 (2018)
7. J. Ablinger, J. Blümlein, A. De Freitas, M. van Hoeij, E. Imamoglu, C.G. Raab, C.-S. Radu, C. Schneider, Iterated elliptic and hypergeometric integrals for Feynman diagrams. J. Math. Phys. **59**, Article 062305 (2018) [arXiv:1706.01299]
8. J. Ablinger, J. Blümlein, C.G. Raab, C. Schneider, Iterated binomial sums and their associated iterated integrals. J. Math. Phys. **55**, Article 112301 (2014) [arXiv:1407.1822]
9. J. Ablinger, J. Blümlein, C.G. Raab, C. Schneider, F. Wißbrock, Calculating massive 3-loop graphs for operator matrix elements by the method of hyperlogarithms. Nucl. Phys. B **885**, 409–447 (2014) [arXiv:1403.1137]
10. J. Ablinger, J. Blümlein, C. Schneider, Harmonic sums and polylogarithms generated by cyclotomic polynomials. J. Math. Phys. **52**, Article 102301 (2011) [arXiv:1105.6063]

11. J. Ablinger, J. Blümlein, C. Schneider, Analytic and algorithmic aspects of generalized harmonic sums and polylogarithms. J. Math. Phys. **54**, Article 082301 (2013). [arXiv:1302.0378]
12. S.A. Abramov, M. Petkovšek, D'Alembertian solutions of linear differential and difference equations, in *Proc. ISSAC'94* (1994), pp. 169–174
13. U. Aglietti, R. Bonciani, Master integrals with 2 and 3 massive propagators for the 2-loop electroweak form factor—planar case. Nucl. Phys. B **698**, 277–318 (2004) [arXiv:hep-ph/0401193]
14. M. Besier, D. van Straten, S. Weinzierl, Rationalizing roots: an algorithmic approach. Commun. Number Theory Phys. **13**, 253–297 (2019) [arXiv:1809.10983]
15. M. Besier, P. Wasser, S. Weinzierl, RationalizeRoots: software package for the rationalization of square roots. Comput. Phys. Commun. **253**, Article 107197 (2020) [arXiv:1910.13251]
16. J. Blümlein, A. De Freitas, C.G. Raab, K. Schönwald, The unpolarized two-loop massive pure singlet Wilson coefficients for deep-inelastic scattering. Nucl. Phys. B **945**, Article 114659 (2019) [arXiv:1903.06155]
17. J. Blümlein, A. De Freitas, C.G. Raab, K. Schönwald, The $O(\alpha^2)$ initial state QED corrections to $e^+e^- \to \gamma^*/Z^*$. Nucl. Phys. B **956**, Article 115055 (2020) [arXiv:2003.14289]
18. J. Blümlein, C.G. Raab, K. Schönwald, The polarized two-loop massive pure singlet Wilson coefficient for deep-inelastic scattering. Nucl. Phys. B **948**, Article 114736 (2019) [arXiv:1904.08911]
19. S.T. Boettner, Mixed transcendental and algebraic extensions for the Risch-Norman algorithm. PhD Thesis, Tulane University, New Orleans, 2010
20. A. Bostan, F. Chyzak, P. Lairez, B. Salvy, Generalized Hermite reduction, creative telescoping and definite integration of D-finite functions, in *Proc. ISSAC'18* (2018), pp. 95–102 [arXiv:1805.03445]
21. D.J. Broadhurst, J. Fleischer, O.V. Tarasov, Two-loop two-point functions with masses: asymptotic expansions and Taylor series, in any dimension. Z. Phys. C Part. Fields **60**, 287–301 (1993) [arXiv:hep-ph/9304303]
22. S. Chen, M. van Hoeij, M. Kauers, C. Koutschan, Reduction-based creative telescoping for fuchsian D-finite functions. J. Symb. Comput. **85**, 108–127 (2018) [arXiv:1611.07421]
23. S. Chen, M. Kauers, C. Koutschan, A generalized Apagodu-Zeilberger algorithm, in *Proc. ISSAC'14* (2014), pp. 107–114 [arXiv:1402.2409]
24. F. Chyzak, An extension of Zeilberger's fast algorithm to general holonomic functions. Discret. Math. **217**, 115–134 (2000)
25. M. Deneufchâtel, G.H.E. Duchamp, V.H.N. Minh, A.I. Solomon, Independence of hyperlogarithms over function fields via algebraic combinatorics, in *Proc. CAI 2011* (2011), pp. 127–139 [arXiv:1101.4497]
26. J. Fleischer, A.V. Kotikov, O.L. Veretin, Analytic two-loop results for self-energy- and vertex-type diagrams with one non-zero mass. Nucl. Phys. B **547**, 343–374 (1999) [arXiv:hep-ph/9808242]
27. A.B. Goncharov, Multiple polylogarithms, cyclotomy and modular complexes. Math. Res. Lett. **5**, 497–516 (1998) [arXiv:1105.2076]
28. C. Hermite, Sur la réduction des intégrales hyperelliptiques aux fonctions de première, de seconde et de troisième espèce. Bull. Sci. Math. Astron. (2e série) **7**, 36–42 (1883)
29. M. Karr, Summation in finite terms. J. Assoc. Comput. Mach. **28**, 305–350 (1981)
30. E.E. Kummer, Ueber die Transcendenten, welche aus wiederholten Integrationen rationaler Formeln entstehen. J. Reine Angew. Math. **21**, 74–90, 193–225, 328–371 (1840)
31. I.A. Lappo-Danilevski, Résolution algorithmique des problèmes réguliers de Poincaré et de Riemann – Mémoire premier: Le problème de Poincaré, concernant la construction d'un groupe monodromie d'un système donné d'équations différentielles linéaires aux integrales régulières. J. Soc. Phys-Math. Léningrade. **2**, 94–120 (1928)
32. C. Mack, Integration of affine forms over elementary functions. Computational Physics Group Report UCP-39, University of Utah, 1976
33. S.-O. Moch, P. Uwer, S. Weinzierl, Nested sums, expansion of transcendental functions, and multiscale multiloop integrals. J. Math. Phys. **43**, 3363–3386 (2002) [arXiv:hep-ph/0110083]

34. A.C. Norman, P.M.A. Moore, Implementing the new Risch integration algorithm, in *Proc. 4th International Colloquium on Advanced Computing Methods in Theoretical Physics* (1977), pp. 99–110
35. H. Poincaré, Sur les groupes des équations linéaires. Acta Math. **4**, 201–312 (1884)
36. C.G. Raab, Definite integration in differential fields. PhD Thesis, Johannes Kepler Univ. Linz, 2012. http://www.risc.jku.at/publications/download/risc_4583/PhD_CGR.pdf
37. C.G. Raab, On the arithmetic of d'Alembertian functions, *Presentation at the 19th Conference on Applications of Computer Algebra (ACA 2013), Málaga*, 2–6 July 2013. Paper in preparation
38. C.G. Raab, Symbolic computation of parameter integrals, in *Proc. ISSAC'16* (2016), pp. 13–15
39. C.G. Raab, G. Regensburger, The fundamental theorem of calculus in differential rings. In preparation
40. R. Ree, Lie elements and an algebra associated to shuffles. Ann. Math. **68**, 210–220 (1958)
41. E. Remiddi, J.A.M. Vermaseren, Harmonic polylogarithms. Int. J. Mod. Phys. A **15**, 725–754 (2000) [arXiv:hep-ph/9905237]
42. R.H. Risch, The problem of integration in finite terms. Trans. Am. Math. Soc. **139**, 167–189 (1969)
43. C. Schneider, A difference ring theory for symbolic summation. J. Symb. Comput. **72**, 82–127 (2016) [arXiv:1408.2776]
44. J.R. Sendra, F. Winkler, S. Pérez-Díaz, *Rational Algebraic Curves – A Computer Algebra Approach* (Springer, Berlin, 2008)
45. M.F. Singer, B.D. Saunders, B.F. Caviness, An extension of Liouville's theorem on integration in finite terms. SIAM J. Comput. **14**, 966–990 (1985)
46. J. van der Hoeven, Constructing reductions for creative telescoping — the general differentially finite case. Appl. Algebra Eng. Commun. Comput. **32**, 575–602 (2021)
47. J.A.M. Vermaseren, Harmonic sums, Mellin transforms, and integrals. Int. J. Mod. Phys. A **14**, 2037–2076 (1999) [arXiv:hep-ph/9806280]
48. G. Wechsung, Functional equations of hyperlogarithms, in *Structural Properties of Polylogarithms* (AMS, Providence, 1991), pp. 171–184
49. D. Zeilberger, A holonomic systems approach to special functions identities. J. Comput. Appl. Math. **32**, 321–368 (1990)

Term Algebras, Canonical Representations and Difference Ring Theory for Symbolic Summation

Carsten Schneider

Abstract A general overview of the existing difference ring theory for symbolic summation is given. Special emphasis is put on the user interface: the translation and back translation of the corresponding representations within the term algebra and the formal difference ring setting. In particular, canonical (unique) representations and their refinements in the introduced term algebra are explored by utilizing the available difference ring theory. Based on that, precise input-output specifications of the available tools of the summation package `Sigma` are provided.

1 Introduction

In the last 40 years exciting results have been accomplished in symbolic summation as elaborated, e.g., in [18, 19, 22, 24, 25, 29, 31, 49, 52, 53, 55, 56, 61, 63, 65–67, 71, 74, 85–87, 90, 92, 93, 97, 98, 108, 114, 116–118, 127, 130, 132–134] that will be sketched in more details below. In most cases, symbolic summation can be subsumed by the following problem description: given an algorithm that computes/represents a sequence, find a simpler algorithm that computes/represents (from a certain point on) the same sequence. Based on the context of a given problem, simpler can have different meanings: e.g., the output algorithm can be represented uniquely (by a canonical form in the sense of [50]), it might be computed more efficiently, or it can be formulated in terms of certain classes of special functions.

Often symbolic summation is subdivided in the following summation paradigms.

C. Schneider (✉)
Johannes Kepler University Linz, Research Institute for Symbolic Computation, Linz, Austria
e-mail: Carsten.Schneider@risc.jku.at

J. Blümlein, C. Schneider (eds.), *Anti-Differentiation and the Calculation of Feynman Amplitudes*, Texts & Monographs in Symbolic Computation,
https://doi.org/10.1007/978-3-030-80219-6_17

- *Telescoping:* Given an algorithm $F(k)$ that computes a sequence, find an algorithm $G(k)$, that is not more complicated than $F(k)$, such that

$$F(k) = G(k+1) - G(k) \tag{1}$$

holds for all $k \in \mathbb{Z}_{\geq 0}$ with $k \geq \delta$ for some $\delta \in \mathbb{Z}_{\geq 0}$. Then summing this equation over k from δ to n yields a simpler way to compute $S(n) = \sum_{k=\delta}^{n} F(k)$, namely

$$\sum_{k=\delta}^{n} F(k) = G(n+1) - G(\delta). \tag{2}$$

- *Zeilberger's creative telescoping [134]:* Given an algorithm $F(n,k)$ that computes a bivariate sequence, find an algorithm $G(n,k)$ that is not more complicated than $F(n,k)$, and algorithms $c_0(n), \ldots, c_d(n)$ (for univariate sequences), such that

$$\begin{aligned} c_0(n)\,F(n,k) + c_1(n)\,F(n+1,k) + \cdots + c_d(n)\,F(n+d,k)& \\ = G(n,k+1) - G(n,k)& \end{aligned} \tag{3}$$

holds for all $n, k \in \mathbb{Z}_{\geq 0}$ with $n, k \geq \delta$ for some $\delta \in \mathbb{Z}_{\geq 0}$. Then summing this equation over k from δ to n yields for the definite sum $S(n) = \sum_{k=\delta}^{n} F(n,k)$ the recurrence

$$c_0(n)\,S(n) + c_1(n)\,S(n+1) + \cdots + c_d(n)\,S(n+d) = H(n) \tag{4}$$

with $H(n) = G(n, n+1) - G(n, \delta) + \sum_{i=1}^{d} c_i(n) \sum_{j=1}^{i} F(n+i, n+j)$. In many cases $H(n)$ collapses to a rather simple "algorithm" and thus (4) yields (together with d initial values and the assumption that $c_d(n)$ is nonzero for $n \geq \delta$) an efficient algorithm to compute the sequence $(S(n))_{n \geq \delta}$.

- *Recurrence solving:* Given a recurrence of the form (4) where the algorithms $c_0(n), \ldots, c_d(n)$ and $H(n)$ can be given by expressions in terms of certain classes of special functions (that can be evaluated accordingly) and given d initial values, say $S(\delta), S(\delta+1), \ldots, S(\delta+d-1)$ which determines the sequence $(S(n))_{n \geq \delta}$, find an expression that computes the sequence $(S(n))_{n \geq \delta}$ in terms of the same class of special functions or an appropriate extension of it.

We emphasize that all of the above summation paradigms are strongly interwoven (as illustrated, e.g., in the book [93]) and they often yield a strong toolbox by combining them in a nontrivial way.

Another natural classification of symbolic summation is based on the input class of algorithms and the focus how they can be formally represented. In most cases they are either given by evaluable expressions in terms of sums/products or linear recurrences accompanied with initial values that uniquely determine/enable one to calculate the underlying sequences. The first breakthrough in this regard has been achieved by Abramov [18, 19] who solved the telescoping problem for a rational function $F(x) \in \mathbb{K}(x)$ and proposed an algorithm for finding all rational solutions of $\mathbb{K}(x)$ of a given linear recurrence of the form (4) with $c_i(x), H(x) \in \mathbb{K}(x)$. In particular, Gosper's telescoping algorithm [61] for hypergeometric products $F(n) = \prod_{k=l}^{n} H(k)$ with $H(x) \in \mathbb{K}(x)$ and Zeilberger's extension to definite sums via his creative telescoping paradigm [52, 53, 85, 90, 93, 134] made symbolic summation highly popular in many areas of sciences; recently also the treatment of contiguous relations has been extensively explored in [86]. In particular, the interplay with Petkovšek's algorithm Hyper [91] or van Hoeij's improvements [127] to find all hypergeometric product solutions enables one to simplify definite hypergeometric products to expressions given in terms of hypergeometric products; first methods are on the way to find even definite sum solutions [92]. More generally, one can use these solvers as subroutines to hunt for all d'Alembertian solutions [22, 24] (solutions that are expressible in terms of indefinite nested sums defined over hypergeometric products) and Liouvillian solutions [63, 94] (incorporating in addition the interlacing operator). This successful story has been pushed forward for indefinite and definite summation problems in terms of q-hypergeometric products and their mixed version [25, 31, 87]. Further generalizations opened up substantially the class of applications, like the holonomic approach [55, 74, 133] dealing with objects that can be described by recurrence systems or the multi-summation approach of (q–)hypergeometric products [29, 130, 132]. Even non-holonomic summation problems [56, 67, 71] involving, e.g., Stirling numbers, can be treated nowadays automatically.

In the following we will focus on the difference ring/field approach. It has been initiated by Karr's telescoping algorithm [65, 66] in $\Pi\Sigma$-fields which can be considered as the discrete analog of Risch's indefinite integration algorithm [48, 95]. This pioneering work has been explored further in [49, 97, 98, 108] and has been pushed forward to a general summation theory in the setting of $R\Pi\Sigma$-ring extensions [114, 116–118] which is the driving engine of the summation package `Sigma` [107, 112]. In this setting, one can deal not only with expressions containing (q–)hypergeometric products and their mixed versions, but also with those containing sums and products that are indefinite nested (that, depending on the ring or field setting, can appear also in the denominator). In particular, it covers a significant class of special functions that arise frequently, e.g., within the calculation of (massive) 2-loop and 3-loop Feynman integrals: harmonic sums [37, 129], generalized harmonic sums [15, 81], cyclotomic sums [2] and binomial sums [4, 58, 131].

Internally, the following construction is performed in `Sigma`.

1. Rephrase the expression in terms of nested sums and products in an appropriate difference ring (built by $\Pi\Sigma$-field and $R\Pi\Sigma$-ring extensions).
2. Solve the summation problems (given above) in this formal difference ring.
3. Translate the obtained solution from the difference ring to the term algebra setting.

The goal of this article is two-fold. First, we will present the existing algorithms in the difference ring setting (step 2) that have been implemented in large part within

Sigma. In particular, we will summarize the available parameterized telescoping algorithms [100, 102, 106, 108–111, 115] (containing telescoping/creative telescoping as special cases), the multiplicative version of telescoping for the representation of products [23, 51, 83, 84, 104, 114, 120] and recurrence solving algorithms [26, 49, 80, 98, 99, 101, 105] which generalize many contributions of the literature mentioned above. In addition, we will comment on further enhancements in order to treat new classes of summation objects, like unspecified sequences [68, 69, 89] and radical objects [70], or to combine the difference field/ring and holonomic approaches yielding a new toolbox for multi-summation [43, 103].

Besides these difference ring algorithms and the underlying difference ring theory (step 2), the translation mechanism between the summation objects and the formal representation (step 1 and 3) will be elaborated in detail. In particular, the summation package Sigma benefits strongly on this stable toolbox: the user can define expressions in terms of symbolic sums and products in a term algebra and obtains simplifications of the expressions by executing the rather technical difference ring/field machinery in the background. However, rigorous input/output specifications on the sum-product level are missing: many of the properties that one can extract on the formal level (step 2) are not properly carried over to the user level. The second main result of the article is a contribution towards closing this gap. In particular, inspired by Nemes and Paule [82] and utilizing ideas from [109, 114, 126] we will show that the difference ring theory implies a canonical simplification in the sense of [50]. We can write the sums and products in a σ-reduced basis (see Definition 4) such that two expressions evaluate to the same sequence iff they are syntactically equal.

In Sect. 2 we will define a term algebra in which we will represent our sequences in terms of indefinite nested sums and products. In particular, we will introduce one of the main features of Sigma given in Problem SigmaReduce: one can represent the expressions of our term algebra in canonical form. In Sect. 3 we will elaborate how this distinguished representation can be accomplished by exploiting the difference ring theory of $R\Pi\Sigma$-extensions. Here we will utilize the interplay (see Fig. 1) between the difference ring of sequences, the term algebra (equipped with an evaluation function) in which the sequences can be introduced by the user and the formal difference ring setting (also equipped with an evaluation function) in which the sequences can be modeled on the computer algebra level. In Sect. 4 we will make this construction precise by using the existing difference ring algorithms. In particular, we will concentrate on refined simplifications, like finding expressions with minimal nesting depth. Finally, we are in the position to specify in Sect. 5 the above introduced summation paradigms of Sigma within the term algebra level. In Sect. 6 we present the main applications of the presented algorithms that support the evaluation of Feynman integrals. We conclude the article in Sect. 7.

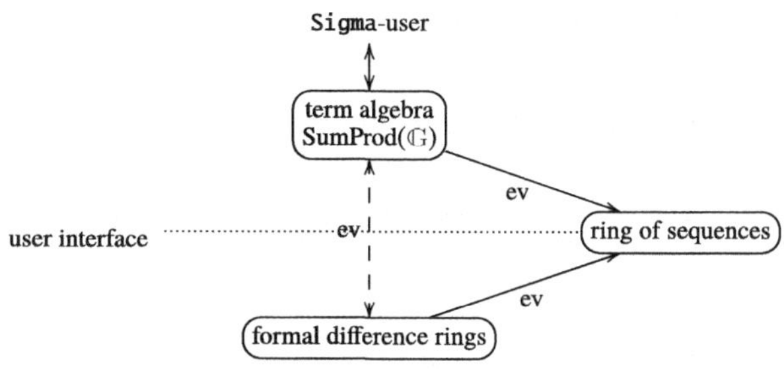

Fig. 1 The symbolic summation framework for difference rings and fields

2 The Term Algebra SumProd($\mathbb{G}$)

Inspired by Nemes and Paule [82] we will refine the construction from [111] to introduce a term algebra for a big class of indefinite nested sums and products.

The basis of our construction (see also [31]) will be the rational function field extension $\mathbb{K} = K(q_1, \ldots, q_v)$ over a field K and on top of it the rational function field extension $\mathbb{G}_m := \mathbb{K}(x, x_1, \ldots, x_v)$ over $\mathbb{K}$. For any element $f = \frac{p}{q} \in \mathbb{G}_m$ with $p, q \in \mathbb{K}[x, x_1, \ldots, x_v]$ where $q \neq 0$ and p, q being coprime we define

$$\mathrm{ev}(f, k) = \begin{cases} 0 & \text{if } q(k, q_1^k, \ldots, q_v^k) = 0 \\ \frac{p(k, q_1^k, \ldots, q_v^k)}{q(k, q_1^k, \ldots, q_v^k)} & \text{if } q(k, q_1^k, \ldots, q_v^k) \neq 0. \end{cases} \tag{5}$$

Note that there is a $\delta \in \mathbb{Z}_{\geq 0}$ with $q(k, q_1^k, \ldots, q_v^k) \neq 0$ for all $k \in \mathbb{Z}_{\geq 0}$ with $k \geq \delta$; for an algorithm that determines δ if one can factorize polynomials over K see [31, Sec. 3.2]. We define $L(f)$ to be the minimal value $\delta \in \mathbb{Z}_{\geq 0}$ such that $q(k, q_1^k, \ldots, q_v^k) \neq 0$ holds for all $k \geq \delta$; further, we define $Z(f) = \max(L(1/p), L(1/q))$ for $f \neq 0$. Later we will call $L : \mathbb{G}_m \to \mathbb{Z}_{\geq 0}$ also an *o-function* and[1] $Z : \mathbb{G}_m^* \to \mathbb{Z}_{\geq 0}$ a *z-function*. $\mathbb{G}_m = \mathbb{K}(x, x_1, \ldots, x_v)$ represents the *multibasic mixed sequences*. The special cases $\mathbb{G}_r = \mathbb{K}(x)$ and $\mathbb{G}_b = \mathbb{K}(x_1, \ldots, x_v)$ represent the *rational* and the *multi-basic sequences*, respectively. If not specified further, $\mathbb{G}$ will stand for one of the three cases $\mathbb{G}_m$, $\mathbb{G}_r$ or $\mathbb{G}_b$.

Now we extend $\mathbb{G}$ to expressions SumProd($\mathbb{G}$) in terms of indefinite nested sums defined over indefinite nested products. For the set of nontrivial roots of unity

$$\mathcal{R} = \{r \in \mathbb{K} \setminus \{1\} \mid r \text{ is a root of unity}\}$$

we introduce the function $\mathrm{ord} : \mathcal{R} \to \mathbb{Z}_{\geq 1}$ with

$$\mathrm{ord}(r) = \min\{n \in \mathbb{Z}_{\geq 1} \mid r^n = 1\}.$$

[1]For a ring $\mathbb{A}$ we denote by $\mathbb{A}^*$ the set of units. If $\mathbb{A}$ is a field, this means $\mathbb{A}^* = \mathbb{A} \setminus \{0\}$.

Let $\circledwedge$, $\oplus$, $\odot$, Sum, Prod and RPow be operations with the signatures

$$\begin{array}{lll}\circledwedge: & \mathrm{SumProd}(\mathbb{G})\times\mathbb{Z} & \to \mathrm{SumProd}(\mathbb{G})\\ \oplus: & \mathrm{SumProd}(\mathbb{G})\times\mathrm{SumProd}(\mathbb{G}) & \to \mathrm{SumProd}(\mathbb{G})\\ \odot: & \mathrm{SumProd}(\mathbb{G})\times\mathrm{SumProd}(\mathbb{G}) & \to \mathrm{SumProd}(\mathbb{G})\\ \mathrm{Sum}: & \mathbb{Z}_{\geq 0}\times\mathrm{SumProd}(\mathbb{G}) & \to \mathrm{SumProd}(\mathbb{G})\\ \mathrm{Prod}: & \mathbb{Z}_{\geq 0}\times\mathrm{SumProd}(\mathbb{G}) & \to \mathrm{SumProd}(\mathbb{G})\\ \mathrm{RPow}: \mathcal{R} & & \to \mathrm{SumProd}(\mathbb{G}).\end{array}$$

In the following we write $^{\circledwedge}$, $\oplus$ and $\odot$ in infix notation, and Sum and Prod in prefix notation. Further, for $(\dots((f_1 \square f_2)\square f_3)\square\dots\square f_r)$ with $\square\in\{\odot,\oplus\}$ and $f_1,\dots,f_r\in \mathrm{SumProd}(\mathbb{G})$ we write $f_1\square f_2\square f_3\square\dots\square f_r$.

More precisely, we define the following chain of set inclusions:

$$\begin{array}{ccccccc} & & \mathrm{Prod}_1(\mathbb{G}) & \subset & \mathrm{SumProd}_1(\mathbb{G}) & & \text{expressions with single nested products}\\ & & \cap & & \cap & & \\ \mathrm{Prod}^*(\mathbb{G}) & \subset & \mathrm{Prod}(\mathbb{G}) & \subset & \mathrm{SumProd}(\mathbb{G}) & & \text{expressions with nested products}\\ \text{power products in products} & & \text{expressions in products} & & \text{expressions in sums and products.} & & \end{array} \tag{6}$$

Here we start with the *set of power products of nested products* $\mathrm{Prod}^*(\mathbb{G})$ which is the smallest set that contains 1 with the following properties:

1. If $r\in\mathcal{R}$ then $\mathrm{RPow}(r)\in\mathrm{Prod}^*(\mathbb{G})$.
2. If $p\in\mathrm{Prod}^*(\mathbb{G})$, $f\in\mathbb{G}^*$, $l\in\mathbb{Z}_{\geq 0}$ with $l\geq Z(f)$ then[2] $\mathrm{Prod}(l, f\odot p)\in\mathrm{Prod}^*(\mathbb{G})$.
3. If $p,q\in\mathrm{Prod}^*(\mathbb{G})$ then $p\odot q\in\mathrm{Prod}^*(\mathbb{G})$.
4. If $p\in\mathrm{Prod}^*(\mathbb{G})$ and $z\in\mathbb{Z}\setminus\{0\}$ then $p^{\circledwedge}z\in\mathrm{Prod}^*(\mathbb{G})$.

Later we will also use the sets

$$\Pi(\mathbb{G}) = \{\mathrm{RPow}(r)\mid r\in\mathcal{R}\}\cup\{\mathrm{Prod}(l, f\odot p)\mid l, f, p \text{ as given in item 2}\}$$
$$\Pi_1(\mathbb{G}) = \{\mathrm{RPow}(r)\mid r\in\mathcal{R}\}\cup\{\mathrm{Prod}(l, f)\mid f\in\mathbb{G}^*, l\in\mathbb{Z}_{\geq 0} \text{ with } l\geq Z(f)\}$$

where $\Pi(\mathbb{G})$ and $\Pi_1(\mathbb{G})$ contains all nested and single nested products, respectively.

Example 1 In $\mathrm{Prod}^*(\mathbb{G})$ with $\mathbb{G}=\mathbb{Q}(q_1)(x, x_1)$ we get, e.g.,

$$P = \underbrace{(\mathrm{Prod}(1,\mathrm{Prod}(1,x)^{\circledwedge}(-2))}_{\in\Pi(\mathbb{G})}{}^{\circledwedge}2)\odot\underbrace{\mathrm{Prod}(1,\tfrac{x_1+x_1^2}{x})}_{\in\Pi_1(\mathbb{G})}\odot\underbrace{\mathrm{RPow}(-1)}_{\Pi_1(\mathbb{G})}\in\mathrm{Prod}^*(\mathbb{G}).$$

[2]We also write p instead of $f\odot p$ if $f=1$; similarly we write f instead of $f\odot p$ if $p=1$.

Finally, we define $\text{SumProd}(\mathbb{G})$ as the smallest set containing $\mathbb{G} \cup \text{Prod}^*(\mathbb{G})$ with the following properties:

1. For all $f, g \in \text{SumProd}(\mathbb{G})$ we have $f \oplus g \in \text{SumProd}(\mathbb{G})$.
2. For all $f, g \in \text{SumProd}(\mathbb{G})$ we have $f \odot g \in \text{SumProd}(\mathbb{G})$.
3. For all $f \in \text{SumProd}(\mathbb{G})$ and $k \in \mathbb{Z}_{\geq 1}$ we have $f^{\circledwedge} k \in \text{SumProd}(\mathbb{G})$.
4. For all $f \in \text{SumProd}(\mathbb{G})$ and $l \in \mathbb{Z}_{\geq 0}$ we have $\text{Sum}(l, f) \in \text{SumProd}(\mathbb{G})$.

$\text{SumProd}(\mathbb{G})$ is also called the *set of expressions in terms of nested sums over nested products*. In addition, we define the following subsets:

1. the *set* $\text{Prod}(\mathbb{G})$ *of expressions in terms of nested products (over* $\mathbb{G}$*)*, i.e., all elements from $\text{SumProd}(\mathbb{G})$ which are free of sums;
2. the *set* $\text{Prod}_1(\mathbb{G})$ *of expressions in terms of depth-1 products (over* $\mathbb{G}$*)*, i.e., all elements from $\text{Prod}(\mathbb{G})$ where the arising products are taken from $\Pi_1(\mathbb{G})$;
3. the *set* $\text{Sum}(\mathbb{G})$ *of expressions in terms of nested sums (over* $\mathbb{G}$*)*, i.e., all elements from $\text{SumProd}(\mathbb{G})$ where no products appear;
4. the *set* $\text{SumProd}_1(\mathbb{G})$ *of expressions in terms of nested sums over depth-1 products (over* $\mathbb{G}$*)*, i.e., all elements from $\text{SumProd}(\mathbb{G})$ with products taken from $\Pi_1(\mathbb{G})$.

In other words, besides the chain of set inclusions given in (6) we also get

$$\text{Sum}(\mathbb{G}) \subset \text{SumProd}_1(\mathbb{G}) \subset \text{SumProd}(\mathbb{G}).$$

Furthermore, we introduce the *set of nested sums over nested products* given by

$$\Sigma(\mathbb{G}) = \{\text{Sum}(l, f) \mid l \in \mathbb{Z}_{\geq 0} \text{ and } f \in \text{SumProd}(\mathbb{G})\},$$

and the *set of nested sums over single nested products* given by

$$\Sigma_1(\mathbb{G}) = \{\text{Sum}(l, f) \mid l \in \mathbb{Z}_{\geq 0} \text{ and } f \in \text{SumProd}_1(\mathbb{G})\}.$$

For convenience we will also introduce the *set* $\Sigma\Pi(\mathbb{G}) = \Sigma(\mathbb{G}) \cup \Pi(\mathbb{G})$ *of nested sums and products* and the *set* $\Sigma\Pi_1(\mathbb{G}) = \Sigma_1(\mathbb{G}) \cup \Pi_1(\mathbb{G})$ *of nested sums and single-nested products*. In short, we obtain the following chain of sets:

$$\begin{array}{ccccccl}
\Pi_1(\mathbb{G}) & \subset & \Sigma\Pi_1(\mathbb{G}) & \supset & \Sigma_1(\mathbb{G}) & & \text{with single nested products} \\
\cap & & \cap & & \cap & & \\
\Pi(\mathbb{G}) & \subset & \Sigma\Pi(\mathbb{G}) & \supset & \Sigma(\mathbb{G}) & & \text{with nested products}
\end{array}$$

products	products and sums over products	sums over products

Example 2 With $\mathbb{G} = \mathbb{K}(x)$ we get, e.g., the following expressions:

$$E_1 = \text{Sum}(1, \text{Prod}(1, x)) \in \Sigma_1(\mathbb{G}) \subset \text{SumProd}_1(\mathbb{G}),$$

$$E_2 = \text{Sum}(1, \tfrac{1}{x+1} \odot \text{Sum}(1, \tfrac{1}{x^3}) \odot \text{Sum}(1, \tfrac{1}{x})) \in \Sigma(\mathbb{G}) \subset \text{Sum}(\mathbb{G}),$$

$$E_3 = (E_1 \oplus E_2) \odot E_1 \in \text{SumProd}_1(\mathbb{G}).$$

Finally, we introduce a function ev (a model of the term algebra) which evaluates a given expression of our term algebra to sequence elements. In addition, we also introduce the depth for our expressions. We start with the evaluation function ev : $\mathbb{G} \times \mathbb{Z}_{\geq 0} \to \mathbb{K}$ given by (5) and the depth function $\mathrm{d} : \mathbb{G} \to \mathbb{Z}_{\geq 0}$ given by

$$\mathrm{d}(f) = \begin{cases} 0 & \text{if } f \in \mathbb{K} \\ 1 & \text{if } f \in \mathbb{G} \setminus \mathbb{K}. \end{cases}$$

Now ev and d are extended recursively from $\mathbb{G}$ to ev: $\text{SumProd}(\mathbb{G}) \times \mathbb{Z}_{\geq 0} \to \text{SumProd}(\mathbb{G})$ and $\mathrm{d} : \text{SumProd}(\mathbb{G}) \to \mathbb{Z}_{\geq 0}$ as follows.

1. For $f, g \in \text{SumProd}(\mathbb{G})$ and $k \in \mathbb{Z} \setminus \{0\}$ ($k > 0$ if $f \notin \text{Prod}^*(\mathbb{G})$) we set

$$\begin{aligned} \text{ev}(f^{\circledwedge}k, n) &:= \text{ev}(f, n)^k, & \mathrm{d}(f^{\circledwedge}k) &:= \mathrm{d}(f), \\ \text{ev}(f \oplus g, n) &:= \text{ev}(f, n) + \text{ev}(g, n), & \mathrm{d}(f \oplus g) &:= \max(\mathrm{d}(f), \mathrm{d}(g)), \\ \text{ev}(f \odot g, n) &:= \text{ev}(f, n)\ \text{ev}(g, n) & \mathrm{d}(f \odot g) &:= \max(\mathrm{d}(f), \mathrm{d}(g)); \end{aligned}$$

2. for $r \in \mathcal{R}$ and $\text{Sum}(l, f), \text{Prod}(\lambda, g) \in \text{SumProd}(\mathbb{G})$ we define

$$\begin{aligned} \text{ev}(\text{RPow}(r), n) &:= \prod_{i=1}^{n} r = r^n, & \mathrm{d}(\text{RPow}(r)) &:= 1, \\ \text{ev}(\text{Sum}(l, f), n) &:= \sum_{i=l}^{n} \text{ev}(f, i), & \mathrm{d}(\text{Sum}(l, f)) &:= \mathrm{d}(f) + 1, \\ \text{ev}(\text{Prod}(\lambda, g), n) &:= \prod_{i=\lambda}^{n} \text{ev}(g, i), & \mathrm{d}(\text{Prod}(\lambda, g)) &:= \mathrm{d}(g) + 1. \end{aligned}$$

Remark 1

(1) Since $\text{ev}(\text{Prod}(r, 1), n) = \text{ev}(\text{RPow}(r), n)$, RPow is redundant. But it will be convenient for the treatment of canonical representations (see Definition 3).
(2) Any evaluation of $\text{Prod}^*(\mathbb{G})$ is well defined and nonzero since the lower bounds of the products are set large enough via the z-function.
(3) $\text{SumProd}_1(\mathbb{G}_r)$ covers as special cases generalized/cyclotomic harmonic sums [2, 15, 37, 81, 129] and binomial sums [4, 58, 131].

In a nutshell, ev *applied to* $f \in \text{SumProd}(\mathbb{G})$ *represents a sequence. In particular, f can be considered as a simple program and* $\text{ev}(f, n)$ *with* $n \in \mathbb{Z}_{\geq 0}$ *executes it (like an interpreter/compiler) yielding the nth entry of the represented sequence.*

Definition 1 For $F \in \text{SumProd}(\mathbb{G})$ and $n \in \mathbb{Z}_{\geq 0}$ we write $F(n) := \text{ev}(F, n)$.

Example 3 For $E_i \in \text{SumProd}(\mathbb{K}(x))$ with $i = 1, 2, 3$ in Example 2 we get $\mathtt{d}(E_i) = 3$ and

$$E_1(n) = \text{ev}(E_1, n) = \sum_{k=1}^{n} \prod_{i=1}^{k} i = \sum_{k=1}^{n} k!,$$

$$E_2(n) = \text{ev}(E_2, n) = \sum_{k=1}^{n} \tfrac{1}{1+k} \Big(\sum_{i=1}^{k} \tfrac{1}{i^3} \Big) \sum_{i=1}^{k} \tfrac{1}{i}$$

and $E_3(n) = (E_1(n) + E_2(n))E_1(n)$. For $P \in \text{SumProd}(\mathbb{K}(x, x_1))$ in Ex. 1 we get

$$P(n) = \text{ev}(P, n) = \Big(\prod_{k=1}^{n} \Big(\prod_{i=1}^{k} i \Big)^{-2} \Big)^2 \Big(\prod_{k=1}^{n} \frac{q^k + q^{2k}}{k} \Big)(-1)^n, \quad \mathtt{d}(P) = 3.$$

Example 4 We show how the expressions of $\text{SumProd}(\mathbb{G})$ with ev are handled in

In[1]:= **<< Sigma.m**

Sigma - A summation package by Carsten Schneider © RISC-JKU

Instead of $F = \text{Sum}(1, \frac{1}{x})$ with $F(n) = \text{ev}(F, n) = \sum_{k=1}^{n} \frac{1}{k}$ we introduce the sum by

In[2]:= **F = SigmaSum[$\frac{1}{k}$, {k, 1, n}]**

Out[2]= $\sum_{k=1}^{n} \frac{1}{k}$

where n is kept symbolically. However, if the user replaces n by a concrete integer, say 5, the evaluation mechanism is carried out and we get $F(5) = \text{ev}(F, 5)$:

In[3]:= **F/.n → 5**

Out[3]= $\frac{137}{60}$

Similarly, we can define E_1 from Example 2 as follows:

In[4]:= **E_1 = SigmaSum[SigmaFactorial[k], {k, 1, n}]**

Out[4]= $\sum_{k=1}^{n} k!$

Here `SigmaFactorial` defines the factorials; its full definition is given by:

In[5]:= **GetFullDefinition[E$_1$]**

$$\text{Out[5]=}\ \sum_{k=1}^{n}\prod_{o_1=1}^{k} o_1$$

Similarly, one can introduce as shortcuts powers, Pochhammer symbols, binomial coefficients, (generalized) harmonic sums [15] etc. with the function calls `SigmaPower`, `SigmaPochhammer`, `SigmaBinomial` or `S`, respectively; analogously q-versions are available. Together with Ablinger's package `HarmonicSums`, also function calls for cyclotomic sums [2] and binomial sums [4] are available.

In the same fashion, we can define $E_2, E_3 \in \text{SumProd}(\mathbb{Q}(x))$ from Example 2 and $P \in \text{SumProd}(\mathbb{Q}(q)(x, x_1))$ with $q = q_1$ from Example 1 by

In[6]:= **E$_2$ = SigmaSum[SigmaSum[1/i, {i, 1, k}]SigmaSum[1/i^3, {i, 1, k}]/(k + 1), {k, 1, n}]**

$$\text{Out[6]=}\ \sum_{k=1}^{n}\frac{\left(\sum_{i=1}^{k}\frac{1}{i^3}\right)\sum_{i=1}^{k}\frac{1}{i}}{1+k}$$

In[7]:= **E$_3$ = (E$_1$ + E$_2$)E$_1$**

$$\text{Out[7]=}\ \Big(\sum_{k=1}^{n}k!\Big)\Big(\sum_{k=1}^{n}k! + \sum_{k=1}^{n}\frac{\left(\sum_{i=1}^{k}\frac{1}{i^3}\right)\sum_{i=1}^{k}\frac{1}{i}}{1+k}\Big)$$

In[8]:= **P = SigmaProduct[SigmaProduct[i, {i, 1, k}]$^{-2}$, {k, 1, n}]^2**
SigmaProduct[(SigmaPower[q, k] + SigmaPower[q, k]^2)/k, {k, 1, n}]
SigmaPower[−1, n]

$$\text{Out[8]=}\ \Big(\prod_{k=1}^{n}\Big(\prod_{i=1}^{k}i\Big)^{-2}\Big)^{2}\Big(\prod_{k=1}^{n}\frac{q^k+(q^k)^2}{k}\Big)(-1)^n$$

Note that within `Sigma` the root of unity product RPow(α) with $\alpha \in \mathcal{R}$ can be either defined by `SigmaPower[`α`,n]` or `SigmaProduct[`α`,{k,1,n}]`. Whenever α is recognized as an element of $\mathcal{R}$, it is treated as the special product RPow(α).

Expressions in SumProd($\mathbb{G}$) (similarly within Mathematica using `Sigma`) can be written in different ways such that they produce the same sequence. In the remaining part of this section we will elaborate on canonical (unique) representations [50].

In a preprocessing step we can rewrite the expressions to a reduced representation; note that the equivalent definition in the ring setting is given in Definition 10.

Definition 2 An expression $A \in \text{SumProd}(\mathbb{G})$ is in *reduced representation* if

$$A = (f_1 \odot P_1) \oplus (f_2 \odot P_2) \oplus \cdots \oplus (f_r \odot P_r) \tag{7}$$

with $f_i \in \mathbb{G}^*$ and

$$P_i = (a_{i,1} \circledwedge z_{i,1}) \odot (a_{i,2} \circledwedge z_{i,2}) \odot \cdots \odot (a_{i,n_i} \circledwedge z_{i,n_i}) \in \text{Prod}^*(\mathbb{G}) \tag{8}$$

for $1 \leq i \leq r$ where

- $a_{i,j} = \text{Sum}(l_{i,j}, f_{i,j}) \in \Sigma(\mathbb{G})$ and $z_{i,j} \in \mathbb{Z}_{\geq 1}$,
- $a_{i,j} = \text{Prod}(l_{i,j}, f_{i,j}) \in \Pi(\mathbb{G})$ and $z_{i,j} \in \mathbb{Z} \setminus \{0\}$, or
- $a_{i,j} = \text{RPow}(f_{i,j})$ with $f_{i,j} \in \mathcal{R}$ and $1 \leq z_{i,j} < \text{ord}(r_{i,j})$

such that the following properties hold:

1. for each $1 \leq i \leq r$ and $1 \leq j < j' < n_i$ we have $a_{i,j} \neq a_{i,j'}$;
2. for each $1 \leq i < i' \leq r$ with $n_i = n_j$ there does not exist a $\sigma \in S_{n_i}$ with $P_{i'} = (a_{i,\sigma(1)} \circledwedge z_{i,\sigma(1)}) \odot (a_{i,\sigma(2)} \circledwedge z_{i,\sigma(2)}) \odot \cdots \odot (a_{i,\sigma(n_i)} \circledwedge z_{i,\sigma(n_i)})$.

We say that $H \in \text{SumProd}(\mathbb{G})$ is in *sum-product reduced representation* (or in *sum-product reduced form*) if it is in reduced representation and for each $\text{Sum}(l, A)$ and $\text{Prod}(l, A)$ that occur recursively in H the following holds: A is in reduced representation as given in (7), $l \geq \max(L(f_1), \ldots, L(f_r))$ (i.e. the first case of (5) is avoided during evaluations) and the lower bound l is greater than or equal to the lower bounds of the sums and products inside of A.

Example 5 In `Sigma` the reduced representation of E_3 is calculated with the call

In[9]:= **CollectProdSum[E3]**

Out[9]= $\left(\sum_{k=1}^{n} k!\right)^2 + \left(\sum_{k=1}^{n} k!\right) \sum_{k=1}^{n} \frac{\left(\sum_{i=1}^{k} \frac{1}{i^3}\right) \sum_{i=1}^{k} \frac{1}{i}}{1+k}$

Before we can state one of `Sigma`'s crucial features we need the following definitions.

Definition 3 Let $W \subseteq \Sigma\Pi(\mathbb{G})$. We define $\text{SumProd}(W, \mathbb{G})$ as the set of elements from $\text{SumProd}(\mathbb{G})$ which are in reduced representation and where the arising sums and products are taken from W. More precisely, $A \in \text{SumProd}(W, \mathbb{G})$ if and only if it is of the form (7) with (8) where $a_{i,j} \in W$. In the following we seek a W with the following properties:

- W is called *shift-closed over* $\mathbb{G}$ if for any $A \in \text{SumProd}(W, \mathbb{G})$, $s \in \mathbb{Z}$ there are $B \in \text{SumProd}(W, \mathbb{G})$ and $\delta \in \mathbb{Z}_{\geq 0}$ such that $A(n + s) = B(n)$ holds for all $n \geq \delta$.

- W is called *shift-stable over* $\mathbb{G}$ if for any product or sum in W the multiplicand or summand is built by sums and products from W.
- W is called *canonical reduced over* $\mathbb{G}$ if for any $A, B \in \text{SumProd}(W, \mathbb{G})$ with $A(n) = B(n)$ for all $n \geq \delta$ for some $\delta \in \mathbb{Z}_{\geq 0}$ the following holds: A and B are the same up to permutations of the operands in $\oplus$ and $\odot$.

The sum-product reduced form is only a minor simplification, but it will be convenient to connect to the difference ring theory below; see Corollary 1. In Lemma 1 we note further that shift-stability implies shift-closure. In particular, the shift operation can be straightforwardly carried out; the proof will be delivered later on page 447.

Lemma 1 *If a finite set $W \subset \Sigma\Pi(\mathbb{G})$ is shift-stable and the elements are in sum-product reduced form,*[3] *then it is also shift-closed. If $\mathbb{K}$ is computable then one can compute for $F \in SumProd(W, \mathbb{G})$ and $\lambda \in \mathbb{Z}$ a $G \in SumProd(W, \mathbb{G})$ such that $F(n+\lambda) = G(n)$ holds for all $n \geq \delta$ for some δ. If one can factor polynomials over $\mathbb{K}$, δ can be determined.*

Based on this observation, we focus on σ-reduced sets which we define as follows.

Definition 4 $W \subseteq \Sigma\Pi(\mathbb{G})$ is called *σ-reduced over* $\mathbb{G}$ if it is canonical reduced, shift-stable and the elements in W are in sum-product reduced form. In particular, $A \in \text{SumProd}(W, \mathbb{G})$ is called *σ-reduced (w.r.t. W)* if W is σ-reduced over $\mathbb{G}$.

More precisely, we are interested in the following problem.

Problem SigmaReduce: Compute a σ-Reduced Representation

Given: $A_1, \dots, A_u \in \text{SumProd}(\mathbb{G})$ with $\mathbb{G} \in \{\mathbb{G}_r, \mathbb{G}_b, \mathbb{G}_m\}$, i,e., $\mathbb{G} = \mathbb{K}(x, x_1, \dots, x_v)$ or $\mathbb{G} = \mathbb{K}(x_1, \dots, x_v)$.
Find: a σ-reduced set $W = \{T_1, \dots, T_e\} \subset \Sigma\Pi(\mathbb{G}')$ in[4] $\mathbb{G}'$, $B_1 \dots, B_u \in \text{SumProd}(W, \mathbb{G}')$ and $\delta_1, \dots, \delta_u \in \mathbb{Z}_{\geq 0}$ such that for all $1 \leq i \leq r$ we get

$$A_i(n) = B_i(n) \quad n \geq \delta_i.$$

[3] The sum-product reduced form is not necessary, but simplifies the proof given on page 447.

[4] In general, we might need a larger field $\mathbb{G}' = \mathbb{K}'(x, x_1, \dots, x_v)$ or $\mathbb{G}' = \mathbb{K}'(x_1, \dots, x_v)$ where the field $\mathbb{K}$ is extended to $\mathbb{K}'$.

Example 6 Consider the following two expressions from SumProd($\mathbb{Q}(x)$):

In[10]:= $\mathbf{A_1 = SigmaSum[SigmaSum[1/i, \{i, 1, k\}]SigmaSum[1/i^3, \{i, 1, k\}]/(k + 1), \{k, 1, n\}]}$

Out[10]= $\sum_{k=1}^{n} \frac{\left(\sum_{i=1}^{k} \frac{1}{i^3}\right) \sum_{i=1}^{k} \frac{1}{i}}{1+k}$

In[11]:= $\mathbf{A_2} = \sum_{i=1}^{n} \frac{1}{i^5} - \frac{\sum_{i=1}^{n} \frac{1}{i^4}}{1+n} - \sum_{j=1}^{n} \frac{\sum_{i=1}^{j} \frac{1}{i^4}}{j} - \sum_{j=1}^{n} \frac{\sum_{i=1}^{j} \frac{1}{i^3}}{j^2} + \frac{\sum_{j=1}^{n} \frac{\sum_{i=1}^{j} \frac{1}{i^3}}{j}}{1+n} - \sum_{j=1}^{n} \frac{\sum_{i=1}^{j} \frac{1}{i}}{j^4} + \frac{\sum_{j=1}^{n} \frac{\sum_{i=1}^{j} \frac{1}{i}}{j^3}}{1+n} + \sum_{k=1}^{n} \frac{\sum_{j=1}^{k} \frac{\sum_{i=1}^{j} \frac{1}{i^3}}{j}}{k} + \sum_{k=1}^{n} \frac{\sum_{j=1}^{k} \frac{\sum_{i=1}^{j} \frac{1}{i}}{j^3}}{k};$

Then we solve Problem SigmaReduce by executing:

In[12]:= $\mathbf{\{B_1, B_2\} = SigmaReduce[\{A_1, A_2\}, n]}$

Out[12]= $\left\{\sum_{k=1}^{n} \frac{\left(\sum_{i=1}^{k} \frac{1}{i^3}\right) \sum_{i=1}^{k} \frac{1}{i}}{1+k}, \sum_{k=1}^{n} \frac{\left(\sum_{i=1}^{k} \frac{1}{i^3}\right) \sum_{i=1}^{k} \frac{1}{i}}{1+k}\right\}$

Since $B_1 = B_2$, it follows $A_1 = A_2$. Note that the set W pops up only implicitly. The set of all sums and products in the output, in our case

$$W_0 = \left\{\sum_{k=1}^{n} \frac{1}{1+k} \left(\sum_{i=1}^{k} \frac{1}{i^3}\right) \sum_{i=1}^{k} \frac{1}{i}\right\} (= \left\{\text{Sum}(1, \tfrac{1}{x+1} \odot \text{Sum}(1, \tfrac{1}{x^3}) \odot \text{Sum}(1, \tfrac{1}{x}))\right\})$$

forms a canonical set in which A_1 and A_2 can be represented by B_1 and B_2 respectively. Adjoining in addition all sums and products that arise inside of the elements in W_0 we get $W = \{\sum_{i=1}^{n} \frac{1}{i}, \sum_{i=1}^{n} \frac{1}{i^3}\} \cup W_0$ which is a σ-reduced set. Internally, `SigmaReduce` parses the arising objects from left to right and constructs the underlying σ-reduced set W in which the input expressions can be rephrased.

Reversing the order of the input elements yields the following result:

In[13]:= $\mathbf{\{B_2, B_1\} = SigmaReduce[\{A_2, A_1\}, n]}$

Out[13]= $\Big\{-\left(\sum_{k=1}^{n} \frac{1}{k^4}\right) \sum_{k=1}^{n} \frac{1}{k} + \frac{\left(\sum_{k=1}^{n} \frac{1}{k^3}\right) \sum_{k=1}^{n} \frac{1}{k}}{1+n} - \sum_{k=1}^{n} \frac{\sum_{k=1}^{k} \frac{1}{k^3}}{k^2} + \sum_{k=1}^{n} \frac{\sum_{k=1}^{k} \frac{1}{k^4}}{k} + \sum_{k=1}^{n} \frac{\left(\sum_{k=1}^{k} \frac{1}{k^3}\right) \sum_{k=1}^{k} \frac{1}{k}}{k},$
$-\left(\sum_{k=1}^{n} \frac{1}{k^4}\right) \sum_{k=1}^{n} \frac{1}{k} + \frac{\left(\sum_{k=1}^{n} \frac{1}{k^3}\right) \sum_{k=1}^{n} \frac{1}{k}}{1+n} - \sum_{k=1}^{n} \frac{\sum_{k=1}^{k} \frac{1}{k^3}}{k^2} + \sum_{k=1}^{n} \frac{\sum_{k=1}^{k} \frac{1}{k^4}}{k} + \sum_{k=1}^{n} \frac{\left(\sum_{k=1}^{k} \frac{1}{k^3}\right) \sum_{k=1}^{k} \frac{1}{k}}{k}\Big\}$

In this case we get the σ-reduced set

$$W = \Big\{\sum_{j=1}^{n}\frac{1}{j^4}, \sum_{j=1}^{n}\frac{1}{j^3}, \sum_{j=1}^{n}\frac{1}{j}, \sum_{j=1}^{n}\frac{\sum_{k=1}^{j}\frac{1}{k^4}}{j}, \sum_{j=1}^{n}\frac{\sum_{k=1}^{j}\frac{1}{k^3}}{j^2}, \sum_{j=1}^{n}\frac{\big(\sum_{k=1}^{j}\frac{1}{k^3}\big)\sum_{k=1}^{j}\frac{1}{k}}{j}\Big\}$$

(expressed in the `Sigma`-language) and since $B_1 = B_2$ we conclude again that $A_1 = A_2$ holds for all $n \geq 0$. To check that $A_1 = A_2$ holds, one can also execute

In[14]:= **SigmaReduce[$\mathbf{A_1 - A_2}$, n]**

Out[14]= $\emptyset$

Here $W = \{\}$ is the σ-reduced set in which we can represent $A_1 - A_2$ by 0.

Such a unique representation (up to trivial permutations) immediately gives rise to the following application: One can compare if two expressions A_1 and A_2 evaluate to the same sequences (from a certain point on): simply check if the resulting B_1 and B_2 in SumProd$(W, \mathbb{G})$ for a σ-reduced W are the same (up to trivial permutations). Alternatively, just check if $A_1 - A_2$ can be reduced to zero. Besides that we will refine the above problem further. E.g., given $A \in$ SumProd$(\mathbb{G})$, one can find an expression $B \in$ SumProd$(W, \mathbb{G})$ and $\delta \in \mathbb{Z}_{\geq 0}$ such that $A(n) = B(n)$ holds for all $n \geq \delta$ and such that B is as simple as possible. Here simple can mean that $\mathtt{d}(B)$ is as small as possible. Other aspects might deal with the task of minimizing the number of elements in the set W. Finally, we want to emphasize that the above considerations can be generalized such that also unspecified/generic sequences can appear. The first important steps towards such a summation theory have been elaborated in [89].

As it turns out, the theory of difference rings provides all the techniques necessary to tackle the above problems. In the next section we introduce all the needed ingredients and will present our main result in Theorem 2 below.

3 The Difference Ring Approach for SumProd($\mathbb{G}$)

In the following we will rephrase expressions $H \in$ SumProd$(\mathbb{G})$ as elements h in a formal difference ring. More precisely, we will design

- a ring $\mathbb{A}$ with $\mathbb{A} \supseteq \mathbb{G} \supseteq \mathbb{K}$ in which H can be represented by $h \in \mathbb{A}$;
- an evaluation function ev : $\mathbb{A} \times \mathbb{Z}_{\geq 0} \to \mathbb{K}$ such that $H(n) = \text{ev}(h, n)$ holds for sufficiently large $n \in \mathbb{Z}_{\geq 0}$;
- a ring automorphism $\sigma : \mathbb{A} \to \mathbb{A}$ which models the shift $H(n+1)$ with $\sigma(h)$.

Example 7 We will rephrase $F = \text{Sum}(1, \frac{1}{x}) \in$ SumProd$(\mathbb{G}_r)$ with $\mathbb{G}_r = \mathbb{K}(x)$ where $\mathbb{K} = \mathbb{Q}$ in a formal ring. Namely, we take the polynomial ring $\mathbb{A} = \mathbb{G}_r[s] =$

$\mathbb{Q}(x)[s]$ (s transcendental over $\mathbb{G}_r$) and extend $\mathrm{ev} : \mathbb{G}_r \times \mathbb{Z}_{\geq 0} \to \mathbb{Q}$ to $\mathrm{ev}' : \mathbb{A} \times \mathbb{Z}_{\geq 0} \to \mathbb{Q}$ as follows: for $h = \sum_{k=0}^{d} f_k\, s^k$ with $f_k \in \mathbb{G}_r$ we set

$$\mathrm{ev}'(h, n) := \sum_{k=0}^{d} \mathrm{ev}(f_k, n)\, \mathrm{ev}'(s, n)^k \tag{9}$$

with

$$\mathrm{ev}'(s, n) = \sum_{i=1}^{n} \frac{1}{i} =: S_1(n)(= H_n); \tag{10}$$

since ev and ev′ agree on $\mathbb{G}_r$, we do not distinguish them anymore. For any

$$H = f_0 \oplus (f_1 \odot (F^{\otimes} 1)) \oplus \cdots \oplus (f_d \odot (F^{\otimes} d))$$

with $d \in \mathbb{Z}_{\geq 0}$ and $f_0, \ldots, f_d \in \mathbb{G}_r$ we can take $h = \sum_{k=0}^{d} f_k s^k \in \mathbb{A}$ and get

$$H(n) = \mathrm{ev}(h, n) \quad \forall n \in \mathbb{Z}_{\geq 0}.$$

Further, we introduce the shift operator acting on the elements in $\mathbb{A}$. For the field $\mathbb{G}_r$ we simply define the field automorphism $\sigma : \mathbb{G}_r \to \mathbb{G}_r$ with $\sigma(f) = f|_{x \mapsto x+1}(= f(x+1))$. Moreover, based on the observation that for any $n \in \mathbb{Z}_{\geq 0}$ we have

$$F(n+1) = \sum_{i=1}^{n+1} \frac{1}{i} = \sum_{i=1}^{n} \frac{1}{i} + \frac{1}{n+1},$$

we extend the automorphism $\sigma : \mathbb{G}_r \to \mathbb{G}_r$ to $\sigma' : \mathbb{A} \to \mathbb{A}$ as follows: for $h = \sum_{k=0}^{d} f_k\, s^k$ with $f_k \in \mathbb{G}_r$ we set $\sigma'(h) := \sum_{k=0}^{d} \sigma(f_k)\sigma'(s)^k$ with $\sigma'(s) = s + \frac{1}{x+1}$; since σ' and σ agree on $\mathbb{G}_r$, we do not distinguish them anymore. We observe that

$$\mathrm{ev}(s, n+1) = \sum_{i=1}^{n+1} \frac{1}{i} = \sum_{i=1}^{n} \frac{1}{i} + \frac{1}{n+1} = \mathrm{ev}(s + \tfrac{1}{x+1}, n) = \mathrm{ev}(\sigma(s), n)$$

holds for all $n \in \mathbb{Z}_{\geq 0}$ and more generally that $\mathrm{ev}(h, n+l) = \mathrm{ev}(\sigma^l(h), n)$ holds for all $h \in \mathbb{A}$, $l \in \mathbb{Z}$ and $n \in \mathbb{Z}_{\geq 0}$ with $n \geq \max(-l, 0)$.

As illustrated in the example above, the following definitions will be relevant.

Definition 5 A *difference ring/difference field* is a ring/field $\mathbb{A}$ equipped with a ring/field automorphism $\sigma : \mathbb{A} \to \mathbb{A}$ which one also denotes by $(\mathbb{A}, \sigma)$. $(\mathbb{A}, \sigma)$ is *difference ring/field extension* of a difference ring/field $(\mathbb{H}, \sigma')$ if $\mathbb{H}$ is a subring/subfield of $\mathbb{A}$ and $\sigma|_{\mathbb{H}} = \sigma'$. For a difference ring $(\mathbb{A}, \sigma)$ and a subfield $\mathbb{K}$ of $\mathbb{A}$ with[5] $\sigma|_{\mathbb{K}} = \text{id}$ we introduce the following functions.

1. A function $\text{ev}\colon \mathbb{A} \times \mathbb{Z}_{\geq 0} \to \mathbb{K}$ is called *evaluation function* for $(\mathbb{A}, \sigma)$ if for all $f, g \in \mathbb{A}$ and $c \in \mathbb{K}$ there exists a $\lambda \in \mathbb{Z}_{\geq 0}$ with the following properties:

$$\forall n \geq \lambda : \text{ev}(c, n) = c, \tag{11}$$

$$\forall n \geq \lambda : \text{ev}(f + g, n) = \text{ev}(f, n) + \text{ev}(g, n), \tag{12}$$

$$\forall n \geq \lambda : \text{ev}(f\, g, n) = \text{ev}(f, n)\, \text{ev}(g, n). \tag{13}$$

   In addition, we require that for all $f \in \mathbb{A}$ and $l \in \mathbb{Z}$ there exists a λ with

$$\forall n \geq \lambda : \text{ev}(\sigma^l(f), n) = \text{ev}(f, n + l). \tag{14}$$

2. A function $L\colon \mathbb{A} \to \mathbb{Z}_{\geq 0}$ is called an *operation-function* (in short *o-function*) for $(\mathbb{A}, \sigma)$ and an evaluation function ev if for any $f, g \in \mathbb{A}$ with $\lambda = \max(L(f), L(g))$ the properties (12) and (13) hold and for any $f \in \mathbb{A}$ and $l \in \mathbb{Z}$ with $\lambda = L(f) + \max(0, -l)$ property (14) holds.
3. Let G be a subgroup of $\mathbb{A}^*$. $Z\colon G \to \mathbb{Z}_{\geq 0}$ is called a *zero-function* (in short *z-function*) for ev and $\mathbb{G}$ if $\text{ev}(f, n) \neq 0$ holds for any $f \in \mathbb{G}$ and integer $n \geq Z(f)$.

We note that a construction of a map $\text{ev} : \mathbb{A} \times \mathbb{Z}_{\geq 0} \to \mathbb{K}$ with the properties (11) and (13) is straightforward. It is property (14) that brings in extra complications: the evaluation of the elements in $\mathbb{A}$ must be compatible with the automorphism σ.

In this article we will always start with the following ground field; see [31].

Example 8 Take the rational function field $\mathbb{G}_m := \mathbb{G} = \mathbb{K}(x, x_1, \dots, x_v)$ over $\mathbb{K} = K(q_1, \dots, q_v)$, $v \geq 0$, with the function (5), together with the functions $L : \mathbb{G}_m \to \mathbb{Z}_{\geq 0}$ and $Z\colon \mathbb{G}_m^* \to \mathbb{Z}_{\geq 0}$ from the beginning of Sect. 2. It is easy to see that $\text{ev} : \mathbb{G}_m \times \mathbb{Z}_{\geq 0} \to \mathbb{K}$ satisfies for all $c \in \mathbb{K}$ and $f, g \in \mathbb{G}$ the property (11) for $L(c) = 0$ and the properties (12) and (13) with $\lambda = \max(L(f), L(g))$. Finally, we take the automorphism $\sigma : \mathbb{G}_m \to \mathbb{G}_m$ defined by $\sigma|_{\mathbb{K}} = \text{id}$, $\sigma(x) = x + 1$ and $\sigma(y_i) = q_i\, y_i$ for $1 \leq i \leq v$. Then one can verify in addition that (14) holds for all $f \in \mathbb{G}_m$ and $l \in \mathbb{Z}$ with $\lambda = \max(-l, L(f))$. Consequently, ev is an evaluation function for $(\mathbb{G}_m, \sigma)$ and L is an o-function for $(\mathbb{G}_m, \sigma)$. In addition, Z is a z-function for ev and $\mathbb{G}_m^*$ by construction. In the following we call $(\mathbb{G}_m, \sigma)$ also a *multibasic mixed difference field*. If $v = 0$, i.e., $\mathbb{G}_r = \mathbb{K}(x) = \mathbb{K}'(x)$, we get the *rational difference field* $(\mathbb{G}_r, \sigma)$, and if we restrict to $\mathbb{G}_b = \mathbb{K}(x_1, \dots, x_v)$, we get the *multibasic difference field* $(\mathbb{G}_b, \sigma)$.

We continue with the convention from above: if we write $(\mathbb{G}, \sigma)$, then it can be replaced by any of the difference rings $(\mathbb{G}_m, \sigma)$, $(\mathbb{G}_r, \sigma)$ or $(\mathbb{G}_b, \sigma)$.

[5]Note that $(\mathbb{A}, \sigma)$ is a difference ring extension of $(\mathbb{K}, \text{id})$.

In the following we look for such a formal difference ring $(\mathbb{A}, \sigma)$ with a computable evaluation function ev and o-function L in which we can model a finite set of expressions $A_1, \ldots, A_u \in \text{SumProd}(\mathbb{G})$ with $a_1, \ldots, a_u \in \mathbb{A}$.

Definition 6 Let $F \in \text{SumProd}(\mathbb{G})$ and $(\mathbb{A}, \sigma)$ be a difference ring extension of $(\mathbb{G}, \sigma)$ equipped with an evaluation function $\text{ev} : \mathbb{A} \times \mathbb{Z}_{\geq 0} \to \mathbb{K}$. We say that $f \in \mathbb{A}$ *models* F if $\text{ev}(f, n) = F(n)$ holds for all $n \geq \lambda$ for some $\lambda \in \mathbb{Z}_{\geq 0}$.

3.1 The Naive Representation in APS-Extensions

As indicated in Example 7 our sum-product expressions will be rephrased in a tower of difference field and ring extensions. We start with the field version which will lead later to $\Pi\Sigma$-fields [65, 66].

Definition 7 A difference field $(\mathbb{F}, \sigma)$ is called a *PS-field extension* of a difference field $(\mathbb{H}, \sigma)$ if $\mathbb{H} = \mathbb{H}_0 \leq \mathbb{H}_1 \leq \cdots \leq \mathbb{H}_e = \mathbb{F}$ is a tower of field extensions where for all $1 \leq i \leq e$ one of the following holds:

- $\mathbb{H}_i = \mathbb{H}_{i-1}(t_i)$ is a rational function field extension with $\frac{\sigma(t_i)}{t_i} \in (\mathbb{H}_{i-1})^*$ (t_i is called a *P-field monomial*);
- $\mathbb{H}_i = \mathbb{H}_{i-1}(t_i)$ is a rational function extension with $\sigma(t_i) - t_i \in \mathbb{H}_{i-1}$ (t_i is called an *S-field monomial*).

Example 9 Following Example 8, $(\mathbb{G}_m, \sigma)$ with $\mathbb{G}_m = \mathbb{K}(x, x_1, \ldots, x_v)$ is a PS-field extension of $(\mathbb{K}, \sigma)$ with the S-field monomial x and the P-monomials $x_1, \ldots, x_v$. Similarly, $(\mathbb{G}_b, \sigma)$ with $\mathbb{G}_b = \mathbb{K}(x_1, \ldots, x_v)$ forms a tower of P-field extensions of $(\mathbb{K}, \sigma)$ and $(\mathbb{G}_r, \sigma)$ with $\mathbb{G}_r = \mathbb{K}(x)$ is an S-field extension of $(\mathbb{K}, \sigma)$.

In addition, we will modify the field version to obtain the following ring version (allowing us to model also products over roots of unity).

Definition 8 A difference ring $(\mathbb{E}, \sigma)$ is called an *APS-extension* of a difference ring $(\mathbb{A}, \sigma)$ if $\mathbb{A} = \mathbb{A}_0 \leq \mathbb{A}_1 \leq \cdots \leq \mathbb{A}_e = \mathbb{E}$ is a tower of ring extensions where for all $1 \leq i \leq e$ one of the following holds:

- $\mathbb{A}_i = \mathbb{A}_{i-1}[t_i]$ is a ring extension subject to the relation $t_i^{\nu} = 1$ for some $\nu > 1$ where $\frac{\sigma(t_i)}{t_i} \in (\mathbb{A}_{i-1})^*$ is a primitive νth root of unity (t_i is called an *A-monomial*, and ν is called the *order of the A-monomial*);
- $\mathbb{A}_i = \mathbb{A}_{i-1}[t_i, t_i^{-1}]$ is a Laurent polynomial ring extension with $\frac{\sigma(t_i)}{t_i} \in (\mathbb{A}_{i-1})^*$ (t_i is called a *P-monomial*);
- $\mathbb{A}_i = \mathbb{A}_{i-1}[t_i]$ is a polynomial ring extension with $\sigma(t_i) - t_i \in \mathbb{A}_{i-1}$ (t_i is called an *S-monomial*).

Depending on the occurrences of the APS-monomials such an extension is also called an *A-/P-/S-/AP-/AS/-/PS-extension.*

Example 10 Take the rational difference ring $(\mathbb{Q}(x), \sigma)$ with $\sigma(x) = x + 1$ and $\sigma|_{\mathbb{Q}} = \text{id}$. Then the difference ring $(\mathbb{Q}(x)[s], \sigma)$ with $\sigma(s) = s + \frac{1}{x+1}$ defined in Example 7 is an S-extension of $(\mathbb{Q}(x), \sigma)$ and s is an S-monomial over $(\mathbb{Q}(x), \sigma)$.

For an APS-extension $(\mathbb{E}, \sigma)$ of a difference ring $(\mathbb{A}, \sigma)$ we will also write $\mathbb{E} = \mathbb{A}\langle t_1\rangle \dots \langle t_e\rangle$. Depending on whether t_i with $1 \le i \le e$ is an A-monomial, a P-monomial or an S-monomial, $\mathbb{G}\langle t_i\rangle$ with $\mathbb{G} = \mathbb{A}\langle t_1\rangle \dots \langle t_{i-1}\rangle$ stands for the algebraic ring extension $\mathbb{G}[t_i]$ with t_i^ν for some $\nu > 1$, for the ring of Laurent polynomials $\mathbb{G}[t_1, t_1^{-1}]$ or for the polynomial ring $\mathbb{G}[t_i]$, respectively.

For such a tower of APS-extensions we can use the following lemma iteratively to construct an evaluation function; for the corresponding proofs see [118, Lemma 5.4].

Lemma 2 *Let $(\mathbb{A}, \sigma)$ be a difference ring with a subfield $\mathbb{K} \subseteq \mathbb{A}$ where $\sigma|_{\mathbb{K}} = \text{id}$ that is equipped with an evaluation function $\text{ev} : \mathbb{A} \times \mathbb{Z}_{\ge 0} \to \mathbb{K}$ and o-function L. Let $(\mathbb{A}\langle t\rangle, \sigma)$ be an APS-extension of $(\mathbb{A}, \sigma)$ with $\sigma(t) = \alpha\, t + \beta$ ($\alpha = 1$, $\beta \in \mathbb{A}$ or $\alpha \in \mathbb{A}^*$, $\beta = 0$). Further, suppose that $\text{ev}(\sigma^{-1}(\alpha), n) \neq 0$ for all $n \ge \mu$ for some $\mu \in \mathbb{Z}_{\ge 0}$. Then the following holds.*

1. *Take $l \in \mathbb{Z}_{\ge 0}$ with $l \ge \max(L(\sigma^{-1}(\alpha), L(\sigma^{-1}(\beta)), \mu)$; if $t^\lambda = 1$ for some $\lambda > 1$ (t is an A-monomial), set $l = 1$. Then $\text{ev}' : \mathbb{A}\langle t\rangle \times \mathbb{Z}_{\ge 0} \to \mathbb{K}$ given by*

$$\text{ev}'(\sum_{i=a}^{b} f_i\, t^i, n) = \sum_{i=a}^{b} \text{ev}(f_i, n)\, \text{ev}'(t, n)^i \quad \forall n \in \mathbb{Z}_{\ge 0} \tag{15}$$

with $f_i \in \mathbb{A}$ for $a \le i \le b$ and[6]

$$\text{ev}'(t, n) = \begin{cases} \prod_{i=l}^{n} \text{ev}(\sigma^{-1}(\alpha), i) & \text{if } \sigma(t) = \alpha\, t \\ \sum_{i=l}^{n} \text{ev}(\sigma^{-1}(\beta), i) & \text{if } \sigma(t) = t + \beta \end{cases} \tag{16}$$

is an evaluation function for $(\mathbb{A}\langle t\rangle, \sigma)$.

2. *There is an o-function $L' : \mathbb{A}\langle t\rangle \to \mathbb{Z}_{\ge 0}$ for ev' defined by*

$$L'(f) = \begin{cases} L(f) & \text{if } f \in \mathbb{A}, \\ \max(l - 1, L(f_a), \dots, L(f_b)) & \text{if } f = \sum_{i=a}^{b} f_i t^i \notin \mathbb{A}\langle t\rangle \setminus \mathbb{A}. \end{cases} \tag{17}$$

Example 11 In Example 7 we followed precisely the construction (1) of the above lemma to construct for $(\mathbb{Q}(x)[s], \sigma)$ an evaluation function. For this ev we can

[6]If t is an A-monomial, we have $\text{ev}(t, n) = \alpha^n$.

now apply also the construction (2) to enhance the o function $L : \mathbb{Q}(x) \to \mathbb{Z}_{\geq 0}$ (given in Example 8 with $v = 0$) to $L : \mathbb{Q}(x)[s] \to \mathbb{Z}_{\geq 0}$ by setting $L(f) = \max(0, L(f_0), \ldots, L(f_b))$ for $f = \sum_{i=0}^{b} f_i s^i$.

More precisely, the main idea is to apply the above lemma iteratively to extend the evaluation function ev from $\mathbb{A}$ to $\mathbb{E}$. However, if one wants to treat, e.g., the next P-monomial t with $\frac{\sigma(t)}{t} = \alpha \in \mathbb{E}^*$, one has to check if there is a $\mu \in \mathbb{Z}_{\geq 0}$ such that $\text{ev}(\sigma^{-1}(\alpha), n) \neq 0$ holds for all $n \geq \mu$. So far, we are not aware of a general algorithm that can accomplish this task. In order to overcome these difficulties, we will restrict APS-extensions further to a subclass which covers all summation problems that we have encountered in concrete problems so far.

Let G be a multiplicative subgroup of $\mathbb{A}^*$. Following [116, 118] we call

$$\{G\}_{\mathbb{A}}^{\mathbb{E}} := \{h\, t_1^{m_1} \ldots t_e^{m_e} \,|\, h \in G \text{ and } m_i \in \mathbb{Z} \text{ where } m_i = 0 \text{ if } t_i \text{ is an } S\text{-monomial}\}$$

the *simple product group* over G and

$$[G]_{\mathbb{A}}^{\mathbb{E}} := \{h\, t_1^{m_1} \ldots t_e^{m_e} \,|\, h \in G \text{ and } m_i \in \mathbb{Z} \text{ where } m_i = 0 \text{ if } t_i \text{ is an } AS\text{-monomial}\}$$

the *basic product group* over G for the nested APS–extension $(\mathbb{E}, \sigma)$ of $(\mathbb{A}, \sigma)$. Note that we have the chain of subgroups $[G]_{\mathbb{A}}^{\mathbb{E}} \leq \{G\}_{\mathbb{A}}^{\mathbb{E}} \leq \mathbb{E}^*$. In the following we will restrict ourselves to the following subclass of APS-extensions.

Definition 9 Let $(\mathbb{A}, \sigma)$ be a difference ring and let G be a subgroup of $\mathbb{A}^*$. Let $(\mathbb{E}, \sigma)$ be an APS-extension of $(\mathbb{A}, \sigma)$ with $\mathbb{E} = \mathbb{A}\langle t_1\rangle \ldots \langle t_e\rangle$.

1. The extension is called *G-basic* if for any P-monomial t_i we have $\frac{\sigma(t_i)}{t_i} \in [G]_{\mathbb{A}}^{\mathbb{A}\langle t_1\rangle \ldots \langle t_{i-1}\rangle}$ and for any A-mon. t_i we have $\alpha_i = \frac{\sigma(t_i)}{t_i} \in G$ with $\sigma(\alpha_i) = \alpha_i$.
2. It is called *G-simple* if for any AP-monomial t_i we have $\frac{\sigma(t_i)}{t_i} \in \{G\}_{\mathbb{A}}^{\mathbb{A}\langle t_1\rangle \ldots \langle t_{i-1}\rangle}$.

If $G = \mathbb{A}^*$, it is also called *basic* (resp. *simple*) instead of $\mathbb{A}^*$-basic (resp. $\mathbb{A}^*$-simple).

By definition any simple APS-extension is also a basic APS-extension. We will start with the more general setting of simple extensions, but will restrict later mostly to basic extensions. For both cases we can supplement Lemma 2 as follows.

Lemma 3 *Let* $(\mathbb{A}, \sigma)$ *be a difference ring with a subfield* $\mathbb{K} \subseteq \mathbb{A}$ *where* $\sigma|_{\mathbb{K}} = \text{id}$ *that is equipped with an evaluation function* ev *and o-function L. Let G be a subgroup of* $\mathbb{A}^*$ *and let* $(\mathbb{A}\langle t\rangle, \sigma)$ *be an APS-extension of* $(\mathbb{A}, \sigma)$ *with* $\sigma(t) = \alpha\, t + \beta$ *with* $\alpha \in G$ *and* $\beta \in \mathbb{A}$. *Suppose that there is in addition a z-function for* ev *and G. Take* $l \in \mathbb{Z}_{\geq 0}$ *with*

$$l \geq \begin{cases} \max(L(\sigma^{-1}(\alpha)), Z(\sigma^{-1}(\alpha))) & \text{if } t \text{ is an } AP\text{-monomial} \\ L(\sigma^{-1}(\beta)) & \text{if } t \text{ is an } S \text{ monomial.} \end{cases} \tag{18}$$

Then we obtain an evaluation function ev′ *and o-function* L' *for* $(\mathbb{A}\langle t\rangle, \sigma)$ *as given in Lemma 2. In addition, we can construct a z-function* Z' *for* $\{G\}_{\mathbb{A}}^{\mathbb{A}\langle t\rangle}$. *If* ev, L *and* Z *are computable,* ev′, L' *and* Z' *are computable.*

Proof For r as defined in (18) the assumptions in Lemma 2 are fulfilled and the ev′ with L' defined in the lemma yield an evaluation function together with an o-function. If t is an S-monomial, $\{G\}_{\mathbb{A}}^{\mathbb{A}\langle t\rangle} = G$ and we can set $Z' := Z$. Otherwise, if t is an AP-monomial, we have $\text{ev}'(t, n) \neq 0$ for all $n \in \mathbb{Z}_{\geq 0}$ by construction. Thus for $f = g\,t^m \in \{G\}_{\mathbb{A}}^{\mathbb{A}\langle t\rangle}$ with $g \in G$ and $m \in \mathbb{Z}$ we have $\text{ev}(f, n) \neq 0$ for all $n \geq Z(g)$. Thus we can define $Z'(f) = Z(g)$. If L and Z are computable, also L' and Z' are computable. In addition, if we can compute ev, then clearly also ev′ is computable. □

In general, suppose that we are given a difference ring $(\mathbb{A}, \sigma)$ with a subfield $\mathbb{K} \subseteq \mathbb{A}$ where $\sigma|_{\mathbb{K}} = \text{id}$. Assume in addition that we are given a (computable) evaluation function $\text{ev} : \mathbb{A} \times \mathbb{Z}_{\geq 0} \to \mathbb{K}$ together with a (computable) o-function $L : \mathbb{A} \to \mathbb{Z}_{\geq 0}$ and a (computable) z-function $Z : \mathbb{A}^* \to \mathbb{Z}_{\geq 0}$. Furthermore, suppose that we are given a simple APS-extension $(\mathbb{E}, \sigma)$ of $(\mathbb{A}, \sigma)$ with $\mathbb{E} = \mathbb{A}\langle t_1\rangle \dots \langle t_e\rangle$. Then we can apply iteratively Lemmas 2 and 3 and get a (computable) evaluation function $\text{ev} : \mathbb{E} \times \mathbb{Z}_{\geq 0} \to \mathbb{K}$ together with a (computable) o-function $L : \mathbb{E} \to \mathbb{Z}_{\geq 0}$ and a (computable) z-function for $\{\mathbb{A}^*\}_{\mathbb{A}}^{\mathbb{A}\langle t_1\rangle\dots\langle t_e\rangle}$; note that $\{\{\mathbb{A}^*\}_{\mathbb{A}}^{\mathbb{H}}\}_{\mathbb{H}}^{\mathbb{H}\langle t_i\rangle} = \{\mathbb{A}^*\}_{\mathbb{A}}^{\mathbb{H}\langle t_i\rangle}$ for all $\mathbb{H} = \mathbb{A}\langle t_1\rangle \dots \langle t_{i-1}\rangle$ with $1 \leq i < e$.

It is natural to define the evaluation function iteratively using Lemma 2 but it is inconvenient to compute the o-function in this iterative fashion. Here the following lemma provides a shortcut for expressions which are given in reduced representation; for the corresponding representation in SumProd($\mathbb{G}$) see Definition 2.

Definition 10 Let $(\mathbb{E}, \sigma)$ be an APS-extension of $(\mathbb{A}, \sigma)$ with $\mathbb{E} = \mathbb{A}\langle t_1\rangle \dots \langle t_e\rangle$. Then we say that $f \in \mathbb{E}$ is in *reduced representation* if it is written in the form

$$f = \sum_{(m_1,\dots,m_e)\in S} f_{(m_1,\dots,m_e)} t_1^{m_1} \dots t_e^{m_e} \tag{19}$$

with $f_{(m_1,\dots,m_e)} \in \mathbb{A}$ and $S \subseteq M_1 \times \dots \times M_e$ finite where

$$M_i = \begin{cases} \{0, \dots, \nu_i - 1\} & \text{if } t_i \text{ is an } A\text{-extension of order } \nu_i, \\ \mathbb{Z} & \text{if } t_i \text{ is a } P\text{-monomial}, \\ \mathbb{Z}_{\geq 0} & \text{if } t_i \text{ is an } S\text{-monomial}. \end{cases}$$

Lemma 4 *Take a difference ring* $(\mathbb{A}, \sigma)$ *with a subfield* $\mathbb{K} \subseteq \mathbb{A}$ *where* $\sigma|_{\mathbb{K}} = \text{id}$ *that is equipped with an evaluation function* $\text{ev} : \mathbb{A} \times \mathbb{Z}_{\geq 0} \to \mathbb{K}$ *together with an o-function* L *and z-function* Z. *Let* $(\mathbb{E}, \sigma)$ *with* $\mathbb{E} = \mathbb{A}\langle t_1\rangle \dots \langle t_e\rangle$ *be a simple* APS*-extension of* $(\mathbb{A}, \sigma)$ *and let* ev *be an evaluation function and* Z *be a z-function* *(using iteratively Lemmas 2 and 3). Here the* $l_i \in \mathbb{Z}_{\geq 0}$ *for* $1 \leq i \leq e$ *are the lower bounds of the corresponding sums/products in* (16) *with* $t = t_i$. *Then for any* $f \in \mathbb{E}$ *with* (19) *where* $f_{(m_1,\dots,m_e)} \in \mathbb{E}$ *and* $S \subseteq \mathbb{Z}^e$ *we have*

$$L(f) = \max(\max_{s \in S} L(f_s), \max_{j \in \mathrm{sup}(f)} l_j - 1)$$

where $\mathrm{sup}(f) = \{1 \leq j \leq e \mid t_j \text{ depends on } f\}$.

Proof We show the statement by induction on e. If $e = 0$, the statement holds trivially. Now suppose that the statement holds for $e \geq 0$ extensions and let $f \in \mathbb{E}\langle t_{e+1}\rangle$ with $\mathbb{E} = \mathbb{G}\langle t_1\rangle \dots \langle t_e\rangle$ where $f = \sum_{i=a}^{b} f_i t_{e+1}^i$ with $f_i \in \mathbb{E}$. If $f \in \mathbb{E}$, then the statement holds by the induction assumption. Otherwise write $f_i = \sum_{(s_1,\dots,s_e)\in S_i} f^{(i)}_{s_1,\dots,s_e} t_1^{s_1} \dots t_e^{s_e}$ with $S_i \subseteq \mathbb{Z}^e$ and $f_s^{(i)}$ for $s \in S_i$ in reduced representation. In particular, we get $f = \sum_{(s_1,\dots,s_{e+1})\in S} h_{(s_1\dots,s_{e+1})} t_1^{s_1} \dots t_{e+1}^{s_{e+1}}$ with $h_{(s_1\dots,s_{e+1})} = f^{(s_{e+1})}_{(s_1,\dots,s_e)}$ and $S = \cup_{a \leq i \leq b}\{(s_1,\dots,s_e,i) \mid (s_1,\dots,s_e) \in S_i\}$. Then by the induction assumption we get $L(f_i) = \max(\max_{s\in S_i} L(f_s^{(i)}), \max_{j\in\mathrm{sup}(f_i)} l_j - 1)$. Thus by the definition in (17) we get

$$\begin{aligned} L(f) &= \max(\max_{a\leq i\leq b} L(f_i), l_{e+1} - 1) \\ &= \max(\max_{s\in S_a} L(f_s^{(a)}), \max_{s\in S_{a+1}} L(f_s^{(a+1)}), \dots, \max_{s\in S_b} L(f_s^{(b)}), \max_{j\in\mathrm{sup}(f)} l_j - 1) \\ &= \max(\max_{s\in S} L(h_s), \max_{j\in\mathrm{sup}(f)} l_j - 1). \end{aligned}$$

□

Utilizing the above constructions with $\mathbb{A} := \mathbb{G}$, we are now ready to show in Lemmas 5 and 6 given below that the representations in SumProd($\mathbb{G}$) and in a basic APS-extension are closely related. Their proofs are rather technical (but not very deep). Still we will present all the details, since this construction will be crucial for further refinements. This will finally lead to a strategy to solve Problem SigmaReduce.

Lemma 5 *Take the difference field* $(\mathbb{G}, \sigma)$ *with* $\mathbb{G} \in \{\mathbb{G}_r, \mathbb{G}_b, \mathbb{G}_m\}$ *with the evaluation function* ev, *o-function* L *and z-function* Z *from Example 8. Let* $(\mathbb{E}, \sigma)$ *with* $\mathbb{E} = \mathbb{G}\langle t_1\rangle \dots \langle t_e\rangle$ *be a basic* APS*-extension of* $(\mathbb{G}, \sigma)$ *and let* ev, L *and* Z *be extended versions for* $(\mathbb{E}, \sigma)$ *(using Lemmas 2 and 3). Then for each* $1 \leq i \leq e$ *one can construct* $T_i \in \Sigma\Pi(\mathbb{G})$ *in sum-product reduced representation with* $\mathrm{ev}(t_i, n) = T_i(n)$ *for all* $n \geq L(t_i)$. *In particular, if* $f \in \mathbb{E} \setminus \{0\}$, *then there is* $0 \neq F \in \mathit{SumProd}(\{T_1, \dots, T_e\}, \mathbb{G})$ *with* $F(n) = \mathrm{ev}(f, n)$ *for all* $n \geq L(f)$.
If $\mathbb{K}$ *is computable and polynomials can be factored over* $\mathbb{K}$, *all components can be computed.*

Proof First suppose that we can construct such $T_i \in \Sigma\Pi(\mathbb{G})$ with $T_i(n) = \mathrm{ev}(t_i, n)$ for all $n \geq L(t_i)$ and $1 \leq i \leq e$. Now take $f \in \mathbb{E}$ in reduced representation, i.e., it is given in the form (19) with $S \subseteq \mathbb{Z}^e$. Now replace each $f_{(m_1,\dots,m_e)} \cdot t_1^{m_1} \dots t_e^{m_e}$

by $f_{(m_1,\ldots,m_e)} \odot (T_1^{\circledwedge} m_1) \odot \cdots \odot (T_e^{\circledwedge} m_e)$ and replace $+$ by $\oplus$ in f yielding $F \in \mathrm{SumProd}(\mathbb{G})$ in reduced representation. Then for each $n \geq L(f)$ we get

$$\begin{aligned}
\mathrm{ev}(f, n) &= \mathrm{ev}\Big(\sum_{(m_1,\ldots,m_e)\in S} f_{(m_1,\ldots,m_e)} t_1^{m_1} \ldots t_e^{m_e}, n\Big)\\
&= \sum_{(m_1,\ldots,m_e)\in S} \mathrm{ev}(f_{(m_1,\ldots,m_e)}, n)\, \mathrm{ev}(t_1, n)^{m_1} \ldots \mathrm{ev}(t_e, n)^{m_e}\\
&= \sum_{(m_1,\ldots,m_e)\in S} \mathrm{ev}(f_{(m_1,\ldots,m_e)}, n)\, \mathrm{ev}(T_1, n)^{m_1} \ldots \mathrm{ev}(T_e, n)^{m_e}\\
&= \mathrm{ev}(F, n).
\end{aligned} \tag{20}$$

Note: if $f \neq 0$, we can find $(m_1, \ldots, m_e) \in S$ with $f_{(m_1,\ldots,m_e)} \in \mathbb{G}^*$ which implies that $F \neq 0$. This shows the second part of statement (1).
Finally we show the existence of the T_i by induction on e. For $e = 0$ nothing has to be shown. Suppose that the statement holds for $e \geq 0$ extensions and consider the APS-monomial t_{e+1} over $\mathbb{E} = \mathbb{G}\langle t_1\rangle \ldots \langle t_e\rangle$. By assumption we can take $T_i \in \mathrm{SumProd}(\mathbb{G})$ in sum-product reduced representation with $T_i(n) = \mathrm{ev}(t_i, n)$ for all $n \geq L(t_i)$ and $1 \leq i \leq e$. Now consider the APS-monomial t_{e+1} with $\sigma(t_{e+1}) = \alpha\, t_{e+1} + \beta$. By assumption we have (16) (ev' replaced by ev) with $l \in \mathbb{Z}_{\geq 0}$ where (18) and L is defined by (17) (L' replaced by L). In particular, we have $l \geq \max(L(\sigma^{-1}(\alpha)), L(\sigma^{-1}(\beta)), \mu)$ with $\mu \geq Z(\sigma^{-1}(\alpha))$, and $L(t_{e+1}) = l - 1$.

***A*-monomial Case** If t_{e+1} is an A-monomial, we have $\sigma(t_{e+1}) = \alpha\, t_{e+1}$ with $\alpha \in \mathcal{R}$. In particular, we have $\mathrm{ev}(t_{e+1}, n) = \alpha^n$. Thus we set $T_{e+1} = \mathrm{RPow}(\alpha)$ and get $\mathrm{ev}(t_{e+1}, n) = T_{e+1}(n)$ for all $n \geq L(t_{e+1}) = 0$.

***S*-Monomial Case** If t_{e+1} is an S-monomial, we have $\sigma(t_{e+1}) = t_{e+1} + \beta$ with $\beta \in \mathbb{E}$. Now take $f = \sigma^{-1}(\beta)$ in reduced representation. Then by construction $l \geq \max(L(\sigma^{-1}(\beta)), 0) = L(f)$. Further, we can take $F \in \mathrm{SumProd}(\mathbb{G})$ as constructed above with (20) for all $n \geq l \geq L(f)$. Thus for $T_{e+1} = \mathrm{Sum}(l, F)$ we get $\mathrm{ev}(t_{e+1}, n) = T_{e+1}(n)$ for all $n \geq l - 1 = L(t_{e+1})$.

***P*-Monomial Case** If t_{e+1} is a P-monomial, we have $\sigma(t_{e+1}) = \alpha\, t_{e+1}$ with $\alpha \in [\mathbb{G}]_{\mathbb{G}}^{\mathbb{E}}$, i.e., $\alpha = g\, t_1^{n_1} \ldots t_e^{n_e}$ with $g \in \mathbb{G}^*$ and $n_1, \ldots, n_e \in \mathbb{Z}$ with $n_i = 0$ if t_i is an AS-monomial. Thus $f = \sigma^{-1}(\alpha) = h\, t_1^{m_1} \ldots t_e^{m_e}$ with $h := f_{(m_1,\ldots,m_e)} \in \mathbb{G}^*$ and $m_1, \ldots, m_e \in \mathbb{Z}$ with $m_i = 0$ if t_i is an AS-monomial. By construction, $l \geq \max(L(f), Z(f)) = \max(L(f), Z(h))$. As above we get $F = h \odot (T_1^{\circledwedge} m_1) \odot \cdots \odot (T_e^{\circledwedge} m_e) \in \mathrm{SumProd}(\mathbb{G})$ such that $\mathrm{ev}(f, n) = F(n)$ holds for all $n \geq L(f)$ and $\mathrm{ev}(f, n) = F(n) \neq 0$ for all $n \geq l$. Thus for $T_{e+1} = \mathrm{Prod}(l, F) \in \mathrm{Prod}(\mathbb{G})$ we get $\mathrm{ev}(t_{e+1}, n) = T_{e+1}(n)$ for all $n \geq l - 1 = L(t_{e+1})$.
We note that in the last two cases T_{e+1} is in sum-product reduced representation: the arising sums and products in F are in sum-product reduced representation by induction, F given by (20) is in reduced representation and we have $l \geq$

$\max_{k\in S} L(f_k)$ where l is larger than all the lower bounds of the sums and product in F due to Lemma 4. This completes the induction step.
If $\mathbb{K}$ is computable and one can factorize polynomials over $\mathbb{K}$, the functions Z and L are computable and thus all the ingredients can be computed. □

Definition 11 Given $(\mathbb{G}, \sigma)$ where $\mathbb{G} \in \{\mathbb{G}_r, \mathbb{G}_b, \mathbb{G}_m\}$ with ev, L and Z from Example 8, let $(\mathbb{E}, \sigma)$ be a basic APS-extension with an evaluation function ev together with L and Z given by iterative application of Lemmas 2 and 3. Let $a \in \mathbb{E}$ be in reduced representation. Then following the construction of Lemma 5 one obtains $A \in \text{SumProd}(\mathbb{G})$ in sum-product reduced representation with $A(n) = \text{ev}(a, n)$ for all $n \geq L(a)$. The derived A is also called the *canonical induced sum-product expression* of a w.r.t. $(\mathbb{A}, \sigma)$ and ev and we write $\text{expr}(a) := A$.

Example 12 (Cont. of Example 7) For $a = x + \frac{x+1}{x}s^4 \in \mathbb{Q}(x)[s]$ with our evaluation function ev we obtain the canonical induced sum-product expression

$$\text{expr}(a) = A = x \oplus \left(\tfrac{x+1}{x} \odot (\text{Sum}(1, \tfrac{1}{x})^{\textcircled{\wedge}} 4)\right) \in \text{Sum}(\mathbb{Q}(x))$$

with $A(n) = \text{ev}(a, n)$ for all $n \geq 1$.

Lemma 6 *Take the difference field* $(\mathbb{G}, \sigma)$ *with* $\mathbb{G} \in \{\mathbb{G}_r, \mathbb{G}_b, \mathbb{G}_m\}$ *with the evaluation function* ev, *o-function L and z-function Z from Example 8. Let* $(\mathbb{H}, \sigma)$ *be a basic APS-extension of* $(\mathbb{G}, \sigma)$ *and let* ev, *L and Z be extended versions for* $(\mathbb{H}, \sigma)$ *(using Lemmas 2 and 3). Let* $A \in \text{SumProd}(\mathbb{G})$*. Then there is an APS-extension* $(\mathbb{E}, \sigma)$ *of* $(\mathbb{H}, \sigma)$ *which forms a basic APS-extension of* $(\mathbb{G}, \sigma)$ *together with the extended functions* ev, *L and Z (using Lemmas 2 and 3) in which one can model A by* $a \in \mathbb{E}$*: i.e.,* $\text{ev}(a, n) = A(n)$ *holds for all* $n \geq \delta$ *for some* $\delta \in \mathbb{Z}_{\geq 0}$.
If $\mathbb{K}$ *is computable and one can factorize polynomials over* $\mathbb{K}$*, all the ingredients can be computed.*

Proof We prove the lemma by induction on the depth of the arising sums (Sum) and products (Prod and RPow) in $A \in \text{SumProd}(\mathbb{G})$. If no sums and products arise in A, then $A \in \mathbb{G}$ and the statement clearly holds. Now suppose that the statement holds for all expressions with sums/products whose depth is smaller than or equal to $d \geq 0$. Take all products and sums $T_1, \ldots, T_r \in \Sigma\Pi(\mathbb{G})$ that arise in A. We proceed stepwise for $i = 1, \ldots, r$ with the starting field $\mathbb{H}$. Suppose that we have constructed an APS-extension $(\mathbb{A}, \sigma)$ of $(\mathbb{H}, \sigma)$ which forms a basic APS-extension of $(\mathbb{G}, \sigma)$. Suppose in addition that we are given an extended evaluation function ev, o-function L and z-function Z function (using Lemmas 2 and 3) in which we find $b_1, \ldots, b_{i-1}$ with $\text{ev}(b_j, n) = T_j(n)$ for all $n \geq L(b_j)$ and all $1 \leq j < i$. Now we consider T_i.

Bookkeeping[7] If T_i has been treated earlier (i.e., by handling sums and products of depth $\leq d$), we get $b_i \in \mathbb{A}$ with $\text{ev}(b_i, n) = T_i(n)$ for all $n \geq L(a_i)$.

RPow-Case If $T_i = \text{RPow}(\alpha)$, we take the A-extension $(\mathbb{A}\langle t\rangle, \sigma)$ of $(\mathbb{A}, \sigma)$ with $\sigma(t) = \alpha t$ of order $\text{ord}(\alpha)$ and extend ev to $\mathbb{A}\langle t\rangle$ by $\text{ev}(t, n) = \alpha^n$. Further, we extend $L : \mathbb{A}\langle t\rangle \to \mathbb{Z}_{\geq 0}$ with (17) and get $L(t) = 0$. Thus we can take $b_i = t$ and get $\text{ev}(b_i, n) = T_i(n)$ for all $n \geq L(b_i) = 0$.

Otherwise, we can write $T_i = \text{Sum}(\lambda, H)$ or $T_i = \text{Prod}(\lambda, H)$ where the sums and products in $H \in \text{SumProd}(\mathbb{G})$ have depth at most d. By assumption we can construct an APS-extension $(\mathbb{A}', \sigma)$ of $(\mathbb{A}, \sigma)$ which is a basic APS-extension of $(\mathbb{G}, \sigma)$ and we can extend ev, L and Z (using Lemmas 2 and 3) and get $h \in \mathbb{A}'$ with $\text{ev}(h, n) = H(n)$ for all $n \geq \delta$ for some $\delta \in \mathbb{Z}_{\geq 0}$ with $\delta \geq L(h)$.

Sum-Case If $T_i = \text{Sum}(\lambda, H)$, we take the S-extension $(\mathbb{A}'\langle t\rangle, \sigma)$ of $(\mathbb{A}', \sigma)$ with $\sigma(t) = t + \sigma(h)$. In addition, we extend ev to $\mathbb{A}'\langle t\rangle$ by $\text{ev}(t, n) = \sum_{k=l}^{n} \text{ev}(h, k)$ with $l = \max(\delta, \lambda)$; note that (18) is satisfied. Further, we extend $L : \mathbb{A}'\langle t\rangle \to \mathbb{Z}_{\geq 0}$ with (17) and get $L(t) = l - 1$. Finally, we set $c = \sum_{k=\lambda}^{l-1} H(k) \in \mathbb{K}$. Then we get $b_i = t + c$ with $\text{ev}(b_i, n) = \sum_{k=\lambda}^{n} H(k) = \text{ev}(\text{Sum}(\lambda, H), n)$ for all $n \geq L(b_i) = l - 1$.

Product-Case If $T_i = \text{Prod}(\lambda, H)$, we take the P-extension $(\mathbb{A}'\langle t\rangle, \sigma)$ of $(\mathbb{A}', \sigma)$ with $\sigma(t) = \sigma(h)t$. In addition we extend ev to $\mathbb{A}'\langle t\rangle$ by $\text{ev}(t, n) = \prod_{k=l}^{n} \text{ev}(h, k)$ with $l = \max(L(h), Z(h), \lambda)$; note that (18) is satisfied. Further, we extend $L : \mathbb{A}'\langle t\rangle \to \mathbb{Z}_{\geq 0}$ with (17) and get $L(t) = l - 1$. Thus we can take $b_i = ct$ with $c = \prod_{k=\lambda}^{l-1} H(k) \in \mathbb{K}^*$ (the product evaluation is nonzero by assumption of $\Pi(\mathbb{G})$) and get $\text{ev}(b_i, n) = \prod_{k=\lambda}^{n} H(k) = \text{ev}(\text{Prod}(\lambda, H), n)$ for all $n \geq L(b_i) = l - 1$.

In all three cases we can follow Lemma 3 and extend the z-function accordingly. After carrying out the steps $i = 1, \ldots, r$ we get a basic APS-extension $(\mathbb{E}, \sigma)$ of $(\mathbb{H}, \sigma)$ together with an evaluation function ev, o-function L and z-function Z (using Lemmas 2 and 3) and $b_1, \ldots, b_r$ such that $T_i(n) = \text{ev}(b_i, n)$ holds for all $1 \leq i \leq r$ and $n \geq L(b_i)$. Finally, let $f_1, \ldots, f_s \in \mathbb{G}$ be all arising elements in A (that do not arise within Prod and Sum). Define $\delta = \max(L(f_1), \ldots, L(f_s)) \in \mathbb{Z}_{\geq 0}$. Then for each $n \in \mathbb{Z}_{\geq 0}$ with $n \geq \delta$ we have that $\text{ev}(A, n)$ can be carried out without catching poles in the second case of (5). Now replace each T_i with $1 \leq i \leq r$ in A by b_i and replace $\oplus, \odot, ^{\wedge}$ by $+$, $\cdot$, and $\hat{}$, respectively. This yields $a \in \mathbb{E}$ which we can write in reduced representation. Note that in a some f_k remain and others are combined by putting elements over a common denominator which lies in $\mathbb{K}[x, x_1, \ldots, x_v]$ (or in $\mathbb{K}[x_1, \ldots, x_v]$). Further, some factors of the denominators might cancel. Thus $L(a) \leq \delta$. In particular, when carrying out the evaluations $\text{ev}(a, n)$ and $\text{ev}(A, n)$ for $n \geq \delta$ no poles arise and thus by the homomorphic

[7]This step is not necessary for the proof, but avoids unneccesary copies of APS-monomials. When we refine this construction later, this step will be highly relevant.

property of the evaluation it follows that $\text{ev}(a, n) = \text{ev}(A, n)$ for all $n \geq \delta$. This completes the induction step.
If $\mathbb{K}$ is computable and one can factorize polynomials over $\mathbb{K}$, then the z- and o-function for $\mathbb{G}$ are computable. Thus all the components of the iterative construction (using Lemmas 2 and 3) are computable. □

As consequence, we can establishe with Lemma 5 above and the following corollary a 1-1 correspondence between basic APS-extensions and shift-stable sets whose expressions are in sum-reduced representation.

Corollary 1 *Let $W = \{T_1, \ldots, T_e\} \subset \Sigma\Pi(\mathbb{G})$ be in sum-product reduced representation and shift-stable. More precisely, for each $1 \leq i \leq e$ the arising sums/products in T_i are contained in $\{T_1, \ldots, T_{i-1}\}$. Then there is a basic APS-extension $(\mathbb{E}, \sigma)$ of $(\mathbb{G}, \sigma)$ with $\mathbb{E} = \mathbb{G}\langle t_1\rangle \ldots \langle t_e\rangle$ equipped with an evaluation function* ev *(using Lemmas 2 and 3) such that $T_i = expr(t_i) \in \Sigma\Pi(\mathbb{G})$ holds for $1 \leq i \leq e$.*

Proof We can treat the elements $T_1, \ldots, T_e$ following the construction of Lemma 6 iteratively. Let us consider the ith step with $T_i = \text{Sum}(\lambda, H)$ or $T_i = \text{Prod}(\lambda, H)$. Since the T_i are in sum-product reduced form it follows from Lemma 4 that within the sum-case (resp. product-case) we can guarantee $l = \lambda$, i.e, $c = 0$ (resp. $c = 1$). Thus $\text{ev}(t_i, n) = T_i(n)$ for all $n \geq l$ and hence $\text{expr}(t_i) = T_i(n)$. □

In addition, we can provide the following simple proof of Lemma 1.

Proof of Lemma 1 Let $W = \{T_1, \ldots, T_e\} \subseteq W$ be shift-stable and the T_i in sum-product reduced form. Take $F \in \text{SumProd}(W, \mathbb{G})$ and $\lambda \in \mathbb{Z}$. W.l.o.g. we may assume that the T_i are given as in Corollary 1. Thus we can take an APS-extension $(\mathbb{G}\langle t_1\rangle \ldots \langle t_e\rangle, \sigma)$ of $(\mathbb{G}, \sigma)$ equipped with an evaluation function ev and o-function L such that $\text{expr}(t_i) = T_i$ for $1 \leq i \leq e$. Then we can take $f \in \mathbb{E}$ with (19) and get $F(n + \lambda) = \text{ev}(F, n + \lambda) = \text{ev}(\sigma^{\lambda}(t_i), n) = G(n)$ for all $n \in L(f) + \max(0, -\lambda)$ with $G(n) := \text{expr}(\sigma^{\lambda}(t_i)) \in \text{SumProd}(W, \mathbb{G})$. Thus W is shift-closed. If $\mathbb{K}$ is computable and one can factor polynomials over $\mathbb{K}$, then one can compute the o-function L and all the above components are computable. □

In short, the naive construction of APS-extensions will not gain any substantial simplification (except a transformation to a sum-product reduced representation). In the next section we will refine this construction further to solve Problem SigmaReduce.

3.2 *The Embedding into the Ring of Sequences and $R\Pi\Sigma$-Extensions*

Let $(\mathbb{A}, \sigma)$ be a difference ring with a subfield $\mathbb{K} \subseteq \mathbb{A}$ where $\sigma|_{\mathbb{K}} = \text{id}$ that is equipped with an evaluation function $\text{ev} : \mathbb{A} \times \mathbb{Z}_{\geq 0} \to \mathbb{K}$. Then ev naturally produces sequences in the commutative ring $\mathbb{K}^{\mathbb{Z}_{\geq 0}}$ with the identity element $\mathbf{1} =$

$(1, 1, 1, \dots)$ with component-wise addition and multiplication. More precisely, we can define the function $\tau \colon \mathbb{A} \to \mathbb{K}^{\mathbb{Z}_{\geq 0}}$ with

$$\tau(f) = (\mathrm{ev}(f, n))_{n\geq 0} = (\mathrm{ev}(f, 0), \mathrm{ev}(f, 1), \mathrm{ev}(f, 2), \dots). \tag{21}$$

Due to (12) and (13) the map τ can be turned to a ring homomorphism by defining the equivalence relation $(f_n)_{n\geq 0} \equiv (g_n)_{n\geq 0}$ with $f_j = g_j$ for all $j \geq \lambda$ for some $\lambda \in \mathbb{Z}_{\geq 0}$; compare [93]. It is easily seen that the set of equivalence classes $[f]$ with $f \in \mathbb{K}^{\mathbb{Z}_{\geq 0}}$ forms with $[f]+[g] := [f+g]$ and $[f][g] := [fg]$ again a commutative ring with the identity element $[\mathbf{1}]$ which we will denote by $S(\mathbb{K})$. In the following we will simply write f instead of $[f]$. In this setting, $\tau \colon \mathbb{A} \to S(\mathbb{K})$ forms a ring homomorphism. In addition the shift operator $S : S(\mathbb{K}) \to S(\mathbb{K})$ defined by

$$S((a_0, a_1, a_2, \dots)) = (a_1, a_2, a_3, \dots)$$

turns to a ring automorphism. In the following we call $(S(\mathbb{K}), S)$ *also the (difference) ring of sequences over* $\mathbb{K}$. Finally, we observe that property (14) implies that

$$\tau(\sigma(f)) = S(\tau(f)) \tag{22}$$

holds for all $f \in \mathbb{A}$, i.e., τ turns to a difference ring homomorphism. Finally, property (11) implies

$$\tau(c) = \boldsymbol{c} = (c, c, c, \dots) \tag{23}$$

for all $c \in \mathbb{K}$. In the following we call a ring homomorphism $\tau : \mathbb{A} \to S(\mathbb{K})$ with (22) and (23) also a $\mathbb{K}$-*homomorphism*.

We can now link these notions to our construction from above with $\mathbb{G} \in \{\mathbb{G}_r, \mathbb{G}_b, \mathbb{G}_m\}$. Let $(\mathbb{E}, \sigma)$ with $\mathbb{E} = \mathbb{G}\langle t_1\rangle \dots \langle t_e\rangle$ be a basic APS-extension of $(\mathbb{G}, \sigma)$ and take an evaluation function $\mathrm{ev} : \mathbb{E} \times \mathbb{Z}_{\geq 0} \to \mathbb{K}$ with o-function L. Such a construction can be accomplished by iterative application of Lemmas 2 and 3. Then the function $\tau : \mathbb{E} \to \mathbb{A}$ with (21) for $f \in \mathbb{E}$ yields a $\mathbb{K}$-homomorphism.

If we find two different elements $a, b \in \mathbb{E}$ with $\tau(a) = \tau(b)$, then we find two different sum-product reduced representations $\mathrm{expr}(a)$ and $\mathrm{expr}(b)$ in terms of the sums and products given in $W = \{\mathrm{expr}(t_1), \dots, \mathrm{expr}(t_e)\} \subseteq \Sigma\Pi(\mathbb{G})$ which evaluates to the same sequence. In short, W is not canonical reduced (and thus not σ-reduced) over $\mathbb{G}$. This shows that a solution of Problem SigmaReduce can be only accomplished if τ is injective.

In this context, the set of constants plays a decisive role.

Definition 12 For a difference ring $(\mathbb{A}, \sigma)$ the *set of constants* is defined by $\mathrm{const}_\sigma \mathbb{A} = \{c \in \mathbb{A} \mid \sigma(c) = c\}$.

In general, $\mathrm{const}_\sigma \mathbb{A}$ is a subring of $\mathbb{A}$. If $\mathbb{A}$ is a field, then $\mathrm{const}_\sigma \mathbb{A}$ itself is a field which one also calls the *constant field of* $(\mathbb{A}, \sigma)$.

With this extra notion we can state now the following remarkable property that is based on results from [118]; compare also [126].

Theorem 1 *Let* $(\mathbb{E}, \sigma)$ *be a basic* APS*-extension of a difference field* $(\mathbb{F}, \sigma)$ *with* $\mathbb{K} = \text{const}_\sigma \mathbb{F}$ *and let* τ *be a* $\mathbb{K}$*-homomorphism. Then* τ *is injective iff* $\text{const}_\sigma \mathbb{E} = \mathbb{K}$.

Proof Suppose that $\text{const}_\sigma \mathbb{E} = \mathbb{K}$. By Theorem [118, Thm 3.3] it follows that $(\mathbb{E}, \sigma)$ is simple (i.e., the only difference ideals in $\mathbb{E}$ are $\{0\}$ or $\mathbb{E}$) and thus by [118, Lemma 5.8] we can conclude that τ is injective. Conversely, if τ is injective, it follows by [118, Lemma 5.13] that $\text{const}_\sigma \mathbb{E} = \mathbb{K}$. □

This result gives rise to the following refined definition of PS-field/APS-extensions.

Definition 13 Let $(\mathbb{F}, \sigma)$ be a PS-field extension of $(\mathbb{H}, \sigma)$ as defined in Definition 7. Then this is called a $\Pi\Sigma$*-field extension* if $\text{const}_\sigma \mathbb{F} = \text{const}_\sigma \mathbb{H}$. The arising P-field and S-field monomials are also called Π*-field and* Σ*-field monomials*, respectively. In particular, we call it a Π-/Σ-/$\Pi\Sigma$*-field extension* if it is built by the corresponding monomials. $(\mathbb{F}, \sigma)$ is called a $\Pi\Sigma$*-field over* $\mathbb{K}$ if $(\mathbb{F}, \sigma)$ is a $\Pi\Sigma$-field extension of $(\mathbb{K}, \sigma)$ and $\text{const}_\sigma \mathbb{K} = \mathbb{K}$.

Example 13 As mentioned in Examples 8 and 9, the difference fields $(\mathbb{G}_r, \sigma)$, $(\mathbb{G}_b, \sigma)$ and $(\mathbb{G}_m, \sigma)$ are PS-field extensions of $(\mathbb{K}, \sigma)$. Using the technologies given in Theorems 5 and 10 below one can show that they are all $\Pi\Sigma$-field extensions. Since $\text{const}_\sigma \mathbb{K} = \mathbb{K}$, they are also $\Pi\Sigma$-fields over $\mathbb{K}$; compare also [83].

Definition 14 Let $(\mathbb{E}, \sigma)$ be an APS-extension of $(\mathbb{A}, \sigma)$ as defined in Definition 8. Then this is called an $R\Pi\Sigma$*-extension* if $\text{const}_\sigma \mathbb{E} = \text{const}_\sigma \mathbb{A}$. The arising A-monomials are also called R*-monomials*, the P-monomials are called Π*-monomials* and the S-monomials are called Σ*-monomials*. In particular, we call it an R-/Π-/Σ-/$R\Pi$-/$R\Sigma$/-$\Pi\Sigma$*-extension* if it is built by the corresponding monomials.

Example 14 (Cont. of Example 10) Consider the difference ring $(\mathbb{Q}(x)[s], \sigma)$ from Example 10. Since $\text{ev} : \mathbb{Q}(x)[s] \to \mathbb{Q}$ defined by (9) and (10) (with $\text{ev}' = \text{ev}$) is an evaluation function of $(\mathbb{Q}(x)[s], \sigma)$ we can construct the $\mathbb{Q}$-homomorphism $\tau : \mathbb{Q}(x)[s] \to \mathbf{S}(\mathbb{Q})$ defined by (21). Since s is a Σ-monomial over $\mathbb{Q}(x)$, we get $\text{const}_\sigma \mathbb{Q}(x)[s] = \mathbb{Q}$. Thus we can apply Theorem 1 and it follows that

$$\tau(\mathbb{Q}(x))[\tau(s)] = \tau(\mathbb{Q}(x))[(\text{ev}(s, n))_{n\geq 0}] = \tau(\mathbb{Q}(x))[(S(n))_{n\geq 0}]$$

with $S = \text{expr}(s) = \text{Sum}(1, \frac{1}{x}) \in \Sigma(\mathbb{Q}(x))$ is isomorphic to the polynomial ring $\mathbb{Q}(x)[s]$. Further, $(S(n))_{n\geq 0}$ with $S(n) = \sum_{k=1}^{n} \frac{1}{k}$ is transcendental over $\tau(\mathbb{Q}(x))$.

Example 14 generalizes as follows. Suppose that we are given a basic $R\Pi\Sigma$-extension $(\mathbb{E}, \sigma)$ of $(\mathbb{G}, \sigma)$ with

$$\mathbb{G}[\rho_1] \dots [\rho_l][p_1, p_1^{-1}] \dots [p_u, p_u^{-1}][s_1] \dots [s_r]$$

where the ρ_i are R-monomials with $\zeta_i = \frac{\sigma(\rho_i)}{\rho_i} \in \mathcal{R}$ being primitive roots of unity, p_i are Π-monomials and the s_i are Σ-monomials. In addition, take an evaluation function ev with o-function L by iterative applications of Lemmas 2 and 3. Here we may assume that

- $\text{ev}(\rho_i, n) = \zeta_i^n$ for all $1 \leq i \leq l$,
- $\text{ev}(p_i, n) = P_i(n)$ with $\text{expr}(p_i) = P_i \in \Pi(\mathbb{G})$ for all $1 \leq i \leq u$, and
- $\text{ev}(s_i, n) = S_i(n)$ with $\text{expr}(s_i) = S_i \in \Sigma(\mathbb{G})$ for all $1 \leq i \leq r$.

Then $\tau : \mathbb{E} \to S(\mathbb{K})$ with (21) is a $\mathbb{K}$-homomorphism. By Theorem 1 it follows that τ is injective and thus

$$\begin{aligned}\tau(\mathbb{E}) &= \tau(\mathbb{G})\,[\tau(\rho_1)]\dots[\tau(\rho_l)]\\ &\qquad\times[\tau(p_1), \tau(p_1)^{-1}]\dots[\tau(p_u), \tau(p_u)^{-1}]\\ &\qquad\times[\tau(s_1)]\dots[\tau(s_r)]\\ &= \tau(\mathbb{G})\,[(\zeta_1^n)_{n\geq 0}]\dots[(\zeta_l^n)_{n\geq 0}]\\ &\qquad\times[(P_1(n))_{n\geq 0}, (\tfrac{1}{P_1(n)})_{n\geq 0}]\dots[(P_u(n))_{n\geq 0}, (\tfrac{1}{P_u(n)})_{n\geq 0}]\\ &\qquad\times[(S_1(n))_{n\geq 0}]\dots[(S_r(n))_{n\geq 0}]\end{aligned}$$

forms a (Laurent) polynomial ring extension over the ring of sequences $R = \tau(\mathbb{G})[(\zeta_1^n)_{n\geq 0}]\dots[(\zeta_l^n)_{n\geq 0}]$. In particular, we conclude that the sequences

$$(P_1(n))_{n\geq 0}, (\tfrac{1}{P_1(n)})_{n\geq 0}, \dots (P_u(n))_{n\geq 0}, (\tfrac{1}{P_u(n)})_{n\geq 0}, (S_1(n))_{n\geq 0}, \dots, (S_r(n))_{n\geq 0}$$

are, up to the trivial relations $(P_i(n))_{n\geq 0}\cdot(\frac{1}{P_i(n)})_{n\geq 0} = 1$ for $1 \leq i \leq u$, algebraically independent among each other over the ring R.

We are now ready to state the main result of this section that connects SumProd($\mathbb{G}$) with difference ring theory.

Theorem 2 *Let* $(\mathbb{E}, \sigma)$ *be a basic* APS*-extension of* $(\mathbb{G}, \sigma)$ *with* $\mathbb{G} \in \{\mathbb{G}_r, \mathbb{G}_b, \mathbb{G}_m\}$ *and* $\mathbb{A} = \mathbb{E}\langle t_1\rangle\dots\langle t_e\rangle$ *equipped with an evaluation function* $\text{ev} : \mathbb{E} \times \mathbb{Z}_{\geq 0} \to \mathbb{K}$ *(using Lemmas 2 and 3). Take the* $\mathbb{K}$*-homomorphism* $\tau : \mathbb{E} \to S(\mathbb{K})$ *with* $\tau(f) = (\text{ev}(f, n))_{n\geq 0}$ *and* $T_i = \text{expr}(t_i) \in \Sigma\Pi(\mathbb{G})$ *for* $1 \leq i \leq e$. *Then the following statements are equivalent.*

1. $(\mathbb{E}, \sigma)$ *is an* $R\Pi\Sigma$*-extension of* $(\mathbb{G}, \sigma)$.
2. τ *is a* $\mathbb{K}$*-isomorphism between* $(\mathbb{E}, \sigma)$ *and* $(\tau(\mathbb{E}), S)$*; in particular all sequences generated by the* $\Pi\Sigma$*-monomials are algebraically independent over the ring given by the sequences of* $\tau(\mathbb{G})$ *adjoined with the sequences generated by* R*-monomials.*
3. $W = \{T_1, \dots, T_e\}$ *is canonical-reduced over* $\mathbb{G}$.
4. *The zero recognition problem is trivial, i.e., for any* $F \in$ *SumProd*$(W, \mathbb{G})$ *the following holds:* $F = 0$ *if and only if* $\text{ev}(F, n) = 0$ *for all* $n \geq \delta$ *for some* $\delta \in \mathbb{Z}_{\geq 0}$.

Proof (1) $\Leftrightarrow$ (2) is an immediate consequence of Theorem 1.
(2) $\Rightarrow$ (3): Let $F, F' \in \text{SumProd}(W, \mathbb{G})$ with $F(n) = F'(n)$ for all $n \geq \delta$ for some $\delta \in \mathbb{Z}_{\geq 0}$. Replace in F, F' any occurrences of $T_i \textcircled{\wedge} z_i$ for $1 \leq i \leq e$ with $z_i \in \mathbb{Z}$ by $t_i^{z_i}$, $\oplus$ by $+$, and $\odot$ by $\cdot$. This yields $f, f' \in \mathbb{E}$ with $\text{ev}(f, n) = F(n)$ for all $n \geq L(f)$ and $\text{ev}(f', n) = F'(n)$ for all $n \geq L(f')$. Hence $\tau(f) = \tau(f')$. Since τ is injective, $f = f'$. But this implies that F and F' are the same up to trivial permutations of the operands in $\odot$ and $\oplus$. Consequently W is canonical reduced.
(3) $\Rightarrow$ (4): Suppose that W is canonical reduced and take $F \in \text{SumProd}(W, \mathbb{G})$ with $F(n) = 0$ for all $n \geq \delta$ for some $\delta \in \mathbb{Z}_{\geq 0}$. Since $\text{ev}(0, n) = 0$ for all $n \geq 0$ and W is canonical reduced, it follows that $F = 0$.
(4) $\Rightarrow$ (1): Suppose that τ is not injective and take $f \in \mathbb{E} \setminus \{0\}$ with $\tau(f) = \mathbf{0}$. By Lemma 5 we can take $0 \neq F \in \text{SumProd}(\{T_1, \ldots, T_e\}, \mathbb{G})$ and $\delta \in \mathbb{Z}_{\geq 0}$ with $\text{ev}(f, n) = F(n) = 0$ for all $n \geq \delta$. Thus statement (4) does not hold. □

In order to derive the equivalences in Theorem 2 we assumed that an APS-extension is given. We can relax this assumption if the set W is shift-stable.

Corollary 2 *Let $W = \{T_1, \ldots, T_e\} \subset \Sigma\Pi(\mathbb{G})$ be in sum-product reduced representation and shift-stable. More precisely, for each $1 \leq i \leq e$ the arising sums/products in T_i are contained in $\{T_1, \ldots, T_{i-1}\}$. Then the following two statements are equivalent:*

1. *There is a basic $R\Pi\Sigma$-extension $(\mathbb{E}, \sigma)$ of $(\mathbb{G}, \sigma)$ with $\mathbb{E} = \mathbb{G}\langle t_1 \rangle \ldots \langle t_e \rangle$ equipped with an evaluation function* ev *(using Lemmas 2 and 3) with $T_i = expr(t_i) \in \Sigma\Pi(\mathbb{G})$ for $1 \leq i \leq e$.*
2. *$W = \{T_1, \ldots, T_e\}$ is σ-reduced over $\mathbb{G}$.*

Proof (1) $\Rightarrow$ (2): By assumption W is sum-product reduced and shift-stable, and by (1) $\Rightarrow$ (3) of Theorem 2 it is canonical-reduced. Thus W is σ-reduced.
(2) $\Rightarrow$ (1): By Corollary 1 we get an APS-extension $(\mathbb{E}, \sigma)$ of $(\mathbb{G}, \sigma)$ with $\mathbb{E} = \mathbb{G}\langle t_1 \rangle \ldots \langle t_e \rangle$ equipped with an evaluation function ev (using Lemmas 2 and 3) with $T_i = \text{expr}(t_i) \in \Sigma\Pi(\mathbb{G})$ for $1 \leq i \leq e$. Since W is canonical reduced, it follows by (3) $\Rightarrow$ (1) of Theorem 2 that $(\mathbb{E}, \sigma)$ is an $R\Pi\Sigma$-extension of $(\mathbb{G}, \sigma)$. □

Corollary 2 yields immediately a strategy (actually the only strategy for shift-stable sets) to solve Problem SigmaReduce.

Strategy to Solve Problem SigmaReduce

Given: $A_1, \ldots, A_u \in \text{SumProd}(\mathbb{G})$ with $\mathbb{G} \in \{\mathbb{G}_r, \mathbb{G}_b, \mathbb{G}_m\}$, i.e., $\mathbb{G} = \mathbb{K}(x, x_1, \ldots, x_v)$ or $\mathbb{G} = \mathbb{K}(x_1, \ldots, x_v)$.
Find: a σ-reduced set $W = \{T_1, \ldots, T_e\} \subset \Sigma\Pi(\mathbb{G}')$ with $B_1 \ldots, B_u \in \text{SumProd}(W, \mathbb{G}')$ and $\delta_1, \ldots, \delta_u \in \mathbb{Z}_{\geq 0}$ such that $A_i(n) = B_i(n)$ holds for all $n \geq \delta_i$ and $1 \leq i \leq r$.

1. Construct an $R\Pi\Sigma$-extension $(\mathbb{E}, \sigma)$ of[8] $(\mathbb{G}', \sigma)$ with $\mathbb{E} = \mathbb{G}'\langle t_1\rangle \dots \langle t_e\rangle$ equipped with an evaluation function $\text{ev} : \mathbb{E} \times \mathbb{Z}_{\geq 0} \to \mathbb{K}'$ and o-function L (using Lemmas 2 and 3) in which $A_1, \dots, A_u$ are modeled by $a_1, \dots, a_u \in \mathbb{E}$. More precisely, for $1 \leq i \leq u$ we compute in addition $\delta_i \in \mathbb{Z}_{\geq 0}$ with $\delta_i \geq L(a_i)$ such that

$$A_i(n) = \text{ev}(a_i, n) \quad \forall n \geq \delta_i. \tag{24}$$

2. Set $W = \{T_1, \dots, T_e\}$ with $T_i := \text{expr}(t_i) \in \Sigma\Pi(\mathbb{G}')$ for $1 \leq i \leq e$.
3. Set $B_i := \text{expr}(a_i) \in \text{SumProd}(W, \mathbb{G}')$ for $1 \leq i \leq u$.
4. Return W, $(B_1, \dots, B_u)$ and $(\delta_1, \dots, \delta_u)$.

What remains open is to enrich this general method with the construction required in step (1). This task will be considered in detail in the next section.

4 The Representation Problem

In this section we will give an overview of the existing algorithms that assist in the task of solving the open subproblem given in step (1) of our general method SigmaReduce. The resulting machinery can be summarized as follows.

Theorem 3 *Given $A_1, \dots, A_u \in SumProd_1(\mathbb{G})$ with $\mathbb{G} \in \{\mathbb{G}_r, \mathbb{G}_b, \mathbb{G}_m\}$ where $\mathbb{K}$ is a rational function field over an algebraic number field. Then one can compute a σ-reduced set $W = \{T_1, \dots, T_e\} \subset \Sigma\Pi_1(\mathbb{G})$ with $B_1 \dots, B_u \in SumProd(W, \mathbb{G})$ and $\delta_1, \dots, \delta_u \in \mathbb{Z}_{\geq 0}$ such that $A_i(n) = B_i(n)$ holds for all $n \geq \delta_i$ and $1 \leq i \leq u$.*

Theorem 4 *Given $A_1, \dots, A_u \in SumProd(\mathbb{K}(x))$ where $\mathbb{K} = \mathcal{A}(y_1, \dots, y_o)$ is a rational function field over an algebraic number field $\mathcal{A}$. Then one can take $\mathbb{K}' = \mathcal{A}'(y_1, \dots, y_o)$ where $\mathcal{A}'$ is an algebraic extension of $\mathcal{A}$ and can compute a σ-reduced set $W = \{T_1, \dots, T_e\} \subset \Sigma\Pi(\mathbb{K}'(x))$ with $B_1 \dots, B_u \in SumProd(W, \mathbb{G}')$ and $\delta_1, \dots, \delta_u \in \mathbb{Z}_{\geq 0}$ such that $A_i(n) = B_i(n)$ holds for all $n \geq \delta_i$ and $1 \leq i \leq u$.*

Here we will start with the problem to represent products in $R\Pi$-monomials (see Sect. 4.1). More precisely, we will show various tactics that enable one to represent expressions of $\text{Prod}_1(\mathbb{G}_r)$, $\text{Prod}_1(\mathbb{G}_b)$, $\text{Prod}_1(\mathbb{G}_m)$ and $\text{Prod}(\mathbb{G}_r)$. Afterwards, we will consider the problem to represent nested sums over such products (i.e., expressions of $\text{Sum}(\mathbb{G})$, $\text{SumProd}_1(\mathbb{G}_r)$, $\text{SumProd}_1(\mathbb{G}_b)$, $\text{SumProd}_1(\mathbb{G}_m)$ and $\text{SumProd}(\mathbb{G}_r)$) in Σ-monomials (see Sect. 4.2).

---

[8] Here we get $\mathbb{G}' = \mathbb{K}'(x, x_1, \dots, x_v)$ or $\mathbb{G}' = \mathbb{K}'(x_1, \dots, x_v)$ where $\mathbb{K}'$ is a field extension of $\mathbb{K}$; if $A_1, \dots, A_u \in \text{SumProd}_1(\mathbb{G})$, one can restrict to the special case $\mathbb{G} = \mathbb{G}'$.

Remark 2 `Sigma` can represent fully algorithmically single nested products in $R\Pi$-extensions; in addition, Ocansey's package `NestedProducts` [83, 84] can deal with the case $\text{Prod}(\mathbb{G}_r)$. Expressions from more general domains (e.g., sums and products that arise nontrivially in denominators) also work with the function call **`SigmaReduce`** of `Sigma`. But for these cases the underlying summation mechanisms (like those given in Lemmas 5 and 6) are only partially developed and the back translation from the difference ring setting to the term algebra might fail.

In general, it suffices in our proposed construction to compute an $R\Pi\Sigma$-extension in which a finite set of sums and products are modeled. However, in some important instances it is possible to perform this constructions stepwise.

Definition 15 Fix X as one of the term algebras $\text{Prod}_1(\mathbb{G})$, $\text{Prod}(\mathbb{G})$, $\text{Sum}(\mathbb{G})$, $\text{SumProd}_1(\mathbb{G})$, $\text{SumProd}(\mathbb{G})$, and let $\mathcal{D}$ be a subclass of basic $R\Pi\Sigma$-extensions of $(\mathbb{G}, \sigma)$. Then $\mathcal{D}$ is called *X-extension-stable* if for any $(\mathbb{H}, \sigma) \in \mathcal{D}$ and any $A \in X$ one can construct an $R\Pi\Sigma$-extension $(\mathbb{E}, \sigma)$ of $(\mathbb{H}, \sigma)$ with $(\mathbb{E}, \sigma) \in \mathcal{D}$ and $a \in \mathbb{E}$ such that one can model A with a.

We note that within such an extension-stable class of $R\Pi\Sigma$-extensions one does not have to treat the arising sums and products in one stroke, but can consider them iteratively. This is in particular interesting, when unforeseen sums and products arise in a later step, that have to be considered in addition. In a nutshell, we will provide a general overview of the existing tools to design basic $R\Pi\Sigma$-extensions. In particular, we will emphasize the available algorithms to construct extension-stable versions.

4.1 Representation of Products in $R\Pi$-Extensions

We start with algorithmic tools that enable one to test if a P-extension forms a Π-extension. Based on these tools we present (without further details) the existing techniques to represent a finite set of products in an $R\Pi$-extension.

4.1.1 Algorithmic Tests

In [109, Theorem 9.1] based on Karr's work [65, 66] a general criterion for Π-field extensions is elaborated. Here we present a more flexible version in the ring setting.

Theorem 5 *Let* $(\mathbb{E}, \sigma)$ *be a* P*-extension of a difference ring* $(\mathbb{H}, \sigma)$ *with* $\mathbb{E} = \mathbb{H}\langle t_1\rangle \dots \langle t_d\rangle$ *and* $f_i = \frac{\sigma(t_i)}{t_i} \in \mathbb{H}^*$ *for* $1 \leq i \leq d$*. Suppose that*

$$\{g \in \mathbb{H} \setminus \{0\} \mid \sigma(g) = u\, g \text{ for some } u \in \mathbb{H}^*\} \subseteq \mathbb{H}^* \tag{25}$$

holds. Then the following statements are equivalent:

1. $(\mathbb{E}, \sigma)$ *is a* Π*-extension of* $(\mathbb{H}, \sigma)$*, i.e.,* $\mathrm{const}_\sigma \mathbb{E} = \mathrm{const}_\sigma \mathbb{H}$.
2. *There do not exist* $g \in \mathbb{H} \setminus \{0\}$ *and* $(z_1, \dots, z_d) \in \mathbb{Z}^d \setminus \{\mathbf{0}\}$ *with*

$$\sigma(g) = f_1^{z_1} \dots f_d^{z_d} g.$$

Proof (1) $\Rightarrow$ (2): Suppose that there is a $g \in \mathbb{H} \setminus \{0\}$ and $(z_1, \dots, z_d) \in \mathbb{Z}^d \setminus \{\mathbf{0}\}$ with $\sigma(g) = f_1^{z_1} \dots f_d^{z_d} g$. Let i be maximal such that $z_i \neq 0$. Then we can take $h = g\, f_1^{-z_1} \dots f_{i-1}^{-z_{i-1}}$ and get $\sigma(h) = f_i^{z_i} h$. With part (2) of Theorem 2.12 in [116] it follows that $(\mathbb{H}\langle t_1 \rangle \dots \langle t_i \rangle, \sigma)$ is not a Π-extension of $(\mathbb{H}\langle t_1 \rangle \dots \langle t_{i-1} \rangle, \sigma)$.
(2) $\Rightarrow$ (1): Let i with $1 \leq i \leq d$ be minimal such that $(\mathbb{H}\langle t_1 \rangle \dots \langle t_i \rangle, \sigma)$ is not a Π-extension of $(\mathbb{H}\langle t_1 \rangle \dots \langle t_{i-1} \rangle, \sigma)$. Then $\sigma(g) = \alpha_i^{z_i} g$ for some $g \in \mathbb{H}\langle t_1 \rangle \dots \langle t_{i-1} \rangle \setminus \{0\}$ and $z_i \in \mathbb{Z} \setminus \{0\}$ by part (2) of Theorem 2.12 in [116]. In particular, with property (25) we can apply Theorem 22 of [116] and it follows that $g = h\, t_1^{-z_1} \dots t_{i-1}^{-z_{i-1}}$ for some $z_i \in \mathbb{Z}$ and $h \in \mathbb{H}^*$. Thus we get $\sigma(h) = \alpha_1^{z_1} \dots \alpha_i^{z_i} h$ with $z_i \neq 0$ which proves statement (1). □

Remark 3

(1) Theorem 10 contains the following special case (see [66] for the field and [116] for the ring case): a P extension $(\mathbb{A}\langle p \rangle, \sigma)$ of $(\mathbb{A}, \sigma)$ with $f := \frac{\sigma(p)}{p} \in \mathbb{A}^*$ is a Π-extension iff there are no $g \in \mathbb{A}$, $m \in \mathbb{Z} \setminus \{0\}$ with $\sigma(g) = f\, g$.
(2) Often Theorem 5 is applied to the special case when the ground ring $(\mathbb{H}, \sigma)$ forms a field. Note that in this particular instance, the assumption (25) trivially holds.

Let $(\mathbb{A}, \sigma)$ be a difference ring and $\boldsymbol{f} = (f_1, \dots, f_d) \in (\mathbb{A}^*)^d$. Then we define

$$M(\boldsymbol{f}, \mathbb{A}) := \{(m_1, \dots, m_d) \in \mathbb{Z}^d \,|\, \sigma(g) = f_1^{m_1} \dots f_d^{m_d}\, g \text{ for some } g \in \mathbb{A} \setminus \{0\}\};$$

see also [65]. Note that Theorem 5 states that the P-extension $(\mathbb{E}, \sigma)$ of the difference ring $(\mathbb{H}, \sigma)$ with $\mathbb{E} = \mathbb{H}\langle t_1 \rangle \dots \langle t_d \rangle$ and $f_i = \frac{\sigma(t_i)}{t_i} \in \mathbb{H}^*$ for $1 \leq i \leq d$ is a Π-extension if and only if $M(\boldsymbol{f}, \mathbb{H}) = \{\mathbf{0}\}$. If $\boldsymbol{f} \in ([\mathbb{F}^*]_{\mathbb{F}}^{\mathbb{H}})^d$ (which holds for $\mathbb{F}^*$-basic P-extensions), this latter property can be checked by utilizing the following result.

Theorem 6 *Let* $(\mathbb{H}, \sigma)$ *be a basic* $R\Pi\Sigma$*-extension of a difference field* $(\mathbb{F}, \sigma)$ *and* $\boldsymbol{f} \in ([\mathbb{F}^*]_{\mathbb{F}}^{\mathbb{H}})^d$*. Then the following holds:*

1. $M(\boldsymbol{f}, \mathbb{H})$ *is a* $\mathbb{Z}$*-module over* $\mathbb{Z}^d$.
2. *If one can compute a basis of* $M(\boldsymbol{h}, \mathbb{F})$ *for any* $\boldsymbol{h} \in (\mathbb{F}^*)^m$ *with* $m \geq 1$*, then one can compute a basis of* $M(\boldsymbol{f}, \mathbb{H})$.

Proof Part (1) follows by Lemma 2.6 and Theorem 2.22 of [116] and part (2) by [116, Theorem 2.23]. □

In other words, we can apply Theorem 5 to test if a basic P-extension over $\mathbb{F}$ is a Π-extension if one can compute a basis of $M(\boldsymbol{h}, \mathbb{F})$ in a difference field $(\mathbb{F}, \sigma)$. In particular, using the algorithms from [65] this is possible if $(\mathbb{F}, \sigma)$ is a $\Pi\Sigma$-field over $\mathbb{K}$ where the constant field satisfies certain algorithmic properties.

Definition 16 A field $\mathbb{K}$ is called *σ-computable* if the following holds:

1. One can factorize multivariate polynomials over $\mathbb{K}$;
2. given $(f_1, \dots, f_d) \in (\mathbb{K}^*)^d$ one can compute for $\{(z_1, \dots, z_d) \in \mathbb{Z}^d \mid f_1^{z_1} \dots f_d^{z_d} = 1\}$ a $\mathbb{Z}$-basis;
3. one can decide if $c \in \mathbb{K}$ is an integer.

More precisely, the following holds if $(\mathbb{F}, \sigma)$ is a $\Pi\Sigma$-field over a σ-computable constant field; special cases are $\mathbb{G}_r$, $\mathbb{G}_b$ or $\mathbb{G}_m$ where $\mathbb{K}$ is σ-computable.

Corollary 3 *Let $(\mathbb{E}, \sigma)$ be a basic $R\Pi\Sigma$-extension of a $\Pi\Sigma$-field $(\mathbb{F}, \sigma)$ over $\mathbb{K}$. If $\mathbb{K}$ is σ-computable, one can compute a basis of $M(\boldsymbol{h}, \mathbb{E})$ for any $\boldsymbol{h} \in ([\mathbb{F}^*]_{\mathbb{F}}^{\mathbb{E}})^d$ with $d \geq 1$. This in particular is the case, if $\mathbb{K} = \mathcal{A}(y_1, \dots, y_o)$ is a rational function over an algebraic number field $\mathcal{A}$.*

Proof If $\mathbb{K}$ is σ-computable, it follows by Karr [65, Theorem 9] that one can compute a basis of $M(\boldsymbol{f}, \mathbb{F})$ for any $\boldsymbol{f} \in (\mathbb{F}^*)^m$ with $m \geq 1$. Thus by part 2 of Theorem 6 one can compute a basis of $M(\boldsymbol{h}, \mathbb{E})$ for any $\boldsymbol{h} \in ([\mathbb{F}^*]_{\mathbb{F}}^{\mathbb{E}})^d$ with $d \geq 1$. In particular, it follows by Schneider [104, Thm. 3.5] (based on the algorithm of [59]) that a rational function field over an algebraic number field is σ-computable. $\square$

Remark 4

(1) By [116, Theorem 2.26] Corollary 3 is also valid for $\boldsymbol{f} \in (\{\mathbb{F}^*\}_{\mathbb{F}}^{\mathbb{E}})^d$ in simple $R\Pi\Sigma$-extension defined over a $\Pi\Sigma$-field. As elaborated in [116, Sect. 2.3.3] (using ideas of [69]) it holds even in the more general setting that $(\mathbb{F}, \sigma)$ is a $\Pi\Sigma$-field extension of a difference field $(\mathbb{F}_0, \sigma)$ where all roots of unity in $\mathbb{F}$ are constants and $(\mathbb{F}_0, \sigma)$ is σ-computable; for the definition of these algorithmic properties we refer to [69, Def. 1]. Further aspects can be also found in [21]. In particular, all these properties hold, if $(\mathbb{F}_0, \sigma)$ is a free difference field [68, 69] (covering generic/unspecified sequences X_n) or is built by radical extensions [70] (covering objects like $\sqrt{n}$). For the underlying implementations enhancing `Sigma` we refer to [69, 70].

(2) Within `Sigma` the case of $\Pi\Sigma$-fields is implemented properly where the constant field is given by a rational function field over the rational numbers. In parts also algebraic numbers work, but here we rely on sub-optimal routines of Mathematica.

4.1.2 Algorithmic Representations

In this section we present several algorithms that provide proofs of Theorems 3 and 4 if one restricts to the cases $\text{Prod}_1(\mathbb{G})$ with $\mathbb{G} \in \{\mathbb{G}_r, \mathbb{G}_b, \mathbb{G}_m\}$ or $\text{Prod}(\mathbb{G}_r)$, i.e., if one drops expressions where sums arise. More precisely, we will introduce several solutions of step (1) for our method SigmaReduce.

First, we treat the case $\text{Prod}_1(\mathbb{G})$. In this setting (where also sums can arise) single-basic $R\Pi\Sigma$-extensions, a subclass of basic $R\Pi\Sigma$-extensions, are sufficient.

Definition 17 An $R\Pi\Sigma$-extension $(\mathbb{E}, \sigma)$ of a difference ring $(\mathbb{A}, \sigma)$ with $\mathbb{E} = \mathbb{A}\langle t_1\rangle \dots \langle t_e\rangle$ is called *single-basic* if for any R-monomial t_i we have $\frac{\sigma(t_i)}{t_i} \in \text{const}_\sigma \mathbb{A}^*$ and for any P-monomial t_i we have $\frac{\sigma(t_i)}{t_i} \in \mathbb{A}^*$.

We will present the following two main strategies.

- *Optimal product representations.* In [120, Theorem 69] we showed that one can construct $R\Pi$-extensions with minimal extension degree and minimal order.

Theorem 7 *Let $\mathbb{G} \in \{\mathbb{G}_r, \mathbb{G}_b, \mathbb{G}_m\}$ and $A_1, \dots, A_u \in \text{Prod}_1(\mathbb{G})$. Then there is a single-basic $R\Pi$-extension $(\mathbb{E}, \sigma)$ of $(\mathbb{G}, \sigma)$ with $\mathbb{E} = \mathbb{G}\langle t_1\rangle \dots \langle t_e\rangle$ together with an evaluation function* ev *and o-function L (based on the construction given in Lemmas 2 and 3) with the following properties:*

1. *$A_1, \dots, A_u$ are modeled by $a_1, \dots, a_u \in \mathbb{E}$, i.e., for all $1 \leq i \leq u$ we have* (24) *for some explicitly given $\delta_i \in \mathbb{Z}_{\geq 0}$ with $\delta_i \geq L(a_i)$.*
2. *There is at most one R-monomial in $\mathbb{E}$. This implies that the order λ is minimal among all such extensions in which one can model $a_1, \dots, a_u$.*
3. *The number of Π-monomials in $\mathbb{E}$ is minimal among all such extensions in which one can model $a_1, \dots, a_u$.*

If the constant field of $(\mathbb{K}, \sigma)$ is a rational function field over an algebraic number field, then the above components are computable.

Example 15 For the following products in $\text{Prod}_1(\mathbb{Q}[\mathbb{i}](x))$ with the imaginary unit $\mathbb{i}$:

$$A_1 = \text{Prod}\big(1, \tfrac{-13122x(1+x)}{(3+x)^3}\big), \qquad A_2 = \text{Prod}\big(1, \tfrac{26244x^2(2+x)^2}{(3+x)^2}\big),$$

$$A_3 = \text{Prod}\big(1, \tfrac{\mathbb{i}k(2+x)^3}{729(5+x)}\big), \qquad A_4 = \text{Prod}\big(1, \tfrac{-162x(2+x)}{5+x}\big),$$

we compute the alternative expressions $B_1 = \frac{5(1+x)^2(2+x)^5(3+x)^8}{52488(4+x)(5+x)} T_1 T_2 T_3^{-2}$, $B_2 = \frac{(4+x)^2(5+x)^2}{400} T_2^2$, $B_3 = \frac{2754990144(4+x)^2(5+x)^2}{25(1+x)^4(2+x)^{10}(3+x)^{16}} T_3^3$ and $B_4 = T_2$ in terms of the σ-reduced set $W = \{T_1, T_2, T_3\}$ with

$$T_1 = \text{RPow}(-1), \quad T_2 = \text{Prod}(1, \tfrac{-162x(2+x)}{5+x}), \quad T_3 = \text{Prod}(1, \tfrac{-\mathbb{i}(3+x)^6}{9x(1+x)^2(2+x)(5+x)});$$

internally, T_1 is modeled by an R-monomial of order 2 and T_2, T_3 are modeled by two Π-monomials. Details on this construction are given in [120, Ex. 70].

We remark that this optimal representation has one essential drawback: if further products have to be treated in a later situation, the existing difference ring cannot be reused, but a completely new difference ring has to be designed.

- *Extension stable representations for completely factorizable constant fields.* In the following we will follow another approach: instead of computing the smallest ring in which one can model a finite set of single nested products, we design a difference ring where the multiplicands are as small as possible such that the constructed difference rings are $\mathrm{Prod}_1(\mathbb{G})$-extension-stable. In order to accomplish this task, we will restrict the constant field $\mathbb{K}$ further as follows.

A *ring R is called completely factorizable* if R is a unique factorization domain (UFD) and all units in R are roots of unity. In particular, any element $a \in R\setminus\{0\}$ can be written in the form $a = u\,a_1^{n_1} \dots a_l^{n_l}$ with a root of unity u, $n_1, \dots, n_l \in \mathbb{Z}_{\geq 1}$ and $a_1, \dots, a_l \in R$ being coprime irreducible elements. In addition, a *field K is called completely factorizable* if it is the quotient field of a completely factorizable ring R. In such a field any element $a \in K^*$ can be written in the form $a = u\,a_1^{n_1} \dots a_l^{n_l}$ with a root of unity u, $n_1, \dots, n_l \in \mathbb{Z}\setminus\{0\}$ and $a_1, \dots, a_l \in R$ being coprime irreducible elements. We call *K completely factorizable of order* $\lambda \in \mathbb{Z}_{\geq 0}$, if the set of roots of unity is finite and the maximal order is λ. We say that *complete factorizations are computable over such a field K* if for any rational function from $K(x_1, \dots, x_r)$ a complete factorization can be computed.

The following lemma allows to lift the property of completely factorizable rings.

Lemma 7 *If a ring (resp. field) $\mathbb{A}$ is completely factorizable, the polynomial ring $\mathbb{A}[x_1, \dots, x_r]$ (resp. rat. function field $\mathbb{A}(x_1, \dots, x_r)$) is completely factorizable.*

Example 16 The ring $\mathbb{Z}$ and the Gaussian ring $\mathbb{Z}[\mathtt{i}]$ with the roots of unity $1, -1$ and $1, -1, \mathtt{i}, -\mathtt{i}$, respectively, are examples of completely factorizable rings. Thus $\mathbb{Z}$, $\mathbb{Z}[\mathtt{i}]$ and, in particular $\mathbb{Z}[x_1, \dots, x_r]$ and $\mathbb{Z}[\mathtt{i}][x_1, \dots, x_r]$ are completely factorizable rings. Furthermore, their quotient fields $\mathbb{Q}$, $\mathbb{Q}[\mathtt{i}]$, $\mathbb{Q}(x_1, \dots, x_r)$ and $\mathbb{Q}[\mathtt{i}](x_1, \dots, x_r)$ are completely factorizable of order 2 or 4, respectively. In particular, one can compute complete factorizations over $\mathbb{Q}$ and $\mathbb{Q}[\mathtt{i}]$.

Definition 18 Let $\mathbb{F}$ be the quotient field of a completely factorizable ring R of order λ. A single-basic $R\Pi\Sigma$-extension $(\mathbb{E}, \sigma)$ of $(\mathbb{F}, \sigma)$ with $\mathbb{E} = \mathbb{F}\langle t_1 \rangle \dots \langle t_e \rangle$ is called *completely factorized* if there is at most one R-monomial ρ with $\frac{\sigma(\rho)}{\rho} \in (\mathrm{const}_\sigma \mathbb{F})^*$ of order λ and for any Π-monomial t_i we have that $\frac{\sigma(t_i)}{t_i} \in R$ is irreducible.

We are now ready to state the following result implemented within `Sigma`; the case $\mathbb{G}_r$ is covered by [114, Theorem 2]; the extension to $\mathbb{G}_b$ and $\mathbb{G}_m$ is straightforward.

Theorem 8 *Let* $\mathbb{G} \in \{\mathbb{G}_r, \mathbb{G}_b, \mathbb{G}_m\}$ *where* $\mathbb{K}$ *is completely factorizable of order* λ. *Then the class of completely factorized* $R\Pi$*-extensions over* $(\mathbb{G}, \sigma)$ *is* $Prod_1(\mathbb{G})$*-extension-stable. More precisely, let* $(\mathbb{H}, \sigma)$ *be a completely factorized* $R\Pi$*-extension of* $(\mathbb{G}, \sigma)$ *equipped with an evaluation function* ev *an o-function* L. *Let* $A \in Prod_1(\mathbb{G})$. *Then there is an* $R\Pi$*-extension* $(\mathbb{E}, \sigma)$ *of* $(\mathbb{H}, \sigma)$ *with an extended evaluation function* ev *and o-function* L *(using Lemmas 2 and 3) with the following properties:*

1. $(\mathbb{E}, \sigma)$ *is a completely factorizable* $R\Pi$*-extension of* $(\mathbb{G}, \sigma)$.
2. A *is modeled by* $a \in \mathbb{E}$, *i.e.,* $A(n) = \text{ev}(a, n)$ *for all* $n \geq \delta$ *for some* $\delta \in \mathbb{Z}_{\geq 0}$.

If complete factorization over $\mathbb{K}$ *can be computed, all components are computable.*

Example 17 Given the products (15), we can split the multiplicands into irreducible factors and get (after some technical details) the product representations $B_1 = \frac{216 T_1^2 T_2 T_3^8}{(n+1)^2(n+2)^3(n+3)^3 T_4}$, $B_2 = \frac{9 T_2^2 T_3^8 T_4^2}{(n+3)^2}$, $B_3 = \frac{15(n+1)^2(n+2)^2 T_1^2 T_4^3}{(n+3)(n+4)(n+5) T_3^6}$ and $B_4 = \frac{60 T_1^2 T_2 T_3^4 T_4}{(n+3)(n+4)(n+5)}$ in terms of the σ-reduced set $W = \{T_1, T_2, T_3, T_4\}$ with

$$T_1 = \text{RPow}(\mathbb{i}), \quad T_2 = \text{Prod}(1, 2), \quad T_3 = \text{Prod}(1, 3), \quad T_4 = \text{Prod}(1, x);$$

internally, T_1 is modeled by an R-monomial of order 4, and T_2, T_3, T_4 are modeled by three Π-monomials.

It would be interesting to see extension-stable difference ring constructions that work in more general settings. A first step in this direction has been elaborated in [83, Theorem 6.2]. Here a toolbox (implemented within `NestedProducts`) is summarized where one tries to follow the above construction of completely factorized $R\Pi$-extensions as much as possible. In this way, a modification of the existing $R\Pi$-extension will arise only for products whose multiplicands are taken from an algebraic number field.

- *Representation of nested products.* We obtained the first algorithm in [84, Theorem 9] (implemented in `NestedProducts`) to represent products from $\text{Prod}(\mathbb{G}_r)$ fully algorithmically in a basic $R\Pi$-extension. This result can be stated as follows.

Theorem 9 *Let* $\mathbb{G} = \mathbb{G}_r = \mathbb{K}(x)$ *where* $\mathbb{K} = \mathcal{A}(y_1, \ldots, y_o)$ *with* $o \geq 0$ *is a rational function field over an algebraic number field* $\mathcal{A}$. *Then for* $A_1, \ldots, A_u \in Prod(\mathbb{G})$ *one can compute a basic* $R\Pi$*-extension* $(\mathbb{E}, \sigma)$ *of* $(\mathbb{G}', \sigma)$ *with an evaluation function* ev *and o-function* L *(using Lemmas 2 and 3) with the following properties:*

1. *The ground field* $\mathbb{G}$ *is extended to* $\mathbb{G}' = \mathbb{K}'(x)$ *where* $\mathbb{K}' = \mathcal{A}'(y_1, \ldots, y_o)$ *with* $\mathcal{A}'$ *being an algebraic field extension of* $\mathcal{A}$.
2. *Within the* $R\Pi$*-monomials in* $(\mathbb{E}, \sigma)$ *there is at most one* R*-monomial.*
3. $A_1, \ldots, A_u$ *are modeled by* $a_1, \ldots, a_u \in \mathbb{E}$, *i.e., for all* $1 \leq i \leq u$ *we have* (24) *for some explicitly given* $\delta_i \in \mathbb{Z}_{\geq 0}$ *with* $\delta_i \geq L(a_i)$.

Remark 5 Theorem 7 holds also for general ground rings $(\mathbb{G}, \sigma)$ with certain algorithmic properties; see [120]. Fascinating structural properties of mixed hypergeometric products (and related objects within the differential case) are presented in [51]. Further simplification aspects within $\Pi\Sigma$-fields (e.g., finding products where the degrees of the top most sum or product in the numerator and denominator of a multiplicand are minimal) are elaborated in [23, 104]. In addition, methods to find algebraic relations of sequences built by products are given in [72, 84, 109, 120, 124].

4.2 *Representation of Sums*

4.2.1 Algorithmic Tests via (Parameterized) Telescoping

We will proceed as in the product case. The additive version of Theorem 5, which is nothing else than parameterized telescoping (see Sect. 5.2), reads as follows.

Theorem 10 ([118, Thm. 7.10]) *Let $(\mathbb{E}, \sigma)$ be an S-extension of a difference ring $(\mathbb{H}, \sigma)$ with $\mathbb{E} = \mathbb{H}\langle t_1\rangle \dots \langle t_d\rangle$ and $f_i = \sigma(t_i) - t_i \in \mathbb{H}$ for $1 \leq i \leq d$. If $\mathbb{K} := \text{const}_\sigma\mathbb{H}$ is a field, then the following statements are equivalent:*

1. *$(\mathbb{E}, \sigma)$ is a Σ-extension of $(\mathbb{H}, \sigma)$, i.e., $\text{const}_\sigma\mathbb{E} = \text{const}_\sigma\mathbb{H}$.*
2. *There do not exist $g \in \mathbb{H}$ and $(c_1, \dots, c_d) \in \mathbb{K}^d \setminus \{\mathbf{0}\}$ with*

$$\sigma(g) - g = c_1 f_1 + \dots + c_d f_d.$$

Note that Theorem 10 contains the following special case (compare [65] for the field case and [116] for the ring case): an S extension $(\mathbb{A}[s], \sigma)$ of $(\mathbb{A}, \sigma)$ with $f := \sigma(s) - s \in \mathbb{A}$ is a Σ-extension if and only if there is no $g \in \mathbb{A}$ such that the telescoping equation $\sigma(g) - g = f$ holds; this property will be crucial for the construction that establishes Theorem 12 given below.

Let $(\mathbb{A}, \sigma)$ be a difference ring with constant field $\mathbb{K}$, $u \in \mathbb{A} \setminus \{0\}$ and $\boldsymbol{f} = (f_1, \dots, f_d) \in \mathbb{A}^d$. Then following [65] we define the *set of solutions of parameterized first-order linear difference equations*:

$$V_1(u, \boldsymbol{f}, \mathbb{A}) = \{(c_1, \dots, c_d, g) \in \mathbb{K}^d \times \mathbb{A} \mid \sigma(g) - u\, g = c_1 f_1 + \dots + c_d f_d\}.$$

With this notion, Theorem 10 can be restated as follows: $(\mathbb{E}, \sigma)$ is a Σ-extension of $(\mathbb{H}, \sigma)$ if and only if $V_1(1, (f_1, \dots, f_d), \mathbb{H}) = \{0\}^d \times \mathbb{K}$. In order to check that this is the case, we can utilize the following theorem.

Theorem 11 *Let $(\mathbb{H}, \sigma)$ be a basic $R\Pi\Sigma$-extension of a difference field $(\mathbb{F}, \sigma)$ with constant field $\mathbb{K}$, $u \in [\mathbb{F}^*]_{\mathbb{F}}^{\mathbb{H}}$ and $\boldsymbol{f} \in \mathbb{H}^d$. Then the following holds:*

1. *$V_1(u, \boldsymbol{f}, \mathbb{H})$ is a $\mathbb{K}$-vector space of dimension $\leq d + 1$.*

2. *If one can compute a basis of* $M(\boldsymbol{h}, \mathbb{F})$ *for any* $\boldsymbol{h} \in (\mathbb{F}^*)^n$ *and a basis of* $V_1(v, \boldsymbol{h}, \mathbb{F})$ *for any* $v \in \mathbb{F}^*$, $\boldsymbol{h} \in \mathbb{F}^n$, *then one can compute a basis of* $V_1(u, \boldsymbol{f}, \mathbb{H})$.

Proof Lemma 2.17 and Thm. 2.22 of [116] gives (1); [116, Thm. 2.23][9] shows (2). □

In particular, we can activate this machinery if $(\mathbb{F}, \sigma)$ is a $\Pi\Sigma$-field over a σ-computable constant field; a special case is, e.g., $\mathbb{F} = \mathbb{G}_m$.

Corollary 4 *Let* $(\mathbb{E}, \sigma)$ *be an* $R\Pi\Sigma$*-extension of a* $\Pi\Sigma$*-field* $(\mathbb{F}, \sigma)$ *over* $\mathbb{K}$. *If* $\mathbb{K}$ *is* σ*-computable, one can compute a basis of* $V_1(1, \boldsymbol{f}, \mathbb{E})$ *for any* $\boldsymbol{f} \in (\mathbb{E}^*)^d$. *This in particular is the case, if* $\mathbb{K}$ *is a rational function field over an algebraic number field.*

Proof If $\mathbb{K}$ is σ-computable, it follows by Karr [65] (or [104]) that one can compute a basis of $V_1(u, \boldsymbol{f}, \mathbb{F})$ for any $u \in \mathbb{F}^*$, $\boldsymbol{f} \in (\mathbb{F}^*)^d$. Thus by part 2 of Theorem 11 one can compute a basis of $V_1(1, \boldsymbol{h}, \mathbb{E})$ for any $\boldsymbol{h} \in (\mathbb{E}^*)^d$. In particular, it follows by Schneider [104, Thm. 3.5] (based on the algorithm of [59]) that a rational function field over an algebraic number field is σ-computable. □

Remark 6 (1) By [116, Thm. 2.26], Corollary 4 is also valid for $\boldsymbol{f} \in (\{\mathbb{F}^*\}_{\mathbb{F}}^{\mathbb{E}})^d$ in simple $R\Pi\Sigma$-extensions over a $\Pi\Sigma$-field. As elaborated in [116, Sect. 2.3.3] it holds even in the more general setting where $(\mathbb{F}, \sigma)$ is a $\Pi\Sigma$-field extension of a difference field $(\mathbb{F}_0, \sigma)$ which is σ^*-computable (see [69, Def. 1]) and one can compute a basis of $V(u, \boldsymbol{f})$ in $(\mathbb{F}_0, \sigma^k)$ for any[10] $k > 0$, $u \in \mathbb{F}^*$ and $\boldsymbol{f} \in \mathbb{F}_0^m$; see also Remark 4.(1).

4.2.2 Basic Representations

The following theorem (based on Theorem 10 and the property that one can solve the telescoping problem (26) given below) enables one to lift the results of $\text{Prod}_1(\mathbb{G})$ and $\text{Prod}(\mathbb{G}_r)$ form Sect. 4.1 to the cases $\text{Sum}(\mathbb{G})$, $\text{SumProd}_1(\mathbb{G})$ and $\text{SumProd}(\mathbb{G}_r)$.

Theorem 12 *Let* $\mathbb{G} \in \{\mathbb{G}_r, \mathbb{G}_b, \mathbb{G}_m\}$ *and* $A_1, \ldots, A_u \in \textit{SumProd}(\mathbb{G})$. *Let* $(\mathbb{H}, \sigma)$ *be a basic* $R\Pi\Sigma$*-extension of* $(\mathbb{G}, \sigma)$ *equipped with an evaluation function* ev *and an o-function* L *where all arising products in* $A_1, \ldots, A_u$ *can be modeled. Then there is a* Σ*-extension* $(\mathbb{E}, \sigma)$ *of* $(\mathbb{H}, \sigma)$ *with an extended evaluation function* ev *and o-function* L *(using Lemmas 2 and 3) such that* $a_1, \ldots, a_u \in \mathbb{E}$ *model* $A_1, \ldots, A_u$, *i.e., for all* $1 \leq i \leq u$ *we have* (24) *for some explicitly given* $\delta_i \in \mathbb{Z}_{\geq 0}$ *with* $\delta_i \geq L(a_i)$.

---

[9] For an alternative algorithm we refer to [118, Section 6].

[10] If the extension is basic, we only need the case $k = 1$.

If $\mathbb{K}$ is σ-computable, and $L : \mathbb{H} \to \mathbb{Z}_{\geq 0}$ and ev $: \mathbb{H} \times \mathbb{Z}_{\geq 0} \to \mathbb{K}$ *are computable, the above components can be computed.*

Proof This result follows from the construction given in [118, pp. 657–658] which can be summarized as follows. We suppose that we have constructed already a basic $R\Pi\Sigma$-extension of $(\mathbb{G}, \sigma)$ equipped with an evaluation function ev and an o-function L where all arising products in $A_1, \ldots, A_u$ can be modeled. Then we can adapt the construction of Lemma 6 and deal with all arising sums and products arising in the $A_1, \ldots, A_u$. Suppose that we have constructed already a Σ-extension $(\mathbb{A}, \sigma)$ of $(\mathbb{G}, \sigma)$ and we are treating now the product or sum T_i. If it is a product, we sort it out in the bookkeeping step and obtain an element $b_i \in \mathbb{H}^* \subseteq \mathbb{E}^*$ that models T_i by assumption. Otherwise, $T_i = \text{Sum}(\lambda, H)$. By induction (on the depth of the arising sums) we can construct a Σ-extension $(\mathbb{A}', \sigma)$ of $(\mathbb{A}, \sigma)$ together with an extended evaluation function ev and o-function L such that we can take $h \in \mathbb{A}'$ with $\text{ev}(h, n) = H(n)$ for all $n \geq L(h)$. Now we enter the sum-case and perform the following extra test. We check if there is a $g \in \mathbb{A}'$ with

$$\sigma(g) = g + f \quad \Leftrightarrow \quad \sigma(g) - g = f \tag{26}$$

for $f := \sigma(h)$. Suppose there is such a g. We define $\delta_i := \max(L(f), L(g), \lambda)$. Then for $b_i := g + \sum_{j=\lambda}^{\delta_i} H(j) - \text{ev}(g, \delta_i) \in \mathbb{A}'$ we get $\text{ev}(b_i, n+1) - \text{ev}(b_i, n) = \text{ev}(g, n+1) - \text{ev}(g, n) = H(n+1)$ and $T_i(n+1) = T_i(n) + H(n+1)$ for all $n \geq \delta_i$. Since $\text{ev}(b_i, \delta_i) = \sum_{j=\lambda}^{\delta_i} F(j) = \text{ev}(T_i, \delta_i)$, we get $\text{ev}(b_i, n) = \text{ev}(T_i, n)$ for all $n \geq \delta_i$.
Otherwise, if there is no such g, we proceed as in the sum-case of Lemma 6: we adjoin the Σ-monomial t to $\mathbb{A}'$ with $\sigma(t) = t + f$ with $f = \sigma(h)$ and get the claimed $b_i = t + c$ with $c \in \mathbb{K}$ such that $\text{ev}(b_i, n) = \text{ev}(T_i, n)$ holds for all $n \geq L(b_i) = \delta_i$.
Summarizing, we can construct a nested Σ-extension in which the elements from SumProd($\mathbb{G}$) can be modeled. If $\mathbb{K}$ is σ-computable, one can decide constructively by Corollary 4 if there exists such a g. Furthermore, if $L : \mathbb{H} \to \mathbb{Z}_{\geq 0}$ and ev $: \mathbb{H} \times \mathbb{Z}_{\geq 0} \to \mathbb{K}$ are computable also their extensions for $(\mathbb{E}, \sigma)$ are computable by recursion. Consequently, all components are computable. □

We get immediately the following result for Sum($\mathbb{G}$)-stable extensions.

Corollary 5 *Let $\mathbb{G} \in \{\mathbb{G}_r, \mathbb{G}_b, \mathbb{G}_m\}$. The class of Σ-extensions over $(\mathbb{G}, \sigma)$ is Sum($\mathbb{G}$)-extension-stable. More precisely, let $(\mathbb{H}, \sigma)$ be a Σ-extension of $(\mathbb{G}, \sigma)$ with an evaluation function* ev *and an o-function L, and let $A \in$ Sum($\mathbb{G}$). Then there is a Σ-extension $(\mathbb{E}, \sigma)$ of $(\mathbb{H}, \sigma)$ with an extended evaluation function* ev *and an o-function L (using Lemmas 2 and 3) together with $a \in \mathbb{E}$ and $\delta \in \mathbb{Z}_{\geq 0}$ with $A(n) = \text{ev}(a, n)$ for all $n \geq \delta$. If $\mathbb{K}$ is σ-computable, these components can be computed.*

Combining Theorems 8 and 12 we get `Sigma`'s main translation mechanism.

Corollary 6 *Let $\mathbb{G} \in \{\mathbb{G}_r, \mathbb{G}_b, \mathbb{G}_m\}$ where $\mathbb{K}$ is completely factorizable of order λ. Then the class of completely factorized $R\Pi\Sigma$-extensions is SumProd$_1(\mathbb{G})$-extension-stable. More precisely, let $(\mathbb{H}, \sigma)$ be a completely factorized $R\Pi\Sigma$-extension of $(\mathbb{G}, \sigma)$ equipped with an evaluation function* ev *an o-function L. Let $A \in$ SumProd$_1(\mathbb{G})$. Then there is an $R\Pi\Sigma$-extension $(\mathbb{E}, \sigma)$ of $(\mathbb{H}, \sigma)$ with an extended evaluation function* ev *and o-function L (using Lemmas 2 and 3) with the following properties:*

1. *$(\mathbb{E}, \sigma)$ is a completely factorizable $R\Pi\Sigma$-extension of $(\mathbb{G}, \sigma)$.*
2. *A is modeled by $a \in \mathbb{E}$, i.e., $A(n) = \mathrm{ev}(a, n)$ for all $n \geq \delta$ for some $\delta \in \mathbb{Z}_{\geq 0}$.*

If $\mathbb{K}$ is σ-computable and complete factorizations over $\mathbb{K}$ can be computed, all the components can be given explicitly.

Proof We can write $\mathbb{H} = \mathbb{G}\langle t_1\rangle \dots \langle t_e\rangle[s_1, \dots, s_u]$ where the t_i are $R\Pi$-monomials and the s_i are Σ-monomials. Take all products that arise in A. Since $(\mathbb{H}_0, \sigma)$ with $\mathbb{H}_0 = \mathbb{G}\langle t_1\rangle \dots \langle t_e\rangle$ is a completely factorized $R\Pi$-extension of $(\mathbb{G}, \sigma)$, we can apply Theorem 8 and get an $R\Pi$-extension $(\mathbb{H}_1, \sigma)$ of $(\mathbb{H}_0, \sigma)$ with $\mathbb{H}_1 = \mathbb{H}_0\langle p_1\rangle \dots \langle p_v\rangle$ together with an extended evaluation function ev and o-function L such that $(\mathbb{H}_1, \sigma)$ is a completely factorized $R\Pi$-extension of $(\mathbb{G}, \sigma)$ and such that all products in A can be modeled in $\mathbb{H}_1$. By Schneider [118, Cor. 6.5] (together with [118, Prop 3.23]) it follows that also $(\mathbb{H}_2, \sigma)$ with $\mathbb{H}_2 = \mathbb{H}\langle p_1\rangle \dots \langle p_v\rangle$ is a Π-extension of $(\mathbb{H}, \sigma)$. In particular, $(\mathbb{H}_2, \sigma)$ is a completely factorized $R\Pi\Sigma$-extension of $(\mathbb{G}, \sigma)$ and we can merge the evaluation functions and o-functions to $\mathrm{ev} : \mathbb{H}_2 \times \mathbb{Z}_{\geq 0} \to \mathbb{K}$ and $L : \mathbb{H}_2 \to \mathbb{Z}_{\geq 0}$. Finally, we apply Theorem 12 and get a Σ-extension $(\mathbb{E}, \sigma)$ of $(\mathbb{H}_2, \sigma)$ with an appropriately extended evaluation function ev and o-function L together with $a \in \mathbb{E}$ and $\delta \in \mathbb{Z}_{\geq 0}$ such that $\mathrm{ev}(a, n) = A(n)$ holds for all $n \geq \delta$. By definition $(\mathbb{E}, \sigma)$ is a completely factorized $R\Pi\Sigma$-extension of $(\mathbb{G}, \sigma)$.
If $\mathbb{K}$ is σ-computable and one can compute complete factorizations over $\mathbb{K}$, Theorems 8 and 12 are constructive and all components can be computed. □

Furthermore, combining Theorems 7 and 12 gives the following result (we omit the optimality properties given in Theorem 7).

Corollary 7 *Let $\mathbb{G} \in \{\mathbb{G}_r, \mathbb{G}_b, \mathbb{G}_m\}$ where $\mathbb{K}$ is built by a rational function field defined over an algebraic number field. Then for $A_1, \dots, A_u \in$ SumProd$_1(\mathbb{G})$ there is a single-basic $R\Pi\Sigma$-extension $(\mathbb{E}, \sigma)$ of $(\mathbb{G}, \sigma)$ together with an extended evaluation function* $\mathrm{ev} : \mathbb{E} \times \mathbb{Z}_{\geq 0} \to \mathbb{K}$ *and o-function* $L : \mathbb{E} \to \mathbb{Z}_{\geq 0}$ *(using Lemmas 2 and 3) with the following properties: $A_1, \dots, A_u$ are modeled by $a_1, \dots, a_u \in \mathbb{E}$, i.e., for all $1 \leq i \leq u$ we have* (24) *for some explicitly given $\delta_i \in \mathbb{Z}_{\geq 0}$ with $\delta_i \geq L(a_i)$.*

In addition, the applications of Theorems 9 and 12 yield the following statement.

Corollary 8 *Let $\mathbb{G}_r = \mathbb{K}(x)$ with $\mathbb{K} = \mathcal{A}(y_1, \dots, y_o)$ be a rational function field over an algebraic number field $\mathcal{A}$. Then for $A_1, \dots, A_u \in$ SumProd$(\mathbb{G})$ there is a basic $R\Pi$-extension $(\mathbb{E}, \sigma)$ of $(\mathbb{G}'_r, \sigma)$ with an evaluation function* $\mathrm{ev} : \mathbb{E} \times \mathbb{Z}_{\geq 0} \to$

$\mathbb{K}'$ and o-function $L : \mathbb{E} \to \mathbb{Z}_{\geq 0}$ (using Lemmas 2 and 3) with the following properties:

1. *The ground field $\mathbb{G}_r$ is extended to $\mathbb{G}'_r = \mathbb{K}'(x)$ where $\mathbb{K}' = \mathcal{A}'(y_1, \ldots, y_o)$ with $\mathcal{A}'$ being an algebraic field extension of $\mathcal{A}$.*
2. *Within the $R\Pi\Sigma$-monomials in $(\mathbb{E}, \sigma)$ there is at most one R-monomial.*
3. *$A_1, \ldots, A_u$ are modeled by $a_1, \ldots, a_u \in \mathbb{E}$, i.e., for all $1 \leq i \leq u$ we have (24) for some explicitly given $\delta_i \in \mathbb{Z}_{\geq 0}$ with $\delta_i \geq L(a_i)$.*

In particular, activating our method **`SigmaReduce`** in combination with Corollaries 6 and 7 establishes Theorems 3 and 3, respectively.

Most of the above results are implemented within the summation package `Sigma` or are available by using in addition the package `NestedProducts`. Further details can be found in the following remark.

Technical Details of the Summation Package `Sigma`

Remark 7

(1) Within `Sigma` the function call `SigmaReduce` follows the method given on page 451. Note that in this construction the σ-reduced set W is constructed by treating stepwise the sums and products that occur in the A_i.

(2) The user can control the σ-reduced set W manually by introducing extra sums and products with the option `Tower`$\to$ $\{S_1, \ldots, S_v\}$ that will be parsed before the arising sums in $A_1, \ldots, A_u$ are considered; as an example we refer to In[20] in Example 19.

(3) `Sigma` is tuned for expressions from $\text{SumProd}_1(\mathbb{G})$ where the constant field $\mathbb{K}$ is a completely factorizable field. In particular for the case that $\mathbb{K}$ is a rational function field over the rational numbers, the machinery given in Corollary 6 is highly robust. `Sigma` also works partially with rational function fields over algebraic number fields; but here it depends on the stability of the subroutines in Mathematica.

(4) For nested products the machinery of `SigmaReduce` works if the objects can be transformed straightforwardly to $R\Pi\Sigma$-extensions. For more complicated situations the objects $\text{SumProd}(\mathbb{G}_r)$ can be handled fully algorithmically in combination with Ocansey's package `NestedProducts`.

Remark 8 As observed in [36] an algebraically independent basis of certain classes of indefinite nested sums can be obtained by exploiting the underlying quasi-shuffle algebra. In [36] this aspect has been utilized for the class of harmonic sums, and it has been enhanced for generalized, cyclotomic and binomial sums in [2, 4, 15]. Later it has been shown in [16] that the relations in the class of cyclotomic harmonic sums produced by difference ring theory (compare Theorem 1) and by the quasi-shuffle algebra are equivalent. As a consequence, the quasi-shuffle algebra of cyclotomic sums induces a canonical representation. We emphasize that many of the above aspects can be carried over to a summation theory of unspecified sequences [89].

4.2.3 Depth-Optimal Representations

In [102, 108] we have refined Karr's definition of $\Pi\Sigma$-field extensions to depth-optimal $\Pi\Sigma$-field extensions and have developed improved telescoping algorithms therein. In this way, we could provide a general toolbox in [111] that can find representations such that the nesting depths of the arising sums are minimal. As it turns out, the underlying telescoping algorithms can be adapted (and even simplified) for $R\Pi\Sigma$-extensions. For the specification of the refined representation (without entering into technical details) we need the following definition.

Definition 19 A finite set $W \subset \Sigma\Pi(\mathbb{G})$ is called *depth-optimal* if for any $G \in \text{SumProd}(W, \mathbb{G})$ and $G' \in \text{SumProd}(\mathbb{G})$ with $G(n) = G'(n)$ for all $n \geq \delta$ for some $\delta \in \mathbb{Z}_{\geq 0}$ it follows that $\delta(G) \leq \delta(G')$ holds.

Then combining the results from Sect. 4.2.2 with the tools from [102, 108, 111] we obtain algorithms that can solve the following problem if $\mathbb{K}$ is σ-computable; for simplicity we skipped the general case SumProd($\mathbb{G}$). Further technical details concerning the implementation in `Sigma` can be found in Remark 7.

Problem DOS: Depth-Optimal SigmaReduce

Given: $A_1, \ldots, A_u \in \text{SumProd}_1(\mathbb{G}_m)$.
Find: a finite σ-reduced depth-optimal set $W \subset \Sigma\Pi_1(\mathbb{G}_m)$ together with $B_1, \ldots, B_u \in \text{SumProd}(W, \mathbb{G}_m)$ and $\delta_1, \ldots, \delta_u \in \mathbb{Z}_{\geq 0}$ such that $A_i(n) = B_i(n)$ holds for all $n \geq \delta_i$ and $1 \leq i \leq u$

Example 18 Given the sums $A_1, A_2, A_3 \in \text{Sum}(\mathbb{Q}(x))$ defined by

In[15]:= $$\{\mathbf{A_1}, \mathbf{A_2}, \mathbf{A_3}\} = \left\{\sum_{k=1}^{n}\frac{\left(\sum_{i=1}^{k}\frac{1}{i^2}\right)\sum_{i=1}^{k}\frac{(-1)^i}{i}}{1+k}, \sum_{k=1}^{n}\frac{\left(\sum_{i=1}^{k}\frac{1}{i^2}\right)\sum_{i=1}^{k}\frac{(-1)^i}{i}}{2+k}, \sum_{k=1}^{n}\frac{\left(\sum_{i=1}^{k}\frac{1}{i^2}\right)\sum_{i=1}^{k}\frac{(-1)^i}{i}}{3+k}\right\};$$

we get the alternative expressions $B_1, B_2, B_3 \in \text{SumProd}(W, \mathbb{Q}(x))$ by executing

In[16]:= $\{\mathbf{B_1}, \mathbf{B_2}, \mathbf{B_3}\} = \mathbf{SigmaReduce}[\{\mathbf{A_1}, \mathbf{A_2}, \mathbf{A_3}\}, \mathbf{n}]$

Out[16]= $$\left\{\sum_{k=1}^{n}\frac{\left(\sum_{i=1}^{k}\frac{1}{i^2}\right)\sum_{i=1}^{k}\frac{(-1)^i}{i}}{1+k}, \sum_{k=1}^{n}\frac{\left(\sum_{i=1}^{k}\frac{1}{i^2}\right)\sum_{i=1}^{k}\frac{(-1)^i}{i}}{2+k},\right.$$
$$\frac{3}{16}+\frac{(-3-2n)(-1)^n}{8(1+n)(2+n)}+\frac{(-1)^n}{2(2+n)}\sum_{i=1}^{n}\frac{1}{i^2}+\frac{1}{2}\sum_{i=1}^{n}\frac{(-1)^i}{i^2}+\frac{-3+2n+2n^2}{4(1+n)(2+n)}\sum_{i=1}^{n}\frac{(-1)^i}{i}$$
$$\left.-\frac{(1+n)(5+2n)}{2(2+n)(3+n)}\left(\sum_{i=1}^{n}\frac{1}{i^2}\right)\sum_{i=1}^{n}\frac{(-1)^i}{i}+\frac{1}{2}\sum_{i=1}^{n}\frac{\left(\sum_{j=1}^{i}\frac{1}{j^2}\right)\sum_{j=1}^{i}\frac{(-1)^j}{j}}{1+i}+\frac{1}{2}\sum_{i=1}^{n}\frac{\left(\sum_{j=1}^{i}\frac{1}{j^2}\right)\sum_{j=1}^{i}\frac{(-1)^j}{j}}{2+i}\right\}$$

with the σ-reduced set

$$W = \Big\{\sum_{k=1}^{n}\frac{1}{k^2}, \sum_{k=1}^{n}\frac{(-1)^k}{k}, \sum_{i=1}^{n}\frac{(-1)^i}{i^2}, \sum_{k=1}^{n}\frac{\Big(\sum_{i=1}^{k}\frac{1}{i^2}\Big)\sum_{i=1}^{k}\frac{(-1)^i}{i}}{1+k}, \sum_{k=1}^{n}\frac{\Big(\sum_{i=1}^{k}\frac{1}{i^2}\Big)\sum_{i=1}^{k}\frac{(-1)^i}{i}}{2+k}\Big\}.$$

Note: instead of A_3 (a sum of nesting depth 3) the simpler sum $\sum_{i=1}^{n}\frac{(-1)^i}{i^2}$ (with nesting depth 2) has been introduced automatically.

Remark 9 Further refined $\Pi\Sigma$-extensions, such as reduced $\Pi\Sigma$-extensions, have been elaborated in [110] (based on improved telescoping algorithms given in [100, 115]).

5 The Summation Paradigms

We have explained in detail how sums and products can be modeled automatically within $R\Pi\Sigma$-extensions. Thus steps 1 and 3 on page (1) are settled and we focus on step 2: We will introduce the summation paradigms in difference rings and fields; further details how these problems are handled in `Sigma` are given below.

5.1 *Refined Telescoping*

As indicated in Sect. 4.2.2, in particular in Theorem 12, the construction of basic $R\Pi\Sigma$-extensions for the representation of SumProd($\mathbb{G}$) is based on algorithms that solve the telescoping problem (26). In particular, the quality of the constructed extensions and the used telescoping algorithms are mutually intertwined. As illustrated for instance in Sect. 4.2.3, the underlying telescoping algorithms could be refined further (using [102, 108, 111]) to compute depth-optimal representations.

In the following we will focus on the available telescoping technologies in `Sigma` (based on [100, 102, 106, 108–111, 115]) that enable one to simplify sums further. For simplicity we will focus on sums from $\Sigma\Pi_1(\mathbb{G}_m)$ and skip, e.g., the case $\Sigma\Pi(\mathbb{G}_r)$.

Problem RT: Refined Telescoping

Given: $F \in \text{SumProd}_1(\mathbb{G}_m)$.
Find: $\delta \in \mathbb{Z}_{\geq 0}$ and a σ-reduced set $W = \{T_1, \ldots, T_e\} \subset \Sigma\Pi_1(\mathbb{G}_m)$ where $\mathrm{d}(T_1) \leq \mathrm{d}(T_2) \leq \cdots \leq \mathrm{d}(T_e)$ together with $F', G \in \text{SumProd}(W, \mathbb{G}_m)$ such that for all $k \geq \delta$ we have $F(k) = F'(k)$ and

$$G(k+1) - G(k) = F'(k).$$

- Refinement 1: W is depth-optimal (by using `SimplifyByExt→MinDepth`).
- Refinement 2: In addition, if $\mathrm{d}(G) = \mathrm{d}(F')+1$, then $\mathrm{d}(T_{e-1}) < \mathrm{d}(T_e) = \mathrm{d}(G)$ and $T_e = \mathrm{Sum}(\delta, H)$ with $H \in \mathrm{SumProd}(\{T_1, \dots, T_i\}, \mathbb{G}_m)$ where i with $1 \leq i < e$ is minimal (by using `SimplifyByExt→DepthNumber`).
- Refinement 3: One can compute, among all possible choices with i minimal, H such that also $\deg_{T_i}$ is minimal (by using `SimplifyByExt→DepthNumberDegree`).

Given such G and $\delta \in \mathbb{Z}_{\geq 0}$ for F we obtain the simplification (2) for all $n \geq \delta$.

Example 19 We start with the following sum:

In[17]:= $\mathbf{mySum1} = \sum_{k=1}^{n}\Big(\sum_{j=1}^{k}\frac{(-1)^j}{j^2}\Big)\Big(\sum_{j=1}^{k}\frac{(-1)^j}{j}\Big)^2;$

Telescoping without any refinements (by setting `SimplifyByExt→None`) does not yield a simplification. However, by activating the first refinement with the option `SimplifyByExt→MinDepth` (which actually is the default option) we get

In[18]:= **SigmaReduce[mySum1, n, SimplifyByExt → MinDepth]**

Out[18]= $$\frac{1}{3}\sum_{i=1}^{n}\frac{(-1)^i}{i^3} + (-1)^{1+n}\Big(\sum_{j=1}^{n}\frac{(-1)^j}{j^2}\Big)\sum_{j=1}^{n}\frac{(-1)^j}{j} + (1+n)\Big(\sum_{j=1}^{n}\frac{(-1)^j}{j^2}\Big)\Big(\sum_{j=1}^{n}\frac{(-1)^j}{j}\Big)^2 - \frac{1}{3}\Big(\sum_{j=1}^{n}\frac{(-1)^j}{j}\Big)^3$$

We illustrate the second refinement with the sum:

In[19]:= $\mathbf{mySum2} = \sum_{k=1}^{n}\Big(\sum_{j=1}^{k}\frac{(-1)^j}{j^2}\Big)\Big(\sum_{j=1}^{k}\frac{(-1)^j}{j}\Big)^3;$

In[20]:= **SigmaReduce[mySum2, n, SimplifyByExt → DepthNumber,**
Tower → $\Big\{\sum_{i=1}^{n}\frac{(-1)^i}{i}, \sum_{i=1}^{n}\frac{(-1)^i}{i^2}\Big\}$
SimpleSumRepresentation → False]

Out[20]= $$\frac{1}{4}\Big(\sum_{j=1}^{n}\frac{(-1)^j}{j^2}\Big)^2 - \frac{3}{2}(-1)^n\Big(\sum_{j=1}^{n}\frac{(-1)^j}{j^2}\Big)\Big(\sum_{j=1}^{n}\frac{(-1)^j}{j}\Big)^2 + (1+n)\Big(\sum_{j=1}^{n}\frac{(-1)^j}{j^2}\Big)\Big(\sum_{j=1}^{n}\frac{(-1)^j}{j}\Big)^3 - \frac{1}{4}\sum_{i=1}^{n}\Big(\frac{1}{i^4} - \frac{6\Big(\sum_{j=1}^{i}\frac{(-1)^j}{j}\Big)^2}{i^2} + \frac{4(-1)^i\Big(\sum_{j=1}^{i}\frac{(-1)^j}{j}\Big)^3}{i}\Big)$$

Namely, within the given extension (specified by `Tower→` $\{\sum_{i=1}^{n}\frac{(-1)^i}{i}, \sum_{i=1}^{n}\frac{(-1)^i}{i^2}\}$, compare Remark 7) we find a sum extension which is free of $\sum_{i=1}^{n}\frac{(-1)^i}{i^2}$. Without the option **SimpleSumRepresentation→False** further simplifications on the found sum (using in addition partial fraction decomposition) are applied and one gets:

In[21]:= **SigmaReduce[mySum2, n, SimplifyByExt → DepthNumber, Tower → $\{\sum_{i=1}^{n}\frac{(-1)^i}{i}, \sum_{i=1}^{n}\frac{(-1)^i}{i^2}\}$]**

Out[21]= $$-\frac{1}{4}\sum_{i=1}^{n}\frac{1}{i^4}+\frac{1}{4}\Big(\sum_{j=1}^{n}\frac{(-1)^j}{j^2}\Big)^2-\frac{3}{2}(-1)^n\Big(\sum_{j=1}^{n}\frac{(-1)^j}{j^2}\Big)\Big(\sum_{j=1}^{n}\frac{(-1)^j}{j}\Big)^2 +(1+n)\Big(\sum_{j=1}^{n}\frac{(-1)^j}{j^2}\Big)\Big(\sum_{j=1}^{n}\frac{(-1)^j}{j}\Big)^3+\frac{3}{2}\sum_{i=1}^{n}\frac{\Big(\sum_{j=1}^{i}\frac{(-1)^j}{j}\Big)^2}{i^2}-\sum_{i=1}^{n}\frac{(-1)^i\Big(\sum_{j=1}^{i}\frac{(-1)^j}{j}\Big)^3}{i}$$

If one changes the order of the extension with the option **Tower→** $\{\sum_{i=1}^{n}\frac{(-1)^i}{i^2}, \sum_{i=1}^{n}\frac{(-1)^i}{i}\}$, no simplification is possible with the option **SimplifyByExt→DepthNumber**. However, using the option **SimplifyByExt→DepthNumberDegree** one finds a sum extension where in the summand the degree w.r.t. $T=\sum_{i=1}^{n}\frac{(-1)^i}{i}$ is minimal. In this case we find

In[22]:= **SigmaReduce[mySum2, n, SimplifyByExt → DepthNumberDegree, Tower → $\{\sum_{i=1}^{n}\frac{(-1)^i}{i^2}, \sum_{i=1}^{n}\frac{(-1)^i}{i}\}$ SimpleSumRepresentation → False]**

Out[22]= $$-\frac{3}{2}(-1)^n\Big(\sum_{j=1}^{n}\frac{(-1)^j}{j^2}\Big)\Big(\sum_{j=1}^{n}\frac{(-1)^j}{j}\Big)^2 + (1+n)\Big(\sum_{j=1}^{n}\frac{(-1)^j}{j^2}\Big)\Big(\sum_{j=1}^{n}\frac{(-1)^j}{j}\Big)^3 - \frac{1}{4}\Big(\sum_{j=1}^{n}\frac{(-1)^j}{j}\Big)^4+\frac{1}{4}\sum_{i=1}^{n}\Big(-\frac{3}{i^4}+\frac{2(-1)^i}{i^2}\sum_{j=1}^{i}\frac{(-1)^j}{j^2}+\frac{4(-1)^i}{i^3}\sum_{j=1}^{i}\frac{(-1)^j}{j}\Big)$$

where in the summand of the found sum the degree w.r.t. T is 1. With the option **SimpleSumRepresentation→True** (which is the standard option) this sum is simplified further (by splitting it into atomics by partial fraction decomposition) and we get:

In[23]:= **SigmaReduce[mySum2, n, SimplifyByExt → DepthNumberDegree, Tower → $\{\sum_{i=1}^{n}\frac{(-1)^i}{i^2}, \sum_{i=1}^{n}\frac{(-1)^i}{i}\}$]**

Out[23]= $$-\frac{1}{2}\sum_{i=1}^{n}\frac{1}{i^4}+\frac{1}{4}\Big(\sum_{j=1}^{n}\frac{(-1)^j}{j^2}\Big)^2-\frac{3}{2}(-1)^n\Big(\sum_{j=1}^{n}\frac{(-1)^j}{j^2}\Big)\Big(\sum_{j=1}^{n}\frac{(-1)^j}{j}\Big)^2 +(1+n)\Big(\sum_{j=1}^{n}\frac{(-1)^j}{j^2}\Big)\Big(\sum_{j=1}^{n}\frac{(-1)^j}{j}\Big)^3-\frac{1}{4}\Big(\sum_{j=1}^{n}\frac{(-1)^j}{j}\Big)^4+\sum_{i=1}^{n}\frac{(-1)^i}{i^3}\sum_{j=1}^{i}\frac{(-1)^j}{j}$$

5.2 Parameterized Telescoping (Including Creative Telescoping)

The summation paradigm of telescoping can be generalized as follows.

Problem PT: Parameterized Telescoping

Given: $F_1, \ldots, F_d \in \text{SumProd}(\mathbb{G})$ with $\mathbb{G} \in \{\mathbb{G}_r, \mathbb{G}_b, \mathbb{G}_m\}$.
Find: Find, if possible, a suitable σ-reduced finite set $W \subset \Sigma\Pi(\mathbb{G}')$ and $\delta \in \mathbb{Z}_{\geq 0}$ with the following properties; as in Problem SigmaReduce, one might have to extend the constant field $\mathbb{K}$ of $\mathbb{G}$ to $\mathbb{K}'$ yielding $\mathbb{G}'$.

- One can take $F'_1, \ldots, F'_d \in \text{SumProd}(W, \mathbb{G}')$ such that for all $1 \leq i \leq d$ and all $k \geq \delta$ we have $F_i(k) = F'_i(k)$;
- one can take $c_1, \ldots, c_d \in \mathbb{K}'$ with $c_1 \neq 0$ and $G \in \text{SumProd}(W, G')$ such that for all $k \geq \delta$ we have

$$G(k+1) - G(k) = c_1\, F'_1(k) + \cdots + c_d\, F'_d(k). \tag{27}$$

Given such $c_1, \ldots, c_d \in \mathbb{K}$, G and $\delta \in \mathbb{Z}_{\geq 0}$ for $F_1, \ldots, F_d$, we obtain

$$c_1 \sum_{k=\delta}^{n} F_1(k) + \cdots + c_d \sum_{k=\delta}^{n} F_d(k) = G(n+1) - G(\delta) \tag{28}$$

for all $n \geq \delta$. In particular, if one is given a bivariate sequence $F(n, k)$ with $F_i(k) = F(n+i-1, k) \in \text{SumProd}(\mathbb{G})$ for $i = 1, \ldots, d$, Eq. (27) turns into (3). In particular, the sum relation (28) can be transformed to the recurrence (4) for the sum $S(n) = \sum_{k=\delta}^{n} F(n, k)$. Summarizing, parameterized telescoping contains creative telescoping [134] as a special case.

A straightforward solution to the above problem can be obtained by the application of Theorem 13. In the context of σ-reduced sets this can be rephrased as follows.

Proposition 1 *Let $W = \{T_1, \ldots, T_e\} \subseteq \Sigma\Pi(\mathbb{G})$ be σ-reduced where for each $1 \leq i \leq e$ the arising sums and products within T_i are contained in $\{T_1, \ldots, T_{i-1}\}$ and are in sum-product reduced form. Let $F'_1, \ldots, F'_d \in$ SumProd$(W, \mathbb{G})$. Then one can compute, in case of existence, $(c_1, \ldots, c_d) \in \mathbb{K}^d$ with $c_1 \neq 0$ together with $G \in$ SumProd$(W, \mathbb{G})$ and $\delta \in \mathbb{Z}_{\geq 0}$ such that* (27) *holds for all $k \geq \delta$.*

Proof By Corollary 2 we get an $R\Pi\Sigma$-extension $(\mathbb{E}, \sigma)$ of $(\mathbb{G}, \sigma)$ with $\mathbb{E} = \mathbb{G}\langle t_1 \rangle \ldots, \langle t_e \rangle$ together with an evaluation function ev and o-function L with $\text{expr}(t_i) = T_i$ for all $1 \leq i \leq e$. In particular, we get $f = (f_1, \ldots, f_d) \in \mathbb{E}^d$ with $\text{ev}(f_i, k) = F'_i(k)$ for all $1 \leq i \leq u$ and all $n \geq L(f_i)$. Note that

$(c_1, \ldots, c_d, G) \in \mathbb{K}^d \times \text{SumProd}(W, \mathbb{G})$ with (27) for all $k \geq \delta$ for some $\delta \in \mathbb{Z}_{\geq 0}$ iff $(c_1, \ldots, c_d, g) \in \mathbb{K}^d \times \mathbb{E}$. By Theorem 13 we can compute a basis $V = V_1(1, \boldsymbol{f}, \mathbb{E})$ and can check if there is $(c_1, \ldots, c_d, g) \in V$ with $c_1 \neq 0$. If this is not the case, then there is no $(c_1, \ldots, c_d, G) \in \mathbb{K}^d \times \text{SumProd}(W, \mathbb{G})$ with $c_1 \neq 0$. Otherwise, we rephrase the result as $(c_1, \ldots, c_d, G) \in \mathbb{K}^d \times \text{SumProd}(W, \mathbb{G})$ such that (27) holds for all $k \geq \delta$ with $\delta = \max(L(F_1'), \ldots, L(F_u'), L(G))$. □

Remark 10 In Proposition 1 we assume that the input expressions from $\text{SumProd}(\mathbb{G})$ can be rephrased directly in an $R\Pi\Sigma$-extension. If this is not the case, the representation machinery has to be applied in a preprocessing step. To support this construction, the user can control the σ-reduced set W as outlined in the Remark 7.(2) above. But this should be done with care in order to avoid useless results. If W contains, e.g., $T_j \in \text{Sum}(l, F_1')$, one gets trivially $G = T_j$ and $(c_1, c_2, \ldots, c_d) = (1, 0, \ldots, 0)$.

Example 20 We activate Proposition 1 to apply Zeilberger's creative telescoping paradigm. Take the summand $F(n, k)$ defined in

In[24]:= $\mathbf{F} = \dfrac{(-1)^k}{k}\dbinom{n}{k}\sum_{i=1}^{k}\dfrac{1}{i}\sum_{j=1}^{i}\dfrac{1}{j+n};$

and define the definite sum

In[25]:= **definiteSum = SigmaSum[F, {k, 1, n}]**

Out[25]= $\sum_{k=1}^{n}\dfrac{(-1)^k}{k}\dbinom{n}{k}\sum_{i=1}^{k}\dfrac{1}{i}\sum_{j=1}^{i}\dfrac{1}{j+n}$

Then we can compute a linear recurrence for `SUM[n]=definiteSum` with the call

In[26]:= **rec = GenerateRecurrence[definiteSum, n, SimplifyByExt → None]**

Out[26]= $$\Big\{(1+n)^3(8+3n)^2\text{SUM}[n]+\big(-1692-4306n-4369n^2-2202n^3-549n^4-54n^5\big)\text{SUM}[n+1] + (7+3n)\big(554+1072n+764n^2+237n^3+27n^4\big)\text{SUM}[n+2] - 2(3+n)^2(5+2n)(5+3n)^2\text{SUM}[n+3] == \frac{808+2008n+2007n^2+1017n^3+261n^4+27n^5}{(2+n)^2(3+n)}\Big\}$$

Here `Sigma` searches for a solution of (3) with $d = 0, 1, 2, \ldots$ and finally computes a solution for $d = 3$. Internally, it takes the shifted versions $F(n+i, k)$ with $i = 0, 1, 2, 3$

In[27]:= **FList = {F, (F/.n → n + 1), (F/.n → n + 2), (F/.n → n + 3)};**

and rewrites the expressions in a σ-reduced representation:

In[28]:= **FListRed = SigmaReduce[FList, k]**

Out[28]= $$\Big\{\frac{(-1)^k}{k}\binom{n}{k}\sum_{i=1}^{k}\frac{1}{i}\sum_{j=1}^{i}\frac{1}{j+n},$$
$$(-1)^k\binom{n}{k}\Big(\frac{1}{(1+n)(1-k+n)(1+k+n)}+\frac{1}{k(-1+k-n)}\sum_{i=1}^{k}\frac{1}{i+n}$$
$$+\frac{(-1-n)}{k(-1+k-n)}\sum_{i=1}^{k}\frac{1}{i}\sum_{j=1}^{i}\frac{1}{n+j}\Big),\dots\Big\}$$

Here we have printed only the first two entries of the output list. Afterwards it activates Proposition 1 by executing the command

In[29]:= **ParameterizedTelescoping[FListRed, n]**

Out[29]= $\{\{\emptyset,\emptyset,\emptyset,\emptyset,1\},\{c_1,c_2,c_3,c_4,G\}\}$

The expressions c_1, c_2, c_3, c_4 and G equal

$$c_1 = -(1+n)^3(8+3n)^2,$$
$$c_2 = 1692 + 4306n + 4369n^2 + 2202n^3 + 549n^4 + 54n^5,$$
$$c_3 = -(7+3n)(554 + 1072n + 764n^2 + 237n^3 + 27n^4),$$
$$c_4 = 2(3+n)^2(5+2n)(5+3n)^2,$$
$$G(n,k) = (-1)^k\binom{n}{k}\Big(Q_1\sum_{i=1}^{k}\frac{1}{i}\sum_{j=1}^{i}\frac{1}{n+j}+Q_2\sum_{i=1}^{k}\frac{1}{i+n}+Q_3\Big)$$

for some $Q_1, Q_2, Q_3 \in \mathbb{Q}(n,k)$. Alternatively, `ParameterizedTelescoping[FList,k]` (without `SigmaReduce` as a preprocessing step) could be used. The same result could be produced with `CreativeTelescoping[definiteSum,n, SimplifyByExt→ None]`.

Finally, summing (3) with $d = 3$ over k from 0 to n yields the recurrence given in Out[26]. Note that the correctness of the solution (c_1, c_2, c_3, c_4, G) of (3) with $d = 4$ can be verified straightforwardly: Since W is σ-reduced, one simply has to plug in the solutions and checks that the left-hand and right-hand sides agree. Thus we have shown rigorously that the definite sum given in In[25] is a solution of Out[26].

In order to introduce refined methods, we need the following definition.

Definition 20 Let $W \subset \Sigma\Pi(\mathbb{G})$ be σ-reduced depth-optimal and $\boldsymbol{F}' = (F'_1,\dots,F'_d) \in \mathrm{SumProd}_1(W,\mathbb{G})^d$. W is called $\boldsymbol{F}'$*-one complete* if the following holds: If there is $(c_1,\dots,c_d,G) \in \mathbb{K}^d \times \mathrm{SumProd}_1(\mathbb{G})$ with $c_1 \neq 0$,

$\mathrm{d}(G) \leq \min(\mathrm{d}(F_1'), \ldots, \mathrm{d}(F_d'))$ such that (27) holds for all n sufficiently large, then there is $G' \in \mathrm{SumProd}_1(W, \mathbb{G})$ with the same c_i such that[11] (27) holds (G replaced by G') for all n sufficiently large.

Using the techniques from [100, 102, 106, 108–111, 115] the following refined parameterized telescoping techniques are available for the class $\mathrm{SumProd}_1(\mathbb{G}_m)$ over a σ-computable field $\mathbb{K}$; for simplicity we skip more general cases, like $\mathrm{SumProd}(\mathbb{G}_r)$.

Problem RPT: Refined Parameterized Telescoping

Given: $F_1, \ldots, F_d \in \mathrm{SumProd}_1(\mathbb{G}_m)$.
Find: $\delta \in \mathbb{Z}_{\geq 0}$ and a depth-optimal σ-reduced set $W = \{T_1, \ldots, T_e\} \subset \Sigma\Pi_1(\mathbb{G}_m)$ with $\mathrm{d}(T_1) \leq \mathrm{d}(T_2) \leq \cdots \leq \mathrm{d}(T_e)$ with the following properties:

- One gets $\boldsymbol{F'} = (F_1', \ldots, F_d') \in \mathrm{SumProd}(W, \mathbb{G}_m)^d$ such that $F_i(k) = F_i'(k)$ holds for all $1 \leq i \leq d$ and $k \geq \delta$.

In addition, based on the refinements given below, one obtains $(c_1, \ldots, c_d, G) \in \mathbb{K}^d \times \mathrm{SumProd}(W, \mathbb{G}_m)$ with $c_1 \neq 1$ such that (27) holds for all $k \geq \delta$.

- Refinement 1: W is $\boldsymbol{F'}$-one complete. Further, one can compute (it exists) such a solution with $\mathrm{d}(G) \leq \mathrm{d}(F_1')$ (by using **SimplifyByExt→MinDepth**).
- Refinement 2: If this is not possible, one gets $\mathrm{d}(G) = \mathrm{d}(F_1') + 1$ with the following extra property: $\mathrm{d}(T_{e-1}) < \mathrm{d}(T_e) = \mathrm{d}(G)$ and $T_e = \mathrm{Sum}(\delta, H)$ with $H \in \mathrm{SumProd}(\{T_1, \ldots, T_i\}, \mathbb{G}_m)$ where i with $1 \leq i < e$ is minimal (by using the option **SimplifyByExt→DepthNumber**).
- Refinement 3: One can compute, among all possible choices with i minimal, H such that also $\deg_{T_i}$ is minimal (by using **SimplifyByExt→ DepthNumberDegree**).

For technical details concerning `Sigma` we refer to Remarks 7 and 10 above.

Example 21 While the standard approach finds for the definite sum given in In[25] only a recurrence of order 3, the refined parameterized telescoping toolbox (refinement 1) computes a recurrence of order 1:

In[30]:= **GenerateRecurrence[definiteSum, n, SimplifyByExt → MinDepth]**

Out[30]= $$\left\{\mathrm{SUM}[n] - \mathrm{SUM}[n+1] == \frac{1}{(1+n)^3} - \frac{1}{2(1+n)^2}\sum_{i=0}^{n}\frac{(-1)^i\binom{n}{i}}{1+i+n} + \frac{1}{1+n}\sum_{i=1}^{n}\frac{(-1)^i\binom{n}{i}\sum_{j=1}^{i}\frac{1}{n+j}}{i}\right\}$$

[11] Since W is depth-optimal, it follows in particular that $\mathrm{d}(G') \leq \mathrm{d}(G)$.

by introducing in addition the sum $\sum_{i=0}^{n}\frac{(-1)^i\binom{n}{i}}{1+i+n}$. The right-hand side is given by definite sums which are simpler than the input sum. In this situation, they can be simplified further to

$$\frac{1}{(1+n)^3}-\frac{1}{2(1+n)^2(1+2n)}\frac{1}{\binom{2n}{n}}+\frac{1}{1+n}\sum_{i=1}^{n}\frac{1}{i^2}-\frac{3}{1+n}\sum_{i=1}^{n}\frac{1}{i^2\binom{2i}{i}}$$

in $\mathrm{SumProd}_1(\mathbb{Q}(x))$ by applying again the creative telescoping paradigm plus recurrence solving (which we will introduce in the next subsection).

This refined version turns out to be highly valuable in concrete applications. First, one can discover in many problems the minimal recurrence relation. Sometimes this enables one even to read off hypergeometric series solutions, like, e.g., in [88]. In addition, the calculation of such recurrences of lower order is more efficient, and the extra time to simplify the more complicated right hand sides is often negligible. In applications from particle physics, like in [13, 14], the standard approach is even out of scope and only our improved methods produced the desired results.

Remark 11

(1) Structural theorems (together with algorithmic versions) that are strongly related to Liouville's theorem of integration [79, 96] can be found in [110].
(2) Based on Theorems 1 and 10 additional aspects of the algebraic independence of indefinite nested sums (related to [62]) are worked out in [109] and [118, Section 7.2]. Namely, if there is no solution of a parameterized telescoping problem (in particular of a creative telescoping problem), then the indefinite sums defined over these parameters are algebraically independent.

5.3 Recurrence Solving

Finally, we turn to difference ring algorithms that solve parameterized higher-order linear difference equations. Let $(\mathbb{A},\sigma)$ be a difference ring with constant field $\mathbb{K}$, $\boldsymbol{a}=(a_0,\dots,a_m)\in\mathbb{A}^{m+1}$ and $\boldsymbol{f}=(f_1,\dots,f_d)\in\mathbb{A}^d$. Then we define [65]

$$V(\boldsymbol{a},\boldsymbol{f},\mathbb{A})=\{(c_1,\dots,c_d,g)\in\mathbb{K}^d\times\mathbb{A}\mid \\ a_m\,\sigma^m(g)+\dots+a_1\,\sigma(g)+a_0\,g=c_1\,f_1+\dots+c_d\,f_d\};$$

note that we have $V((-u,1),\boldsymbol{f},\mathbb{A})=V_1(u,\boldsymbol{f},\mathbb{A})$.

In `Sigma` algorithms are available to solve parameterized linear difference equations that are based on the following theorem.

Theorem 13 *Let* $(\mathbb{E}, \sigma)$ *be a basic* $R\Pi\Sigma$*-extension of a* $\Pi\Sigma$*-field* $(\mathbb{F}, \sigma)$ *over* $\mathbb{K}$, $\mathbf{0} \neq \boldsymbol{a} = (a_0, \ldots, a_m) \in \mathbb{F}^{m+1}$ *and* $\boldsymbol{f} \in \mathbb{E}^d$. *Then the following holds:*

1. $V(\boldsymbol{a}, \boldsymbol{f}, \mathbb{E})$ *is a* $\mathbb{K}$*-vector space of dimension* $\leq m + d$.
2. *If* $\mathbb{K}$ *is* σ*-computable, then one can compute a basis of* $V(\boldsymbol{a}, \boldsymbol{f}, \mathbb{E})$.

Proof (1) follows by a slight variant of [98, Prop 3.1.1] and [57, Thm. XII (page 272)]. By Abramov et al. [26, Theorem 9] (based on [20, 49, 91, 99, 101, 105]) the statement (2) holds for $\boldsymbol{f} \in \mathbb{F}^n$. Thus with [17] statement (2) holds also for $\boldsymbol{f} \in \mathbb{E}^n$. □

In addition, `Sigma` contains a solver that finds all hypergeometric solutions in the setting of $\Pi\Sigma$-fields. This result follows by Theorems 9 and 10 of [26], which can be considered as the differential version of Singer's celebrated algorithm [123] that finds Liouvillian solutions of linear differential equations with Liouvillian coefficients.

Theorem 14 *Let* $(\mathbb{F}, \sigma)$ *be a* $\Pi\Sigma$*-field over a* σ*-computable* $\mathbb{K}$. *Let* $a_0, \ldots, a_m \in \mathbb{F}$ *with* $a_0 a_m \neq 0$. *Then one can compute a P-extension* $(\mathbb{E}, \sigma)$ *of* $(\mathbb{F}, \sigma)$ *with* $\mathbb{E} = \mathbb{F}\langle t_1\rangle \ldots \langle t_e\rangle$ *and* $\frac{\sigma(t_i)}{t_i} \in \mathbb{F}^*$ *and finite sets* $\emptyset \neq S_i \subset \mathbb{F}^*$ *for* $1 \leq i \leq e$ *as follows.*

1. *For any* $1 \leq i \leq e$ *and any* $h \in S_i$ *it follows that* $g = h t_i$ *is a solution of*

$$a_m \sigma^m(g) + \cdots + a_1 \sigma(g) + a_0 g = 0. \tag{29}$$

2. *For any difference ring extension* $(\mathbb{H}, \sigma)$ *of* $(\mathbb{F}, \sigma)$ *with* $\mathbb{H} = \mathbb{F}\langle p_1\rangle \ldots \langle p_u\rangle$ *and* $\frac{\sigma(p_i)}{p_i} \in \mathbb{F}^*$ *and any solution* $g \in \mathbb{H}$ *of* (29) *with* $\alpha = \frac{\sigma(g)}{g} \in \mathbb{F}^*$ *one can take* $i \in \{1, \ldots, e\}$ *with* $f_1, \ldots, f_l \in S_i$ *and* $c_1, \ldots, c_l \in \mathbb{K}^*$ *such that* $\frac{\sigma(g')}{g'} = \alpha$ *holds for* $g' = (c_1 h_1 + \cdots + c_l h_l) t_i$.

We note that the obtained solver of hypergeometric solutions covers the special cases $\mathbb{G}_r$ (see [91, 127]), $\mathbb{G}_b$ with $v = 1$ (see [25]) and $\mathbb{G}_m$ (see [31]).

Remark 12 Theorems 13 and 14 hold in the more general setting where $(\mathbb{F}, \sigma)$ is a $\Pi\Sigma$-field extension of a difference field $(\mathbb{F}_0, \sigma)$ where certain properties are satisfied (see [26, Def. 7]). In addition, there is a generalization of Theorem 13 given in [17] (based on [26, 118]) where the a_i (with some extra properties) can be taken from the ring $\mathbb{E}$; the implementation can be found in the Mathematica package `PLDESolver`.

Based on [22, 24] we obtain the following result to find all d'Alembertian solutions, a subclass of Liouvillian solutions [63]. The solver relies on [43, Cor 2.1] and [98, Alg. 4.5.3] and the algorithmic machinery of Theorems 14 and 13.

Problem PLDE: Solving Parameterized Linear Difference Equations

Given: $a_0, \ldots, a_m \in \mathbb{G}$ with $a_m \neq 0$ and $F_1, \ldots, F_d \in \text{SumProd}_1(\mathbb{G})$ with $\mathbb{G} \in \{\mathbb{G}_r, \mathbb{G}_b, \mathbb{G}_m\}$, i.e., $\mathbb{G} = \mathbb{K}(x, x_1, \ldots, x_v)$ (or $\mathbb{G} = \mathbb{K}(x_1, \ldots, x_v)$) where $\mathbb{K} = \mathcal{A}(y_1, \ldots, y_o)(q_1, \ldots, q_v)$ is a rational function field over an algebraic number field $\mathcal{A}$.

Find: $\delta \in \mathbb{Z}_{\geq 0}$, a finite σ-reduced set $W \subset \Sigma\Pi_1(\mathbb{G}')$ and $B = \{(c_{i,1}, \ldots, c_{i,d}, G_i)\}_{1 \leq i \leq \nu} \subseteq \mathbb{K}^d \times \text{SumProd}(W, \mathbb{G}')$ such that

$$a_m(n)\, G_i(n+m) + \cdots + a_0(n) G_i(n) = c_{i,1} F_1(n) + \cdots + c_{i,d} F_d(n)$$

holds for all $n \geq \delta$ with $1 \leq i \leq \nu$; here $\mathbb{G}' = \mathbb{K}'(x, x_1, \ldots, x_v)$ (or $\mathbb{G}' = \mathbb{K}'(x_1, \ldots, x_v)$) with $\mathbb{K}' = \mathcal{A}'(y_1, \ldots, y_o)(q_1, \ldots, q_v)$ where $\mathcal{A}'$ is an algebraic field extension of $\mathcal{A}$.

In addition, the following properties hold:

1. Completeness: For any $\mathbb{G}'' = \mathbb{K}''(x, x_1, \ldots, x_v)$ (or $\mathbb{G}'' = \mathbb{K}''(x_1, \ldots, x_v)$) with $\mathbb{K}'' = \mathcal{A}''(y_1, \ldots, y_o)(q_1, \ldots, q_v)$ where $\mathcal{A}''$ is an algebraic extension of $\mathcal{A}$ and $(c_1, \ldots, c_d, G) \in \mathbb{K}^d \times \text{SumProd}_1(\mathbb{G}'')$ with

$$(a_m(n)\, G(n+m) + \cdots + a_0(n) G(n))_{n \geq 0} = (c_1 F_1(n) + \cdots + c_d F_d(n))_{n \geq 0}$$

there is a $(\kappa_1, \ldots, \kappa_\nu) \in (\mathbb{K}'')^\nu$ with

$$(c_1, \ldots, c_d) = \kappa_1(c_{1,1}, \ldots, c_{1,d}) + \cdots + \kappa_\nu(c_{\nu,1}, \ldots, c_{\nu,d}),$$
$$(G(n))_{n \geq 0} = (\kappa_1 G_1(n) + \cdots + \kappa_\nu G_\nu(n))_{n \geq 0}.$$

2. Linear independence: If there is a $(\kappa_1, \ldots, \kappa_\nu) \in (\mathbb{K}')^\nu$ with

$$\kappa_1(c_{1,1}, \ldots, c_{1,d}) + \cdots + \kappa_\nu(c_{\nu,1}, \ldots, c_{\nu,d}) = 0,$$
$$(\kappa_1 G_1(n) + \cdots + \kappa_\nu G_\nu(n))_{n \geq 0} = \mathbf{0},$$

then $(\kappa_1, \ldots, \kappa_\nu) = \mathbf{0}$.

Remark 13 Right from the start the case $d = 1$ was available (and fully solved with [26]) in `Sigma` with the function call **`SolveRecurrence`**`[`$a_0 G[n] + \cdots + a_m G[n+m] == F_1, G[n]$`]`. The case $d > 1$ has been incorporated in `Sigma` only recently. It can be carried out with **`SolveRecurrence`**`[`$a_0 G[n] + \cdots + a_m G[n+m] == \{F_1, \ldots, F_d\}, G[n]$`]` or **`SolveRecurrenceList`**`[`$\{a_0, \ldots, a_m\}, \{F_1, \ldots, F_d\}, n$`]`. It works also for nested products, i.e., $F_1, \ldots, F_d \in \text{SumProd}(\mathbb{G})$, if the F_i can be expressed straightforwardly in an $R\Pi\Sigma$-extension.

Using in addition the package `NestedProducts` this toolbox works also fully algorithmically for the case SumProd($\mathbb{G}_r$).

Example 22 [Cont. of Ex. 20] We proceed with the calculations given in Example 20. We apply our solver in `Sigma` to the already computed recurrence Out[26] and get

In[31]:= **recSol = SolveRecurrence[rec[[1]], SUM[n]]**

Out[31]= $$\Big\{\{0,1\},\{0,\sum_{i=1}^{n}\frac{1}{i}\},\{0,\frac{4}{9}\sum_{i=1}^{n}\frac{i!^2}{i^3(2i)!}+\frac{4}{3}\Big(\sum_{i=1}^{n}\frac{1}{i}\Big)\sum_{i=1}^{n}\frac{i!^2}{i^2(2i)!}-\frac{4}{3}\sum_{i=1}^{n}\frac{i!^2\sum_{j=1}^{i}\frac{1}{j}}{i^2(2i)!}\},\{1,-\sum_{i=1}^{n}\frac{\sum_{j=1}^{i}\frac{1}{j^2}}{i}\}\Big\}$$

The first three entries provide three linearly independent solutions of the homogeneous version of the recurrence and the last entry gives a particular solution of the recurrence itself.

Remark 14

(1) By default the found solutions are represented in a depth-optimal σ-reduced set W to keep the nesting depth of the solutions as small as possible.
(2) Since all components (i.e., a_i, F_i, G_i) can be represented in the given σ-reduced set W, the correctness of the solutions G_i can be verified by plugging them into the recurrence and checking if the left-hand and right-hand sides are equal.
(3) If one finds m linearly independent solutions of the homogeneous version together with a particular solution, the solution space is fully determined. In particular, any sequence, which is a solution of the recurrence, can be represented by SumProd($\mathbb{G}$): simply combine the found solutions accordingly (which is always possible from a certain point on) such that the evaluation of the expression agrees with the first m initial values.

Example 23 (Cont. of Ex. 22) In Example 22 we found all solutions of the recurrence. Since also the definite sum given in In[25] is a solution of the recurrence, we can combine the solutions accordingly and get an alternative solution of the input sum:

In[32]:= **sol = FindLinearCombination[recSol, definiteSum, n, 3]**

Out[32]= $$3\Big(\sum_{i=1}^{n}\frac{1}{i}\Big)\sum_{i=1}^{n}\frac{i!^2}{i^2(2i)!}-\sum_{i=1}^{n}\frac{\sum_{j=1}^{i}\frac{1}{j^2}}{i}-3\sum_{i=1}^{n}\frac{i!^2\sum_{j=1}^{i}\frac{1}{j}}{i^2(2i)!}+\sum_{i=1}^{n}\frac{i!^2}{i^3(2i)!}$$

Finally, we can rewrite the result in terms of the central binomial coefficient with

In[33]:= **sol = SigmaReduce[sol, n, Tower → {SigmaBinomial[2n, n]}]**

Out[33]= $3\Big(\sum_{i=1}^{n}\frac{1}{i}\Big)\sum_{i=1}^{n}\frac{1}{i^2\binom{2i}{i}}+\sum_{i=1}^{n}\frac{1}{i^3\binom{2i}{i}}-\sum_{i=1}^{n}\frac{1}{i}\sum_{j=1}^{i}\frac{1}{j^2}-3\sum_{i=1}^{n}\frac{1}{i^2\binom{2i}{i}}\sum_{j=1}^{i}\frac{1}{j}$

Summarizing we have discovered and proved the identity

$$\sum_{k=1}^{n}\frac{(-1)^k\binom{n}{k}}{k}\sum_{i=1}^{k}\frac{\sum_{j=1}^{i}\frac{1}{j+n}}{i}=3\Big(\sum_{i=1}^{n}\frac{1}{i}\Big)\sum_{i=1}^{n}\frac{1}{i^2\binom{2i}{i}}$$
$$-\sum_{i=1}^{n}\frac{\sum_{j=1}^{i}\frac{1}{j^2}}{i}-3\sum_{i=1}^{n}\frac{\sum_{j=1}^{i}\frac{1}{j}}{i^2\binom{2i}{i}}+\sum_{i=1}^{n}\frac{1}{i^3\binom{2i}{i}}.$$

Example 24 More generally, using the algorithms from [26] we can solve recurrences where the coefficients are represented within a $\Pi\Sigma$-field. E.g., for the recurrence

In[34]:= **recFactorial = −F[n+2]+(1+n)(8+9n+2n²)n!F[n+1]−2(1+n)³(3+n)n!²F[n] = 0;**

where the coefficients are taken from $\mathrm{SumProd}_1(\mathbb{Q}(x))$, we can find all its solutions (in this instance, they are again from $\mathrm{SumProd}_1(\mathbb{Q}(x))$) by executing the `Sigma`-call

In[35]:= **SolveRecurrence[recFactorial, F[n]]**

Out[35]= $\Big\{\{0,\prod_{i=1}^{n}i!\},\{0,-2^n n!\prod_{i=1}^{n}i!+\frac{3}{2}\prod_{i=1}^{n}i!\sum_{i=1}^{n}2^i i!\}\Big\}$

6 Application: Evaluation of Feynman Integrals

The elaborated summation tools from above contributed to highly nontrivial applications, e.g., in the research areas of combinatorics, number theory and particle physics. Here we emphasize the following striking aspects that are most relevant for the treatment of Feynman integrals.

Multi-Summation In order to support the user for the evaluation of definite multi-sums to expressions in $\mathrm{SumProd}(\mathbb{G})$, the package `EvaluateMultiSums` [112, 113]

In[36]:= **<< EvaluateMultiSums.m**

EvaluateMultiSums by Carsten Schneider © RISC-JKU

has been developed to tackle definite sums in one stroke. It uses as backbone `Sigma` with all the available tools introduced above. E.g., by executing

In[37]:= **EvaluateMultiSum**$[\sum_{k=1}^{n}\frac{(-1)^k}{k}\binom{n}{k}\sum_{i=1}^{k}\frac{1}{i}\sum_{j=1}^{i}\frac{1}{j+n}, \{\}, \{n\}, \{1\}, \{\infty\}]$

Out[37]= $$3\Big(\sum_{i=1}^{n}\frac{1}{i}\Big)\sum_{i=1}^{n}\frac{i!^2}{i^2(2i)!}-\sum_{i=1}^{n}\frac{\sum_{j=1}^{i}\frac{1}{j^2}}{i}-3\sum_{i=1}^{n}\frac{i!^2\sum_{j=1}^{i}\frac{1}{j}}{i^2(2i)!}+\sum_{i=1}^{n}\frac{i!^2}{i^3(2i)!}$$

we reproduce the identity given in Example 23. In particular, it can tackle definite multi-sums by zooming from inside to outside and, in case that this is possible, transforming stepwise the sums to expressions in $\mathrm{SumProd}_1(\mathbb{G})$. In this way we could treat highly complicated massive 3-loop Feynman integrals. More precisely, using techniques described in [42] these integrals can be transformed to several thousands of multiple sums with summands from $\mathrm{Prod}_1(\mathbb{G}_r)$. Afterwards, the package **SumProduction** [40, 113] is applied. It combines these sums to few (but large) sums tailored for our summation toolbox. Afterwards the command **EvaluateMultiSum** can be applied (without any further interaction) to treat the obtained sums. In the course of these calculations, we treated up to sevenfold multi-sums [8] or fourfold sums with up to 1GB of size [13, 14]. In addition, this package helped significantly to solve problems from combinatorics [75, 119, 121].

In addition, the difference field/ring approach described in this article has been united with important parts of the holonomic approach [55, 133] in [103]. While its first main application arose in combinatorics [28], this combined toolbox has been improved further in [43] and enabled us to tackle various multi-sums coming from particle physics [3, 5, 6]. In addition, these improved tools have been applied in [122] to complicated multi-sums that arose in the context of irrationality proofs of $\zeta(4)$. We remark further, that also other multi-sum and integral techniques from [1, 8, 42] have been explored; for further technologies see also [39] and the references therein.

Solving Coupled Systems Using integration-by-parts methods [54, 76] one can represent physical expressions in terms of master integrals which can be calculated by solving recursively defined coupled systems of linear differential equations. Most of these master integrals can be represented in terms of power series. Utilizing the techniques from above, this gives rise to two general tactics to compute the physical expressions in terms of known special functions (in case that this is possible).

Uncoupling and Solving the Underlying Recurrences In the first approach we uncouple iteratively the systems of linear differential equation using Gerhold's package `OreSys` [60] and reduce the problem to solving scalar linear differential equations of each master integral $I(x) = \sum_{n=0}^{\infty} F(n)x^n$. In a first step, each linear differential equation can be transformed to a linear recurrence. Applying `Sigma`'s recurrence solver in a second step enables one to decide constructively if the coefficient $F(n)$ can be expressed in terms of $\mathrm{SumProd}_1(\mathbb{G})$. If this is possible for each master integral, one can express also the physical expressions in $\text{SumProd}_1(\mathbb{G})$. Using these technologies implemented in the package `SolveCoupledSystem` [7, 9] (using `Sigma`) we could treat highly nontrivial problems of particle physics as given in [3, 10, 32–34]. Note that there are also other methods available [64, 78] that can solve certain classes of systems. Furthermore, in ongoing investigations nontrivial methods are developed to solve the coupled systems directly without recourse to uncoupling methods; see [30, 80, 128] and the literature therein.

The Large Moment Method The second highly successful approach is based on the technology [38, 46] implemented within the package `SolveCoupledSystem`. It enables one to produce for the master integrals the first coefficients $F(n)$ with $n = 0, \dots, \mu$; so far we encountered cases where $\mu = 10.000$ was necessary. Here one does not solve the arising recurrences as proposed above, but uses them to produce a large number of sequence values; as starting point one needs in addition a few initial values that can be produced by our summation tools or procedures like `Mincer` [77] or `MATAD` [125]. A significant feature of the large moment method is that one can avoid complicated function spaces (either nested sums with high weight or new classes, like nested sums over, e.g., elliptic functions [11, 27, 45, 47]) during the calculation. Only in the last step, one combines all the calculations and gets large moments of the physical expressions. Then one can use, e.g., the package `ore_algebra` [73] in Sage to guess recurrences (so far up to order 40) that specify precisely the different components of the physical problem. Finally, one can decide algorithmically if the physical problem (or individual subexpressions) can be represented within the class $\text{SumProd}_1(\mathbb{G}_r)$. In this way we could compute, e.g., the 3-loop splitting functions [10], the polarized 3-loop anomalous dimensions [35] and the massive 2- and 3-loop form factor [12, 44]; for another case study see, e.g., [41].

7 Conclusion

We presented two different layers to treat the class of indefinite nested sums defined over nested products in the context of symbolic summation. First, the term algebra layer $\text{SumProd}(\mathbb{G})$ (covering the rational case $\mathbb{G} = \mathbb{G}_r$, the multibasic case $\mathbb{G} = \mathbb{G}_b$ and the mixed multibasic case $\mathbb{G} = \mathbb{G}_m$) equipped with an evaluation function $\text{ev} : \text{SumProd}(\mathbb{G}) \times \mathbb{Z}_{\geq 0} \to \mathbb{K}$ has been introduced. There the user can define, evaluate and manipulate the class of nested sums and products conveniently. In particular, we illustrated how this user interface is implemented within the summation package `Sigma`.

Second, the formal difference ring/field layer has been elaborated. Here the elements of $\text{SumProd}(\mathbb{G})$ are rephrased in a ring $\mathbb{E}$ that is built by (Laurent) polynomial ring extensions. More precisely, the adjoined variables (in some instances factored out by ideals) represent the summation objects with two extra ingredients: a ring automorphism $\sigma : \mathbb{E} \to \mathbb{E}$ that describes the action of the shift operator on the ring elements and an evaluation function ev : $\mathbb{E} \times \mathbb{Z}_{\geq 0} \to \mathbb{K}$ that allows one to evaluate the formal ring elements to sequences. In this formal setting one can develop and implement not only complicated summation algorithms, but also set up a summation theory that enables one to embed the formal ring extensions into the ring of sequences (see Theorem 1).

One of the secrets of `Sigma`'s success within, e.g., particle physics, combinatorics and number theory is the smooth interaction between these two different layers: as illustrated in Fig. 1 on page 427 one can represent the objects from the two worlds so that their interpretation with the corresponding evaluation function agrees. In this article, we worked out in detail this algorithmic translation back and forth between the user-friendly term algebra and the complicated difference ring setting. To gain a better understanding of `Sigma`'s capabilities we established a precise input-output specification of the available summation tools using the introduced term algebra language. Special emphasis has been put on the canonical form representation (and its relation to the difference ring theory) for the class SumProd($\mathbb{G}$).

Acknowledgments This article got inspired during the workshop *Antidifferentiation and the Calculation of Feynman Amplitudes* at Zeuthen, Germany. In this regard, I would like to thank DESY for its hospitality and the Wolfgang Pauli Centre (WPC) for the financial support. In particular, I would like to thank the referee for the very thorough reading and valuable suggestions which helped me to improve the presentation.

This project has received funding from the European Union's Horizon 2020 research and innovation programme under the Marie Skłodowska–Curie grant agreement No. 764850, SAGEX and from the Austrian Science Foundation (FWF) grant SFB F50 (F5005-N15) in the framework of the Special Research Program "Algorithmic and Enumerative Combinatorics".

References

1. J. Ablinger, Extensions of the AZ-algorithm and the package MultiIntegrate, in *Anti-Differentiation and the Calculation of Feynman Amplitudes*, ed. by J. Blümlein, C. Schneider, Texts and Monographs in Symbolic Computuation (Springer, Berlin, 2021). arXiv:2101.11385.
2. J. Ablinger, J. Blümlein, C. Schneider, Harmonic sums and polylogarithms generated by cyclotomic polynomials. J. Math. Phys. **52**(10), 1–52 (2011). arXiv:1007.0375
3. J. Ablinger, A. Behring, J. Blümlein, A. De Freitas, A. von Manteuffel, C. Schneider, The 3-loop pure singlet heavy flavor contributions to the structure function $F_2(x, Q^2)$ and the anomalous dimension. Nucl. Phys. B **890**, 48–151 (2014). arXiv:1409.1135
4. J. Ablinger, J. Blümlein, C.G. Raab, C. Schneider, Iterated binomial sums and their associated iterated integrals. J. Math. Phys. **55**(112301), 1–57 (2014). arXiv:1407.1822
5. J. Ablinger, J. Blümlein, A. De Freitas, A. Hasselhuhn, A. von Manteuffel, M. Round, C. Schneider, F. Wissbrock, The transition matrix element $A_{gq}(N)$ of the variable flavor number scheme at $O(\alpha_s^3)$. Nucl. Phys. B **882**, 263–288 (2014). arXiv:1402.0359
6. J. Ablinger, A. Behring, J. Blümlein, A. De Freitas, A. Hasselhuhn, A. von Manteuffel, M. Round, C. Schneider, F. Wißbrock, Heavy flavour corrections to polarised and unpolarised deep-inelastic scattering at 3-loop order, in *Proceedings of QCD Evolution 2016*, vol. PoS(QCDEV2016)052 (2016), pp. 1–16. arXiv:1611.01104

7. J. Ablinger, A. Behring, J. Blümlein, A. De Freitas, C. Schneider, Algorithms to solve coupled systems of differential equations in terms of power series, in *Proceedings of Loops and Legs in Quantum Field Theory - LL 2016*, ed. by J. Blümlein, P. Marquard, T. Riemann, vol. PoS(LL2016)005 (2016), pp. 1–15. arXiv:1608.05376
8. J. Ablinger, A. Behring, J. Blümlein, A. De Freitas, A. von Manteuffel, C. Schneider, Calculating three loop ladder and V-topologies for massive operator matrix elements by computer algebra. Comput. Phys. Commun. **202**, 33–112 (2016). arXiv:1509.08324
9. J. Ablinger, J. Blümlein, A. De Freitas, C. Schneider, A toolbox to solve coupled systems of differential and difference equations, in *Proceedings of the 13th International Symposium on Radiative Corrections (Applications of Quantum Field Theory to Phenomenology)*, vol. PoS(RADCOR2015)060 (2016), pp. 1–13. arXiv:1601.01856
10. J. Ablinger, A. Behring, J. Blümlein, A. De Freitas, A. von Manteuffel, C. Schneider, The three-loop splitting functions $P_{qg}^{(2)}$ and $P_{gg}^{(2,N_F)}$. Nucl. Phys. B **922**, 1–40 (2017). arxiv:1705.01508
11. J. Ablinger, J. Blümlein, A.De Freitas, M. van Hoeij, E. Imamoglu, C. Raab, C.-S. Radu, C. Schneider, Iterated elliptic and hypergeometric integrals for Feynman diagrams. J. Math. Phys. **59**(062305), 1–55 (2018). arXiv:1706.01299
12. J. Ablinger, J. Blümlein, P. Marquard, N. Rana, C. Schneider, Heavy quark form factors at three loops in the planar limit. Phys. Lett. B **782**, 528–532 (2018). arXiv:1804.07313
13. J. Ablinger, J. Blümlein, A. De Freitas, A. Goedicke, C. Schneider, K. Schönwald, The two-mass contribution to the three-loop gluonic operator matrix element $A_{gg,Q}^{(3)}$. Nucl. Phys. B **932**, 129–240 (2018). arXiv:1804.02226
14. J. Ablinger, J. Blümlein, A. De Freitas, M. Saragnese, C. Schneider, K. Schönwald, The three-loop polarized pure singlet operator matrix element with two different masses. Nucl. Phys. B **952**(114916), 1–18 (2020). arXiv:1911.11630
15. J. Ablinger, J. Blümlein, C. Schneider, Analytic and algorithmic aspects of generalized harmonic sums and polylogarithms. J. Math. Phys. **54**(8), 1–74 (2013). arXiv:1302.0378
16. J. Ablinger, C. Schneider, Algebraic independence of sequences generated by (cyclotomic) harmonic sums. Ann. Combin. **22**(2), 213–244 (2018). arXiv:1510.03692
17. J. Ablinger, C. Schneider, Solving linear difference equations with coefficients in rings with idempotent representations, in *Proceedings of the 2021 International Symposium on Symbolic and Algebraic Computation (Proc. ISSAC 21)*, ed. by M. Mezzarobba (2021), pp. 27–34. arXiv:2102.03307.
18. S.A. Abramov, On the summation of rational functions. Zh. vychisl. mat. Fiz. **11**, 1071–1074 (1971)
19. S.A. Abramov, The rational component of the solution of a first-order linear recurrence relation with a rational right-hand side. U.S.S.R. Comput. Maths. Math. Phys. **15**, 216–221 (1975). Transl. from Zh. vychisl. mat. mat. fiz. 15, pp. 1035–1039 (1975)
20. S.A. Abramov, Rational solutions of linear differential and difference equations with polynomial coefficients. U.S.S.R. Comput. Math. Math. Phys. **29**(6), 7–12 (1989)
21. S.A. Abramov, M. Bronstein, Hypergeometric dispersion and the orbit problem, in *Proceedings of ISSAC'00*, ed. by C. Traverso (ACM Press, New York, 2000)
22. S.A. Abramov, M. Petkovšek, D'Alembertian solutions of linear differential and difference equations, in *Proceedings of ISSAC'94*, ed. by J. von zur Gathen (ACM Press, New York, 1994), pp. 169–174
23. S.A. Abramov, M. Petkovšek, Polynomial ring automorphisms, rational (w, σ)-canonical forms, and the assignment problem. J. Symb. Comput. **45**(6), 684–708 (2010)
24. S.A. Abramov, E.V. Zima, D'Alembertian solutions of inhomogeneous linear equations (differential, difference, and some other), in *Proceedings of ISSAC'96* (ACM Press, New York, 1996), pp. 232–240
25. S.A. Abramov, P. Paule, M. Petkovšek, q-Hypergeometric solutions of q-difference equations. Discrete Math. **180**(1–3), 3–22 (1998)

26. S.A. Abramov, M. Bronstein, M. Petkovšek, C. Schneider, On rational and hypergeometric solutions of linear ordinary difference equations in $\Pi\Sigma^*$-field extensions. J. Symb. Comput. **107**, 23–66 (2021). arXiv:2005.04944 [cs.SC]
27. L. Adams, C. Bogner, A. Schweitzer, S. Weinzierl, The kite integral to all orders in terms of elliptic polylogarithms. J. Math. Phys. **57**(12), 122302 (2016)
28. G.E. Andrews, P. Paule, C. Schneider, Plane partitions VI: Stembridge's TSPP theorem. Adv. Appl. Math. **34**(4), 709–739 (2005)
29. M. Apagodu, D. Zeilberger, Multi-variable Zeilberger and Almkvist-Zeilberger algorithms and the sharpening of Wilf-Zeilberger theory. Adv. Appl. Math. **37**, 139–152 (2006)
30. M. Barkatou, On rational solutions of systems for linear differential equations. J. Symb. Comput. **28**, 547–567 (1999)
31. A. Bauer, M. Petkovšek, Multibasic and mixed hypergeometric Gosper-type algorithms. J. Symb. Comput. **28**(4–5), 711–736 (1999)
32. A. Behring, J. Blümlein, A. De Freitas, A. Hasselhuhn, A. von Manteuffel, C. Schneider, The $O(\alpha_s^3)$ heavy flavor contributions to the charged current structure function $xF_3(x, Q^2)$ at large momentum transfer. Phys. Rev. D **92**(114005), 1–19 (2015). arXiv:1508.01449
33. A. Behring, J. Blümlein, A. De Freitas, A. von Manteuffel, C. Schneider, The 3-loop non-singlet heavy flavor contributions to the structure function $g_1(x, Q^2)$ at large momentum transfer. Nucl. Phys. B **897**, 612–644 (2015). arXiv:1504.08217
34. A. Behring, J. Blümlein, G. Falcioni, A. De Freitas, A. von Manteuffel, C. Schneider, The asymptotic 3-loop heavy flavor corrections to the charged current structure functions $F_L^{W^+-W^-}(x, Q^2)$ and $F_2^{W^+-W^-}(x, Q^2)$.Phys. Rev. D **94**(11), 1–19 (2016). arXiv:1609.06255
35. A. Behring, J. Blümlein, A. De Freitas, A. Goedicke, S. Klein, A. von Manteuffel, C. Schneider, K. Schönwald, The polarized three-loop anomalous dimensions from on-shell massive operator matrix elements. Nucl. Phys. B **948**(114753), 1–41 (2019). arXiv:1908.03779
36. J. Blümlein, Algebraic relations between harmonic sums and associated quantities. Comput. Phys. Commun. **159**(1), 19–54 (2004). arXiv:hep-ph/0311046
37. J. Blümlein, S. Kurth, Harmonic sums and Mellin transforms up to two-loop order. Phys. Rev. **D60**, 014018 (1999)
38. J. Blümlein, C. Schneider, The method of arbitrarily large moments to calculate single scale processes in quantum field theory. Phys. Lett. B **771**, 31–36 (2017). arXiv:1701.04614
39. J. Blümlein, C. Schneider, Analytic computing methods for precision calculations in quantum field theory. Int. J. Mod. Phys. A **33**(1830015), 1–35 (2018). arXiv:1809.02889
40. J. Blümlein, A. Hasselhuhn, C. Schneider, Evaluation of multi-sums for large scale problems, in *Proceedings of RADCOR 2011*, vol. PoS(RADCOR2011)32 (2012), pp. 1–9. arXiv:1202.4303
41. J. Blümlein, M. Kauers, S. Klein, C. Schneider, Determining the closed forms of the $O(a_s^3)$ anomalous dimensions and Wilson coefficients from Mellin moments by means of computer algebra. Comput. Phys. Commun. **180**, 2143–2165 (2009). arXiv:0902.4091
42. J. Blümlein, S. Klein, C. Schneider, F. Stan, A symbolic summation approach to Feynman integral calculus. J. Symb. Comput. **47**, 1267–1289 (2012). arXiv:1011.2656
43. J. Blümlein, M. Round, C. Schneider, Refined holonomic summation algorithms in particle physics, in *Advances in Computer Algebra. WWCA 2016.*, ed. by E. Zima, C. Schneider. Springer Proceedings in Mathematics & Statistics, vol. 226 (Springer, Berlin, 2018), pp. 51–91. arXiv:1706.03677
44. J. Blümlein, P. Marquard, N. Rana, C. Schneider, The heavy fermion contributions to the massive three loop form factors. Nucl. Phys. B **949**(114751), 1–97 (2019). arXiv:1908.00357
45. J. Blümlein, P. Paule, C. Schneider (Eds.), *Elliptic Integrals, Elliptic Functions and Modular Forms in Quantum Field Theory*. Texts & Monographs in Symbolic Computation (Springer, Berlin, 2019)

46. J. Blümlein, P. Marquard, C. Schneider, A refined machinery to calculate large moments from coupled systems of linear differential equations, in *14th International Symposium on Radiative Corrections (RADCOR2019)*, ed. by D. Kosower, M. Cacciari, POS(RADCOR2019)078 (2020), pp. 1–13. arXiv:1912.04390
47. J. Broedel, C. Duhr, F. Dulat, L. Tancredi, Elliptic polylogarithms and iterated integrals on elliptic curves. II. An application to the sunrise integral. Phys. Rev. **D97**, 116009 (2018)
48. M. Bronstein, *Symbolic Integration I, Transcendental Functions* (Springer, Berlin, 1997)
49. M. Bronstein, On solutions of linear ordinary difference equations in their coefficient field. J. Symb. Comput. **29**(6), 841–877 (2000)
50. B. Buchberger, R. Loos, Algebraic simplification, in *Computer Algebra* (Springer, Vienna, 1983), pp. 11–43
51. S. Chen, R. Feng, G. Fu, Z. Li, On the structure of compatible rational functions, in *Proceedings of ISSAC 2011* (2011), pp. 91–98
52. W.Y.C. Chen, Q.-H. Hou, H.-T. Jin, The Abel-Zeilberger algorithm. Electron. J. Combin. **18**(2), Paper 17 (2011)
53. S. Chen, M. Jaroschek, M. Kauers, M.F. Singer, Desingularization explains order-degree curves for ore operators, in *Proceedingsd of ISSAC'13*, ed. by M. Kauers (2013), pp. 157–164
54. K.G. Chetyrkin, F.V. Tkachov, Integration by parts: the algorithm to calculate beta functions in 4 loops. Nucl. Phys. B **192**, 159–204 (1981)
55. F. Chyzak, An extension of Zeilberger's fast algorithm to general holonomic functions. Discrete Math. **217**, 115–134 (2000)
56. F. Chyzak, M. Kauers, B. Salvy, A non-holonomic systems approach to special function identities, in *Proceedings of ISSAC'09*, ed. by J. May (2009), pp. 111–118
57. R.M. Cohn, *Difference Algebra* (Wiley, Hoboken, 1965)
58. A.I. Davydychev, M.Yu. Kalmykov, Massive Feynman diagrams and inverse binomial sums. Nucl. Phys. B **699**, 3–64 (2004)
59. G. Ge, Algorithms related to the multiplicative representation of algebraic numbers. Ph.D. Thesis, Univeristy of California at Berkeley (1993)
60. S. Gerhold, Uncoupling systems of linear ore operator equations. Master's Thesis, RISC, J. Kepler University Linz (2002)
61. R.W. Gosper, Decision procedures for indefinite hypergeometric summation. Proc. Nat. Acad. Sci. U.S.A. **75**, 40–42 (1978)
62. C. Hardouin, M.F. Singer, Differential Galois theory of linear difference equations. Math. Ann. **342**(2), 333–377 (2008)
63. P.A. Hendriks, M.F. Singer, Solving difference equations in finite terms. J. Symb. Comput. **27**(3), 239–259 (1999)
64. J.M. Henn, Multiloop integrals in dimensional regularization made simple. Phys. Rev. Lett. **110**, 251601 (2013)
65. M. Karr, Summation in finite terms. J. ACM **28**, 305–350 (1981)
66. M. Karr, Theory of summation in finite terms. J. Symb. Comput. **1**, 303–315 (1985)
67. M. Kauers, Summation algorithms for stirling number identities. J. Symb. Comput. **42**(10), 948–970 (2007)
68. M. Kauers, C. Schneider, Application of unspecified sequences in symbolic summation, in *Proceedings of ISSAC'06.*, ed. by J. Dumas (ACM Press, New York, 2006), pp. 177–183
69. M. Kauers, C. Schneider, Indefinite summation with unspecified summands. Discrete Math. **306**(17), 2021–2140 (2006)
70. M. Kauers, C. Schneider, Symbolic summation with radical expressions, in *Proceedings of ISSAC'07*, ed. by C. Brown (2007), pp. 219–226
71. M. Kauers, C. Schneider, Automated proofs for some stirling number identities. Electron. J. Combin. **15**(1), 1–7 (2008). R2
72. M. Kauers, B. Zimmermann, Computing the algebraic relations of c-finite sequences and multisequences. J. Symb. Comput. **43**(11), 787–803 (2008)

73. M. Kauers, M. Jaroschek, F. Johansson, Ore polynomials in Sage, in *Computer Algebra and Polynomials*, ed. by J. Gutierrez, J. Schicho, M. Weimann. Lecture Notes in Computer Science (2014), pp. 105–125
74. C. Koutschan, Creative telescoping for holonomic functions, in *Computer Algebra in Quantum Field Theory: Integration, Summation and Special Functions*, ed. by C. Schneider, J. Blümlein. Texts and Monographs in Symbolic Computation (Springer, Berlin, 2013), pp. 171–194. arXiv:1307.4554
75. C. Krattenthaler, C. Schneider, Evaluation of binomial double sums involving absolute values, in *Algorithmic Combinatorics: Enumerative Combinatorics, Special Functions and Computer Algebra, in Honour of Peter Paule on his 60th Birthday*, ed. by V. Pillwein, C. Schneider. Texts and Monographs in Symbolic Computuation (Springer, Berlin, 2020), pp. 249–295. arXiv:1607.05314
76. S. Laporta, High precision calculation of multiloop Feynman integrals by difference equations. Int. J. Mod. Phys. **A15**, 5087–5159 (2000)
77. S.A. Larin, F. Tkachov, J.A.M. Vermaseren, The FORM version of Mincer. Technical Report NIKHEF-H-91-18, NIKHEF, Netherlands (1991)
78. R.N. Lee, Reducing differential equations for multiloop master integrals. JHEP **04**, 108 (2015)
79. J. Liouville, Mémoire sur l'intégration d'une classe de fonctions transcendantes. J. Reine Angew. Math. **13**, 93–118 (1835)
80. J. Middeke, C. Schneider, Denominator bounds for systems of recurrence equations using $\Pi\Sigma$-extensions, in *Advances in Computer Algebra: In Honour of Sergei Abramov's 70th Birthday*, ed. by C. Schneider, E. Zima. Springer Proceedings in Mathematics & Statistics, vol. 226 (Springer, Berlin, 2018), pp. 149–173. arXiv:1705.00280.
81. S.O. Moch, P. Uwer, S. Weinzierl, Nested sums, expansion of transcendental functions, and multiscale multiloop integrals. J. Math. Phys. **6**, 3363–3386 (2002)
82. I. Nemes, P. Paule, A canonical form guide to symbolic summation, in *Advances in the Design of Symbolic Computation Systems*, ed. by A. Miola, M. Temperini. Texts & Monographs in Symbolic Computation (Springer, Wien-New York, 1997), pp. 84–110
83. E.D. Ocansey, C. Schneider, Representing (q-)hypergeometric products and mixed versions in difference rings, in *Advances in Computer Algebra. WWCA 2016.*, C. Schneider, E. Zima. Springer Proceedings in Mathematics & Statistics, vol. 226 (Springer, Berlin, 2018), pp. 175–213. arXiv:1705.01368
84. E.D. Ocansey, C. Schneider, Representation of hypergeometric products of higher nesting depths in difference rings. RISC Report Series 20-19, Research Institute for Symbolic Computation (RISC), Johannes Kepler University Linz, Schloss Hagenberg, 4232 Hagenberg, Austria (2020). arXiv:2011.08775
85. P. Paule, Greatest factorial factorization and symbolic summation. J. Symb. Comput. **20**(3), 235–268 (1995)
86. P. Paule, Contiguous relations and creative telescoping, in *Anti-Differentiation and the Calculation of Feynman Amplitudes*, ed. by J. Blümlein, C. Schneider. Texts and Monographs in Symbolic Computuation (Springer, Berlin, 2021)
87. P. Paule, A. Riese, A Mathematica q-analogue of Zeilberger's algorithm based on an algebraically motivated approach to q-hypergeometric telescoping, in *Special Functions, q-Series and Related Topics*, ed. by M. Ismail, M. Rahman, vol. 14 (AMS, Providence, 1997), pp. 179–210
88. P. Paule, C. Schneider, Computer proofs of a new family of harmonic number identities. Adv. Appl. Math. **31**(2), 359–378 (2003). Preliminary version online
89. P. Paule, C. Schneider, Towards a symbolic summation theory for unspecified sequences, in *Elliptic Integrals, Elliptic Functions and Modular Forms in Quantum Field Theory*, ed. by J. Blümlein, P. Paule, C. Schneider. Texts and Monographs in Symbolic Computation (Springer, Berlin, 2019), pp. 351–390. arXiv:1809.06578
90. P. Paule, M. Schorn, A Mathematica version of Zeilberger's algorithm for proving binomial coefficient identities. J. Symb. Comput. **20**(5–6), 673–698 (1995)

91. M. Petkovšek, Hypergeometric solutions of linear recurrences with polynomial coefficients. J. Symb. Comput. **14**(2–3), 243–264 (1992)
92. M. Petkovšek, Definite sums as solutions of linear recurrences with polynomial coefficients (2018). arXiv:1804.02964 [cs.SC]
93. M. Petkovšek, H.S. Wilf, D. Zeilberger, $A = B$ (A. K. Peters, Wellesley, 1996)
94. M. Petkovšek, H. Zakrajšek, Solving linear recurrence equations with polynomial coefficients, in *Computer Algebra in Quantum Field Theory: Integration, Summation and Special Functions*, ed. by C. Schneider, J. Blümlein. Texts and Monographs in Symbolic Computation (Springer, Berlin, 2013), pp. 259–284
95. R. Risch, The problem of integration in finite terms. Trans. Am. Math. Soc. **139**, 167–189 (1969)
96. M. Rosenlicht, Liouville's theorem on functions with elementary integrals. Pac. J. Math. **24**, 153–161 (1968)
97. C. Schneider, An implementation of Karr's summation algorithm in Mathematica. Sem. Lothar. Combin. **S43b**, 1–10 (2000)
98. C. Schneider, Symbolic summation in difference fields. Technical Report 01-17, RISC-Linz, J. Kepler University, Ph.D. Thesis (2001)
99. C. Schneider, A collection of denominator bounds to solve parameterized linear difference equations in $\Pi\Sigma$-extensions. An. Univ. Timişoara Ser. Mat.-Inform. **42**(2), 163–179 (2004). Extended version of Proc. SYNASC'04
100. C. Schneider, Symbolic summation with single-nested sum extensions, in *Proceedings of ISSAC'04*, ed. by J. Gutierrez (ACM Press, New York, 2004), pp. 282–289
101. C. Schneider, Degree bounds to find polynomial solutions of parameterized linear difference equations in $\Pi\Sigma$-fields. Appl. Algebra Engrgy Comm. Comput. **16**(1), 1–32 (2005)
102. C. Schneider, Finding telescopers with minimal depth for indefinite nested sum and product expressions, in *Proceedings of ISSAC'05*, ed. by M. Kauers (ACM, New York, 2005)
103. C. Schneider, A new Sigma approach to multi-summation. Adv. Appl. Math. **34**(4), 740–767 (2005)
104. C. Schneider, Product representations in $\Pi\Sigma$-fields. Ann. Combin. **9**(1), 75–99 (2005)
105. C. Schneider, Solving parameterized linear difference equations in terms of indefinite nested sums and products. J. Differ. Equ. Appl. **11**(9), 799–821 (2005)
106. C. Schneider, Simplifying sums in $\Pi\Sigma$-extensions.J. Algebra Appl. **6**(3), 415–441 (2007)
107. C. Schneider, Symbolic summation assists combinatorics. Sém. Lothar. Combin. **56**, 1–36 (2007). Article B56b
108. C. Schneider, A refined difference field theory for symbolic summation. J. Symb. Comput. **43**(9), 611–644 (2008). arXiv:0808.2543v1
109. C. Schneider, Parameterized telescoping proves algebraic independence of sums. Ann. Comb. **14**, 533–552 (2010). arXiv:0808.2596; for a preliminary version see FPSAC 2007
110. C. Schneider, Structural theorems for symbolic summation. Appl. Algebra Engrgy Comm. Comput. **21**(1), 1–32 (2010)
111. C. Schneider, A symbolic summation approach to find optimal nested sum representations, in *Motives, Quantum Field Theory, and Pseudodifferential Operators*, ed. by A. Carey, D. Ellwood, S. Paycha, S. Rosenberg. Clay Mathematics Proceedings, vol. 12 (American Mathematical Society, Providence, 2010), pp. 285–308. arXiv:0808.2543
112. C. Schneider, Simplifying multiple sums in difference fields, in *Computer Algebra in Quantum Field Theory: Integration, Summation and Special Functions*, ed. by C. Schneider, J. Blümlein. Texts and Monographs in Symbolic Computation (Springer, Berlin, 2013), pp. 325–360. arXiv:1304.4134
113. C. Schneider, Modern summation methods for loop integrals in quantum field theory: the packages Sigma, EvaluateMultiSums and SumProduction, in *Proceedings of ACAT 2013*. Journal of Physics: Conference Series, vol. 523(012037) (2014), pp. 1–17. arXiv:1310.0160

114. C. Schneider, A streamlined difference ring theory: indefinite nested sums, the alternating sign and the parameterized telescoping problem, in *Symbolic and Numeric Algorithms for Scientific Computing (SYNASC), 2014 15th International Symposium*, ed. by F. Winkler, V. Negru, T. Ida, T. Jebelean, D. Petcu, S. Watt, D. Zaharie (IEEE Computer Society, Washington, 2014), pp. 26–33. arXiv:1412.2782
115. C. Schneider, Fast algorithms for refined parameterized telescoping in difference fields, in *Computer Algebra and Polynomials*, ed. by M.W.J. Guitierrez, J. Schicho. Lecture Notes in Computer Science (LNCS), vol. 8942 (Springer, Berlin, 2015), pp. 157–191. arXiv:1307.7887.
116. C. Schneider, A difference ring theory for symbolic summation. J. Symb. Comput. **72**, 82–127 (2016). arXiv:1408.2776
117. C. Schneider, Symbolic summation in difference rings and applications, in *Proceedings of ISSAC 2016*, ed. by M. Rosenkranz (2016), pp. 9–12
118. C. Schneider, Summation theory II: characterizations of $R\Pi\Sigma$-extensions and algorithmic aspects. J. Symb. Comput. **80**(3), 616–664 (2017). arXiv:1603.04285
119. C. Schneider, The absent-minded passengers problem: a motivating challenge solved by computer algebra. Math. Comput. Sci. (2020), https://doi.org/10.1007/s11786-020-00494-w. arXiv:2003.01921
120. C. Schneider, Minimal representations and algebraic relations for single nested products. Program. Comput. Softw. **46**(2), 133–161 (2020). arXiv:1911.04837
121. C. Schneider, R. Sulzgruber, Asymptotic and exact results on the complexity of the Novelli–Pak–Stoyanovskii algorithm. Electron. J. Combin. **24**(2), 1–33 (2017). #P2.28, arXiv:1606.07597
122. C. Schneider, W. Zudilin, A case study for $\zeta(4)$, in *Proceedings of the Conference 'Transient Transcendence in Transylvania'*. Proceedings in Mathematics & Statistics (Springer, Berlin, 2021). arXiv:2004.08158
123. M.F. Singer, Liouvillian solutions of linear differential equations with Liouvillian coefficients. J. Symb. Comput. **11**(3), 251–274 (1991)
124. M.F. Singer, Algebraic and algorithmic aspects of linear difference equations, in *Galois Theories of Linear Difference Equations: An Introduction*, ed. by C. Hardouin, J. Sauloy, M.F. Singer. Mathematical Surveys and Monographs, vol. 211 (AMS, Providence, 2016)
125. M. Steinhauser, MATAD: a program package for the computation of MAssive TADpoles. Comput. Phys. Commun. **134**, 335–364 (2001)
126. M. van der Put, M. Singer, *Galois Theory of Difference Equations*. Lecture Notes in Mathematics, vol. 1666 (Springer, Berlin, 1997)
127. M. van Hoeij, Finite singularities and hypergeometric solutions of linear recurrence equations. J. Pure Appl. Algebra **139**(1–3), 109–131 (1999)
128. M. van Hoeij, M. Barkatou, J. Middeke, A family of denominator bounds for first order linear recurrence systems. Technical Report (2020). arXiv:2007.02926
129. J.A.M. Vermaseren, Harmonic sums, Mellin transforms and integrals. Int. J. Mod. Phys. **A14**, 2037–2976 (1999)
130. K. Wegschaider, Computer generated proofs of binomial multi-sum identities. Master's Thesis, RISC, J. Kepler University (1997)
131. S. Weinzierl, Expansion around half integer values, binomial sums and inverse binomial sums. J. Math. Phys. **45**, 2656–2673 (2004)
132. H. Wilf, D. Zeilberger, An algorithmic proof theory for hypergeometric (ordinary and "q") multisum/integral identities. Invent. Math. **108**, 575–633 (1992)
133. D. Zeilberger, A holonomic systems approach to special functions identities. J. Comput. Appl. Math. **32**, 321–368 (1990)
134. D. Zeilberger, The method of creative telescoping. J. Symb. Comput. **11**, 195–204 (1991)

Expansion by Regions: An Overview

Vladimir A. Smirnov

Abstract A short review of expansion by regions is presented. It is a well-known strategy to obtain an expansion of a given multiloop Feynman integral in a given limit where some kinematic invariants and/or masses have certain scaling measured in powers of a given small parameter. Prescriptions of this strategy are formulated in a simple geometrical language and are illustrated through simple examples.

1 Historiographical Notes

Expansion by regions is a universal strategy to obtain an expansion of a given Feynman integral in a given limit, where kinematic invariants and/or masses essentially differ in scale. For simplicity, let us consider a Feynman integral $G_\Gamma(q^2, m^2)$ depending on two scales, for example, q^2 and m^2, and let the limit be $t = -m^2/q^2 \to 0$. Experience tells us that the expansion at $t \to 0$ has the form

$$G_\Gamma(t, \varepsilon) \sim (-q^2)^\omega \sum_{n=n_0}^{\infty} \sum_{k=0}^{2h} c_{n,k}(\varepsilon)\, t^n \, \log^k t \;, \tag{1}$$

where $\omega = 4h - 2\sum a_i$ is the degree of divergence, with a_l powers of the propagators, h is the number of loops and $\varepsilon = (4-d)/2$ is the parameter of dimensional regularization. The expansion is often called asymptotic, in the sense that the remainder of expansion has the order $o(t^N)$ after keeping terms up to t^N. However, every power series at a power of logarithm in expansions in various limits of momenta and masses has a non-zero radius of convergence which is determined usually by the nearest threshold.

V. A. Smirnov (✉)
Skobeltsyn Institute of Nuclear Physics of Moscow State University, Moscow, Russia
e-mail: smirnov@theory.sinp.msu.ru

J. Blümlein, C. Schneider (eds.), *Anti-Differentiation and the Calculation of Feynman Amplitudes*, Texts & Monographs in Symbolic Computation,
https://doi.org/10.1007/978-3-030-80219-6_18

There can be different reasons to consider some limit and the corresponding expansion. Typically, different scaling of kinematic invariants and/or masses involved is dictated by a phenomenological situation. Moreover, experience obtained when expanding Feynman integrals in some limit can show a way to construct the corresponding effective theory. At the level of individual Feynman integrals, expanding a complicated Feynman integral in some limit can approximately substitute the analytic evaluation of the integral.

One can use various techniques in order to obtain an expansion of a given Feynman integral in some limit: one can start with a parametric representation, or apply the method of Mellin–Barnes representation, or obtain an expansion within the method of differential equations. However, the *general* strategy of expansion by regions provides the possibility to write down a result for the expansion immediately once relevant regions are known. Such a result looks similar to (1) but now exponents of the expansion parameter depending linearly on ε are not yet expanded in ε,

$$G_\Gamma(t,\varepsilon) \sim (-q^2)^\omega \sum_{n=n_0}^{\infty} \sum_{k=0}^{h} \sum_{j=0}^{h} c'_{n,j,k}(\varepsilon) t^{n-j\varepsilon} \log^k t \qquad (2)$$

and the coefficients in the expansion can be represented in terms of integrals over loop momenta or over Feynman parameters. These integrals on the right-hand side of the expansion are constructed according to certain rules starting from the Feynman integral or a parametric integral for the initial Feynman integrals G_Γ. This means that expansion by regions reduces the problem to the evaluation of integrals present in (2).

Logarithms in (2) within dimensional regularization do not appear in limits typical of Euclidean space such as the off-shell large momentum limit and the large mass limit. Rather, they are typical for limits typical of Minkowski space such as the Regge limit and various versions of the Sudakov limit. In fact, one can avoid such logarithms by introducing an auxiliary analytic regularization which can be introduced as additional complex numbers in the exponents of the propagators. One can say that, after this, the various scales in the problem become separated so that the expansion becomes only in powers of the expansion parameter. After turning off this regularization, spurious poles in the auxiliary analytic parameters cancel giving rise to the logarithms, and this happens to be an important consistency check. A lot of examples illustrating this phenomenon can be found, e.g., in [1]. We will come back to this point in Section 2 when discussing the geometrical formulation of expansion by regions.

According to the first formulation of expansion by regions [2] one analyzes various regions in a given integral over loop momenta and, in every region, expands the integrand in parameters which are there small. Then the integration in the integral with so expanded propagators is extended to the whole domain of the loop momenta and, finally, one obtains an expansion of the given integral as the corresponding sum over the regions. Although these recipes were formulated in a

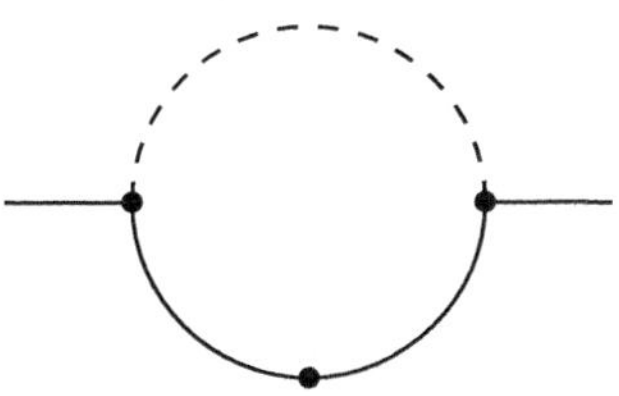

Fig. 1 A one-loop graph

suspicious mathematical language, expansion by regions was successfully applied in numerous calculations.

A very simple example is given by the Feynman integral corresponding to the graph depicted in Fig. 1,

$$G(q^2, m^2; d) = \int \frac{\mathrm{d}^d k}{(k^2 - m^2)^2 (q - k)^2} \tag{3}$$

in the limit $m^2/q^2 \to 0$.

The relevant regions are the region of small loop momenta, $k \sim m$, and the region of large loop momenta, $k \sim q$. According to the above prescriptions, in the first region, the first propagator is unexpanded and the second propagator is expanded in a Taylor series in k. In the second region, the first propagator is expanded in a Taylor series in m the second propagator is unexpanded. The leading terms of expansion give

$$G(q^2, m^2; d) \sim \int \frac{\mathrm{d}^d k}{(k^2)^2 (q - k)^2} + \frac{1}{q^2} \int \frac{\mathrm{d}^d k}{(k^2 - m^2)^2} + \ldots \tag{4}$$

The integrals involved can be evaluated by Feynman parameters, with the following result

$$G(q^2, m^2; d) \sim i\pi^{d/2} \left(\frac{\Gamma(1-\varepsilon)^2 \Gamma(\varepsilon)}{\Gamma(1-2\varepsilon)(-q^2)^{1+\varepsilon}} + \frac{\Gamma(\varepsilon)}{q^2 (m^2)^{\varepsilon}} + \ldots \right) \tag{5}$$

Although the initial Feynman integral is finite at $d = 4$, there are simple poles above: an infrared pole in the first term and an ultraviolet term in the second term. They are successfully canceled, with the following result

$$i\pi^{d/2} \left(\log \left(\frac{-q^2}{m^2} \right) + \ldots \right) . \tag{6}$$

Such an interplay of various divergences is a typical feature of expansions in momenta and masses. Only in rare situations, such as an expansion in the small momentum limit of a Feynman integral without massless threshold in the corresponding channel, there is no such phenomenon. Let me also point out that the first term in (4) is convergent at $\mathrm{Re}(\varepsilon) < 0$ while the second term in (4) is convergent at $\text{Re}(\varepsilon) > 0$. This can be seen from an analysis of convergence of the corresponding integrals over Feynman parameters. Thus, there is no domain in the complex plane of ε where both terms are given by convergent integrals. In fact, using auxiliary subtraction operators, it is possible to write down the result of expansion in such a way that both terms on the right-hand side will be convergent in some domain of ε. However. I prefer to follow the prescription which is implied in practice: to evaluate every term in the result for expansion in a domain of ε where it is convergent and then analytically continue the corresponding result to some desirable domain.

Expansion by regions has the status of experimental mathematics. Usually, when studying a given limit, one starts from one-loop examples, checks results by independent methods and, finally, one understands which regions are relevant to the limit and that one obtains reliable expansion within this strategy. Beneke provided a one-parametric example showing explicitly how expansion by regions works. The example was used in Chapter 3 of [1]. Guided by this example, Jantzen [3] provided detailed explanations of how this strategy works in several two-loop examples by starting from regions determined by some inequalities and covering the whole integration space of the loop momenta, then expanding the integrand and then extending integration and analyzing all the pieces which are obtained, with the hope that 'readers would be convinced that the expansion by regions is a well-founded method'.

However, there is an important class of limits for which there is a mathematical proof. These are limits typical of Euclidean space: for example, the off-shell large momentum limit and the large mass limit. In [4] (see also Appendix B of [1]) that the remainder of such expansion constructed with the help of an operator which has the structure of the R-operation (i.e. renormalization at the diagrammatical level) has the desirable order with respect to the parameter of expansion. This proof was for a general h-loop graph. It was similar to proofs of results on the R-operation and was based on sector decompositions and a resolution of singularities in parametric integrals, with power counting of sector variables.

For this class of limit, the expansion of a given Feynman integral corresponding to a graph Γ is given [4–6] (see also [7] and Chapter 9 of [8]) by the following simple formula:

$$G_\Gamma \sim \sum_\gamma G_{\Gamma/\gamma} \circ \mathcal{T}_{q_\gamma, m_\gamma} G_\gamma \ . \tag{7}$$

which is written for the off-shell large-momentum limit, i.e. where a momentum Q is considered large and momenta q_i as well as the masses m_j are small. The sum runs over subgraphs γ of Γ which can be called asymptotically irreducible (AI): they are one-particle irreducible after identifying the two external vertices associated with the large external momentum Q. Moreover, $\mathcal{T}$ is the operator of Taylor expansion in internal masses and external momenta of a subgraph γ, the symbol $\circ$ means the insertion of the polynomial obtained after this Taylor expansion into the vertex of the reduced graph Γ/γ to which γ is reduced.

In the case of limits typical of Euclidean space, there is a natural one-to-one correspondence between AI subgraphs and regions in the description of the expansion within expansion by regions, so that we obtain an indirect justification of expansion by regions for such limits. The set of relevant regions exactly corresponds to the set of AI subgraphs. There are two kind of regions for each loop momentum: small and large. For a given AI subgraph γ, the corresponding region is defined by considering each loop momentum of γ as large and the rest of the loop momenta of Γ (i.e. loop momenta of Γ/γ) as small. For example, two subgraphs are AI for Fig. 1: the graph Γ and the subgraph consisting of the massless line. As a result, we obtain the same contributions as above.

For limits typical of Minkowski space, to reveal the set of relevant regions is not so simple. For example, for the threshold limit in the case where the threshold in the q channel is at $q^2 = 4m^2$ and the small expansion parameter is introduced by $y = m^2 - q^2/4 \to 0$, the following four kind of regions for a loop momentum are relevant [2]:

$$
\begin{aligned}
\text{(hard)},\ & k_0 \sim \sqrt{q^2}\,,\ \mathbf{k} \sim \sqrt{q^2}\,,\\
\text{(soft)},\ & k_0 \sim \sqrt{y}\,,\ \mathbf{k} \sim \sqrt{y}\,,\\
\text{(potential)},\ & k_0 \sim y/\sqrt{q^2}\,,\ \mathbf{k} \sim \sqrt{y}\,,\\
\text{(ultrasoft)},\ & k_0 \sim y/\sqrt{q^2}\,,\ \mathbf{k} \sim y/\sqrt{q^2}\,.
\end{aligned}
$$

where $q = (q_0, \mathbf{0})$.

An alternative version of expansion by regions was formulated and illustrated via examples in [9] within the well-known Feynman parametric representation. This representation in the case of propagators with $-k^2$ propagators with general indices a_i (powers of the propagators) is

$$
G(q_1, \ldots, q_n; d) = \left(i\pi^{d/2}\right)^h \frac{\Gamma(\sum a - hd/2)}{\prod_i \Gamma(a_i)}
$$

$$
\times \int_0^\infty \ldots \int_0^\infty \delta\left(\sum x_i - 1\right) \prod x_i^{a_i - 1}\, U^{a-(h+1)d/2} F^{hd/2-a} \mathrm{d}x_1 \ldots \mathrm{d}x_n \quad (8)
$$

where n is the number of lines (edges), $a = \sum a_i$, h is the number of loops of the graph,

$$
F = -V + U \sum m_l^2 x_l \,, \quad (9)
$$

and U and V are two basic functions (Symanzik polynomials, or graph polynomials) for the given graph,

$$
U = \sum_{T \in T^1} \prod_{l \notin T} x_l \,, \quad (10)
$$

$$V = \sum_{T \in T^2} \prod_{l \notin T} x_l \left(q^T\right)^2 . \tag{11}$$

In (10), the sum runs over trees of the given graph, and, in (11), over *2-trees*, i.e. subgraphs that do not involve loops and consist of two connectivity components; $\pm q^T$ is the sum of the external momenta that flow into one of the connectivity components of the 2-tree T. The products of the Feynman parameters involved are taken over the lines that do not belong to a given tree or a 2-tree T. As is well known, one can choose the sum in the argument of the delta-function over any subset of lines. In particular, one can choose just one Feynman parameter, x_l, and then the integration will be over the other parameters at $x_l = 1$. The functions U and V are homogeneous with respect to Feynman parameters, with the homogeneity degrees h and $h + 1$, respectively.

One can consider quite general limits for a Feynman integral which depends on external momenta q_i and masses and is a scalar function of kinematic invariants and squares of masses, s_i, and assume that each s_i has certain scaling ρ^{κ_i} where ρ is a small parameter.

An algorithmic way to reveal regions relevant to a given limit was found in [10]. It is based on the geometry of polytopes connected with the basic functions U and F in (8). This was a real breakthrough, both in theoretical and practical sense because, on the one hand, it became possible to formulate expansion by regions in an unambiguous mathematical language and, on the other hand, the authors of [10] presented also a public code `asy.m` which was later successfully applied in various problems with Feynman integrals.

Ironically, this algorithm and the code didn't find, in this first version, the potential region for the threshold expansion. Later, this algorithm was updated and, in its current version, it can reveal potential region as well as Glauber region. This was done by introducing an additional decomposition of the integration domain and introducing new variables. Consider, for example, one-loop diagram with two massive lines in the threshold limit $y = m^2 - q^2/4 \to 0$

$$G(q^2, y) = i\pi^{d/2}\,\Gamma(\varepsilon) \int_0^\infty \int_0^\infty \frac{(x_1 + x_2)^{2\varepsilon-2}\, \delta\left(x_1 + x_2 - 1\right)\, dx_1 dx_2}{\left[\frac{q^2}{4}(x_1 - x_2)^2 + y(x_1 + x_2)^2 - i0\right]^\varepsilon} . \tag{12}$$

The code `asy.m` in its first version revealed only the contribution of the hard region, i.e. $x_i \sim y^0$. To make the potential region visible, let us decompose integration over $x_1 \le x_2$ and $x_2 \le x_1$, with equal contributions. In the first domain, let us turn to new variables by $x_1 = x_1'/2,\ x_2 = x_2' + x_1'/2$ and arrive at

$$i\pi^{d/2}\,\frac{\Gamma(\varepsilon)}{2} \int_0^\infty \int_0^\infty \frac{(x_1 + x_2)^{2\varepsilon-2}\, \delta\left(x_1 + x_2 - 1\right)\, dx_1 dx_2}{\left[\frac{q^2}{4}x_2^2 + y(x_1 + x_2)^2 - i0\right]^\varepsilon} .$$

Now we observe two regions with the scalings (0, 0) and (0, 1/2). The second one, with $x_1 \sim y^0, x_2 \sim \sqrt{y}$, gives

$$i\pi^{d/2}\frac{\Gamma(\varepsilon)}{2}\int_0^\infty \frac{\mathrm{d}x_2}{\left(\frac{q^2}{4}x_2^2+y\right)^\varepsilon} = i\pi^{d/2}\frac{1}{2}\Gamma(\varepsilon-1/2)\sqrt{\frac{\pi y}{q^2}}y^{-\varepsilon} .$$

Taking into account that we have two identical contributions after the above decomposition, we obtain a result for the potential contribution equal to the previous expression with omitted 1/2.

Observe that the expression for the function F in the Feynman parametric representation is non-negatively defined and only some individual terms are negative but this brings problems when looking for potential contributions. In the current version [11] of the code `asy.m`, one can get rid of the negative terms due to additional decompositions and introduction of new variable. Let me emphasize that this code can work successfully also in situations with a function F not positively defined even without additional decompositions—see, e.g. [12, 13].

For completeness, let me refer to [14, 15] where two specific ways of dealing with expansion by region were applied.

Let us realize that the very word 'region' is used within the strategy under discussion in a physical rather mathematical way. By region, we mean some scaling behaviour of parameters involved. I will present expansion by regions in a mathematical language in the next section using another form of parametric representation, rather than (8) and illustrate it through simple examples.

2 Geometrical Formulation

Lee and Pomeransky [16] have recently derived another form of parametric representation which turns out to be preferable in certain situations

$$G(q_1,\ldots,q_n;d) = \left(i\pi^{d/2}\right)^h \frac{\Gamma(d/2)}{\Gamma((h+1)d/2-a)\prod_i\Gamma(a_i)} \times \int_0^\infty \ldots \int_0^\infty \prod_i x_i^{a_i-1} P^{-\delta}\mathrm{d}x_1\ldots\mathrm{d}x_n , \qquad (13)$$

where $\delta = 2-\varepsilon$ and $P = U+F$. One can obtain (8) from (13) by [16] inserting $1 = \int \delta(\sum_i x_i - \eta)\mathrm{d}\eta$, scaling $x \to \eta x$ and integrating over η.

The parametric representation takes now a very simple form: up to general powers of the integration variables, there is only one polynomial raised to a general complex power. I believe that the fact that this function is the sum of the two basic functions in Feynman parametric representation is not crucial and expansion by regions holds for any polynomial.

Let us formulate, following [17], expansion by regions for integral (13) with a polynomial with positive coefficients in the case of limits with two kinematic invariants and/or masses of essentially different scale, where one introduces one parameter, t, which is the ratio of two scales and is considered small. These can be such limits typical of Minkowski space as the Regge limit, with $t \ll s$ and various versions of the Sudakov limit. Then the polynomial in Eq. (13) is a function of Feynman parameters and t,

$$P(x_1, \ldots, x_n, t) = \sum_{w \in S} c_w x_1^{w_1} \ldots x_n^{w_n} t^{w_{n+1}} , \tag{14}$$

where S is a finite set of points $w = (w_1, \ldots, w_{n+1})$ and $c_w > 0$.

By definition, the Newton polytope $\mathcal{N}_P$ of P is the convex hull of the points w in the $n+1$-dimensional Euclidean space $\mathbb{R}^{n+1}$ equipped with the scalar product $v \cdot w = \sum_{i=1}^{n+1} v_i w_i$. A facet of P is a face of maximal dimension, i.e. n.

The Main Conjecture (Expansion by Regions) The expansion of (13) in the limit $t \to +0$ is given by

$$G(t, \varepsilon) \sim \sum_{\gamma} \int_0^\infty \ldots \int_0^\infty \left[\mathcal{M}_\gamma \left(P(x_1, \ldots, x_n, t) \right)^{-\delta} \right] \mathrm{d}x_1 \ldots \mathrm{d}x_n , \tag{15}$$

where the sum runs over facets of the Newton polytope $\mathcal{N}_P$ of P, for which the normal vectors $r^\gamma = (r_1^\gamma, \ldots, r_n^\gamma, r_{n+1}^\gamma)$, oriented inside the polytope have $r_{n+1}^\gamma > 0$. Let us normalize these vectors by $r_{n+1}^\gamma = 1$. Let us call these facets *essential*.

The contribution of a given essential facet is defined by the change of variables $x_i \to t^{r_i^\gamma} x_i$ in the integral (13) and expanding the resulting integrand in powers of t. Let us write this procedure explicitly. For a given essential facet γ, the polynomial P is transformed into

$$P^\gamma(x_1, \ldots, x_n, t) = P(t^{r_1^\gamma} x_1, \ldots, t^{r_n^\gamma} x_n, t) \equiv \sum_{w \in S} c_w x_1^{w_1} \ldots x_n^{w_n} t^{w \cdot r^\gamma} . \tag{16}$$

The scalar product $w \cdot r^\gamma$ is proportional to the projection of the point w on the vector r^γ. For $w \in S$, it takes a minimal value for all the points belonging to the considered facet $w \in S \cap \gamma$. Let us denote it by $L(\gamma)$.

The polynomial (16) can be represented as

$$t^{L(\gamma)} \left(P_0^\gamma(x_1, \ldots, x_n) + P_1^\gamma(x_1, \ldots, x_n, t) \right) , \tag{17}$$

where

$$P_0^\gamma(x_1, \ldots, x_n) = \sum_{w \in S \cap \gamma} c_w x_1^{w_1} \ldots x_n^{w_n} , \tag{18}$$

$$P_1^\gamma(x_1,\ldots,x_n,t) = \sum_{w\in S\setminus\gamma} c_w x_1^{w_1}\ldots x_n^{w_n} t^{w\cdot r^\gamma - L(\gamma)} \,. \tag{19}$$

The polynomial P_0^γ is independent of t while P_1^γ can be represented as a linear combination of positive rational powers of t with coefficients which are polynomials of x.

For a given facet γ, let us define the operator

$$\begin{aligned} M_\gamma\left(P(x_1,\ldots,x_n,t)\right)^{-\delta} &= t^{\sum_{i=1}^n r_i^\gamma - L(\gamma)\delta}\mathcal{T}_t\left(P_0^\gamma(x_1,\ldots,x_n) + P_1^\gamma(x_1,\ldots,x_n,t)\right)^{-\delta} \\ &= t^{\sum_{i=1}^n r_i^\gamma - L(\gamma)\delta}\left(P_0^\gamma(x_1,\ldots,x_n)\right)^{-\delta} + \ldots \end{aligned} \tag{20}$$

where $\mathcal{T}_t$ performs an expansion in powers of t at $t = 0$.

Comments

- An operator M_γ can equivalently be defined by introducing a parameter ρ_γ, replacing x_i by $\rho^{r_i^\gamma} x_i$, pulling an overall power of ρ_γ, expanding in ρ_γ and setting $\rho_\gamma = 1$ in the end.
- The leading order term of a given facet γ corresponds to the leading order of the operator M_γ^0:

$$\begin{aligned} &\int_0^\infty \ldots \int_0^\infty \left[M_\gamma^0\left(P(x_1,\ldots,x_n,t)\right)^{-\delta}\right] \mathrm{d}x_1\ldots\mathrm{d}x_n \\ &= t^{-L(\gamma)\delta + \sum_{i=1}^n r_i^\gamma} \int_0^\infty \ldots \int_0^\infty \left(P_0^\gamma(x_1,\ldots,x_n)\right)^{-\delta} \mathrm{d}x_1\ldots\mathrm{d}x_n \,. \end{aligned} \tag{21}$$

- In fact, with the above definitions, we can write down the equation of the hyperplane generated by a given facet γ as follows

$$w_{n+1} = -\sum_{i=1}^n r_i^\gamma w_i + L(\gamma) \,. \tag{22}$$

- Let us agree that the action of an operator M_γ on an integral reduces to the action of M_γ on the integrand described above. Then we can write down the expansion in a shorter way,

$$G(t,\varepsilon) \sim \sum_\gamma M_\gamma G(t,\varepsilon) \tag{23}$$

- In the usual Feynman parametrization (8), the expansion by regions in terms of operators M_γ is formulated in a similar way, and this is exactly how it is implemented in the code `asy.m` [10]. The expansion can be written in the same form (23) but the operators M_γ act on the product of the two basic polynomials U

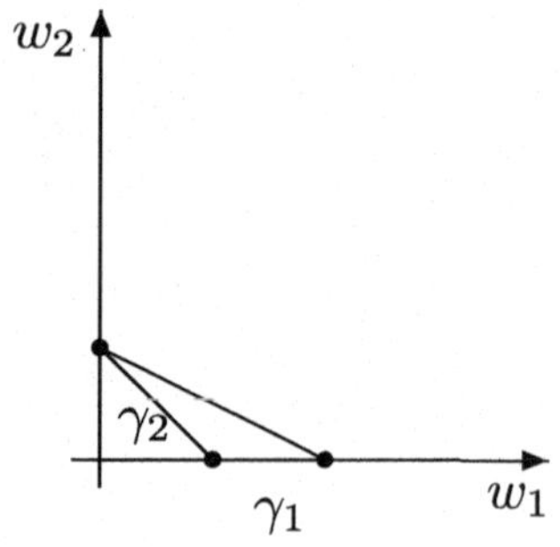

Fig. 2 The Newton polytope for (24)

and F raised to certain powers present in (8). Now, each of the two polynomials is decomposed in the form (17) and so on.

- Of course, prescriptions based on representation (13) are algorithmically preferable because the degree of the sum of the two basic polynomials is smaller than the degree of their product UF (previously used in `asy.m`) so that looking for facets of the corresponding Newton polytope becomes a simpler procedure.[1] Therefore, the current version of the code `asy.m` included in `FIESTA` [18] (called with the command `SDExpandAsy` is now based on this more effective procedure.
- It is well known that dimensional regularization might be insufficient to regularize individual contributions to the asymptotic expansion. As it was explained in the discussion after Eq. (2), the natural way to overcome this problem is to introduce an auxiliary analytic regularization, i.e. to introduce additional exponents λ_i to powers of the propagators. This possibility exists in the code `asy.m` [10] included in `FIESTA` [18]. One can choose these additional parameters in some way and obtain a result in terms of an expansion in λ_i followed by an expansion in ε. If an initial integral can be well defined as a function of ε then the cancellation of poles in λ_i in the sum of contributions of different regions serves as a good check of the calculational procedure, so that in the end one obtains a result in terms of a Laurent expansion in ε up to a desired order.

To illustrate the above prescriptions let us consider a very simple example of the integral

$$G(t,\varepsilon) = \int_0^\infty (x^2 + x + t)^{\varepsilon-1} \mathrm{d}x \tag{24}$$

in the limit $t \to 0$. The polynomial involved is $P(x,t) = \sum_{(w_1,w_2)\in S} c_{(w_1,w_2)} x^{w_1} t^{w_2}$. The corresponding Newton polytope (triangle) is shown in Fig. 2.

[1]In fact, this step is performed within `asy.m` with the help of another code `qhull`. It is most time-consuming and can become problematic in higher-loop calculations.

There are two essential facets γ_1 and γ_2 with the corresponding normal vectors $r_1 = (0, 1)$ and $r_2 = (1, 1)$. For the facet γ_1, we obtain the contribution given by expanding the integrand in t. In the leading order, we have

$$\int_0^\infty (x^2 + x)^{\varepsilon-1} \mathrm{d}x = \frac{\Gamma(1-2\varepsilon)\Gamma(\varepsilon)}{\Gamma(1-\varepsilon)} . \tag{25}$$

For the facet γ_2, we obtain t times the integral of the integrand with $x \to tx$ expanded in powers of t. In the leading order, we have

$$t^\varepsilon \int_0^\infty (x+1)^{\varepsilon-1} \mathrm{d}x = -\frac{t^\varepsilon}{\varepsilon} . \tag{26}$$

The sum of the two contributions in the leading order gives

$$G(t, \varepsilon) \sim -\log t + O(\varepsilon) . \tag{27}$$

Let us now consider again the example of Fig. 1. The two basic functions of Feynman parameters are

$$F = x_1(t(x_1 + x_2) + x_2) , \quad U = x_1 + x_2 . \tag{28}$$

The set S involved in the definition (14) consists of the vertices

$$A(2, 0, 1), B(1, 1, 1), C(1, 1, 0), D(1, 0, 0), E(0, 1, 0)$$

of the Newton polytope for the polynomial $P = U + F$, as it is shown in Fig. 3.

There are two essential facets. The first one is CDE which belongs to the plane $w_3 = 0$ and has the normal vector $(0, 0, 1)$. It gives the contribution obtained by expanding the integrand in t.

The second essential facet is ACD which belongs to the plane $w_1 - w_3 = 1$ and has the normal vector $(-1, 0, 1)$. It gives $t^{-\varepsilon}$ times the integral

$$\frac{\Gamma(2-\varepsilon)}{\Gamma(1-2\varepsilon)} \int_0^\infty \int_0^\infty x_1 \left[x_1 + x_1^2 + x_1 x_2 + t x_2 + t x_1 x_2\right]^{\varepsilon-2} \mathrm{d}x_1 \mathrm{d}x_2$$

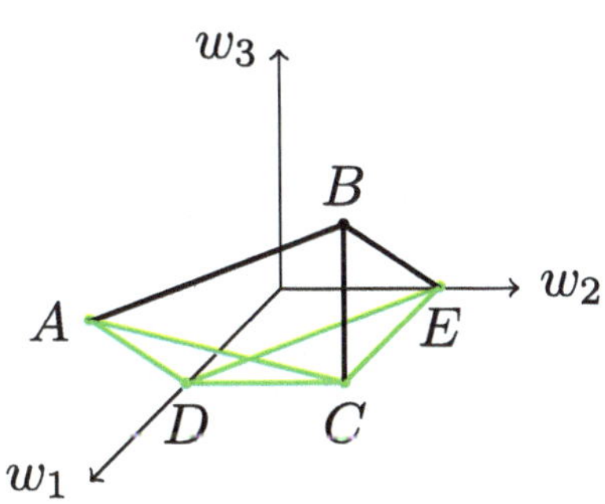

Fig. 3 The Newton polytope for Fig. 1

with the integrand expanded in t. Taking the leading orders in both contributions we reproduce (5).

3 Conclusion

As it was argued in [17], the more general parametric representation (13), with a general polynomial not necessarily related to Feynman integrals, looks mathematically more natural for the proof of expansion by regions. Moreover, first steps of analysis of convergence of integrals (13) were made and expansion by regions was proven in a partial case in the leading order of expansion. Hopefully, expansion by regions will be sooner or later mathematically justified in the case of a general polynomial P.

Practically, expansion by regions is a very important strategy which is successfully applied for several purposes. Let me, finally, point out that one can use expansion by regions in various ways.

- One can apply the code `asy.m` included in `FIESTA` [18] (i.e. the command `SDExpandAsy`) to obtain an expansion in some limit treating all the involved parameters numerically. In particular, one can check analytic results.
- One can use `SDExpandAsy` with the option `OnlyPrepareRegions = True` in order to reveal relevant regions and to construct contributions to the expansion as parametric integrals which can then analytically be evaluated. Here the method of Mellin-Barnes representation can serve as an appropriate additional technique.
- One can study expansion in multiscale limits, applying `asy.m` several times, in various orders.

Acknowledgments I would like to thank the organizers of the Paris Winter Workshop "The Infrared in QFT" (2–6 March 2020, Paris) and the workshop "Antidifferentiation and the Calculation of Feynman Amplitudes" (DESY Zeuthen,4–9 October 2020) for the possibility to present this talk. Supported by the Russian Science Foundation, agreement no. 21-71-30003.

References

1. V.A. Smirnov, Applied asymptotic expansions in momenta and masses. Springer Tracts Mod. Phys. **177**, 1–262 (2002)
2. M. Beneke, V.A. Smirnov, Nucl. Phys. **B522**, 321 (1998). [hep-ph/9711391]
3. B. Jantzen, JHEP **1112**, 076 (2011). [arXiv:1111.2589 [hep-ph]].
4. V.A. Smirnov, Commun. Math. Phys. **134**, 109 (1990)
5. K.G. Chetyrkin, Theor. Math. Phys. **75**, 346 (1988); **76**, 809 (1988)
6. S.G. Gorishnii, Nucl. Phys. B **319**, 633 (1989)
7. V.A. Smirnov, Mod. Phys. Lett. A **10**, 1485 (1995). [arXiv:hep-th/9412063 [hep-th]]

8. V.A. Smirnov, Analytic tools for Feynman integrals. Springer Tracts Mod. Phys. **250**, 1–296 (2012)
9. V.A. Smirnov, Phys. Lett. B **465**, 226 (1999). [hep-ph/9907471]
10. A. Pak, A.V. Smirnov, Eur. Phys. J. C **71**, 1626 (2011). [arXiv:1011.4863 [hep-ph]]
11. B. Jantzen, A.V. Smirnov, V.A. Smirnov, Eur. Phys. J. C **72**, 2139 (2012). [arXiv:1206.0546 [hep-ph]]
12. J.M. Henn, K. Melnikov, V.A. Smirnov, JHEP **05**, 090 (2014). [arXiv:1402.7078 [hep-ph]]
13. F. Caola, J.M. Henn, K. Melnikov, V.A. Smirnov, JHEP **09**, 043 (2014). [arXiv:1404.5590 [hep-ph]]
14. B. Ananthanarayan, A. Pal, S. Ramanan, R. Sarkar, Eur. Phys.J . C **79** 1, 57 (2019). [arXiv:1810.06270 [hep-ph]]
15. G. Mishima, JHEP **02**, 080 (2019). [arXiv:1812.04373 [hep-ph]]
16. R.N. Lee, A.A. Pomeransky, JHEP **1311**, 165 (2013). [arXiv:1308.6676 [hep-ph]]
17. T.Yu. Semenova, A.V. Smirnov, V.A. Smirnov, Eur. Phys. J. C **79**, 2, 136 (2019). [arXiv:1809.04325 [hep-th]]
18. A.V. Smirnov, Comput. Phys. Commun. **204**, 189 (2016). [arXiv:1511.03614 [hep-ph]]

Some Steps Towards Improving IBP Calculations and Related Topics

J.A.M. Vermaseren

Abstract A number of aspects of IBP reductions are discussed, indicating some of the major bottlenecks. Potential future developments are indicated, including some in the realm of mathematics and computer algebra.

1 Introduction

We all know the principle of the IBP relations. They are based on the equation

$$\int d^D p \frac{\partial}{\partial p^\mu} I^\mu = 0 \tag{1}$$

and working out the derivative. This sounds simple, but when applied to Feynman diagrams it can lead to great complications. This working out was first done in a nontrivial way by Chetyrkin and Tkachov [1], but it took about 10 years before their algorithms could actually be used in an automated way to compute the $O(\alpha_S^3)$ contribution [2] to the reaction $e^+e^- \rightarrow hadrons$. This is a principle we are still struggling with: there is much time involved in going from an idea to obtaining physics results. In particular more and more mathematics and computer science is needed.

The general problem is of course how to compute reactions by whatever means. The 'classical' way is to write down all Feynman diagrams and try to work them out. Before the days of LHC data relatively few people were involved at the technological edge of this approach, but the need of accurate theoretical results for very large numbers of reactions has changed this and nowadays this is a very active field with very many very smart people making steady progress.

J.A.M. Vermaseren (✉)
Nikhef Theory Group, Amsterdam, The Netherlands
e-mail: t68@nikhef.nl

J. Blümlein, C. Schneider (eds.), *Anti-Differentiation and the Calculation of Feynman Amplitudes*, Texts & Monographs in Symbolic Computation,
https://doi.org/10.1007/978-3-030-80219-6_19

The development knows two main branches:

- many legs and few loops
- few legs and many loops.

There are of course more subdivisions. Inclusion of mass parameters creates horrendous complications, just to name one.

Personally I work in the 'few legs and many loops' branch and in addition I try to keep the particles massless. There are some similarities here with the 'many legs and few loops' approach, because in principle the equations have the same structure, but in for instance massless propagator-type diagrams the equations can/should be reduced until only the dimension parameter remains in the coefficients, which I call a *complete reduction*, while otherwise the reduction steps finish with still other parameters remaining, resulting in very complicated final answers. In many cases one is already happy with numerical results.

Deriving and applying the IBP relations cannot be done without the use of sophisticated computer programs, either of the generic computer algebra type or more dedicated programs. One computer algebra program that has been developed with this kind of work in mind is Form [3, 4]. As mentioned later in this talk, it is still being extended.

There are different approaches to solving the IBP relations.

The oldest method is to work the integrals down to simpler and simpler integrals until the IBP relations cannot go any further as is done in the Mincer [5–7] and Forcer [8] programs. This requires a rather lengthy development time for creating good programs. Currently the attempt is to automate this approach.

In Laporta-style [9] methods one starts with the easiest integrals and uses the IBP relations to create more and more complicated integrals until one has all the integrals that are needed for a given problem. This generates usually many more integrals than needed and hence needs much storage. Its advantage is that it can (and has been) automated much easier. This has been done in programs like FIRE [10], Reduze [11, 12] and Kira [13]. In principle they could work at any number of loops and legs, provided enough computer resources are available, but in practise a 4-loop gluon propagator with all powers of the gauge parameter might be an enormous challenge. To our knowledge this has not been tried yet.

In this talk I will concentrate on the first method. It should be noted that also for this there are more methods than the one described here. The first was the method of Baikov [14], used for (among others) the first calculation of the 5-loop beta function in QCD [15]. In addition Roman Lee developed LiteRed [16, 17] which is an automatically constructed program that could in principle do the same things as Forcer, except for that (a) it is written in Mathematica and (b) it does not have the tailored tricks and methods that make Forcer so fast.

Of course, to get a full answer for a Feynman diagram involves more than 'just' solving a set of IBP equations. There are also the master integrals. These have become a specialization by themselves, attracting also attention from the more mathematically minded scientists as we will see in this talk.

In this talk I will highlight what I see as several paths that are followed in the 'few legs and many loops' branch to make it more powerful. Next I will discuss some mathematical aspects in the field of multiple zeta values (MZV's) and finally some new features in Form to support this.

There will not be very many formulas in this talk. In this subject it is either not many or too many.

2 Few Legs and Many Loops

The program with which this all started was Mincer [5, 6], originally developed for Schoonschip [18, 19] in the 1980s in Russia, and later reprogrammed into Form [7] at Nikhef. It managed to reduce 3-loop massless propagator-type diagrams into two master integrals plus convolutions of well known one-loop integrals. With ever increasing computer power it could be used for many more useful physics calculations.

The most complicated topology in Mincer was what was called the NO topology (for non planar) (Fig. 1):

The reduction scheme can be worked out by hand, but you would not want it to be much more complicated than it is. The diagram in which all denominators have one power is a master diagram. Originally it was worked out to its finite term as [1]

$$NO = 20\zeta_5 + O(\epsilon) \tag{2}$$

and a few improvements later it was worked out all the way to the ϵ^7 term (Roman Lee and the Smirnovs [20]).

This topology shows the beginning of a very annoying effect. Whereas there are 8 propagators/edges, there are 9 variables (not considering Q^2 which is used for scaling). One of these can only occur in the numerator. There are various possibilities to choose this variable, but all run into the problem that much of the reduction scheme deals with reducing its power to zero. Once this has been done, the remaining part is much simpler.

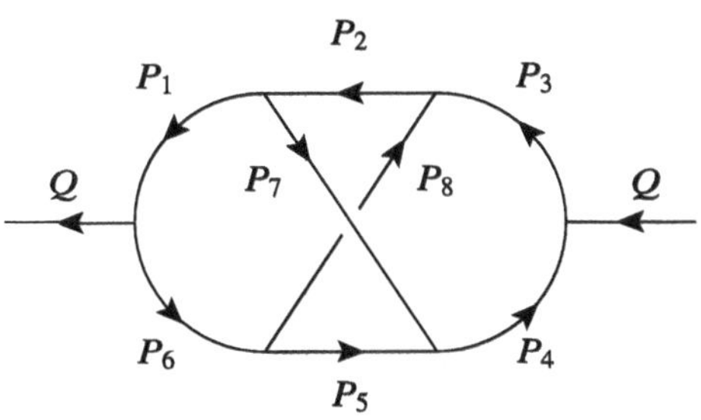

Fig. 1 The non-planar or NO topology

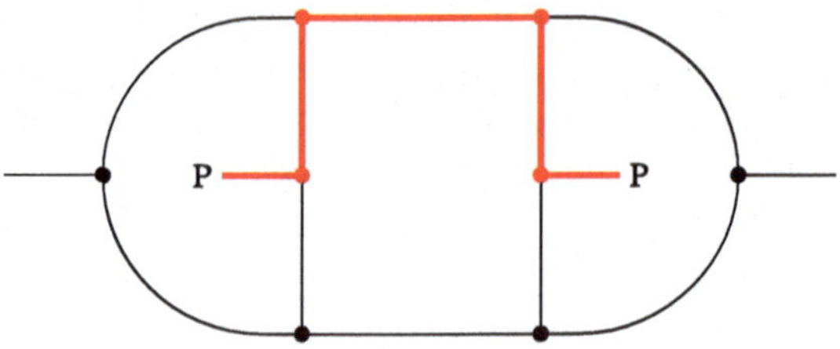

Fig. 2 The LA78 topology

One may wonder why we need reductions of higher powers of the denominators. Normally diagrams give only one power, and the diagram with all propagators at one is a master integral. Well, there are at least two cases for which higher powers occur:

First: When we want to make gauge checks gluon propagators look like

$$-i(g_{\mu\nu} - \xi q_\mu q_\nu / q^2)/q^2 \tag{3}$$

and hence we can obtain two powers of $1/q^2$ (Fig. 2).

Second: When we study Mellin moments, we have diagrams with 4 legs, two of which have momentum P. This P flows through the diagram and we expand in P. This means that for the N-th moment we have to take N derivatives in P, after which we set P to zero. This can give up to $N + 2$ extra powers of denominators in the remaining propagator-type diagram. The $+2$ comes from what remains after putting P to zero.

When we go to four loops things become much more complicated. Now there are 14 variables and at most 11 propagators. This gives very problematic reduction schemes. Some of the worst ones are (Fig. 3):

This was all implemented in the Forcer program. In the Forcer program actually the worst topology was (Fig. 4).

More recent research has shown that the scheme for this topology can be vastly improved by making a different selection for its 5 numerator variables. This brings us to the crucial question, relevant for all IBP solving systems: what is the optimal set of variables and how do we find it?

I have been involved in trying to find such variables in a systematic way. This is by no means easy for several reasons:

- There are many ways to select an independent set of variables. At the four-loop level there can already be hundreds and sometimes even thousands.
- Once you have a set there are many orders in which you can eliminate variables.
- Many orders of elimination become very slow and obtain rational polynomial coefficients that can crash the program.
- Some nasties that I will mention below.

Considering the above restrictions a nearly exhaustive search program has been executed only for one topology until now (Fig. 5):

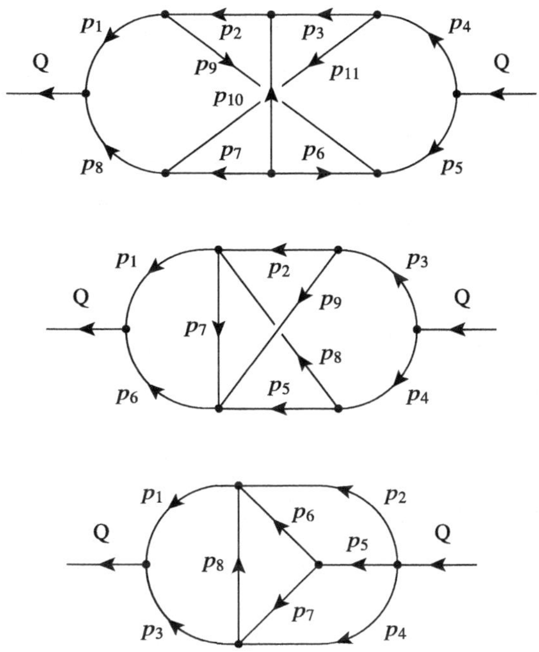

Fig. 3 The NO2, NONO and BEBE topologies

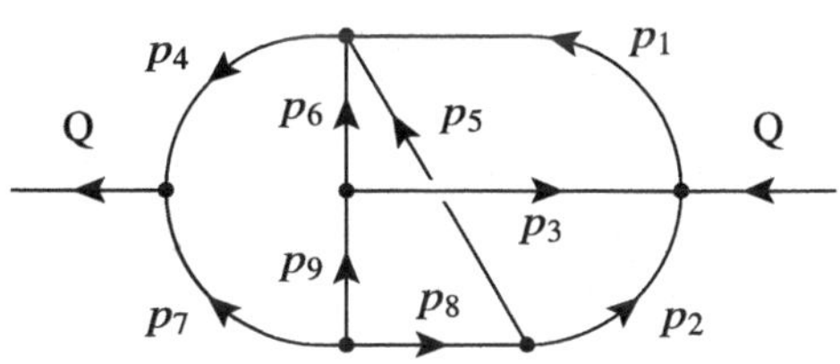

Fig. 4 The BUBU topology

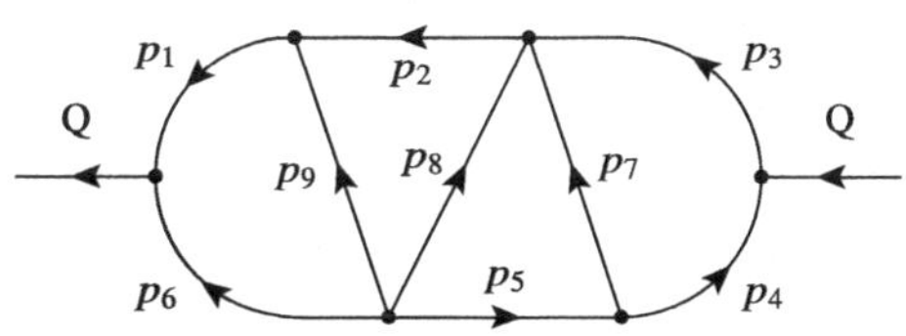

Fig. 5 The LALA topology

It did give indeed a much better reduction scheme than what was in Form. The more important BEBE topology however has still resisted a complete investigation due to extra complications which reduce the number of available equations and hence make most selections run into very long execution times (to be mentioned when we define complexity).

Programs like the above depend very much on built in smartness to reject hopeless cases as soon as possible. Such rules have to be found heuristically and which involves much time.

Of course one can derive reduction schemes automatically without such optimizations, but the result will be that the actual calculations will be extremely slow due to the inefficiency of the scheme.

The worst reduction equations are typically about halfway the reduction, when about half the number of parameters is still present, but already the powers of the monomials can become rather high. This is comparable to final results in the 'many legs and few loops' case where there are many physical parameters remaining. In the case of the 'few legs and many loops' diagrams and higher powers of numerators or denominators, one may have to apply such reduction formulas many times, hence it is important to keep such formulas as short as possible.

The Forcer program has been used already for a number of calculations. Some of them are

- Two different ways to compute the five-loop beta function.
- Higgs decay.
- A number of Mellin moments of the splitting functions in DIS at the 4-loop level. This is needed for the 3-loop Higgs production at the LHC.
- Even some Mellin moments of the splitting functions in DIS at the 5-loop level.

Currently the splitting functions for the non-singlet are being run at $N = 10$ and the singlet at $N = 8$. These are enormous tasks with individual diagrams that can take weeks on a computer with 16 cores. Sven Moch [21] will tell us more about it.

The hope is that when all topologies can be optimized some more Mellin moments can be computed, giving a higher precision in the determination of the PDF's at the LHC.

Having a four loop program, the next question is: what about five loops? The Forcer program as it exists was derived by hand guided computer programs. Reduction schemes were needed for 21 topologies and took three people several months to derive. For five loops there are about 10 times as many of such topologies, and most have 20 variables that need to be reduced. This is only feasible with an even higher level of automatization.

The flow from topology to topology is not so much the problem. For Forcer this gluing together of the topologies was done with a Python program, but the newest version of Form has the Kaneko diagram generator [22] built in, making it at least as easy to do this in Form itself. For a five (or more) loop program most of this has now been set up in Form. There are however a few unresolved issues. The most important are:

- The reduction scheme of the 'nontrivial' topologies.
- The selection of the notation for such topologies.

In the Forcer program it turns out that often most of the time is spent when transitioning from one topology to simpler topologies (with one fewer edge). They may have different notations. This is shown by (Fig. 6):

If we eliminate either of the lines 1,4,5,8 we obtain identical topologies, but the transitions to the notation of this topology are different. The same holds when we

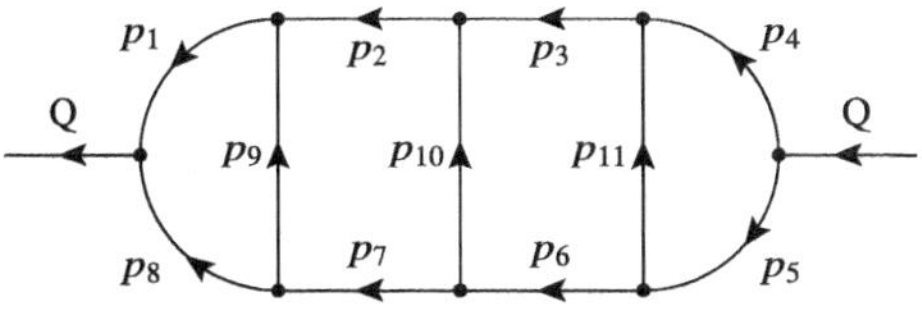

Fig. 6 The 4-loop ladder topology

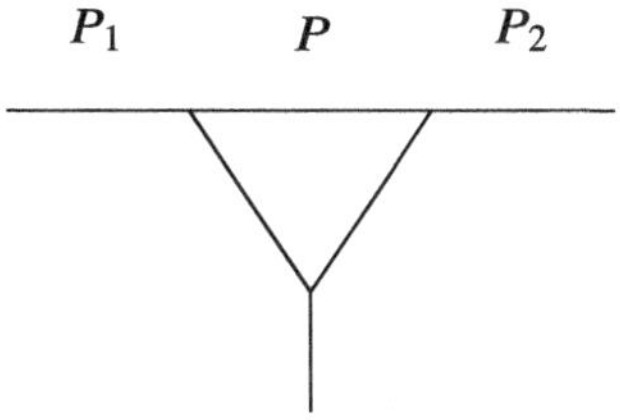

Fig. 7 The triangle diagram

eliminate either of the lines 2,3,6,7, and similar when we eliminate one of the lines 9 or 11.

It means that the numerators have to be rewritten, unless we have a reduction scheme for each notation that can be encountered. This would need for instance 34 different reduction schemes for the bebe topology. It has already been noticed that not all of those schemes would be very efficient. Hence one should create a number of efficient schemes, and rely for the others on a minimal amount of rewriting. It should be clear that designing a program that can do all this automatically is not trivial. Some progress has been made though.

We did find a much better bebe reduction scheme than the one in Forcer. Replacing the Forcer routine was not easy because it was all handwork to change the notations at the input and the output. The results was:

- A much faster overal running time.
- Fewer spurious poles (the equivalent of better numerical stability).

What are spurious poles?

IBP reduction algorithms have the tendency to create poles in ϵ whenever a line/edge is eliminated. The result is that diagrams with for instance the 3-loop ladder topology at the three loop level will give individual terms with up to 6 powers of $1/\epsilon$ during the reduction, while, once all terms are added, the leading divergence is at most $1/\epsilon^3$. This is the normal situation. Spurious poles can occur when there are powers of numerators. Let us take a close look at the most common reduction equation which is called the triangle relation (Fig. 7):

$$\begin{aligned}
&I(n,\alpha_0,\beta_1,\beta_2,\alpha_1,\alpha_2) = \Big(\\
&+\beta_1(I(n,\alpha_0-1,\beta_1+1,\beta_2,\alpha_1,\alpha_2) - I(n,\alpha_0,\beta_1+1,\beta_2,\alpha_1-1,\alpha_2)) \\
&+\beta_2(I(n,\alpha_0-1,\beta_1,\beta_2+1,\alpha_1,\alpha_2) - I(n,\alpha_0,\beta_1,\beta_2+1,\alpha_1,\alpha_2-1)) \\
&\Big)/(n+4-2\epsilon-2\alpha_0-\beta_1-\beta_2). \qquad (4)
\end{aligned}$$

The parameter n is the number of times the vector P occurs in the numerator. The parameters α_1 and α_2 are the powers of the two lines outside the triangle with the momenta P_1 and P_2, while β_1 and β_2 are the powers of the adjacent lines inside the triangle.

If all denominators have power 1 and there are no numerators, we see indeed that eliminating one line gives us a power of $1/\epsilon$. But we can obtain more than one power when there are more powers of numerators and denominators because now $(n + 4 - 2\alpha_0 - \beta_1 - \beta_2)$ can pass through zero more than once during the full reduction.

If one works with rational polynomials in ϵ this is not so much of a problem, but this is rather slow. It is faster to work with finite expansions in ϵ, thereby avoiding rational polynomial arithmetic. This is similar to rational numbers versus floating point numbers which have a finite accuracy and are sensitive to numerical instabilities.

In the case of the Mincer program the spurious poles were avoided by a summation of the recursion [23] and properly adding all resulting terms. This is however either not feasible or not practical when more loops are involved. Hence, whenever powers of numerators are involved we have to be prepared for extra powers of $1/\epsilon$. It can be rather difficult to predict how many. There are two solutions around this:

1. Work with exact rational coefficients.
2. Work with a cut off series but maybe run more than once to get an idea how many powers are needed cq. sufficient.

The first solution turned out to be rather expensive. Hence we worked with the second solution and eventually we got a heuristic formula to how many powers in ϵ we had to go to be able to trust the results. This formula depended linearly on the number of the Mellin moment.

As it turns out, with the new bebe routine this dependence disappeared. Let us see how that can be, because this will be important for the automated derivation of reduction schemes.

In order to have a reduction scheme one needs an ordering of the integrals in the form of a *complexity*. If each equation in the scheme lowers this complexity, and below a certain level at least one line/edge vanishes creating a simpler topology, eventually the scheme will terminate. The most important parameter is of course the number of edges. We cannot possibly accept relations that increase the number of edges. We do not include this parameter in our definition of complexity.

We have to proceed as follows:

1. The most important part of the complexity is the sum of the deviations from their minimal value for all parameters combined. We call this the sum-complexity.
2. Next we can lower individual parameters , but this may go at the cost of raising others, provided the sum complexity is not increasing.
3. Then we have to find an ordering of these reductions that does not create loops in the scheme.

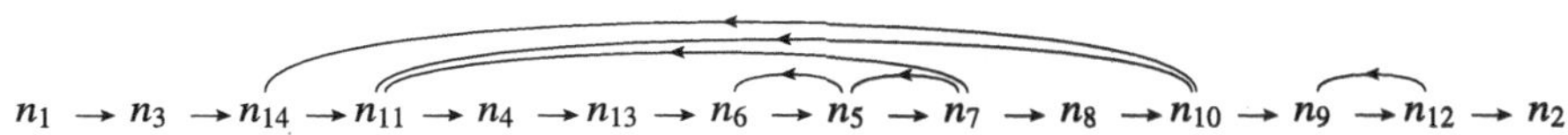

Fig. 8 A kickback scheme

The above three steps are in principle sufficient, but not necessarily efficient.

For the next step we need the concept of *kickbacks*. When we lower the sum complexity by one, it may well be that we lower several variables at the cost of several others. And if one of those that are raised is at the top of the scheme, we have to start all over again, be it with one sum complexity less. We call this a large kickback. The better schemes have as few of such kickbacks as possible, and the ones they have are only very few steps back. In particular, if whatever exists in kickbacks does not involve numerators, the number of spurious poles will be severely limited. Also the rational coefficients will be comparatively mild and so will be the number of terms at the end of the reduction (Fig. 8).

Hence the ordering of the reduction of the variables should take these kickbacks into account.

This is exactly what happened with the bebe routine. The old bebe routine, created by handguided computing, but without much experience, had a few rather bad kickbacks. The new one, designed with much more automation, gave already shorter code, and by exchanging the order of some of the variables by hand, the code became only marginally longer, but kickback-free. This made a big difference, both in speed and in the number of terms in the final answer.

The bebe reduction has another problem. The nasties mentioned before. The structure of the equations is such that the only solutions that we could find start off with eliminating two variables, but each step increases the sum complexity by one. This had to be done before the equations were generated that are used for the other variables, thereby avoiding kickbacks and hence infinite loops. It does however make the number of useful relations for the other reductions significantly smaller.

Such approaches are not needed for the other topologies, but can be used anyway in such a way that the results are improved. This is still being studied.

Unfortunately the situation with the other topologies that still need to be cleaned up is slightly more complicated. Also, it is not easy to replace topology routines inside Forcer, because of the changes in notation. A routine like bebe can be called in 34 different ways, and creates output in 8 different ways. This has to be changed by hand, because the original Python code had some global optimizations in which tiny changes anywhere, could change all numberings and notations of diagrams in the complete program, invalidating all partially built up databases. The future Form program should not have that problem. The hope is that with the new computer we just got we can make a new attempt at the other topologies.

Of course at the five loop level this will be even worse, because then 20 variables have to be eliminated, rather than 14, and in addition the equations will be lengthier. In the next section we will see how one can still obtain some five-loop results.

3 A Little Bit About Rstar

When calculating only the UV divergences of a diagram there are a few theorems [24–27] about extracting this divergence and rearranging external lines. They make the calculation feasible with a program that can solve diagrams with one loop fewer.

In the case of propagator diagrams one may extract the divergence by taking enough derivatives of the external momentum to make the result dimensionless.

If a propagator diagram is dimensionless, we may as well set the momentum to zero, or move external legs to other points in the diagram. This does not change the ultraviolet divergence, but it may affect the IR divergences. This way we can change the diagram to one in which we can do one integral and reduce the problem to an integral that can be done at a simpler level (Fig. 9).

Integrals with only one line between the two Q's we call carpet integrals and they can be done 'trivially'. But because now $p_1^2 = p_8^2$ we have introduced a potential infrared divergence. This is the main reason why it took so many years to do the four-loop QCD beta function and later the five-loop QCD beta function. Cannot we hunt down those IR divergences?

For the subtraction of divergences there exists a procedure, called Rstar. This has been used a lot in ϕ^4 theories. In QCD it is more complicated because the gluons introduce dotproducts in the numerators and hence complicated tensors. The whole is a very tricky interplay of determining subgraphs, locally eliminating tensors in the numerators by derivation and keeping the orders of limits and derivations correct. Details are in the paper by Herzog and Ruijl [28].

One practical problem is that the tensors need projectors to take them from inside the derivatives and the counterterm subtractions to outside of these.

Effectively one needs a projection operator $T^{\mu_1\cdots\mu_n}_{\nu_1\cdots\nu_n}$ that projects out products of Kronecker deltas like $\delta^{\mu_1}_{\nu_1}\cdots\delta^{\mu_n}_{\nu_n}$. As it turns out, it is an exercise in cosets of the symmetric group S_n in which each coset has a coefficient that is a rational

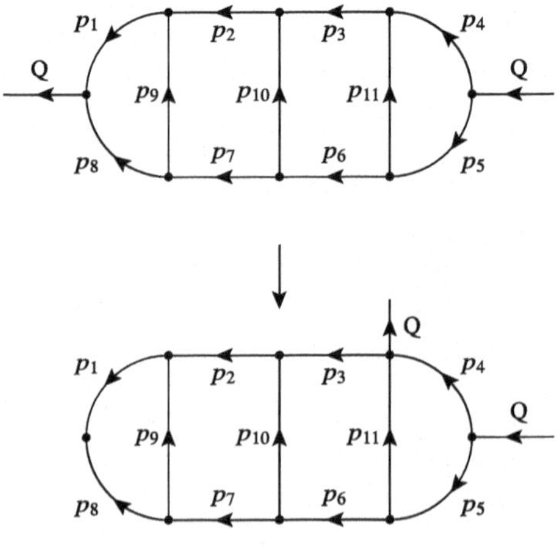

Fig. 9 Moving an external line to create a carpet integral

polynomial in the dimension D. And it happens that the generalised Kronecker delta dd_ in Form is ideal for this kind of work because, when the indices are contracted with momenta, of which there are only a few different ones, Form gets the combinatorics factors right without having to generate the same term many times. We managed to determine these coefficients all the way to 16 pairs of indices (in the process of being written up).

```
Vector p1,p2,p3;
Local F = dd_(p1,p2,p3,p1,p2,p3,p1,p2,p3,p1,p2,p3);
Print +f +s;
.end

Time =        0.00 sec    Generated terms =        15
                F         Terms in output =        15
                          Bytes used      =       868

   F =
       + 1728*p1.p1*p1.p2*p1.p3*p2.p2*p2.p3*p3.p3
       + 1152*p1.p1*p1.p2*p1.p3*p2.p3^3
       + 216*p1.p1*p1.p2^2*p2.p2*p3.p3^2
       + 864*p1.p1*p1.p2^2*p2.p3^2*p3.p3
       + 864*p1.p1*p1.p3^2*p2.p2*p2.p3^2
       + 216*p1.p1*p1.p3^2*p2.p2^2*p3.p3
       + 216*p1.p1^2*p2.p2*p2.p3^2*p3.p3
       + 27*p1.p1^2*p2.p2^2*p3.p3^2
       + 72*p1.p1^2*p2.p3^4
       + 1152*p1.p2*p1.p3^3*p2.p2*p2.p3
       + 864*p1.p2^2*p1.p3^2*p2.p2*p3.p3
       + 1728*p1.p2^2*p1.p3^2*p2.p3^2
       + 1152*p1.p2^3*p1.p3*p2.p3*p3.p3
       + 72*p1.p2^4*p3.p3^2
       + 72*p1.p3^4*p2.p2^2
      ;
```

To do all possible diagrams in QCD one also needs to deal with indices on gamma matrices. It is still an outstanding problem when there is a mixture of Kronecker delta's and gamma matrices. The gamma matrices can be written in the antisymmetric 'σ' basis, but it ceases to be a simple application of the symmetric group. This is currently under study.

Also here the continuous changing of (sub)topologies and hence notations is a serious problem. In the local Rstar operation this is done term by term making such operations very time consuming, unless specialized tables are constructed in advance. Part of this is already present in Form by means of the Kaneko diagram generator, but the canonicalization of the notations still has to be built in. Using external code is rather slow.

4 Mathematical Aspects

In the sequel we ignore color factors. Each color channel can be seen as a separate object for the purpose of the current discussion.

Evaluating master integrals is a science by itself. In the case that we have massless propagator graphs, the answer should be 'just a number'. This holds also for beta functions. What numbers are involved?

Until now Multiple Zeta Values have been sufficient. There are indications that at more loops these may not be sufficient. Thus far this has been shown for individual diagrams cq. topologies. This is not yet a proof that it will hold for complete physical results. Personally I have encountered two cases in which the more complicated objects cancelled each other.

In the first computation of the four-loop beta function in QCD there were master integrals that we could not evaluate at the time, but they dropped out when the diagrams were added. Here the method was to blame. The infrared regularization was based on masses and this gave rather hard integrals. When done with a program like Forcer there are no mass parameters and this effect does not occur. Also not at five loops.

In the computation of all Mellin moments of the coefficient functions in three-loop DIS, one sum occurred that was not a simple harmonic sum. We just gave it a name when it survived in individual diagrams, but when all diagrams were added this object vanished (Fig. 10).

It is of course likely that with extra loops such miracles do not occur.

When we have integrals or sums that end up in terms of MZV's or Euler sums there are usually very many different ones. There exist however many relations between them due to shuffle and stuffle relations. Such relations can be combined to reduce all different MZV's and Euler sums to an independent set. This set is usually rather restricted.

Example of a stuffle relation, which comes from the sum property of MZV's:

```
CFunction S;
L F = S(2,3)*S(5);
Stuffle S-;
Print +s;
.end
```

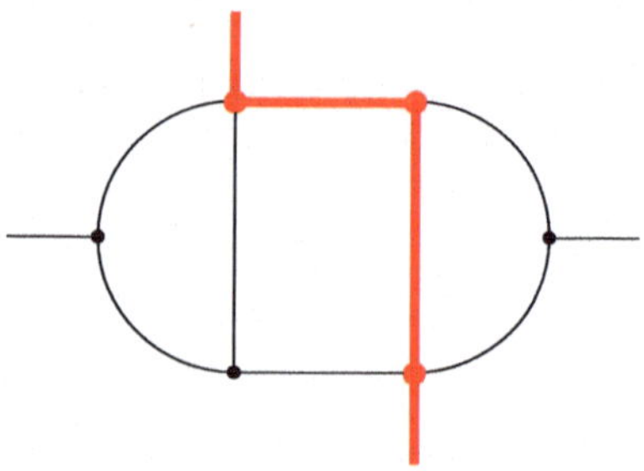

Fig. 10 A nontrivial but cancelling integral

```
F =
    + S(2,3,5) + S(2,5,3)
    - S(2,8) + S(5,2,3) - S(7,3);
```

Example of a shuffle relation, which comes from the integral property of MZV's:

```
 CFunction H;
 L F = H(2,3)*H(5);
 Transform H,ToIntegralNotation(1,last);
 Print;
 .sort
F =
   H(0,0,0,0,1)*H(0,1,0,0,1);

 Shuffle H;
 Transform H,ToSumNotation(1,last);
 Print +s;
 .end
F =
    + H(2,3,5) + 3*H(2,4,4) + 7*H(2,5,3) + 15*H(2,6,2)
    + 30*H(2,7,1) + 2*H(3,3,4) + 8*H(3,4,3) + 20*H(3,5,2)
    + 40*H(3,6,1) + 6*H(4,3,3) + 18*H(4,4,2) + 36*H(4,5,1)
    + 5*H(5,2,3) + 12*H(5,3,2) + 24*H(5,4,1) + 10*H(6,1,3)
    + 5*H(6,2,2) + 10*H(6,3,1);
```

By writing down all shuffle and stuffle relations and 'solving' them, one can determine a basis and express all MZV's in terms of this basis.

Complete reduction relations have been worked out in the MZV datamine [29] and go to weight 12 for the Euler sums and weight 22 for the MZV's. This was the limit of what could be reached with the existing computer resources at the time (2009). Recently I got a new computer with 64 cores and 1 Tbyte of memory. I managed to compute the weight 13 Euler sum reductions. The run took 11 days in which it used 637 days of CPU time (a pseudo-efficiency of 58). The run for weight 12 took 5 hours, using 161 hours of CPU time. It shows that it is not realistic to expect much of an extension to the datamine. Actually the situation can become much worse when the alphabet has more than 3 elements.

As Broadhurst [30] showed, at a given moment some not very high loop diagrams may result in sums in which the alphabet involves the sixth root of unity. Let us consider how many finite sums exist for a given alphabet as a function of the number m of its elements. For the MZV's the alphabet contains the two elements 0 and 1, while for the Euler sums the alphabet contains the elements 0, 1 and -1. The generic formula is $(m-1)^2 m^{N-2}$, possibly divided roughly by two when there are duality considerations. In general one needs to solve only for about one half the number of variables (Table 1).

To do this for such a large number of variables requires special programs, as is explained in the paper about the datamine. One complication is that the rational coefficients become rather bad. Their complexity seems to be dictated by the number of variables. Of course one can try to solve the systems over a finite field, like modulus a 31 bit prime number and after doing this several times, one can reconstruct the coefficients. This was also done for the MZV's at weight 22, but the

Table 1 Number of sums for various sizes of the alphabet

Weight	2	3	7
2	1	4	36
3	2	12	252
4	4	36	1764
5	8	108	12,348
6	16	324	86,436
7	32	972	605,052
8	64	2916	4,235,364
9	128	8748	
10	256	26,244	
11	512	78,732	
12	1024	236,196	
13	2048	708,588	
22	1,048,576		

result was that the combined running time of the modulus programs took more time than working directly with the rational numbers.

If weight 13 for the alternating sums is already barely feasible, it should be clear that we can never reach such weights for the sixth root of unity. Hence another approach is called for, which is based on a paper by Francis Brown [31] and a paper by Borwein, Bradley and Broadhurst [32]. Most advanced in applying this is Oliver Schnetz [33]. The way it works is as follows:

Based on the theory of motivic multiple zeta values one can construct algebraic operators that can map any MZV onto elements in a space that has the same dimension d as a basis for the given weight w. Because this is all linear, one can determine whether, if one chooses d MZV's, these form a basis for this weight. This needs however already expressions for weights lower or equal to $w-3$, which means that the whole algorithm is effectively recursive. Once a basis is known one can obtain the coefficients for each individual MZV in terms of this basis with the exception of one coefficient: the coefficient of the depth 1 term, which, when the weight is even, can be expressed as a power of ζ_2.

This is where the Broadhurst paper comes in. He explains how one can determine any MZV numerically to very high precision. If necessary many thousands of digits. Having routines for this allows then to determine the last coefficient numerically in terms of a high precision floating point number, which then can be converted to a fraction.

The above method is explained for the MZV's but can also be applied to Euler sums or to alphabets based on higher roots of unity. In this way Oliver Schnetz [33] has created already programs that can determine alternating sums to weight 21 and MZV's to weight 32. In addition he has already some programs for higher roots of unity, like up to weight 13 for the sixth roots of unity and up to weight 17 for the third and fourth roots of unity.

For the higher weights the programs can become a bit slow, because they will need to do calculations with many thousands of bits accuracy, and the evaluation of the MZV's needs as many steps as there are bits in the accuracy.

Currently those programs are mainly interesting from the mathematical viewpoint, but it is inevitable that once more loops can be dealt with, physics will need them as well.

5 Computer Algebra

It should be clear that when diagrams or MZV's are calculated the easiest would be if it can all be done in a single program. Unfortunately version 4 of Form cannot deal with floating point numbers. Hence for the necessary type of operations version 5 will be equipped with some new functions that allow access to the arbitrary precision floating point facilities of the GMP library. In addition there will be some built in functions like mzv_ and euler_ to evaluate these sums to a user defined precision. More functions (sqrt_, ln_, etc.) still need to be programmed. And of course there are some commands that allow conversions between rational and floating point numbers. This was not completely trivial, because this is rather specialized stuff that should not be in the way of the regular operations of Form.

Let me give an example: Assume we want to know the expression for the

```
L  F = mzv_(2,5,3);
```

The Francis Brown part of the program would give us

```
L   Fa = +20*mzv_(2)*mzv_(3)*mzv_(5)
         -34*mzv_(3)*mzv_(7)
         -447/14*mzv_(5)^2
          -22/7*mzv_(7,3);
```

but the coefficient of `mzv_(10)` or `mzv_(2)^5` is missing. The interesting part of the program becomes now

```
*
* Program to test the reconstruction of the depth 1
*   term for MZV's.
* We take a relation from the datamine. For example:
*
*Fill mzv10(0,1,0,0,0,0,1,0,0,1)=
*   38686/13475*z2^5+20*z2*z3*z5
*   -34*z3*z7-447/14*z5^2-22/7*z7z3;
*
#StartFloat 100
*
Off Statistics;
L F = mzv_(2,5,3);
L Fa = +20*mzv_(2)*mzv_(3)*mzv_(5)
      -34*mzv_(3)*mzv_(7)
      -447/14*mzv_(5)^2
```

```
          -22/7*mzv_(7,3);
 L Fb = mzv_(2)^5;
 Evaluate mzv_;
 Print +f +s;
 .sort

F=
   +1.401734757585418072158234569434e-02
   ;

Fa=
   -3.456140777152661943477201458825e+01
   ;

Fb=
   +1.204321598200656133921253731802e+01
   ;
 L FF = (F-Fa)/Fb;
 Print +f +s FF;
 .sort

FF=
   +2.870946196660482374768089053803e+00
   ;
 ToRational;
 Print +f +s FF;
 .end

FF=
   +38686/13475
   ;
```

Checking with the datamine gives indeed this fraction.

The major work now is to figure out what is a good accuracy for the floating point numbers. This will be a function of the type of the sums and the maximal weight we would like to go to. This needs some experimentation and extrapolation. For now we can use the datamine of course.

For other roots of unity the experimental method is called for. If there is not enough accuracy, the conversion to a fraction will give rather ‘unphysical’ results. One can start with too much accuracy for a number of cases and then work it down to see how long the proper fractions are reconstructed. Then give it some extra accuracy and one is ready to go. And there is always the check of the stuffle and shuffle relations.

Of course, much speed can be gained if we tabulate the numerical value of the basis elements. This has one problem. These basis elements have different numbers of indices and tables have typically a fixed number of indices. Hence Form has now been extended with tables of which the number of indices does not have to be fixed. This gives yet another complication. Tables in Form can have arguments in addition to indices, and the number of arguments follows the rules of wildcarding, and hence the number of arguments can also not be fixed. Therefore the notation becomes:

```
Table,sparse,T1(<=5,arguments);
Table,sparse,T2(<=10,arguments);
```

in which ‘arguments’ represents a potential field of arguments as in the definition of regular tables. It needs a maximum number, because space is reserved for this maximum.

When such tables are used, the first index should be a number that tells how many real indices there are, because as mentioned before, the number of arguments does not have to be constant when `?a` style wildcards are involved, which pick up a whole field of arguments.

With the above we can make one single table for all basis elements of which we would like to store the numerical value. Such a table can be stored for future programs and its size is only a small fraction of the size of complete tables.

6 Outlook

To make progress in the field of few legs and many loops we need either completely new methods, or a number of very powerful new automated programs to break down the integrals to master integrals.

It is worth to put effort in this, because accurate calculations for reactions at the LHC need reliable 4-loop splitting functions.

Steady progress is made and more is needed, both in the fields of mathematics and computer algebra.

Dedicated programs for specific tasks may help, but have mainly been constructed for the ‘many legs and few loops’ calculations.

References

1. K.G. Chetyrkin, F.V. Tkachov, Integration by parts: The algorithm to calculate beta functions in 4 loops. Nucl. Phys. **B192**, 159 (1981)
2. S.G. Gorishny, A.L. Kataev, S.A. Larin, The $O(\alpha_s^3)$-corrections to $\sigma_{tot}(e^+e^- \rightarrow hadrons)$ and $\Gamma(\tau \rightarrow \nu_\tau + hadrons)$ in QCD. Phys. Lett. **B212**, 238 (1988); ibid. **B259**, 144 (1991)
3. J.A.M. Vermaseren, New features of FORM. arXiv:math-ph/0010025
4. B.Ruijl, T. Ueda, J. Vermaseren, FORM version 4.2, e-print arXiv:1707.06453(hep-ph)
5. S.G. Gorishny, S.A. Larin, F.V. Tkachov, INR preprint P-0330 (Moscow, 1984)
6. S.G. Gorishny, S.A. Larin, L.R. Surguladze, F.V. Tkachov, Mincer: Program for multiloop calculations in quantum field theory for the schoonschip system. Comput. Phys. Commun. **55**, 381 (1989)
7. S.A. Larin, F.V. Tkachov, J.A.M. Vermaseren, The FORM version of MINCER, Nikhef report NIKHEF-H-91-18
8. B. Ruijl, T. Ueda, J.A.M. Vermaseren, Forcer, a FORM program for the parametric reduction of four-loop massless propagator diagrams. Comput. Phys. Commun. **253**, 107198 (2020)

9. S. Laporta, High precision calculation of multiloop Feynman integrals by difference equations. Int. J. Mod. Phys. A **15**, 5087–5159 (2000)
10. A.V. Smirnov, Algorithm FIRE – Feynman integral REduction. JHEP **10**, 107 (2008)
11. C. Studerus, Reduze-Feynman integral reduction in C++. Comput. Phys. Commun. **181**, 1293–1300 (2010)
12. A. von Manteuffel, C. Studerus, Reduze 2 - Distributed Feynman integral reduction. e-Print: arXiv:1201.4330 [hep-ph]
13. P. Maierher, J. Usovitsch, P. Uwer, Kira-A Feynman integral reduction program. Comput. Phys. Commun. **230**, 99–112 (2018)
14. P.A. Baikov, K.G. Chetyrkin, J.H. Kuhn, Order alpha**4(s) QCD corrections to Z and tau decays. Phys. Rev. Lett. **101**, 012002 (2008)
15. P.A. Baikov, K.G. Chetyrkin, J.H. Kühn, Five-loop running of the QCD coupling constant. Phys. Rev. Lett. **118**(8), 082002 (2017
16. R.N. Lee, Presenting LiteRed: a tool for the loop InTEgrals REDuction. e-Print: arXiv:1212.2685 [hep-ph]
17. R.N. Lee, LiteRed 1.4: a powerful tool for reduction of multiloop integrals. J. Phys. Conf. Ser. **523**, 012059 (2014)
18. M. Veltman, An IBM-7090 Program for Symbolic Evaluation Of Algebraic Expressions, Especially Feynman Diagrams. CERN preprint (1965)
19. H. Strubbe, 'Manual for schoonschip: A CDC 6000/7000 program for symbolic evaluation of algebraic expressions. Comput. Phys. Commun. **8**, 1 (1974)
20. R.N. Lee, A.V. Smirnov, V.A. Smirnov, Master integrals for four-loop massless propagators up to transcendentality weight twelve. Nucl. Phys. **B856**, 95–110 (2012)
21. S.O. Moch, Talk in this workshop
22. T. Kaneko, A Feynman graph generator for any order of coupling constants. Comput. Phys. Commun. **92**, 127–152 (1995)
23. F.V. Tkachov, An algorithm for calculating multiloop integrals. Teor. Mat. Fiz. **56**, 350 (1983)
24. K.G. Chetyrkin, F.V. Tkachov, Infrared R operation and ultraviolet counterterms in the MS scheme. Phys. Lett. **B114**, 340 (1982)
25. K.G. Chetyrkin, V.A. Smirnov, R* operation corrected. Phys. Lett. **B144**, 419 (1984)
26. V.A. Smirnov, K.G. Chetyrkin, R* operation in the minimal subtraction scheme. Theor. Math. Phys. **63**, 462 (1985)
27. K.G. Chetyrkin, Combinatorics of R-, R-1-, and R*-operations and asymptotic expansions of feynman integrals in the limit of large momenta and masses. arXiv:1701.08627
28. F. Herzog, B. Ruijl, The R*-operation for Feynman graphs with generic numerators. JHEP **05**, 037 (2017)
29. J. Blümlein, D. Broadhurst, J.A.M. Vermaseren, The multiple zeta value data mine. Comput. Phys. Commun. **181**, 582–625 (2010). e-Print: arXiv:0907.2557 [math-ph]
30. D.J. Broadhurst, Massive three - loop Feynman diagrams reducible to SC* primitives of algebras of the sixth root of unity. Eur. Phys. J. C **8**, 311–333 (1999). e-Print: arXiv:hep-th/9803091 [hep-th]
31. F. Brown, On the decomposition of motivic multiple zeta values. e-Print: arXiv:1102.1310 [math.NT]
32. J.M. Borwein, D.M. Bradley, D.J. Broadhurst, Special values of multiple polylogarithms. Trans. Am. Math. Soc. **353**, 907–941 (2001). e-Print: arXiv:math/9910045 [math.CA]
33. O. Schnetz: Private communication

Iterated Integrals Related to Feynman Integrals Associated to Elliptic Curves

Stefan Weinzierl

Abstract This talk reviews Feynman integrals, which are associated to elliptic curves. The talk will give an introduction into the mathematics behind them, covering the topics of elliptic curves, elliptic integrals, modular forms and the moduli space of n marked points on a genus one curve. The latter will be important, as elliptic Feynman integrals can be expressed as iterated integrals on the moduli space $\mathcal{M}_{1,n}$, in same way as Feynman integrals which evaluate to multiple polylogarithms can be expressed as iterated integrals on the moduli space $\mathcal{M}_{0,n}$. With the right language, many methods from the genus zero case carry over to the genus one case. In particular we will see in specific examples that the differential equation for elliptic Feynman integrals can be cast into an ε-form. This allows to systematically obtain a solution order by order in the dimensional regularisation parameter.

1 Introduction

In this talk we review Feynman integrals associated to elliptic curves and the mathematics behind them. It has become common practice to call these Feynman integrals "elliptic Feynman integrals". Elliptic Feynman integrals and closely related integrals in string theory have received considerable attention in recent years [1–63].

We call a Feynman integral elliptic, if it can be expressed as a linear combination of iterated integrals on a covering space of the moduli space $\mathcal{M}_{1,n}$ of a genus one curve with n marked points with integrands having only simple poles. "Ordinary" Feynman integrals, which evaluate to multiple polylogarithms, can be expressed as a linear combination of iterated integrals on a covering space of the moduli space

S. Weinzierl (✉)
PRISMA Cluster of Excellence, Institut für Physik, Johannes Gutenberg-Universität Mainz, Mainz, Germany
e-mail: weinzierl@uni-mainz.de

J. Blümlein, C. Schneider (eds.), *Anti-Differentiation and the Calculation of Feynman Amplitudes*, Texts & Monographs in Symbolic Computation,
https://doi.org/10.1007/978-3-030-80219-6_20

$\mathcal{M}_{0,n}$ of a genus zero curve with n marked points, again with integrands having only simple poles.

This definition already uses some mathematical terminology, which we explain in the sequel. As a rough guide, elliptic Feynman integrals are the next-to-easiest Feynman integrals, with Feynman integrals evaluating to multiple polylogarithms being the easiest Feynman integrals. Of course, there are also more complicated Feynman integrals beyond these two categories [64–68]. These more complicated integrals are not the topic of this talk.

2 Background from Mathematics

We review the background from mathematics. The material presented in this section is probably well-known to mathematicians. It might help physicists as a starting guide into this topic. Textbooks on elliptic curves are Du Val [69] and Silverman [70], textbooks on modular forms are Stein [71], Miyake [72] and Diamond and Shurman [73].

We start with the definition of an algebraic curve. As ground field we take the complex numbers $\mathbb{C}$. An algebraic curve in $\mathbb{C}^2$ is defined by the zero set of a polynomial $P(x, y)$ in two variables x and y:

$$P(x, y) = 0 \tag{1}$$

It is more common to consider algebraic curves not in the affine space $\mathbb{C}^2$, but in the projective space $\mathbb{CP}^2$. Let $[x : y : z]$ be homogeneous coordinates of $\mathbb{CP}^2$. An algebraic curve in $\mathbb{CP}^2$ is defined by the zero set of a homogeneous polynomial $P(x, y, z)$ in the three variables x, y and z:

$$P(x, y, z) = 0 \tag{2}$$

The requirement that $P(x, y, z)$ is a homogeneous polynomial is necessary to have a well-defined zero set on $\mathbb{CP}^2$.

We usually work in the chart $z = 1$. In this chart Eq. (2) reduces to

$$P(x, y, 1) = 0. \tag{3}$$

If d is the degree of the polynomial $P(x, y, z)$, the arithmetic genus of the algebraic curve is given by

$$g = \frac{1}{2}(d-1)(d-2). \tag{4}$$

For a smooth curve the arithmetic genus equals the geometric genus, therefore just using "genus" is unambiguous in the smooth case. Let's look at an example: The equation

$$y^2 z - x^3 - x z^2 = 0 \tag{5}$$

defines a smooth algebraic curve of genus 1.

Please note that the geometric genus of a singular curve is given by a more complicated formula. For example, the geometric genus of an algebraic curve with k irreducible components, s nodes (i.e. double points, where the curve crosses transversely) and no other singularities is given by $g_{\text{geom}} = g_{\text{arithm}} + k - s - 1$.

Let us now turn to elliptic curves: An elliptic curve over $\mathbb{C}$ is a smooth algebraic curve in $\mathbb{CP}^2$ of genus one with one marked point. It is common practice to work in the chart $z = 1$ and to take as the marked point the "point at infinity". Equation (5) reads in the chart $z = 1$

$$y^2 - x^3 - x = 0, \tag{6}$$

The point at infinity, which is not contained in this chart, is given by $[x : y : z] = [0 : 1 : 0]$.

Over the complex numbers $\mathbb{C}$ any elliptic curve can be cast into the Weierstrass normal form. In the chart $z = 1$ the Weierstrass normal form reads

$$y^2 = 4x^3 - g_2 x - g_3. \tag{7}$$

A second important example is to define an elliptic curve by a quartic polynomial in the chart $z = 1$:

$$y^2 = (x - x_1)(x - x_2)(x - x_3)(x - x_4). \tag{8}$$

If all roots of the quartic polynomial on the right-hand side are distinct, this defines a smooth elliptic curve. (The attentive reader may ask, how this squares with the genus formula above. The answer is that the elliptic curve in $\mathbb{CP}^2$ is not given by the homogenisation $y^2 z^2 = (x - x_1 z)(x - x_2 z)(x - x_3 z)(x - x_4 z)$. The latter curve is singular at infinity. However, there is a smooth elliptic curve, which in the chart $z = 1$ is isomorphic to the affine curve defined by Eq. (8).)

As one complex dimension corresponds to two real dimensions, we may consider a smooth algebraic curve (i.e. an object of complex dimension one) also as a real surface (i.e. an object of real dimension two). The latter objects are called Riemann surfaces, as the real surface inherits the structure of a complex manifold. We may therefore view an elliptic curve either as a complex one-dimensional smooth algebraic curve in $\mathbb{CP}^2$ with one marked point or as a real Riemann surface of genus one with one marked point. This is shown in Fig. 1.

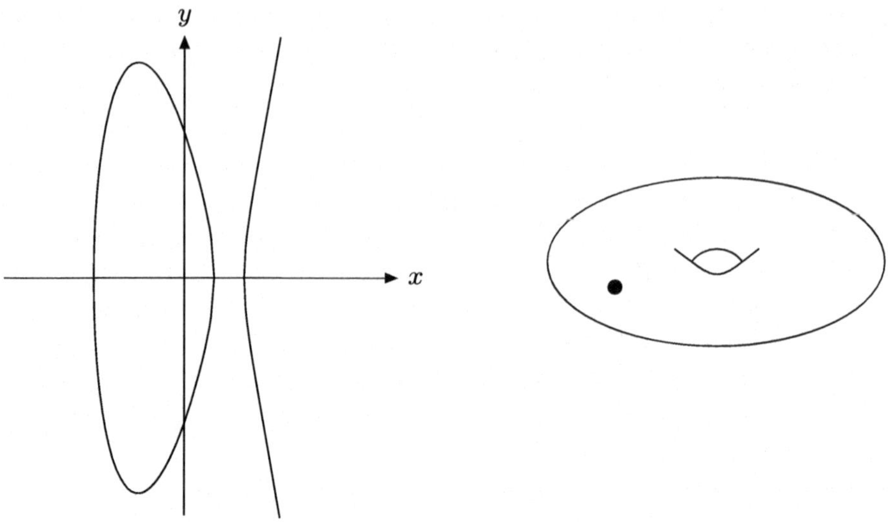

Fig. 1 The left picture shows the real part of an elliptic curve in the Weierstrass normal form $y^2 = 4x^3 - g_2x - g_3$. The marked point is at infinity. The right part shows a real Riemann surface of genus one with one marked point

Let us now turn to a second topic: Periodic functions. We consider a non-constant meromorphic function f of a complex variable z. A period ω of the function f is a constant such that for all z:

$$f(z + \omega) = f(z) \tag{9}$$

The set of all periods of f forms a lattice, which is either

- trivial (i.e. the lattice consists of $\omega = 0$ only),
- a simple lattice, generated by one period $\omega : \Lambda = \{n\omega \mid n \in \mathbb{Z}\}$,
- a double lattice, generated by two periods ω_1, ω_2 with $\text{Im}(\omega_2/\omega_1) \neq 0$:

$$\Lambda = \{n_1\omega_1 + n_2\omega_2 \mid n_1, n_2 \in \mathbb{Z}\}. \tag{10}$$

It is common practice to order these two periods such that $\text{Im}(\omega_2/\omega_1) > 0$.

There cannot be more possibilities: Assume that there is a third period ω_3, which is not an element of the lattice Λ spanned by ω_1 and ω_2. In this case we may construct arbitrary small periods as linear combinations of ω_1, ω_2 and ω_3 with integer coefficients. In the next step one shows that this implies that the derivative of $f(z)$ vanishes at any point z, hence $f(z)$ is a constant. This contradicts our assumption that f is a non-constant function.

An example for a singly periodic function is given by

$$\exp(z). \tag{11}$$

In this case the simple lattice is generated by $\omega = 2\pi i$.

Double periodic functions are called elliptic functions. An example for a doubly periodic function is given by Weierstrass's $\wp$-function. Let Λ be the lattice generated by ω_1 and ω_2. Then

$$\wp(z) = \frac{1}{z^2} + \sum_{\omega \in \Lambda \backslash \{0\}} \left(\frac{1}{(z+\omega)^2} - \frac{1}{\omega^2} \right). \tag{12}$$

$\wp(z)$ is periodic with periods ω_1 and ω_2.

Of particular interest are also the corresponding inverse functions. These are in general multivalued functions. In the case of the exponential function $x = \exp(z)$, the inverse function is given by

$$z = \ln(x). \tag{13}$$

The inverse function to Weierstrass's elliptic function $x = \wp(z)$ is an elliptic integral given by

$$z = \int\limits_x^\infty \frac{dt}{\sqrt{4t^3 - g_2 t - g_3}} \tag{14}$$

with

$$g_2 = 60 \sum_{\omega \in \Lambda \backslash \{0\}} \frac{1}{\omega^4}, \qquad g_3 = 140 \sum_{\omega \in \Lambda \backslash \{0\}} \frac{1}{\omega^6}. \tag{15}$$

The standard elliptic integrals are classified as complete or incomplete elliptic integrals and as integrals of the first, second or third kind. Table 1 shows the definition of the six standard elliptic integrals. The complete elliptic integrals are a special case of the incomplete elliptic integrals and obtained from the incomplete elliptic integrals by setting the variable z to one.

The classification of elliptic integrals as integrals of the first, second or third kind follows the classification of Abelian differentials: An Abelian differential $f(z)dz$ is called Abelian differential of the first kind, if $f(z)$ is holomorphic. It is called an Abelian differential of the second kind, if $f(z)$ is meromorphic, but with all residues vanishing. It is called an Abelian differential of the third kind, if $f(z)$ is meromorphic with non-zero residues.

So far we introduced elliptic curves and elliptic integrals. The link between the two is provided by the periods of an elliptic curve. An elliptic curve has one

Table 1 The six standard elliptic integrals. They are classified as complete or incomplete elliptic integrals and as integrals of the first, second or third kind

	Complete	Incomplete
First kind	$K(x) = \int\limits_0^1 \frac{dt}{\sqrt{(1-t^2)(1-x^2t^2)}}$	$F(z, x) = \int\limits_0^z \frac{dt}{\sqrt{(1-t^2)(1-x^2t^2)}}$
Second kind	$E(x) = \int\limits_0^1 dt \frac{\sqrt{1-x^2t^2}}{\sqrt{1-t^2}}$	$E(z, x) = \int\limits_0^z dt \frac{\sqrt{1-x^2t^2}}{\sqrt{1-t^2}}$
Third kind	$\Pi(v, x) = \int\limits_0^1 \frac{dt}{(1-vt^2)\sqrt{(1-t^2)(1-x^2t^2)}}$	$\Pi(v, z, x) = \int\limits_0^z \frac{dt}{(1-vt^2)\sqrt{(1-t^2)(1-x^2t^2)}}$

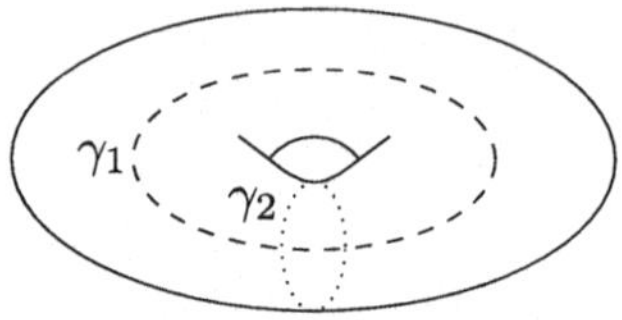

Fig. 2 A genus one Riemann surface, where the two independent cycles γ_1 and γ_2 are indicated

holomorphic differential (i.e. one Abelian differential of the first kind). If we view the elliptic curve as a genus one Riemann surface (i.e. a torus), we see that there are two independent cycles γ_1 and γ_2, as shown in Fig. 2. A period of an elliptic curve is the integral of the holomorphic differential along a cycle. As there are two independent cycles, there are two independent periods. Let's study this for an elliptic curve in the Legendre form

$$y^2 = x(x-1)(x-\lambda), \tag{16}$$

where λ is a parameter not equal to 0, 1 or infinity. The periods are

$$\omega_1 = 2\int\limits_0^\lambda \frac{dx}{y} = 4K\left(\sqrt{\lambda}\right), \quad \omega_2 = 2\int\limits_1^\lambda \frac{dx}{y} = 4iK\left(\sqrt{1-\lambda}\right). \tag{17}$$

The elliptic curve $y^2 = x(x-1)(x-\lambda)$ depends on a parameter λ, and so do the periods $\omega_1(\lambda)$ and $\omega_2(\lambda)$. We may now ask: How do the periods change, if we change λ? The variation is governed by a second-order differential equation: With $t = \sqrt{\lambda}$ we have

$$\left[t\left(1-t^2\right)\frac{d^2}{dt^2} + \left(1-3t^2\right)\frac{d}{dt} - t\right]\omega_j = 0, \qquad j = 1, 2. \tag{18}$$

The differential operator

$$t\left(1-t^2\right)\frac{d^2}{dt^2} + \left(1-3t^2\right)\frac{d}{dt} - t \tag{19}$$

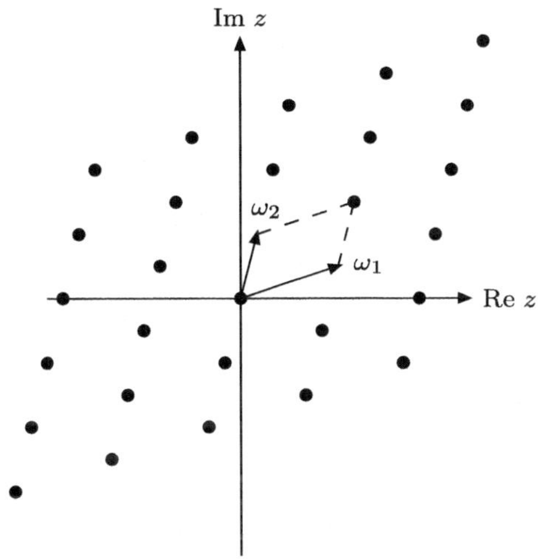

Fig. 3 $\mathbb{C}/\Lambda$, where Λ is a double lattice generated by ω_1 and ω_2. Points inside the fundamental parallelogram correspond to points on the elliptic curve. A point outside the fundamental parallelogram can always be shifted inside the fundamental parallelogram through the addition of some lattice vector

is called the Picard-Fuchs operator of the elliptic curve $y^2 = x(x-1)(x-\lambda)$.

There is a third possibility to represent an elliptic curve: We may also represent an elliptic curve as $\mathbb{C}/\Lambda$, where Λ is the double lattice generated by ω_1 and ω_2. This is shown in Fig. 3. Points, which differ by a lattice vector are considered to be equivalent. The different equivalence classes are represented by the points inside the fundamental parallelogram, as shown in Fig. 3. They correspond to points on the elliptic curve. Before we go into the details, let us first remark that this is not too surprising: If we start from the representation of an elliptic curve as a genus one Riemann surface and cut open this surface along the two cycles γ_1 and γ_2 shown in Fig. 2, we obtain a parallelogram.

Let's now fill in the technical detail: We would like to map a point on an elliptic curve, defined by a polynomial P, to a point in $\mathbb{C}/\Lambda$ and vice versa. For simplicity we assume that the elliptic curve is given in the Weierstrass normal form $y^2 - 4x^3 + g_2 x + g_3 = 0$.

We start with the direction from the Weierstrass normal form to $\mathbb{C}/\Lambda$: Given a point (x, y) with $y^2 - 4x^3 + g_2 x + g_3 = 0$ the corresponding point $z \in \mathbb{C}/\Lambda$ is given by

$$z = \int\limits_x^\infty \frac{dt}{\sqrt{4t^3 - g_2 t - g_3}}. \tag{20}$$

Let's now consider the reverse direction from $z \in \mathbb{C}/\Lambda$ to a point on the curve defined by the Weierstrass normal form. Given a point $z \in \mathbb{C}/\Lambda$ the corresponding point (x, y) on $y^2 - 4x^3 + g_2 x + g_3 = 0$ is given by

$$(x, y) = \left(\wp(z), \wp'(z)\right). \tag{21}$$

$\wp(z)$ denotes Weierstrass's $\wp$-function.

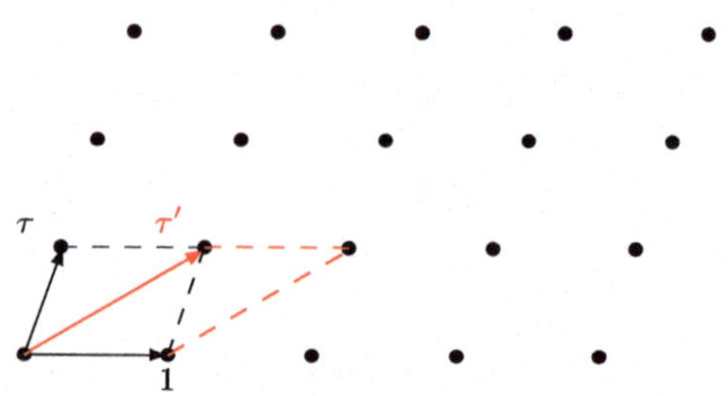

Fig. 4 The generators τ and 1 generate the same lattice as the generators τ' and 1

Let us now introduce some additional notation and conventions: It is common practise to normalise one period to one: $(\omega_2, \omega_1) \to (\tau, 1)$, where

$$\tau = \frac{\omega_2}{\omega_1}. \tag{22}$$

In addition one requires $\mathrm{Im}(\tau) > 0$. This is always possible: If $\mathrm{Im}(\tau) < 0$ simply exchange ω_1 and ω_2 and proceed as above. The possible values of τ lie therefore in the complex upper half-plane, defined by

$$\mathbb{H} = \{\tau \in \mathbb{C} | \mathrm{Im}(\tau) > 0\}. \tag{23}$$

Let us now turn to modular transformations: We have seen that we may represent an elliptic curve as $\mathbb{C}/\Lambda$, where Λ is a double lattice generated by ω_1 and ω_2. As only the lattice Λ matters, but not the specific generators, we may consider a different pair of periods (ω_2', ω_1'), which generate the same lattice Λ. An example is shown in Fig. 4: The generators τ and 1 generate the same lattice as the generators τ' and 1.

Let's return to the general case and consider a change of basis from the pair of periods (ω_2, ω_1) to the pair of periods (ω_2', ω_1'). The new pair of periods (ω_2', ω_1') is again a pair of lattice vectors, so it can be written as

$$\begin{pmatrix} \omega_2' \\ \omega_1' \end{pmatrix} = \begin{pmatrix} a & b \\ c & d \end{pmatrix} \begin{pmatrix} \omega_2 \\ \omega_1 \end{pmatrix}, \tag{24}$$

with $a, b, c, d \in \mathbb{Z}$. The transformation should be invertible and (ω_2, ω_1) and (ω_2', ω_1') should generate the same lattice Λ. This implies

$$\begin{pmatrix} a & b \\ c & d \end{pmatrix} \in \mathrm{SL}_2(\mathbb{Z}). \tag{25}$$

In terms of τ and τ' we have

$$\tau' = \frac{a\tau + b}{c\tau + d}. \tag{26}$$

A transformation of the form as in Eq. (26) is called a modular transformation.

We may then look at functions $f(\tau)$, which transform under modular transformations in a particular way. This leads to modular forms. A meromorphic function $f : \mathbb{H} \to \mathbb{C}$ is a modular form of modular weight k for $\mathrm{SL}_2(\mathbb{Z})$ if

1. f transforms under modular transformations as

$$f\left(\frac{a\tau + b}{c\tau + d}\right) = (c\tau + d)^k \cdot f(\tau) \qquad \text{for } \gamma = \begin{pmatrix} a & b \\ c & d \end{pmatrix} \in \mathrm{SL}_2(\mathbb{Z}), \tag{27}$$

2. f is holomorphic on $\mathbb{H}$,
3. f is holomorphic at $i\infty$.

The prefactor $(c\tau + d)^k$ in Eq. (27) is called automorphic factor and equals

$$(c\tau + d)^k = \left(\frac{\omega_1'}{\omega_1}\right)^k. \tag{28}$$

It is convenient to introduce the $|_k\gamma$ operator, defined by

$$(f|_k\gamma)(\tau) = (c\tau + d)^{-k} \cdot f(\gamma(\tau)). \tag{29}$$

With the help of the $|_k\gamma$ operator we may rewrite Eq. (27) as

$$(f|_k\gamma) = f \qquad \text{for } \gamma \in \mathrm{SL}_2(\mathbb{Z}) \tag{30}$$

A meromorphic function $f : \mathbb{H} \to \mathbb{C}$, which only satisfies Eq. (27) (or equivalently only Eq. (30)) is called weakly modular of weight k for $\mathrm{SL}_2(\mathbb{Z})$.

Apart from $\mathrm{SL}_2(\mathbb{Z})$ we may also look at congruence subgroups. The standard congruence subgroups are defined by

$$\Gamma_0(N) = \left\{ \begin{pmatrix} a & b \\ c & d \end{pmatrix} \in \mathrm{SL}_2(\mathbb{Z}) : c \equiv 0 \bmod N \right\},$$

$$\Gamma_1(N) = \left\{ \begin{pmatrix} a & b \\ c & d \end{pmatrix} \in \mathrm{SL}_2(\mathbb{Z}) : a, d \equiv 1 \bmod N,\ c \equiv 0 \bmod N \right\},$$

$$\Gamma(N) = \left\{ \begin{pmatrix} a & b \\ c & d \end{pmatrix} \in \mathrm{SL}_2(\mathbb{Z}) : a, d \equiv 1 \bmod N,\ b, c \equiv 0 \bmod N \right\}.$$

$\Gamma(N)$ is called the principle congruence subgroup of level N. The principle congruence subgroup $\Gamma(N)$ is a normal subgroup of $\mathrm{SL}_2(\mathbb{Z})$. In general, a subgroup Γ of $\mathrm{SL}_2(\mathbb{Z})$ is called a congruence subgroup, if there exists an N such that

$$\Gamma(N) \subseteq \Gamma. \tag{31}$$

The smallest such N is called the level of the congruence subgroup.

We may now define modular forms for a congruence subgroup Γ, by relaxing the transformation law in Eq. (27) to hold only for modular transformations from the subgroup Γ, plus holomorphicity on $\mathbb{H}$ and at the cusps. In detail: A meromorphic function $f : \mathbb{H} \to \mathbb{C}$ is a modular form of modular weight k for the congruence subgroup Γ if

1. f transforms as

$$(f|_k\gamma) = f \qquad \text{for} \;\; \gamma \in \Gamma, \tag{32}$$

2. f is holomorphic on $\mathbb{H}$,
3. $f|_k\gamma$ is holomorphic at $i\infty$ for all $\gamma \in \mathrm{SL}_2(\mathbb{Z})$.

For a congruence subgroup Γ of $\mathrm{SL}_2(\mathbb{Z})$ we denote by $\mathcal{M}_k(\Gamma)$ the space of modular forms of weight k. From the inclusions

$$\Gamma(N) \subseteq \Gamma_1(N) \subseteq \Gamma_0(N) \subseteq \mathrm{SL}_2(\mathbb{Z}) \tag{33}$$

follow the inclusions

$$\mathcal{M}_k(\mathrm{SL}_2(\mathbb{Z})) \subseteq \mathcal{M}_k(\Gamma_0(N)) \subseteq \mathcal{M}_k(\Gamma_1(N)) \subseteq \mathcal{M}_k(\Gamma(N)). \tag{34}$$

For a given N, the space $\mathcal{M}_k(\Gamma(N))$ of modular forms of weight k for the principal congruence subgroup $\Gamma(N)$ is the largest one among the spaces listed in Eq. (34). By definition we have for $f \in \mathcal{M}_k(\Gamma(N))$ and $\gamma \in \Gamma(N)$

$$f|_k\gamma = f, \qquad\qquad \gamma \in \Gamma(N).$$

We may ask what happens if we transform by a $\gamma \in \mathrm{SL}_2(\mathbb{Z})$, which does not belong to the congruence subgroup $\Gamma(N)$. One may show that in this case we have

$$f|_k\gamma \in \mathcal{M}_k(\Gamma(N)), \qquad\qquad \gamma \in \mathrm{SL}_2(\mathbb{Z})\backslash\Gamma(N),$$

i.e. $f|_k\gamma$ is again a modular form of weight k for $\Gamma(N)$, although not necessarily identical to f. The proof relies on the fact that $\Gamma(N)$ is a normal subgroup of $\mathrm{SL}_2(\mathbb{Z})$. This is essential: If Γ is a non-normal congruence subgroup of $\mathrm{SL}_2(\mathbb{Z})$ one has in general $f|_k\gamma \notin \mathcal{M}_k(\Gamma)$.

Modular forms of $\mathrm{SL}_2(\mathbb{Z})$ are invariant under $\tau' = \tau + 1$, since

$$\begin{pmatrix} 1 & 1 \\ 0 & 1 \end{pmatrix} \in \mathrm{SL}_2(\mathbb{Z}). \tag{35}$$

In other words, they are periodic with period 1: $f(\tau + 1) = f(\tau)$.

It is convenient to introduce for $\tau \in \mathbb{H}$ and $z \in \mathbb{C}$

$$\bar{q} = \exp(2\pi i \tau), \quad \bar{w} = \exp(2\pi i z). \tag{36}$$

Throughout this review we use the notation $\bar{q}$ to denote the nome squared. The bar is used to distinguish this quantity from the nome, usually denoted by $q = \exp(i\pi\tau)$. The notation for z and w we will be useful in the next section. $\bar{q} = \exp(2\pi i\tau)$ maps the complex upper half-plane $\tau \in \mathbb{H}$ to the unit disk $|\bar{q}| < 1$.

The maps in Eq. (36) trivialise periodicity with period 1, i.e.

$$\bar{w}(z+1) = \bar{w}(z). \tag{37}$$

On the other hand, shifts by τ correspond to multiplication with $\bar{q}$:

$$\bar{w}(z+\tau) = \bar{w}(z) \cdot \bar{q}(\tau). \tag{38}$$

We now introduce iterated integrals of modular forms. Let $f_1, \dots, f_n$ be modular forms. We set

$$\begin{aligned} I(f_1, f_2, \dots, f_n; \tau) &= \\ &= (2\pi i)^n \int\limits_{\tau_0}^{\tau} d\tau_1 f_1(\tau_1) \int\limits_{\tau_0}^{\tau_1} d\tau_2 f_2(\tau_2) \dots \int\limits_{\tau_0}^{\tau_{n-1}} d\tau_n f_n(\tau_n). \end{aligned} \tag{39}$$

As basepoint we usually take $\tau_0 = i\infty$. Please note that an integral over a modular form is in general not a modular form. This is not surprising if we consider the following analogy: An integral over a rational function is in general not a rational function.

We usually like iterated integrals appearing in solutions of Feynman integrals to have at worst simple poles. Let's study iterated integrals of modular forms. As modular forms are holomorphic in the complex upper half-plane, there are no poles there. So the only interesting points are the cusps. Let's focus on modular forms $f \in \mathcal{M}_k(\mathrm{SL}_2(\mathbb{Z}))$, so the only cusp is at $\tau = i\infty$. By definition a modular form $f(\tau)$ is holomorphic at the cusp and has a $\bar{q}$-expansion

$$f(\tau) = a_0 + a_1\bar{q} + a_2\bar{q}^2 + \dots, \qquad \bar{q} = \exp(2\pi i\tau). \tag{40}$$

The transformation $\bar{q} = \exp(2\pi i\tau)$ transforms the point $\tau = i\infty$ to $\bar{q} = 0$ and we have

$$2\pi i\ f(\tau)d\tau = \frac{d\bar{q}}{\bar{q}}\left(a_0 + a_1\bar{q} + a_2\bar{q}^2 + \dots\right). \tag{41}$$

Thus a modular form non-vanishing at the cusp $\tau = i\infty$ has a simple pole at $\bar{q} = 0$.

3 Moduli Spaces

This section gives an introduction into moduli spaces.

Let X be a topological space. The configuration space of n ordered points in X is

$$\text{Conf}_n\left(X\right) = \left\{ \left(x_1, \dots, x_n\right) \in X^n \middle| x_i \neq x_j \text{ for } i \neq j \right\}. \tag{42}$$

Please note that we require that the points are distinct: $x_i \neq x_j$. As a simple example consider the configuration space of 2 ordered points in $\mathbb{R}$:

$$\text{Conf}_2\left(\mathbb{R}\right) = \left\{ \left(x_1, x_2\right) \in \mathbb{R}^2 \middle| x_1 \neq x_2 \right\}. \tag{43}$$

$\text{Conf}_2(\mathbb{R})$ is the plane $\mathbb{R}^2$ with the diagonal $x_1 = x_2$ removed.

As a second example consider the configuration space of 2 ordered points in the complex projective space $\mathbb{CP}^1$ (i.e. the Riemann sphere):

$$\text{Conf}_2\left(\mathbb{CP}^1\right) = \left\{ \left(z_1, z_2\right) \in \left(\mathbb{CP}^1\right)^2 \middle| z_1 \neq z_2 \right\}. \tag{44}$$

This is a two-dimensional space. A Möbius transformation

$$z' = \frac{az+b}{cz+d} \tag{45}$$

transforms the Riemann sphere into itself. These transformations form a group $\text{PSL}\left(2, \mathbb{C}\right)$. Usually we are not interested in configurations

$$\left(z_1, \dots, z_n\right) \in \text{Conf}_n\left(\mathbb{CP}^1\right) \text{ and } \left(z'_1, \dots, z'_n\right) \in \text{Conf}_n\left(\mathbb{CP}^1\right), \tag{46}$$

which differ only by a Möbius transformation. This brings us to the definition of the moduli space of the Riemann sphere with n marked points:

$$\mathcal{M}_{0,n} = \text{Conf}_n\left(\mathbb{CP}^1\right) / \text{PSL}\left(2, \mathbb{C}\right). \tag{47}$$

We may use the freedom of Möbius transformations to fix three points (usually 0, 1 and ∞). Therefore

$$\begin{aligned} \dim\left(\text{Conf}_n\left(\mathbb{CP}^1\right)\right) &= n, \\ \dim\left(\mathcal{M}_{0,n}\right) &= n-3. \end{aligned} \tag{48}$$

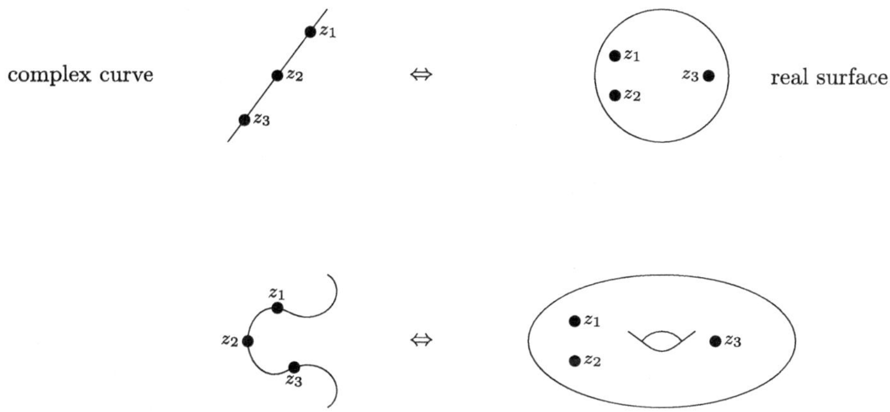

Fig. 5 The upper left figure shows a configuration of three marked points on a complex curve of genus zero, the upper right figure shows the corresponding configuration when the complex curve is viewed as a real Riemann surface. The lower figures show the analogous situation for a complex curve of genus one

Let's generalise this: We are interested in the situation, where the topological space X is a smooth algebraic curve C in $\mathbb{CP}^2$. This implies that there exists a homogeneous polynomial $P(z_1, z_2, z_3)$ such that

$$C = \left\{ [z_1 : z_2 : z_3] \in \mathbb{CP}^2 \middle| P(z_1, z_2, z_3) = 0 \right\}. \tag{49}$$

If d is the degree of the polynomial $P(z_1, z_2, z_3)$, the genus g of C is given by Eq. (4). Alternatively we may view C as a Riemann surface of genus g (Fig. 5).

Let us now consider a smooth curve C of genus g with n marked points. Two such curves $(C; z_1, \dots, z_n)$ and $(C'; z'_1, \dots, z'_n)$ are isomorphic if there is an isomorphism

$$\phi : C \to C' \qquad \text{such that } \phi(z_i) = z'_i. \tag{50}$$

The moduli space

$$\mathcal{M}_{g,n} \tag{51}$$

is the space of isomorphism classes of smooth curves of genus g with n marked points. For $g \geq 1$ the isomorphism classes do not only depend on the positions of the marked points, but also on the "shape" of the curve. For $g = 0$ there is only one "shape", the Riemann sphere.

The dimension of $\mathcal{M}_{g,n}$ is

$$\dim\left(\mathcal{M}_{g,n}\right) - 3g + n - 3, \tag{52}$$

for $g = 0$ this formula agrees with the previous result in Eq. (48).

In this talk we are mainly interested in the moduli spaces $\mathcal{M}_{0,n}$ and $\mathcal{M}_{1,n}$. Let us work out natural choices for coordinates on $\mathcal{M}_{0,n}$ and $\mathcal{M}_{1,n}$.

- We start with genus 0. We have $\dim \mathcal{M}_{0,n} = n - 3$. As mentioned above, the sphere has a unique shape. We may use Möbius transformations to fix three points, say $z_{n-2} = 1$, $z_{n-1} = \infty$, $z_n = 0$. This leaves

$$(z_1, \ldots, z_{n-3}) \tag{53}$$

as coordinates on $\mathcal{M}_{0,n}$.
- We now turn to genus 1. From Eq. (52) we have $\dim \mathcal{M}_{1,n} = n$. We need one coordinate to describe the shape of the elliptic curve (or the shape of the torus or the shape of the parallelogram). We may take τ as defined in Eq. (22) for this. We may use translation to fix one marked point, say $z_n = 0$. This gives

$$(\tau, z_1, \ldots, z_{n-1}) \tag{54}$$

as coordinates on $\mathcal{M}_{1,n}$.

We then consider iterated integrals on $\mathcal{M}_{0,n}$ and $\mathcal{M}_{1,n}$. In general, iterated integrals are defined as follows: Let $\omega_1, \ldots, \omega_k$ be differential 1-forms on a manifold M and $\gamma : [0, 1] \to M$ a path. We write for the pull-back of ω_j to the interval [0, 1]

$$f_j(\lambda)\, d\lambda = \gamma^* \omega_j. \tag{55}$$

Iterated integral are defined by Chen [74]

$$I_\gamma(\omega_1, \ldots, \omega_k; \lambda) = \int\limits_0^{\lambda} d\lambda_1 f_1(\lambda_1) \int\limits_0^{\lambda_1} d\lambda_2 f_2(\lambda_2) \ldots \int\limits_0^{\lambda_{k-1}} d\lambda_k f_k(\lambda_k). \tag{56}$$

Let us now specialise to iterated integrals on $\mathcal{M}_{0,n}$. We are interested in differential one-forms, which have only simple poles. We therefore consider

$$\omega^{\mathrm{mpl}} = \frac{dy}{y - z_j}. \tag{57}$$

The iterated integrals constructed from these differential one-forms are the multiple polylogarithms:

$$G(z_1, \ldots, z_k; y) = \int\limits_0^{y} \frac{dy_1}{y_1 - z_1} \int\limits_0^{y_1} \frac{dy_2}{y_2 - z_2} \ldots \int\limits_0^{y_{k-1}} \frac{dy_k}{y_k - z_k}, \qquad z_k \neq 0. \tag{58}$$

We may slightly enlarge the set of functions by setting

$$G(\underbrace{0, \dots, 0}_{k}; y) = \frac{1}{k!} \ln^k (y) \tag{59}$$

and for $(z_1, z_2, \dots, z_k) \neq (0, 0, \dots, 0)$

$$G(z_1, z_2, \dots, z_k; y) = \int_0^y \frac{dy_1}{y_1 - z_1} G(z_2, \dots, z_k; y_1). \tag{60}$$

This allows trailing zeros. We say that the multiple polylogarithm $G(z_1, \dots, z_k; y)$ has a trailing zero if $z_k = 0$. Using the shuffle product we may convert any multiple polylogarithm with trailing zeros into multiple polylogarithm without trailing zeros and powers of $\ln(y)$.

Let's now consider iterated integrals on $\mathcal{M}_{1,n}$. We recall that we may take $(\tau, z_1, \dots, z_{n-1})$ as coordinates on $\mathcal{M}_{1,n}$. We may decompose an arbitrary integration path into pieces along $d\tau$ (with $z_1 = \dots = z_{n-1} =$ const) and pieces along the dz_j's (with $\tau =$ const). Thus we obtain two classes of standardised iterated integrals: Iterated integrals on $\mathcal{M}_{1,n}$ with integration along $d\tau$ and iterated integrals on $\mathcal{M}_{1,n}$ with integration along the dz_j's.

In addition we have to specify the differential one-forms we want to integrate. The differential one-forms which we want to consider in the case of $\mathcal{M}_{1,n}$ are derived from the Kronecker function. The Kronecker function $F(x, y, \tau)$ is defined in terms of the first Jacobi theta function by

$$F(x, y, \tau) = \pi \theta_1' (0, q) \frac{\theta_1 (\pi (x + y), q)}{\theta_1 (\pi x, q) \theta_1 (\pi y, q)}, \tag{61}$$

where $q = \exp(\pi i \tau)$ and θ_1' denotes the derivative with respect to the first argument. The first Jacobi theta function $\theta_1(z, q)$ is defined by

$$\theta_1 (z, q) = -i \sum_{n=-\infty}^{\infty} (-1)^n q^{\left(n+\frac{1}{2}\right)^2} e^{i(2n+1)z}, \qquad q = e^{i\pi\tau}. \tag{62}$$

Please note that in order to make contact with the standard notation for the Jacobi theta functions we used here the nome $q = \exp(\pi i \tau)$ and not the nome squared $\bar{q} = q^2 = \exp(2\pi i \tau)$. The definition of the Kronecker function is cleaned up if we define

$$\bar{\theta}_1 (z, \bar{q}) = \theta_1 \left(\pi z, \bar{q}^{\frac{1}{2}}\right). \tag{63}$$

Then

$$F(x, y, \tau) = \bar{\theta}_1'(0, \bar{q}) \frac{\bar{\theta}_1(x + y, \bar{q})}{\bar{\theta}_1(x, \bar{q}) \bar{\theta}_1(y, \bar{q})}. \tag{64}$$

It is obvious from the definition that the Kronecker function is symmetric in x and y. We are interested in the Laurent expansion in one of these variables. We define functions $g^{(k)}(z, \tau)$ through

$$F(z, \alpha, \tau) = \sum_{k=0}^{\infty} g^{(k)}(z, \tau) \alpha^{k-1}. \tag{65}$$

We are primarily interested in the coefficients $g^{(k)}(z, \tau)$ of the Kronecker function. Let us recall some of their properties [39, 75, 76].

1. When viewed as a function of z, the function $g^{(k)}(z, \tau)$ has only simple poles. More concretely, the function $g^{(1)}(z, \tau)$ has a simple pole with unit residue at every point of the lattice. For $k > 1$ the function $g^{(k)}(z, \tau)$ has a simple pole only at those lattice points that do not lie on the real axis.
2. The (quasi-) periodicity properties are

$$\begin{aligned} g^{(k)}(z + 1, \tau) &= g^{(k)}(z, \tau), \\ g^{(k)}(z + \tau, \tau) &= \sum_{j=0}^{k} \frac{(-2\pi i)^j}{j!} g^{(k-j)}(z, \tau). \end{aligned} \tag{66}$$

   We see that $g^{(k)}(z, \tau)$ is invariant under translations by 1, but not by τ.
3. The functions $g^{(k)}(z, \tau)$ have the symmetry

$$g^{(k)}(-z, \tau) = (-1)^k g^{(k)}(z, \tau). \tag{67}$$

4. Under modular transformations the functions $g^{(k)}(z, \tau)$ transform as

$$g^{(k)}\left(\frac{z}{c\tau + d}, \frac{a\tau + b}{c\tau + d}\right) = (c\tau + d)^k \sum_{j=0}^{k} \frac{(2\pi i)^j}{j!} \left(\frac{cz}{c\tau + d}\right)^j g^{(k-j)}(z, \tau). \tag{68}$$

5. The $\bar{q}$-expansion of the $g^{(k)}(z,\tau)$ functions is given by (with $\bar{q} = \exp(2\pi i \tau)$ and $\bar{w} = \exp(2\pi i z)$)

$$g^{(0)}(z,\tau) = 1,$$
$$g^{(1)}(z,\tau) = -2\pi i \left[\frac{1+\bar{w}}{2(1-\bar{w})} + \overline{\mathrm{E}}_{0,0}(\bar{w}; 1; \bar{q}) \right],$$
$$g^{(k)}(z,\tau) = -\frac{(2\pi i)^k}{(k-1)!} \left[-\frac{B_k}{k} + \overline{\mathrm{E}}_{0,1-k}(\bar{w}; 1; \bar{q}) \right], \qquad k > 1, \quad (69)$$

where B_k denote the k-th Bernoulli number, defined by

$$\frac{x}{e^x - 1} = \sum_{k=0}^{\infty} \frac{B_k}{k!} x^k, \quad (70)$$

and

$$\overline{\mathrm{E}}_{n;m}(\bar{u}; \bar{v}; \bar{q}) = \mathrm{ELi}_{n;m}(\bar{u}; \bar{v}; \bar{q}) - (-1)^{n+m} \mathrm{ELi}_{n;m}\left(\bar{u}^{-1}; \bar{v}^{-1}; \bar{q}\right),$$
$$\mathrm{ELi}_{n;m}(\bar{u}; \bar{v}; \bar{q}) = \sum_{j=1}^{\infty} \sum_{k=1}^{\infty} \frac{\bar{u}^j}{j^n} \frac{\bar{v}^k}{k^m} \bar{q}^{jk}. \quad (71)$$

Having defined the functions $g^{(k)}(z,\tau)$, we may now state the differential one forms which we would like to integrate on $\mathcal{M}_{1,n}$. To keep the discussion simple, we focus on $\mathcal{M}_{1,2}$ with coordinates (τ, z). (The general case $\mathcal{M}_{1,n}$ is only from a notational perspective more cumbersome.) We consider

$$\omega_k^{\mathrm{Kronecker}} = (2\pi i)^{2-k} \left[g^{(k-1)}\left(z - c_j, \tau\right) dz + (k-1)\, g^{(k)}\left(z - c_j, \tau\right) \frac{d\tau}{2\pi i} \right], \quad (72)$$

with c_j being a constant. The differential one-form $\omega_k^{\mathrm{Kronecker}}$ is closed

$$d\omega_k^{\mathrm{Kronecker}} = 0. \quad (73)$$

For the integration along dz (i.e. $\tau = \mathrm{const}$) the part

$$\omega_k^{\mathrm{Kronecker},z} = (2\pi i)^{2-k}\, g^{(k-1)}\left(z - c_j, \tau\right) dz \quad (74)$$

is relevant. The iterated integrals of the differential one-forms in Eq. (74) along a path γ from $z = 0$ to z are the elliptic multiple polylogarithms $\widetilde{\Gamma}$, as defined in ref. [31]:

$$\widetilde{\Gamma}\left(\begin{smallmatrix} n_1 & \dots & n_r \\ c_1 & \dots & c_r \end{smallmatrix} ; z; \tau \right) =$$
$$(2\pi i)^{n_1+\dots+n_r-r} I_\gamma \left(\omega^{\mathrm{Kronecker},z}_{n_1+1}(c_1,\tau), \dots, \omega^{\mathrm{Kronecker},z}_{n_r+1}(c_r,\tau); z \right). \quad (75)$$

It is not possible that the differential one-forms ω entering the definition of elliptic multiple polylogarithms have at the same time the following three properties: (i) ω is double-periodic, (ii) ω is meromorphic and (iii) ω has only simple poles. We can only require two of these three properties. The definition of the $\widetilde{\Gamma}$-functions selects meromorphicity and simple poles. The differential one-forms are not double-periodic. (This is spoiled by the quasi-periodicity of $g^{(k)}(z,\tau)$ in τ.) However, this is what physics (i.e. the evaluation of Feynman integrals) dictates us to choose. The integrands are then either multi-valued functions on $\mathcal{M}_{1,n}$ or single-valued functions on a covering space, in the same way as $\ln(z)$ is a multi-valued function on $\mathbb{C}^\times$ or a single-valued function on a covering space of $\mathbb{C}^\times$. Of course, in mathematics one might also consider alternative definitions, which prioritise other properties. A definition of elliptic multiple polylogarithms, which implements properties (i) and (ii), but gives up property (iii) can be found in [77], a definition, which implements properties (i) and (iii), but gives up (ii) can be found in [76]. It is a little bit unfortunate that these different function are all named elliptic multiple polylogarithms. The reader is advised to carefully check what is meant by the name "elliptic multiple polylogarithm".

Let us now consider the integration along $d\tau$ (i.e. $z =$ const). Here, the part

$$\begin{aligned} \omega^{\mathrm{Kronecker},\tau}_k &= (2\pi i)^{2-k} (k-1)\, g^{(k)}\left(z - c_j, \tau\right) \frac{d\tau}{2\pi i} \\ &= \frac{(k-1)}{(2\pi i)^k} g^{(k)}\left(z - c_j, \tau\right) \frac{d\bar{q}}{\bar{q}} \end{aligned} \quad (76)$$

is relevant. This is supplemented by z-independent differential one-forms constructed from modular forms: Let $f_k(\tau)$ be a modular form of weight k. We set

$$\omega^{\mathrm{modular}}_k = (2\pi i)\, f_k(\tau)\, d\tau \; = \; f_k(\tau) \frac{d\bar{q}}{\bar{q}}. \quad (77)$$

Let ω_{k_j} be as in Eq. (76) or as in Eq. (77) and γ the path from $\tau = i\infty$ to τ, corresponding in $\bar{q}$-space to a path from $\bar{q} = 0$ to $\bar{q}$. We then consider in $\bar{q}$-space the iterated integrals

$$I_\gamma\left(\omega_{k_1}, \dots, \omega_{k_r}; \bar{q}\right). \quad (78)$$

The integrands have no poles in $0 < |\bar{q}| < 1$. A simple pole at $\bar{q} = 0$ is possible and allowed. If ω_{k_r} has simple pole $\bar{q} = 0$ we say that the iterated integral has a trailing zero. We may split ω_{k_r} into a part proportional to $d\bar{q}/\bar{q}$ and a regular remainder. The singular part of a trailing zero can be treated in exactly the same way as we did in the case of multiple polylogarithms.

4 Physics

After reviewing the mathematical background let us now turn to physics, and here in particular to the computation of Feynman integrals.

Integration-by-parts identities [78, 79] and differential equations [24, 80–89] are standard tools for the computation of Feynman integrals. In essence, integration-by-parts identities allow us to express a Feynman integral from a large set of Feynman integrals as a linear combination of Feynman integrals from a smaller set. The Feynman integrals in the smaller set are called master integrals and we may think of the master integrals as a basis of an (abstract) vector space. We denote the number of master integrals by $N_F = N_{\text{Fibre}}$ and the master integrals by $I = (I_1, \dots, I_{N_F})$. The notation used in relation with Feynman integrals is summarised in Table 2. Public available computer programs based on the Laporta algorithm [90] like `REDUZE` [91], `FIRE` [92] or `KIRA` [93] can be used to perform the reduction to the master integrals.

For the master integrals one derives (again by using integration-by-parts identities) differential equations in the external invariants or internal masses. We denote the number of kinematic variables by $N_B = N_{\text{Base}}$ and the kinematic variables by $x = (x_1, \dots, x_{N_B})$. The system of differential equations for the master integrals can be written as

$$(d + A)\, I = 0, \tag{79}$$

where $A(\varepsilon, x)$ is a matrix-valued one-form

$$A = \sum_{i=1}^{N_B} A_i dx_i. \tag{80}$$

Table 2 The notation used in connection with Feynman integrals: N_F denotes the number of master integrals, N_B the number of kinematic variables the master integrals depend on and N_L the number linearly independent differential one-forms appearing in the ε-form of the differential equation

$N_F = N_{\text{Fibre}}$:	Number of master integrals, master integrals denoted by	$I = (I_1, \dots, I_{N_F})$.
$N_B = N_{\text{Base}}$:	Number of kinematic variables, kinematic variables denoted by	$x = (x_1, \dots, x_{N_B})$.
$N_L = N_{\text{Letters}}$:	Number of letters, differential one-forms denoted by	$\omega = (\omega_1, \dots, \omega_{N_L})$.

The $A_i(\varepsilon, x)$'s are matrices of size $N_F \times N_F$, whose entries are rational functions in the dimensional regularisation parameter ε and the kinematic variables x. The matrix-valued one-form A satisfies the integrability condition

$$dA + A \wedge A = 0. \tag{81}$$

Geometrically we have a vector bundle with a fibre of dimension N_F spanned by $I_1, \dots I_{N_F}$ and a base space of dimension N_B with local coordinates $x_1, \dots, x_{N_B}$. The matrix-valued one-form A defines a flat connection.

Up to this point everything is general and applies to any Feynman integral. In particular, computing a Feynman integral is reduced to the problem of solving a system of differential equations as in Eq. (79). The solution of a system of differential equations requires in addition boundary values. The boundary values correspond to simpler Feynman integrals, where some kinematic variables have special values or vanish. Therefore at this stage the boundary values can be considered to be known (otherwise one would first set up a system of differential equations for the boundary values).

The system of differential equations is particular simple [86], if A is of the form

$$A = \varepsilon \sum_{j=1}^{N_L} C_j \, \omega_j, \tag{82}$$

where

- the only dependence on the dimensional regularisation parameter ε is given by the explicit prefactor,
- the C_j's are $N_F \times N_F$-matrices, whose entries are numbers $r_1 + i r_2$ with $r_1, r_2 \in \mathbb{Q}$,
- the differential one-forms ω_j have only simple poles (and depend only on x).

We denote by $N_L = N_{\text{Letters}}$ the number of letters, i.e. the number of $\mathbb{Q}[i]$-linear independent differential one-forms ω_j. The set of letters is denoted by $\omega = (\omega_1, \dots, \omega_{N_L})$.

Let us now discuss the possibilities to transform a generic system of differential equations as in Eq. (79) into the simple form of Eq. (82). On the one hand we may change the basis of the master integrals

$$I' = UI, \tag{83}$$

where $U(\varepsilon, x)$ is a $N_F \times N_F$-matrix. The new connection matrix is

$$A' = UAU^{-1} + UdU^{-1}. \tag{84}$$

On the other hand, we may perform a coordinate transformation on the base manifold:

$$x_i' = f_i(x), \qquad 1 \le i \le N_B. \tag{85}$$

The connection transforms as

$$A = \sum_{i=1}^{N_B} A_i dx_i \qquad \Rightarrow \qquad A' = \sum_{i,j=1}^{N_B} A_i \frac{\partial x_i}{\partial x_j'} dx_j'. \tag{86}$$

Let us consider some examples of elliptic Feynman integrals. The most prominent example is the two-loop sunrise integral. The two-loop sunrise integral is defined by

$$S_{\nu_1\nu_2\nu_3}(\varepsilon, x) = (-1)^{\nu_{123}} e^{2\gamma_E\varepsilon} \left(m_3^2\right)^{\nu_{123}-D} \int \frac{d^D k_1}{i\pi^{\frac{D}{2}}} \frac{d^D k_2}{i\pi^{\frac{D}{2}}} \frac{1}{D_1^{\nu_1} D_2^{\nu_2} D_3^{\nu_3}}, \tag{87}$$

with the propagators

$$D_1 = k_1^2 - m_1^2, \quad D_2 = (k_1 - k_2)^2 - m_2^2, \quad D_3 = (p - k_2)^2 - m_3^2 \tag{88}$$

and $\nu_{123} = \nu_1 + \nu_2 + \nu_3$. γ_E denotes Euler's constant. It is convenient to consider this Feynman integral in $D = 2 - 2\varepsilon$ space-time dimensions. With the help of dimensional shift relations [94, 95] the result in $D = 2 - 2\varepsilon$ dimensions is easily related to the corresponding Feynman integrals in $D = 4 - 2\varepsilon$ dimensions.

The simplest example for an elliptic Feynman integral is the equal mass sunrise integral. In the equal mass case we have $m_1 = m_2 = m_3 = m$. In this case we have 3 master integrals and one kinematic variable, which we may take originally as $x = p^2/m^2$. This corresponds to $N_F = 3$ and $N_B = 1$. In mathematical terms we are looking at a rank 3 vector bundle over $\mathcal{M}_{1,1}$.

The first question which we should address is how to obtain the elliptic curve associated to this integral. For the sunrise integral there are two possibilities, we may either obtain an elliptic curve from the Feynman graph polynomial or from the maximal cut. The sunrise integral has three propagators, hence we need three Feynman parameters, which we denote by $\alpha_1, \alpha_2, \alpha_3$. The second graph polynomial defines an elliptic curve

$$E^{\text{Feynman}} : -\alpha_1\alpha_2\alpha_3 x + (\alpha_1 + \alpha_2 + \alpha_3)(\alpha_1\alpha_2 + \alpha_2\alpha_3 + \alpha_3\alpha_1) = 0, \tag{89}$$

in $\mathbb{CP}^2$, with $[\alpha_1 : \alpha_2 : \alpha_3]$ being the homogeneous coordinates of $\mathbb{CP}^2$. The elliptic curve varies with the kinematic variable x. In general, the Feynman parameter space can be viewed as $\mathbb{CP}^{n-1}$, with n being the number of propagators of the Feynman integral. It is clear that this approach does not generalise in a straightforward way to other elliptic Feynman integrals with more than three propagators. (For an elliptic curve we want the zero set of a single polynomial in $\mathbb{CP}^2$).

We therefore turn to the second method of obtaining the elliptic curve, which generalises easily: From the maximal cut of the sunrise integral we obtain the elliptic curve as a quartic polynomial $P(w, z) = 0$:

$$E^{\text{cut}} : w^2 - z\,(z+4)\left[z^2 + 2\,(1+x)\,z + (1-x)^2\right] = 0. \tag{90}$$

Also this elliptic curve varies with the kinematic variable x. Please note that these two elliptic curves E^{Feynman} and E^{cut} are not isomorphic, but only isogenic. Let ω_1 and ω_2 be two periods of this elliptic curve E^{cut} with $\text{Im}(\omega_2/\omega_1) > 0$ and set $\tau = \omega_2/\omega_1$. We denote the Wronskian by

$$W = \omega_1 \frac{d}{dx}\omega_2 - \omega_2 \frac{d}{dx}\omega_1. \tag{91}$$

In order to bring the system of differential equations for the equal mass sunrise integral into the simple form of Eq. (82) we perform a change of the basis of the master integrals from a pre-canonical basis $(S_{110}, S_{111}, S_{211})$ to

$$\begin{aligned}
J_1 &= 4\varepsilon^2\, S_{110}\,(\varepsilon, x)\,, \\
J_2 &= \varepsilon^2 \frac{\pi}{\omega_1}\, S_{111}\,(\varepsilon, x)\,, \\
J_3 &= \frac{1}{\varepsilon}\frac{\omega_1^2}{2\pi i W}\frac{d}{dx} J_2 + \frac{\omega_1^2}{2\pi i W}\frac{\left(3x^2 - 10x - 9\right)}{2x\,(x-1)\,(x-9)} J_2.
\end{aligned} \tag{92}$$

This transformation is not rational or algebraic in x, as can be seen from the prefactor $1/\omega_1$ in the definition of J_2. The period ω_1 is a transcendental function of x. In addition we change the kinematic variable from x to τ (or $\bar{q}$). Again, this is a non-algebraic change of variables. One obtains

$$(d + A)\, J = 0 \tag{93}$$

with

$$A = 2\pi i\, \varepsilon \begin{pmatrix} 0 & 0 & 0 \\ 0 & \eta_2\,(\tau) & \eta_0\,(\tau) \\ \eta_3\,(\tau) & \eta_4\,(\tau) & \eta_2\,(\tau) \end{pmatrix} d\tau, \tag{94}$$

where $\eta_k(\tau)$ denotes a modular form of modular weight k for $\Gamma(6)$. The differential equation for the equal mass sunrise system is now in ε-form and the kinematic variable matches the standard coordinate on $\mathcal{M}_{1,1}$. With the additional information of a boundary value, the differential equation is now easily solved order by order in ε in terms of iterated integrals of modular forms. One finds for example

$$J_2 = \left[3\,\mathrm{Cl}_2\left(\frac{2\pi}{3}\right) + I\left(\eta_0, \eta_3; \tau\right) \right] \varepsilon^2 + O\left(\varepsilon^3\right). \tag{95}$$

The Clausen value $\mathrm{Cl}_2(2\pi/3)$ comes from the boundary value.

Let us also consider an example where the kinematic space is $\mathcal{M}_{1,n}$ with $n > 1$. We don't have to go very far, the unequal mass sunrise integral provides an example. We now take the three masses squared m_1^2, m_2^2 and m_3^2 in Eq. (88) to be pairwise distinct. We now have 7 master integrals and 3 kinematic variables. As original kinematic variables we use $x = p^2/m_3^2$, $y_1 = m_1^2/m_3^2$, $y_2 = m_2^2/m_3^2$. This corresponds to $N_F = 7$ and $N_B = 3$. In mathematical terms we are looking at a rank 7 vector bundle over $\mathcal{M}_{1,3}$.

Finding the elliptic curve proceeds exactly in the same way as discussed in the equal mass case. In the next step we would like to change the kinematic variables from (x, y_1, y_2) to the standard coordinates (τ, z_1, z_2) on $\mathcal{M}_{1,3}$. This raises the question: How to express the new coordinates in terms of the old coordinates (or vice versa)? For τ the answer is straightforward: τ is a again the ratio of the two periods $\tau = \omega_2/\omega_1$, and ω_1 and ω_2 are functions of x, y_1 and y_2.

Also for z_1 and z_2 there is a simple geometric interpretation: In the Feynman parameter representation there are two geometric objects of interest: the domain of integration σ (the simplex $\alpha_1, \alpha_2, \alpha_3 \geq 0$, $\alpha_1 + \alpha_2\alpha_3 \leq 1$) and the elliptic curve E^{Feynman} (the zero set X of the second graph polynomial). X and σ intersect at three points, as shown in Fig. 6. The images of these three points in $\mathbb{C}/\Lambda$ are $0, z_1, z_2$, where we used a translation transformation to fix one point at 0.

The system of differential equations can again be transformed into the simple form of Eq. (82) by a redefinition of the master integrals and a change of coordinates

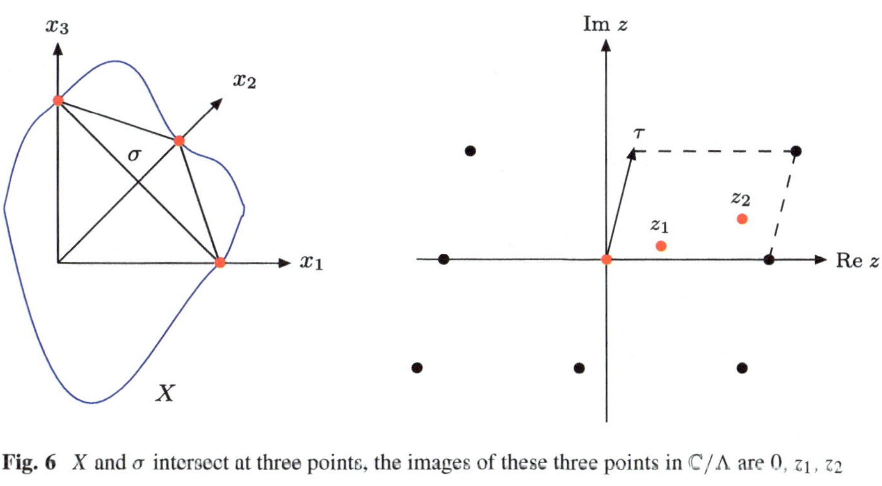

Fig. 6 X and σ intersect at three points, the images of these three points in $\mathbb{C}/\Lambda$ are $0, z_1, z_2$

from (x, y_1, y_2) to (τ, z_1, z_2). The explicit formula for the fibre transformation is a little bit lengthy and can be found in the literature [48, 55]. Doing so, one finds

$$A = \varepsilon \sum_{j=1}^{N_L} C_j \, \omega_j, \qquad \text{with } \omega_j \text{ having only simple poles,} \tag{96}$$

where ω_j is either

$$2\pi i \; f_k(\tau) \, d\tau, \tag{97}$$

where $f_k(\tau)$ is a modular form, or of the form

$$\omega_k \left(L\left(z\right), \tau\right) = \\ \left(2\pi i\right)^{2-k} \left[g^{(k-1)}\left(L\left(z\right), \tau\right) dL\left(z\right) + \left(k-1\right) g^{(k)}\left(L\left(z\right), \tau\right) \frac{d\tau}{2\pi i} \right], \tag{98}$$

with $L(z)$ being a linear function of z_1 and z_2:

$$L\left(z\right) = \sum_{j=1}^{2} \alpha_j z_j + \beta, \tag{99}$$

and α_1, α_2 and β being constants.

With the additional information of a boundary value, the differential equation in Eq. (96) is now easily solved order by order in ε in terms of iterated integrals as discussed in Sect. 3. We are free to choose a suitable point in kinematic space for the boundary value and to integrate the differential equation from the chosen boundary point to the kinematic point of interest. We are free to choose any path (as long as the path avoids branch cuts). An arbitrary path will involve integrations along $d\tau$ and the dz_j's. It is advantageous to use as boundary condition the values on the hypersurface $\tau = i\infty$. There the elliptic curve degenerates, i.e. the geometric genus equals zero, and the sought-after boundary values of the Feynman integrals are expressible in terms of multiple polylogarithms. We may then integrate the differential equation only along $d\tau$. This avoids integrations along the dz_j's, the analytic expressions tend to be shorter and the final expressions are easier to evaluate numerically. This approach also avoids poles and branch cuts along the integration path. The only problem, which might occur, are a slow convergence of the $\bar{q}$-expansion of the final result in the case $|\bar{q}| \lesssim 1$. This can be avoided by using in this kinematic region a different choice of periods ω_1 and ω_2, related to the original ones by a modular transformation [55]. It is therefore always possible to achieve

$$|\bar{q}| \le e^{-\pi\sqrt{3}} \approx 0.0043, \tag{100}$$

which guarantees a fast convergence of the $\bar{q}$-expansion of the final result.

5 Conclusions

Feynman integrals are important in many areas of physics and indispensable for precision calculations within perturbation theory beyond the leading order. Feynman integrals, which evaluate to multiple polylogarithms are by now well understood. Multiple polylogarithms are iterated integrals on the moduli space $\mathcal{M}_{0,n}$. From two loops onwards, there is a class of Feynman integrals related to elliptic curves, which evaluate to iterated integrals on the moduli space $\mathcal{M}_{1,n}$. These integrals were the main topic of this talk. We discussed the mathematical background of elliptic curves, elliptic functions, modular forms and the moduli space of n marked points on a smooth curve of genus one. The investment in the mathematical foundations pays off, as with the right language we may transfer methods known from the genus zero case to the genus one case. In particular we may achieve through a redefinition of the master integrals and a change of the kinematic variables that the differential equation is transformed to

$$A = \varepsilon \sum_{k=1}^{N_L} C_k \, \omega_k, \qquad \text{with } \omega_k \text{ having only simple poles.} \tag{101}$$

This form can be reached for many Feynman integrals evaluating to multiple polylogarithms and—as we have seen in this talk—also for non-trivial elliptic Feynman integrals.

References

1. D.J. Broadhurst, J. Fleischer, O. Tarasov, Z. Phys. **C60**, 287 (1993). arXiv:hep-ph/9304303
2. S. Laporta, E. Remiddi, Nucl. Phys. **B704**, 349 (2005). hep-ph/0406160
3. D.H. Bailey, J.M. Borwein, D. Broadhurst, M.L. Glasser, J. Phys. **A41**, 205203 (2008). arXiv:0801.0891
4. S. Müller-Stach, S. Weinzierl, R. Zayadeh, Commun. Num. Theor. Phys. **6**, 203 (2012). arXiv:1112.4360
5. L. Adams, C. Bogner, S. Weinzierl, J. Math. Phys. **54**, 052303 (2013). arXiv:1302.7004
6. S. Bloch, P. Vanhove, J. Numb. Theor. **148**, 328 (2015). arXiv:1309.5865
7. E. Remiddi, L. Tancredi, Nucl. Phys. **B880**, 343 (2014). arXiv:1311.3342
8. L. Adams, C. Bogner, S. Weinzierl, J. Math. Phys. **55**, 102301 (2014). arXiv:1405.5640
9. L. Adams, C. Bogner, S. Weinzierl, J. Math. Phys. **56**, 072303 (2015). arXiv:1504.03255
10. L. Adams, C. Bogner, S. Weinzierl, J. Math. Phys. **57**, 032304 (2016). arXiv:1512.05630
11. S. Bloch, M. Kerr, P. Vanhove, Adv. Theor. Math. Phys. **21**, 1373 (2017). arXiv:1601.08181
12. L. Adams, S. Weinzierl, Commun. Num. Theor. Phys. **12**, 193 (2018). arXiv:1704.08895
13. C. Bogner, A. Schweitzer, S. Weinzierl, Nucl. Phys. **B922**, 528 (2017). arXiv:1705.08952
14. L. Adams, S. Weinzierl, Phys. Lett. **B781**, 270 (2018). arXiv:1802.05020
15. I. Hönemann, K. Tempest, S. Weinzierl, Phys. Rev. **D98**, 113008 (2018). arXiv:1811.09308
16. S. Bloch, M. Kerr, P. Vanhove, Compos. Math. **151**, 2329 (2015). arXiv:1406.2664
17. M. Søgaard, Y. Zhang, Phys. Rev. **D91**, 081701 (2015). arXiv:1412.5577
18. L. Tancredi, Nucl. Phys. **B901**, 282 (2015). arXiv:1509.03330

19. A. Primo, L. Tancredi, Nucl. Phys. **B916**, 94 (2017). arXiv:1610.08397
20. E. Remiddi, L. Tancredi, Nucl. Phys. **B907**, 400 (2016). arXiv:1602.01481
21. L. Adams, C. Bogner, A. Schweitzer, S. Weinzierl, J. Math. Phys. **57**, 122302 (2016). arXiv:1607.01571
22. R. Bonciani et al., J. High Energy Phys. **12**, 096 (2016). arXiv:1609.06685
23. A. von Manteuffel, L. Tancredi, J. High Energy Phys. **06**, 127 (2017). arXiv:1701.05905
24. L. Adams, E. Chaubey, S. Weinzierl, Phys. Rev. Lett. **118**, 141602 (2017). arXiv:1702.04279.
25. J. Ablinger et al., J. Math. Phys. **59**, 062305 (2018). arXiv:1706.01299
26. A. Primo, L. Tancredi, Nucl. Phys. **B921**, 316 (2017). arXiv:1704.05465
27. G. Passarino, Eur. Phys. J. C **77**, 77 (2017). arXiv:1610.06207
28. E. Remiddi, L. Tancredi, Nucl. Phys. **B925**, 212 (2017). arXiv:1709.03622
29. J.L. Bourjaily, A.J. McLeod, M. Spradlin, M. von Hippel, M. Wilhelm, Phys. Rev. Lett. **120**, 121603 (2018). arXiv:1712.02785
30. M. Hidding, F. Moriello, J. High Energy Phys. **01**, 169 (2019). arXiv:1712.04441
31. J. Broedel, C. Duhr, F. Dulat, L. Tancredi, J. High Energy Phys. **05**, 093 (2018). arXiv:1712.07089
32. J. Broedel, C. Duhr, F. Dulat, L. Tancredi, Phys. Rev. **D97**, 116009 (2018). arXiv:1712.07095
33. J. Broedel, C. Duhr, F. Dulat, B. Penante, L. Tancredi, J. High Energy Phys. **08**, 014 (2018). arXiv:1803.10256
34. R.N. Lee, A.V. Smirnov, V.A. Smirnov, J. High Energy Phys. **03**, 008 (2018). arXiv:1709.07525
35. R.N. Lee, A.V. Smirnov, V.A. Smirnov, J. High Energy Phys. **07**, 102 (2018). arXiv:1805.00227
36. L. Adams, E. Chaubey, S. Weinzierl, Phys. Rev. Lett. **121**, 142001 (2018). arXiv:1804.11144
37. L. Adams, E. Chaubey, S. Weinzierl, J. High Energy Phys. **10**, 206 (2018). arXiv:1806.04981
38. L. Adams, Ph.D. thesis, University of Mainz, 2018
39. J. Broedel, C. Duhr, F. Dulat, B. Penante, L. Tancredi, J. High Energy Phys. **01**, 023 (2019). arXiv:1809.10698
40. J.L. Bourjaily, A.J. McLeod, M. von Hippel, M. Wilhelm, J. High Energy Phys. **08**, 184 (2018). arXiv:1805.10281
41. M. Besier, D. Van Straten, S. Weinzierl, Commun. Num. Theor. Phys. **13**, 253 (2019). arXiv:1809.10983
42. P. Mastrolia, S. Mizera, J. High Energy Phys. **02**, 139 (2019). arXiv:1810.03818.
43. J. Ablinger, J. Blümlein, P. Marquard, N. Rana, C. Schneider, Nucl. Phys. **B939**, 253 (2019). arXiv:1810.12261
44. H. Frellesvig et al., J. High Energy Phys. **05**, 153 (2019). arXiv:1901.11510
45. J. Broedel, C. Duhr, F. Dulat, B. Penante, L. Tancredi, J. High Energy Phys. **05**, 120 (2019). arXiv:1902.09971
46. J. Blümlein, (2019). arXiv:1905.02148
47. J. Broedel, A. Kaderli, J. Phys. A **53**, 245201 (2020). arXiv:1906.11857
48. C. Bogner, S. Müller-Stach, S. Weinzierl, Nucl. Phys. B **954**, 114991 (2020). arXiv:1907.01251
49. B.A. Kniehl, A.V. Kotikov, A.I. Onishchenko, O.L. Veretin, Nucl. Phys. **B948**, 114780 (2019). arXiv:1907.04638
50. J. Broedel et al., J. High Energy Phys. **09**, 112 (2019). arXiv:1907.03787
51. S. Abreu, M. Becchetti, C. Duhr, R. Marzucca, J. High Energy Phys. **02**, 050 (2020). arXiv:1912.02747
52. C. Duhr, L. Tancredi, J. High Energy Phys. **02**, 105 (2020). arXiv:1912.00077
53. M. Leitner, (2019). arXiv:1908.11815
54. M. Walden, S. Weinzierl, (2020). arXiv:2010.05271
55. S. Weinzierl, (2020). arXiv:2011.07311
56. J. Campert, F. Moriello, A. Kotikov, (2020). arXiv:2011.01904
57. M. Bezuglov, A. Onishchenko, O. Veretin, (2020). arXiv:2011.13337
58. J. Broedel, C.R. Mafra, N. Matthes, O. Schlotterer, J. High Energy Phys. **07**, 112 (2015). arXiv:1412.5535

59. J. Broedel, N. Matthes, O. Schlotterer, J. Phys. **A49**, 155203 (2016). arXiv:1507.02254
60. J. Broedel, N. Matthes, G. Richter, O. Schlotterer, J. Phys. **A51**, 285401 (2018). arXiv:1704.03449
61. E. D'Hoker, M.B. Green, Ö. Gürdogan, P. Vanhove, Commun. Num. Theor. Phys. **11**, 165 (2017). arXiv:1512.06779
62. S. Hohenegger, S. Stieberger, Nucl. Phys. **B925**, 63 (2017). arXiv:1702.04963
63. J. Broedel, O. Schlotterer, F. Zerbini, J. High Energy Phys. **01**, 155 (2019). arXiv:1803.00527
64. F. Brown, O. Schnetz, Duke Math. J. **161**, 1817 (2012). arXiv:1006.4064
65. J.L. Bourjaily, A.J. McLeod, M. von Hippel, M. Wilhelm, Phys. Rev. Lett. **122**, 031601 (2019). arXiv:1810.07689
66. J.L. Bourjaily et al., J. High Energy Phys. **01**, 078 (2020). arXiv:1910.01534
67. A. Klemm, C. Nega, R. Safari, J. High Energy Phys. **04**, 088 (2020). arXiv:1912.06201
68. K. Bönisch, F. Fischbach, A. Klemm, C. Nega, R. Safari, (2020). arXiv:2008.10574
69. P. Du Val, *Elliptic Functions, Elliptic Curves*. London Mathematical Society Lecture Note Series (Cambridge University Press, Cambridge, 1973)
70. J. Silverman, *The Arithmetic of Elliptic Curves* (Springer, Berlin, 1986)
71. W.A. Stein, *Modular Forms and a Computational Approach* (American Mathematical Society, Providence, 2007)
72. T. Miyake, *Modular Forms* (Springer, Berlin, 1989)
73. F. Diamond, J. Shurman, *A First Course in Modular Forms* (Springer, Berlin, 2005)
74. K.-T. Chen, Bull. Am. Math. Soc. **83**, 831 (1977)
75. D. Zagier, Invent. Math. **104**, 449 (1991)
76. F. Brown, A. Levin, (2011). arXiv:1110.6917
77. A. Levin, G. Racinet, (2007). arXiv:math/0703237
78. F.V. Tkachov, Phys. Lett. **B100**, 65 (1981)
79. K.G. Chetyrkin, F.V. Tkachov, Nucl. Phys. **B192**, 159 (1981)
80. A.V. Kotikov, Phys. Lett. **B254**, 158 (1991)
81. A.V. Kotikov, Phys. Lett. **B267**, 123 (1991)
82. E. Remiddi, Nuovo Cim. **A110**, 1435 (1997). hep-th/9711188
83. T. Gehrmann, E. Remiddi, Nucl. Phys. **B580**, 485 (2000). hep-ph/9912329
84. M. Argeri, P. Mastrolia, Int. J. Mod. Phys. **A22**, 4375 (2007). arXiv:0707.4037
85. S. Müller-Stach, S. Weinzierl, R. Zayadeh, Commun.Math.Phys. **326**, 237 (2014). arXiv:1212.4389
86. J.M. Henn, Phys. Rev. Lett. **110**, 251601 (2013). arXiv:1304.1806
87. J.M. Henn, J. Phys. **A48**, 153001 (2015). arXiv:1412.2296
88. J. Ablinger et al., Comput. Phys. Commun. **202**, 33 (2016). arXiv:1509.08324
89. J. Bosma, K.J. Larsen, Y. Zhang, Phys. Rev. **D97**, 105014 (2018). arXiv:1712.03760
90. S. Laporta, Int. J. Mod. Phys. **A15**, 5087 (2000). hep-ph/0102033
91. A. von Manteuffel, C. Studerus, (2012). arXiv:1201.4330
92. A.V. Smirnov, Comput. Phys. Commun. **189**, 182 (2015). arXiv:1408.2372
93. P. Maierhöefer, J. Usovitsch, P. Uwer, Comput. Phys. Commun. **230**, 99 (2018). arXiv:1705.05610
94. O.V. Tarasov, Phys. Rev. **D54**, 6479 (1996). hep-th/9606018
95. O.V. Tarasov, Nucl. Phys. **B502**, 455 (1997). hep-ph/9703319

Printed by Printforce, the Netherlands